CORE MATHS
ADVANCED LEVEL
3rd Edition

CORE MATHS
ADVANCED LEVEL
3rd Edition

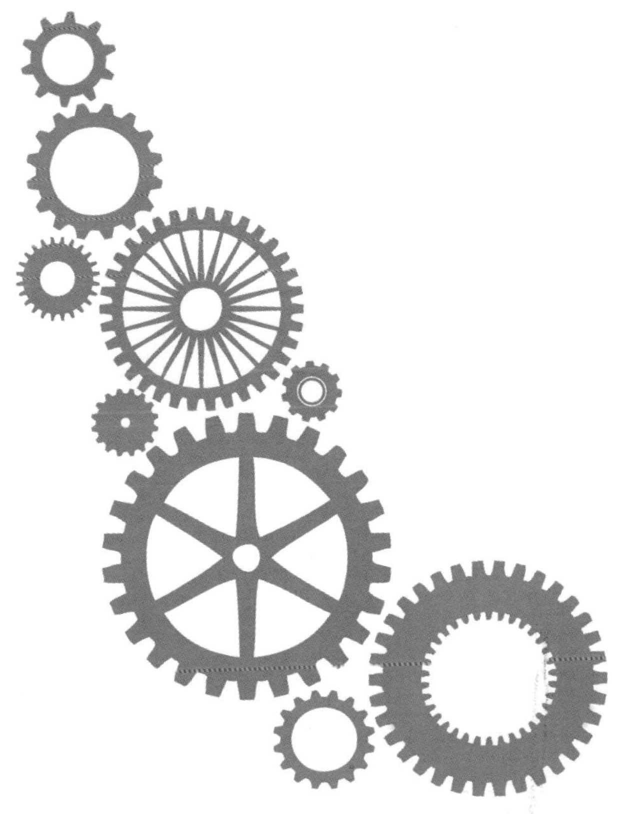

L. BOSTOCK
S. CHANDLER

OXFORD
UNIVERSITY PRESS

Great Clarendon Street, Oxford, OX2 6DP, United Kingdom

Oxford University Press is a department of the University of Oxford. It furthers the University's objective of excellence in research, scholarship, and education by publishing worldwide. Oxford is a registered trade mark of Oxford University Press in the UK and in certain other countries

First published by Stanley Thornes (Publishers) Ltd in 1990
Second edition published in 1994
Third edition published in 2000
Fourth edition published by Nelson Thornes Ltd in 2013
This edition published by Oxford University Press in 2015

British Library Cataloguing in Publication Data
Data available

978-1-4085-2228-8

10 9 8 7 6 5

Printed in India by Multivista Global Pvt. Ltd

Acknowledgements

Page make-up: Tech-Set Ltd

Although we have made every effort to trace and contact all copyright holders before publication this has not been possible in all cases. If notified, the publisher will rectify any errors or omissions at the earliest opportunity.

INTRODUCTION

This edition of *Core Maths for Advanced Level* covers the pure mathematics for AS and A Levels in Mathematics based on subject criteria. The material is dealt with in a progressive and logical way so that the knowledge and skills acquired for AS stage are used and developed in later topics for A2. Because this book is not limited to any specific scheme, it also covers material that some boards may not examine. Each specification needs to be looked at in order to identify topics that are optional. Each topic in a specification is mapped to the pages in this book. These maps are available to download for free at www.nelsonthornes.com\coremaths-alevel

As a starting point, the book assumes the minimum level of success on the national curriculum for access to A-level. Many of you will have reached a higher level, particularly in algebraic skills and so will find an overlap between some work in this book and what you already know. For you the early chapters provide useful revision but they also contain some work that you are unlikely to have covered. We suggest using the mixed exercises at the ends of these chapters to identify unfamiliar topics, so that you can, if you wish, restrict your study to these sections.

All too many students regard A-level mathematics as being intrinsically difficult. We strongly disagree with this opinion. Part of the reason for this myth may be that students, at an early stage in their course, tackle problems that are too sophisticated. The exercises in this book are designed to overcome this problem, all starting with straightforward questions. There are many A-level examination questions at regular intervals throughout the book. These extensive exercises are intended for use at a later date, to give practice in examination questions when confidence and sophistication have been developed. The summary sections also include a brief recap of the work in preceding chapters and a set of multiple choice questions, which are useful for self-testing even if they do not form part of the examination to be taken.

There are many computer programs that help with the understanding of mathematics. In particular, good graph drawing software is invaluable for investigating graphical aspects of functions. Graphics calculators are also invaluable. However, as their use is forbidden in certain examination papers, you should not get into the habit of relying on them too much.

Another very valuable aid, particularly for investigating sequences and analysing data, is a computer spreadsheet. In a few places we have indicated where such aids can be used effectively, but this should be regarded as a minimum indication of the possible use of technology.

We would like to thank Alison Gee for her thorough work in checking the book.

We are grateful to the following examination boards for giving their permission to reproduce questions from past examination papers.

London Examinations, a division of Edexcel Foundation (Edexcel)
Oxford, Cambridge and RSA Examinations (OCR)
Assessment and Qualifications Alliance (AQA)
Welsh Joint Education Committee (WJEC)

S. Chandler
2013

CONTENTS

CHAPTER 1

Algebra 1

Skill in manipulating algebraic expressions is essential in any mathematics course beyond GCSE and needs to be almost as instinctive as the ability to manipulate simple numbers. This and the next two chapters present the facts and provide practice necessary for the development of these skills.

Multiplication of Algebraic Expressions

The multiplication sign is usually omitted, so that, for example,

$2q$ means $2 \times q$

and $x \times y$ can be simplified to xy

Remember also that if a string of numbers and letters are multiplied, the multiplication can be done in any order, for example

$$2p \times 3q = 2 \times p \times 3 \times q$$
$$= 6pq$$

Powers can be used to simplify expressions such as $x \times x$,

i.e. $x \times x = x^2$

and $x \times x^2 = x \times x \times x = x^3$

But remember that a power refers only to the number or letter it is written above, for example

$2x^2$ means that x is squared, but 2 is not.

EXAMPLE 1A

Simplify **a** $(4pq)^2 \times 5$ **b** $\dfrac{ax^2}{y} \div \dfrac{x}{ay^2}$

a $(4pq)^2 \times 5 = 4pq \times 4pq \times 5$
$= 80p^2q^2$

b $\dfrac{ax^2}{y} \div \dfrac{x}{ay^2} = \dfrac{ax^{\not 2}}{\not y} \times \dfrac{ay^{\not 2}}{\not x}$
$= a^2xy$

EXERCISE 1A

Simplify

1 $3 \times 5x$

2 $x \times 2x$

3 $(2x)^2$

4 $5p \times 2q$

5 $4x \times 2x$

6 $2pq \times 5pr$

7 $(3a)^2$

8 $7u \times 9b$

9 $8t \times 3st$

10 $2a^2 \times 4a$

11 $25x^2 \div 15x$

12 $12m^2 \div 6m$

13 $b^2 \times 4ab$

14 $25x^2y \div 5x$

15 $(7pq)^2 \times (2p)^2$

16 $\dfrac{22ab}{11b}$

17 $\dfrac{18ax^2}{3x}$

18 $\dfrac{36xy}{18y}$

19 $\dfrac{72ab^2}{40a^2b}$

20 $\dfrac{2}{5} \div \dfrac{1}{x}$

21 $\dfrac{x^2}{y} \div \dfrac{y}{x}$

Addition and Subtraction of Expressions

The *terms* in an algebraic expression are the parts separated by a plus or minus sign.

Like terms contain the same combination of letters; like terms can be added or subtracted.

For example, $2ab$ and $5ab$ are like terms and can be added,

i.e. $\quad 2ab + 5ab = 7ab$

Unlike terms contain different combinations of letters; they cannot be added or subtracted. For example, ab and ac are unlike terms and $ab + ac$ cannot be simplified; similarly $x^2 + x^3$ cannot be simplified.

EXAMPLE 1B

Simplify $5x - 3(4 - x)$

$$5x - 3(4 - x) = 5x - 12 + 3x$$
$$= 8x - 12$$

Note that $-3(4 - x)$ means 'take away 3 times everything inside the bracket': remember that $(-3) \times (-x) = +3x$.

EXERCISE 1B

Simplify

1 $2x^2 - 4x + x^2$

2 $5a - 4(a + 3)$

3 $2y - y(x - y)$

4 $8pq - 9p^2 - 3pq$

5 $4xy - y(x - y)$

6 $x^3 - 2x^2 + x^2 - 4x + 5x + 7$

7 $t^2 - 4t + 3 - 2t^2 + 5t + 2$

8 $2(a^2 - b) - a(a + b)$

9 $3 - (x - 4)$
Note that $-(x - 4)$ means $-1(x - 4)$.

10 $5x - 2 - (x + 7)$

11 $3x(x + 2) + 4(3x - 5)$

12 $a(b - c) - c(a - b)$

13 $2cT(3 - T) + 5T(c - 11T)$

14 $x^2(x + 7) - 3x^3 + x(x^2 - 7)$

15 $(3y^2 + 4y - 2) - (7y^2 - 20y + 8)$

16 $6RS + 5RF - R(R + S)$

Coefficients

We can identify a particular term in an expression by using the letter, or combination of letters, involved,

for example $2x^2$ is 'the term in x^2',
$3xy$ is 'the term in xy'.

The number (including its sign) in front of the letters is called the *coefficient*,

for example in the term $2x^2$, 2 is the coefficient of x^2
in the term $3xy$, 3 is the coefficient of xy.

If no number is written in front of a term, the coefficient is 1 or -1, depending on the sign of the term.

Consider the expression $x^3 + 5x^2y - y^3$

the coefficient of x^3 is 1

the coefficient of x^2y is 5

the coefficient of y^3 is -1

There is no term in x^2, so the coefficient of x^2 is zero.

EXERCISE 1C

1 Write down the coefficient of x in $x^2 - 7x + 4$.

2 What is the coefficient of xy^2 in the expression $y^3 + 2xy^2 - 7xy$?

3 For the expression $x^2 - 5xy - y^2$ write down the coefficient of

a x^2 **b** xy **c** y^2

4 For the expression $x^3 - 3x + 7$ write down the coefficient of

a x^3 **b** x^2 **c** x

Expansion of Two Brackets

Expanding an expression means multiplying it out.

To expand $(2x + 4)(x - 3)$ each term in the first bracket is multiplied by each term in the second bracket. To make sure that nothing is missed out, it is sensible to follow the same order every time.

The order used in this book is:

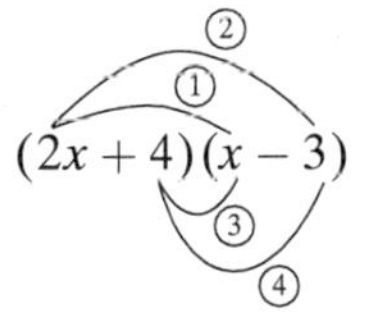

$$(2x + 4)(x - 3) = 2x^2 - 6x + 4x - 12$$
$$= 2x^2 - 2x - 12$$

Use the next exercise to practice expanding and to develop the confidence to go straight to the simplified form.

EXERCISE 1D

Expand and simplify

1 $(x + 2)(x + 4)$

2 $(x + 5)(x + 3)$

3 $(a + 6)(a + 7)$

4 $(t + 8)(t + 7)$

5 $(s + 6)(s + 11)$

6 $(2x + 1)(x + 5)$

7 $(5y + 3)(y + 5)$

8 $(2a + 3)(3a + 4)$

9 $(7t + 6)(5t + 8)$

10 $(11s + 3)(9s + 2)$

11 $(x - 3)(x - 2)$

12 $(y - 4)(y - 1)$

13 $(a - 3)(a - 8)$

14 $(b - 8)(b \quad 9)$

15 $(p - 3)(p - 12)$

16 $(2y - 3)((y - 5)$

17 $(x - 4)(3x - 1)$

18 $(2r - 7)(3r - 2)$

19 $(4x - 3)(5x - 1)$

20 $(2a - b)(3a - 2b)$

21 $(x - 3)(x + 2)$

22 $(a - 7)(a + 8)$

23 $(y + 9)(y - 7)$

24 $(s - 5)(s + 6)$

25 $(q - 5)(q + 13)$

26 $(2t - 5)(t + 4)$

27 $(x + 3)(4x - 1)$

28 $(2q + 3)(3q - 5)$

29 $(x + y)(x - 2y)$

30 $(s + 2t)(2s - 3t)$

Difference of Two Squares

Consider the expansion of $(x-4)(x+4)$,

$$\begin{aligned}(x-4)(x+4) &= x^2-4x+4x-16\\ &= x^2-16\end{aligned}$$

EXERCISE 1E Expand and simplify

1 $(x-2)(x+2)$

2 $(5-x)(5+x)$

3 $(x+3)(x-3)$

4 $(2x-1)(2x+1)$

5 $(x+8)(x-8)$

6 $(x-a)(x+a)$

Questions 1 to 6 show clearly that an expansion of the form $(ax+b)(ax-b)$ can be written down directly,

i.e $\qquad (ax+b)(ax-b) = a^2x^2-b^2$

Use this result to expand the following brackets.

7 $(x-1)(x+1)$

8 $(3b+4)(3b-4)$

9 $(2y-3)(2y+3)$

10 $(ab+6)(ab-6)$

11 $(5x+1)(5x-1)$

12 $(xy+4)(xy-4)$

Squares

$(2x+3)^2$ means $(2x+3)(2x+3)$

$$\begin{aligned}\therefore\qquad (2x+3)^2 &= (2x+3)(2x+3)\\ &= (2x)^2+(2)(2x)(3)+(3)^2\\ &= 4x^2+12x+9\end{aligned}$$

In general, $\quad (ax+b)^2 = a^2x^2+(2)(ax)(b)+b^2$

$$= a^2x^2+2abx+b^2$$

and $\qquad (ax-b)^2 = a^2x^2-2abx+b^2$

EXERCISE 1F Use the results above to expand

1 $(x+4)^2$

2 $(x+2)^2$

3 $(2x+1)^2$

4 $(3x+5)^2$

5 $(2x+7)^2$

6 $(x-1)^2$

7 $(x-3)^2$

8 $(2x-1)^2$

9 $(4x-3)^2$

10 $(5x-2)^2$

11 $(3t-7)^2$

12 $(x+y)^2$

13 $(2p+9)^2$

14 $(3q-11)^2$

15 $(2x-5y)^2$

Important Expansions

The results from the last two section should be memorised. They are summarised here.

$$(ax + b)^2 = a^2x^2 + 2abx + b^2$$
$$(ax - b)^2 = a^2x^2 - 2abx + b^2$$
$$(ax + b)(ax - b) = a^2x^2 - b^2$$

The next exercise contains a variety of expansions including some of the forms given above.

EXAMPLE 1G

Expand $(4p+5)(3-2p)$

$$\begin{aligned}(4p+5)(3-2p) &= (5+4p)(3-2p)\\ &= 15+2p-8p^2\end{aligned}$$

EXERCISE 1G

Expand

1 $(2x-3)(4-x)$

2 $(x-7)(x+7)$

3 $(6-x)(1-4x)$

4 $(7p+2)(2p-1)$

5 $(3p-1)^2$

6 $(5t+2)(3t-1)$

7 $(4-p)^2$

8 $(4t-1)(3-2t)$

9 $(x+2y)^2$

10 $(4x-3)(4x+3)$

11 $(3x+7)^2$

12 $(R+3)(5-2R)$

13 $(a-3b)^2$

14 $(2x-5)^2$

15 $(7a+2b)(7a-2b)$

16 $(3a+5b)^2$

17 Write down the coefficients of x^2 and x in the expansion of

a $(2x-4)(3x-5)$

b $(5x+2)(3x+5)$

c $(2x-3)(7x-5)$

d $(9x+1)^2$

Harder Expansions

Consider the product $(x-2)(x^2-x+5)$

Expansions like this should be done in a systematic way.

First multiply each term of the quadratic by x, writing down the separate results as they are found. Then multiply each term of the quadratic by -2. Do not attempt to simplify at this stage.

$$\begin{aligned}&(x-2)(x^2-x+5)\\ &\quad = x^3-x^2+5x-2x^2+2x-10\end{aligned}$$

Now simplify

$$= x^3-3x^2+7x-10$$

EXAMPLE 1H

Expand $(x+2)(2x-1)(x+4)$

First we expand the last two brackets.

$$\begin{aligned}(x+2)(2x-1)(x+4) &= (x+2)(2x^2+7x-4)\\ &= 2x^3+7x^2-4x+4x^2+14x-8\\ &= 2x^3+11x^2+10x-8\end{aligned}$$

EXERCISE 1H Expand and simplify

1 $(x-2)(x^2+x+1)$

2 $(3x-2)(x^2-x-1)$

3 $(2x-1)(2x^2-3x+5)$

4 $(x-1)(x^2-x-1)$

5 $(2x+3)(x^2-6x-3)$

6 $(x+1)(x+2)(x+3)$

7 $(x+4)(x-1)(x+1)$

8 $(x-2)(x-3)(x+1)$

9 $(x+1)(2x+1)(x+2)$

10 $(x+2)(x+1)^2$

11 $(2x-1)^2(x+2)$

12 $(3x-1)^3$

13 $(4x+3)(x+1)(x-4)$

14 $(x-1)(2x-1)(2x+1)$

15 $(2x+1)(x+2)(3x-1)$

16 $(x+1)^3$

17 $(x-2)(x+2)(x+1)$

18 $(x+3)(2x+3)(x-1)$

19 $(3x-2)(2x+5)(4x-1)$

20 $2(x-7)(2x+3)(x-5)$

21 Expand and simplify $(x-2)^2(3x-4)$. Write down the coefficients of x^2 and x.

22 Find the coefficients of x^3 and x^2 in the expansion of $(x-4)(2x+3)(3x-1)$.

23 Expand and simplify $(x+y)^3$.

24 Expand and simplify $(x+y)^4$.

Pascal's Triangle

We sometimes need to expand expressions such as $(x+y)^4$ but the multiplication is tedious when the power is three or more.

We now describe a far quicker way of obtaining such expansions.

Consider the following expansions,

$$(x+y)^1 = x+y$$
$$(x+y)^2 = x^2+2xy+y^2$$
$$(x+y)^3 = x^3+3x^2y+3xy^2+y^3$$
$$(x+y)^4 = x^4+4x^3y+6x^2y^2+4xy^3+y^4$$

The first thing to notice is that the powers of x and y in the terms of each expansion form a pattern. Looking at the expansion of $(x+y)^4$ we see that the first term is x^4 and then the power of x decreases by 1 in each succeeding term while the power of y increases by 1. For all the terms, the sum of the powers of x and y is 4 and the expansion ends with y^4. There is a similar pattern in the other expansions.

Now consider just the coefficients of the terms. Writing these as a triangular array gives

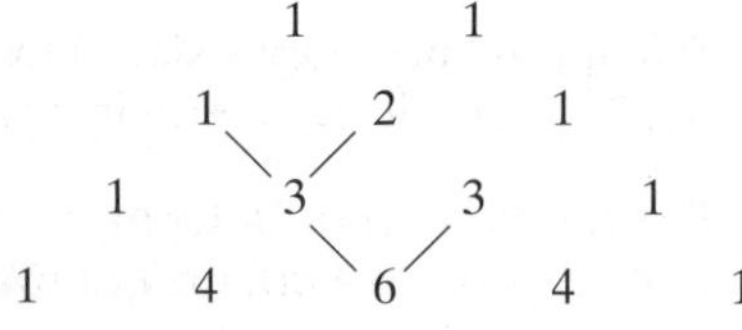

This array is called *Pascal's Triangle* and it has a pattern. Each row starts and ends with 1 and each other number is the sum of the two numbers in the row above it, as shown. When the pattern is known, Pascal's triangle can be written down to as many rows as needed. Using Pascal's triangle to expand $(x+y)^6$, for example, we go as far as row 6:

1 1

1 2 1

1 3 3 1

1 4 6 4 1

1 5 10 10 5 1

1 6 15 20 15 6 1

We then use our knowledge of the pattern of the powers, together with row 6 of the array, to fill in the coefficients,

i.e. $(x+y)^6 = x^6 + 6x^5y + 15x^4y^2 + 20x^3y^3 + 15x^2y^4 + 6xy^5 + y^6$

The following worked examples show how expansions of other brackets can be found.

EXAMPLES 1I

1 Expand $(x+5)^3$

From Pascal's triangle $(x+y)^3 = x^3 + 3x^2y + 3xy^2 + y^3$

Replacing y by 5 gives $(x+5)^3 = x^3 + 3x^2(5) + 3x(5)^2 + (5)^3$

$= x^3 + 15x^2 + 75x + 125$

2 Expand $(2x-3)^4$

From Pascal's triangle, $(x+y)^4 = x^4 + 4x^3y + 6x^2y^2 + 4xy^3 + y^4$

Replacing x by $2x$ and y by -3 gives

$(2x-3)^4 = (2x)^4 + 4(2x)^3(-3) + 6(2x)^2(-3)^2 + 4(2x)(-3)^3 + (-3)^4$

$= 16x^4 - 96x^3 + 216x^2 - 216x + 81$

EXERCISE 1I

Expand

1 $(x+3)^3$

2 $(x-2)^4$

3 $(x+1)^4$

4 $(2x+1)^3$

5 $(x-3)^5$

6 $(p-q)^4$

7 $(2x+3)^3$

8 $(x-4)^5$

9 $(3x-1)^4$

10 $(1+5a)^4$

11 $(2a-b)^6$

12 $(2x-5)^3$

Factorising Quadratic Expressions

An expression of the form $ax+b$, where a and b are numbers, is called a linear expression in x.

When two linear expressions in x are multiplied, the result usually contains three terms: a term in x^2, a term in x and a number.

Expressions of this form, i.e. ax^2+bx+c where a, b and c are numbers and $a \neq 0$, are called *quadratic expressions in* x.

Since the product of two linear brackets is quadratic, we might expect to be able to reverse this process. For instance, given a quadratic such as $x^2 - 5x + 6$, we could try to find two linear expressions in x whose product is $x^2 - 5x + 6$. To be able to do this we need to appreciate the relationship between what is inside the brackets and the resulting quadratic.

Consider the examples

$$(2x + 1)(x + 5) = 2x^2 + 11x + 5 \quad [1]$$

$$(3x - 2)(x - 4) = 3x^2 - 14x + 8 \quad [2]$$

$$(x - 5)(4x + 2) = 4x^2 - 18x - 10 \quad [3]$$

The first thing to notice about the quadratic in each example is that

the coefficient of x^2 is the product of the coefficients of x in the two brackets,

the number is the product of the numbers in the two brackets,

we get the coefficient of x by adding the coefficients formed by multiplying the x term in one bracket by the number term in the other bracket.

The next thing to notice is the relationship between the signs.

Positive signs throughout the quadratic come from positive signs in both brackets, as in [1].

A positive number term and a negative coefficient of x in the quadratic come from a negative sign in each bracket, as in [2].

A negative number term in the quadratic comes from a negative sign in one bracket and a positive sign in the other, as in [3].

EXAMPLES 1J

1 Factorise $x^2 - 5x + 6$.

> The x term in each bracket is x as x^2 can only be $x \times x$.
> The sign in each bracket is $-$, so $x^2 - 5x + 6 = (x - \quad)(x - \quad)$.
> The numbers in the brackets could be 6 and 1 or 2 and 3.
> Checking the middle term tells us that the numbers must be 2 and 3.

$x^2 - 5x + 6 = (x - 2)(x - 3)$

> Mentally expanding the brackets checks that they are correct.

2 Factorise $x^2 - 3x - 10$.

> The x term in each bracket is x so $x^2 - 3x - 10 = (x - \quad)(x + \quad)$.
> The numbers could be 10 and 1 or 5 and 2.
> Checking the middle term shows that they are 5 and 2.

$x^2 - 3x - 10 = (x - 5)(x + 2)$

> Mentally expanding the brackets confirms that they are correct.

EXERCISE 1J Factorise

1 $x^2 + 8x + 15$
2 $x^2 + 11x + 28$
3 $x^2 + 7x + 6$
4 $x^2 + 7x + 12$
5 $x^2 - 10x + 9$
6 $x^2 - 6x + 9$
7 $x^2 + 8x + 12$
8 $x^2 - 9x + 8$
9 $x^2 + 5x - 14$
10 $x^2 + x - 12$
11 $x^2 - 4x - 5$
12 $x^2 - 10x - 24$
13 $x^2 + 9x + 14$
14 $x^2 - 2x + 1$
15 $x^2 - 9$
16 $x^2 + 5x - 24$
17 $x^2 + 4x + 4$
18 $x^2 - 1$
19 $x^2 - 3x - 18$
20 $x^2 + 10x + 25$
21 $x^2 - 16$
22 $4 + 5x + x^2$
23 $2x^2 - 3x + 1$
24 $3x^2 + 4x + 1$
25 $9x^2 - 6x + 1$
26 $6x^2 - x - 1$
27 $9 + 6x + x^2$
28 $4x^2 - 9$
29 $x^2 + 2ax + a^2$
30 $x^2y^2 - 2xy + 1$

Harder Factorising

When the number of possible combinations of terms for the brackets increases, common sense considerations can help to reduce the possibilities.

For example, if the coefficient of x in the quadratic is odd, then there must be an even number in one bracket and an odd number in the other.

EXAMPLE 1K Factorise $12 - x - 6x^2$.

The x terms in the brackets could be $6x$ and x, or $3x$ and $2x$, one positive and the other negative.

The number terms could be 12 and 1 or 3 and 4 (not 6 and 2 because the coefficient of x in the quadratic is odd).

Now we try various combinations until we find the correct one.

$$12 - x - 6x^2 = (3 + 2x)(4 - 3x)$$

EXERCISE 1K Factorise

1 $6x^2 + x - 12$
2 $4x^2 - 11x + 6$
3 $4x^2 + 3x - 1$
4 $3x^2 - 17x + 10$
5 $4x^2 - 12x + 9$
6 $3 - 5x - 2x^2$
7 $25x^2 - 16$
8 $3 - 2x - x^2$
9 $5x^2 - 61x + 12$
10 $9x^2 + 30x + 25$
11 $3 + 2x - x^2$
12 $12 + 7x - 12x^2$
13 $1 - x^2$
14 $9x^2 + 12x + 4$
15 $x^2 + 2xy + y^2$
16 $1 - 4x^2$
17 $4x^2 - 4xy + y^2$
18 $9 - 4x^2$
19 $36 + 12x + x^2$
20 $40x^2 - 17x - 12$
21 $7x^2 - 5x - 150$
22 $36 - 25x^2$
23 $x^2 - y^2$
24 $81x^2 - 36xy + 4y^2$
25 $49 - 84x + 36x^2$
26 $25x^2 - 4y^2$
27 $36x^2 + 60xy + 25y^2$
28 $4x^2 - 4xy - 3y^2$
29 $6x^2 + 11xy + 4y^2$
30 $49p^2q^2 - 28pq + 4$

Common Factors

If the terms in a quadratic equation have a common factor it should be taken out first,

e.g. $4x^2 + 8x + 4 = 4(x^2 + 2x + 1)$

The quadratic inside the bracket now has smaller coefficients and can be factorised more easily:

$$\begin{aligned}4x^2 + 8x + 4 &= 4(x+1)(x+1)\\ &= 4(x+1)^2\end{aligned}$$

Not all quadratics factorise.

Consider $3x^2 - x + 5$.

The options we can try are $(3x-5)(x-1)$ [1]

$(3x-1)(x-5)$ [2]

From [1], $(3x-5)(x-1) = 3x^2 - 8x + 5$

From [2], $(3x-1)(x-5) = 3x^2 - 16x + 5$

As neither of the possible pairs of brackets expand to give $3x^2 - x + 5$, we conclude that $3x^2 - x + 5$ has no factors of the form $ax + b$ where a and b are integers.

EXAMPLE 1L Factorise $2x^2 - 8x + 16$.

$2x^2 - 8x + 16 = 2(x^2 - 4x + 8)$

> The possible brackets are $(x-1)(x-8)$ and $(x-2)(x-4)$.
> Neither pair expands to $x^2 - 4x + 8$, so there are no further factors.

EXERCISE 1L Factorise where possible

1 $x^2 + x + 1$

2 $2x^2 + 4x + 2$

3 $x^2 + 3x + 2$

4 $3x^2 + 12x - 15$

5 $x^2 + 4$

6 $x^2 - 4x - 6$

7 $x^2 + 3x + 1$

8 $2x^2 - 8x + 8$

9 $3x^2 - 3x - 6$

10 $2x^2 - 6x + 8$

11 $3x^2 - 6x - 24$

12 $x^2 - 4x - 12$

13 $x^2 + 1$

14 $4x^2 - 100$

15 $5x^2 - 25$

16 $7x^2 + x + 4$

17 $10x^2 - 39x - 36$

18 $x^2 + xy + y^2$

Factor Theorem

It is not easy to spot the factors of an expression such as $x^3 - 3x^2 + 7x - 10$. But we can say that if $(x - a)$ is a factor, then $x^3 - 3x^2 + 7x - 10$ can be expressed as $(x - a) \times$ (a quadratic),

i.e. $x^3 - 3x^2 + 7x - 10 = (x - a) \times \text{(a quadratic)}$ [1]

Substituting a for x in [1] we get

$$a^3 - 3a^2 + 7a - 10 = (0) \times (\text{quadratic})$$

i.e. $$a^3 - 3a^2 + 7a - 10 = 0$$

The same argument can be applied to any expression containing powers of x and it means that

if $(x - a)$ is a factor of an expression containing powers of x, when we substitute a for x in that expression the result is zero.

Conversely, if the result is not zero, $x - a$ is not a factor.

This result is called the *factor theorem* and we can use it to help find factors of cubic and higher order expressions.

Consider again $x^3 - 3x^2 + 7x - 10$.
The possible factors of this expression are $(x \pm 1)$, $(x \pm 2)$ and $(x \pm 5)$.
We can try each of these in turn using the factor theorem.

Try $(x - 1)$: substitute 1 for x in $x^3 - 3x^2 + 7x - 10$

$$\Rightarrow (1)^3 - 3(1)^2 + 7(1) - 10 = -5 \neq 0$$

so $(x - 1)$ is not a factor.

Try $(x + 1)$: substitute -1 for x in $x^3 - 3x^2 + 7x - 10$

$$\Rightarrow (-1)^3 - 3(-1)^2 + 7(-1) - 10 = -21 \neq 0$$

so $(x + 1)$ is not a factor.

Try $(x - 2)$: substitute 2 for x in $x^3 - 3x^2 + 7x - 10$

$$\Rightarrow (2)^3 - 3(2)^2 + 7(2) - 10 = 0$$

so $(x - 2)$ is a factor.

Now we know that $x^3 - 3x^2 + 7x - 10 = (x - 2)(ax^2 + bx + c)$

The values of a and c can be written down directly from observation;

$$x^3 - 3x^2 + 7x - 10 = (x - 2)(ax^2 + bx + c)$$

> x^3 comes from the product of x and ax^2, so $a = 1$,
> -10 comes from the product of -2 and c, so $c = 5$.

The value of b is found by checking the pairs of products that give $7x$ (or $-3x^2$)

i.e. $$x^3 - 3x^2 + 7x - 10 = (x - 2)(x^2 + bx + 5)$$

shows that $7x = 5x - 2bx$ so $b = -1$.

$\therefore$ $$x^3 - 3x^2 + 7x - 10 = (x - 2)(x^2 - x + 5)$$

The Factors of $a^3 - b^3$ and $a^3 + b^3$

$a^3 - b^3 = 0$ when $a = b$, so $a - b$ is a factor of $a^3 - b^3$

therefore

$$\mathbf{a^3 - b^3 = (a - b)(a^2 + ab + b^2)} \qquad [1]$$

$a^3 + b^3 = 0$ when $a = -b$, so $a + b$ is a factor of $a^3 + b^3$

therefore

$$\mathbf{a^3 + b^3 = (a + b)(a^2 - ab + b^2)} \qquad [2]$$

These two facts can be used to factorise similar forms,
e.g. $x^3 - 8 = x^3 - 2^3$, so replacing a by x and b by 2 in [1] gives
$x^3 - 8 = (x - 2)(x^2 + 2x + 4)$

EXAMPLES 1M

1 Find the value of a for which $2x - 1$ is a factor of $4x^3 - 2x^2 + ax - 4$.

> Using the factor theorem we know that the value of the expression is zero when $x = \frac{1}{2}$ (the value of x for which $2x - 1 = 0$).

As $2x - 1$ is a factor of $4x^3 - 2x^2 + ax - 4$,

$4(\frac{1}{2})^3 - 2(\frac{1}{2})^2 + a(\frac{1}{2}) - 4 = 0$

i.e. $\frac{1}{2} - \frac{1}{2} + \frac{1}{2}a - 4 = 0 \quad \therefore \quad a = 8$

2 Factorise $x^4 - 16$.

> Using the factor theorem will find factors if they exist, but it is not always the quickest method. Look for forms that can be recognised. In this case we see that $x^4 - 16$ is the difference of two squares.

$$x^4 - 16 = (x^2 - 4)(x^2 + 4)$$
$$= (x - 2)(x + 2)(x^2 + 4)$$

EXERCISE 1M

1 Find whether $x - 1$ is a factor of $x^3 - 7x + 6$.

2 Is $x + 1$ a factor of $x^3 - 2x^2 + 1$?

3 Show that $2x - 1$ is a factor of $2x^4 - x^3 + 6x^2 - x - 1$.

4 Determine whether $x - 3$ and/or $2x + 1$ are factors of $4x^3 - 7x + 9$.

5 Show that $x - 3$ is a factor of $x^3 - 7x - 6$.

6 Factorise fully

a $x^3 + 2x^2 - x - 2$

b $x^3 - x^2 - x - 2$

c $2x^3 - x^2 + 2x - 1$

d $x^4 - 81$

e $x^3 + 27$

f $x^4 + x^3 - 3x^2 - 4x - 4$

7 $x-4$ is a factor of $x^3-ax+16$. Find a.

8 Find one factor of $2x^3+x^2+9x-5$. Hence express $2x^3+x^2+9x-5$ in the form $(ax+b)(px^2+qx+r)$ giving the values of a, b, p, q and r.

9 $(x+1)$ and $(x+2)$ are both factors of $2x^3+bx^2-5x+c$. Find the values of b and c.

10 x^3-4x^2-25 has a factor $(x-a)$. Find the value of a.

Simplification of Fractions

The value of a fraction is unaltered if we multiply or divide both the *numerator* and the *denominator* by the same number,

e.g. $$\frac{3}{6}=\frac{1}{2}=\frac{2}{4}=\frac{7}{14}=\dots$$

and $$\frac{ax}{ay}=\frac{x}{y}=\frac{3x}{3y}=\frac{x(a+b)}{y(a+b)}=\dots$$

A fraction can be simplified by multiplying or dividing *top* and *bottom* by a factor which is common to both.

Sometimes the numerator and/or the denominator of a fraction themselves contain fractions. If they do, get rid of these fractions first, then factorise the numerator and denominator, remembering to look for common factors.

EXAMPLES 1N

1 Simplify $\dfrac{2a^2-2ab}{6ab-6b^2}$.

First factorise top and bottom.

$$\frac{2a^2-2ab}{6ab-6b^2}=\frac{2a(a-b)}{{}_{3}6b(a-b)}$$

$$=\frac{a}{3b}$$

2 Simplify $\dfrac{\frac{1}{2}x^2-2}{\frac{1}{4}y^2+3}$.

Remove fractions from top and bottom.

$$\frac{\frac{1}{2}x^2-2}{\frac{1}{4}y^2+3}=\frac{2x^2-8}{y^2+12}$$

$$=\frac{2(x^2-4)}{y^2+12}$$

$$=\frac{2(x-2)(x+2)}{y^2+12}$$

EXERCISE 1N Simplify where possible

1 $\dfrac{x-2}{4x-8}$

2 $\dfrac{2x+4}{3x-6}$

3 $\dfrac{2a+8}{3a+12}$

4 $\dfrac{3p-3q}{5p-5q}$

5 $\dfrac{x^2+xy}{xy+y^2}$

6 $\dfrac{x-3p}{2x+p}$

7 $\dfrac{a-4}{a-2}$

8 $\dfrac{x^2y+xy^2}{y^2+\frac{2}{5}xy}$

9 $\dfrac{\frac{1}{3}a-b}{a+\frac{1}{6}b}$

10 $\dfrac{2x(b-4)}{6x^2(b+4)}$

11 $\dfrac{(x-4)(x-3)}{x^2-16}$

12 $\dfrac{4y^2+3}{y^2-9}$

13 $\dfrac{\frac{1}{3}(x-3)}{x^2-9}$

14 $\dfrac{x^2-x-6}{2x^2-5x-3}$

15 $\dfrac{(x-2)(x+2)}{x^2+x-2}$

16 $\dfrac{\frac{1}{2}(a+5)}{a^2-25}$

17 $\dfrac{3p+9q}{p^2+6pq+9q^2}$

18 $\dfrac{a^2+2a+4}{a^2+7a+10}$

19 $\dfrac{x^2+2x+1}{3x^2+12x+9}$

20 $\dfrac{4(x-3)^2}{(x+1)(x^2-2x-3)}$

Multiplication and Division

Fractions are multiplied by taking the product of the numerators and the product of the denominators,

e.g. $$\frac{x}{a}\times\frac{y}{b}=\frac{x\times y}{a\times b}=\frac{xy}{ab}$$

To divide by a fraction, we multiply by the reciprocal of that fraction, for example

$$\frac{x}{a}\div\frac{y}{b}=\frac{x}{a}\times\frac{b}{y}=\frac{xb}{ay}$$

EXAMPLE 1P Simplify $\dfrac{2\pi x^2}{7y^2}\div 4\pi x$.

> Think of $4\pi x$ as $\dfrac{4\pi x}{1}$

$$\frac{2\pi x^2}{7y^2}\div 4\pi x=\frac{\cancel{2}\cancel{\pi}x^{\cancel{2}}}{7y^2}\times\frac{1}{{}_2\cancel{4}\cancel{\pi}\cancel{x}}$$

$$=\frac{x}{14y^2}$$

EXERCISE 1P Simplify

1 $\dfrac{4x}{y} \times \dfrac{x}{6y}$

2 $2st \times \dfrac{3t}{s^2}$

3 $\dfrac{4uv}{3} \div \dfrac{u}{2v}$

4 $\dfrac{4\pi r^2}{3} \div 2\pi r$

5 $x(2-x) \div \dfrac{2-x}{3}$

6 $\dfrac{3x^2}{2y} \Big/ \dfrac{6xy}{9}$

7 $\dfrac{\pi x^3}{3} \div 8\pi x$

8 $\dfrac{1}{a^2+ab} \Big/ \dfrac{1}{2}$

9 $\dfrac{1}{x^2-1} \div \dfrac{1}{x-1}$

10 $\left(\dfrac{1}{a}\right)^2 \times \tfrac{1}{2}a$

11 $(x+1) \times \dfrac{1}{x^2-1}$

12 $\dfrac{x-4}{x+3} \div \dfrac{2(x-4)}{3}$

13 $\dfrac{a^2}{3} \times \left(\dfrac{a}{3}\right)^2$

14 $\dfrac{x^2}{6} \Big/ \left(\dfrac{x}{2}\right)^2$

15 $\dfrac{2r^3}{3} \times \left(\dfrac{1}{rs}\right)^2$

16 $\dfrac{3x^2}{2y} \times \dfrac{y}{y-2}$

17 $\dfrac{ab}{c} \div \dfrac{ac}{b}$

18 $\dfrac{x^2+4x+3}{5} \times \dfrac{10}{x+1}$

19 $\dfrac{x^2+x-12}{3} \Big/ (x+4)$

20 $\dfrac{4x^2-9}{(x-1)^2} \div \dfrac{2x+3}{x(x-1)}$

Addition and Subtraction of Fractions

Before fractions can be added or subtracted, they must be expressed with the same denominator, i.e. we have to find a common denominator. Then the numerators can be added or subtracted,

e.g. $$\frac{2}{p}+\frac{3}{q}=\frac{2q}{pq}+\frac{3p}{pq}=\frac{2q+3p}{pq}$$

EXAMPLE 1Q Simplify $x-\dfrac{1}{x}$.

$$x-\frac{1}{x}=\frac{x}{1}-\frac{1}{x}=\frac{x^2}{x}-\frac{1}{x}=\frac{x^2-1}{x}$$
$$=\frac{(x-1)(x+1)}{x}$$

EXERCISE 1Q Simplify

1 $\dfrac{1}{a}-\dfrac{1}{b}$

2 $\dfrac{1}{3x}+\dfrac{1}{5x}$

3 $\dfrac{1}{p}-\dfrac{1}{q}$

4 $\dfrac{1}{2x}+\dfrac{3}{5x}$

5 $x+\dfrac{1}{x}$

6 $\dfrac{x}{y}-\dfrac{y}{x}$

7 $2p-\dfrac{1}{p}$

8 $\dfrac{x}{3}+\dfrac{x+1}{4}$

9 $\tfrac{1}{2}(x-1)+\tfrac{1}{3}(x+1)$

10 $\frac{x+2}{5}-\frac{2x-1}{3}$

11 $\frac{1}{\sin A}+\frac{1}{\sin B}$

12 $\frac{1}{\cos A}+\frac{1}{\sin A}$

13 $3x+\frac{1}{4x}$

14 $x-\frac{2}{2x+1}$

15 $x+1+\frac{1}{x+1}$

16 $1+\frac{1}{x}+\frac{1}{2x}$

17 $1-x+\frac{1}{x}$

18 $\frac{1}{n}+\frac{1}{n^2}$

19 $\frac{x}{a^2}+\frac{x}{b^2}$

20 $1+\frac{1}{a}+\frac{1}{a+1}$

EXAMPLE 1R Simplify $\frac{2}{x+2}-\frac{x-4}{2x^2+x-6}$

$$\begin{aligned}\frac{2}{x+2}-\frac{x-4}{2x^2+x-6} &= \frac{2}{x+2}-\frac{x-4}{(x+2)(2x-3)}\\ &= \frac{2(2x-3)}{(x+2)(2x-3)}-\frac{x-4}{(x+2)(2x-3)}\\ &= \frac{2(2x-3)-(x-4)}{(x+2)(2x-3)}\\ &= \frac{4x-6-x+4}{(x+2)(2x-3)}\\ &= \frac{3x-2}{(x+2)(2x-3)}\end{aligned}$$

EXERCISE 1R Simplify

1 $\frac{1}{x+1}+\frac{1}{x-1}$

2 $\frac{1}{x+1}+\frac{1}{x-2}$

3 $\frac{4}{x+2}+\frac{3}{x+3}$

4 $\frac{1}{x^2-1}+\frac{1}{x+1}$

5 $\frac{2}{a^2-1}-\frac{3}{a-1}$

6 $\frac{1}{x^2+2x+1}+\frac{1}{x+1}$

7 $\frac{3}{4x^2+4x+1}-\frac{2}{2x+1}$

8 $\frac{2}{x^2+5x+4}-\frac{3}{x+1}$

9 $\frac{4}{(x+1)^2}+\frac{2}{x+1}$

10 $\frac{3}{(x+2)^2}-\frac{1}{x+4}$

11 $\dfrac{1}{2(x-1)}+\dfrac{2}{3(x+4)}$

12 $\dfrac{7}{5(x+2)}-\dfrac{2}{x+4}$

13 $\dfrac{4}{3(x+2)}-\dfrac{3}{2(3x-5)}$

14 $\dfrac{3}{x+1}-\dfrac{2}{x-2}+\dfrac{4}{x+3}$

15 $\dfrac{1}{x+1}-\dfrac{2}{x+2}+\dfrac{3}{x+3}$

16 $\dfrac{x+2}{(x+1)^2}-\dfrac{1}{x}$

17 $\dfrac{4t}{t^2+2t+1}+\dfrac{3}{t+1}$

18 $\dfrac{2t}{t^2+1}-\dfrac{t^2+1}{t^2-1}$

19 $\dfrac{1}{y^2-x^2}+\dfrac{3}{y+x}$

20 $1+\dfrac{1}{n}+\dfrac{1}{n+1}+\dfrac{1}{n+2}$

Algebraic Division

A fraction where both the numerator and the denominator are algebraic expressions, is *proper* if the highest power of x in the numerator is less than the highest power of x in the denominator, e.g. $\dfrac{x+1}{x^2+x+2}$.

If this is not the case the fraction is *improper* and we can divide the top by the bottom using long division.

The following example shows how you divide $x+2$ into x^2+3x-7.

(x^2+3x-7 is called the *dividend* and $x+2$ is called the *divisor*.)

$$\begin{array}{r|l} x\ +1 & \\ x+2\,\overline{)x^2+3x-7} & \text{Start by dividing } x \text{ into } x^2\text{; it goes } x \text{ times.} \\ x^2+2x & \text{Multiply } x+2 \text{ by } x \text{ and then subtract this from } x^2+3x. \\ \hline x-7 & \text{Bring down the } -7\text{, divide } x \text{ into } x \text{ and repeat the process.} \\ x+2 & \\ \hline -9 & \text{No more division by } x \text{ can be done, so we stop here.} \end{array}$$

$x+1$ is called the *quotient* and the *remainder* is -9.

This compares with dividing, say, 32 by 5, where the quotient is 6 and the remainder is 2.

But writing $\frac{32}{5}$ as a mixed number gives $6+\frac{2}{5}$.

Similarly $\dfrac{x^2+3x-7}{x+2}\equiv x+1-\dfrac{9}{x+2}$

where $x+1$ is a linear function and $\dfrac{9}{x+2}$ is a proper fraction.

All the examples above are very simple. You may meet a long division where the dividend has a 'missing' term, e.g. $2x^3-x+5$ has no term in x^2. In this case you should put in a term $0x^2$ (or at least leave a space where that term should be). By doing this you ensure that the terms always line up.

EXAMPLE 1S Divide $2x^3 - x + 5$ by $x + 3$

$$\begin{array}{r l}
2x^2 - 6x + 17 & \\
x + 3\overline{)2x^3 + 0x^2 - x + 5} & x \text{ into } 2x^3 \text{ goes } 2x^2 \text{ times.} \\
2x^3 + 6x^2 & \text{Multiply } 2x^2 \text{ by } x + 3 \text{ and subtract.} \\
-6x^2 - x & \text{Bring down } -x. \\
-6x^2 - 18x & x \text{ into } -6x^2 \text{ goes } -6x \text{ times.} \\
17x + 5 & \\
17x + 51 & \\
-46 &
\end{array}$$

The quotient is $2x^2 - 6x + 17$ and the remainder is -46.

EXERCISE 1S Carry out each of the following divisions, giving the quotient and the remainder.

1 $(2x^2 + 5x - 3) \div (x + 2)$

2 $(x^2 - x + 4) \div (x + 1)$

3 $(4x^3 + x - 1) \div (2x - 1)$

4 $(2x^3 - x^2 + 2) \div (x - 2)$

MIXED EXERCISE 1

1 Find the coefficient of x in the expansion of $(3x - 7)(5x + 4)$.

2 Expand $5(3x - 2)(3 - 7x)$.

3 Write down the coefficient of y^2 in the expansion of $(2y + 9)^3$.

4 Factorise $3x^2 - 9x + 6$.

5 Write down the coefficient of x^3 in the expansion of $(x - 5)^5$.

6 Factorise $4x^2 - 36$.

7 Expand $x(2x - 1)^2$.

8 Find the factors of $25 + x^2 - 10x$.

9 Find the coefficient of x in the expansion of $(x - 4)(3 - x)$.

10 Expand $(3a - 2b)(5a + 3b)$.

11 Expand $(x - 2)(x^2 - 5x + 3)$.

12 Expand $(2x + 1)(x + 3)(3x - 2)$.

13 $(2x - 3)(x^2 + ax - 2) = 2x^3 - 13x^2 + 11x + 6$. Find the value of a.

14 Show that $x - 3$ is a factor of $2x^3 - 9x^2 + 10x - 3$.

15 Factorise $2x^3 - x^2 - 2x + 1$.

16 Show that $x^3 - 2x^2 - 9x + 4 = (x - 4)(ax^2 + bx + c)$ and find the values of a, b and c

17 Find the value of k given that $x - 2$ is a factor of $x^3 - x^2 + kx + 8$.

18 Factorise $x^4 - 6x^3 + 7x^2 + 6x - 8$.

19 Factorise

a $x^3 - 1$ **b** $x^3 + y^3$ **c** $8x^3 + 27$

20 Given that $(x - 1)$ and $(x + 2)$ are factors of $x^3 + ax^2 + bx - 6$, find the values of a and b.

21 Simplify

a $\dfrac{x^2 - 9}{2x - 6}$ **b** $\dfrac{1}{x^2 - 9} \div \dfrac{1}{x + 3}$

22 Simplify

a $\left(\dfrac{2p}{r}\right)^2 \times \dfrac{ar}{p^3}$ **b** $\dfrac{2p}{r} - \dfrac{3}{p}$

23 Simplify

a $\dfrac{2n - 4}{3} \div (n^2 - 4)$

b $\dfrac{1}{x + 1} + \dfrac{1}{2x - 1} + \dfrac{1}{x}$

24 Simplify

a $\dfrac{4x^2 - 25}{4x^2 + 20x + 25}$

b $\dfrac{2t}{t^2 + 1} \div \dfrac{t^2 - 1}{t^2 + 1}$

25 Simplify

a $\left(\dfrac{x - 1}{x + 1}\right)^2 \times (x^2 - 1)$

b $\dfrac{1}{a} + \dfrac{1}{b} + \dfrac{1}{c}$

26 Divide $2x^2 + 3x + 4$ by $x - 2$, giving the quotient and the remainder.

27 Express $\dfrac{3x^2 - 5x + 1}{x + 3}$ as a linear expression plus a proper fraction.

28 Divide $x^3 - 4x^2 + 5$ by $x - 1$, giving the quotient and the remainder.

Hence express $\dfrac{x^3 - 4x^2 + 5}{x - 1}$ as a quadratic expression plus a proper fraction.

CHAPTER 2

Surds, indices and logarithms

Square Roots

When we express a number as the product of two equal factors, that factor is called the *square root* of the number, for example

$$4 = 2 \times 2 \quad \Rightarrow \quad 2 \text{ is the square root of } 4.$$

This is written $2 = \sqrt{4}$

Now -2 is also a square root of 4, as $4 = -2 \times -2$ but we do *not* write $\sqrt{4} = -2$.

The symbol $\sqrt{\ }$ is used *only for the positive square root*.
So, although $x^2 = 4 \quad \Rightarrow \quad x = \pm 2$, the only value of $\sqrt{4}$ is 2.

The negative square root of 4 would be written as $-\sqrt{4}$ and, when both square roots are wanted, we write $\pm\sqrt{4}$.

Cube Roots

When a number can be expressed as the product of three equal factors, that factor is called the *cube root* of the number.

e.g. $\quad 27 = 3 \times 3 \times 3 \quad$ so 3 is the cube root of 27.

This is written $\sqrt[3]{27} = 3$.

Other Roots

The notation used for square and cube roots can be extended to represent fourth roots, fifth roots, etc,

e.g. $\quad 16 = 2 \times 2 \times 2 \times 2 \quad \Rightarrow \quad \sqrt[4]{16} = 2$

and $\quad 243 = 3 \times 3 \times 3 \times 3 \times 3 \quad \Rightarrow \quad \sqrt[5]{243} = 3$

In general, if a number, n, can be expressed as the product of p equal factors then each factor is called the pth root of n and is written $\sqrt[p]{n}$.

Rational and Irrational Numbers

A number which is either an integer, or a fraction whose numerator and denominator are both integers, is called a *rational number*.

The square roots of certain numbers are rational.

e.g. $\quad \sqrt{9} = 3, \quad \sqrt{25} = 5, \quad \sqrt{\frac{4}{49}} = \frac{2}{7}$

This is not true of all square roots however, $\sqrt{2}$, $\sqrt{5}$ and $\sqrt{11}$, for example, are not rational numbers. Such square roots can be given to as many decimal places as are required, for example

$$\sqrt{3} = 1.73 \qquad \text{correct to 2 d.p.}$$
$$\sqrt{3} = 1.732\,05 \qquad \text{correct to 5 d.p.}$$

but they can never be expressed exactly as a decimal. They are called *irrational numbers*, and cannot be written as $\dfrac{a}{b}$ where a and b are integers.

The only way to give an exact answer when such irrational numbers are involved is to leave them in the form $\sqrt{2}$, $\sqrt{7}$ etc; in this form they are called *surds*. At this level of mathematics *answers should be given exactly unless an approximate answer is asked for*, e.g. give your answer correct to 3 s.f.

Surds arise in many topics and you need to be able to manipulate them.

Simplifying Surds

Consider $\sqrt{18}$.

One of the factors of 18 is 9, and 9 has an exact square root,

i.e. $\sqrt{18} = \sqrt{(9 \times 2)} = \sqrt{9} \times \sqrt{2}$

But $\sqrt{9} = 3$, therefore $\sqrt{18} = 3\sqrt{2}$

$3\sqrt{2}$ is the simplest possible surd form for $\sqrt{18}$.

Similarly $\sqrt{\dfrac{2}{25}} = \dfrac{\sqrt{2}}{\sqrt{25}} = \dfrac{\sqrt{2}}{5}$

EXERCISE 2A

Express in terms of the simplest possible surd.

1 $\sqrt{12}$ **4** $\sqrt{50}$ **7** $\sqrt{162}$ **10** $\sqrt{48}$

2 $\sqrt{32}$ **5** $\sqrt{200}$ **8** $\sqrt{288}$ **11** $\sqrt{500}$

3 $\sqrt{27}$ **6** $\sqrt{72}$ **9** $\sqrt{75}$ **12** $\sqrt{20}$

Multiplying Surds

Consider $(4 - \sqrt{5})(3 + \sqrt{2})$.

The multiplication is carried out in the same way and order as when multiplying two linear brackets.

i.e
$$\begin{aligned}(4 - \sqrt{5})(3 + \sqrt{2}) &= (4)(3) + (4)(\sqrt{2}) - (3)(\sqrt{5}) - (\sqrt{5})(\sqrt{2}) \\ &= 12 + 4\sqrt{2} - 3\sqrt{5} - \sqrt{5}\sqrt{2} \\ &= 12 + 4\sqrt{2} - 3\sqrt{5} - \sqrt{10}\end{aligned}$$

In this example there are no like terms to collect but when the same surd occurs in each bracket the expansion can be simplified.

EXAMPLES 2B

1 Expand and simplify $(2+2\sqrt{7})(5-\sqrt{7})$.

$$\begin{aligned}(2+2\sqrt{7})(5-\sqrt{7}) &= (2)(5)-(2)(\sqrt{7})+(5)(2\sqrt{7})-(2\sqrt{7})(\sqrt{7})\\ &= 10-2\sqrt{7}+10\sqrt{7}-14\\ &= 8\sqrt{7}-4\end{aligned}$$

2 Expand and simplify $(4-\sqrt{3})(4+\sqrt{3})$.

$$\begin{aligned}(4-\sqrt{3})(4+\sqrt{3}) &= 4^2-(\sqrt{3})^2\\ &= 16-3 = 13\end{aligned}$$

The example above is a special case because the result is a rational number. This is because the two given brackets are of the form $(x-a)(x+a)$, i.e. the factors of x^2-a^2.

The product of any two brackets of the type $(p-\sqrt{q})(p+\sqrt{q})$ is, similarly, $p^2-(\sqrt{q})^2 = p^2-q$, which is always rational.

This property has an important application later on.

EXERCISE 2B

Expand and simplify where this is possible.

1 $\sqrt{3}(2-\sqrt{3})$

2 $\sqrt{2}(5+4\sqrt{2})$

3 $\sqrt{5}(2+\sqrt{75})$

4 $\sqrt{2}(\sqrt{32}-\sqrt{8})$

5 $(\sqrt{3}+1)(\sqrt{2}-1)$

6 $(\sqrt{3}+2)(\sqrt{3}+5)$

7 $(\sqrt{5}-1)(\sqrt{5}+1)$

8 $(2\sqrt{2}-1)(\sqrt{2}-1)$

9 $(\sqrt{5}-3)(2\sqrt{5}-4)$

10 $(4+\sqrt{7})(4-\sqrt{7})$

11 $(\sqrt{6}-2)^2$

12 $(2+3\sqrt{3})^2$

Multiply by a bracket which will make the product rational.

13 $(4-\sqrt{5})$

14 $(\sqrt{11}+3)$

15 $(2\sqrt{3}-4)$

16 $(\sqrt{6}-\sqrt{5})$

17 $(3-2\sqrt{3})$

18 $(2\sqrt{5}-\sqrt{2})$

Rationalising a Denominator

A fraction whose denominator contains a surd is more awkward to deal with than one where a surd occurs only in the numerator.

There is a technique for transferring the surd expression from the denominator to the numerator; it is called *rationalising the denominator* (i.e. making the denominator into a rational number).

EXAMPLES 2C

1 Rationalise the denominator of $\dfrac{2}{\sqrt{3}}$.

> The square root in the denominator can be removed if we multiply it by another $\sqrt{3}$. If this is done we must also multiply the numerator by $\sqrt{3}$, otherwise the value of the fraction is changed.

$$\frac{2}{\sqrt{3}} = \frac{2\sqrt{3}}{(\sqrt{3})(\sqrt{3})} = \frac{2\sqrt{3}}{3}$$

2 Rationalise the denominator and simplify $\dfrac{3\sqrt{2}}{5-\sqrt{2}}$.

> We saw in Example 2b number 2, that a product of the type $(a-\sqrt{b})(a+\sqrt{b})$ is wholly rational so in this question we multiply numerator and denominator by $5+\sqrt{2}$

$$\frac{3\sqrt{2}}{5-\sqrt{2}} = \frac{3\sqrt{2}(5+\sqrt{2})}{(5-\sqrt{2})(5+\sqrt{2})}$$

$$= \frac{15\sqrt{2}+3(\sqrt{2})(\sqrt{2})}{25-(\sqrt{2})(\sqrt{2})}$$

$$= \frac{15\sqrt{2}+6}{23}$$

EXERCISE 2C

Rationalise the denominator, simplifying where possible.

1 $\dfrac{3}{\sqrt{2}}$

2 $\dfrac{1}{\sqrt{7}}$

3 $\dfrac{2}{\sqrt{11}}$

4 $\dfrac{3\sqrt{2}}{\sqrt{5}}$

5 $\dfrac{1}{\sqrt{27}}$

6 $\dfrac{\sqrt{5}}{\sqrt{10}}$

7 $\dfrac{1}{\sqrt{2}-1}$

8 $\dfrac{3\sqrt{2}}{5+\sqrt{2}}$

9 $\dfrac{2}{2\sqrt{3}-3}$

10 $\dfrac{5}{2-\sqrt{5}}$

11 $\dfrac{1}{\sqrt{7}-\sqrt{3}}$

12 $\dfrac{4\sqrt{3}}{2\sqrt{3}-3}$

13 $\dfrac{3-\sqrt{5}}{\sqrt{5}+1}$

14 $\dfrac{2\sqrt{3}-1}{4-\sqrt{3}}$

15 $\dfrac{\sqrt{5}-1}{\sqrt{5}-2}$

16 $\dfrac{3}{\sqrt{3}-\sqrt{2}}$

17 $\dfrac{3\sqrt{5}}{2\sqrt{5}+1}$

18 $\dfrac{\sqrt{2}+1}{\sqrt{2}-1}$

19 $\dfrac{2\sqrt{7}}{\sqrt{7}+2}$

20 $\dfrac{\sqrt{5}-1}{3-\sqrt{5}}$

21 $\dfrac{1}{\sqrt{11}-\sqrt{7}}$

22 $\dfrac{4-\sqrt{3}}{3-\sqrt{3}}$

23 $\dfrac{1-3\sqrt{2}}{3\sqrt{2}+2}$

24 $\dfrac{1}{3\sqrt{2}-2\sqrt{3}}$

25 $\dfrac{\sqrt{3}}{\sqrt{2}(\sqrt{6}-\sqrt{3})}$

26 $\dfrac{1}{\sqrt{3}(\sqrt{21}+\sqrt{7})}$

27 $\dfrac{\sqrt{2}}{\sqrt{3}(\sqrt{5}-\sqrt{2})}$

Indices

Base and Index

In an expression such as 3^4, the *base* is 3 and the 4 is called the *power* or *index* (the plural is *indices*).

Working with indices involves using some properties which apply to any base, so we express these laws in terms of a general base a (i.e. a stands for any number).

Law 1 Because a^3 means $a \times a \times a$ and a^2 means $a \times a$ it follows that

$$a^3 \times a^2 = (a \times a \times a) \times (a \times a) = a^5$$

i.e. $a^3 \times a^2 = a^{3+2}$

Similar examples with different powers all indicate the general law that

$$\mathbf{a^p \times a^q = a^{p+q}}$$

Law 2 Now dealing with division we have

$$a^7 \div a^4 = \frac{\not{a} \times \not{a} \times \not{a} \times \not{a} \times a \times a \times a}{\not{a} \times \not{a} \times \not{a} \times \not{a}} = a^3$$

i.e. $a^7 \div a^4 = a^{7-4}$

Again this is just one example of the general law

$$\mathbf{a^p \div a^q = a^{p-q}}$$

When this law is applied to certain fractions some interesting cases arise.

Consider $a^3 \div a^5$

$$\frac{a^3}{a^5} = \frac{\not{a} \times \not{a} \times \not{a}}{\not{a} \times \not{a} \times \not{a} \times a \times a} = \frac{1}{a^2}$$

But from Law 2 we have $a^3 \div a^5 = a^{3-5} = a^{-2}$

Therefore a^{-2} means $\frac{1}{a^2}$.

In general $\mathbf{a^{-p} = \frac{1}{a^p}}$

i.e. **a^{-p} means 'the reciprocal of a^p'**

Now consider $a^4 \div a^4$.

$$\frac{a^4}{a^4} = \frac{\not{a} \times \not{a} \times \not{a} \times \not{a}}{\not{a} \times \not{a} \times \not{a} \times \not{a}} = 1$$

From Law 2, $\frac{a^4}{a^4} = a^{4-4} = a^0$

Therefore $\mathbf{a^0 = 1}$

i.e. **any base to the power zero is equal to 1.**

Law 3 $$(a^2)^3 = (a \times a)^3$$
$$= (a \times a) \times (a \times a) \times (a \times a)$$
$$= a^6$$

i.e. $(a^2)^3 = a^{2\times 3}$

In general $\mathbf{(a^p)^q = a^{pq}}$

Law 4 This law explains the meaning of a fractional index.

From the first law we have

$$a^{1/2} \times a^{1/2} = a^{1/2+1/2} = a^1 = a$$

i.e. $a = a^{1/2} \times a^{1/2}$

But $a = \sqrt{a} \times \sqrt{a}$

Therefore $a^{1/2}$ means $\sqrt{a}$, i.e. the positive square root of a.

Similarly $a^{1/3} \times a^{1/3} \times a^{1/3} = a^{1/3+1/3+1/3} = a^1 = a$

But $\sqrt[3]{a} \times \sqrt[3]{a} \times \sqrt[3]{a} = a$

Therefore $a^{1/3}$ means $\sqrt[3]{a}$, i.e. the cube root of a.

In general $\mathbf{a^{1/p} = \sqrt[p]{a}}$**, i.e. the pth root of a.**

For a more general fraction index, $\frac{p}{q}$, the third law shows that

$\mathbf{a^{p/q} = (a^p)^{1/q}}$ **or** $\mathbf{(a^{1/q})^p}$

For example

$$a^{3/4} = (a^3)^{1/4} \text{ or } (a^{1/4})^3$$
$$= \sqrt[4]{a^3} \text{ or } (\sqrt[4]{a})^3$$

i.e. $a^{3/4}$ represents either 'the fourth root of a^3' or 'the cube of the fourth root of a'.

All the general laws can be applied to simplify a wide range of expressions containing indices *provided that the terms all have the same base.*

EXAMPLES 2D

1 Simplify **a** $\dfrac{2^3 \times 2^7}{4^3}$ **b** $(x^2)^7 \times x^{-3}$ **c** $\sqrt[3]{a^4b^5} \times b^{1/3}/a$

a First we express all the terms to a base 2.

$$\frac{2^3 \times 2^7}{4^3} = \frac{2^3 \times 2^7}{(2^2)^3}$$
$$= \frac{2^{3+7}}{2^{2\times 3}} = \frac{2^{10}}{2^6} = 2^4$$

b $$(x^2)^7 \times x^{-3} = x^{2\times 7} \times \frac{1}{x^3}$$
$$= x^{14} \times \frac{1}{x^3} = x^{11}$$

c $$\sqrt[3]{a^4b^5} \times \frac{b^{1/3}}{a} = (a^{4/3})(b^{5/3})(b^{1/3})(a^{-1})$$
$$= (a^{4/3-1})(b^{5/3+1/3}) = a^{1/3}b^2$$

2 Evaluate **a** $(64)^{-1/3}$ **b** $\left(\frac{25}{9}\right)^{-3/2}$

a $$(64)^{-1/3} = \frac{1}{(64)^{1/3}} = \frac{1}{\sqrt[3]{64}} = \frac{1}{4}$$

b $$\left(\frac{25}{9}\right)^{-3/2} = \left(\frac{9}{25}\right)^{3/2} = \left(\sqrt{\frac{9}{25}}\right)^3 = \left(\frac{3}{5}\right)^3 = \frac{27}{125}$$

Note that $\left(\frac{9}{25}\right)^{3/2}$ could have been expressed as $\sqrt{\left(\frac{9}{25}\right)^3}$ but this form involves much bigger numbers.

EXERCISE 2D

Simplify

1 $\dfrac{2^4}{2^2 \times 4^3}$

2 $4^{1/2} \times 2^{-3}$

3 $(3^3)^{1/2} \times 9^{1/4}$

4 $\dfrac{x^{1/3} \times x^{4/3}}{x^{-1/3}}$

5 $\dfrac{p^{1/2} \times p^{-3/4}}{p^{-1/4}}$

6 $(\sqrt{t})^3 \times (\sqrt{t^5})$

7 $(y^2)^{3/2} \times y^{-3}$

8 $(16)^{5/4} \div 8^{4/3}$

9 $\dfrac{y^{1/2}}{y^{-3/4}} \times \sqrt{y^{1/2}}$

10 $x^2 \times x^{5/2} \div x^{-1/2}$

11 $\dfrac{y^{1/6} \times y^{-2/3}}{y^{1/4}}$

12 $(p^{1/3})^2 \times (p^2)^{1/3} \div \sqrt[3]{p}$

Evaluate

13 $\left(\frac{1}{3}\right)^{-1}$

14 $\left(\frac{1}{4}\right)^{5/2}$

15 $(8)^{-1/3}$

16 $\dfrac{1}{(16)^{-1/4}}$

17 $\left(\frac{1}{9}\right)^{-3/2}$

18 $\left(\frac{27}{8}\right)^{2/3}$

19 $\left(\frac{100}{9}\right)^0$

20 $\dfrac{1}{4^{-2}}$

21 $(0.64)^{-1/2}$

22 $\left(-\frac{1}{5}\right)^{-1}$

23 $(121)^{3/2}$

24 $\left(\frac{125}{27}\right)^{-1/3}$

25 $18^{1/2} \times 2^{1/2}$

26 $3^{-3} \times 2^0 \times 4^2$

27 $\dfrac{8^{1/2} \times 32^{1/2}}{(16)^{1/4}}$

28 $5^{1/3} \times 25^0 \times 25^{1/3}$

29 $27^{1/4} \times 3^{1/4} \times (\sqrt{3})^{-2}$

30 $\dfrac{9^{1/3} \times 27^{-1/2}}{3^{-1/6} \times 3^{-2/3}}$

Logarithms

Consider the statement $10^2 = 100$

We can read this as

the base 10 raised to the power 2 gives 100

Now this relationship can be rearranged to give the same information, but with a different emphasis, i.e.

2 is the power to which the base 10 must be raised to give 100.

In this form the power is called a logarithm (log).

The whole relationship can then be abbreviated to read

2 is the logarithm to the base 10 of 100

or $2 = \log_{10} 100$

In the same way, $2^3 = 8 \Rightarrow 3 = \log_2 8$

and $3^4 = 81 \Rightarrow 4 = \log_3 81$

Similarly $\log_5 25 = 2 \Rightarrow 25 = 5^2$

and $\log_9 3 = \frac{1}{2} \Rightarrow 3 = 9^{1/2}$

Although we have so far used only certain bases, the base of a logarithm can be any positive number, or even an unspecified number represented by a letter, for example

$$b = a^c \iff \log_a b = c$$

Note that the symbol $\iff$ means that each of these facts implies the other.

Natural Logarithms

We saw on page 21 that some numbers, such as $\sqrt{2}$, can never be expressed as exact decimals and are called irrational numbers.

The constant π is a well known irrational number: it is equal to 3.141 592 ...

There is another constant which by the end of this course will be as familiar as π. It is, like π, an irrational number, i.e. a non-repeating and never-ending decimal.
It is denoted by e and is equal to 2.718 28 ...

Later in this course we find that the number e plays a vital role in the modelling of population growth and decay. Because e appears in many other unrelated areas of mathematics, it is called a *natural number*.

This constant was first named e by Euler who showed that, as x gets larger and larger, $\left(1 + \frac{1}{x}\right)^x$ approaches the value e.

Another interesting appearance of e was discovered by Newton who found that the sum $1 + \frac{1}{1} + \frac{1}{1 \times 2} + \frac{1}{1 \times 2 \times 3} + \frac{1}{1 \times 2 \times 3 \times 4} + \dots$ approaches e as more and more terms are added.

When e is used as the base for logarithms, they are called *natural logarithms*. To avoid confusion with other bases, $\log_e x$ is written as $\ln x$.

$\ln x$ means $\log_e x$
so $\ln x = y \iff e^y = x$

The power of a positive number always gives a positive result,
e.g. $4^2 = 16$, $4^{-2} = \frac{1}{64}, \ldots$

This means that, if $\log_a b = c$, i.e. $b = a^c$, then b must be positive, so logs of positive numbers exist, but

the logarithm of a negative number does not exist.

EXAMPLE 2E

a Write $\log_2 64 = 6$ in index form.

b Write $5^3 = 125$ in logarithmic form.

c Complete the statement $2^{-3} = ?$ and then write it in logarithmic form.

d Write $\ln x = 2$ in index form.

a If $\log_2 64 = 6$ then the base is 2, the number is 64 and the power (i.e. the log) is 6.

$\log_2 64 = 6 \quad \Rightarrow \quad 64 = 2^6$

b If $5^3 = 125$ then the base is 5, the log (i.e. the power) is 3 and the number is 125.

$5^3 = 125 \quad \Rightarrow \quad 3 = \log_5 125$

c $2^{-3} = \frac{1}{8}$

The base is 2, the power (log) is −3 and the number is $\frac{1}{8}$.

$2^{-3} = \frac{1}{8} \quad \Rightarrow \quad -3 = \log_2\left(\frac{1}{8}\right)$

d $\ln x$ means that the base is e

$\ln x = 2 \quad \Rightarrow \quad x = e^2$

EXERCISE 2E

Convert each of the following facts to logarithmic form.

1 $10^3 = 1000$

2 $2^4 = 16$

3 $10^4 = 10\,000$

4 $3^2 = 9$

5 $4^2 = 16$

6 $5^2 = 25$

7 $10^{-2} = 0.01$

8 $9^{1/2} = 3$

9 $5^0 = 1$

10 $4^{1/2} = 2$

11 $12^0 = 1$

12 $8^{1/3} = 2$

13 $p = q^2$

14 $x^y = 2$

15 $p^q = r$

16 $e^x = 4$

17 $e^2 = y$

18 $e^a = b$

Convert each of the following facts to index form.

19 $\log_{10} 100\,000 = 5$

20 $\log_4 64 = 3$

21 $\log_{10} 10 = 1$

22 $\log_2 4 = 2$

23 $\log_2 32 = 5$

24 $\log_{10} 1000 = 3$

25 $\log_5 1 = 0$

26 $\log_3 9 = 2$

27 $\log_4 16 = 2$

28 $\log_3 27 = 3$

29 $\log_{36} 6 = \frac{1}{2}$

30 $\log_a 1 = 0$

31 $\log_x y = z$

32 $\log_a 5 = b$

33 $\log_p q = r$

34 $\ln x = 4$

35 $\ln 0.5 = a$

36 $\ln a = b$

Evaluating Logarithms

It is generally easier to solve a simple equation in index form than in log form so we often use an index equation in order to evaluate a logarithm. For example to evaluate $\log_{49} 7$ we can say

if $\quad x = \log_{49} 7$ then $49^x = 7 \quad \Rightarrow \quad x = \frac{1}{2}$

therefore $\quad \log_{49} 7 = \frac{1}{2}$

In particular, for any base b,

if $\quad x = \log_b 1$ then $b^x = 1 \quad \Rightarrow \quad x = 0$

i.e. **the logarithm of 1 to any base is zero.**

Using a Calculator

A scientific calculator can be used to find the values of logarithms with a base of e or with a base of 10.

The value of ln x for various values of x can be obtained by using the button marked [ln].

The button marked [log] gives the value of a logarithm with a base of 10.

Powers of e, such as e^2, $e^{0.15}$, can also be obtained from a calculator using 'e^x' which is usually above the ln button.

EXERCISE 2F

Evaluate

1 $\log_2 4$

2 $\log_{10} 1\,000\,000$

3 $\log_2 64$

4 $\log_3 81$

5 $\log_8 64$

6 $\log_4 64$

7 $\log_9 3$

8 $\log_{1/2} 4$

9 $\log_{10} 0.1$

10 $\log_{121} 11$

11 $\log_5 1$

12 $\log_2 2$

13 $\log_{64} 4$

14 $\log_{99} 1$

15 $\log_{27} 3$

16 $\log_a a^3$

17 $\ln e$

18 $\ln e^2$

19 $\log_b b^3$

20 $\ln e^{1.5}$

21 Use a calculator to find, correct to 3 significant figures, the value of

a e^2 **b** $e^{1.5}$ **c** e^{-2} **d** $e^{0.05}$

22 Use a calculator to evaluate, correct to 3 significant figures,

a $\ln 3$ **b** $\ln 2.4$ **c** $\ln 0.201$ **d** $\ln 17.3$ **e** $\log_{10} 5.6$ **f** $\log_{10} 250$

The Laws of Logarithms

When working with indices earlier in this chapter we found that powers obey certain rules in the multiplication and division of numbers. Because logarithm is just another word for index or power, it is to be expected that logarithms too obey certain laws and we are now going to investigate these.

Consider $\quad x = \log_a b$ and $y = \log_a c$

$\Rightarrow \quad a^x = b$ and $a^y = c$

Now $\quad bc = (a^x)(a^y)$

$\Rightarrow \quad bc = a^{x+y}$

Therefore $\quad \log_a bc = x + y$

i.e. $\quad \log_a bc = \log_a b + \log_a c$

This is the first law of logarithms and, as a can represent *any* base, this law applies to the log of *any* product *provided that the same base is used for all the logarithms in the formula*.

Using x and y again, a law for the log of a fraction can be found.

$$\frac{b}{c} = \frac{a^x}{a^y} \quad \Rightarrow \quad \frac{b}{c} = a^{x-y}$$

Therefore $\quad \log_a\left(\dfrac{b}{c}\right) = x - y$

i.e $\quad \log_a\left(\dfrac{b}{c}\right) = \log_a b - \log_a c$

A third law allows us to deal with an expression of the type $\log_a b^n$

Using $\quad x = \log_a b^n \quad \Rightarrow \quad a^x = b^n$

i.e. $\quad a^{x/n} = b$

Therefore $\quad \dfrac{x}{n} = \log_a b \quad \Rightarrow \quad x = n\log_a b$

i.e. $\quad \log_a b^n = n\log_a b$

So we now have the three most important laws of logarithms. Because they are true for *any* base it is unnecessary to include a base in the formula but

in each of these laws every logarithm must be to the same base.

$$\log bc = \log b + \log c$$

$$\log \frac{b}{c} = \log b - \log c$$

$$\log b^n = n\log b$$

EXAMPLES 2G

1 Express $\log pq^2\sqrt{r}$ in terms of $\log p$, $\log q$ and $\log r$.

$$\log pq^2\sqrt{r} = \log p + \log q^2 + \log\sqrt{r}$$
$$= \log p + 2\log q + \tfrac{1}{2}\log r$$

2 Simplify $3\log p + n\log q - 4\log r$

$$3\log p + n\log q - 4\log r = \log p^3 + \log q^n - \log r^4$$
$$= \log \frac{p^3 q^n}{r^4}$$

3 Express $\ln \dfrac{x+1}{x^2}$ as separate logarithms.

$$\ln \frac{x+1}{x^2} = \ln(x+1) - \ln x^2 = \ln(x+1) - 2\ln x$$

4 Express $\ln(x+1) + \ln 4 - \frac{1}{2}\ln x$ as a single logarithm.

$$\ln(x+1) + \ln 4 - \tfrac{1}{2}\ln x = \ln 4(x+1) - \ln\sqrt{x}$$
$$= \ln \frac{4(x+1)}{\sqrt{x}}$$

EXERCISE 2G

Express in terms of $\log p$, $\log q$, and $\log r$

1 $\log pq$

2 $\log pqr$

3 $\log \dfrac{p}{q}$

4 $\log \dfrac{pq}{r}$

5 $\log \dfrac{p}{qr}$

6 $\log p^2 q$

7 $\log \dfrac{q}{r^2}$

8 $\log p\sqrt{q}$

9 $\log \dfrac{p^2 q^3}{r}$

10 $\log \sqrt{\dfrac{q}{r}}$

11 $\log q^n$

12 $\log p^n q^m$

Simplify

13 $\log p + \log q$

14 $2\log p + \log q$

15 $\log q - \log r$

16 $3\log q + 4\log p$

17 $n\log p - \log q$

18 $\log p + 2\log q - 3\log r$

Express as the sum or difference of the simplest possible logarithms.

19 $\ln 5x$

20 $\ln 5x^2$

21 $\ln 3(x+1)$

22 $\ln \dfrac{x}{x+1}$

23 $\ln \dfrac{2x}{x-1}$

24 $\ln xy^2$

25 $\ln \sqrt{x+1}$

26 $\ln x(x+4)$

27 $\ln(x^2-1)$

28 $\ln x^2(x+y)$

29 $\ln \mathrm{e}x$

30 $\ln \mathrm{e}^2 x(x-\mathrm{e})$

Express as a single logarithm

31 $\ln 2 + \ln x$

32 $\ln 3 - \ln x$

33 $2\ln x - \ln 4$

34 $\ln x - 2\ln(1-x)$

35 $1 - \ln x$

36 $2 + \ln x$

37 $2\ln x - \frac{1}{2}\ln(x-1)$

38 $2\ln x - 3\ln y$

MIXED EXERCISE 2

1 Simplify

a $\sqrt{84}$ **b** $\sqrt{300}$ **c** $\sqrt{45}$

2 Expand and simplify

a $(3+\sqrt{2})(4-2\sqrt{2})$

b $(\sqrt{5}-\sqrt{2})^2$

3 Multiply by a bracket that will make the product rational

a $(7-\sqrt{3})$ **b** $(2\sqrt{2}+1)$

c $(\sqrt{7}-\sqrt{5})$

4 Rationalise the denominator and simplify where possible

a $\dfrac{5}{\sqrt{7}}$ **b** $\dfrac{3}{\sqrt{13}-2}$

c $\dfrac{4}{\sqrt{3}-\sqrt{2}}$ **d** $\dfrac{\sqrt{3}-1}{\sqrt{3}+1}$

5 Simplify

a $\dfrac{2^3 \times 4^{-2}}{2^{-1}}$ **b** $(x^3)^{-2} \times (x^2)^3$

6 Evaluate

a $(64)^{-1/3}$ **b** $\left(\dfrac{49}{16}\right)^{-1/2}$

c $\left(\dfrac{8}{27}\right)^{3/2}$

7 Simplify

a $8^{1/6} \times 2^0 \times 2^{-1/2}$

b $(\sqrt{5})^{-2} \times 75^{1/2} \times 25^{-1/4}$

8 Evaluate

a $\log_2 128$ **b** $\log_{25} 5$

c $\log_{12} 1$ **d** $\ln e^5$

9 Express in terms of $\log a$, $\log b$ and $\log c$

a $\log \dfrac{a^3}{(bc^2)}$ **b** $\log \dfrac{a^n}{b}$

c $\log \dfrac{ab}{c}$ **d** $\log a\sqrt{1+b}$

10 Simplify

a $3\log a - \log b$

b $\log \dfrac{1}{a} + \log 1$

11 Express as a single logarithm

a $\ln x - \ln y$

b $2 + \ln(x+1)$

c $\ln A + \ln x$

d $\ln x - \ln xy + \ln y^2$

***12** You borrow £1 from a loan company. The company charges interest at the rate of 100% per annum.
After one year,
when this interest is added yearly,
you owe 200% of £1 = £2,
when the interest is added half yearly,
you owe 150% of 150% of £1 = £$(1.5)^2$ = £2.25,
when the interest is added each quarter (three monthly),
you owe 125% of 125% of 125% of 125% of £1 = £$(1.25)^4$ = £2.44

a If the interest could be added continuously, have a guess at what you would owe after one year.

b Work out what you would owe if the interest is added
i daily **ii** hourly **iii** by the second.

c Now repeat part **a**.

CHAPTER 3

Equations 1

Expressions, Identities and Equations

It is important to know the difference between expressions, identities and equations. $2x - 4$ is an *expression*; it is not linked to anything else.

$2x - 4 \equiv 2(x - 2)$ is an *identity*, $2(x - 2)$ is just a different form of the expression $2x - 4$. The symbol '$\equiv$' means 'is identical to', and $2x - 4 \equiv 2(x - 2)$ is true for all values of x.

In general, an *identity* is the relationship between two different forms of the same expression.

$2x - 4 = x$ is an *equation*; the equality is true when $x = 4$, but not true for any other value of x.

In general, an *equation* is the equality of two different expressions. This equality is true only for a number of distinct values (or none) of the unknown quantity.

The process of finding these values is called solving the equation.

EXERCISE 3A

State which sign, $=$ or $\equiv$, connects the two expressions.

1 $x^2 - 3;\ 2$

2 $x^2 - 9;\ (x-3)(x+3)$

3 $\dfrac{1}{x} - \dfrac{1}{x+1};\ \dfrac{1}{x^2+x}$

4 $p^2 + 2p - 3;\ 3 - 2p - p^2$

5 $(x+y)(x-y);\ x^2 - y^2$

6 $\dfrac{1}{x-1} + \dfrac{1}{x+1};\ \dfrac{2}{x^2-1}$

7 $\log 2 + \log 3x;\ \log 6x$

8 $\ln x - \ln 2;\ \ln(x-2)$

9 $y - 1;\ \dfrac{1}{y}$

Quadratic Equations

When a quadratic expression has a particular value we have a quadratic equation, for example

$$2x^2 - 5x + 1 = 0$$

Using *a*, *b* and *c* to stand for any numbers, any quadratic equation can be written in the general form

$$ax^2 + bx + c = 0$$

Solution by Factorising

Consider the quadratic equation $x^2 - 3x + 2 = 0$

The quadratic expression on the left-hand side can be factorised,

i.e. $$x^2 - 3x + 3 \equiv (x - 2)(x - 1)$$

Therefore the given equation becomes

$$(x - 2)(x - 1) = 0 \qquad [1]$$

Now if the product of two quantities is zero then one, or both, of those quantities must be zero.

Applying this fact to equation [1] gives

$$x - 2 = 0 \quad \text{or} \quad x - 1 = 0$$

i.e. $$x = 2 \quad \text{or} \quad x = 1$$

This is the solution of the given equation.
The values 2 and 1 are called the *roots* of that equation.

This method of solution can be used for any quadratic equation in which the quadratic expression factorises.

EXAMPLE 3B

Find the roots of the equation $x^2 + 6x - 7 = 0$

$$x^2 + 6x - 7 = 0$$

$\Rightarrow$ $$(x - 1)(x + 7) = 0$$

$\therefore$ $$x - 1 = 0 \quad \text{or} \quad x + 7 = 0$$

$\therefore$ $$x = 1 \quad \text{or} \quad x = -7$$

The roots of the equation are 1 and −7.

EXERCISE 3B

Solve the equations.

1 $x^2 + 5x + 6 = 0$

2 $x^2 + x - 6 = 0$

3 $x^2 - x - 6 = 0$

4 $x^2 + 6x + 8 = 0$

5 $x^2 - 4x + 3 = 0$

6 $x^2 + 2x - 3 = 0$

7 $2x^2 + 3x + 1 = 0$

8 $4x^2 - 9x + 2 = 0$

9 $x^2 + 4x - 5 = 0$

10 $x^2 + x - 72 = 0$

Find the roots of the equations.

11 $x^2 - 2x - 3 = 0$

12 $x^2 + 5x + 4 = 0$

13 $x^2 - 6x + 5 = 0$

14 $x^2 + 3x - 10 = 0$

15 $x^2 - 5x - 14 = 0$

16 $x^2 - 9x + 14 = 0$

Rearranging the Equation

The terms in a quadratic equation are not always given in the order $ax^2 + bx + c = 0$. When they are given in a different order they should be rearranged into the standard form.

For example $x^2 - x = 4$ becomes $x^2 - x - 4 = 0$

$3x^2 - 1 = 2x$ becomes $3x^2 - 2x - 1 = 0$

$x(x-1) = 2$ becomes $x^2 - x = 2 \quad \Rightarrow \quad x^2 - x - 2 = 0$

It is usually best to collect the terms on the side where the x^2 term is positive, for example

$2 - x^2 = 5x$ becomes $0 = x^2 + 5x - 2$

i.e. $x^2 + 5x - 2 = 0$

EXAMPLE 3C

Solve the equation $4x - x^2 = 3$

$$4x - x^2 = 3$$
$$\Rightarrow \quad 0 = x^2 - 4x + 3$$
$$\Rightarrow \quad x^2 - 4x + 3 = 0 \Rightarrow (x-3)(x-1) = 0$$
$$\Rightarrow \quad x - 3 = 0 \quad \text{or} \quad x - 1 = 0$$
$$\Rightarrow \quad x = 3 \text{ or } x = 1$$

Losing a Solution

Quadratic equations sometimes have a common factor containing the unknown quantity. It is very tempting in such cases to divide by the common factor, but doing this results in the loss of part of the solution, as the following example shows.

Correct solution

$x^2 - 5x = 0$

$x(x-5) = 0$

$\therefore \quad x = 0$ or $x - 5 = 0$

$\Rightarrow \quad x = 0$ or 5

Faulty solution

$x^2 - 5x = 0$

$x - 5 = 0$ (Dividing by x.)

$\therefore \quad x = 5$

The solution $x = 0$ has been lost.

Although dividing an equation by a numerical common factor is correct and sensible, dividing by a common factor containing the unknown quantity results in the loss of a solution.

EXERCISE 3C

Solve each equation, making sure that you give all the roots.

1 $x^2 + 10 - 7x = 0$

2 $15 - x^2 - 2x = 0$

3 $x^2 - 3x = 4$

4 $12 - 7x + x^2 = 0$

5 $2x - 1 + 3x^2 = 0$

6 $x(x+7) + 6 = 0$

7 $2x^2 - 4x = 0$

8 $x(4x+5) = -1$

9 $2 - x = 3x^2$

10 $6x^2 + 3x = 0$

11 $x^2 + 6x = 0$

12 $x^2 = 10x$

13 $x(4x+1) = 3x$

14 $20 + x(1-x) = 0$

15 $x(3x-2) = 8$

16 $x^2 - x(2x-1) + 2 = 0$

17 $x(x+1) = 2x$

18 $4 + x^2 = 2(x+2)$

19 $x(x-2) = 3$

20 $1 - x^2 = x(1+x)$

Solution by Completing the Square

When there are no obvious factors, another method is needed to solve the equation. One such method involves adding a constant to the x^2 term and the x term, to make a perfect square. This technique is called *completing the square*.

Consider $x^2 - 2x$

Adding 1 gives $x^2 - 2x + 1$

Now $x^2 - 2x + 1 \equiv (x-1)^2$ which is a perfect square.

Adding the number 1 was not a guess, it was found by using the fact that

$$x^2 + 2ax + \boxed{a^2} \equiv (x+a)^2$$

We see from this that the number to be added is always (half the coefficient of x)^2.

Hence $x^2 + 6x$ requires 3^2 to be added to make a perfect square,

i.e. $\quad x^2 + 6x + 9 \equiv (x+3)^2$

To complete the square when the coefficient of x^2 is not 1, we first take out the coefficient of x^2 as a factor,

e.g. $\quad 2x^2 + x \equiv 2(x^2 + \frac{1}{2}x)$

Now we add $(\frac{1}{2} \times \frac{1}{2})^2$ inside the bracket, giving

$$2(x^2 + \tfrac{1}{2}x + \tfrac{1}{16}) \equiv 2(x + \tfrac{1}{4})^2$$

Take extra care when the coefficient of x^2 is negative

e.g. $\quad -x^2 + 4x \equiv -(x^2 - 4x)$

Then $\quad -(x^2 - 4x + 4) \equiv -(x-2)^2, \quad \therefore \quad -x^2 + 4x - 4 \equiv -(x-2)^2$

EXAMPLES 3D

1 Solve the equation $x^2 - 4x - 2 = 0$, giving the solution in surd form.

$x^2 - 4x - 2 = 0 \quad \Rightarrow \quad x^2 - 4x = 2$

> No factors can be found so we isolate the two terms with x in.

Add $\{\frac{1}{2} \times (-4)\}^2$ to *both* sides, i.e. $x^2 - 4x + 4 = 2 + 4$

$\Rightarrow \quad (x-2)^2 = 6$, i.e. $x - 2 = \pm\sqrt{6}$

$\therefore \quad x = 2 + \sqrt{6}$ or $x = 2 - \sqrt{6}$

2 Find in surd form the roots of the equation $2x^2 - 3x - 3 = 0$.

$2x^2 - 3x - 3 = 0$

$2(x^2 - \frac{3}{2}x) = 3 \quad \Rightarrow \quad x^2 - \frac{3}{2}x = \frac{3}{2}$

$x^2 - \frac{3}{2}x + \frac{9}{16} = \frac{3}{2} + \frac{9}{16} \quad \Rightarrow \quad (x - \frac{3}{4})^2 = \frac{33}{16}$

$\therefore \quad x - \frac{3}{4} = \pm\sqrt{\frac{33}{16}} = \pm\frac{1}{4}\sqrt{33} \quad \Rightarrow \quad x = \frac{3}{4} \pm \frac{1}{4}\sqrt{33}$

The roots of the equations are $\frac{1}{4}(3 + \sqrt{33})$ and $\frac{1}{4}(3 - \sqrt{33})$

EXERCISE 3D Add a number to each expression so that the result contains a perfect square as a factor.

1 x^2-4x **5** $2x^2-4x$ **9** $2x^2-40x$

2 x^2+2x **6** x^2+5x **10** x^2+x

3 x^2-6x **7** $3x^2-48x$ **11** $3x^2-2x$

4 x^2+10x **8** x^2+18x **12** $2x^2+3x$

Solve the equations by completing the square, giving the solutions in surd form.

13 $x^2+8x=1$ **17** $x^2+3x+1=0$ **21** $2x^2+4x=7$

14 $x^2-2x-2=0$ **18** $2x^2-x-2=0$ **22** $x^2-x=3$

15 $x^2+x-1=0$ **19** $x^2+4x=2$ **23** $4x^2+x-1=0$

16 $2x^2+2x=1$ **20** $3x^2+x-1=0$ **24** $2x^2-3x-4=0$

The Formula for Solving a Quadratic Equation

Solving a quadratic equation by completing the square is rather tedious. If the method is applied to a general quadratic equation, a formula can be derived which can then be used to solve any particular equation.

Using a, b and c to represent any numbers we have the general quadratic equation

$$ax^2+bx+c=0$$

Using the method of completing the square for this equation gives

$$ax^2+bx=-c$$

i.e. $$a\left(x^2+\frac{b}{a}x\right)=-c$$

$\Rightarrow$ $$x^2+\frac{b}{a}x=-\frac{c}{a}$$

$\therefore$ $$x^2+\frac{b}{a}x+\left(\frac{b}{2a}\right)^2=\left(\frac{b}{2a}\right)^2-\frac{c}{a}$$

$\therefore$ $$\left(x+\frac{b}{2a}\right)^2=\frac{b^2}{4a^2}-\frac{c}{a}=\frac{b^2-4ac}{4a^2}$$

$\Rightarrow$ $$x+\frac{b}{2a}=\pm\sqrt{\frac{b^2-4ac}{4a^2}}$$

$\Rightarrow$ $$x=-\frac{b}{2a}\pm\frac{\sqrt{b^2-4ac}}{2a}$$

i.e. $$\mathbf{x=\frac{-b\pm\sqrt{b^2-4ac}}{2a}}$$

EXAMPLE 3E

Find, by using the formula, the roots of the equation $2x^2 - 7x - 1 = 0$ giving them correct to 3 decimal places.

$2x^2 - 7x - 1 = 0$

Comparing with $ax^2 + bx + c = 0$ gives $a = 2, b = -7, c = -1$

$$x = \frac{-b \pm \sqrt{b^2 - 4ac}}{2a} = \frac{7 \pm \sqrt{49 - 4(2)(-1)}}{4}$$

Therefore, in surd form, $x = \dfrac{7 \pm \sqrt{57}}{4}$

Correct to 3 d.p. the roots are 3.637 and −0.137.

EXERCISE 3E

Solve the equations by using the formula. Give the solutions in surd form.

1 $x^2 + 4x + 2 = 0$

2 $2x^2 - x - 2 = 0$

3 $x^2 + 5x + 1 = 0$

4 $2x^2 - x - 4 = 0$

5 $x^2 + 1 = 4x$

6 $2x^2 - x = 5$

7 $1 + x - 3x^2 = 0$

8 $3x^2 = 1 - x$

Find, correct to 3 d.p., the roots of the equations.

9 $5x^2 + 9x + 2 = 0$

10 $2x^2 - 7x + 4 = 0$

11 $4x^2 - 7x - 1 = 0$

12 $3x = 5 - 4x^2$

13 $4x^2 + 3x = 5$

14 $1 = 5x - 5x^2$

15 $8x - x^2 = 1$

16 $x^2 - 3x = 1$

Cubic Equations

Equations such as $(x-2)(2x+1)(x+4) = 0$ can be solved directly. As the product of the three brackets is zero, one or more of them must be zero, i.e.
either $x - 2 = 0 \Rightarrow x = 2$ or $2x + 1 = 0 \Rightarrow x = -\frac{1}{2}$
or $x + 4 = 0 \Rightarrow x = -4$.

When the factors of an equation are not known, we can use the factor theorem. For example, to solve the equation $x^3 - 5x^2 - x + 5 = 0$, we see that possible factors of the LHS are $x \pm 1, x \pm 5$.

Try $x - 1$: $1^3 - 5(1)^2 - 1 + 5 = 0$, so $x - 1$ is a factor.

Now we can factorise fully:

$$x^3 - 5x^2 - x + 5 = (x-1)(x^2 - 4x - 5) = (x-1)(x+1)(x-5)$$

$$\therefore\ x^3 - 5x^2 - x + 5 = 0 \Rightarrow (x-1)(x+1)(x-5) = 0$$

$$\Rightarrow x = 1 \text{ or } x = -1 \text{ or } x = 5$$

It is important to realise that not all cubic equations can be solved by factorisation; some may have no linear factors with rational coefficients and others may have only one such linear factor.

The ideas demonstrated here can be applied also to equations containing x^4 or higher powers of x.

EXAMPLE 3F

Use the factor theorem to show that $2x-3$ is a factor of $2x^3-5x^2+x+3$. Hence find all the roots of the equation $2x^3-5x^2+x+3=0$, giving your answers in surd form.

> To show that $2x-3$ is a factor, we need to substitute into the cubic expression the value of x that makes $2x-3$ zero.

When $x=\frac{3}{2}$, $2x^3-5x^2+x+3=2(\frac{3}{2})^3-5(\frac{3}{2})^2+\frac{3}{2}+3=\frac{27}{4}-\frac{45}{4}+\frac{3}{2}+3=0$

$\therefore \quad 2x^3-5x^2+x+3=(2x-3)(x^2-x-1)$

so $\quad 2x^3-5x^2+x+3=0 \Rightarrow (2x-3)(x^2-x-1)=0$

$\therefore \quad x=\frac{3}{2}$ or $x^2-x-1=0 \Rightarrow x=\dfrac{1\pm\sqrt{1+4}}{2}=\dfrac{1+\sqrt{5}}{2}$ or $\dfrac{1-\sqrt{5}}{2}$

EXERCISE 3F

Solve the equations, giving the roots in surd form where necessary.

1 $(x-3)(2x-5)(3x+1)=0$

2 $(2x+5)(x^2+3x-4)=0$

3 $x^3-6x^2+11x-6=0$

4 $3x^3-22x^2+37x-10=0$

5 $2x^3-5x^2-14x+8=0$

6 $x^3-4x^2+8=0$

7 $x^3-3x+2=0$

8 $x^3+6x^2+10x+5=0$

9 One root of the equation $x^3-ax+2=0$ is 1. Find the value of a and hence find the other roots of the equation.

Simultaneous Equations

When only one unknown quantity has to be found, only one equation is needed for a solution.

If two unknown quantities are involved in a problem we need two equations connecting them. Then, between the two equations we can eliminate one of the unknowns, producing just one equation containing just one unknown which we can find.

Solution of Three Linear Equations

For three unknown quantities we need three connections, i.e. three equations. Then one unknown at a time can be eliminated. One way to eliminate an unknown quantity is to add or subtract two of the equations and then go on to eliminate the second unknown in a similar way.

EXAMPLES 3G

1 Solve the equations $\begin{cases} x+y-z=4 \\ 2x+z=7 \\ 3x-2y=5 \end{cases}$

$$x+y-z=4 \qquad [1]$$

$$2x+z=7 \qquad [2]$$

$$3x-2y=5 \qquad [3]$$

> As z appears only in equations [1] and [2] we can eliminate z from these two equations – in this case by adding.

[1] + [2] gives $3x+y=11$ [4]

> Now bring in [3]

$$3x-2y=5 \qquad [3]$$

[4] – [3] gives $3y=6$

$$\Rightarrow \quad y=2$$

> Use $y=2$ in [3]

$$3x-4=5$$

$$\Rightarrow \quad 3x=9$$

$$\Rightarrow \quad x=3$$

> Now use $x=3$ in [2]

$$6+z=7 \quad \Rightarrow \quad z=1$$

Therefore the solution of the three simultaneous equations is

$x=3,\ y=2,\ z=1$

It is not always easy to eliminate the first of the unknown quantities. If all three unknowns occur in all three equations it is necessary to eliminate the same unknown from each of two different pairs of equations.

2 Solve the equations $\begin{cases} x - y + 2z = 6 \\ 2x + y + z = 3 \\ 3x - y + z = 6 \end{cases}$

$$x - y + 2z = 6 \qquad [1]$$

$$2x + y + z = 3 \qquad [2]$$

$$3x - y + z = 6 \qquad [3]$$

> The easiest letter to eliminate from two pairs of equations is y.

[1] + [2] gives $\quad 3x + 3z = 9$

Dividing by 3 gives $\quad x + z = 3 \qquad [4]$

[2] + [3] gives $\quad 5x + 2z = 9 \qquad [5]$

> Now we eliminate either x or z from [4] and [5].

$5 \times [4] - [5]$ gives $\quad 3z = 6$

$$\Rightarrow \quad z = 2$$

Using $z = 2$ in [4] gives $x + 2 = 3$

$$\Rightarrow x = 1$$

Then using $x = 1$ and $z = 2$ in [2] gives

$$2 + y + 2 = 3$$

$$\Rightarrow y = -1$$

Therefore the solution is $x = 1,\ y = -1,\ z = 2$

EXERCISE 3G

Solve the following sets of equations.

Remember first to look for a letter which occurs in only two equations because it can be eliminated completely in one step.

1 $x + 2y = 4$
$x + 3z = 5$
$2y - z = 1$

2 $y - z = 3$
$x - 2y + z = -4$
$x + 2y = 11$

3 $x + y + 3z = 6$
$2x - y = 3$
$4x - z = 2$

4 $2x - y - z = 5$
$4y + 3z = 5$
$x + 2y = 7$

5 $x + y + 4z = 15$
$x - y + z = 2$
$x + 2y - 3z = -4$

6 $2x - 3y + z = 13$
$x + y - 2z = -1$
$3x - y + 2z = 17$

Solution of One Linear and One Quadratic Equation

Another way to eliminate an unknown quantity from two equations is by substitution. From the linear equation we can express one unknown in terms of the other, and then substitute in the quadratic equation.

EXAMPLE 3H

Solve the equations $x - y = 2$
$2x^2 - 3y^2 = 15$

$$x - y = 2 \quad [1]$$

$$2x^2 - 3y^2 = 15 \quad [2]$$

Equation [1] is linear so we use it for the substitution,

i.e. $x = y + 2$

Substituting $y + 2$ for x in [2] gives

$$2(y+2)^2 - 3y^2 = 15$$

$$\Rightarrow \quad 2(y^2 + 4y + 4) - 3y^2 = 15$$

$$\Rightarrow \quad 2y^2 + 8y + 8 - 3y^2 = 15$$

Collecting terms on the side where y^2 is positive gives

$$0 = y^2 - 8y + 7$$

$$\Rightarrow \quad 0 = (y-7)(y-1), \qquad \therefore \quad y = 7 \text{ or } 1$$

Now we use $x = y + 2$ to find corresponding values of x.

y	7	1
x	9	3

$\therefore$ either $x = 9$ and $y = 7$ or $x = 3$ and $y = 1$

Note that the values of x and y must be given in *corresponding pairs*. It is incorrect to write the answer as $y = 7$ or 1 and $x = 9$ or 3 because $\begin{cases} y = 7 \text{ with } x = 3 \\ y = 1 \text{ with } x = 9 \end{cases}$ are *not* solutions.

EXERCISE 3H

Solve the following pairs of equations.

1 $x^2 + y^2 = 5$
$y - x = 1$

2 $y^2 - x^2 = 8$
$x + y = 2$

3 $3x^2 - y^2 = 3$
$2x - y = 1$

4 $y = 4x^2$
$y + 2x = 2$

5 $y^2 + xy = 3$
$2x + y = 1$

6 $x^2 - xy = 14$
$y = 3 - x$

7 $xy = 2$
$x + y - 3 = 0$

8 $2x - y = 2$
$x^2 - y = 5$

9 $y - x = 4$
$y^2 - 5x^2 = 20$

10 $x + y^2 = 10$
$x - 2y = 2$

11 $4x + y = 1$
$4x^2 + y = 0$

12 $3xy - x = 0$
$x + 3y = 2$

13 $x^2 + 4y^2 = 2$
$2y + x + 2 = 0$

14 $x + 3y = 0$
$2x + 3xy = 1$

15 $3x - 4y = 1$
$6xy = 1$

16 $x^2 + 4y^2 = 2$
$x + 2y = 2$

17 $xy = 9$
$x - 2y = 3$

18 $4x + y = 2$
$4x + y^2 = 8$

19 $1 + 3xy = 0$
$x + 6y = 1$

20 $x^2 - xy = 0$
$x + y = 1$

21 $xy + y^2 = 2$
$2x + y = 3$

22 $xy + x = -3$
$2x + 5y = 8$

MIXED EXERCISE 3

In questions 1 to 10 use any method to find the roots of the equations, giving any irrational roots in surd form.

1 $x^2 - 5x - 6 = 0$

2 $x^2 - 6x - 5 = 0$

3 $2x^2 + 3x = 1$

4 $5 - 3x^2 = 4x$

5 $x(2 - x) = 1$

6 $4x^2 - 3 = 11x$

7 $(x - 1)(x + 2) = 1$

8 $x^2 + 4x + 4 = 16$

9 $x^2 + 2x = 2$

10 $2(x^2 + 2) = x(x - 4)$

In questions 11 to 16, solve the equations giving all possible solutions.

11 $x(x - 2) = 0$

12 $x(x + 3) = 4$

13 $(x - 4)(x^2 + 5x + 6) = 0$

14 $x^2 + 5x + 2 = 2(2x + 1)$

15 $x(x - 5) = 2(x + 5)$

16 $x^3 - 1 = 0$

17 $(x + 3)(x - 5)(3x + 1) = 0$

18 $x^3 + x^2 - 11x + 10 = 0$

In questions 19 and 20 solve the pair of equations. (Choose your substitution carefully to keep the amount of squaring to a minimum.)

19 $2x^2 - y^2 = 7$
$x + y = 9$

20 $2x = y - 1$
$x^2 - 3y + 11 = 0$

21 Use the formula to solve the equation $3x^2 - 17x + 10 = 0$.

a Are the roots of the equation rational or irrational?
b What does your answer to part **a** tell you about the LHS of the equation?

CHAPTER 4 Equations 2

Properties of the Roots of a Quadratic Equation

A number of interesting facts can be deduced by examining the formula used for solving a quadratic equation, especially when it is written in the form

$$x = -\frac{b}{2a} \pm \frac{\sqrt{b^2 - 4ac}}{2a}$$

The Sum of the Roots

The separate roots are $-\frac{b}{2a} + \frac{\sqrt{b^2 - 4ac}}{2a}$ and $-\frac{b}{2a} - \frac{\sqrt{b^2 - 4ac}}{2a}$

When the roots are added, the terms containing the square root disappear giving

sum of roots $= -\dfrac{b}{a}$

This fact is very useful as a check on the accuracy of roots that have been calculated.

The Nature of the Roots

In the formula there are two terms. The first of these, $-\frac{b}{2a}$, can always be found for any values of a and b.

The second term however, i.e. $\frac{\sqrt{b^2 - 4ac}}{2a}$, is not so straightforward as there are three different cases to consider.

1 If $b^2 - 4ac$ is positive, its square root can be found and, whether it is a whole number, a fraction or a decimal, it is a number of the type we are familiar with – it is called a *real* number.

The two square roots, i.e. $\pm\sqrt{b^2 - 4ac}$ have different (or distinct) values giving two different real values of x. So the equation has *two different real roots*.

2 If $b^2 - 4ac$ is zero then its square root also is zero and $x = -\frac{b}{2a} - \frac{\sqrt{b^2 - 4ac}}{2a}$ gives

$$x = -\frac{b}{2a} + 0 \text{ and } x = -\frac{b}{2a} - 0$$

i.e. there is just one value of x that satisfies the equation.

An example of this case is $x^2 - 2x + 1 = 0$

From the formula we get $x = -\frac{(-2)}{2} \pm 0$, i.e. $x = 1$ or 1

By factorising we can see that there are two equal roots,

i.e. $(x - 1)(x - 1) = 0 \Rightarrow x = 1$ or $x = 1$

This type of equation can be said to have a *repeated root*.

3 If $b^2 - 4ac$ is negative we cannot find its square root because there is no real number whose square is negative. In this case the equation has *no real roots*.

From these three deductions we see that the roots of a quadratic equation can be

either real and different
or real and equal
or not real

and that it is the value of $b^2 - 4ac$ which determines the nature of the roots.

$b^2 - 4ac$ is called the discriminant.

Condition	**Nature of Roots**
$b^2 - 4ac > 0$	**Real and different**
$b^2 - 4ac = 0$	**Real and equal**
$b^2 - 4ac < 0$	**Not real**

Sometimes it matters only that the roots are real, in which case the first two conditions can be combined to give

if $b^2 - 4ac \geqslant 0$, the roots are real.

EXAMPLES 4A

1 Determine the nature of the roots of the equation $x^2 - 6x + 1 = 0$.

$x^2 - 6x + 1 = 0$

$a = 1,\ b = -6,\ c = 1$

$b^2 - 4ac = (-6)^2 - 4(1)(1) = 32$

$b^2 - 4ac > 0$ so the roots are real and different.

2 If the roots of the equation $2x^2 - px + 8 = 0$ are equal, find the value of p.

$2x^2 - px + 8 = 0$

$a = 2,\ b = -p,\ c = 8$

The roots are equal so $b^2 - 4ac = 0$,

i.e. $(-p)^2 - 4(2)(8) = 0$

$\Rightarrow \quad p^2 - 64 = 0 \quad \Rightarrow \quad p^2 = 64 \quad \therefore \quad p = \pm 8$

3 Prove that the equation $(k-2)x^2 + 2x - k = 0$ has real roots whatever the value of k.

$(k-2)x^2 + 2x - k = 0$

$a = k - 2,\ b = 2,\ c = -k$

$$b^2 - 4ac = 4 - 4(k-2)(-k) = 4 + 4k^2 - 8k$$
$$= 4k^2 - 8k + 4 = 4(k^2 - 2k + 1) = 4(k-1)^2$$

Now $(k-1)^2$ cannot be negative whatever the value of k, so $b^2 - 4ac$ cannot be negative. Therefore the roots are always real.

EXERCISE 4A

Without solving the equation, write down the sum of its roots.

1 $x^2 - 4x - 7 = 0$

2 $3x^2 + 5x + 1 = 0$

3 $2 + x - x^2 = 0$

4 $3x^2 - 4x - 2 = 0$

5 $x^2 + 3x + 1 = 0$

6 $7 + 2x - 5x^2 = 0$

Without solving the equation, determine the nature of its roots.

7 $x^2 - 6x + 4 = 0$

8 $3x^2 + 4x + 2 = 0$

9 $2x^2 - 5x + 3 = 0$

10 $x^2 - 6x + 9 = 0$

11 $4x^2 - 12x - 9 = 0$

12 $4x^2 + 12x + 9 = 0$

13 $x^2 + 4x - 8 = 0$

14 $x^2 + ax + a^2 = 0$

15 $x^2 - ax - a^2 = 0$

16 $x^2 + 2ax + a^2 = 0$

17 If the roots of $3x^2 + kx + 12 = 0$ are equal, find k.

18 If $x^2 - 3x + a = 0$ has equal roots, find a.

19 The roots of $x^2 + px + (p - 1) = 0$ are equal. Find p.

20 Prove that the roots of the equation $kx^2 + (2k + 4)x + 8 = 0$ are real for all values of k.

21 Show that the equation $ax^2 + (a + b)x + b = 0$ has real roots for all values of a and b.

22 Find the relationship between p and q if the roots of the equation $px^2 + qx + 1 = 0$ are equal.

Summary

Methods for solving quadratic equations

1 Collect the terms in the order $ax^2 + bx + c = 0$, then factorise the left-hand side.

2 Arrange in the form $ax^2 + bx = -c$, then complete the square on the left-hand side, adding the appropriate number to *both* sides.

3 Use the formula $x = \dfrac{-b \pm \sqrt{b^2 - 4ac}}{2a}$

Note Roots that are not rational should be given in surd form (i.e. the exact form) unless an approximate form (such as correct to 3 s.f.) is specifically asked for.

Properties of Roots

The nature of the roots depends on the value of the discriminant, i.e. on the value of $b^2 - 4ac$.

$b^2 - 4ac > 0 \quad \Rightarrow \quad$ real different roots

$b^2 - 4ac = 0 \quad \Rightarrow \quad$ real equal roots

$b^2 - 4ac \geqslant 0 \quad \Rightarrow \quad$ real roots

$b^2 - 4ac < 0 \quad \Rightarrow \quad$ no real roots

$$\text{Sum of roots} = -\frac{b}{a}$$

Equations containing Logarithms or x as a Power

When x forms part of an index, first look to see if the value of x is obvious, for example, when $4^x = 16$, it is clear that $x = 2$ since $4^2 = 16$.
Slightly less obvious is the equation $4^x = 32$,
in this case, 4 and 32 can both be expressed as powers of 2;

i.e. $4^x = 32 \Rightarrow (2^2)^x = 2^5$, i.e. $2^{2x} = 2^5$,

from which we have $2x = 5$, i.e. $x = 2.5$.

When the value of the unknown is not so obvious, taking logs will often transform the index into a factor.

For example, when $5^x = 10$, taking logs of both sides gives

$$x \ln 5 = \ln 10 \quad \Rightarrow \quad x = \frac{\ln 10}{\ln 5} = 1.43 \text{ correct to 3 s.f.}$$

> We choose to use logs to the base e (we could also use logs to the base 10) so that we can use a calculator. Note that $\frac{\ln 10}{\ln 5}$ is NOT equal to $\ln \frac{10}{5}$.

When an equation contains logs involving x, first look to see if there is an obvious solution, for example, when $\log_a x = \log_a 4$, it is clear that $x = 4$.
When the solution is not so obvious, the best policy is to express the log terms as a single log and then remove the logarithms.

For example, when $2 \log_2 x - \log_2 8 = 1$,

using the laws of logs gives $\log_2 \frac{x^2}{8} = 1$

then removing the logs gives $\frac{x^2}{8} = 2^1 \quad \Rightarrow \quad x^2 = 16 \quad \Rightarrow \quad x = 4$

($x = -4$ is not a solution because $\log_2 (-4)$ does not exist.)
This shows that it is essential, when solving equations involving logs or indices, that all roots are checked in the original equation.

EXERCISE 4B

Solve the equations. Give answers that are not exact correct to 3 significant figures.

1 $3^x = 9$

2 $3^x = \frac{1}{9}$

3 $9^x = 27$

4 $3^x = 6$

5 $2^{2x} = 5$

6 $5^x = 4$

7 $3^{x-1} = 7$

8 $4^{2x+1} = 8$

9 $\log_2 x = \log_2 (2x - 1)$

10 $\log_4 x = 2$

11 $\log x = 2 \log (x - 2)$

12 $\ln 2 + 2 \ln x = \ln (x + 3)$

13 Express $\log_x 5 - 2\log_x 3$ as a single log term.
Hence find the value of x when $\log_x 5 - 2\log_x 3 = 2$.

14 Solve the equation $\ln 4 - 2\ln(x+1) = \ln x$.

15 Express $\log_3 y - 2\log_3 x$ as a single logarithm.
Hence express y in terms of x when $\log_3 y - 2\log_3 x = 1$.

16 Given that $y = 2^x$, express 2^{2x} in terms of y.
By substituting y for 2^x, solve the equation $2^{2x} - 2^x - 2 = 0$.

Disguised Quadratic Equations

Some equations do not, at first sight, appear to be quadratic but can be reduced to a quadratic equation.
The most obvious are those with two terms containing x where one involves the square of the other:

e.g. $x^6 - 3x^3 + 2 = 0 \quad \Rightarrow \quad (x^3)^2 - 3(x^3) + 2 = 0$

so replacing x^3 by y

gives $y^2 - 3y + 2 = 0 \quad \Rightarrow \quad (y-2)(y-1) = 0 \quad \Rightarrow \quad y = 1$ or 2.

$\therefore \quad x^3 = 1$ or $2 \quad \Rightarrow \quad x = 1$ or $\sqrt[3]{2}$

Similarly, $x^{\frac{2}{5}} - 31x^{\frac{1}{5}} - 32 = 0 \quad \Rightarrow \quad (x^{\frac{1}{5}})^2 - 31(x^{\frac{1}{5}}) - 32 = 0$

$\Rightarrow \quad y^2 - 31y - 32 = 0$ where $y = x^{\frac{1}{5}}$,

and $e^{2x} + 3e^x - 4 = 0 \quad \Rightarrow \quad (e^x)^2 + 3(e^x) - 4 = 0$

$\Rightarrow \quad y^2 + 3y - 4 = 0$ where $y = e^x$,

and $2\ln x = \ln(2x+3) \quad \Rightarrow \quad \ln x^2 = \ln(2x+3) \quad \Rightarrow \quad x^2 = 2x + 3$

Another type involves fractions whose denominators contain x, e.g. $x + \dfrac{1}{x} = 3$.

Multiplying by the common denominator to eliminate fractions may result in a quadratic equation which we can solve provided that we exclude any solutions for which the common denominator is zero.

For $x + \dfrac{1}{x} = 3$, multiplying by x, provided $x \neq 0$ (if $x = 0$ then $\dfrac{1}{x}$ is meaningless),

gives $x^2 + 1 = 3x \quad \Rightarrow \quad x^2 - 3x + 1 = 0 \quad \Rightarrow \quad x = \dfrac{3 \pm \sqrt{5}}{2}$.

When an equation contains a square root involving x, the square root needs to be eliminated by squaring. However, when both sides of an equation are squared an extra equation, and hence extra solutions, are introduced.

For example, if $x = \sqrt{x+2}$, then squaring gives $x^2 = x + 2$

$\Rightarrow \quad x = 2$ or $x = -1$.

But when $x = -1$, $\sqrt{x+2} = 1$ so $x = -1$ does not satisfy the equation, so $x = 2$ is the only solution. This example illustrates again that

it is essential that, when solving any equation, all roots are checked in the original equation.

EXERCISE 4C

Solve the equations, giving answers in exact form.

1 $x^{10} - 31x^5 - 32 = 0$

2 $e^{2x} + 3e^x - 4 = 0$

3 $2\ln x = \ln(2x+3)$

4 $x^4 - 12x^2 + 27 = 0$

5 $\frac{2x}{x+1} = \frac{1}{x}$

6 $\sqrt{x} = 2x - 1$

7 $x^{4/3} - 5x^{2/3} + 4 = 0$

8 $\frac{1}{x-1} = 2x$

9 $4^x + 2^x - 6 = 0$

10 $x^{\frac{2}{5}} - x^{\frac{1}{5}} - 2 = 0$

11 $\sqrt{2x+1} = x - 1$

12 $9^x + 2(3^x) + 1 = 0$

MIXED EXERCISE 4

1 For each equation, first find the value of $-\frac{b}{a}$, then use any method to find the roots of the equation and finally find the sum of the roots and check that it is equal to $-\frac{b}{a}$.

a $x^2 - 6x + 8 = 0$

b $4x^2 + 5x = 3$

2 Determine the nature of the roots of the equations

a $x^2 + 3x + 7 = 0$

b $3x^2 - x - 5 = 0$

c $ax^2 + 2ax + a = 0$

d $2 + 9x - x^2 = 0$

3 For what values of p does the equation $px^2 + 4x + (p-3) = 0$ have equal roots?

4 Show that the equation $2x^2 + 2(p+1)x + p = 0$ always has real roots.

5 The equation $x^2 + kx + k = 1$ has equal roots. Find k.

6 Show that, when $x = 4$, $x^3 - 3x^2 - 3x - 4 = 0$. Hence show that the equation $x^3 - 3x^2 - 3x - 4 = 0$ has only one root.

7 Show that $x - 2$ is a factor of $x^4 - 2x^3 - x + 2$. Hence show that the equation $x^4 - 2x^3 - x + 2 = 0$ has only two roots and give the other one.

In questions **8** to **19**, solve the equations.

8 $5^x = 125$

9 $\ln 2x = \ln(x+2)$

10 $3^x = 10$

11 $\ln 2 - 2\ln(x-1) = \ln x$

12 $e^{4x} - 2 = e^{2x}$

13 $x^{2/3} - 3x^{1/3} + 2 = 0$

14 $3\sqrt{x} = x$

15 $\frac{1}{x+1} = 1 - \frac{1}{x-1}$

16 $2\ln(x+1) = \ln(x+2)$

17 $4^x = 2^x + 12$

18 $2^{2x+1} - 5(2^x) + 2 = 0$

19 $3^{2x+1} - 26(3^x) - 9 = 0$

20 Express $\log_4(x-1) - \frac{1}{2}\log_4 y$ as a single logarithm. Hence express y in terms of x when $\log_4(x-1) - \frac{1}{2}\log_4 y = -\frac{1}{2}$
Given also that $y = 2x$, find the values of x and y.

21 Solve the equations $\log_x y = 2$ and $xy = 8$ simultaneously.

22 Solve the equations $2\ln y = \ln 2 + \ln x$ and $2^y = 4^x$ simultaneously.

23 Find the values of x and y when $xy = 16$ and $\log_2 x - 2\log_2 y = 1$.

SUMMARY A

TERMS AND COEFFICIENTS

In an algebraic expression, terms are separated by plus or minus signs. An individual term is identified by the combination of letters involved. The coefficient of a term is the number in the term, e.g. ②x^2y.

EXPANSION OF BRACKETS

Important results are

$(ax+b)^2 = a^2x^2 + 2abx + b^2$

$(ax-b)^2 = a^2x^2 - 2abx + b^2$

$(ax+b)(ax-b) = a^2x^2 - b^2$

PASCAL'S TRIANGLE

$$\begin{array}{ccccccccc} & & & 1 & & 1 & & & \\ & & 1 & & 2 & & 1 & & \\ & 1 & & 3 & & 3 & & 1 & \\ 1 & & 4 & & 6 & & 4 & & 1 \end{array}$$

The 1st, 2nd, 3rd, ... rows in this array give the coefficients in the expansion of $(1+x)^1, (1+x)^2, (1+x)^3, \ldots$

FACTOR THEOREM

$(x-a)$ is a factor of an expression if and only if the expression is equal to zero when a is substituted for x.

INDICES

$a^n \times a^m = a^{n+m}$

$a^n \div a^m = a^{n-m}$ $\quad (a^n)^m = a^{nm}$

$\sqrt[n]{a} = a^{1/n}$

$a^0 = 1$

LOGARITHMS

$\log_a b = c \iff a^c = b$

$\log_a b + \log_a c = \log_a bc$

$\log_a b - \log_a c = \log_a b/c$

$\log_a b^n = n\log_a b$

$\ln x$ means $\log_e x$ where $e = 2.71\ldots$

QUADRATIC EQUATIONS

The general quadratic equation is $ax^2+bx+c = 0$.

The roots of this equation can be found by factorising when this is possible, or completing the square, or by using the formula

$$x = \frac{-b \pm \sqrt{b^2-4ac}}{2a}$$

When $b^2-4ac > 0$, the roots are real and different.

When $b^2-4ac = 0$, the roots are real and equal.

When $b^2-4ac < 0$, the roots are not real.

MULTIPLE CHOICE EXERCISE A

In questions 1 to 19 write the letter or letters corresponding to a correct answer.

1 The roots of the equation $x^2-3x+2 = 0$ are

A 2, 1 **C** −3, 2 **E** not real

B −2, −1 **D** 0, $\frac{2}{3}$

2 The coefficient of xy in the expansion of $(x-3y)(2x+y)$ is

A 1 **B** 6 **C** 5 **D** 0 **E** −5

3 The value of $\log_5 0.04$ is

A 4 **B** 5 **C** $\frac{1}{2}$ **D** −2 **E** 0.25

4 $\dfrac{1-\sqrt{2}}{1+\sqrt{2}}$ is equal to

A 1 **C** $3-\sqrt{2}$

B −1 **D** $2\sqrt{2}-3$

5 Expanding $(1+\sqrt{2})^3$ gives

A $3+3\sqrt{2}$ **D** $3+\sqrt{6}$

B $7+5\sqrt{2}$ **E** $1+2\sqrt{2}$

C $1+3\sqrt{2}$

6 $x^3-7x^2+7x+15$ has a factor

A $x+1$ **C** $2x-1$

B $x+15$ **D** $x-3$

7 x^3+8 has a factor

A $x-2$ **C** $x-8$

B $x+2$ **D** x^2+2x+4

8 If $x^2+px+6=0$ has equal roots and $p>0$, p is

A $\sqrt{48}$ **C** $\sqrt{6}$ **E** $\sqrt{24}$

B 0 **D** 3

9 If $x^2+4x+p \equiv (x+q)^2+1$, the values of p and q are

A $p=5, q=2$ **D** $p=-1, q=5$

B $p=1, q=2$ **E** $p=0, q=-1$

C $p=2, q=5$

10 $\dfrac{p^{-1/2}\times p^{3/4}}{p^{-1/4}}$ simplifies to

A 1 **C** $p^{3/4}$ **E** $p^{1/2}$

B $p^{-1/2}$ **D** p

11 In the expansion of $(a-2b)^3$ the coefficient of b^2 is

A $-2a^2$ **C** $12a$ **E** -12

B $-8a$ **D** $-4a$

12 If $\log_x y=2$ then

A $x=2y$ **C** $x^2=y$ **E** $y=\sqrt{x}$

B $x=y^2$ **D** $y=2x$

13 $2\ln x+\frac{1}{2}\ln y=$

A $\ln\dfrac{x^2}{2y}$ **C** $\ln(\sqrt{y}\times x^2)$

B $\ln x^2+\ln\sqrt{y}$ **D** $\ln 2x+\ln\frac{1}{2}y$

14 $\log_{10}5-2\log_{10}2+\frac{3}{2}\log_{10}16$ is equal to

A $\log_{10}80$ **D** $2\log_{10}12$

B 10 **E** $1+\log_{10}8$

C 0

15 When $(3-5x)^4$ is expanded

A the coefficient of x^4 is 1.

B the coefficient of x is -540.

C there are four terms after all simplification.

16 $y=\ln x-\ln 4$

A $y=\dfrac{x}{4}$ **B** $x=4e^y$ **C** $y=\ln\dfrac{x}{4}$

17 $x^2-2x+2=$

A $(x-1)^2+1$ **C** $(x-1)^2$

B 0 when $x=1$

18 $\frac{1}{2}\log_4 16-1$

A is equal to zero.

B is equal to $\log_4 7$.

19 $\dfrac{2\sqrt{3}-2}{2\sqrt{3}+2}$

A can be expressed as a fraction with a rational denominator.

B is an irrational number.

C is equal to -1.

In questions 20 to 23, write T if the statement is true and F if the statement is false.

20 If $x-a$ is a factor of x^2+px+q, the equation $x^2+px+q=0$ has a root equal to a.

21 $3\log x+1=\log 10x^3$ is an equation.

22 In the expansion of $(1+x)^6$ the coefficient of x is 6.

23 As $\sqrt{x}=4$ gives $x^2=16$, the equation $\sqrt{x}=4$ has two solutions.

EXAMINATION QUESTIONS A

1 Determine the value of the rational number p for which

$$\frac{3^{1/4} \times 3 \times 3^{1/6}}{\sqrt{3}} = 3^p$$ (OCR)

2 Show that the elimination of x from the simultaneous equations

$$x - 2y = 1$$
$$3xy - y^2 = 8$$

produces the equation $5y^2 + 3y - 8 = 0$.
Solve this equation and hence find the pairs (x, y) for which the simultaneous equations are satisfied. (Edexcel)

3 Express $(a^4)^{-1/2}$ as an algebraic fraction in simplified form. (OCR)

4 Solve the simultaneous equations

$x + y = 1, \quad x^2 - xy + y^2 = 7.$ (OCR)

5 Express $\frac{1}{(\sqrt{a})^{\frac{4}{3}}}$ in the form a^n, stating the value of n. (OCR)

6 Solve the simultaneous equations

$2x + y = 3, \quad 2x^2 - xy = 10.$ (OCR)

7 $y = 5x^3 + 24x^2 + 29x + 2$.

a Use the factor theorem to find one factor of y.

b Hence write y in the form

$$(x + k)(ax^2 + bx + c),$$

giving the value of each of the constants k, a, b, and c.

c Hence find the exact solutions to the equation $y = 0$. (AQA)

8 It is given that $(x + 2)$ is a factor of $x^4 - 4x^2 + 2x + a$. Find the value of the constant a. (OCR)

9 Express each of the following in the form $p + q\sqrt{7}$ where p and q are rational numbers.

a $(2 + 3\sqrt{7})(5 - 2\sqrt{7})$ **b** $\frac{(5 + \sqrt{7})}{(3 - \sqrt{7})}$ (AQA)

10 a Use an algebraic method to solve the simultaneous equations

$y = x^2 - 3x + 2$ and $y = 3x - 7$.

b Interpret your answer geometrically. (AQA)

11 Express $\frac{5(x-3)(x+1)}{(x-12)(x+3)} - \frac{3(x+1)}{x-12}$ as a single fraction in its simplest form. (Edexcel)

12 Use the laws of logarithms to express

$3 \ln 4 - \ln 24 + \frac{1}{2} \ln 2.25$

as a single logarithm in its simplest form. Show all your working. (OCR)

13 a Write down the exact value of x given that $4^x = 8$.
b Use logarithms to find y, correct to 3 decimal places, when $5^y = 10$. (OCR)

14 a Expand $(1+x^2)(1+x^3)$, arranging your answers in ascending powers of x.
b Find, as a decimal number, the exact value of $(1+x^2)(1+x^3)$ for $x = 10^{-3}$. (Edexcel)

15 Write $\ln x^3 + \ln xy - \ln y^3$ as single term.
Hence obtain an expression for y in terms of x if $\ln x^3 + \ln xy - \ln y^3 = 0$ (AQA)

16 Express

$$x^2 - 8x - 3$$

in completed square form.
Hence, or otherwise, find the exact solutions of the equation $x^2 - 8x - 3 = 0$. (AQA)

17 Solve the equation $4x^3 + 12x^2 + 5x - 6 = 0$. (WJEC)

18 One root of the equation $x^3 + kx + 11 = 0$, where k is a constant, is 1. Find the value of k.
Hence find the other two roots of the equation, giving your answer in an exact form. (OCR)

19 Simplify $\dfrac{2+\sqrt{2}}{2-\sqrt{2}}$ expressing your answer in surd form. (WJEC)

20 Given that $p = e^x$ and $q = e^y$, express each of

a e^{x+y} **b** e^{2x-y}

in terms of p and q. Your answers must not involve either logarithms or powers of e. (OCR)

21 Given that $y = \log_b 45 + \log_b 25 - 2\log_b 75$, express y as a single logarithm in base b.
In the case when $b = 5$, state the value of y. (AQA)

22 Express $\log_2(x+2) - \log_2 x$ as a single logarithm.
Hence solve the equation $\log_2(x+2) - \log_2 x = 3$ (OCR)

23 Given that $y = 10^x$, show that

a $y^2 = 100^x$, **b** $\dfrac{y}{10} = 10^{x-1}$

c Using the results from **a** and **b** write the equation

$$100^x - 10\,001(10^{x-1}) + 100 = 0$$

as an equation in y.
d By first solving the equation in y, find the values of x which satisfy the given equation in x. (Edexcel)

24 It is given that

$$(x+a)(x^2+bx+2) \equiv x^3 - 2x^2 - x - 6,$$

where a and b are constants. Find the value of a and the value of b. (OCR)

25 The cubic polynomial $x^3 - 2x^2 - 2x + 4$ has a factor $(x-a)$, where a is an integer.
i Use the factor theorem to find the value of a.
ii Hence find exactly all three roots of the cubic equation $x^3 - 2x^2 - 2x + 4 = 0$. (OCR)

26 Explaining each step clearly, express

$$\log_2(8\sqrt{3}) - \tfrac{1}{3}\log_2\tfrac{9}{16}$$

in the form $p + q\log_2 3$, where p and q are rational numbers to be found. (Edexcel)

27 $y = 2x^3 + 5x^2 - 8x - 15$.

a Show that when $x = -3$, $y = 0$.
b Hence factorise $2x^3 + 5x^2 - 8x - 15$.
c Find, to 2 decimal places, the two other values of x for which $y = 0$. (Edexcel)

28 Prove that $x^2 + px + q$ is a perfect square if, and only if, $p^2 = 4q$. (Edexcel)

29 Use the factor theorem to find one of the factors of the cubic

$$2x^3 - 9x^2 + 7x + 6.$$

Hence factorise the cubic into its linear factors. (WJEC)

30 Given that for all values of x,

$$3x^2 + 12x + 5 \equiv p(x+q)^2 + r,$$

a find the values of p, q and r.
b Hence, or otherwise, find the minimum value of $3x^2 + 12x + 5$.
c Solve the equation $3x^2 + 12x + 5 = 0$, giving your answers to one decimal place. (Edexcel)

31 a Given that a and b are positive numbers, show that

$$\ln(ab) = \ln a + \ln b.$$

b Express $\ln 6 - \ln 4 + \ln 8 - \ln 3$ as a single logarithm. (AQA)

32 Express $(2x+1)(x-2) - 3$ as a product of linear factors. (OCR)

33 Solve the simultaneous equations

$$x + y = 2, \quad x^2 + 2y^2 = 11$$ (OCR)

34 Given that k is a real constant such that $0 < k < 1$, show that the roots of the equation

$$kx^2 + 2x + (1-k) = 0$$

are **a** always real **b** always negative. (Edexcel)

35 The polynomial $x^3 + ax + b$ has $x - 1$ and $x - 3$ as two of its factors. Use this fact to write down two equations for a and b, and solve them. Hence find the third factor. (OCR)

36 Given that $(x-2)$ and $(x+2)$ are each factors of $x^3 + ax^2 + bx - 4$, find the values of a and b. For these values of a and b, find the other linear factor of $x^3 + ax^2 + bx - 4$. (OCR)

37 Show that both $(x - \sqrt{3})$ and $(x + \sqrt{3})$ are factors of $x^4 + x^3 - x^2 - 3x - 6$. Hence write down one quadratic factor of $x^4 + x^3 - x^2 - 3x - 6$, and find a second quadratic factor of this polynomial. (OCR)

38 a Assuming that $a = e^{\ln a}$ where $a > 0$, prove that

$$\ln(a^n) = n \ln a$$

b Find, correct to three decimal places, the value of y given that

$$2^{y+1} = 3 \times 5^y$$ (WJEC)

CHAPTER 5 Reasoning and proof

The Need for Proof

You may have accepted many results in mathematics on the basis of a demonstration that they are true in a few particular cases. For example, acceptance of the fact that the sum of the interior angles of a triangle is 180° may be based on the measured angles of some particular triangles having this property. This may be reinforced by not being able to find a triangle whose angles have a different sum but that does not rule out the possibility that such a triangle exists. However, it is not satisfactory to assume that a fact is *always* true without *proving* that it is, because one fact can be used to produce another and results deduced from an assumption cannot be reliable.

In mathematics, results are accepted because they are deduced by logical steps from already accepted facts. For example, the law that states that $\log_a bc = \log_a b + \log_a c$ is accepted as the truth because it is deduced from previously accepted facts about the behaviour of indices.

Later in this chapter and further on in this book we look at some of the ways in which results can be proved, but first we look at some of the concepts and symbols that can be used in a proof.

Reasoning

Consider the two statements $x = 3$ and $x^2 = 9$ where x can be any real number.

We do not know whether either of these statements is true, but we can say that

if $x = 3$ *then* $x^2 = 9$.

This is an implication which, from our knowledge of numbers, we know is correct and which we can write symbolically as

$$x = 3 \quad \Rightarrow \quad x^2 = 9$$

There are other ways of writing this implication,

e.g. $x = 3$ implies that $x^2 = 9$,
$x = 3$ therefore $x^2 = 9$,
$x = 3$ is a sufficient condition for $x^2 = 9$,
$x = 3$ only if $x^2 = 9$.

An implication is untrue when a false conclusion is drawn from the first statement, e.g. it is obviously not true that $x = 3 \quad \Rightarrow \quad x = 2$.

If we swap the order of the statements in an implication, we get the *converse*,
e.g. starting with $x + 1 = 4 \quad \Rightarrow \quad x = 3$,
the converse is $x = 3 \quad \Rightarrow \quad x + 1 = 4$.

We also get the converse by reversing the arrow,

e.g. starting with $x+1=4 \quad \Rightarrow \quad x=3$

the converse is $x+1=4 \quad \Leftarrow \quad x=3$

which can be read as $x+1=4$ is implied by $x=3$

or $x+1=4$ if $x=3$

or $x+1=4$ because $x=3$

or $x+1=4$ is a necessary condition for $x=3$.

Note that $x+1=4 \quad \Rightarrow \quad x=3$ and $x+1=4 \quad \Leftarrow \quad x=3$ are both true. This is not always the case, however, e.g. $x=3 \quad \Rightarrow \quad x^2=9$ is true, but $x=3 \quad \Leftarrow \quad x^2=9$ is not, because x can also be equal to -3.

When an implication and its converse are both true, we use the symbol $\Leftrightarrow$ to link them,

e.g. $x+1=4 \quad \Leftrightarrow \quad x=3$

Other ways of writing this are

$x+1=4$ implies and is implied by $x=3$

$x+1=4$ if and only if $x=3$

$x+1=4$ is a necessary and sufficient condition for $x=3$.

EXERCISE 5A

In this exercise, x can take any value unless it is stated otherwise.

In questions 1 to 10, determine whether the implication is true.
For those that are not, give a reason.

1 $(x+1)(x-2)=0 \quad \Rightarrow \quad x=-1$ or $x=2$

2 $ax^2+bx+c=0$ has real roots $\quad \Leftrightarrow \quad b^2-4ac \geqslant 0$

3 $\ln(x+y)=1 \quad \Rightarrow \quad \ln x+\ln y=1$

4 $x=4 \Leftarrow x^2=16$

5 $x \neq 0 \Leftarrow x^2=1$

6 In $\triangle$ABC, AB = AC $\quad \Leftrightarrow \quad$ In $\triangle$ABC, $\angle$B = $\angle$C

7 $x=1 \quad \Leftrightarrow \quad x^2=1$

8 x is an even integer $\quad \Rightarrow \quad x^2$ is an even integer

9 $(x+a)$ is a factor of $x^3+1 \quad \Rightarrow \quad (a)^3+1=0$

10 In $\triangle$ABC, $\angle$A = 90° $\quad \Leftrightarrow \quad$ $BC^2 = AB^2 + AC^2$

In questions 11 to 20, insert the symbol $\Rightarrow$, $\Leftarrow$ or $\Leftrightarrow$ which fully represents the link between the two statements.

11 'x is rational' 'x^2 is rational'

12 '$x+2=4$' '$x=2$'

13 'P(x, y) is any point on the line through the origin with gradient 1.'
'$y=x$'

14 '$x^2>0$' '$x>0$'

15 'In an octagon, all the interior angles are equal.'
'An octagon is regular.'

16 '$x=2$' '$x^2=4$'

17 '$\dfrac{x}{x+1}=0$' '$x=0$'

18 '(a, b) is a point on $y=2x-1$'
'$b=2a-1$'

19 'In ABCD, $\angle$A = $\angle$B = $\angle$C = $\angle$D.'
'ABCD is a square.'

20 '$x>1$' '$2x>2$'

Proving a Result by Direct Deduction

This is the most familiar form of proof; we use correct implications from the starting point using known (and proved) facts to arrive at the required result.

EXAMPLES 5B

1 Prove that $x^2+4x+p=0$ has two distinct real roots if and only if $p<4$.

We start with $x^2+4x+p=0$

$\Rightarrow \quad (x+2)^2+p-4=0$ (completing the square)

$\Rightarrow \quad (x+2)^2=4-p$

$\Rightarrow \quad x+2=\pm\sqrt{4-p}$

$\Rightarrow \quad x=-2\pm\sqrt{4-p}$

$\therefore \quad x$ has two distinct values $\Leftrightarrow \sqrt{4-p}$ exists and is not zero,

$\Leftrightarrow \quad 4-p>0$, i.e. $p<4$.

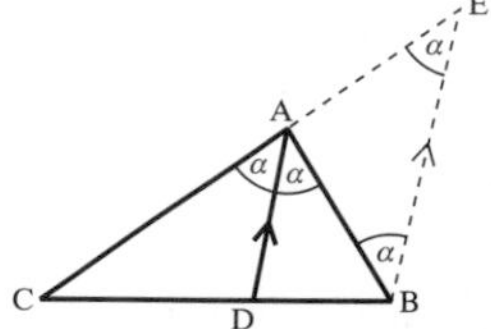

2 Prove that the line bisecting an angle of any triangle divides the side opposite to that angle in the ratio of the sides containing the angle.

In $\triangle$ABC, AD bisects $\angle$A, so we have to prove that CD : DB = CA : AB. Drawing BE parallel to DA to cut CA produced at E, we have

$\angle$BEA = $\angle$DAC (corresponding angles)

$\angle$EBA = $\angle$BAD (alternate angles)

$\therefore$ $\triangle$BEA is isosceles $\Rightarrow$ EA = AB

and $\triangle$s CAD and CEB are similar (equiangular)

$\Rightarrow$ CD : CB = CA : CE $\Rightarrow$ CD : DB = CA : AE

$\therefore$ CD : DB = CA : AB.

This result is called the *angle bisector theorem* and it should be known.

Straight Line Geometry

The following definitions and theorems should be known; they may be needed in the following exercise and elsewhere in the course.

Definitions

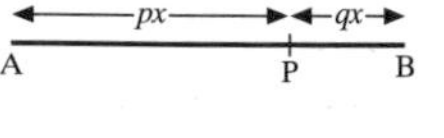

A point P divides a line AB *internally* in the ratio $p:q$
$\Rightarrow$ P is between A and B and AP : PB = $p:q$.

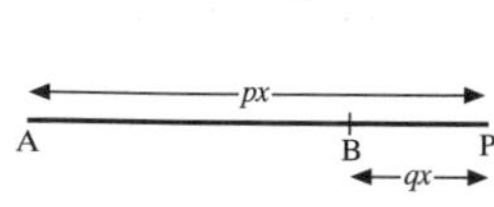

A point P divides a line AB *externally* in the ratio $p:q$ $(p>q)$
$\Rightarrow$ P is on AB produced and AP : PB = $p:q$.

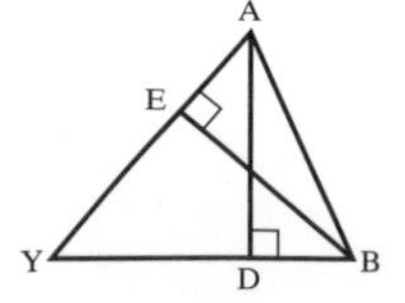

An *altitude* of a triangle is a line drawn from a vertex to the opposite side and perpendicular to that side. In the diagram,

AD is the altitude through A

and BE is the altitude through B.

A *median* of a triangle is a line drawn from a vertex to the midpoint of the opposite side. In the diagram,

XP is the median through X

and YQ is the median through Y.

Theorems

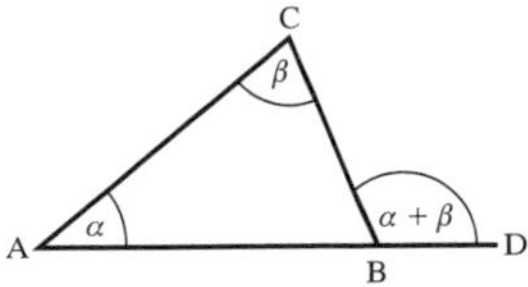

An *exterior angle* of any triangle is equal to the sum of the two interior opposite angles, i.e. in the diagram, $\angle\text{CBD} = \angle\text{A} + \angle\text{C}$

Pythagoras' theorem and its converse state that,
ABC is any triangle in which $\angle\text{B} = 90^\circ \quad \Leftrightarrow \quad \text{AC}^2 = \text{AB}^2 + \text{BC}^2$

Similar triangles (i.e. one triangle is an enlargement of the other)
Two triangles are similar $\Leftrightarrow$ The three angles of one triangle are equal to the three angles of the other triangle.

and

Two triangles are similar $\Leftrightarrow$ The corresponding sides of two triangles are in the same ratio. (To prove that two triangles are similar, we need to show *either* that they are equiangular *or* that their corresponding sides are in the same ratio.)

EXERCISE 5B

1 Prove that if $x^2 - 3x + 2 = 0$ then $x = 1$ or $x = 2$.

2

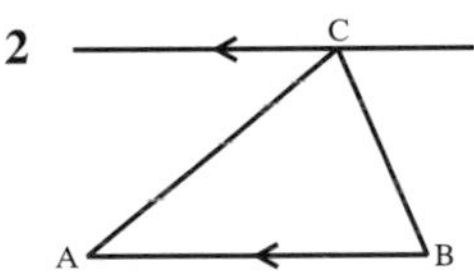

Use a copy of this diagram to prove that the sum of the angles in any triangle is 180°.

3 Prove that if $x = \sqrt{2}$ then

$$\frac{2+x}{2-x} = 3 + 2\sqrt{2}.$$

4

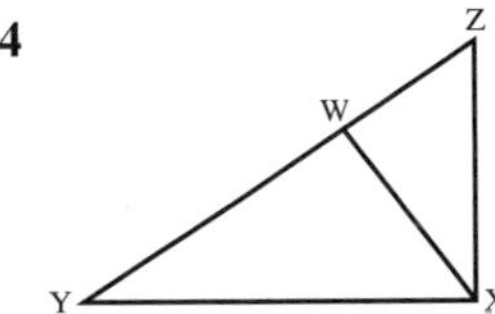

XYZ is a triangle with $\angle\text{X} = 90^\circ$. XW is an altitude of $\triangle\text{XYZ}$. Show that triangles XYZ, XWZ and XYW are all similar.

5 Prove that $\log_a x + \log_a y = \log_a xy$.

6 Prove that if n is an odd integer, then n^2 is also an odd integer. (Start with 'n is an odd integer $\Rightarrow n = 2k + 1$ where k is any integer'.)

7 Prove that $x^2 + bx + c$ is a perfect square if and only if $b^2 - 4c = 0$.

8 Show that if a and b are positive numbers where $a > b$, then $a^2 > b^2$.
(Start with '$a > b \quad \Rightarrow \quad a - b > 0$'
and
'$a > 0$ and $b > 0 \quad \Rightarrow \quad a + b > 0$'.)

9 $y = ax^3 + bx^2 + cx + d$

Prove that $y = 0$ when $x = 1$ if and only if $(x - 1)$ is a factor of $ax^3 + bx^2 + cx + d$.

10 Prove that $\ln x - \ln y = \ln \frac{x}{y}$.

CHAPTER 6 Coordinate geometry 1

Location of a Point in a Plane

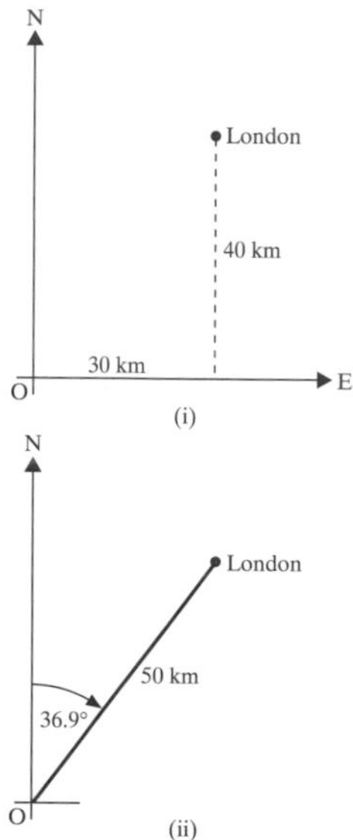

Graphical methods lend themselves particularly well to the investigation of the geometric properties of many curves and surfaces. At this stage we will deal only with rectilinear plane figures (i.e. two dimensional figures bounded by straight lines) and we need a simple and unambiguous way of describing the position of a point.

Consider the problem of describing the location of a city, London say.

There are many ways in which this can be done, but they all require reference to at least one known place and known directions. This is called a *frame of reference*. Within this frame of reference, two measurements are needed to locate the city precisely. These measurements are called coordinates.

The position of London is described in two alternative ways in the diagrams.

In (i) the frame of reference is a fixed point O and the directions due east and due north from O. The coordinates of London are 30 km east of O and 40 km north of O.

In (ii) the frame of reference is a fixed point O and the direction due north from O. The coordinates of London are 50 km from O and a bearing of 036.9°.

For graphical work we use the first of these systems.

Cartesian Coordinates

This system of reference uses a fixed point O, called *the origin*, and a pair of perpendicular lines through O. One of these lines is drawn horizontally and is called the x-axis. The other line is drawn vertically and is called the y-axis.

Coordinate Geometry

Coordinate geometry is the name given to the graphical analysis of geometric properties. For this analysis we need to refer to three types of points:

1) fixed points whose coordinates are known, e.g. the point (1, 2).

2) fixed points whose coordinates are not known numerically. These points are referred to as (x_1, y_1), (x_2, y_2), ... etc. or (a, b), etc.

3) points which are not fixed. We call these general points and we refer to them as (x, y), (X, Y), etc.

It is conventional to use the letters A, B, ... for fixed points and the letters P, Q, ... for general points.

It is also conventional to graduate the axes using identical scales. This avoids distorting the shape of figures.

The Length of a Line Joining Two Points

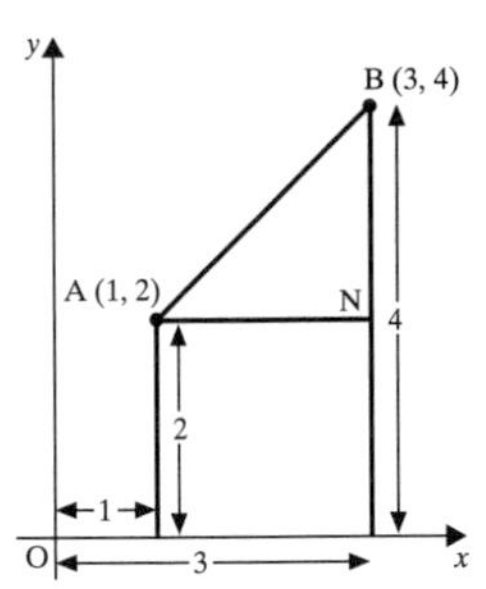

Consider the line joining the points A(1, 2) and B(3, 4).
The length of the line joining A and B can be found by using Pythagoras' theorem,

i.e.
$$\begin{aligned} AB^2 &= AN^2 + BN^2 \\ &= (3-1)^2 + (4-2)^2 \\ &= 8 \end{aligned}$$

Therefore $AB = \sqrt{8} = 2\sqrt{2}$

In the same way the length of the line joining any two points $A(x_1, y_1)$ and $B(x_2, y_2)$ can be found.

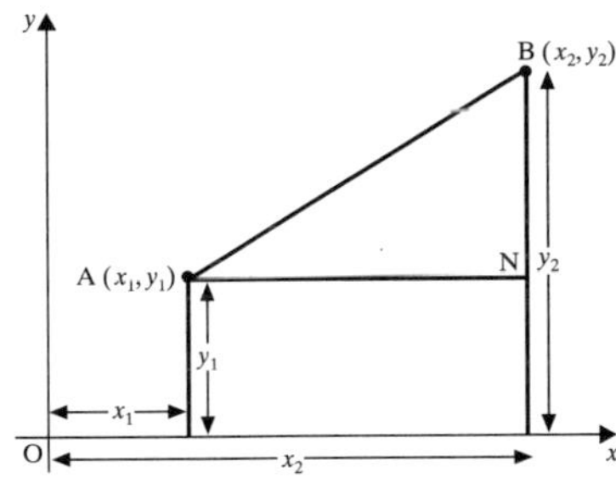

From the diagram,
$$\begin{aligned} AB^2 &= AN^2 + BN^2 \\ &= (x_2 - x_1)^2 + (y_2 - y_1)^2 \end{aligned}$$
$$\Rightarrow \quad AB = \sqrt{(x_2 - x_1)^2 + (y_2 - y_1)^2}$$

i.e.

the length of the line joining $A(x_1, y_1)$ to $B(x_2, y_2)$ is given by

$$AB = \sqrt{(x_2 - x_1)^2 + (y_2 - y_1)^2}$$

This formula still holds when some, or all, of the coordinates are negative. This is illustrated in the next worked example.

EXAMPLES 6A

1 Find the length of the line joining A(−2, 2) to B(3, −1).

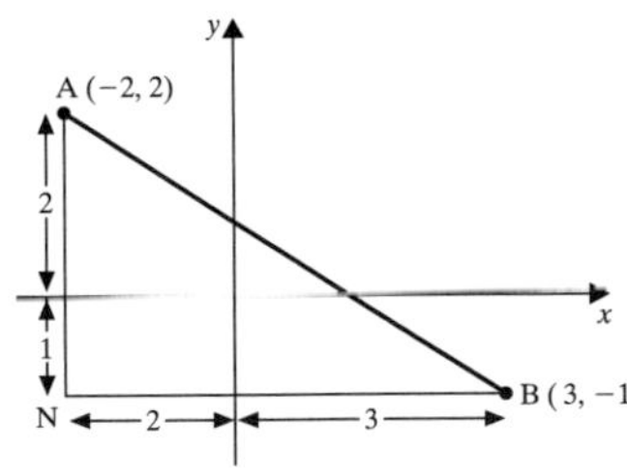

$$\begin{aligned} AB &= \sqrt{(x_2 - x_1)^2 + (y_2 - y_1)^2} \\ &= \sqrt{(3 - \{-2\})^2 + (-1-2)^2} \\ &= \sqrt{5^2 + (-3)^2} \\ &= \sqrt{34} \end{aligned}$$

From the diagram, $BN = 3 + 2 = 5$ and $AN = 2 + 1 = 3$

$\Rightarrow \quad AB^2 = 5^2 + 3^2 = 34 \quad \Rightarrow \quad AB = \sqrt{34}$

This confirms that the formula used above is valid when some of the coordinates are negative.

The Midpoint of the Line Joining Two Given Points

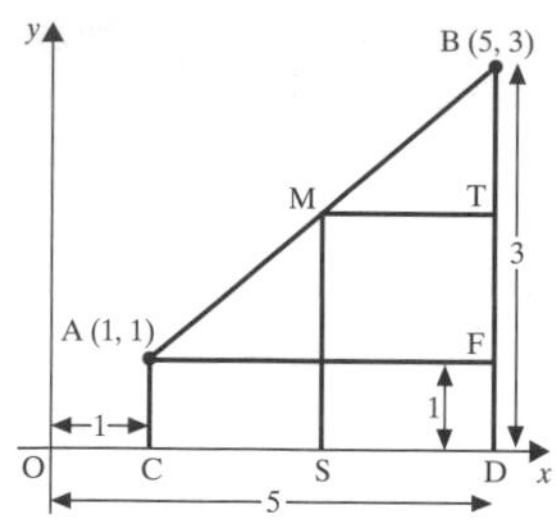

Consider the line joining the points A(1, 1) and B(5, 3).

Using the intercept theorem, we see that if M is the midpoint of AB then S is the midpoint of CD.

Therefore the x-coordinate of M is given by OS, where

$$\text{OS} = \text{OC} + \tfrac{1}{2}\text{CD} = 1 + \tfrac{1}{2}(5-1) = 3$$

Similarly, T is the midpoint of BF, so the y-coordinate of M is given by SM (= DT), where

$$\text{DT} = \text{DF} + \tfrac{1}{2}\text{FB} = 1 + \tfrac{1}{2}(3-1) = 2$$

Therefore M is the point (3, 2).

In general, if $A(x_1, y_1)$ and $B(x_2, y_2)$ are two points, then the coordinates of M, the midpoint of AB, can be found in the same way.

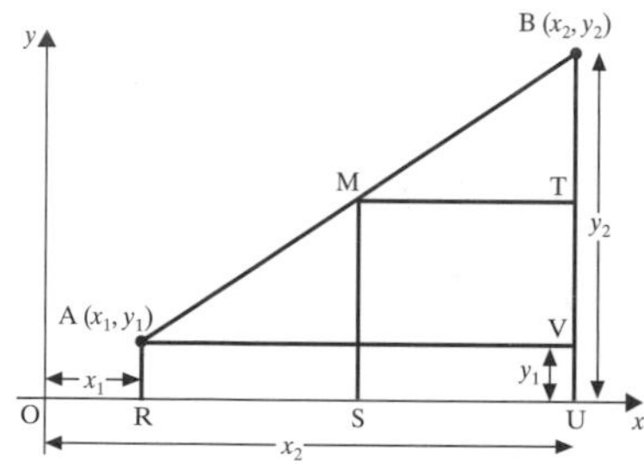

At M,
$$x = \text{OS} = \text{OR} + \tfrac{1}{2}\text{RU} = x_1 + \tfrac{1}{2}(x_2 - x_1) = \tfrac{1}{2}(x_1 + x_2)$$

and
$$y = \text{SM} = \text{UT} = \text{UV} + \tfrac{1}{2}\text{BV} = y_1 + \tfrac{1}{2}(y_2 - y_1) = \tfrac{1}{2}(y_1 + y_2)$$

The coordinates of the midpoint of the line joining $A(x_1, y_1)$ and $B(x_2, y_2)$ are $\left[\frac{1}{2}(x_1 + x_2), \frac{1}{2}(y_1 + y_2)\right]$

These coordinates are easy to remember as the average of the coordinates of A and B.

The next worked example shows that this formula holds when some of the coordinates are negative.

EXAMPLES 6A (continued)

2 Find the coordinates of the midpoint of the line joining A(−3, −2) and B(1, 3).

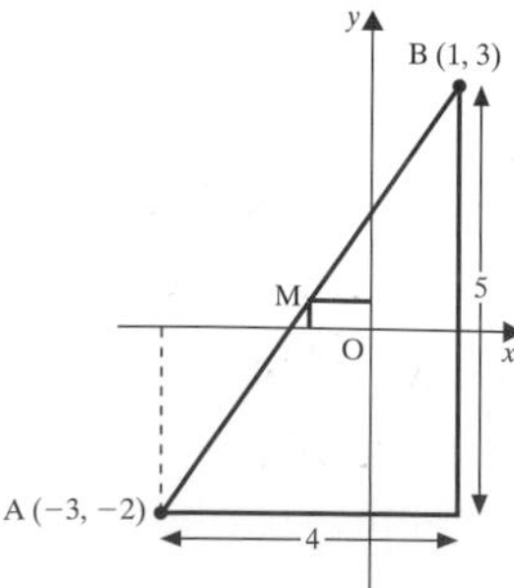

The coordinates of M are $\left[\frac{1}{2}(x_1 + x_2), \frac{1}{2}(y_1 + y_2)\right]$

$$= \left[\tfrac{1}{2}(-3+1), \tfrac{1}{2}(-2+3)\right] = \left(-1, \tfrac{1}{2}\right)$$

Alternatively, from the diagram, M is half-way from A to B horizontally and vertically,

i.e. at M $\quad x = -3 + \frac{1}{2}(4) = -1 \quad$ and $\quad y = -2 + \frac{1}{2}(5) = \frac{1}{2}$

This confirms that the formula works when some of the coordinates are negative.

The Coordinates of a Point Dividing a Line in a Given Ratio

To find the coordinates of a point that divides the line joining two points in a given ratio, a clear diagram together with some simple mental arithmetic is all that is needed.

EXAMPLES 6A (continued)

3 Find the coordinates of the point which divides the line joining A(−2, 5) and B(4, 2) in the ratio 2 : 1 (a) internally at P (b) externally at Q.

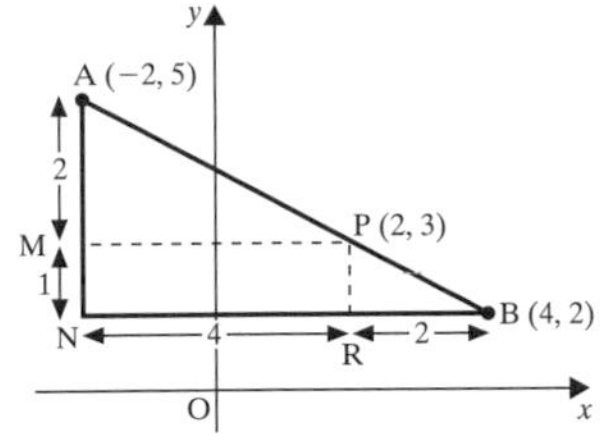

P is between A and B.
As AP : PB = 2 : 1, AM : MN = NR : RB = 2 : 1
P is the point (2, 3).

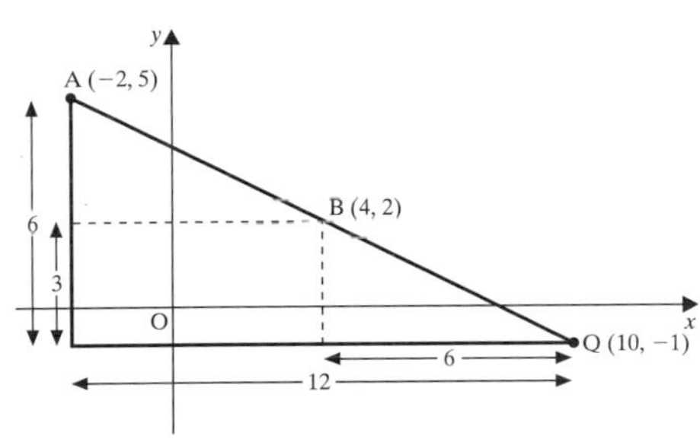

Q is on AB produced, where AQ : BQ = 2 : 1
Q is the point (10, −1).

EXERCISE 6A

1 Find the length of the line joining

a A(1, 2) and B(4, 6)

b C(3, 1) and D(2, 0)

c J(4, 2) and K(2, 5)

2 Find the coordinates of the midpoints of the lines joining the points in Question 1.

3 Find (i) the length,
(ii) the coordinates of the midpoint,
of the line joining

a A(−1, −4), B(2, 6)

b S(0, 0), T(−1, −2)

c E(−1, −4), F(−3, −2)

4 Find the distance from the origin to the point (7, 4).

5 Find the length of the line joining the point (−3, 2) to the origin.

6 Find the coordinates of the midpoint of the line from the point (4, −8) to the origin.

7 Show, by using Pythagoras' Theorem, that the lines joining A(1, 6), B(−1, 4) and C(2, 1) form a right-angled triangle.

8 A, B and C are the points (7, 3), (−4, 1) and (−3, −2) respectively.

a Show that △ABC is isosceles.

b Find the midpoint of BC.

c Find the area of △ABC.

9 The vertices of a triangle are A(0, 2), B(1, 5) and C(−1, 4). Find

a the perimeter of the triangle

b the coordinates of D such that AD is a median of △ABC

c the length of AD.

10 Show that the lines OA and OB are perpendicular where A and B are the points (4, 3) and (3, −4) respectively.

11 M is the midpoint of the line joining A to B. The coordinates of A and M are (5, 7) and (0, 2) respectively. Find the coordinates of B.

12 Find the coordinates of the point that divides the line joining A(2, 4) to B(−3, 9) internally in the ratio 1 : 4.

13 Find the coordinates of the point that divides the line joining L(−3, −4) to M(3, 5) externally in the ratio 3 : 1.

Gradient

The gradient of a straight line is a measure of its slope with respect to the x-axis. Gradient is defined as

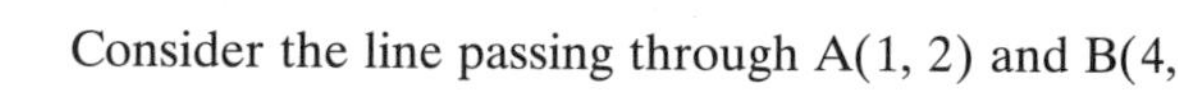
the increase in y divided by the increase in x between one point and another point on the line.

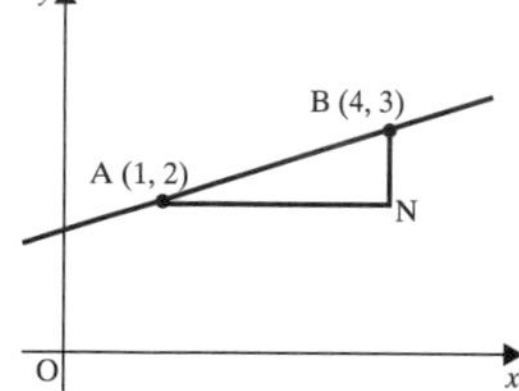

Consider the line passing through A(1, 2) and B(4, 3).

From A to B, the increase in y is 1,
the increase in x is 3.

Therefore the gradient of AB is $\frac{1}{3}$.

The gradient of a line may be found from *any* two points on the line.

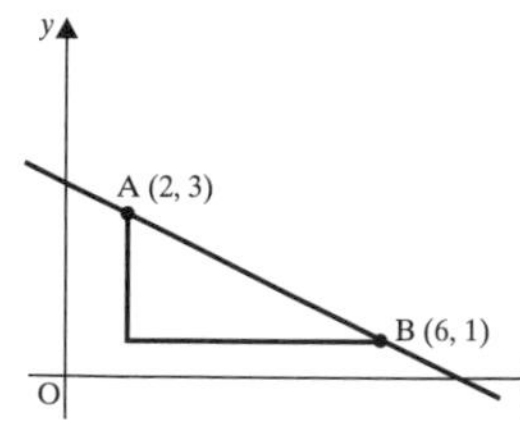

Now consider the line through the points A(2, 3) and B(6, 1).

Moving from A to B $\quad \dfrac{\text{increase in } y}{\text{increase in } x} = \dfrac{-2}{4} = -\dfrac{1}{2}$

Alternatively, moving from B to A $\quad \dfrac{\text{increase in } y}{\text{increase in } x} = \dfrac{2}{-4} = -\dfrac{1}{2}$

This shows that it does not matter in which order the two points are considered, provided that they are considered in the *same* order when calculating the increases in x and in y.

From these two examples we see that the gradient of a line may be positive or negative.

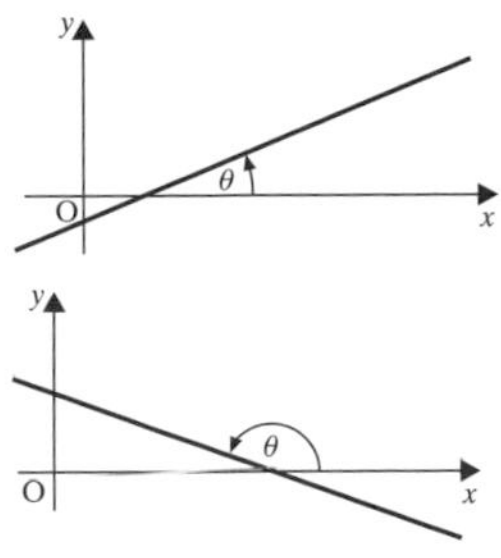

A positive gradient indicates an 'uphill' slope with respect to the positive direction of the x-axis, i.e. the line makes an acute angle with the positive sense of the x-axis.

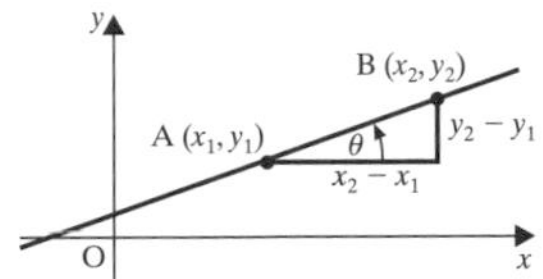

A negative gradient indicates a 'downhill' slope with respect to the positive direction of the x-axis, i.e. the line makes an obtuse angle with the positive sense of the x-axis.

In general,

the gradient of the line passing through $A(x_1, y_1)$ and $B(x_2, y_2)$ is

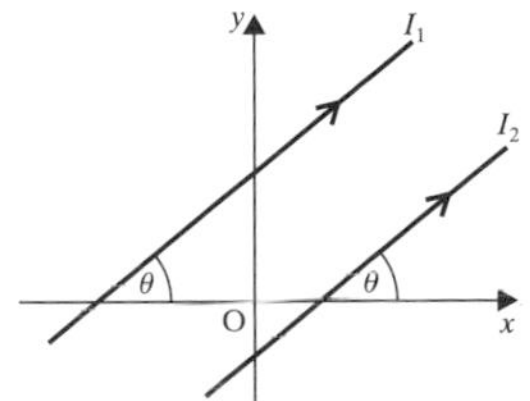

$$\frac{\text{the increase in } y}{\text{the increase in } x} = \frac{y_2 - y_1}{x_2 - x_1}$$

As the gradient of a straight line is the increase in y divided by the increase in x from one point on the line to another,

gradient measures the increase in y per unit increase in x,
i.e. the rate of increase of y with respect to x.

Parallel Lines

If l_1 and l_2 are parallel lines, they are equally inclined to the positive direction of the x-axis,

i.e. **parallel lines have equal gradients.**

Perpendicular lines

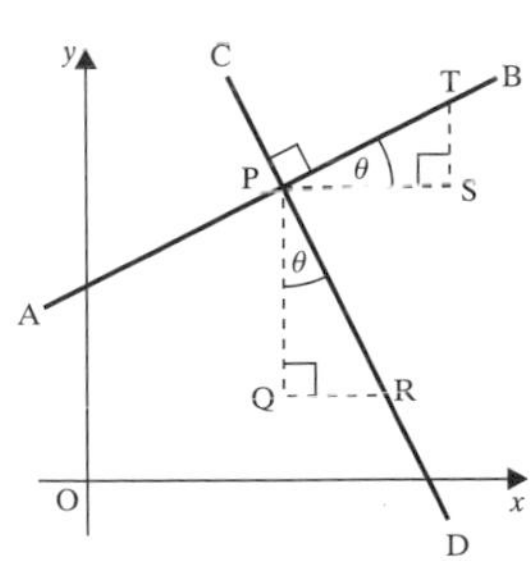

Consider the perpendicular lines AB and CD whose gradients are m_1 and m_2 respectively.

If AB makes an angle θ with the x-axis then CD makes an angle θ with the y-axis. Therefore triangles PQR and PST are similar.

Now the gradient of AB is $\dfrac{\text{ST}}{\text{PS}} = m_1$

and the gradient of CD is $\dfrac{-\text{PQ}}{\text{QR}} = m_2$, i.e. $\dfrac{\text{PQ}}{\text{QR}} = -m_2$

But $\dfrac{\text{ST}}{\text{PS}} = \dfrac{\text{QR}}{\text{PQ}}$ ($\triangle$s PQR and PST are similar)

therefore $m_1 = -\dfrac{1}{m_2}$ or $m_1 m_2 = -1$

The product of the gradients of perpendicular lines is -1,

i.e. if one line has gradient m, any perpendicular line has gradient $-\dfrac{1}{m}$.

EXAMPLE 6B

Determine, by comparing gradients, whether the following three points are collinear (i.e. lie on the same straight line).

$$A(\tfrac{2}{3}, 1),\ B(1, \tfrac{1}{2}),\ C(2, -1)$$

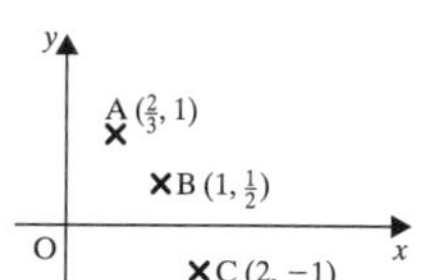

The gradient of AB is $\dfrac{1 - \frac{1}{2}}{\frac{2}{3} - 1} = -\dfrac{3}{2}$

The gradient of BC is $\dfrac{-1 - \frac{1}{2}}{2 - 1} = -\dfrac{3}{2}$

As the gradients of AB and BC are the same, A, B and C are collinear.

> The diagram, although not strictly necessary, gives a check that the answer is reasonable.

EXERCISE 6B

1 Find the gradient of the line through the pair of points.

a (0, 0), (1, 3)

b (1, 4), (3, 7)

c (5, 4), (2, 3)

d (−1, 4), (3, 7)

e (−1, −3), (−2, 1)

f (−1, −6), (0, 0)

g (−2, 5), (1, −2)

h (3, −2), (−1, 4)

i (h, k), (0, 0)

2 Determine whether the given points are collinear.

a (0, −1), (1, 1), (2, 3)

b (0, 2), (2, 5), (3, 7)

c (−1, 4), (2, 1), (−2, 5)

d (0, −3), (1, −4), $(-\frac{1}{2}, -\frac{5}{2})$

3 Determine whether AB and CD are parallel, perpendicular or neither.

a A(0, −1), B(1, 1), C(1, 5), D(−1, 1)

b A(1, 1), B(3, 2), C(−1, 1), D(0, −1)

c A(3, 3), B(−3, 1), C(−1, −1), D(1, −7)

d A(2, −5), B(0, 1), C(−2, 2), D(3, −7)

e A(2, 6), B(−1, −9), C(2, 11), D(0, 1)

Problems in Coordinate Geometry

This chapter ends with a miscellaneous selection of problems on coordinate geometry. A clear and reasonably accurate diagram showing all the given information will often suggest the most direct method for solving a particular problem.

EXAMPLE 6C

The vertices of a triangle are the points A(2, 4), B(1, −2) and C(−2, 3) respectively. The point H(a, b) lies on the altitude through A. Find a relationship between a and b.

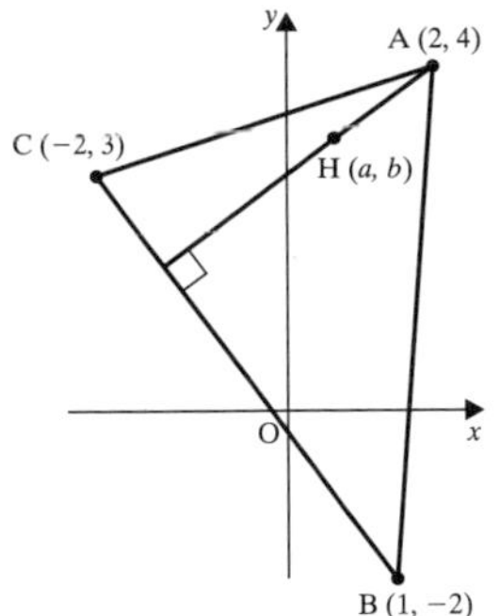

As H is on the altitude through A, AH is perpendicular to BC.

The gradient of AH is $\frac{4-b}{2-a}$,

the gradient of BC is $\frac{3-(-2)}{-2-1} = -\frac{5}{3}$

> The product of the gradients of perpendicular lines is −1.

Therefore $\left(\frac{4-b}{2-a}\right)\left(-\frac{5}{3}\right) = -1$

$\Rightarrow \quad \frac{-20+5b}{6-3a} = -1$

$\Rightarrow \quad 5b = 3a + 14$

EXERCISE 6C

1 A(1, 3), B(5, 7), C(4, 8), D(a, b) form a rectangle ABCD. Find a and b.

2 The triangle ABC has its vertices at the points A(1, 5), B(4, −1) and C(−2, −4).

a Show that △ABC is right-angled.

b Find the area of △ABC.

3 Show that the point $(-\frac{32}{3}, 0)$ is on the altitude through A of the triangle whose vertices are A(1, 5), B(1, −2) and C(−2, 5).

4 Show that the triangle whose vertices are (1, 1), (3, 2), (2, −1) is isosceles.

5 Find, in terms of a and b, the length of the line joining (a, b) and ($2a$, $3b$).

6 The point (1, 1) is the centre of a circle whose radius is 2. Show that the point (1, 3) is on the circumference of this circle.

7 A circle, radius 2 and centre the origin, cuts the x-axis at A and B and cuts the positive y-axis at C. Prove that ∠ACB = 90°.

8 Find in terms of p and q, the coordinates of the midpoint of the line joining C(p, q) and D(q, p). Hence show that the origin is on the perpendicular bisector of the line CD.

9 The point (a, b) is on the circumference of the circle of radius 3 whose centre is at the point (2, 1). Find a relationship between a and b.

10 ABCD is a quadrilateral where A, B, C and D are the points (3, −1), (6, 0), (7, 3) and (4, 2). Prove that the diagonals bisect each other at right angles and hence find the area of ABCD.

11 The vertices of a triangle are at the points A(a, 0), B(0, b) and C(c, d) and ∠B = 90°. Find a relationship between a, b, c and d.

12 A point P(a, b) is equidistant from the y-axis and from the point (4, 0). Find a relationship between a and b.

CHAPTER 7 Triangles

Trigonometric Ratios of Acute Angles

The sine, cosine and tangent of an acute angle A in a right-angled triangle are defined in terms of the sides of the triangle as

$$\cos A = \frac{\text{adjacent}}{\text{hypotenuse}}$$

$$\sin A = \frac{\text{opposite}}{\text{hypotenuse}}$$

$$\tan A = \frac{\text{opposite}}{\text{adjacent}}$$

If any of these trig ratios is given as a fraction, the lengths of two of the sides of the right-angled triangle can be marked. Then the third side can be calculated by using Pythagoras' theorem.

EXAMPLE 7A

Given that $\sin A = \frac{3}{5}$ find $\cos A$ and $\tan A$.

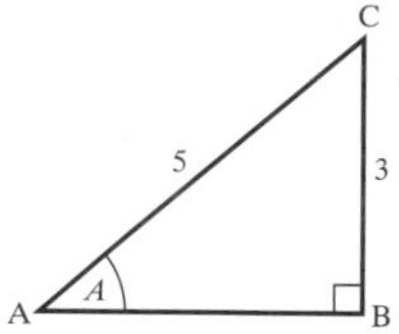

Because $\sin A = \frac{\text{opp}}{\text{hyp}}$, we can draw a right-angled triangle with the side opposite to angle A of length 3 units and a hypotenuse of length 5 units.

Applying Pythagoras' theorem to $\triangle$ABC gives

$(\text{AB})^2 + 3^2 = 5^2 \quad \Rightarrow \quad \text{AB} = 4$

Then $\quad \cos A = \frac{\text{adj}}{\text{hyp}} = \frac{4}{5}$

and $\quad \tan A = \frac{\text{opp}}{\text{adj}} = \frac{3}{4}$

EXERCISE 7A

If any of the square roots in this exercise are not integers, leave them in surd form.

1 If $\tan A = \frac{12}{5}$ find $\sin A$ and $\cos A$.

2 Given that $\cos X = \frac{4}{5}$ find $\tan X$ and $\sin X$.

3 If $\sin P = \frac{40}{41}$ find $\cos P$ and $\tan P$.

4 If $\tan A = 1$ find $\sin A$ and $\cos A$.

5 If $\cos Y = \frac{2}{3}$ find $\sin Y$ and $\tan Y$.

6 Given that $\sin A = \frac{1}{2}$ what is $\cos A$? Use your calculator to find the size of angle A.

7 If $\sin X = \frac{?}{25}$ and $\tan X = \frac{7}{?}$ find $\cos X$.

In each question from 8 to 10, use $\sin X = \frac{3}{5}$.

8 Find $\cos X$ and hence calculate $\cos^2 X - \sin^2 X$. Use a calculator to determine the value of angle X and hence find $\cos 2X$ correct to 2 s.f. What conclusion can you draw?

9 Find $\cos^2 X + \sin^2 X$.

10 Evaluate $2 \sin X \cos X$ as a decimal. Find correct to 2 s.f. the value of $\sin 2X$ and draw any conclusion that you can.

Trigonometric Ratios of Obtuse Angles

The Cosine of an Obtuse Angle

Although the definition of the cosine that we have used so far applies only to an acute angle, larger angles do have cosines. These values are stored in calculators, for example:

θ	0	30°	45°	60°	90°	120°	135°	150°	180°
$\cos\theta$ (to 2 d.p.)	1	0.87	0.71	0.50	0	−0.50	−0.71	−0.87	−1

Plotting a graph of these figures gives a shape that is called a cosine curve. The symbol θ is the most commonly used symbol for a varying angle.

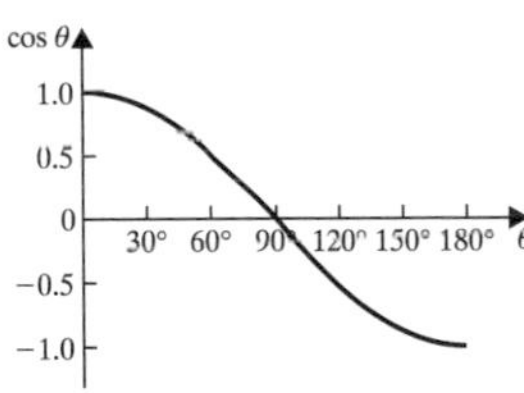

The table of values and the graph show clearly that an acute angle has a positive cosine while the cosine of an obtuse angle is negative.

From the graph or the table it can be seen that

$$\cos 60° = 0.5$$

and $\quad \cos 120° = -0.5$

i.e. $\quad \cos 120° = -\cos 60° \quad (120° + 60° = 180°)$

also $\quad \cos 45° = 0.71$

and $\quad \cos 135° = -0.71$

i.e. $\quad \cos 135° = -\cos 45° \quad (135° + 45° = 180°)$

You can find more pairs of angles which suggest the relationship

$$\mathbf{\cos\theta = -\cos(180° - \theta)}$$

It also appears that the graph has rotational symmetry about the point (90, 0).

The Sine of an Obtuse Angle

As we saw in the case of cosines, sine ratios are not limited to acute angles. The sines of larger angles are given by a calculator, e.g. for angles from 0 to 180° we have

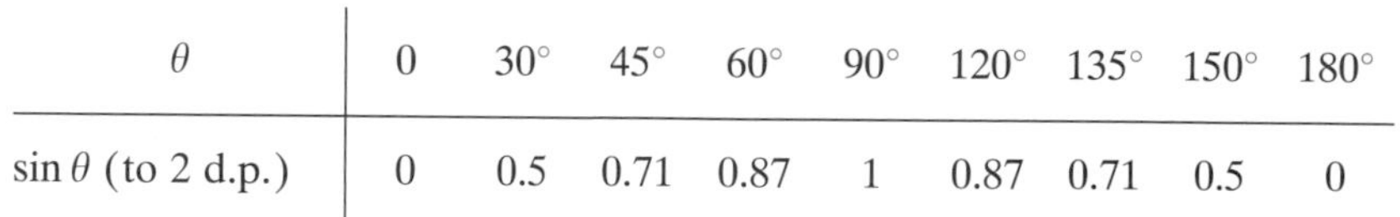

θ	0	30°	45°	60°	90°	120°	135°	150°	180°
$\sin\theta$ (to 2 d.p.)	0	0.5	0.71	0.87	1	0.87	0.71	0.5	0

Plotting these values gives this graph which is called a sine curve.

Again relationships can be seen between the sines of pairs of angles, for example

$\sin 30° = 0.5$ and $\sin 150° = 0.5$

i.e. $\sin 150° = \sin 30°$ $(150° + 30° = 180°)$

also $\sin 60° = 0.87$ and $\sin 120° = 0.87$

i.e. $\sin 120° = \sin 60°$ $(120° + 60° = 180°)$

This time it looks as if

$$\sin(180° - \theta) = \sin\theta$$

and it looks as though the curve is symmetrical about the line $\theta = 90°$.

Similar reasoning shows that

$$\tan(180° - \theta) = -\tan\theta$$

EXAMPLES 7B

1 If $\sin\theta = \frac{1}{5}$ find two possible values for θ.

As given by a calculator, the angle with a sine of 0.2 is 11.5°.

But $\sin\theta = \sin(180° - \theta)$ so $\sin 11.5° = \sin(180° - 11.5°)$, therefore when $\sin\theta = \frac{1}{5}$, two values of θ are 11.5° and 168.5°.

2 Use the information in the diagram to find $\cos\theta$.

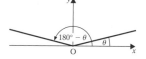

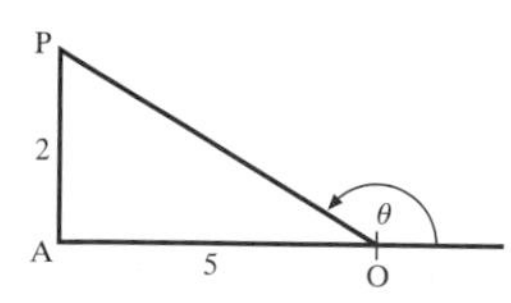

In $\triangle OPA$ $\quad OP^2 = 4 + 25 = 29$ (Pythagoras)

and $\quad A\hat{O}P = (180° - \theta)$

$$\cos(180° - \theta) = \frac{OA}{OP} = \frac{5}{\sqrt{29}}$$

$$\cos\theta = -\cos(180° - \theta) = -\frac{5}{\sqrt{29}}$$

EXERCISE 7B

In each question from 1 to 4, find $\sin\theta$, $\cos\theta$ and $\tan\theta$, giving unknown lengths in surd form when necessary.

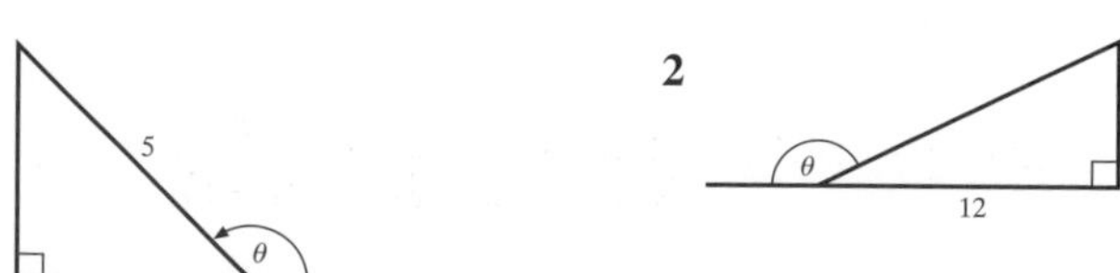

3 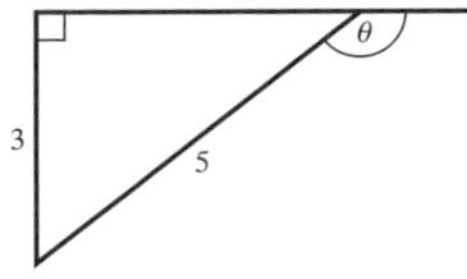

4 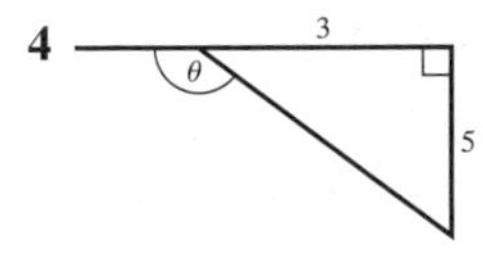

In each question from 5 to 12, find X where X is an angle from 0 to 180°.

5 $\sin X = \sin 80°$

6 $\cos X = -\cos 75°$

7 $\sin X = \sin 128°$

8 $\cos 30° = -\cos X$

9 $\sin X = \sin 81°$

10 $-\cos 123° = \cos X$

11 $\sin 90° = \sin X$

12 $\cos 91° = -\cos X$

The unknown angles in questions 13 to 17 are in the range 0 to 180°.

13 If $\cos X = -\frac{12}{13}$ find $\sin X$.

14 If $\sin\theta = \frac{4}{5}$ find, to the nearest degree, two possible values of θ.

15 Given that $\sin A = 0.5$ and $\cos A = -0.8660$, find $\angle$A.

16 Is there an angle X for which

a $\cos X = 0$ and $\sin X = 1$

b $\sin X = 0$ and $\cos X = 1$

c $\cos X = 0$ and $\sin X = -1$?

17 If $\cos A = -\cos B$, what is the relationship between $\angle$A and $\angle$B?

Finding Unknown Sides and Angles in a Triangle

Triangles are involved in many practical measurements (e.g. surveying) so it is important to be able to make calculations from limited data about a triangle.

Although a triangle has three sides and three angles, it is not necessary to know all of these in order to define a particular triangle. If enough information about a triangle is known, the remaining sides and angles can be calculated by using a suitable formula. This is called *solving* the triangle.

The two relationships that are used most frequently are the sine rule and the cosine rule.

The Sine Rule

When working with a triangle ABC the side opposite to $\angle$A is denoted by a, the side opposite to $\angle$B by b and so on.

In a triangle ABC, $\dfrac{a}{\sin A} = \dfrac{b}{\sin B} = \dfrac{c}{\sin C}$

This rule applies whether or not one of the angles is obtuse.

Proof
Consider a triangle ABC in which there is no right angle.

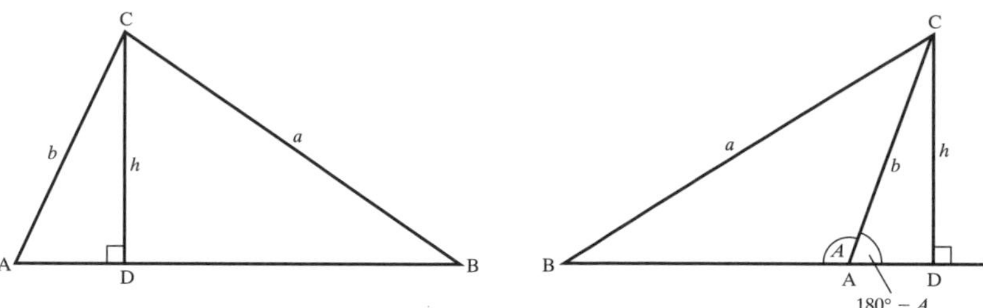

A line drawn from C, perpendicular to AB produced if necessary, divides triangle ABC into two right-angled triangles, CDA and CDB.

In $\triangle$CDA $\quad \sin A = h/b \quad \Rightarrow \quad h = b\sin A \qquad (\sin(180 - A) = \sin A)$

In $\triangle$CDB $\quad \sin B = h/a \quad \Rightarrow \quad h = a\sin B$

Therefore $\quad a\sin B = b\sin A$

i.e. $$\frac{a}{\sin A} = \frac{b}{\sin B}$$

We can equally well divide $\triangle$ABC into two right-angled triangles by drawing the perpendicular from A to BC (or from B to AC). This gives a similar result,

i.e. $$\frac{b}{\sin B} = \frac{c}{\sin C}$$

By combining the two results we produce the sine rule,

$$\frac{a}{\sin A} = \frac{b}{\sin B} = \frac{c}{\sin C}$$

Using the Sine Rule

The sine rule is made up of three separate fractions, only two of which can be used at a time. We select the two which contain three known quantities and only one unknown. It follows that, to use the sine rule, we must know the values of

either two sides and the angle opposite one of them
or two angles and a side.

Note that, when the sine rule is being used to find an unknown angle, we can use it in the form $\dfrac{\sin A}{a} = \dfrac{\sin B}{b} = \dfrac{\sin C}{c}$

EXAMPLES 7C

1 In $\triangle$ABC, BC = 5 cm, $A = 43°$ and $B = 61°$. Find the length of AC.

> ∠A, ∠B and *a* are known and *b* is required, so the two fractions we select from the sine rule are $\dfrac{a}{\sin A} = \dfrac{b}{\sin B}$

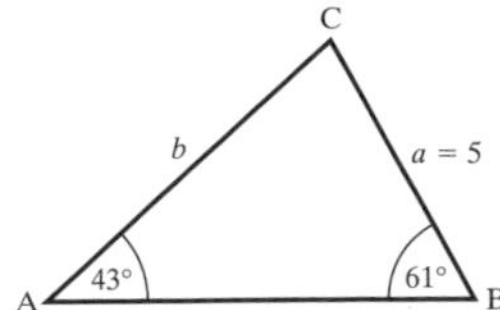

$$\frac{5}{\sin 43°} = \frac{b}{\sin 61°}$$

$$\Rightarrow \quad b = \frac{5 \sin 61°}{\sin 43°} = 6.412\ldots$$

Therefore AC = 6.41 cm correct to 3 s.f.

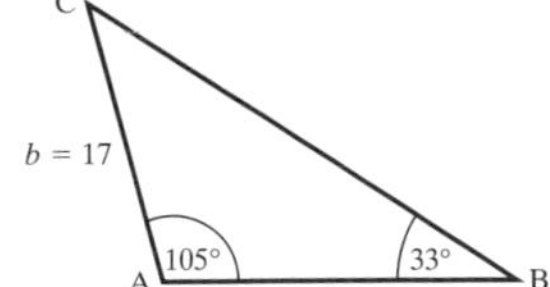

2 In ABC, AC = 17 cm, $\angle A = 105°$ and $\angle B = 33°$. Find AB.

> The two sides involved are b and c, so before the sine rule can be used we must find $\angle C$.

$\angle A + \angle B + \angle C = 180° \quad \Rightarrow \quad \angle C = 42°$

From the sine rule $\dfrac{b}{\sin B} = \dfrac{c}{\sin C}$

$$\Rightarrow \quad \frac{17}{\sin 33°} = \frac{c}{\sin 42°} \quad \Rightarrow \quad c = \frac{17 \times \sin 42°}{\sin 33°} = 20.88\ldots$$

Therefore AB = 20.9 cm correct to 3 s.f.

EXERCISE 7C

1 In $\triangle$ABC, AB = 9 cm, $\angle A = 51°$ and $\angle C = 39°$. Find BC.

2 In $\triangle$PQR, $\angle R = 52°$, $\angle Q = 79°$ and PR = 12.7 cm. Find PQ.

3 In $\triangle$DEF, DE = 174 cm, $\angle D = 48°$ and $\angle F = 56°$. Find EF.

4 In $\triangle$XYZ, $\angle X = 130°$, $\angle Y = 21°$ and XZ = 53 cm. Find YZ.

5 In $\triangle$PQR, $\angle Q = 37°$, $\angle R = 101°$ and PR = 4.3 cm. Find PQ.

6 In $\triangle$XYZ, XY = 92 cm, $\angle X = 59°$ and $\angle Y = 81°$ Find XZ.

7 In $\triangle$PQR, $\angle P = 78°$, $\angle R = 38°$ and PR = 15 cm. Find QR.

8 In $\triangle$ABC, AB = 10 cm, BC = 9.1 cm and AC = 17 cm. Can you use the sine rule to find $\angle A$? If you answer YES, write down the two parts of the sine rule that you would use. If you answer NO, give your reason.

The Ambiguous Case

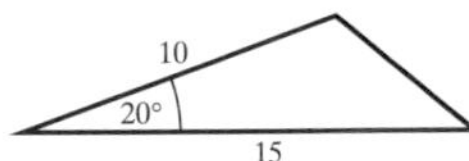

Consider a triangle specified by two sides and one angle.

If the angle is between the two sides there is only one possible triangle.

If, however, the angle is not between the two given sides it is sometimes possible to draw two triangles from the given data.

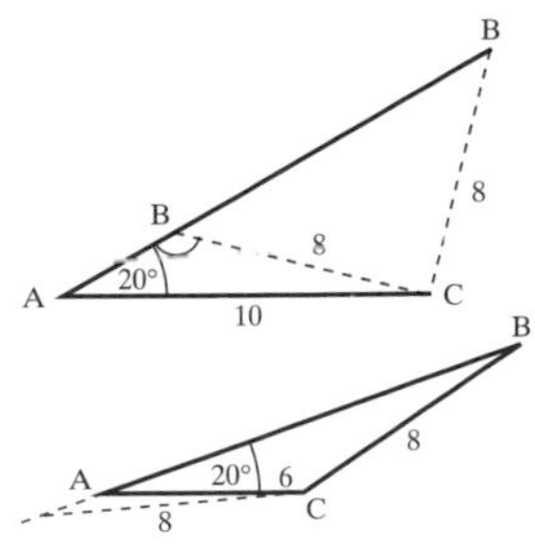

Consider, for example, a triangle ABC in which $\angle A = 20°$, $b = 10$ and $a = 8$.

The two triangles with this specification are shown in the diagram; in one of them B is an acute angle, while in the other one B is obtuse.

Do not assume that there are always two possible triangles however, e.g. if $A = 20°$, $b = 6$ and $a = 8$, only one triangle fits the given data.

So, when using the sine rule to find an angle in a triangle,

it is essential to check whether the obtuse angle is possible.

EXAMPLES 7D

1 In the triangle ABC, find C given that AB = 5 cm, BC = 3 cm and $A = 35°$.

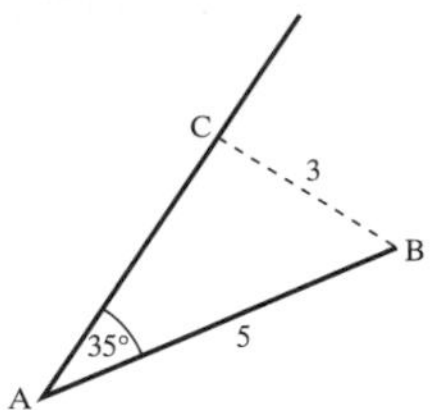

We know a, c and ∠A so the sine rule can be used to find ∠C.

As we are looking for an angle, the form we use is

$$\frac{\sin A}{a} = \frac{\sin C}{c} \quad \Rightarrow \quad \frac{\sin 35°}{3} = \frac{\sin C}{5}$$

Hence $\quad \sin C = \dfrac{5 \times \sin 35°}{3} = 0.959\ldots$

One angle whose sine is 0.959... is 73° (to the nearest degree) but there is also an obtuse angle with the same sine, i.e. 107° (to the nearest degree).

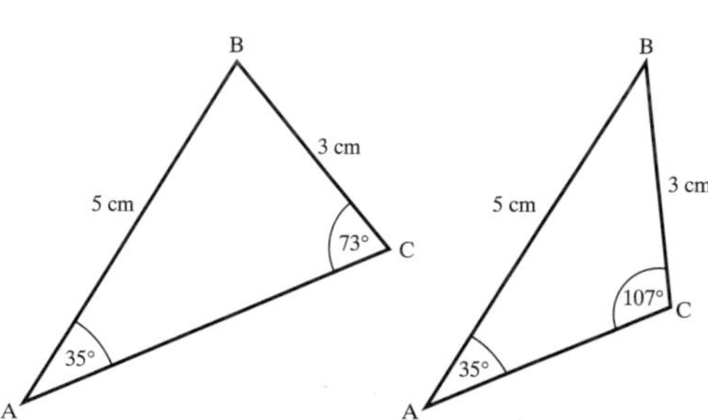

If $C = 107°$, then $A + C = 107° + 35° = 142°$

$\Rightarrow \quad B = 180° - 142° = 38°$

So there is a triangle in which ∠C = 107° and we have two possible triangles.

Therefore ∠C is either 73° or 107°.

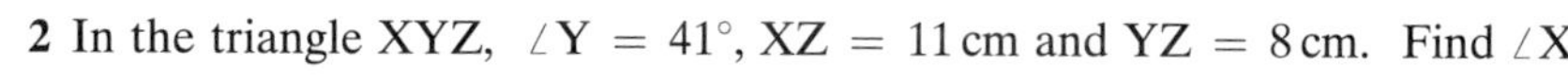

2 In the triangle XYZ, ∠Y = 41°, XZ = 11 cm and YZ = 8 cm. Find ∠X.

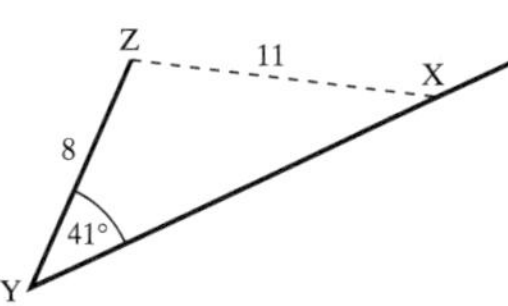

Using the part of the sine rule that involves x, y, ∠X and ∠Y we have

$$\frac{\sin X}{x} = \frac{\sin Y}{y} \quad \Rightarrow \quad \frac{\sin X}{8} = \frac{\sin 41°}{11}$$

Hence $\quad \sin X = \dfrac{8 \times \sin 41°}{11} = 0.4771\ldots$

The two angles with a sine of 0.4771... are 28° and 152° (to the nearest degree).

Checking to see whether 152° is a possible value for ∠X we see that

∠X + ∠Y = 152° + 41° = 193°.

This is greater than 180°, so it is not possible for the angle X to have the value 152°.

In this case then, there is only one possible triangle containing the given data, i.e. the triangle in which ∠X = 28°.

It is interesting to notice how the two different situations that arose in Examples 1 and 2 above can be illustrated by the construction of the triangles with the given data.

When AB = 5 cm, BC = 3 cm and ∠A = 35° we have

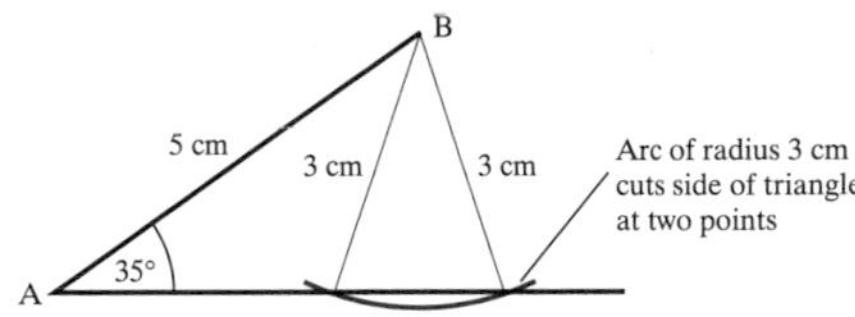

When XZ = 11 cm, YZ = 8 cm and ∠Y = 41° we have

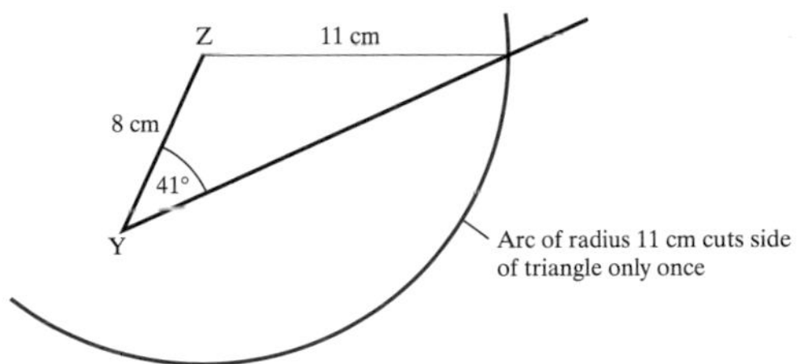

EXERCISE 7D

In each of the following questions, find the angle indicated by the question mark, giving two values in those cases where there are two possible triangles. Illustrate your solution to each question.

	AB	BC	CA	A	B	C
1		2.9 cm	6.1 cm	?	40°	
2	5.7 cm		2.3 cm		20°	?
3	21 cm	36 cm		29.5°		?
4		2.7 cm	3.8 cm	?	54°	
5	4.6 cm		7.1 cm		?	33°
6	9 cm	7 cm		?		40°

The Cosine Rule

The sine rule can be used to solve a triangle only when certain information is given (see p. 72). If, for example, only the three sides are given, the sine rule fails and we need another method. The one we use is called the *cosine rule* and it states that

$$a^2 = b^2 + c^2 - 2bc \cos A$$

This formula is valid whether A is acute or obtuse.

Proof

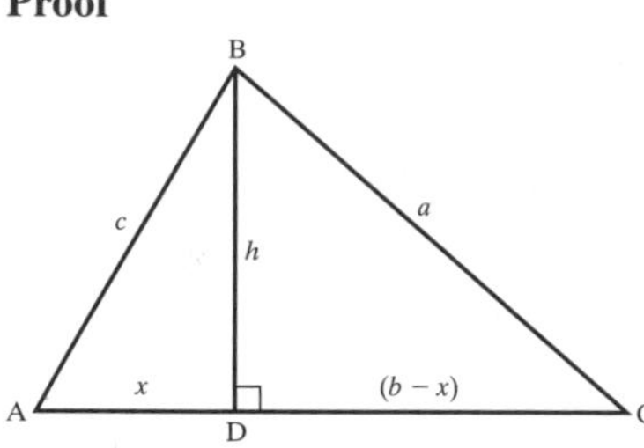

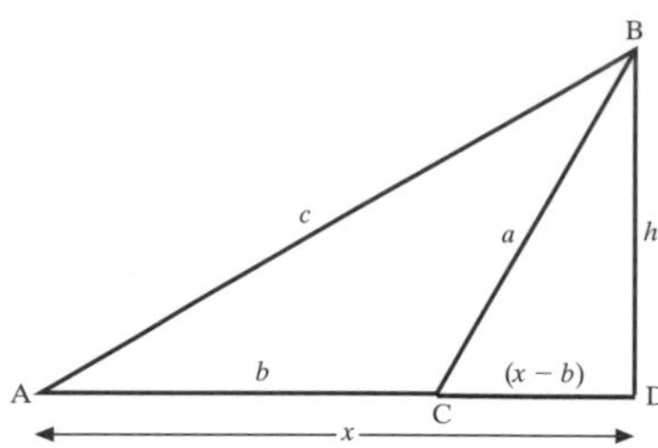

Let ABC be a non-right-angled triangle in which BD is drawn perpendicular to AC produced if necessary. Taking x as the length of AD, the length of CD is $(b-x)$ or $(x-b)$. Then, using h as the length of BD, we can use Pythagoras' theorem to find h in each of the right-angled triangles BDA and BDC, i.e.

$h^2 = c^2 - x^2$ and $h^2 = a^2 - (b-x)^2$ (or $h^2 = a^2 - (x-b)^2$)

Therefore $\quad c^2 - x^2 = a^2 - (b-x)^2$

$\Rightarrow \quad c^2 - x^2 = a^2 - b^2 + 2bx - x^2$

$\Rightarrow \quad a^2 = b^2 + c^2 - 2bx$

But $x = c\cos A$

Therefore $\quad a^2 = b^2 + c^2 - 2bc\cos A$

When the altitude is drawn from A or from C similar expressions for the other sides of a triangle are obtained, i.e.

$\mathbf{b^2 = c^2 + a^2 - 2ca\cos B}$ and $\mathbf{c^2 = a^2 + b^2 - 2ab\cos C}$

EXAMPLES 7E

1 In $\triangle$ABC, BC $= 7$ cm, AC $= 9$ cm and $C = 61°$. Find AB.

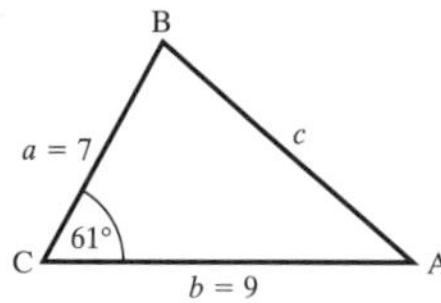

Using the cosine rule, starting with c^2, we have

$$c^2 = a^2 + b^2 - 2ab\cos C$$

$$\Rightarrow \quad c^2 = 7^2 + 9^2 - (2)(7)(9)\cos 61°$$

$$\Rightarrow \quad c = 8.301\ldots$$

Hence AB $= 8.30$ cm correct to 3 s.f.

2 XYZ is a triangle in which $\angle Y = 121°$, XY $= 14$ cm and YZ $= 26.9$ cm. Find XZ.

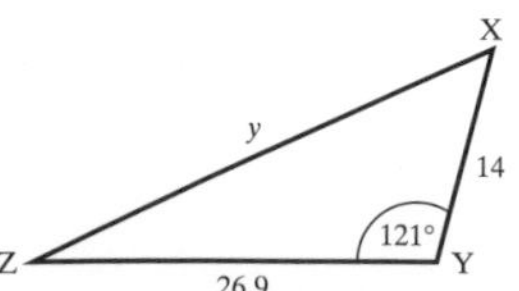

Using $y^2 = z^2 + x^2 - 2zx\cos Y$ gives

$$y^2 = (14)^2 + (26.9)^2 - (2)(14)(26.9)\cos 121°$$

Hence $\quad y^2 = 1307.53\ldots \quad \Rightarrow \quad y = 36.15\ldots$

Therefore XZ $= 36.2$ cm correct to 3 s.f.

EXERCISE 7E In each question use the data given for $\triangle$PQR to find the length of the third side.

	PQ	QR	RP	P	Q	R
1		8 cm	4.6 cm			39°
2	11.7 cm		9.2 cm	75°		
3	29 cm	37 cm			109°	
4		2.1 cm	3.2 cm			97°
5	135 cm		98 cm	48°		
6	4.7 cm	8.1 cm			138°	
7		44 cm	62 cm			72°
8	19.4 cm		12.6 cm	167°		

Using the Cosine Rule to Find an Angle

So far the cosine rule has been used to find an unknown side of a triangle. The formula can be rearranged when we want to find an unknown angle, e.g.

$c^2 = a^2 + b^2 - 2ab\cos C$ can be written $\cos C = \dfrac{a^2 + b^2 - c^2}{2ab}$

with similar expressions for $\cos A$ and $\cos B$.

This form is fairly easy to remember, as the side opposite to the angle being found appears only once in the formula and that is in the last term. On the other hand you may prefer to work from the basic cosine formula for all calculations.

EXAMPLES 7F

1 If, in $\triangle$ABC, $a = 9, b = 16$ and $c = 11$, find, to the nearest degree, the largest angle in the triangle.

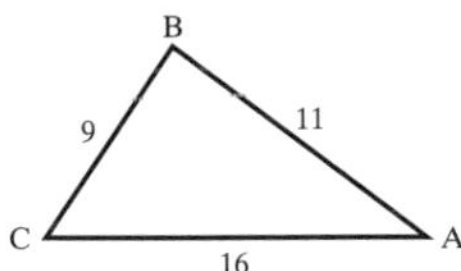

The largest angle in a triangle is opposite to the longest side, so in this question we are looking for angle B and we use

$$\cos B = \frac{c^2 + a^2 - b^2}{2ca}$$

$$= \frac{121 + 81 - 256}{(2)(11)(9)}$$

$$= -0.2727\ldots$$

The negative sign shows that $\angle$B is obtuse.

Hence $B = 106°$ and this is the largest angle in $\triangle$ABC.

2 The sides a, b, c of a triangle ABC are in the ratio 3 : 6 : 5. Find the smallest angle in the triangle.

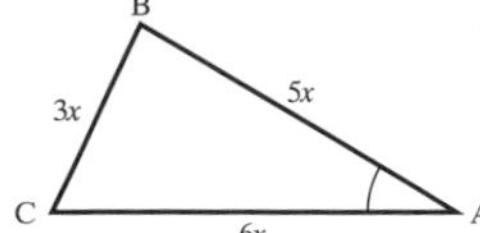

> The actual lengths of the sides are not necessarily 3, 6 and 5 units so we represent them by $3x$, $6x$ and $5x$. The smallest angle is A (opposite to the smallest side).

$$\cos A = \frac{b^2 + c^2 - a^2}{2bc}$$

$$= \frac{36x^2 + 25x^2 - 9x^2}{60x^2} = \frac{52}{60} = 0.8666\ldots$$

Therefore the smallest angle in $\triangle$ABC is 29.9°.

EXERCISE 7F

1 In $\triangle$XYZ, XY = 34 cm, YZ = 29 cm and ZX = 21 cm. Find the smallest angle in the triangle.

2 In $\triangle$PQR, PQ = 1.3 cm, QR = 1.8 cm and RP = 1.5 cm. Find $\angle$Q.

3 In $\triangle$ABC, AB = 51 cm, BC = 37 cm and CA = 44 cm. Find $\angle$A.

4 Find the largest angle in $\triangle$XYZ given that $x = 91$, $y = 77$ and $z = 43$.

5 In $\triangle$ABC, $a = 13$, $b = 18$ and $c = 7$. What is the size of

a the smallest angle

b the largest angle?

6 In $\triangle$PQR the sides PQ, QR and RP are in the ratio 2 : 1 : 2. Find $\angle$P.

7 ABCD is a quadrilateral in which AB = 5 cm, BC = 8 cm, CD = 11 cm, DA = 9 cm and angle ABC = 120°. Find the length of AC and the size of the angle ADC.

General Triangle Calculations

If three independent facts are given about the sides and/or angles of a triangle and further facts are required, a choice must be made between using the sine rule or the cosine rule for the first step.

As the sine rule is easier to work out, use it whenever the given facts make this possible, i.e. whenever an angle and the opposite side are known. (Remember that if two angles are given, then the third angle is also known.)

The cosine rule is used only when the sine rule is not suitable and it is never necessary to use it more than once in solving a triangle.

Suppose, for example, that the triangle PQR is to be solved.

Only one angle is known and the side opposite to it is not given. We must therefore use the cosine rule first to find the length of PR.

Once we know q as well as $\angle$Q, the sine rule can be used to find either of the remaining angles, the third angle then follows from the sum of the angles in the triangle.

EXERCISE 7G Each of the following questions refers to a triangle ABC. Fill in the blank spaces in the table.

	A	B	C	a	b	c
1		80°	50°			68 cm
2			112°	15.7 cm	13 cm	
3	41°	69°		12.3 cm		
4	58°				131 cm	87 cm
5		49°	94°		206 cm	
6	115°		31°			21 cm
7	59°	78°		17 cm		
8		48°	80°		31.3 cm	
9	77°				19 cm	24 cm
10		125°		14 cm		20 cm

11 A tower stands on level ground. From a point P on the ground, the angle of elevation of the top of the tower is 26°. Another point Q is 3 m vertically above P and from this point the angle of elevation of the top of the tower is 21°. Find the height of the tower.

12 A survey of a triangular field, bounded by straight fences, found the three sides to be of lengths 100 m, 80 m and 65 m. Find the angles between the boundary fences.

The Area of a Triangle

The simplest way to find the area of a triangle is to use the formula

$$\text{Area} = \tfrac{1}{2}\ \text{base} \times \text{perpendicular height}$$

If the perpendicular height is not known we can use the formula

area of triangle ABC $= \frac{1}{2}bc\sin A$

with similar expressions for angles B and C,
i.e. area $= \frac{1}{2}ab\sin C = \frac{1}{2}ac\sin B$

Each of these formulae can be expressed in the 'easy to remember' form

$$\text{area} = \tfrac{1}{2}\ \text{product of two sides} \times \text{sine of included angle}$$

EXAMPLE 7H

Find the area of triangle PQR, given that $P = 65°$, $Q = 79°$ and PQ = 30 cm.

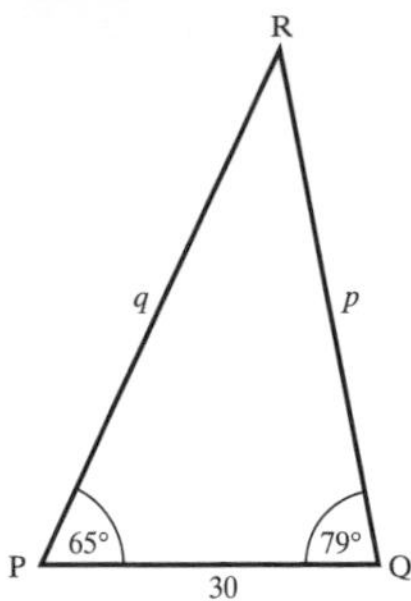

> The given facts do not involve two sides and the included angle so we must first find another side. To do this the sine rule can be used and we need angle R.

$\angle R = 180° - 65° - 79° = 36°$

From the sine rule, $\dfrac{p}{\sin P} = \dfrac{r}{\sin R}$

$\Rightarrow \qquad p = \dfrac{30 \times \sin 65°}{\sin 36°} = 46.25\ldots$

i.e. QR = 46.3 cm (correct to 3 s.f.)

Now we can use area PQR $= \frac{1}{2} pr \sin 79° = 681.1\ldots$

So the area of triangle PQR is 681 cm^2 (correct to 3 s.f.)

EXERCISE 7H

Find the area of each triangle given in Questions 1 to 5.

1 $\triangle$XYZ; XY = 180 cm, YZ = 145 cm, $\angle$Y = 70°.

2 $\triangle$ABC; AB = 75 cm, AC = 66 cm, $\angle$A = 62°.

3 $\triangle$PQR; QR = 69 cm, PR = 49 cm, $\angle$R = 85°.

4 $\triangle$XYZ; $x = 30$, $y = 40$, $\angle Z = 49°$.

5 $\triangle$PQR; $p = 9$, $r = 11$, $\angle Q = 120°$.

6 In triangle ABC, AB = 6 cm, BC = 7 cm and CA = 9 cm. Find $\angle$A and the area of the triangle.

7 $\triangle$PQR is such that $\angle$P = 60°, $\angle$R = 50° and QR = 12 cm. Find PQ and the area of the triangle.

8 In $\triangle$XYZ, XY = 150 cm, YZ = 185 cm and the area is 11 000 cm^2. Find $\angle$Y and XZ.

9 The area of triangle ABC is 36.4 cm^2. Given that AC = 14 cm and $\angle$A = 98°, find AB.

10 In $\triangle$ABC, BD is perpendicular to AC. Using h as the length of BD, find an expression for h in $\triangle$ABD. Hence prove that the area of $\triangle$ABC is $\frac{1}{2} bc \sin A$.

Problems

Many practical problems which involve distances and angles can be illustrated by a diagram. Often, however, this diagram contains too many lines, dimensions, etc. to be clear enough to work from. In these cases we can draw a second figure by extracting a triangle (or triangles) in which three facts about sides and/or angles are known. The various methods given in this chapter can then be used in this triangle.

EXAMPLES 7I

1 Two boats, P and Q, are 300 m apart. The base, A, of a lighthouse is in line with PQ. From the top, B, of the lighthouse the angles of depression of P and Q are found to be 35° and 48°. Write down the values of the angles BQA, PBQ and BPQ and find, correct to the nearest metre, the height of the lighthouse.

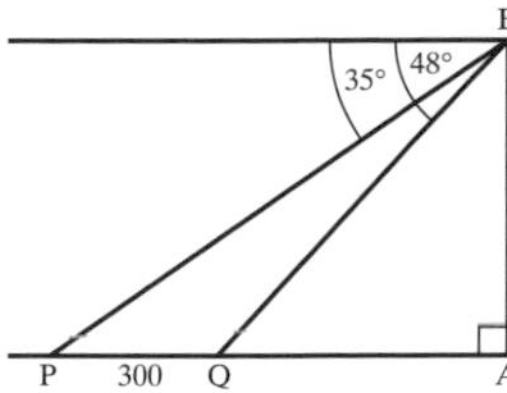

$\angle BQA = 48°$ (alternate angles)

$\angle PBQ = 48° - 35° = 13°$

$\angle BPQ = 35°$ (alternate angles)

> Now we can extract $\triangle PBQ$, knowing two angles and a side.

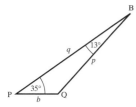

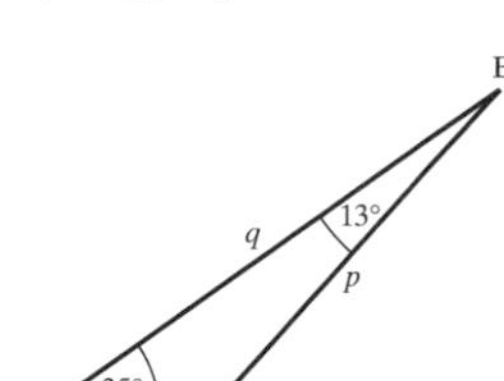

From the sine rule,

$$\frac{p}{\sin P} = \frac{b}{\sin B}$$

$$\therefore \quad p = \frac{(300)(\sin 35°)}{\sin 13°}$$

$$= 764.9\ldots$$

> We can now use the right-angled triangle ABQ.

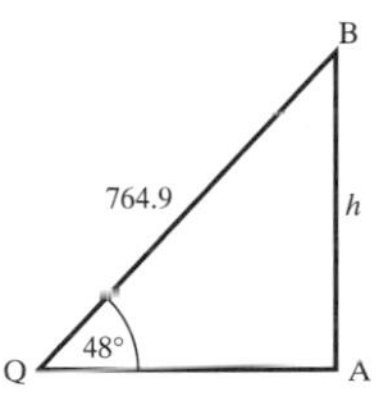

$$h = p \sin 48°$$
$$= (764.9\ldots)(\sin 48°)$$
$$= 568.4\ldots$$

Correct to the nearest metre the height of the lighthouse is 568 m.

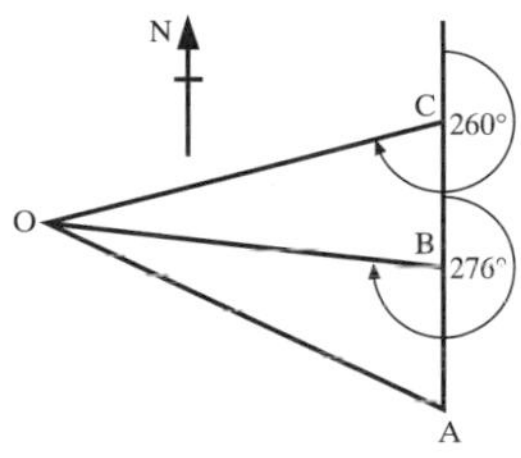

2 A traveller pitches camp in a desert. He knows that there is an oasis in the distance, but cannot see it. Wishing to know how far away it is, he measures 250 m due north from his starting point, A, to a point B where he can see the oasis, O, and finds that its bearing is 276°. He then measures a further 250 m due north to point C from which the bearing of the oasis is 260°. Find how far from the oasis he has camped.

$\angle OCB = 260° - 180° = 80°$ and $\angle OBC = 360° - 276° = 84°$

> As two angles and a side are known, $\triangle OBC$ can be used.

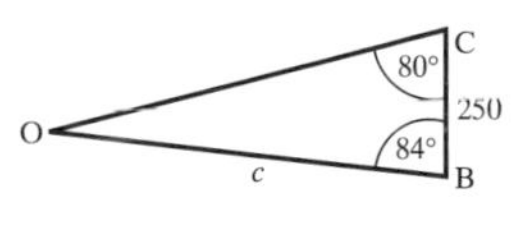

$\angle BOC = 180° - 80° - 84° = 16°$

From the sine rule, $\dfrac{c}{\sin C} = \dfrac{o}{\sin O}$

$$\Rightarrow \quad c = \frac{250 \times \sin 80°}{\sin 16°} = 893.2$$

> Now in $\triangle ABO$, $\angle ABO = 276° - 180° = 96°$ and we also know OB and AB. As two sides and the included angle are known, we must use the cosine rule.

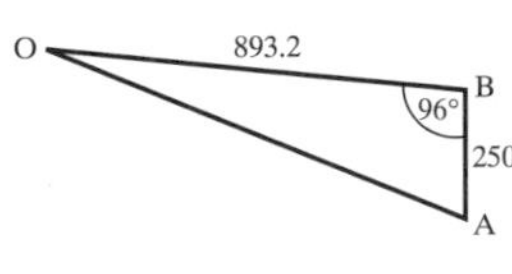

$$OA^2 = OB^2 + AB^2 - 2 \times OB \times AB \times \cos A\hat{B}O$$
$$= (893.2\ldots)^2 + (250)^2 - 2 \times 893.2\ldots \times 250 \times \cos 96°$$
$$= 907\,006.3\ldots$$
$$\Rightarrow \quad OA = 952.3\ldots$$

To the nearest metre the initial distance from the oasis is 952 m.

EXERCISE 7I

1 In a quadrilateral PQRS, PQ = 6 cm, QR = 7 cm, RS = 9 cm, ∠PQR = 115° and ∠PRS = 80°. Find the length of PR. Considering it as split into two separate triangles, find the area of the quadrilateral PQRS.

2 A light aircraft flies from an airfield, A, a distance of 50 km on a bearing of 049° to a town, B. The pilot then changes course and flies on a bearing of 172° to a landing strip, C, 68 km from B. How far is the landing strip from the airfield?

3 In a surveying exercise, P and Q are two points on land which is inaccessible. To find the distance PQ, a line AB of length 300 metres is marked out so that P and Q are on opposite sides of AB. The directions of P and Q relative to the line AB are then measured and are shown in the diagram. Calculate the length of PQ. (*Hint* Find AP and AQ.)

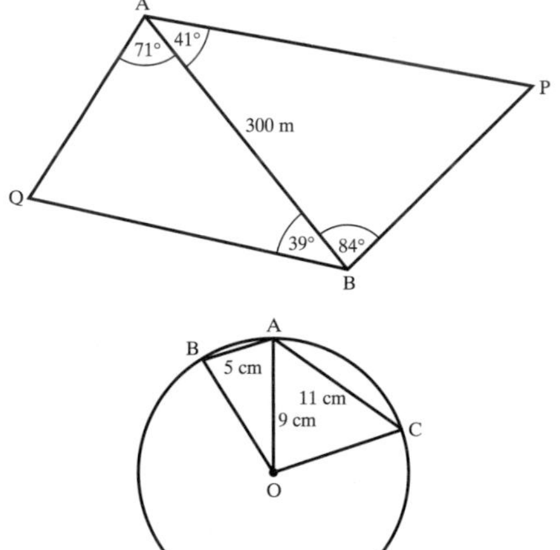

4

AB, of length 5 cm, and AC, of length 11 cm, are two chords of a circle with centre O and radius 9 cm. Find each of the angles BAO and CAO and hence calculate the area of the triangle ABC.

5

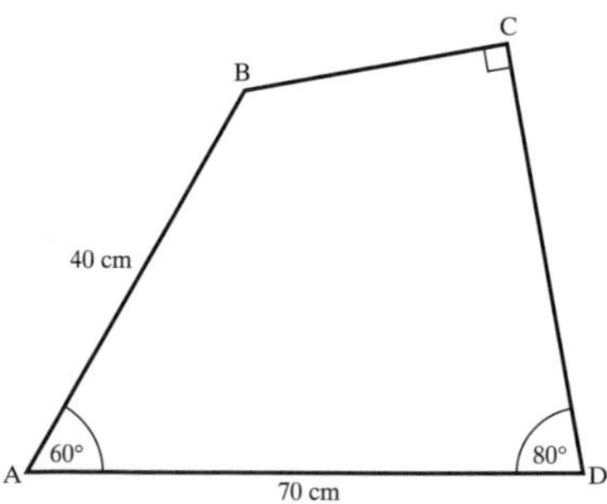

The diagram shows the cross section of a beam of length 2 m. Calculate

a the length of BD
b the angle ADB
c the length of CD
d the area of the cross section
e the volume of the beam.

MIXED EXERCISE 7

1 Find the value of $\angle A$, if A is between $0°$ and $180°$, when

a $\cos A = -\cos 64°$

b $\sin 94° = \sin A$.

2 If $\angle X$ is acute and $\sin X = \frac{7}{25}$, find $\cos(180° - X)$.

3 Given that $\sin A = \frac{5}{8}$, find $\tan A$ in surd form if

a $\angle A$ is acute **b** $\angle A$ is obtuse.

4 Find, in surd form, $\sin\theta$ and $\cos\theta$, given

a

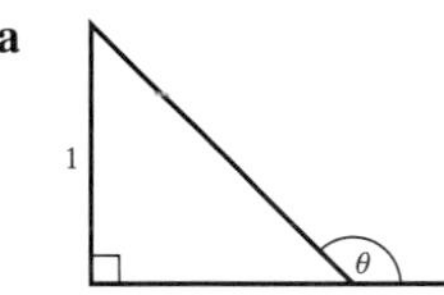

b

5 Given that $\sin X = \frac{12}{13}$ and X is obtuse, find $\cos X$.

6 In $\triangle ABC$, $BC = 11\text{ cm}$, $\angle B = 53°$ and $\angle A = 76°$, find AC.

7 In $\triangle PQR$, $p = 3, q = 5$ and $R = 69°$, find r.

8 In $\triangle XYZ$, $XY = 8\text{ cm}$, $YZ = 7\text{ cm}$ and $ZX = 10\text{ cm}$, find $\angle Y$.

9 In $\triangle ABC$, $AB = 7\text{ cm}$, $BC = 6\text{ cm}$ and $\angle A = 44°$, find all possible values of $\angle C$.

10 Find the angles of a triangle whose sides are in the ratio 2:4:5.

11 Use the cosine formula,

$$\cos A = \frac{b^2 + c^2 - a^2}{2bc}, \text{ to show that}$$

a $\angle A$ is acute if $a^2 < b^2 + c^2$

b $\angle A$ is obtuse if $a^2 > b^2 + c^2$.

12 In $\triangle PQR$, $PQ = 11\text{ cm}$, $PR = 14\text{ cm}$ and $\angle QPR = 100°$. Find the area of the triangle.

13 The area of ABC is 9 cm^2. If $AB = AC = 6\text{ cm}$, find $\sin A$. Are there two possible triangles? Give a reason for your answer.

14

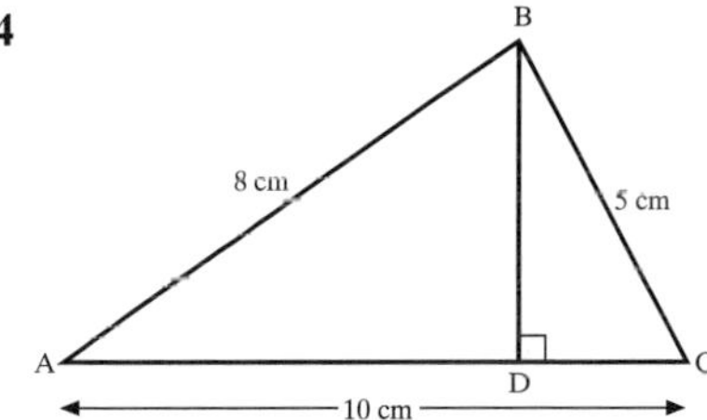

Given the information in the diagram,

a find $\angle ABC$

b find the area of $\triangle ABC$

c *hence* find the length of BD.

Coordinate geometry 2

The Meaning of Equations

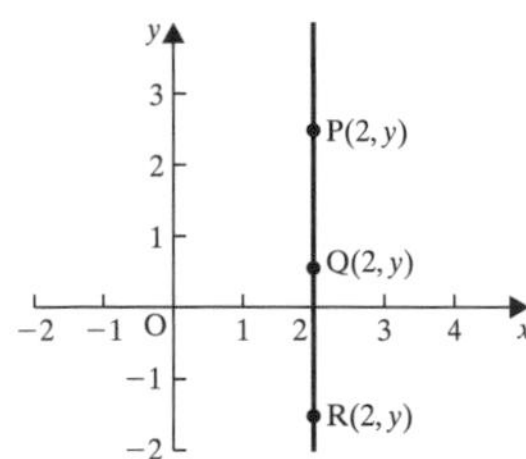

The Cartesian frame of reference provides a means of defining the position of any point in a plane. This plane is called the *xy*-plane.

In general x and y are independent variables. This means that they can each take any value independently of the value of the other unless some restriction is placed on them.

If however, the value of x is restricted to 2, say, but the value of y is not restricted, the condition gives a set of points which form a straight line parallel to the y-axis and passing through P, Q and R as shown.

In the *xy*-plane, the equation $x = 2$ defines the line shown in the diagram; $x = 2$ is called *the equation of this line* and we can refer briefly to *the line* $x = 2$.

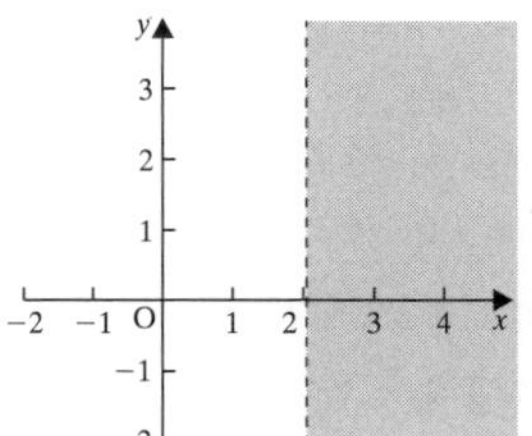

Now consider the set of points for which the condition is $x > 2$.

All the points to the right of the line $x = 2$ have an x-coordinate that is greater than 2.

So the inequality $x > 2$ defines the shaded region of the *xy*-plane shown. Similarly, the inequality $x < 2$ defines the region left unshaded in the diagram.

Note that the region defined by $x > 2$ does not include the line $x = 2$.

When a region does not include a boundary line this is drawn as a broken line. When the points on a boundary *are* included in a region, this boundary is drawn as a solid line.

The Equation of a Straight Line

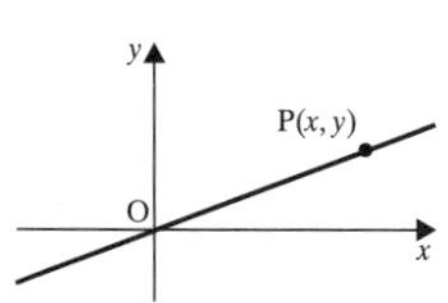

A straight line may be defined in many ways; for example, a line passes through the origin and has a gradient of $\frac{1}{2}$.

The point $P(x,y)$ is on this line if and only if the gradient of OP is $\frac{1}{2}$.

In terms of x and y, the gradient of OP is $\frac{y}{x}$, so the statement above can be written in the form

$$P(x,y) \text{ is on the line if and only if } \frac{y}{x} = \frac{1}{2}, \text{ i.e. } 2y = x.$$

Therefore the coordinates of points on the line satisfy the relationship $2y = x$, and the coordinates of points that are not on the line do not satisfy this relationship.

$2y = x$ is called the equation of the line.

The equation of a line (straight or curved) is a relationship between the x and y–coordinates of all points on the line and which is not satisfied by any other point in the plane.

EXAMPLES 8A

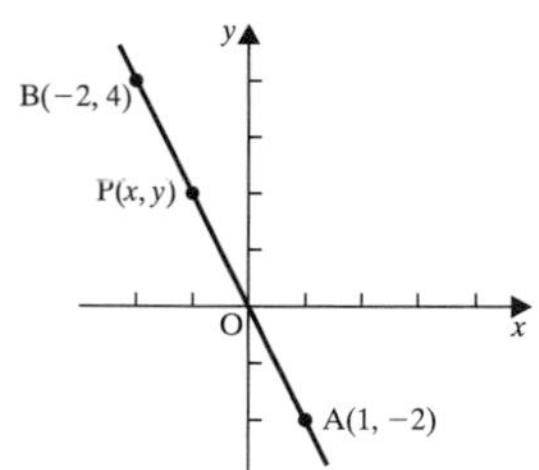

1 Find the equation of the line through the points $(1, -2)$ and $(-2, 4)$.

$P(x, y)$ is on the line if and only if the gradient of PA is equal to the gradient of AB (or PB).

The gradient of PA is $\dfrac{y-(-2)}{x-1} = \dfrac{y+2}{x-1}$

The gradient of AB is $\dfrac{-2-4}{1-(-2)} = -2$

Therefore the coordinates of P satisfy the equation $\dfrac{y+2}{x-1} = -2$

i.e. $\quad y + 2x = 0$

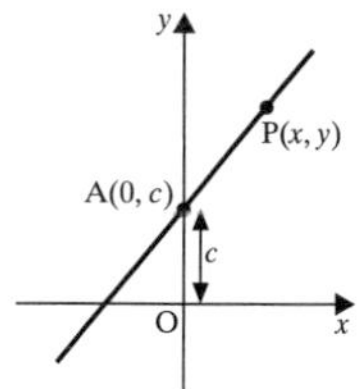

Consider the more general case of a line whose gradient is m and which cuts the y-axis at a directed distance c from the origin.
Note that c is called the *intercept on the y-axis.*

Now $P(x, y)$ is on this line if and only if the gradient of AP is m.

Therefore the coordinates of P satisfy the equation $\dfrac{y-c}{x-0} = m$

i.e. $\quad y = mx + c.$

This is the *standard form* for the equation of a straight line.

An equation of the form $y = mx + c$ represents a straight line with gradient m and intercept c on the y-axis.

Because the value of m and/or c may be fractional, this equation can be rearranged and expressed as $ax + by + c = 0$, i.e.

$ax + by + c = 0$ where a, b and c are constants,
is the equation of a straight line.

Note that in this form c is *not* the intercept.

EXAMPLES 8A (continued)

2 Write down the gradient of the line $3x - 4y + 2 = 0$ and find the equation of the line through the origin which is perpendicular to the given line.

Rearranging $3x - 4y + 2 = 0$ in the standard form gives $y = \frac{3}{4}x + \frac{1}{2}$

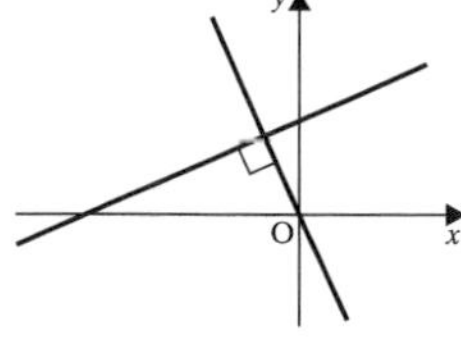

Comparing with $y = mx + c$ we can read off the gradient (m) and the intercept on the y-axis.

The gradient of the line is $\frac{3}{4}$.

The gradient of the perpendicular line is $-\dfrac{1}{m}$, i.e. $-\dfrac{4}{3}$ and it passes through the origin so the intercept on the y-axis is 0.

Therefore its equation is $y = -\frac{4}{3}x + 0 \quad \Rightarrow \quad 3y + 4x = 0$

3 Sketch the line $x - 2y + 3 = 0$.

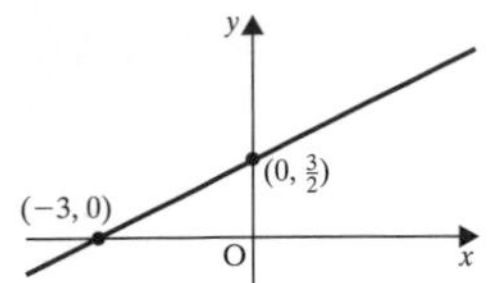

This line can be located accurately in the xy-plane when we know two points on the line. We will use the intercepts on the axes as these can be found easily (i.e. $x = 0 \Rightarrow y = \frac{3}{2}$ and $y = 0 \Rightarrow x = -3$).

Notice that the diagrams in the worked examples are sketches, not accurate plots, but they show reasonably accurately the position of the lines in the plane.

EXERCISE 8A

1 Write down the equation of the line through the origin and with gradient

a 2 **b** -1 **c** $\frac{1}{3}$

d $-\frac{1}{4}$ **e** 0 **f** ∞

Draw a sketch showing all these lines on the same set of axes.

2 Write down the equation of the line passing through the given point and with the given gradient.

a $(0, 1), \frac{1}{2}$ **b** $(0, 0), -\frac{2}{3}$

c $(-1, -4), 4$

Draw a sketch showing all these lines on the same set of axes.

3 Write down the equation of the line passing through the points

a $(0, 0), (2, 1)$ **b** $(1, 4), (3, 0)$

c $(-1, 3), (-4, -3)$

4 Write down the equation of the line passing through the origin and perpendicular to

a $y = 2x + 3$

b $3x + 2y - 4 = 0$

c $x - 2y + 3 = 0$

5 Write down the equation of the line passing through $(2, 1)$ and perpendicular to

a $3x + y - 2 = 0$

b $2x - 4y - 1 = 0$

Draw a sketch showing all four lines on the same set of axes.

6 Write down the equation of the line passing through $(3, -2)$ and parallel to

a $5x - y + 3 = 0$

b $x + 7y - 5 = 0$

7 A$(1, 5)$ and B$(4, 9)$ are two adjacent vertices of a square. Find the equation of the line on which the side BC of the square lies. How long are the sides of this square?

Finding the Equation of a Straight Line

Straight lines play a major role in graphical analysis and it is important to be able to find their equations easily. This section gives two of the commonest ways in which the equation of a straight line can be found. Each leads to a formula that can then be used to write down the equation of a particular line.

The equation of a line with gradient m and passing through the point (x_1,y_1)

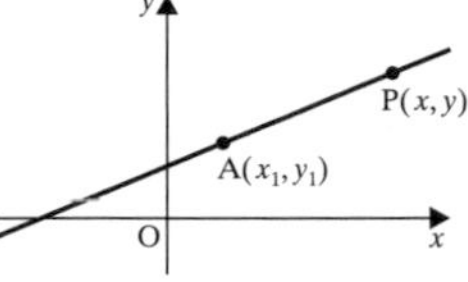

P (x,y) is a point on the line if and only if the gradient of AP is m

i.e. $$\frac{y-y_1}{x-x_1} = m$$

$$\Rightarrow \quad y - y_1 = m(x - x_1) \qquad [1]$$

In a particular case it is often simpler to work from a diagram than to apply a formula.

The equation of the line passing through (x_1,y_1) and (x_2,y_2)

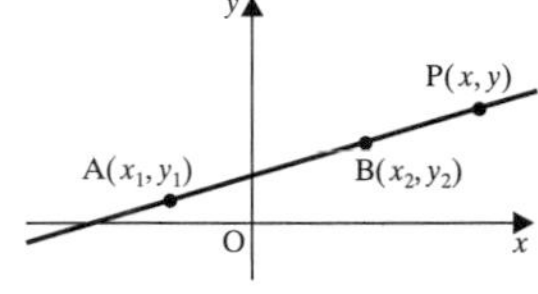

In the equation $y = mx + c$, m is the gradient of AB, i.e. $m = \dfrac{y_2 - y_1}{x_2 - x_1}$

So the equation of the line through A and B is

$$y - y_1 = \left[\frac{y_2 - y_1}{x_2 - x_1}\right](x - x_1) \qquad [2]$$

EXAMPLES 8B

1 Find the equation of the line with gradient $-\frac{1}{3}$ and passing through $(2, -1)$.

Using [1] with $m = -\frac{1}{3}$, $x_1 = 2$ and $y_1 = -1$ gives

$$y - (-1) = -\tfrac{1}{3}(x - 2)$$

$$\Rightarrow \quad x + 3y + 1 = 0$$

> *Alternatively* the equation of this line can be found from the standard form of the equation of a straight line, i.e. $y = mx + c$

Using $y = mx + c$ and $m = -\frac{1}{3}$ we have $y = -\frac{1}{3}x + c$

The point $(2, -1)$ lies on this line so its coordinates satisfy the equation,

i.e. $-1 = -\frac{1}{3}(2) + c \Rightarrow c = -\frac{1}{3}$

Therefore $y = -\frac{1}{3}x - \frac{1}{3}$

$$\Rightarrow \quad x + 3y + 1 = 0$$

2 Find the equation of the line through the points $(1, -2)$, $(3, 5)$

Using formula [2] with $x_1 = 1$, $y_1 = -2$, $x_2 = 3$ and $y_2 = 5$ gives

$$y - (-2) = \frac{5 - (-2)}{3 - 1}(x - 1)$$

$$\Rightarrow \quad 7x - 2y - 11 = 0$$

The worked examples in this book necessarily contain a lot of explanation but do not think that your solutions must be equally long. Avoid 'overworking' a problem, particularly in the case of problems that are basically simple. With practice, you can use any of the methods illustrated to write down the equation of a straight line directly.

3 Find the equation of the line through $(1,2)$ which is perpendicular to the line $3x - 7y + 2 = 0$.

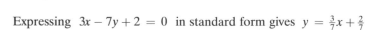

Expressing $3x - 7y + 2 = 0$ in standard form gives $y = \frac{3}{7}x + \frac{2}{7}$

Hence the given line has gradient $\frac{3}{7}$.

So the required line has a gradient of $-\frac{7}{3}$ and it passes through $(1,2)$.

Using $y - y_1 = m(x - x_1) \quad \Rightarrow \quad y - 2 = \frac{-7}{3}(x - 1) \quad \Rightarrow \quad 7x + 3y - 13 = 0$

In the last example note that the line perpendicular to $3x - 7y + 2 = 0$
has equation $7x + 3y - 13 = 0$

i.e. the coefficients of x and y have been transposed and the sign between the x and y terms has changed. This is a particular example of the general fact that

given a line with equation $ax + by + c = 0$ then the equation of any perpendicular line is $bx - ay + k = 0$

This property of perpendicular lines can be used to shorten the working of problems, e.g. to find the equation of the line passing through $(2, -6)$ which is perpendicular to the line $5x - y + 3 = 0$, we can say that the required line has an equation of the form $x + 5y + k = 0$ and then use the fact that the coordinates $(2, -6)$ satisfy this equation to find the value of k.

The Angle between a Straight Line and the *x*-axis

When the equation of a straight line is written in standard form, i.e. $y = mx + c$, then m is the gradient of the line and c is its intercept on the y-axis.

When m is positive, the line makes an acute angle with the positive direction of the x-axis, and when m is negative, the line makes an obtuse angle with the positive direction of the x-axis.

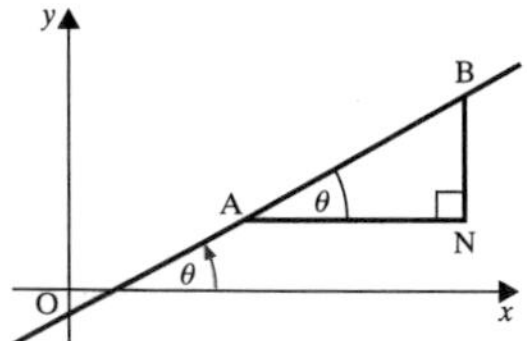

When θ is acute, gradient of AB $= \dfrac{\text{BN}}{\text{AN}} = \tan\theta$

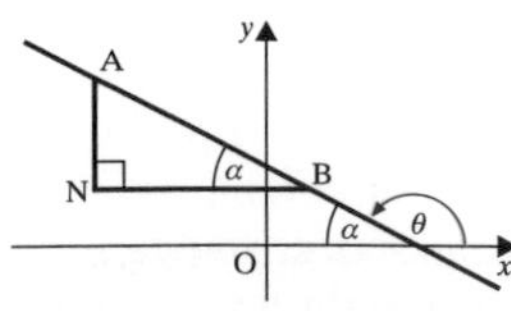

When θ is obtuse, gradient of AB $= \dfrac{-\text{AN}}{\text{BN}} = -\tan\alpha = \tan\theta$

In both cases the gradient of AB $= \tan\theta$.

Therefore the gradient of a line is equal to the tangent of the angle between the line and the positive direction of the x-axis. i.e. $\boldsymbol{m = \tan\theta}$

EXERCISE 8B

1 Find the equation of the line with the given gradient and passing through the given point.

a $3, (4,9)$ b $-5, (2,-4)$
c $\frac{1}{4}, (4,0)$ d $0, (-1,5)$
e $-\frac{2}{5}, (\frac{1}{2}, 4)$ f $-\frac{3}{8}, (\frac{22}{5}, -\frac{5}{2})$

2 Find the equation of the line passing through the points

a $(0,1), (2,4)$ b $(-1,2), (1,5)$
c $(3,-1), (3,2)$

3 Determine which of the following pairs of lines are perpendicular.

a $x-2y+4=0$ and $2x+y=3$
b $x+3y=6$ and $3x+y+2=0$
c $x+3y-2=0$ and $y=3x+2$
d $y+2x+1=0$ and $x=2y-4$

4 Find the equation of the line through the point $(5,2)$ and perpendicular to the line $x-y+2=0$.

5 Find the equation of the perpendicular bisector of the line joining

a $(0,0), (2,4)$
b $(3,-1), (-5,2)$
c $(5,-1), (0,7)$

6 Find the equation of the line through the origin which is parallel to the line $4x+2y-5=0$.

7 The line $4x-5y+20=0$ cuts the x-axis at A and the y-axis at B. Find the equation of the median through O of $\triangle$OAB.

8 Find the equation of the altitude through O of the triangle OAB defined in Question 7.

9 Find the equation of the perpendicular from $(5,3)$ to the line $2x-y+4=0$.

10 The points A$(1,4)$ and B$(5,7)$ are two adjacent vertices of a parallelogram ABCD. The point C$(7,10)$ is another vertex of the parallelogram. Find the equation of the side CD.

11 Find the equation of the line that cuts the y-axis where $y=2$ and that is inclined at 45° to the positive direction of the x-axis.

12 Find the equation of the line that passes through $(2,-1)$ and is inclined at 135° to the positive direction of the x-axis.

13 Find, to the nearest degree, the angle between the lines $y=3x-1$ and $2x-5y+2=0$.

14 Find the equations of the two lines through the origin which are inclined at 45° to the line $2y+x=0$.

Intersection

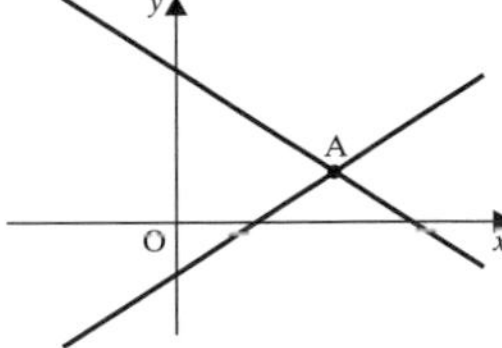

The point where two lines (or curves) cut is called a point of intersection.

If A is the point of intersection of the lines $y-3x+1=0$ [1]

and $y+x-2=0$ [2]

then the coordinates of A satisfy both of these equations. A can be found by solving [1] and [2] simultaneously, i.e.

[2] – [1] $\Rightarrow$ $4x-3=0$ $\Rightarrow$ $x=\frac{3}{4}$ and $y=\frac{5}{4}$

Therefore $(\frac{3}{4}, \frac{5}{4})$ is the point of intersection.

Note that the coordinates of A can also be found using a graphics calculator or graph drawing software on a computer.

EXAMPLE 8C

A circle has radius 4 and its centre is the point C(5, 3).

a Show that the points A(5, −1) and B(1, 3) are on the circumference of the circle.

b Prove that the perpendicular bisector of AB goes through the centre of the circle.

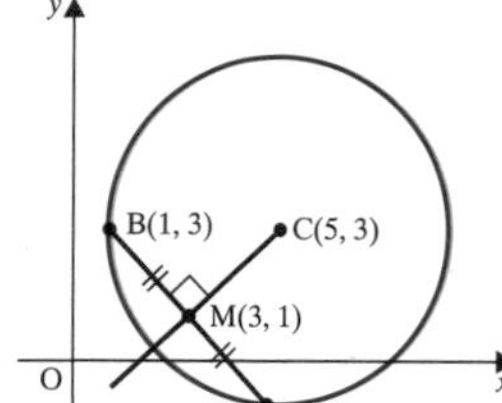

a From the diagram, BC = 4

$\therefore$ B is on the circumference.

Similarly AC = 4,

$\therefore$ A is on the circumference.

b The midpoint, M, of AB is $\left[\dfrac{5+1}{2}, \dfrac{-1+3}{2}\right]$ i.e. (3, 1)

The gradient of AB is $\dfrac{-1-3}{5-1} = -1$

If l is the perpendicular bisector of AB, its gradient is 1 and it goes through (3, 1).

$\therefore$ the equation of l is $y - 1 = 1(x-3) \Rightarrow y = x - 2$ [1]

In equation [1], when $x = 5, y = 3$

$\therefore$ the point (5, 3) is on l.

i.e. the perpendicular bisector of AB goes through C.

EXERCISE 8C

1 Show that the triangle whose vertices are (1, 1), (3, 2) and (2, −1) is isosceles.

2 Find the area of the triangular region enclosed by the x and y axes and the line $2x - y - 1 = 0$.

3 Find the coordinates of the triangular region enclosed by the lines $y = 0$, $y = x + 5$ and $x + 2y - 6 = 0$.

4 Write down the equation of the perpendicular bisector of the line joining the points (2, −3) and $(-\frac{1}{2}, 3\frac{1}{2})$.

5 Find the equation of the line through A(5, 2) which is perpendicular to the line $y = 3x - 5$. Hence find the coordinates of the foot of the perpendicular from A to the line.

6 Find, in terms of a and b, the coordinates of the foot of the perpendicular from the point (a, b) to the line $x + 2y - 4 = 0$.

7 The coordinates of a point P are $(t+1, 2t-1)$. Sketch the position of P when $t = -1, 0, 1$ and 2. Show that these points are collinear and write down the equation of the line on which they lie.

8 Write down the equation of the line which goes through (7, 3) and which is inclined at 45° to the positive direction of the x-axis.

9 Find the equation of the perpendicular bisector of the line joining the points (a, b) and $(2a, -3b)$.

10 The centre of a circle is at the point C(3, 7) and the point A(5, 3) is on the circumference of the circle. Find

a the radius of the circle,

b the equation of the line through A that is perpendicular to AC.

11 The equations of two sides of a square are $y = 3x - 1$ and $x + 3y - 6 = 0$. If $(0, -1)$ is one vertex of the square find the coordinates of the other vertices.

12 The lines $y = 2x$, $2x + y - 12 = 0$ and $y = 2$ enclose a triangular region of the xy-plane. Find

a the coordinates of the vertices of this region,

b the area of this region.

Reduction of a Relationship to a Linear Law

In this part of the chapter we look at a practical application of the equation $y = mx + c$.

If it is thought that a certain relationship exists between two variable quantities, this hypothesis can be tested by experiment, i.e. by giving one variable certain values and measuring the corresponding values of the other variable. The experimental data collected can then be displayed graphically. If the graph shows points that lie approximately on a straight line (allowing for experimental error) then this indicates a linear relationship between the variables (i.e. a relationship of the form $Y = mX + c$). Further, the gradient of the line (m) and the vertical axis intercept (c) provide the values of the constants.

EXAMPLES 8D

1 An elastic string is fixed at one end and a variable weight is hung on the other end. It is believed that the length of the string is related to the weight by a linear law. Use the following experimental data to confirm this belief and find the particular relationship between the length of the string and the weight.

Weight (W) in newtons	1	2	3	4	5	6	7	8
Length (l) in metres	0.33	0.37	0.4	0.45	0.5	0.53	0.56	0.6

If l and W are related by a linear law then, allowing for experimental error, we expect that the points will lie on a straight line. Plotting l against W gives the following graph.

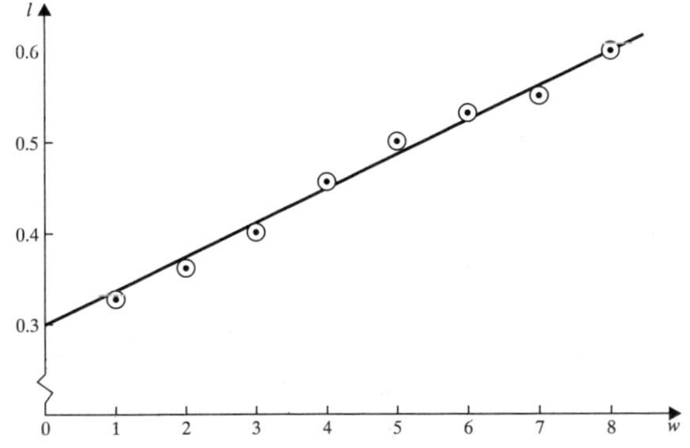

These points do lie fairly close to a straight line.

From the graph, l and W are connected by a linear relationship, i.e. a relationship of the form $l = aW + b$.

Now we draw the line of 'closest fit'. This is the line that has the points distributed above and below it as evenly as possible; it is not necessarily the line which goes through the most points.

By measurement from the graph the gradient $= 0.04$
the intercept on the vertical axis $= 0.3$.

So comparing $l = aW + b$
with $Y = mX + c$ } we have $a = 0.04, b = 0.3$

i.e. within the limits of experimental and graphical accuracy

$l = 0.04W + 0.3.$

When the gradient of a line is found from a graph, the increase in a quantity is measured from the *scale used for that quantity* and it is worth noting that the scales used for the two quantities are *not* usually the same.

The values of the constants found from calculating the gradient and intercept from a drawn graph are approximate. Apart from experimental error in the data, selecting the line of best fit is a personal judgement and so is subject to slight variations which affect the values obtained.

There are methods for calculating the equation of the line of best fit; these are called regression lines and computer programs exist which will give these equations from the data. Using such a program, the values of a and b in the last example are given as $a = 0.039$ and $b = 0.291$.

If the relationship is not of a linear form, the points on the graph will lie on a section of a curve. It is very difficult to identify the equation of a curve from a section of it, so the form of a non-linear relationship can rarely be verified in this way.

Non-linear relationships, however, can often be reduced to a linear form. The following examples illustrate some of the relationships that can be verified by plotting experimental data in a form which gives a straight line.

Relationships of the Form $y = ax^n$

A relationship of the form $y = ax^n$ where a is a constant can be reduced to a linear relationship by taking logarithms, since

$$y = ax^n \iff \ln y = n \ln x + \ln a$$

(Although any base can be used, it is sensible to use either e or 10 as these are built into most calculators.)

Comparing $\ln y = n \ln x + \ln a$

with $Y = mX + c$

we see that plotting values of $\ln y$ against values of $\ln x$ gives a straight line whose gradient is n and whose intercept on the vertical axis is $\ln a$.

EXAMPLES 8D (continued)

2 The following data, collected from an experiment is believed to obey a law of the form $p = aq^n$. Verify this graphically and find the values of a and n.

q	1	2	3	4	5	6
p	0.5	0.63	0.72	0.8	0.85	0.9

If the relationship $p = aq^n$ is correct, then $\ln p = n \ln q + \ln a$
comparing with $y = mx + c$

we expect that $\ln p$ and $\ln q$ will be related by a linear law.

First a table of values of $\ln p$ and $\ln q$ is needed.

$\ln q$	0	0.69	1.10	1.39	1.61	1.79
$\ln p$	−0.69	−0.46	−0.33	−0.22	−0.16	−0.11

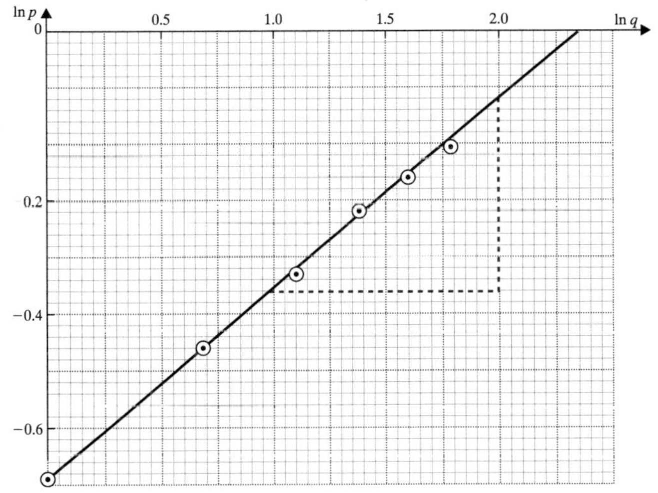

The points lie on a straight line confirming that there is a linear relationship between $\ln q$ and $\ln p$.

From the graph, the gradient of the line is 0.34, $\Rightarrow \quad n = 0.34$
and the intercept on the vertical axis is −0.69, so

$\ln a = -0.69 \quad \Rightarrow \quad a = 0.5$

Therefore the data does obey a law of the form $p = aq^n$, where $a \approx 0.5$ and $n \approx 0.34$.

(Using the tabulated values of $\ln q$ and $\ln p$ and a computer program, gives $n = 0.327$ and $\ln a = -0.687$)

Relationships of the Form $y = ab^x$

A relationship of the form $y = ab^x$ where a and b are constant can be reduced to a linear relationship by taking logs, since

$$y = ab^x \iff \log y = x \log b + \log a$$

Comparing $\log y = x \log b + \log a$

with $Y = mX + c$

we see that plotting values of $\log y$ against corresponding values of x gives a straight line whose gradient is $\log b$ and whose intercept on the vertical axis is $\log a$.

EXAMPLES 8D (continued)

3 In an experiment, the mass, y grams, of a substance is measured at various times, x seconds.

The results are shown in the table below. It is believed that x and y are related by a law of the form $2y + 10 = ab^{(x-3)}$.

x	10	12	15	20	21	24
y	37.5	90	320	2440	3700	4200

a Confirm this graphically, showing that one result does not conform to this law.

b Find approximate values of a and b.

c Explain, with reasons, whether it is sensible to use these results to predict the mass when $x = 30$.

a If $2y + 10 = ab^{(x-3)}$, taking logs of both sides gives

$\log(2y + 10) = (x - 3)\log b + \log a$

which is of the form $Y = mX + c$

where $Y = \log(2y + 10)$, $X = x - 3$ and $m = \log b$, $c = \log a$
i.e. $[\log(2y + 10)]$ and $[x - 3]$ obey a linear law.

So we need to tabulate corresponding values of $(x - 3)$ and $\log(2y + 10)$ from the given values of x and y.

$x - 3$	7	9	12	17	18	21
$\log(2y + 10)$	1.9	2.3	2.8	3.7	3.9	3.9

Then plotting $\log(2y + 10)$ against $x - 3$ gives the graph.

The straight line shows that, except where $x - 3 = 21$, there is a linear relationship between $\log(2y + 10)$ and $x - 3$, confirming that $2y + 10 = ab^{x-3}$ for values of $x - 3$ from 7 to 18, i.e. for values of x from 10 to 21.

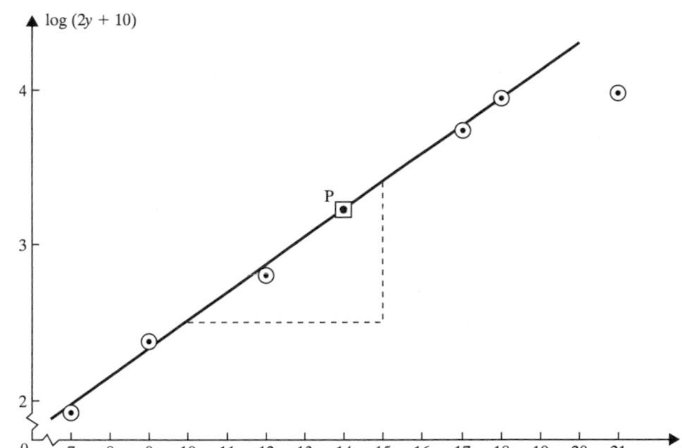

b From the graph, the gradient is 0.175.

$\therefore \quad \log b \approx 0.175 \quad \Rightarrow \quad b \approx 1.49$

Using the point P(14, 3.18) and $m = 0.175$ then $Y = mX + c$ gives

$3.18 = (0.175)(14) + c \quad \Rightarrow \quad c = 0.73$

i.e. $\quad \log a \approx 0.73 \quad \Rightarrow \quad a \approx 5.37$

c From the evidence given, the relationship between x and y applies for $10 \leqslant x \leqslant 21$. It clearly does not apply when $x = 24$; this may be because this is a rogue result, i.e. an error, or because the conditions for the relationship no longer apply so it is not sensible to use the graph to predict the mass for a value of x above 21.

EXERCISE 8D

In Questions 1–3, the table gives sets of values for the related variables and the law which relates the variables. By drawing a straight line graph find approximate values for a and b.

1 $y = ax + ab$

x	3	5	7	10
y	−2	2	6	12

2 $s = ab^{-t}$

t	1	2	3	4
s	1.5	0.4	0.1	0.02

3 $ay = b^x$

x	5	6	7	8
y	1.07	2.13	4.27	8.53

4 The variables x and y are believed to satisfy a relationship of the form $y = k(x + 1)^n$. Show that the experimental values shown in the table do satisfy the relationship. Find approximate values for k and n.

x	4	8	15	19	24
y	4.45	4.60	4.80	4.89	5.00

5 Two variables s and t are related by a law of the form $s = ke^{-nt}$. The values in the table were obtained from an experiment. Show graphically that these values do verify the relationship and use the graph to find approximate values of k and n.

t	1	1.5	2	2.5	3
s	1230	590	260	140	60

6 It is thought that the value of a second-hand car reduces exponentially with age. Its value £y after time x years can be modelled by the relationship

$$y - 2000 = ab^{-x}$$

a If the initial value of a car was £12 000 and its value 2 years later was £8 500, find the values of a and b.

b Reduce the relationship to a linear law and represent it graphically for $0 \leqslant x \leqslant 10$.

c The table gives the values of another car for various values of x.

x	0	2	4	6	8	10
y	12 000	8800	7300	6100	5100	4300

Plot these points on your graph. Explain why the given model does not predict the changing values of this car.

7 When doing an emergency stop, it is thought that the distance, s metres, travelled by a car from the time the brakes are applied until it stops can be modelled by the equation

$$s = ku^2$$

where u km/h is the initial speed of the car and k is a constant.

a A car initially moving at 60 km/h covers 120 m before it stops. Assuming the model is valid, find the value of k.

b Reduce the relationship to a linear law, and using the value of k found in **a**, represent this graphically for $0 \leqslant u \leqslant 200$.

In the same car on another day and under different conditions, the following information was recorded.

u	50	100	130	150	200
s	100	210	310	780	620

c Show that these values do not satisfy a relationship of the form $s = ku^2$, even allowing for error.

d Given that there is one rogue value, identify it.

CHAPTER 9 Circles

Parts of a Circle

We start this chapter with a reminder of the names used for parts of a circle.

Part of the circumference is called an *arc*.

If the arc is less than half the circumference it is called a *minor arc*; if it is greater than half the circumference it is called a *major arc*.

A straight line which cuts a circle in two distinct points is called a *secant*. The part of the line inside the circle is called a *chord*. AB is a secant, CD is a chord.

The area enclosed by two radii and an arc is called a *sector*.

The area enclosed by a chord and an arc is called a *segment*. If the segment is less than half a circle it is called a *minor segment*; if it is greater than half a circle it is called a *major segment*.

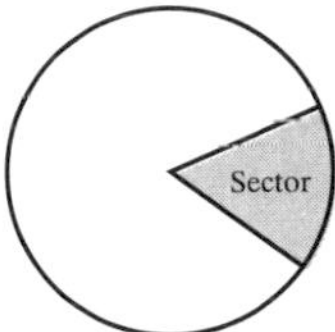

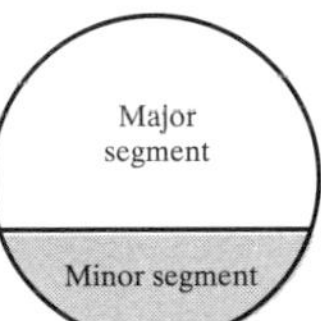

The Angle Subtended by an Arc

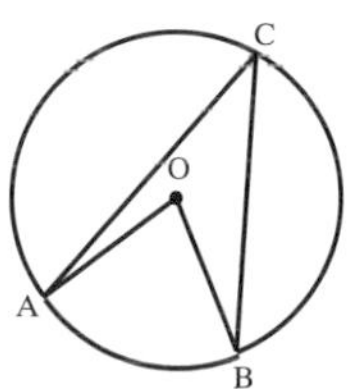

Consider the points A, B and C on the circumference of a circle whose centre is O. We say that $\angle$ACB stands on the minor arc AB.

The minor arc AB *subtends* the angle ACB at the circumference (and the angle is *subtended* by the arc).

The arc AB also subtends the angle AOB at the centre of the circle.

EXAMPLE 9A

A circle of radius 2 units which has its centre at the origin, cuts the x-axis at the points A and B and cuts the y-axis at the point C. Prove that $\angle ACB = 90°$.

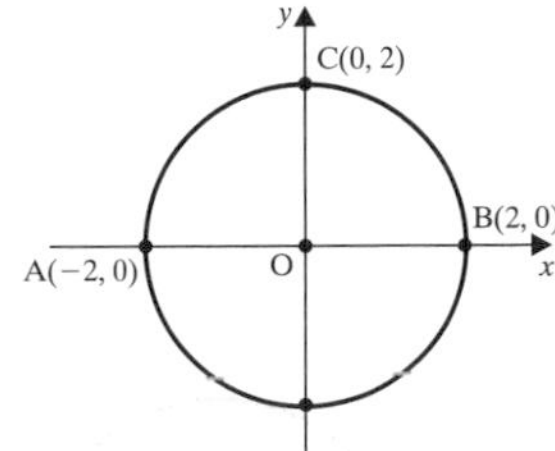

> All the information given in the question, and gleaned from the known properties of the figure, can be marked in the diagram as shown. The gradients we need can then be seen.

From the diagram, the gradient of AC is $\dfrac{2-0}{0-(-2)} = 1$

and the gradient of BC is $\dfrac{2-0}{0-2} = -1$

$\therefore$ (gradient of AC) $\times$ (gradient of BC) $= -1$

i.e. AC is perpendicular to BC $\Rightarrow$ $\angle ACB = 90°$

EXERCISE 9A

1 Name the angles subtended

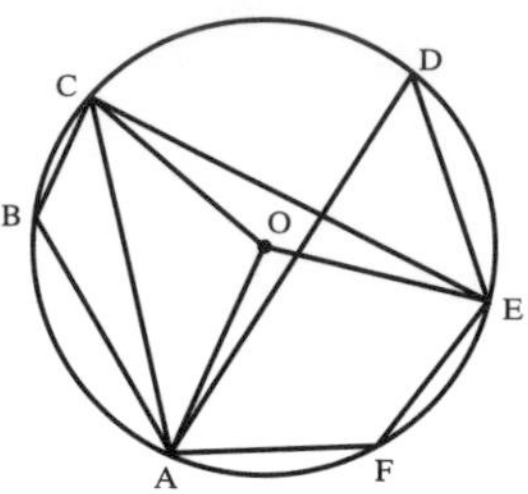

a at the circumference by the minor arc AE

b at the circumference by the major arc AE

c at the centre by the minor arc AC

d at the circumference by the major arc AC

e at the centre by the minor arc CE

f at the circumference by the minor arc CD

g at the circumference by the minor arc BC.

2 AB is a chord of a circle, centre O, and M is its midpoint. The radius from O is drawn through M. Prove that OM is perpendicular to AB.

3 C(5, 3) is the centre of a circle of radius 5 units.

a Show that this circle cuts the x-axis at A(1, 0) and B(9, 0).

b Prove that the radius that is perpendicular to AB goes through the midpoint of AB.

c Find the angle subtended at C by the minor arc AB.

d The point D is on the major arc AB and DC is perpendicular to AB. Find the coordinates of D and hence find the angle subtended at D by the minor arc AB.

4 A and B are two points on the circumference of a circle centre O. C is a point on the major arc AB. Draw the lines AC, BC, AO, BO and CO, extending the last line to a point D inside the sector AOB. Prove that $\angle$AOD is twice $\angle$ACO and that $\angle$BOD is twice angle $\angle$BCO. Hence show that the angle subtended by the minor arc AB at the centre of the circle is twice the angle that it subtends at the circumference of the circle.

Angles in a Circle

There are several important facts about circles that are useful in solving problems (some of them were used in the previous exercise).

The perpendicular bisector of a chord of a circle goes through the centre of the circle.

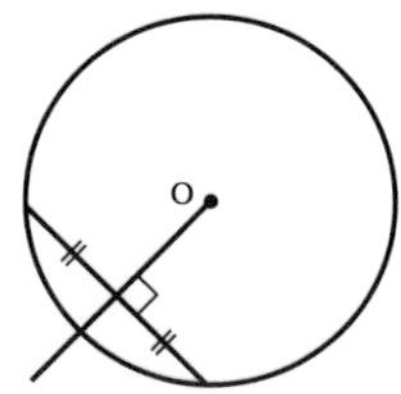

The angle subtended at the centre of a circle by an arc, is twice the angle subtended at the circumference by the same arc.

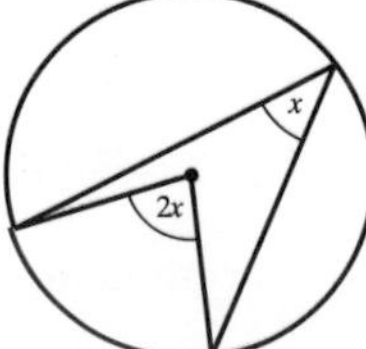

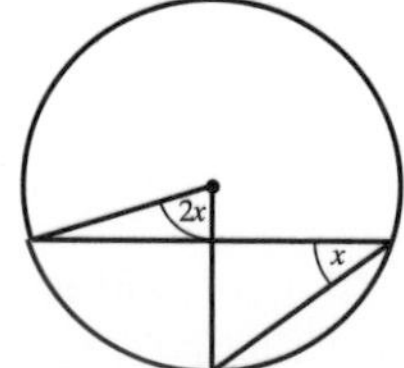

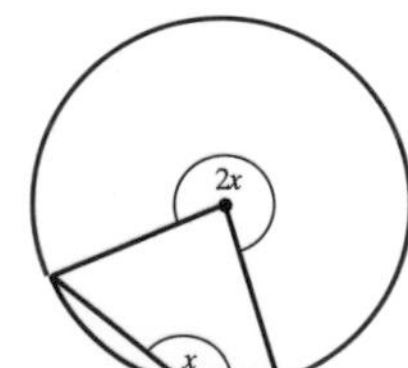

All angles subtended at the circumference by the same arc are equal.

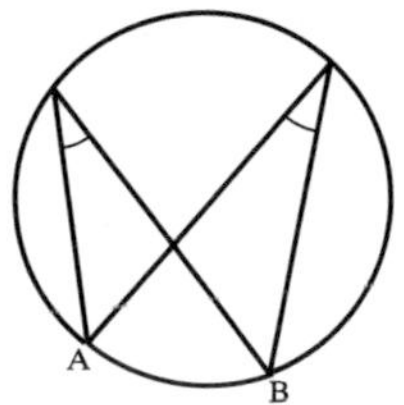

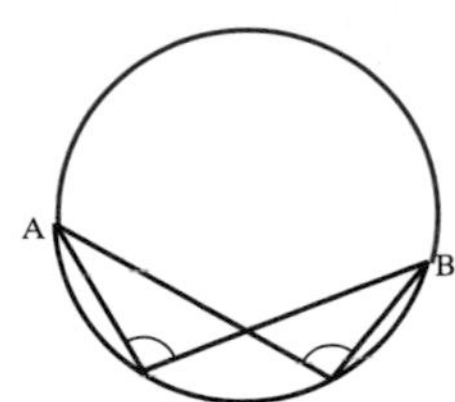

An angle in a semicircle is 90°.

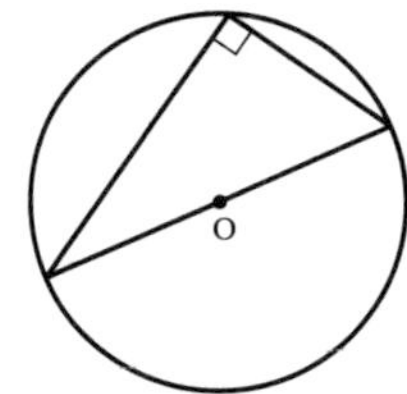

If the four points P, Q, R and S all lie on a circle, PQRS is called a cyclic quadrilateral.

The opposite angles of a cyclic quadrilateral are supplementary (i.e. add up to 180°).

$p + r = 180°, q + s = 180°$

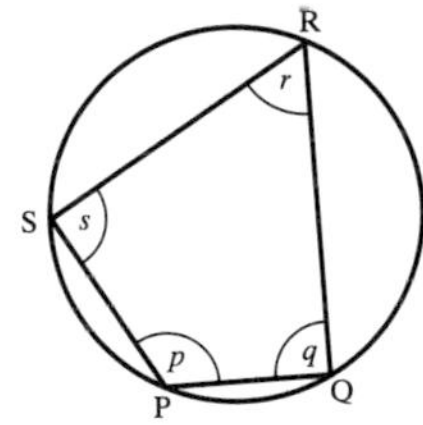

EXAMPLE 9B

A circle circumscribes a triangle whose vertices are at the points A(0, 4), B(2, 3) and C(−2, −1). Find the centre of the circle.

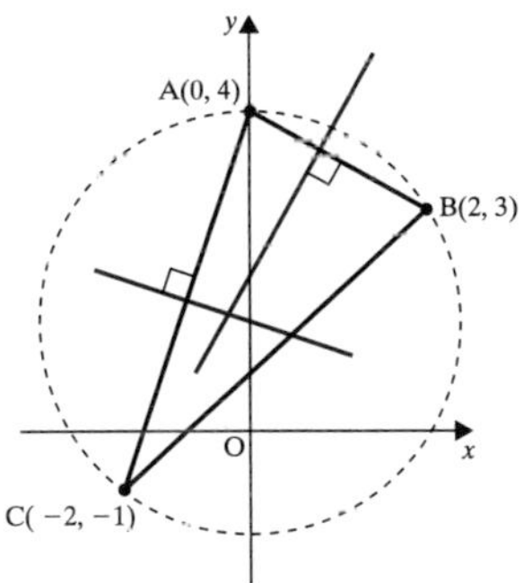

> When a circle circumscribes a figure, it passes through the vertices of the figure. The centre of the circle lies at the point of intersection of the perpendicular bisectors of two chords.

The midpoint of AC is $\left[\frac{0-2}{2}, \frac{4-1}{2}\right] \Rightarrow (-1, \frac{3}{2})$,

and the gradient of AC is $\frac{4-(-1)}{0-(-2)} = \frac{5}{2}$

$\therefore$ the gradient of the perpendicular bisector of AC is $-\frac{2}{5}$ and its equation is

$y = -\frac{2}{5}x + \frac{11}{10} \Rightarrow 4x + 10y - 11 = 0$ [1]

Similarly the midpoint of AB is $(1, \frac{7}{2})$ and its gradient is $-\frac{1}{2}$

$\therefore$ the gradient of the perpendicular bisector of AB is 2 and its equation is

$y = 2x + \frac{3}{2} \Rightarrow 4x - 2y + 3 = 0$ [2]

Solving equations [1] and [2] simultaneously gives

$12y - 14 = 0 \Rightarrow y = \frac{7}{6}$ and $x = -\frac{1}{6}$

Therefore the centre of the circle is the point $(-\frac{1}{6}, \frac{7}{6})$.

EXERCISE 9B

1 Find the size of each marked angle,

a

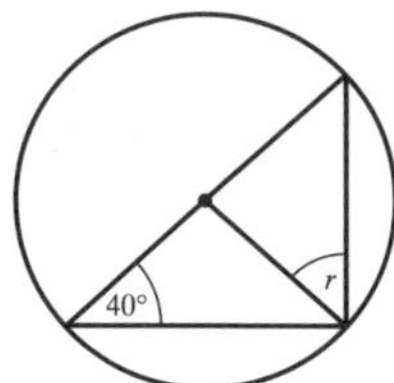

b

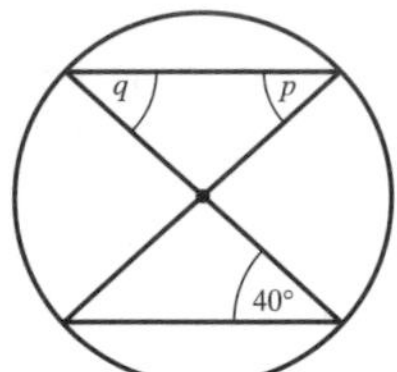

c

d

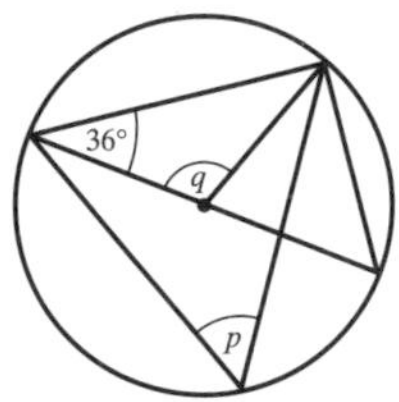

2 O is the centre of a circle that passes through A, B, C and D. ∠A $= x$ and ∠C $= y$. Write down the values of the obtuse angle DOB and the reflex angle DOB. Hence prove that the opposite angles of a cyclic quadrilateral are supplementary.

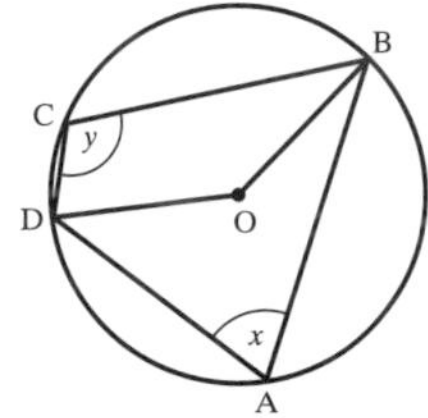

3 AB is a diameter of a circle centre O. C is a point on the circumference. D is a point on AC such that OD bisects ∠AOC. Prove that OD is parallel to BC.

4 A triangle has its vertices at the points A(1, 3), B(5, 1) and C(7, 5). Prove that △ABC is right-angled and hence find the coordinates of the centre of the circumcircle of △ABC.

5 AB and CD are two chords of a circle that cut at E. (E is not the centre of the circle.) Show that △s ACE and BDE are similar.

6 A circle with centre O circumscribes an equilateral triangle ABC. The radius drawn through O and the midpoint of AB meets the circumference at D. Prove that △ADO is equilateral.

7 The line joining A(5, 3) and B(4, −2) is a diameter of a circle. If P(a, b) is a point on the circumference find a relationship between a and b.

8 ABCD is a cyclic quadrilateral. The side CD is produced to a point E outside the circle. Show that ∠ABC = ∠ADE.

9 A triangle has its vertices at the points A(1, 3), B(−2, 5) and C(4, −2). Find the coordinates of the centre and, correct to 3 s.f., the radius of the circle that circumscribes △ABC.

10 In the diagram, O is the centre of the circle and CD is perpendicular to AB. If ∠CAB $= 30°$ find the size of each marked angle.

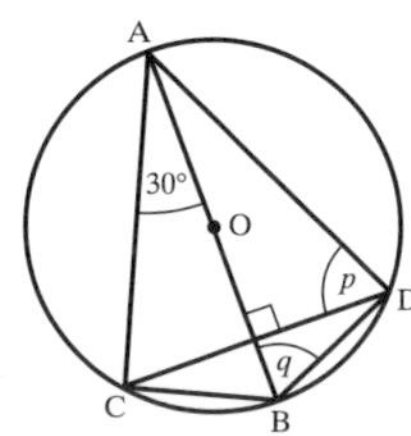

Tangents to Circles

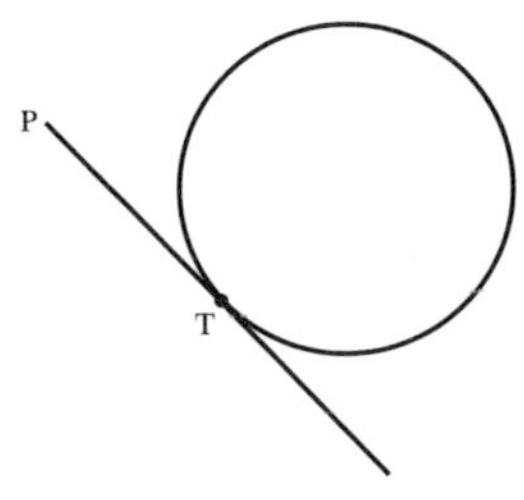

If a line and a circle are drawn in a plane then there are three possibilities for the position of the line in relation to the circle. The line can miss the circle, or it can cut the circle in two distinct points, or it can touch the circle at one point. In the last case the line is called a *tangent* and the point at which it touches the circle is called the *point of contact*.

The length of a tangent drawn from a point to a circle is the distance from that point to the point of contact.

T is the point of contact.

PT is the length of the tangent from P.

Properties of Tangents to a Circle

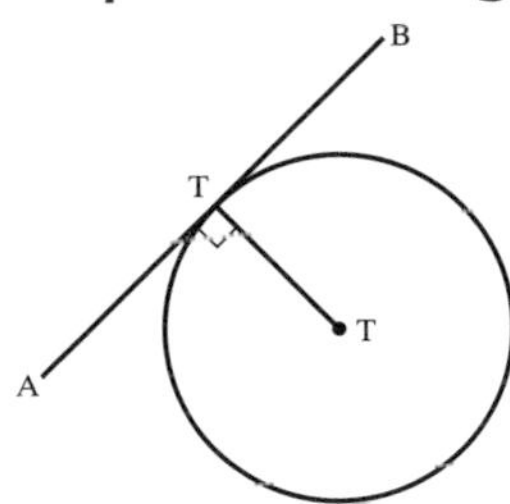

A tangent to a circle is perpendicular to the radius drawn from the point of contact, i.e. AB is perpendicular to OT.

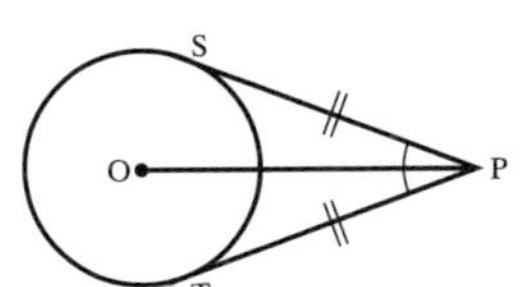

Two tangents drawn to a circle from the same point P, are equal in length, i.e. PS = PT.

Also the line joining P to the centre O bisects the angle between the tangents, i.e. $\angle SPO = \angle TPO$.

EXAMPLES 9C

1 A circle of radius 10 units is circumscribed by a right-angled isosceles triangle. Find the lengths of the sides of the triangle.

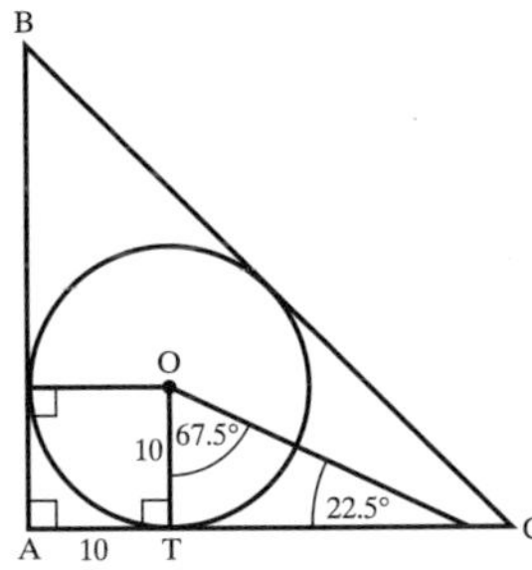

> A circle is *circumscribed* by a figure when all the sides of the figure touch the circle (and the circle is *inscribed* in the figure).

CO bisects $\angle ACB \Rightarrow \angle OCT = 22.5°$

$\Rightarrow \angle TOC = 67.5°$

In $\triangle OTC$, $TC = 10\tan 67.5° = 24.14\ldots$

$AT = 10$

$\therefore\quad AC = 34.14\ldots = AB$

In $\triangle ABC$, $BC = \sqrt{34.14^2 + 34.14^2}$ (Pythagoras)

$= 48.28\ldots$

$\therefore$ correct to 3 s.f. the lengths of the sides of the triangle are 34.1 units, 34.1 units and 48.3 units.

2 The centre of a circle of radius 3 units is the point C(2, 5). The equation of a line, l, is $x + y - 2 = 0$.

a Find the equation of the line through C, perpendicular to l.

b Find the distance of C from l and hence determine whether l is a tangent to the circle.

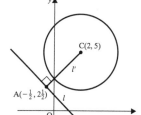

a The line l' is perpendicular to $x + y - 2 = 0$

so its equation is $x - y + k = 0$

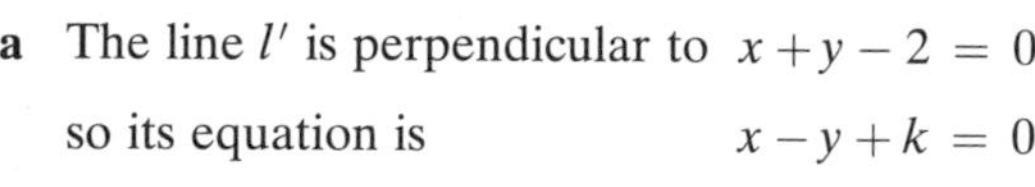

The point (2, 5) lies on l'

$\therefore\ 2 - 5 + k = 0 \quad \Rightarrow \quad k = 3$

$\therefore$ the equation of l' is $x - y + 3 = 0$

b To find the distance of C from l we need the coordinates of A, the point of intersection of l and l'.

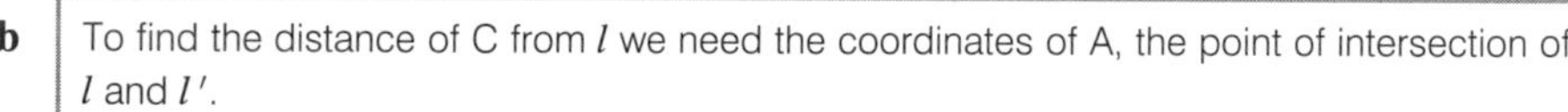

Solving the equations of l and l' simultaneously gives $2x + 1 = 0$

$\Rightarrow \quad x = -\frac{1}{2}$ and $y = \frac{5}{2}$

so A is the point $(-\frac{1}{2}, \frac{5}{2})$

$\therefore\ \text{CA} = \sqrt{\{2 - (-\frac{1}{2})\}^2 + \{5 - \frac{5}{2}\}^2} = 3.54$ to 2 s.f.

For the line to be a tangent, CA would have to be 3 units exactly (i.e. equal to the radius).

CA > 3, therefore l is not a tangent.

EXERCISE 9C

1 PS and PT are two tangents drawn from a point P to a circle whose centre is O. Join PO and prove that PT = PS.

2 The two tangents from a point to a circle of radius 12 units are each of length 20 units. Find the angle between the tangents.

3 Two circles with centres C and O have radii 6 units and 3 units respectively and the distance between O and C is less than 9 units. AB is a tangent to both circles, touching the larger circle at A and the smaller circle at B, where AB is of length 4 units. Find the length of OC.

4 The two tangents from a point A to a circle touch the circle at S and at T. Find the angle between one of the tangents and the chord ST given that the radius of the circle is 5 units and that A is 13 units from the centre of the circle.

5 An equilateral triangle of side 25 cm circumscribes a circle. Find the radius of the circle.

6 AB is a diameter of the circle and C is a point on the circumference. The tangent to the circle at A makes an angle of 30° with the chord AC. Find the angles in $\triangle$ABC.

7 The centre of a circle is at the point C(4, 8) and its radius is 3 units. Find the length of the tangents from the origin to the circle.

8 A circle touches the y-axis at the origin and goes through the point A(8, 0). The point C is on the circumference. Find the greatest possible area of △OAC.

9

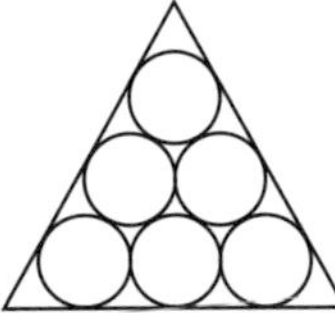

A triangular frame is made to enclose six identical spheres as shown. Each sphere has a radius of 2 cm. Find the lengths of the sides of the frame.

10 The line $y = 3x - 4$ is a tangent to the circle whose centre is the point (5, 2). Find the radius of the circle.

11 A circle of radius 6 units has its centre at the point (9, 0). If the two tangents from the origin to the circle are inclined to the x-axis at angles α and β, find $\tan\alpha$ and $\tan\beta$.

12 A, B and C are three points on the circumference of a circle. The tangent to the circle at A makes an angle α with the chord AB. The diameter through A cuts the circle again at D and D is joined to B.
Prove that $\angle ACB = \alpha$.

13 The equations of the sides of a triangle are $y = 3x$, $y + 3x = 0$ and $3y - x + 12 = 0$. Find the coordinates of the centre of the circumcircle of this triangle.

14 The line $x - 2y + 4 = 0$ is a tangent to the circle whose centre is the point C(−1, 2).

a Find the equation of the line through C that is perpendicular to the line $x - 2y + 4 = 0$.

b Hence find the coordinates of the point of contact of the tangent and the circle.

15 The point A(6, 8) is on the circumference of a circle whose centre is the point C(3, 5). Find the equation of the tangent that touches the circle at A.

CHAPTER 10

Circular measure

Angle Units

An angle is a measure of rotation and the units we have used up to now are the revolution and the degree. The reason why the number of degrees in a revolution is 360 is that the Babylonians believed that the length of the solar year was 360 days so they divided a complete revolution into 360 parts, one for each day as they thought. We now know they did not have the length of the year quite right but the number they used, 360, remains as the number of degrees in one revolution.

Part of an angle smaller than a degree is given as a decimal part although some use is still made of a system that divides a degree into 60 minutes ($60'$) and each minute into 60 seconds ($60''$).

The Radian

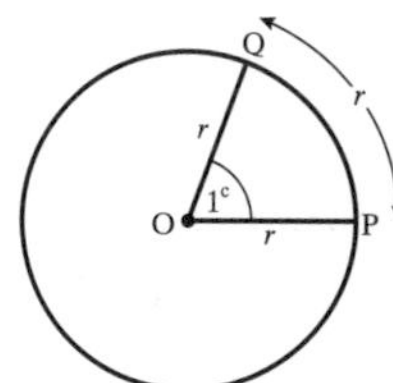

Now we consider a different unit of rotation.

If O is the centre of a circle and an arc PQ is drawn so that its length is equal to the radius of the circle then the angle POQ is called a *radian* (one radian is written 1 rad or 1^c).

An arc equal in length to the radius of a circle subtends an angle of 1 radian at the centre.

It follows that the number of radians in a complete revolution is the number of times the radius divides into the circumference.

Now the circumference of a circle is of length $2\pi r$, so the number of radians in a revolution is $2\pi r \div r$ which is 2π,

i.e. **2π radians = 360°**

Further $\quad \pi$ radians $= 180°$ and $\frac{1}{2}\pi$ radians $= 90°$

(Note that $\frac{1}{2}\pi$ can be written as $\frac{\pi}{2}$, $\frac{2}{3}\pi$ as $\dfrac{2\pi}{3}$, and so on. Both ways of writing a fraction of π should be recognised.)

When an angle is given in terms of π we usually omit the radian symbol, i.e. we write $180° = \pi$ (not $180° = \pi$ rad).

If an angle is a simple fraction of 180°, it can easily be given in terms of π,

e.g. $\quad 60° = \frac{1}{3}$ of $180° = \frac{1}{3}\pi = \dfrac{\pi}{3}$ and $135° = \frac{3}{4}$ of $180° = \frac{3}{4}\pi = \dfrac{3\pi}{4}$

Conversely, $\dfrac{7\pi}{6} = \frac{7}{6}\pi = \frac{7}{6}$ of $180° = 210°$

and $\qquad \frac{2}{3}\pi = \frac{2}{3}$ of $180° = 120°$

Angles that are not simple fractions of 180°, or π, can be converted by using the relationship $\pi = 180°$, taking the value of π from a calculator,

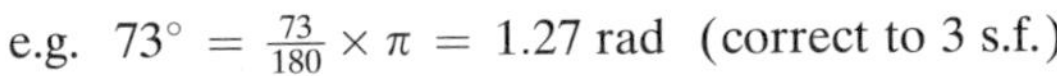

e.g. $73° = \frac{73}{180} \times \pi = 1.27$ rad (correct to 3 s.f.)

and $2.36 \text{ rad} = \dfrac{236}{\pi} \times 180° = 135°$ (correct to the nearest degree)

1^c

Now $1 \text{ rad} = \dfrac{1}{\pi} \times 180° = 57°$ (correct to 2 s.f.),

i.e. 1 radian is just a little less than 60°. This helps to visualise the size of a radian.

EXAMPLES 10A

1 Express 75° in radians in terms of π.

> $180° = \pi$ radians $\Rightarrow$ $1° = \dfrac{\pi}{180}$ radians; so to convert degrees to radians, multiply by $\dfrac{\pi}{180}$.

$$75° = 75 \times \frac{\pi}{180} \text{ radians} = \frac{5\pi}{12} \text{ radians}$$

2 Express $\frac{1}{16}\pi$ radians in degrees.

> π radians $= 180°$ radians $\Rightarrow$ $1^c = \dfrac{180°}{\pi}$; so to convert radians to degrees, multiply by $\dfrac{180}{\pi}$.

$$\tfrac{1}{16}\pi \text{ radians} = \frac{\pi}{16} \times \frac{180°}{\pi} = \frac{45°}{4} = 11\tfrac{1}{4}°$$

EXERCISE 10A

1 Express each of the following angles in radians as a fraction of π.

a 45° **b** 150° **c** 30°
d 90° **e** 270° **f** 120°
g 60° **h** 22.5° **i** 240°
j 300° **k** 315° **l** 135°
m 210° **n** 225°

2 Express each of the following angles in degrees.

a $\frac{1}{6}\pi$ **b** π **c** $\frac{1}{10}\pi$
d $\dfrac{\pi}{3}$ **e** $\frac{5}{6}\pi$ **f** $\frac{1}{12}\pi$
g $\dfrac{7\pi}{6}$ **h** $\frac{3}{4}\pi$ **i** $\dfrac{\pi}{9}$
j $\frac{3}{2}\pi$ **k** $\frac{4}{9}\pi$ **l** $\frac{1}{4}\pi$
m $\dfrac{3\pi}{5}$ **n** $\frac{1}{8}\pi$

3 Express each of the following angles in radians correct to 2 decimal places.

a 35° **b** 47.2° **c** 93°
d 233° **e** 14.1° **f** 117°
g 370°

4 Express each of the following angles in degrees correct to 1 decimal place.

a 1.7 rad **b** 3.32 rad

c 1 rad **d** 2.09 rad

e 5 rad **f** 6.283 19 rad

5 Write down the value of

a $\sin\frac{1}{3}\pi$ **b** $\sin\frac{\pi}{6}$ **c** $\cos\frac{1}{2}\pi$

d $\cos\frac{\pi}{3}$ **e** $\tan\frac{1}{4}\pi$ **f** $\tan\pi$

g $\sin\frac{\pi}{2}$ **h** $\cos\pi$ **i** $\tan\frac{3\pi}{4}$

j $\cos\frac{2}{3}\pi$

6 Write down, as a fraction of π, the possible values of x in the range $0 \leqslant x \leqslant 2\pi$ for which

a $\cos x = 1$ **b** $\tan x = 1$

c $\sin x = \frac{1}{2}$ **d** $\cos x = \frac{1}{2}$

e $\sin x = -1$ **f** $\cos x = -1$

g $\sin x = 1$ **h** $\tan x = -1$

i $\sin x = -\frac{1}{2}$ **j** $\tan x = 0$

k $\cos x = 0$ **l** $\sin x = -\frac{1}{\sqrt{2}}$

7 Use your calculator to find

a $\sin 1.2$ rad **b** $\cos 0.35$ rad

c $\tan 1.47$ rad **d** $\cos 2.5$ rad

There is no need to change the angle to degrees: set the angle mode on your calculator to radians then $\sin\theta$ rad can be keyed in directly. Similarly with the mode in radians, a calculator will give the angle in radians for which, say, $\sin\theta = 0.7$.

8 Use your calculator to find, in radians, the acute angle for which

a $\sin x = 0.28$

b $\tan x = 1.339$

c $\cos x = 0.7997$

d $\sin x = 0.0226$

The Length of an Arc

Consider an arc that subtends an angle θ at the centre of a circle, *where θ is measured in radians.*

From the definition of a radian, the arc that subtends an angle of 1 radian at the centre of the circle is of length r. Therefore if an arc subtends an angle of θ radians at the centre, the length of the arc is $r\theta$.

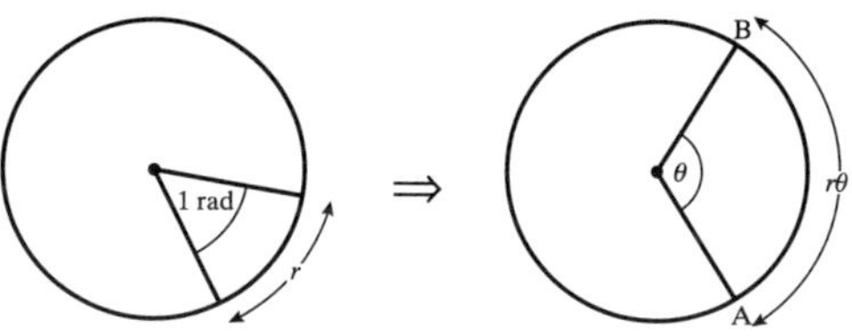

The length of arc AB = $r\theta$

EXAMPLES 10B

1 An arc subtends an angle of $\frac{\pi}{3}$ at the centre of a circle with radius 4.5 cm. Find the length of the arc in terms of π.

$$\text{Length of arc} = r\theta$$
$$= 4.5 \times \frac{\pi}{3} = 1.5\pi$$

2 Find, in radians, the angle subtended at the centre of a circle of radius 4 cm by an arc of length 8 mm.

> Remember that, when using any formula, the units must be consistent. In this case we will use millimetres for both the length of the arc and the radius.
> Remember also that using 'Length of arc $= r\theta$' gives θ in radians.

$$\text{Length of arc} = r\theta$$
$$\therefore \quad 8 = 40\theta \quad \Rightarrow \quad \theta = 0.2 \text{ rad}$$

Note that if an angle is given, or required, in degrees, we can use

$$\frac{\text{length of arc}}{2\pi r} = \frac{\theta}{360}$$

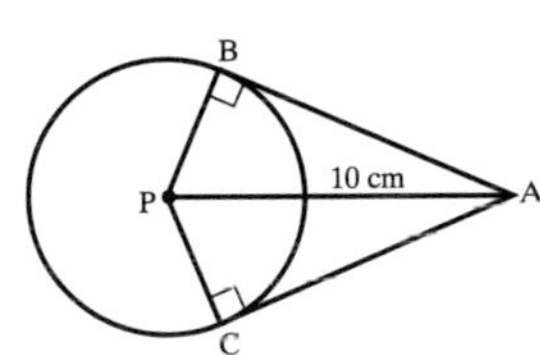

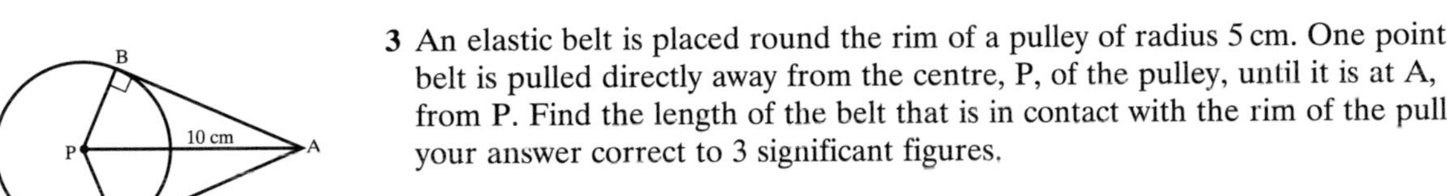

3 An elastic belt is placed round the rim of a pulley of radius 5 cm. One point on the belt is pulled directly away from the centre, P, of the pulley, until it is at A, 10 cm from P. Find the length of the belt that is in contact with the rim of the pulley. Give your answer correct to 3 significant figures.

> The belt leaves the pulley at B and C. At these two points the belt is a tangent to the rim, so AB is perpendicular to the radius BP and AC is perpendicular to PC.
> We have to find the length of the major arc BC, so we need the reflex angle at P. It is easier to find angles APB and APC first.

In $\triangle$ABP BP $= 5$ cm, AP $= 10$ cm and $\angle$ABP $= 90°$

Therefore $\cos \text{APB} = \frac{5}{10} = \frac{1}{2}$

$\Rightarrow \quad \angle\text{APB} = 60° = \frac{1}{3}\pi$

Similarly $\quad \angle\text{APC} = \frac{1}{3}\pi$

$\therefore \quad \angle\text{BPC} = \frac{2}{3}\pi$

Hence the angle subtended at P by the major arc BC is $2\pi - \frac{2}{3}\pi = \frac{4}{3}\pi$

The length of major arc BC is $5 \times \frac{4}{3}\pi = 20.94\ldots$ (Using length of arc $= r\theta$)

i.e. the length of belt in contact with the pulley is 20.9 cm (correct to 3 s.f.)

EXERCISE 10B

1 Find, in terms of π, the length of the arc that subtends an angle of $\frac{1}{6}\pi$ radians at the centre of a circle of radius 4 cm.

2 An arc subtends an angle of $\frac{5}{4}\pi$ radians at the centre of a circle of radius 10 cm. Find, in terms of π, the length of the arc.

3 Find, in radians, the angle subtended at the centre of a circle of radius 5 cm by an arc of length 12 cm.

4 What is the size of the angle subtended at the centre of a circle of radius 65 mm by an arc of length 45 mm. Give your answer in radians.

5 Find the radius of a circle in which an arc of length 15 cm subtends an angle of π radians at the centre.

6 An arc of length 20 cm subtends an angle of $\frac{4}{5}\pi$ radians at the centre of a circle. Find the radius of the circle.

7 Find, in terms of π, the length of the arc that subtends an angle of 60° at the centre of a circle of radius 12 cm.

8 An arc of length 15 cm subtends an angle of 45° at the centre of a circle. Find, in terms of π, the radius of the circle.

9 Find, in degrees, the angle subtended at the centre of a circle of radius a cm by an arc of length $2a$ cm. Give your answer in terms of π.

10 Calculate, in degrees, the angle subtended at the centre of a circle of radius 2.7 cm by an arc of length 6.9 cm.

11 A company logo is made by removing a sector containing an angle of 1 radian from a circle of radius 25 mm. Find the perimeter of the logo.

12 A curve in the track of a railway line is a circular arc of length 400 m and radius 1200 m. Through what angle does the direction of the track turn?

13 Two discs, of radii 5 cm and 12 cm, are placed, partly overlapping, on a table. If their centres are 13 cm apart find the perimeter of the 'figure-eight' shape.

The Area of a Sector

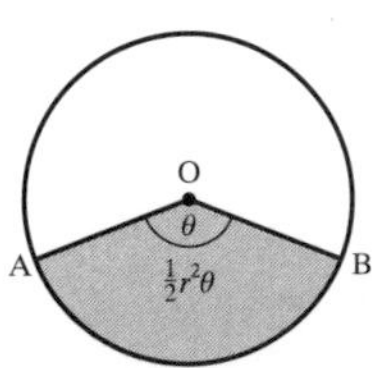

We know that $\dfrac{\text{area of sector}}{\text{area of circle}} = \dfrac{\text{angle contained in the sector}}{\text{complete angle at the centre}}$

Consider a sector containing an angle of θ radians at the centre of the circle.

The complete angle at the centre of the circle is 2π, hence

$$\frac{\text{Area of sector}}{\text{Area of circle}} = \frac{\theta}{2\pi}$$

The area of the circle is πr^2

$$\Rightarrow \quad \text{area of sector} = \frac{\theta}{2\pi} \times \pi r^2$$

$$= \tfrac{1}{2}r^2\theta$$

The area of sector AOB = $\boldsymbol{\frac{1}{2}r^2\theta}$

EXAMPLES 10C

1 Find, in terms of π, the area of the sector of a circle of radius 3 cm that contains an angle of $\frac{\pi}{5}$.

$$\text{Area of sector} = \tfrac{1}{2}r^2\theta = \tfrac{1}{2}(3)^2\left(\frac{\pi}{5}\right)\text{ cm}^2 = \frac{9\pi}{10}\text{ cm}^2$$

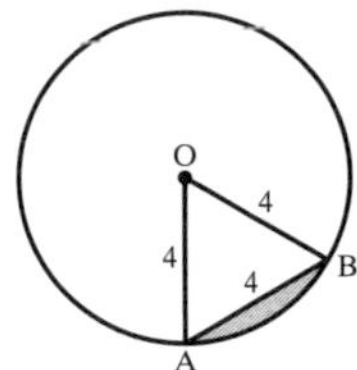

2 AB is a chord of a circle with centre O and radius 4 cm. AB is of length 4 cm and divides the circle into two segments. Find, correct to two decimal places, the area of the minor segment.

ABC is an equilateral triangle, so each angle is 60°, i.e. $\frac{\pi}{3}$ rad.

> To use the formula for the area of a sector, the angle must be in radians. To find the area of the minor segment, we subtract the area of $\triangle$AOB from the area of sector AOB.

$$\text{Area of sector AOB} = \tfrac{1}{2}r^2\theta = \tfrac{1}{2}(4^2)(\tfrac{1}{3}\pi) = 8.3775\ldots$$

$$\text{Area of }\triangle\text{AOB} = \tfrac{1}{2}r^2\sin\theta = \tfrac{1}{2}(4)(4)(\sin 60°) = 6.9282\ldots$$

$$\text{Area of minor segment} = \text{area of sector AOB} - \text{area of }\triangle\text{AOB}$$

$$= 8.3775\ldots - 6.9282\ldots = 1.449\ldots$$

The area of the minor segment is 1.45 cm^2 correct to 2 d.p.

EXERCISE 10C

1 A sector of a circle of radius 4 cm contains an angle of 30°. Find the area of the sector.

2 A sector of a circle of radius 8 cm contains an angle of 135°. Find the area of the sector.

3 The area of a sector of a circle of radius 2 cm is π cm^2. Find the angle contained by the sector.

4 The area of a sector of a circle of radius 5 cm is 12 cm^2. Find the angle contained by the sector.

5 A sector of a circle of radius 10 cm contains an angle of $\frac{5\pi}{6}$.
Find the area of the sector.

6 An arc of length 15 cm subtends an angle π at the centre of a circle. Find the radius of the circle and hence the area of the sector containing the angle π.

7 A sector of a circle has an area 3π cm^2 and contains an angle $\frac{1}{6}\pi$. Find the radius of the circle.

8 A sector of a circle has an area 6π cm^2 and contains an angle of 45°. Find the radius of the circle.

9

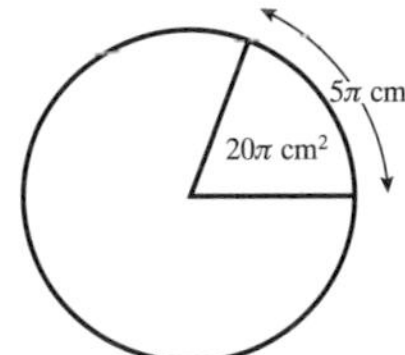

An arc of a circle is of length 5π cm and the sector it bounds has an area of 20π cm^2. Find the radius of the circle.

10 Calculate, in radians, the angle at the centre of a circle of radius 83 mm contained in a sector of area 974 mm^2.

11 In a circle with centre O and radius 5 cm, AB is a chord of length 8 cm. Find

a the area of triangle AOB

b the area of the sector AOB.

12 A chord of length 10 mm divides a circle of radius 7 mm into two segments. Find the area of each segment.

13 A chord PQ, of length 12.6 cm, subtends an angle of $\frac{2}{3}\pi$ at the centre of a circle. Find

a the length of the arc PQ

b the area of the minor segment cut off by the chord PQ.

14 Two circles, each of radius 14 cm, are drawn with their centres 20 cm apart. Find the length of their common chord. Find also the area common to the two circles.

Problems using Arcs and Sectors

When you use the formulae 'arc length $= r\theta$' and 'area of sector $= \frac{1}{2}r^2\theta$', remember that the angle is in radians and units must be consistent.

EXAMPLES 10D

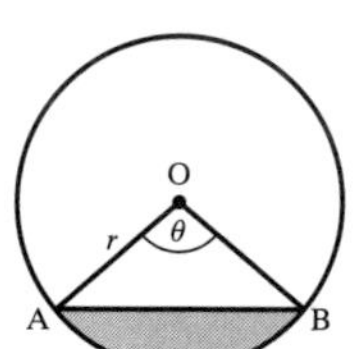

1 The diagram shows a circle centre O, radius r.
Angle AOB $= \theta$ radians.
The area of $\triangle$AOB is twice the area of the shaded segment.
Show that $\theta = \frac{3}{2}\sin\theta$.

The area of $\triangle\text{AOB} = \frac{1}{2}r^2\sin\theta$

The area of sector AOB $= \frac{1}{2}r^2\theta$

Therefore the area of the shaded segment is equal to $\frac{1}{2}r^2\theta - \frac{1}{2}r^2\sin\theta$

We are told that (area of $\triangle$AOB) $= 2 \times$ (area of the shaded segment) so we use this to find a relationship between the expressions for these areas.

Therefore $\frac{1}{2}r^2\sin\theta = 2 \times (\frac{1}{2}r^2\theta - \frac{1}{2}r^2\sin\theta)$

This can be simplified by multiplying both sides by 2 and dividing both sides by r^2.

$\Rightarrow \quad \sin\theta = 2\theta - 2\sin\theta$

$\Rightarrow \quad 3\sin\theta = 2\theta$

i.e. $\quad \theta = \frac{3}{2}\sin\theta$

2 The diagram shows two discs that are fixed with their centres 18 cm apart and connected by a drive belt. The radii of the discs are 5 cm and 8 cm. Find the length of the drive belt.

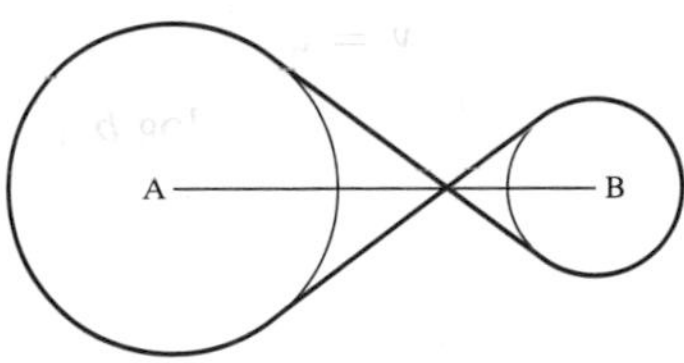

> The diagram is symmetrical about the line AB so we can draw a simpler diagram showing half the belt. The measurements are all in centimetres, so we will omit the units from the diagram and working.

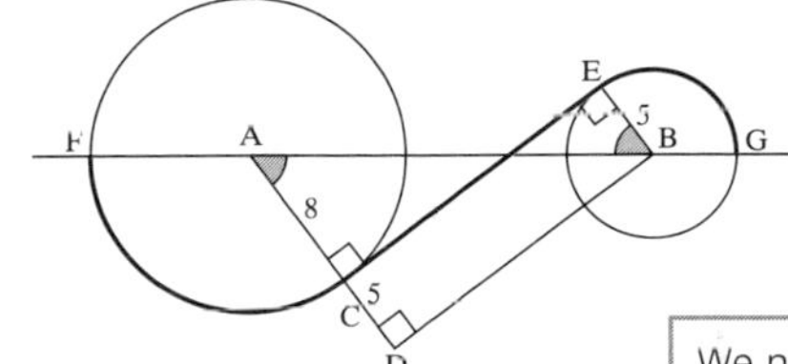

The part of the belt between the two discs is a tangent to both circles, so $\angle ACE = \angle BEC = 90°$

> We need to find the lengths of the common tangent CE and the arcs FC, EG. Adding the line BD, drawn parallel to CE to cut AC extended at D, gives us the single triangle ADB to work with. This avoids having to consider the two sections of CE in two different triangles.

In $\triangle ABD$, $DB^2 = AB^2 - AD^2$

$$= 18^2 - 13^2 = 155$$

$\therefore\ DB = 12.449\ldots$ so $CE = 12.449\ldots$

> To find the length of arc FC, we need $\angle FAC$; $\angle FAC = \pi - \angle BAC$ and $\angle BAC$ can be found from $\triangle ABD$.

In $\triangle ABD$, $\cos \angle BAD = \frac{13}{18} \quad \Rightarrow \quad \angle BAD = 0.7637\ldots$ rad

$\therefore \quad \angle FAC = \pi - 0.7637\ldots$ rad $= 2.3778\ldots$ rad

Hence arc $FC = 8 \times 2.3778\ldots = 19.0224\ldots$

$\angle EBA = \angle BAC$, $\therefore\ \angle EBG = \angle FAC = 2.3778\ldots$ rad

Hence arc $EG = 5 \times 2.3778\ldots = 11.8890\ldots$

Arc FC + arc EG + CE $= 19.022\ldots + 11.889\ldots + 12.449\ldots = 43.36\ldots$

Therefore the length of the drive belt is $2 \times 43.36\ldots$ cm

$= 86.7$ cm correct to 3 s.f.

EXERCISE 10D

1 A chord of a circle subtends an angle of θ radians at the centre of the circle. The area of the minor segment cut off by the chord is one eighth of the area of the circle. Prove that $4\theta = \pi + 4\sin\theta$.

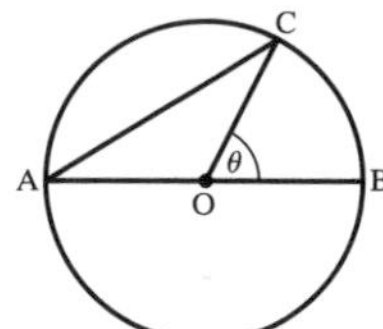

2 AB is the diameter of a circle, centre O. C is a point on the circumference such that $\angle COB = \theta$ radians. The area of the minor segment cut off by AC is equal to twice the area of the sector BOC. Show that $3\theta = \pi - \sin\theta$.

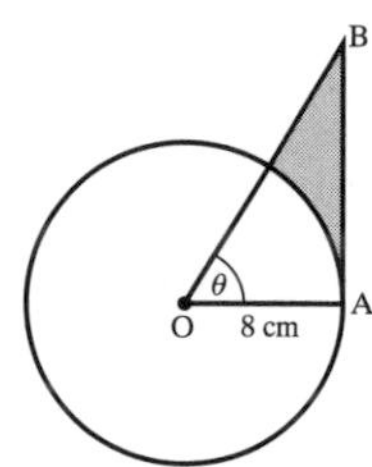

3 The diagram shows a sector of a circle, centre O, containing an angle θ radians. Find the area of the shaded region of the diagram, giving your answer in terms of θ.

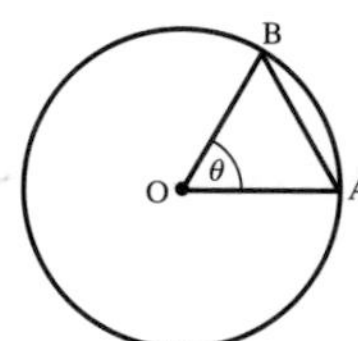

4 O is the centre of a circle of radius r cm. A chord AB subtends an angle of θ radians at O.

a Show that the area of the minor segment cut off by AB is equal to

$$\tfrac{1}{2}r^2(\theta - \sin\theta)$$

b The area of the circle is twenty times the area of the minor segment. Show that

$$\sin\theta = \theta - \frac{\pi}{10}.$$

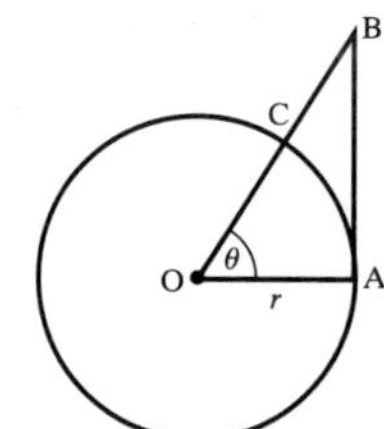

5 AB is a tangent to the circle centre O. $\angle AOC = \theta$ radians. Show that the perimeter of the section bounded by the lines AB, BC and the arc AC is given by

$$r\left(\tan\theta + \frac{1}{\cos\theta} + \theta - 1\right)$$

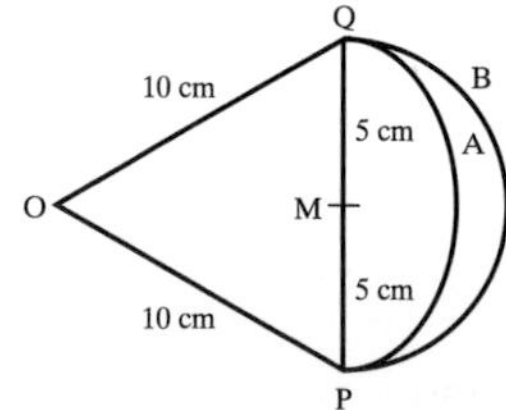

6 The diagram shows two arcs, A and B. Arc A is part of the circle, centre O and radius OP. Arc B is part of the circle, centre M and radius PM, where M is the midpoint of PQ.

a Write down the angle POQ in radians.

b Show that the area enclosed by the two arcs is equal to

$$25\left(\sqrt{3} - \frac{\pi}{6}\right). \qquad \left[\text{Use } \sin\frac{\pi}{3} = \frac{\sqrt{3}}{2}\right]$$

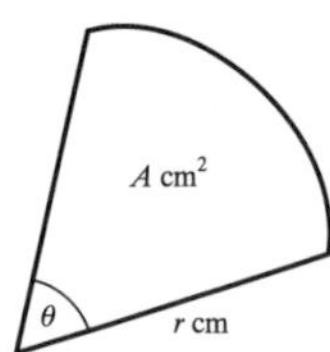

7 The diagram shows a sector of a circle of radius r cm containing an angle θ radians. The area of the sector is A cm^2 and the perimeter of the sector is 50 cm.

a Find θ in terms of r.

b Show that $A = 25r - r^2$.

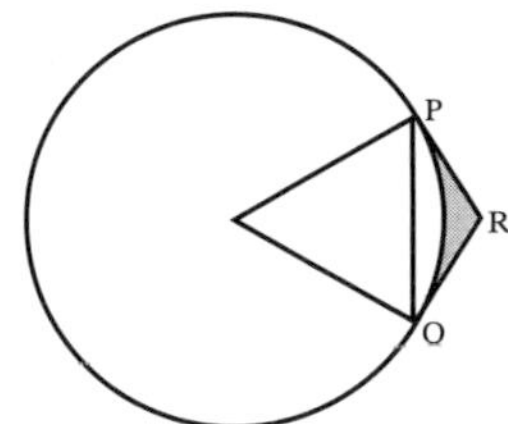

8 A chord PQ of length 6 cm is drawn in a circle of radius 10 cm. The tangents to the circle at P and Q meet at R. Find the area enclosed by PR, QR and the minor arc PQ.

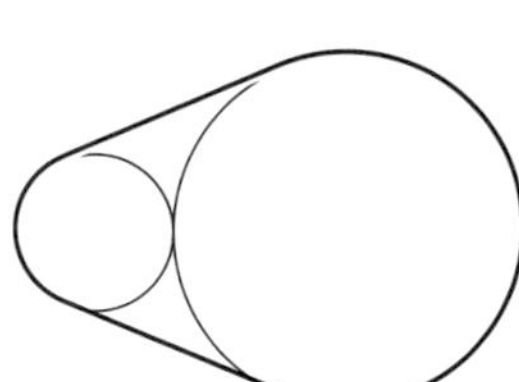

9 Two discs are placed, in contact with each other, on a table. Their radii are 4 cm and 9 cm. An elastic band is stretched round the pair of discs. Calculate

a the angle subtended at the centre of the smaller disc by the arc that is in contact with the elastic band

b the length of the part of the band that is in contact with the smaller disc

c the length of the part of the band that is in contact with the larger disc

d the total length of the stretched band.

(*Hint* The straight parts of the stretched band are common tangents to the two circles.)

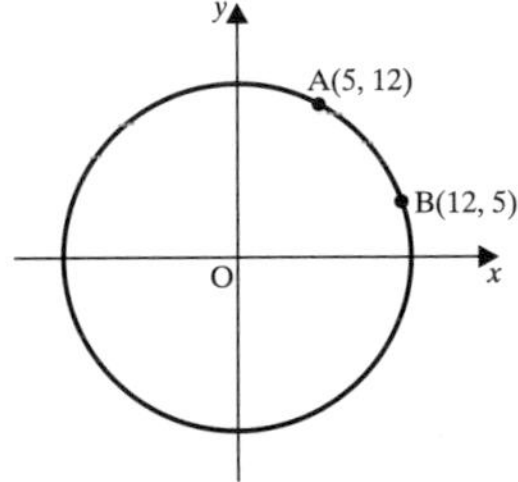

10 The diagram shows a circle with radius 13 cm whose centre is at the origin. The points A(5, 12) and B(12, 5) are on the circumference of the circle. Find the length of the arc AB.

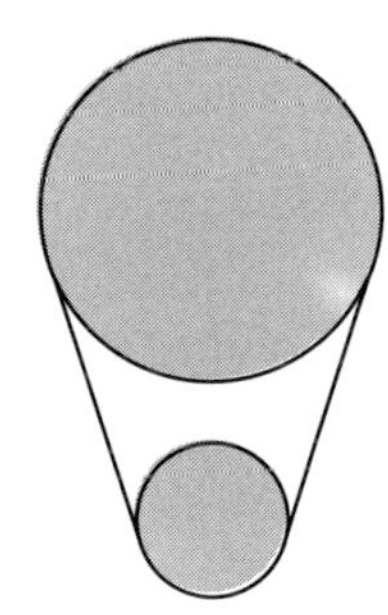

11 The diagram shows a drive belt round two pulleys whose centres are 40 cm apart. The radius of the smaller pulley is 12 cm and the radius of the larger pulley is 20 cm. Find the length of the drive belt.

SUMMARY B

PLANE GEOMETRY

Pythagoras' Theorem

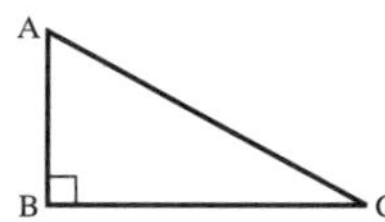

In $\triangle ABC$, $\angle B = 90^\circ \iff AC^2 = AB^2 + BC^2$

Similar Triangles

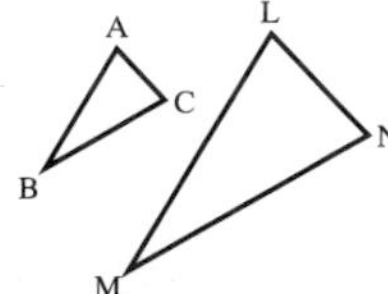

$\angle A = \angle L$, $\angle B = \angle M$, $\angle C = \angle N$
$\iff AB : LM = BC : MN = AC : LN$

Angle Bisector Theorem

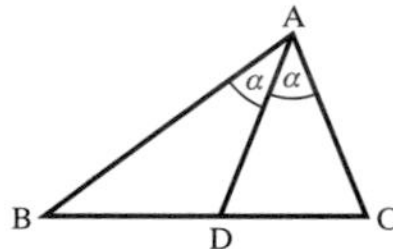

AD bisects $\angle A \iff BD : DC = AB : AC$

Circle Theorem

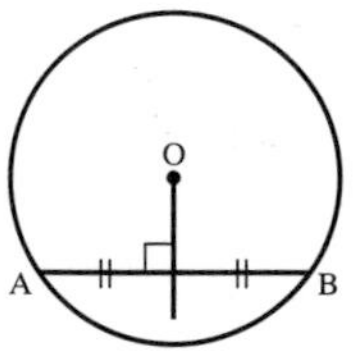

The perpendicular bisector of any chord goes through O, and conversely.

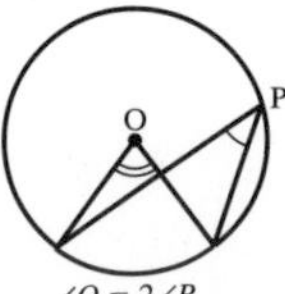

$\angle O = 2\angle P$

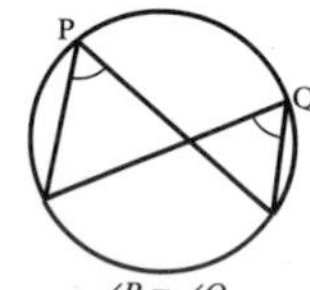

$\angle P = \angle Q$

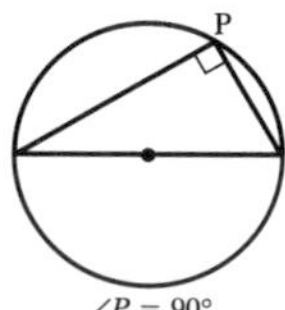

$\angle P = 90^\circ$

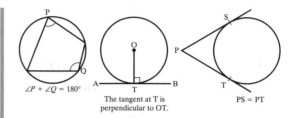

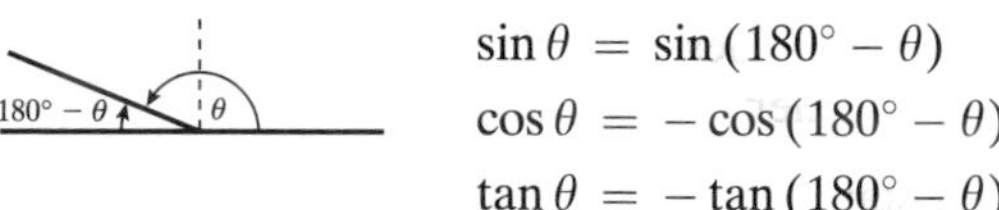

$\angle P + \angle Q = 180^\circ$

The tangent at T is perpendicular to OT.

PS = PT

TRIGONOMETRY

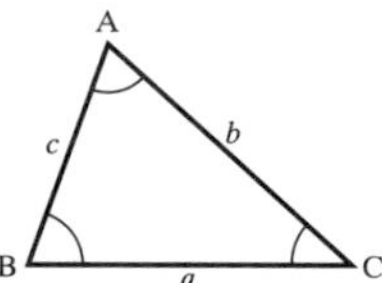

$$\sin\theta = \sin(180^\circ - \theta)$$
$$\cos\theta = -\cos(180^\circ - \theta)$$
$$\tan\theta = -\tan(180^\circ - \theta)$$

In any triangle ABC,

$$\frac{a}{\sin A} = \frac{b}{\sin B} = \frac{c}{\sin C} \quad \text{(sine rule)}$$

$$\left.\begin{aligned} a^2 &= b^2 + c^2 - 2bc\cos A \\ b^2 &= a^2 + c^2 - 2ac\cos B \\ c^2 &= a^2 + b^2 - 2ab\cos C \end{aligned}\right\} \quad \text{(cosine rule)}$$

The area of $\triangle ABC$ is
$\frac{1}{2}ab\sin C$, or $\frac{1}{2}bc\sin A$, or $\frac{1}{2}ac\sin B$

CIRCULAR MEASURE

One radian (1^c) is the size of the angle subtended at the centre of a circle by an arc equal in length to the radius of the circle.

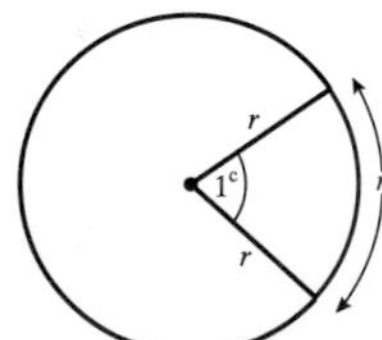

The length of arc AB is $r\theta$.

The area of sector AOB is $\frac{1}{2}r^2\theta$.

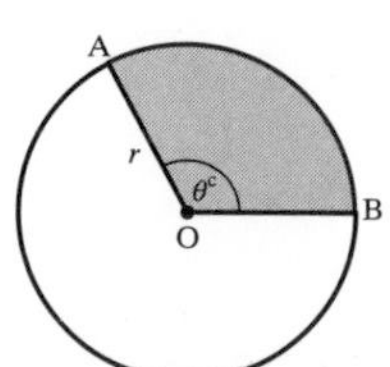

COORDINATE GEOMETRY

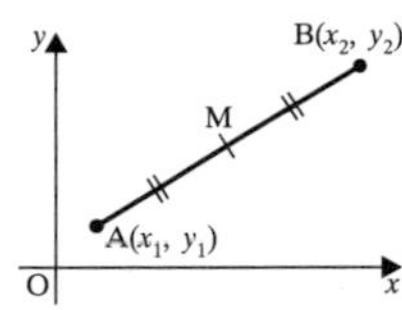

Length of AB is $\sqrt{(x_2 - x_1)^2 + (y_2 - y_1)^2}$

Midpoint, M, of AB is $[\frac{1}{2}(x_1 + x_2), \frac{1}{2}(y_1 + y_2)]$

Gradient of AB is $\dfrac{y_2 - y_1}{x_2 - x_1}$

Parallel lines have equal gradients.

When two lines are perpendicular, the product of their gradient is -1.

The standard equation of a straight line is $y = mx + c$, where m is its gradient and c its intercept on the y-axis.

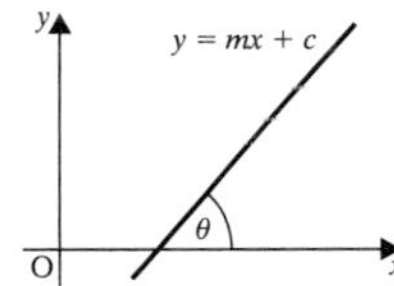

When a line with gradient m makes an angle θ with the positive x-axis, then $m = \tan\theta$.

Any equation of the form $ax + by + c = 0$ gives a straight line.

The equation of a line passing through (x_1, y_1) and with gradient m is

$$y - y_1 = m(x - x_1)$$

Given a line with equation $ax + by + c = 0$, then any perpendicular line has equation $bx - ay + k = 0$.

Reduction of Relationships to Linear Form

When a non-linear law, containing two unknown constants, connects two variables, the relationship can often be reduced to linear form. The aim is to produce an equation in which one term is constant and another term does not contain a constant. The law can then be expressed in the form

$$Y = mX + C$$

Some common conversions are

$y = ax^n$; take logs: $\ln y = \ln a + n \ln x$; use $Y = \ln y$ and $X = \ln x$

$y = ab^x$; take logs: $\ln y = \ln a + x \ln b$; use $Y = \ln y$ and $X = x$

MULTIPLE CHOICE EXERCISE B

In Questions 1 to 23 write down the letter or letters corresponding to a correct answer.

1 A line AB is 10 cm long. P divides AB externally in the ratio 9:4. The length of PB is

A 18 cm **C** 8 cm **E** 4 cm

B $\frac{40}{13}$ cm **D** 40 cm

2 In △ABC, $a = 3$ cm, $b = 4$ cm and $c = 5$ cm,

A A = 90° **D** B = 90°

B C = 60° **E** C = 90°

C B = 45°

3 In the diagram, O is the centre of the circle. Angle P is

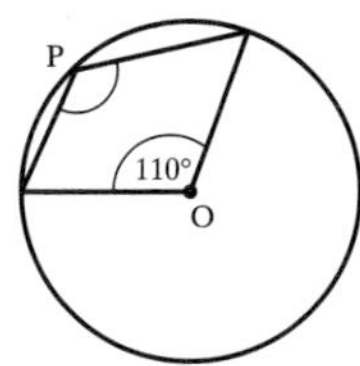

A 110° **C** 125° **E** 220°

B 90° **D** 55°

4 120° in radians is

A $\frac{\pi}{2}$ **C** 2π **E** $\frac{2\pi}{3}$

B 2 **D** $\frac{\pi}{3}$

5 The sign that links the statements 'In $\triangle ABC$, $\angle A = 90°$' and 'In $\triangle ABC$, $BC^2 = AB^2 + AC^2$' could be

A $\Rightarrow$ **B** $\Leftarrow$ **C** $\Leftrightarrow$

6 The length of the line joining $(3, -4)$ to $(-7, 2)$ is

A $2\sqrt{13}$ **C** $2\sqrt{34}$ **E** 6

B 16 **D** $2\sqrt{5}$

7 The midpoint of the line joining $(-1, -3)$ to $(3, -5)$ is

A $(1, 1)$ **C** $(2, -8)$ **E** $(1, -1)$

B $(0, 0)$ **D** $(1, -4)$

8 The gradient of the line joining $(1, 4)$ and $(-2, 5)$ is

A $\frac{1}{3}$ **C** 3 **E** 1.3

B $-\frac{1}{3}$ **D** -3

9 The gradient of the line perpendicular to the join of $(-1, 5)$ and $(2, -3)$ is

A $\frac{3}{8}$ **C** $\frac{1}{2}$ **E** $2\frac{2}{3}$

B $-2\frac{2}{3}$ **D** 2

10 The line joining $(1, 3)$ to (a, b) has unit gradient.

A $b - a = 2$ **D** $b - a = 4$

B $a - b = 2$ **E** $a - b = 4$

C $a + b = 2$

11 A line l goes through the origin and is perpendicular to $3x - 2y = 0$.

A $\Rightarrow$ The equation of l is $2x + 3y = 0$.

B $\Leftarrow$ The equation of l is $2x + 3y = 0$.

C $\Leftrightarrow$ The gradient of l is $-\frac{2}{3}$.

12 The equation of the line with gradient 1 and passing through the point (h, k) is

A $y = x + k - h$

B $y = \dfrac{k}{h}x + 1$

C $y = x + h - k$

D $ky = hx - 1$

E $y + x = k - h$

13 The two lines $x + y = 0$ and $2x - y + 3 = 0$ intersect at the point

A $(-\frac{1}{3}, \frac{1}{3})$ **D** $(-1, 1)$

B $(1, -1)$ **E** $(3, -3)$

C $(-3, 3)$

14 An angle of 1 radian is equivalent to:

A 90° **C** 67.3° **E** 45°

B 60° **D** 57.3°

15 An arc PQ subtends an angle of 60° at the centre of a circle of radius 1 cm. The length of the arc PQ is

A 60 cm **C** $\frac{1}{6}\pi$ cm **E** $\frac{1}{18}\pi^2$ cm

B 30 cm **D** $\frac{1}{3}\pi$ cm

16 The relationship between x and y is $y = ab^x$.

A When values of $\log y$ are plotted against values of $\log x$, the points lie on a straight line.

B When values of y are plotted against values of x, the points lie on a straight line.

C When values of $\log y$ are plotted against values of x, the points lie on a straight line.

D When values of $\log y$ are plotted against values of $x \log b$, the points lie on a straight line.

17 In $\triangle ABC$

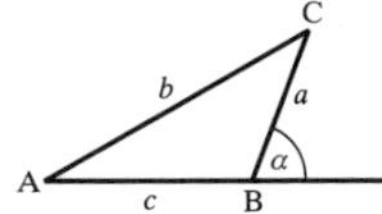

A $\dfrac{a}{\sin A} = \dfrac{b}{\sin \alpha}$

B $b^2 = a^2 + c^2 + 2ac\cos\alpha$

C $a^2 = b^2 + c^2 + 2bc\cos A$

18 The symbol that links 'In $\triangle$ABC, $\angle$A = $\angle$B' and 'In $\triangle$ABC, AB = BC = AC' can be

A $\Rightarrow$ **B** $\Leftarrow$ **C** $\Leftrightarrow$

19 A and B are two points with coordinates $(3, 4)$, $(-1, 6)$.

A Gradient of AB is $-\frac{1}{2}$.

B Midpoint of AB is $(2, 5)$.

C Length of AB is $2\sqrt{5}$.

20 A, B and C are the points $(5, 0)$, $(-5, 0)$, $(2, 3)$

A AB and BC are perpendicular.

B Area of $\triangle$ABC is 15 square units.

C A, B and C are collinear.

21 A, B, C are the points $(0, 13)$, $(0, -13)$, $(5, -12)$.

A A, B, C lie on the circumference of a circle, centre the origin.

B The equation of AC is $5x + y - 13 = 0$.

C The midpoint of BC is the origin.

22 The equation of a line l is $y = 2x - 1$

A The line through the origin perpendicular to l is $y + 2x = 0$.

B The line through $(1, 2)$ parallel to l is $y = 2x - 3$.

C l passes through $(1, 1)$

D $\Leftrightarrow$ $(a, 2a - 1)$ lies on the line for all values of a.

23 The equation of a line l is $7x - 2y + 4 = 0$.

A l has a gradient of $3\frac{1}{2}$.

B l is parallel to $7x + 2y - 3 = 0$.

C l is perpendicular to $2x + 7y - 5 = 0$.

In Questions 24 to 30 a single statement is made. Write T if it is true, F if it is false.

24 If an arc of a circle of radius 0.5 cm subtends an angle of 60° at the centre of the circle, then the length of the arc is 30 cm.

25 If in triangle ABC, angle A is 30° then $\sin B = b/2a$.

26

If BD = DC then AD bisects angle A.

27 The line joining $(0, 0)$ and $(1, 3)$ is equal in length to the line joining $(0, 1)$ and $(3, 0)$.

28 If a line has gradient m and intercept d on the x-axis, its equation is $y = mx - md$.

29 The line passing through $(3, 1)$ and $(-2, 5)$ is perpendicular to the line $4y = 5x - 3$.

30 π radians = 360°

EXAMINATION QUESTIONS B

1 a i Find the gradient of the straight line $2x + 3y = -11$.

ii Find the equation of the line through $(9, -1)$ perpendicular to $2x + 3y = -11$.

b Calculate the coordinates of the point where these two lines meet. (OCR)

2

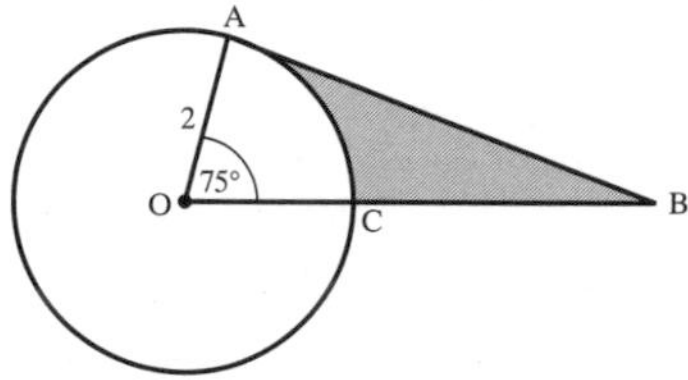

The diagram shows a circle of centre O and of radius 2 cm (not drawn to scale). The line AB is the tangent to the circle at A and the line OB cuts the circle at C. The angle AOB $= 75°$. Calculate the area of the shaded region correct to three significant figures. (AQA)

3 The points A, B and C have coordinates $(5, -3)$, $(7, 8)$ and $(-3, 4)$ respectively. The midpoint of BC is M.

a Write down the coordinates of M.

b Find the equation of the straight line which passes through the points A and M. (AQA)

4

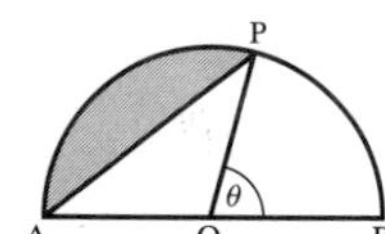

The diagram shows a semicircle APB on AB as diameter. The midpoint of AB is O. The point P on the semicircle is such that the area of the sector POB is equal to twice the area of the shaded segment. Given that angle POB is θ radians, show that

$$3\theta = 2(\pi - \sin\theta).$$

(OCR)

5 The vertices of the triangle ABC are A$(-3, 1)$, B$(10, -8)$ and C$(1, 4)$. Find an equation of the line passing through A and B, giving your answer in the form $px + qy + r = 0$, where p, q and r are integers. Show by calculation that CA and CB are perpendicular. (OCR)

6 In triangle ABC, angle A $= 42°$, AB $= 6.0$ cm and BC $= 4.5$ cm. Calculate the two possible values of angle C. (OCR)

7 The points A$(-2, 4)$, B$(6, -2)$ and C$(5, 5)$ are the vertices of triangle ABC and D is the midpoint of AB.

a Find the equation of the line passing through A and B in the form $y = mx + c$, where the constants m and c are to be found.

b Show that CD is perpendicular to AB. (Edexcel)

8 Two circles with centres A and B, each of radius 13 cm, lie in a plane with their centres 24 cm apart. The circles intersect at the points C and D.

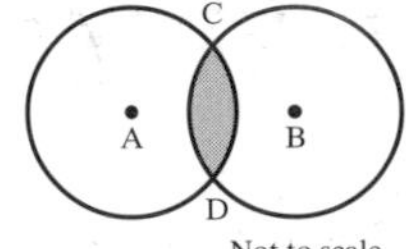

a Determine the length of CD.

b Calculate the size of angle CAD in radians, giving your answer to four significant figures.

c Calculate the area of the shaded region common to both circles, giving your answer to the nearest 0.1 cm^2. (AQA)

9 A, B, C are the points $(4, 3)$, $(2, 2)$ and $(5, -4)$ respectively.

a Show that the lines AB and BC are perpendicular.

b A point D is such that ABCD is a rectangle. Find the equation of the line AD and the equation of the line CD. Hence, or otherwise, find the coordinates of D.

c Find the area of rectangle ABCD. (WJEC)

10 The diagram shows the points $A(-2, 4)$, $B(6, -2)$ and $C(5, 5)$.

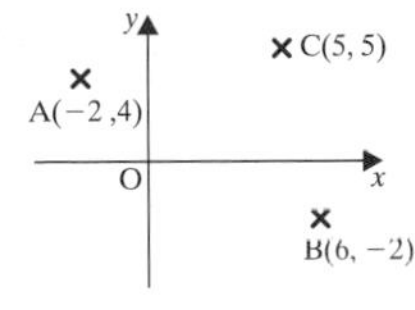

a Find the equation of the line passing through the points A and B giving your answer in the form $y = mx + c$ where the values of m and c are to be found.

b The point D is the midpoint of AB. Prove that CD is perpendicular to AB.

c Show that the line through C parallel to AB has equation $3x + 4y = 35$. (AQA)

11 The straight line passing through the point $P(2, 1)$ and $Q(k, 11)$ has gradient $-\frac{5}{12}$.

a Find the equation of the line in terms of x and y only.

b Determine the value of k.

c Calculate the length of the line segment PQ. (Edexcel)

12

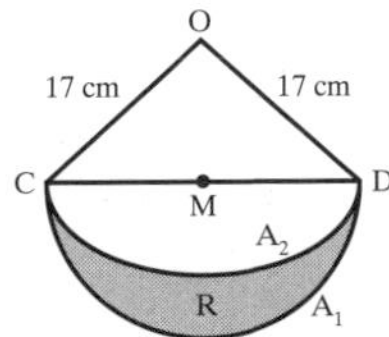

The figure shows the triangle OCD with OC = OD = 17 cm and CD = 30 cm. The midpoint of CD is M. With centre M, a semicircular arc A_1 is drawn on CD as diameter. With centre O and radius 17 cm, a circular arc A_2 is drawn from C to D. The shaded region R is bounded by the arcs A_1 and A_2. Calculate, giving answers to 2 decimal places,

a the area of the triangle OCD

b the angle COD in radians

c the area of the shaded region R. (Edexcel)

13 a Find an equation of the line l which passes through the points A$(1, 0)$ and B$(5, 6)$.

The line m with equation $2x + 3y = 15$ meets l at the point C.

b Determine the coordinates of C.

The point P lies on m and has x-coordinate -3.

c Show, by calculation, that PA = PB. (Edexcel)

14 The left edge of the shaded crescent-shaped region, shown in the figure below, consists of an arc of a circle of radius r cm with centre O. The angle AOB $= \frac{2}{3}\pi$ radians.

The right edge of the shaded region is a circular arc with centre X, where OX $= r$ cm.

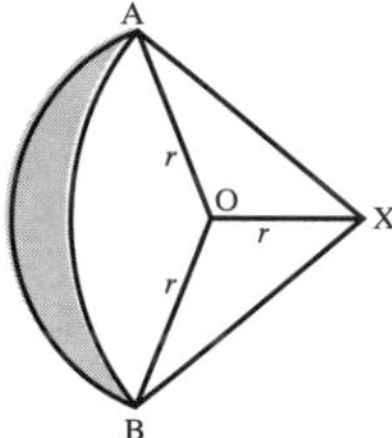

a Show that angle AXB $= \frac{1}{3}\pi$ radians.

b Show that AX $= r\sqrt{3}$ cm.

c Calculate, in terms of r, π and $\sqrt{3}$, the area of the shaded region. (OCR)

15 The straight line p passes through the point $(10, 1)$ and is perpendicular to the line r with equation $2x + y = 1$. Find the equation of p. Find also the coordinates of the point of intersection of p and r, and deduce the perpendicular distance from the point $(10, 1)$ to the line r. (OCR)

16 The figure shows a circle with centre O and radius r. Points A, B and C lie on the circle such that AB is a diameter. Angle BAC $= \theta$ radians.

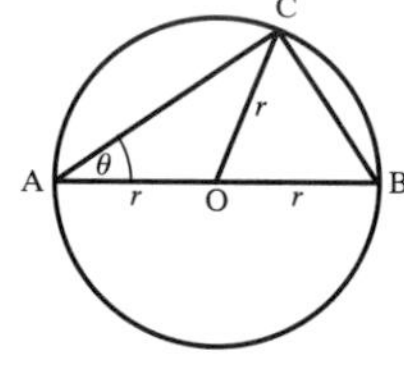

a Write down the size of angle AOC in terms of θ.

b Use the cosine rule in triangle AOC to express AC^2 in terms of r and θ.

c By considering right-angled triangle ABC, write down the length of AC in terms of r and θ. Deduce that $\cos 2\theta = 2\cos^2\theta - 1$. (OCR)

17 The points A and B have coordinates $(3, -1)$ and $(6, 3)$ respectively. The points C and D are each distant 4 units from A and 6 units from B, as shown in the diagram.

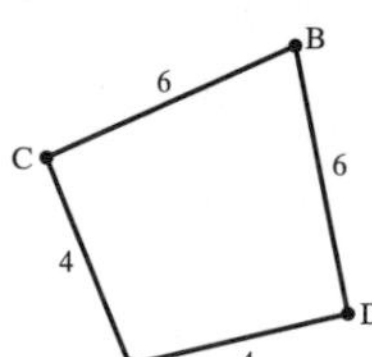

a Calculate the length AB.

b Calculate the cosine of angle CAB.

c Show that the length of CD is $3\sqrt{7}$. (AQA)

18

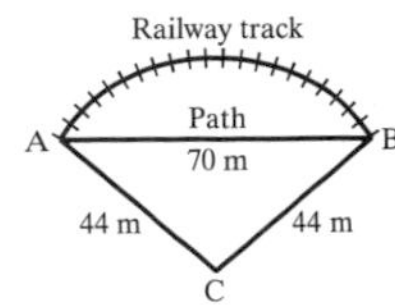

There is a straight path of length 70 m from the point A to the point B. The points are joined also by a railway track in the form of an arc of the circle whose centre is C and whose radius is 44 m, as shown in the figure.

a Show that the size, to 2 decimal places, of ∠ACB is 1.84 radians.

Calculate

b the length of the railway track

c the shortest distance from C to the path

d the area of the region bounded by the railway track and the path. (Edexcel)

19

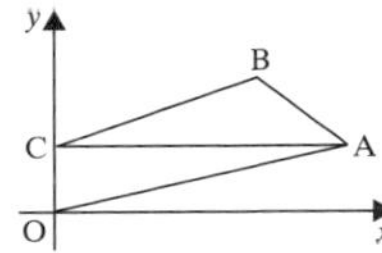

The diagram, not drawn to scale, shows a trapezium OABC with OA parallel to CB. Given that B is the point $(4, 3)$, C is the point $(0, 2)$ and the diagonal CA is parallel to the x-axis, **calculate** the coordinates of A. (OCR)

20

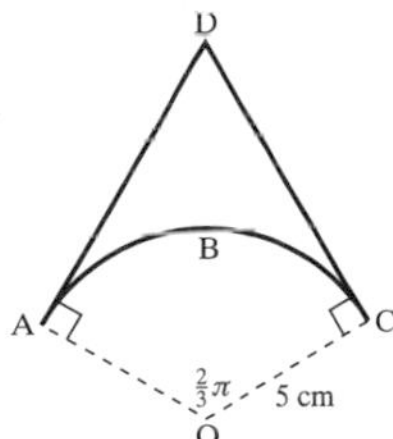

In the diagram, ABC is an arc of a circle with centre O and radius 5 cm. The lines AD and CD are tangents to the circle at A and C respectively. Angle AOC $= \frac{2}{3}\pi$ radians. Calculate the area of the region enclosed by AD, DC and the arc ABC, giving your answer correct to 2 significant figures. (OCR)

21 The line L passes through the points A$(1, 3)$ and B$(-19, -19)$.

a Calculate the distance between A and B.

b Find an equation of L in the form $ax + by + c = 0$, where a, b and c are integers. (Edexcel)

22 a Find an equation of the straight line passing through the points with coordinates $(-1, 5)$ and $(4, -2)$, giving your answer in the form $ax + by + c = 0$, where a, b and c are integers.

The line crosses the x-axis at the point A and the y-axis at the point B, and O is the origin.

b Find the area of △OAB. (Edexcel)

23

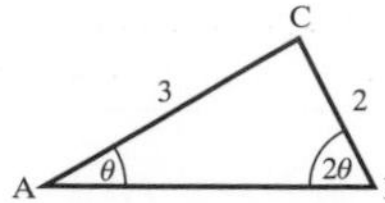

In the triangle ABC, AC = 3 cm, BC = 2 cm, $\angle$BAC = θ and $\angle$ABC = 2θ.
Calculate the value of θ correct to the nearest tenth of a degree.

Hence find the size of the angle ACB and, without further calculation, explain why the length of AB is greater than 2 cm. (AQA)

24 Find the equation of the straight line that passes through the points $(3, -1)$ and $(-2, 2)$, giving your answer in the form $ax + by + c = 0$. Hence find the coordinates of the point of intersection of the line and the x-axis. (OCR)

25 The diagram shows the cross-section of a tunnel.
The cross-section has the shape of a major segment of a circle, and the point O is the centre of the circle.
The radius of the circle is 4 m, and the size of angle AOB is 1.5 radians.
Calculate the perimeter of the cross-section.

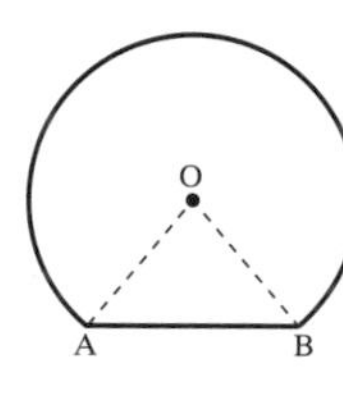

(OCR)

26 The points P, Q and R have coordinates $(2, 4)$, $(7, -2)$ and $(6, 2)$ respectively. Find the equation of the straight line l which is perpendicular to the line PQ and which passes through the midpoint of PR. (AQA)

27 The points A and B have coordinates $(8, 7)$ and $(-2, 2)$ respectively. A straight line l passes through A and B and meets the coordinate axes at the points C and D.

a Find, in the form $y = mx + c$, the equation of l.

b Find the length CD, giving your answer in the form $p\sqrt{q}$, where p and q are integers and q is prime. (Edexcel)

28 The line l has equation $2x - y - 1 = 0$.

The line m passes through the point A$(0, 4)$ and is perpendicular to the line l.

a Find an equation of m and show that the lines l and m intersect at the point P$(2, 3)$.

The line n passes through the point B$(3, 0)$ and is parallel to the line m.

b Find an equation of n and hence find the coordinates of the point Q where the lines l and n intersect.

c Prove that AP = BQ = PQ. (Edexcel)

29 The figure shows a straight line graph of $\ln y$ against $\ln x$.
The line crosses the axes at A$(0, 3)$ and B$(3.5, 0)$.

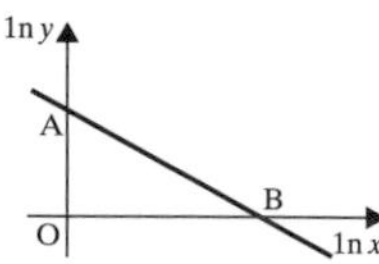

a Find an equation relating $\ln y$ and $\ln x$.

b Hence, or otherwise, express y in the form px^q, giving the values of the constants p and q to 3 significant figures. (Edexcel)

30 The figure shows a graph of ln y against x for two sets of observations x and y. The line of best fit is shown and it crosses the x-axis at $x = 4\frac{1}{3}$ and the (ln y)-axis at ln $y = 3$

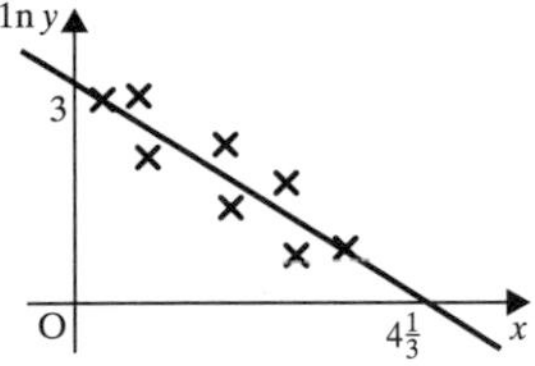

a Find an equation for the line of best fit.

b Express y as a function of x in the form $y = ab^x$ where a and b are constants. Write down the values of a and b to 1 significant figure. (Edexcel)

31 A zoo keeps an official record of the mass, x kg, and the average daily food intake, y kg, of each adult animal. Four selected pairs of values of x and y are given in the table below.

Animal	Cheetah	Deer	Rhinoceros	Hippopotamus
x	40	170	1500	3000
y	1.8	5.0	33.2	50.0

Show, by drawing a suitable linear graph on a sheet of 2 mm graph paper, that these values are approximately consistent with a relationship between x and y of the form

$y = ax^m$,

where a and m are constants.

Use your linear graph to find an estimate of the value of m.

The zoo has a bear with mass 500 kg. Assuming that this animal's food intake and mass conform to the relationship mentioned above, indicate a point on your linear graph corresponding to the bear and estimate the bear's average daily food intake. (AQA)

32 The variables x and y satisfy a relationship of the form $y = ax^b$, where a and b are constants.

Measurements of y for given values of x gave the following results.

x	2	3	4	5	6
y	6.32	7.24	7.98	8.60	9.12

a Plot ln y against ln x and draw the line of best fit to the plotted points.

b Use your line to estimate

i the value of x when $y = 7.50$ giving your answer correct to two significant figures.

ii the values of a and b, giving your answers to an appropriate degree of accuracy. (AQA)

CHAPTER 11

Functions 1

Mappings

If, on a calculator, the number 2 is entered and then the x^2 button is pressed, the display shows the number 4.

We say that 2 is mapped to 2^2 or $2 \to 2^2$.

Under the same rule, i.e. squaring the input number,
$3 \to 9$, $25 \to 625$, $0.5 \to 0.25$, $-4 \to 16$ and in fact,

(any real number) $\to$ (the square of that number)

The last statement can be expressed more briefly as

$$x \to x^2$$

where x is any real number.

This mapping can be represented graphically by plotting values of x^2 against values of x.

The graph, and knowledge of what happens when we square a number, show that one input number gives just one output number.

Now consider the mapping which maps a number to its square roots; the rule by which, for example, $4 \to +2$ and -2.

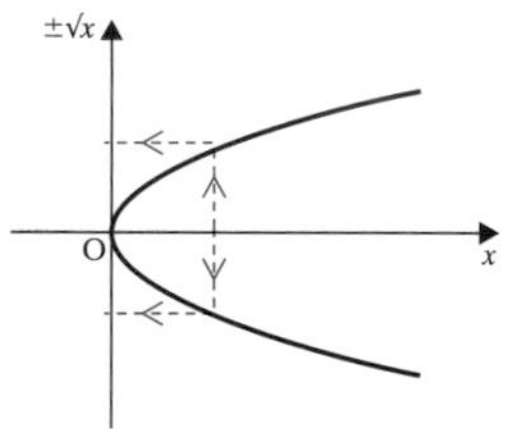

This rule gives a real output only if the input number is greater than zero (negative numbers do not have real square roots). This mapping can now be written in general terms as

$$x \to \pm\sqrt{x} \quad \text{for} \quad x \geqslant 0$$

The graphical representation of this mapping is shown here.

This time we notice that one input value gives two output values.

From these two examples, we can see that a mapping is a rule for changing a number to another number or numbers.

Functions

Under the first mapping, $x \to x^2$, one input number gives one output number. However, for the second mapping, $x \to \pm\sqrt{x}$, one input number gives two output numbers.

We use the word *function* for any rule that gives the same kind of result as the first mapping, i.e. one input value gives one output value.

A function is a rule that maps a single number to another single number.

The second mapping does not satisfy this condition so we cannot call it a function.

Consider again what we can now call the function such that $x \rightarrow x^2$.

Using f for 'function' and the symbol : to mean 'such that', we can write $f: x \rightarrow x^2$ to mean 'f is the function such that x maps to x^2'.

We use the notation $f(x)$ to represent the output values of the function,

e.g.

$$\text{for } f: x \rightarrow x^2 \text{ we have } f(x) = x^2.$$

EXAMPLES 11A

1 Determine whether these mappings are functions,

a $x \rightarrow \frac{1}{x}$ **b** $x \rightarrow y$ where $y^2 - x = 0$

a For any value of x, except $x = 0$, $\frac{1}{x}$ has a single value,

therefore $x \rightarrow \frac{1}{x}$ is a function provided that $x = 0$ is excluded.

> Note that $\frac{1}{0}$ is meaningless, so to make this mapping a function we have to exclude 0 as an input value. The function can be described by $f(x) = \frac{1}{x}, x \neq 0$

b If, as an example, we input $x = 4$, then the output is the value of y given by $y^2 - 4 = 0$ i.e. $y = 2$ and $y = -2$.

Therefore an input gives more than one value for the output, so $x \rightarrow y$ where $y^2 - x = 0$ is not a function.

2 If $f(x) = 2x^2 - 5$, find $f(3)$ and $f(-1)$.

> As $f(x)$ is the output of the mapping, $f(3)$ is the output when 3 is the input, i.e. $f(3)$ is the value of $2x^2 - 5$ when $x = 3$.

$f(3) = 2(3)^2 - 5 = 13$

$f(-1) = 2(-1)^2 - 5 = -3$

EXERCISE 11A

1 Determine which of these mappings are functions.

a $x \rightarrow 2x - 1$

b $x \rightarrow x^3 + 3$

c $x \rightarrow \frac{1}{x - 1}$

d $x \rightarrow t$ where $t^2 = x$

e $x \rightarrow \sqrt{x}$

f $x \rightarrow$ the length of the line from the origin to $(0, x)$.

g $x \rightarrow$ the greatest integer less than or equal to x.

h $x \rightarrow$ the height of a triangle whose area is x.

2 If $f(x) = 5x - 4$ find $f(0)$, $f(-4)$.

3 If $f(x) = 3x^2 + 25$ find $f(0)$, $f(8)$.

4 If $f(x) =$ the value of x correct to the nearest integer, find $f(1.25)$, $f(-3.5)$, $f(12.49)$.

5 If $f(x) = \sin x$, find $f(\frac{1}{2}\pi)$, $f(\frac{2}{3}\pi)$.

Domain and Range

We have assumed that we can use any real number as an input for a function unless some particular numbers have to be excluded because they do not give real numbers as output.

The set of inputs for a function is called the *domain* of the function.

The domain does not have to contain all possible inputs; it can be as wide, or as restricted, as we choose to make it. Hence to define a function fully, the domain must be stated.

If the domain is not stated, we assume that it is the set of all real numbers ($\mathbb{R}$).

Consider the mapping $x \to x^2 + 3$

We can define a function f for this mapping over any domain we choose. Some examples, together with their graphs are given.

1 $f(x) = x^2 + 3$ for $x \in \mathbb{R}$

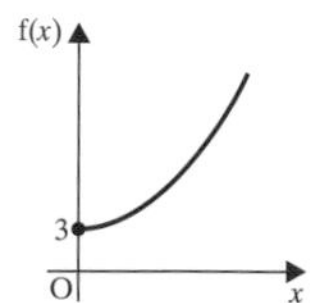

2 $f(x) = x^2 + 3$ for $x \geqslant 0$

Note that the point on the curve where $x = 0$ is included and we denote this on the curve by a solid circle.

If the domain were $x > 0$, then the point would not be part of the curve and we indicate this fact by using an open circle.

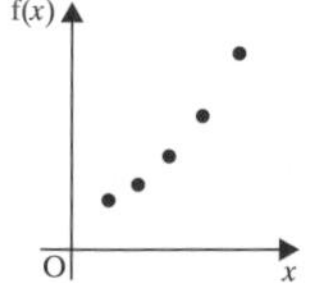

3 $f(x) = x^2 + 3$ for $x = 1, 2, 3, 4, 5$

This time the graphical representation consists of just five discrete (i.e. separate) points.

For each domain, there is a corresponding set of output numbers.

The set of output numbers is called the *range* of the function.

Thus for the function defined in **(1)** above, the range is $f(x) \geqslant 3$ and for the function given in **(2)**, the range is also $f(x) \geqslant 3$. For the function defined in **(3)**, the range is the set of numbers $4, 7, 12, 19, 28$.

Sometimes a function can be made up from more than one mapping, where each mapping is defined for a different domain. This is illustrated in the next worked example.

EXAMPLE 11B

The function, f, is defined by $f(x) = x^2$ for $x \leqslant 0$
and $f(x) = x$ for $x > 0$

a Find $f(4)$ and $f(-4)$. **b** Sketch the graph of f. **c** Give the range of f.

a For $x > 0, f(x) = x,$

$\therefore \quad f(4) = 4$

For $x \leqslant 0, f(x) = x^2,$

$\therefore \quad f(-4) = (-4)^2 = 16$

b To sketch the graph of a function, we can use what we know about lines and curves in the *xy*-plane. In this way we can interpret $f(x) = x$ for $x > 0$, as that part of the line $y = x$ which corresponds to positive values of x, and $f(x) = x^2$ for $x \leqslant 0$ as the part of the parabola $y = x^2$ that corresponds to negative values of x.

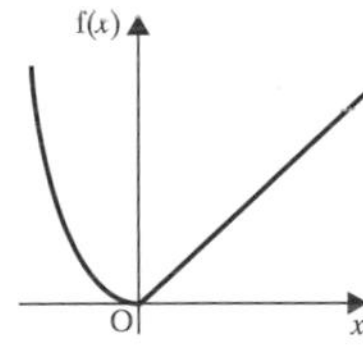

c The range of f is $f(x) \geqslant 0$.

EXERCISE 11B

1 Find the range of each of the following functions.

a $f(x) = 2x - 3$ for $x \geqslant 0$

b $f(x) = x^2 - 5$ for $x \leqslant 0$

c $f(x) = 1 - x$ for $x \leqslant 1$

d $f(x) = 1/x$ for $x \geqslant 2$

2 Draw a sketch graph of each function given in Question 1.

3 The function f is such that $f(x) = -x$ for $x < 0$
and $f(x) = x$ for $x \geqslant 0$

a Find the value of $f(5), f(-4), f(-2)$ and $f(0)$.

b Sketch the graph of the function.

4 The function f is such that $f(x) = x$ for $0 \leqslant x \leqslant 5$
and $f(x) = 5$ for $x > 5$

a Find the value of $f(0)$, $f(2)$, $f(4)$, $f(5)$ and $f(7)$.

b Sketch the graph of the function.

c Give the range of the function.

5 In Utopia, the tax on earned income is calculated as follows. The first £20 000 is tax free and remaining income is taxed at 20%.

a Find the tax payable on an earned income of £15 000 and of £45 000.

b Taking x as the number of pounds earned income and y as the number of pounds of tax payable, define a function f such that $y = f(x)$. Draw a sketch of the function and state the domain and range.

Curve Sketching

When functions have similar definitions they usually have common properties and graphs of the same form. If the common characteristics of a group of functions are known, the graph of any one particular member of the group can be sketched without having to plot points.

Quadratic Functions

The general form of a quadratic function is

$f(x) = ax^2 + bx + c$ for $x \in \mathbb{R}$

where a, b and c are constants and $a \neq 0$.

When a graphics calculator, or a computer, is used to draw the graphs of quadratic functions for a variety of values of a, b and c, the basic shape of the curve is always the same. This shape is called a *parabola*.

Every parabola has an axis of symmetry which goes through the vertex, i.e. the point where the curve turns back upon itself.

If the coefficient of x^2 is positive, i.e. $a > 0$, then $f(x)$ has a least value, and the parabola looks like this.

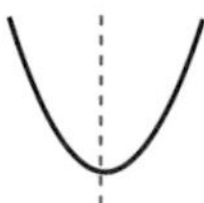

If the coefficient of x^2 is negative, i.e. $a < 0$, then $f(x)$ has a greatest value and the curve is this way up.

These properties of the graph of a quadratic function can be proved algebraically.

For $f(x) = ax^2 + bx + c$, 'completing the square' on the RHS and simplifying, gives

$$f(x) = \left[\frac{4ac - b^2}{4a}\right] + a\left[x + \frac{b}{2a}\right]^2 \qquad [1]$$

Now the first bracket is constant and, as the second bracket is squared, its value is zero when $x = -\frac{b}{2a}$ and greater than zero for all other values of x. Hence

when *a* is positive,

$f(x) = ax^2 + bx + c$ has a least value when $x = -\frac{b}{2a}$

and when *a* is negative,

$f(x) = ax^2 + bx + c$ has a greatest value when $x = -\frac{b}{2a}$

Further, taking values of x that are symmetrical about $x = -\frac{b}{2a}$,

e.g. $x = \pm k - \frac{b}{2a}$, we see from [1] that

$$f\left(k - \frac{b}{2a}\right) = f\left(-k - \frac{b}{2a}\right) = \left[\frac{4ac - b^2}{4a}\right] + ak^2$$

i.e. **the value of $f(x)$ is symmetrical about $x = -\frac{b}{2a}$**

These properties can now be used to draw *sketches* of the graphs of quadratic functions.

EXAMPLES 11C

1 Find the greatest or least value of the function given by $f(x) = 2x^2 - 7x - 4$ and hence sketch the graph of $f(x)$.

$f(x) = 2x^2 - 7x - 4 \quad \Rightarrow \quad a = 2, b = -7$ and $c = -4$.

As $a > 0$, $f(x)$ has a least value and this occurs when $x = -\frac{b}{2a} = \frac{7}{4}$

$\therefore$ the least value of $f(x)$ is $f(\frac{7}{4}) = 2(\frac{7}{4})^2 - 7(\frac{7}{4}) - 4$

$= -\frac{81}{8}$

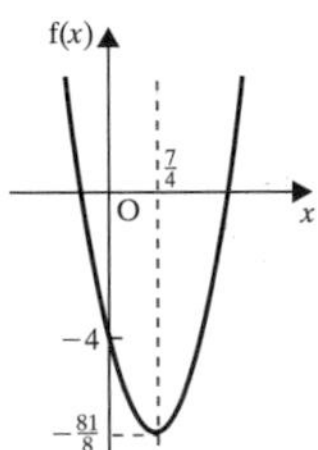

We now have one point on the graph of f(x) and we know that the curve is symmetrical about this value of x. However, to locate the curve more accurately we need another point and we use f(0) as it is easy to find.

$f(0) = -4$

2 Draw a quick sketch of the graph of $f(x) = (1-2x)(x+3)$.

The coefficient of x^2 is negative, so $f(x)$ has a greatest value.

The curve cuts the x-axis when $f(x) = 0$.

When $f(x) = 0$, $(1-2x)(x+3) = 0 \Rightarrow x = \frac{1}{2}$ or -3

The average of these values is $-\frac{5}{4}$, so the curve is symmetrical about $x = -\frac{5}{4}$.

We now have enough information to draw a quick sketch, but note that this method is suitable only when the quadratic function factorises.

EXERCISE 11C

1 Find the greatest or least value of $f(x)$ where $f(x)$ is

a $x^2 - 3x + 5$

b $2x^2 - 4x + 5$

c $3 - 2x - x^2$

2 Find the range of f where $f(x)$ is

a $7 + x - x^2$

b $x^2 - 2$

c $2x - x^2$

3 Sketch the graph of each of the following quadratic functions, showing the greatest or least value and the value of x at which it occurs.

a $x^2 - 2x + 5$ **b** $x^2 + 4x - 8$

c $2x^2 - 6x + 3$ **d** $4 - 7x - x^2$

e $x^2 - 10$ **f** $2 - 5x - 3x^2$

4 Draw a quick sketch of each of the following functions.

a $(x-1)(x-3)$ **d** $(1+x)(2-x)$

b $(x+2)(x-4)$ **e** $x^2 - 9$

c $(2x-1)(x-3)$ **f** $3x^2$

Cubic Functions

The general form of a cubic function is

$$f(x) = ax^3 + bx^2 + cx + d$$

where *a*, *b*, *c* and *d*, are constants and $a \neq 0$.

Investigating the curve $y = ax^3 + bx^2 + cx + d$ for a variety of values of a, b, c and d shows that the shape of the curve is

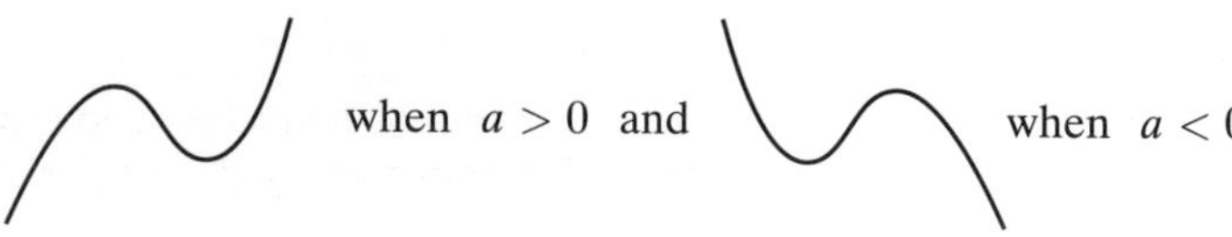

Sometimes there are no turning points and the curve looks like this

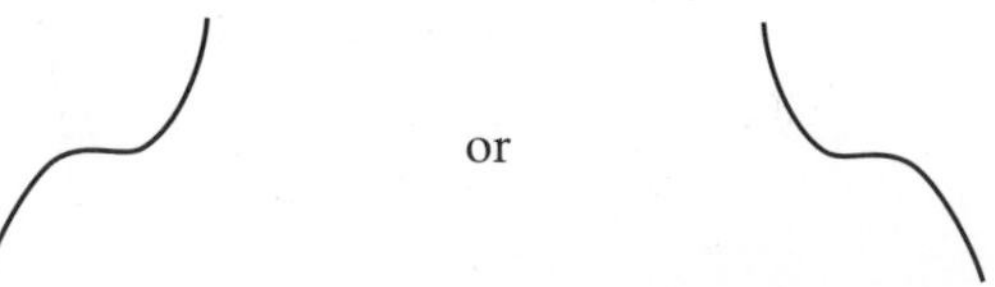

Polynomial Functions

The general form of a polynomial function is

$$f(x) = a_n x^n + a_{n-1}x^{n-1} + \ldots + a_2x^2 + a_1x + a_0$$

where $a_n, a_{n-1}, \ldots, a_0$ are constants, n is a positive integer and $a_n \neq 0$.

Examples of polynomials are

$$f(x) = 3x^4 - 2x^3 + 5, \quad f(x) = x^5 - 2x^3 + x, \quad f(x) = x^2$$

The *order* of a polynomial is the highest power of x in the function.
So $x^4 - 7$ has order 4, and $2x - 1$ has order 1.

We have already investigated the graphs of polynomials of order 1.

e.g. $f(x) = 2x - 1$ which gives a straight line.

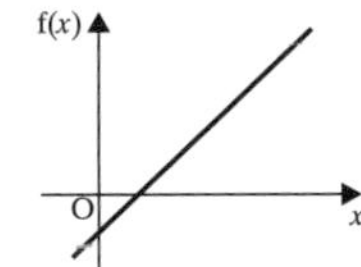

and of order 2,

e.g. $f(x) = x^2 - 4$ which gives a parabola.

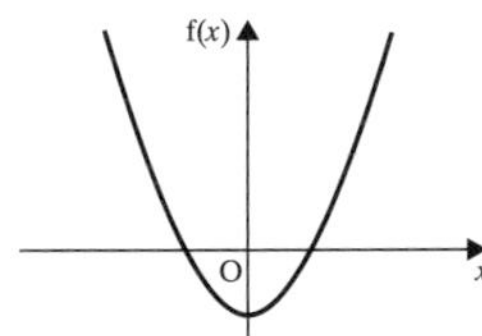

and of order 3,

e.g. $f(x) = x^3 - 2x + 1$ which gives a cubic curve.

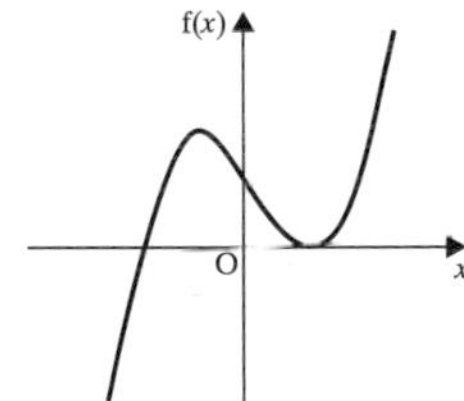

The shape of the curve
$f(x) = x^4 - 3x + 2x^2 + 1$ looks like this.

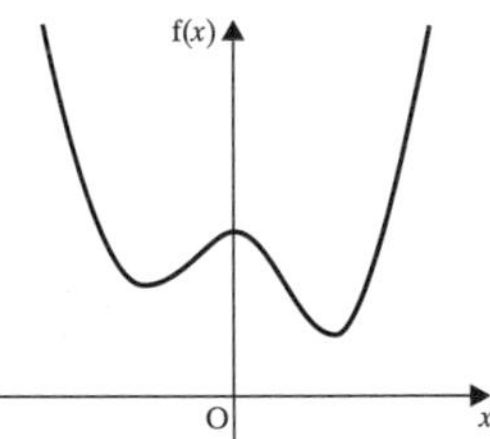

Experimenting with the curves of other polynomial functions of order 4 shows that, in general, the curve has three turning points although some or all of these may merge, e.g.

Rational Functions

A rational function is one in which both numerator and denominator are polynomial.

Examples of rational functions of x are

$$\frac{1}{x},\ \frac{x}{x^2-1},\ \frac{3x^2+2x}{x-1}$$

Now consider the familiar function $f(x) = \dfrac{1}{x}$ and its graph. From its form we can infer various properties of $f(x)$.

1 As the value of x increases, the value of $f(x)$ gets closer to zero,

e.g. when $x = 100, f(x) = \frac{1}{100}$
and when $x = 1000, f(x) = \frac{1}{1000}$

We write this as 'when $x \to \infty, f(x) \to 0$'

Similarly as the value of x decreases, i.e. as $x \to -\infty$, the value of $f(x)$ again gets closer to zero, i.e. when $x \to -\infty, f(x) \to 0$

2 $f(x)$ does not exist when $x = 0$, so this value of x must be excluded from the domain of f.

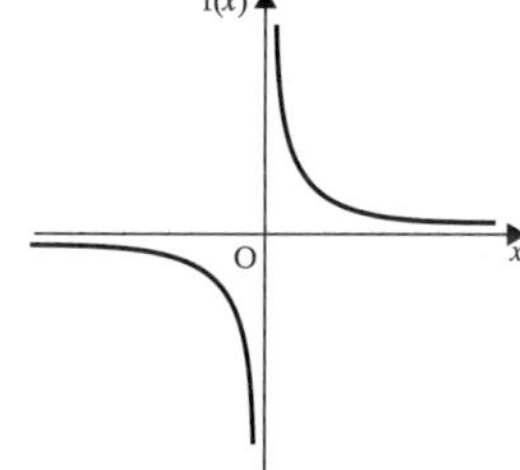

x can get as close as we like to zero however, and can approach zero in two ways.

If $x \to 0$ from above (i.e. from positive values, $\overset{}{\underset{0}{+}}\!\longleftarrow$) then $f(x) \to \infty$.
If $x \to 0$ from below (i.e. from negative values, $\longrightarrow\!\underset{0}{+}$) then $f(x) \to -\infty$.

Notice that, as $x \to \pm\infty$, the curve gets closer to the x-axis but does not cross it. Also, as $x \to 0$, the curve approaches the y-axis but again does not cross it.

We say that the x-axis and the y-axis are *asymptotes* to the curve.

Exponential Functions

Exponent is another word for index or power.

An exponential function is one where the variable is in the index.

For example, 2^x, 3^{-x}, 10^{x+1} are exponential functions of x.

Consider the function $f(x) = 2^x$ for which a table of corresponding values of x and $f(x)$ and a graph are given.

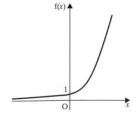

x	$-\infty \leftarrow \ldots$	-10	-1	$-\frac{1}{10}$	0	$\frac{1}{10}$	1	10	$\ldots \to \infty$
$f(x)$	$0 \leftarrow \ldots$	$\frac{1}{1024}$	$\frac{1}{2}$	0.93	1	1.07	2	1024	$\ldots \to \infty$

The graph and the table show that

1 2^x has real values for all real values of x and 2^x is positive for all values of x, i.e. the range of f is $f(x) > 0$.

2 As $x \to -\infty$, $f(x) \to 0$, i.e. the x-axis is an asymptote.

3 As x increases, $f(x)$ increases at a rapidly accelerating rate.

Note also that the curve crosses the y-axis at $(0, 1)$, i.e. $f(0) = 1$. In fact for any function of the form $f(x) = a^x$, where a is a constant greater than 1, $f(0) = 1$, and the curve representing it is similar in shape to that for 2^x.

EXAMPLE 11D Sketch the graph of the function given by $f(x) = \dfrac{1}{2-x}$.

$f(x) = \dfrac{1}{2-x}$ does not exist when $x = 2$, so the curve $y = f(x)$ does not cross the line $x = 2$.

As $x \to 2$ from above, $2 - x$ is negative and approaches zero,

so $$\frac{1}{2-x} \to -\infty$$

As $x \to 2$ from below, $2 - x$ is positive and approaches zero,

so $$\frac{1}{2-x} \to \infty$$

Therefore the line $x = 2$ is an asymptote.

As $x \to \infty$, $\dfrac{1}{2-x} \to 0$ from below

and as $x \to -\infty$, $\dfrac{1}{2-x} \to 0$ from above

$\therefore$ the x-axis is an asymptote.

As this is similar to $f(x) = \dfrac{1}{x}$, we now have enough information to sketch the graph.

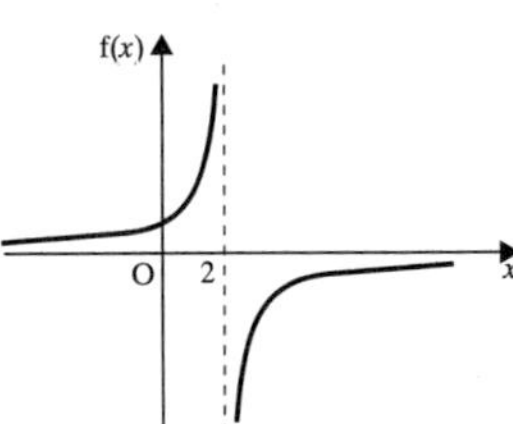

EXERCISE 11D

1 Draw sketch graphs of the following functions.

a 3^x b $\dfrac{1}{2x}$ c 4^{2x} d $\dfrac{1}{x-3}$

2 Write down the values of $f(x) = (\frac{1}{2})^x$ corresponding to $x = -4, -3, -2, -1, 0, 1, 2, 3$ and 4. From these values deduce the behaviour of $f(x)$ as $x \to \pm\infty$ and hence sketch the graph of the function.

3 What value of x must be excluded from the domain of $f(x) = \dfrac{1}{x+2}$?

Describe the behaviour of $f(x)$ as x approaches this value from above and from below. Describe also the behaviour of $f(x)$ as $x \to \pm\infty$. Use this information to sketch the graph of $f(x)$.

4 By following a procedure similar to that given in Question 3 draw sketch graphs of the following functions.

a $-\dfrac{1}{x}$ **b** $\dfrac{1}{1-2x}$ **c** $\dfrac{2}{x+1}$ **d** $1+\dfrac{1}{x}$

5 Find the values of x where the curve $y = f(x)$ cuts the x-axis and sketch the curve when

a $f(x) = x(x-1)(x+1)$ **b** $f(x) = x(x-1)(x+1)(x-2)$

c $f(x) = (x^2-1)(2-x)$ **d** $f(x) = (x^2-1)(4-x^2)$

Simple Transformation of Curves

Transformations of curves are best appreciated if they can be 'seen', so this section starts with an investigative approach using a graphics calculator or graph-drawing software. This exercise is not essential and all the necessary conclusions are drawn analytically in the next part of the text.

EXERCISE 11E

You will need a graphics calculator or computer with appropriate software for this exercise.

1 a On the screen, draw the graph of $y = 2^x$. Superimpose the graphs of $y = 2^x + 2$ and $y = 2^x - 1$. Clear the screen and again draw the graph of $y = 2^x$. This time superimpose the graph of $y = 2^x + c$ for a variety of values of c.

b Describe the transformation that maps the graph of $f(x) = 2^x$ to the graph of $g(x) = 2^x + c$.

c Repeat **a** and **b** for other simple functions,

e.g. x^2, x^3, $1/x$.

2 Use a procedure similar to that described in Question 1 to investigate the relationship between the graphs of $f(x)$ and $f(x+c)$,

e.g. draw the graph of $y = x^2$ and then superimpose the graph of $y = (x+c)^2$ for a variety of values of c.

3 Investigate the relationship between the graphs of

a $f(x)$ and $-f(x)$ **b** $f(x)$ and $f(-x)$

Translations

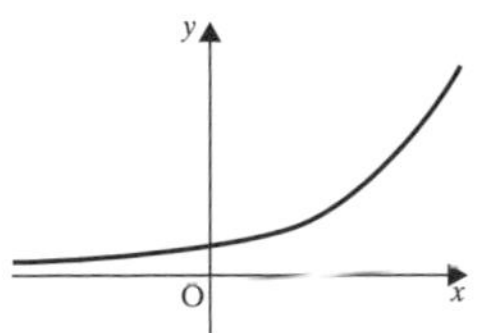

Consider the function f where $f(x) = 2^x$

The graph of this function is the curve $y = 2^x$

1 Now consider the function g where $g(x) = f(x) + 2$.

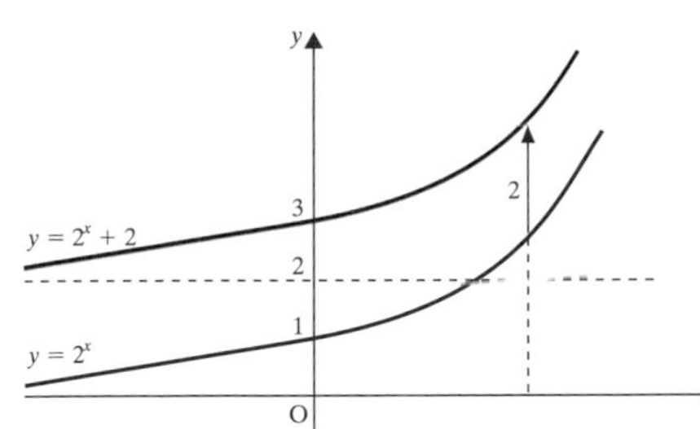

Comparing $f(x) = 2^x$ with $g(x) = 2^x + 2$ we can see that for a particular value of x, the value of $g(x)$ is 2 units greater than the corresponding value of $f(x)$.
Therefore, for equal values of x, points on the curve $y = g(x)$ are two units above points on the curve $y = f(x)$,
i.e. the curve $y = 2^x + 2$ is a translation of the curve $y = 2^x$ by two units in the positive direction of the y-axis.

In general, for any function f, the curve $y = f(x) + c$ is the translation of the curve $y = f(x)$ by c units parallel to the y-axis.

2 Consider the function $g(x) = 2^{x-2}$.

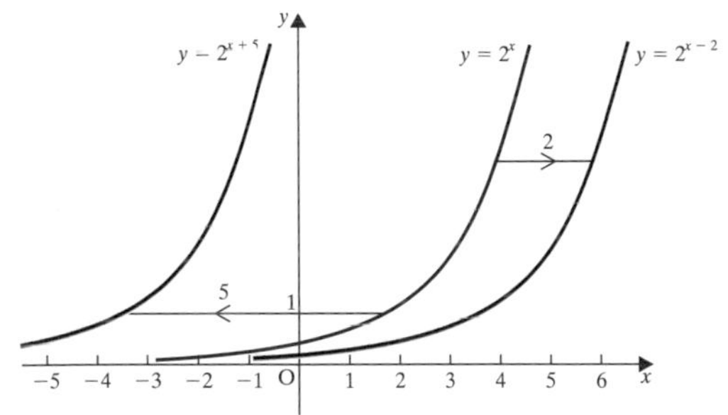

Comparing $f(x) = 2^x$ with $g(x) = 2^{x-2}$ we can see that the values of $f(x)$ and $g(x)$ are the same when the input value to $g(x)$ is 2 units greater than the input value for $f(x)$, i.e. $f(a) = g(a + 2)$.

Therefore, for equal values of y, points on the curve $y = 2^{x-2}$ are 2 units to the *right* of points on the curve $y = 2^x$, i.e. the curve $y = 2^{x-2}$ is a translation of the curve $y = 2^x$ by 2 units in the positive direction of the x-axis.

Similarly, considering $h(x) = 2^{x+5}$, the values of $f(x)$ and $h(x)$ are the same when the input to $h(x)$ is five units less than the input to $f(x)$. Thus for equal values of y, points on the curve $y = 2^{x+5}$ are 5 units to the *left* of points on the curve $y = 2^x$.

In general, the curve $y = f(x + c)$ is a translation of the curve $y = f(x)$ by c units parallel to the x-axis.

If $c > 0$, the translation is in the negative direction of the x-axis and if $c < 0$, the translation is in the positive direction of the x-axis.

Reflections

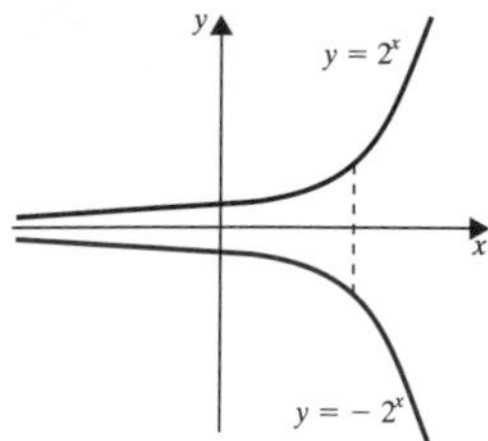

1 Consider the function $g(x) = -f(x)$.

For a given value of x, $g(x)$ is equal to $-f(x)$. Therefore for equal values of x, points on the curve $y = -2^x$ are the reflection in the x-axis of points on the curve $y = 2^x$, i.e. the curve $y = -f(x)$ is the reflection in the x-axis of the curve $y = f(x)$.

In general, the curve $y = -f(x)$ is the reflection of the curve $y = f(x)$ in the x-axis.

2 Consider the function $g(x) = 2^{-x}$.

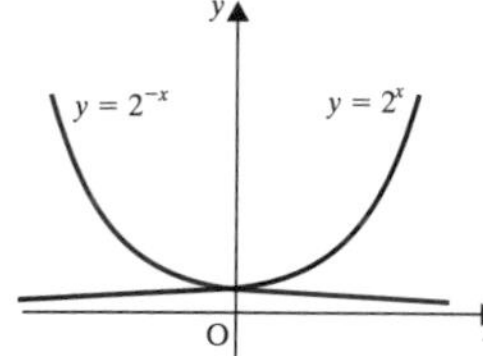

Comparing $f(x) = 2^x$ with $g(x) = 2^{-x}$ we see that $f(x)$ and $g(x)$ have the same value when the inputs to $g(x)$ and $f(x)$ are equal in value but opposite in sign, i.e. $g(a) = f(-a)$.

Therefore points with the same y-coordinates on the curves $y = 2^{-x}$ and $y = 2^x$, are symmetrical about $x = 0$, i.e. the curve $y = 2^{-x}$ is the reflection of the curve $y = 2^x$ in the y-axis.

In general, the curve $y = f(-x)$ is the reflection of the curve $y = f(x)$ in the y-axis.

EXAMPLE 11F

Sketch the curve $y = (x-1)^3$.

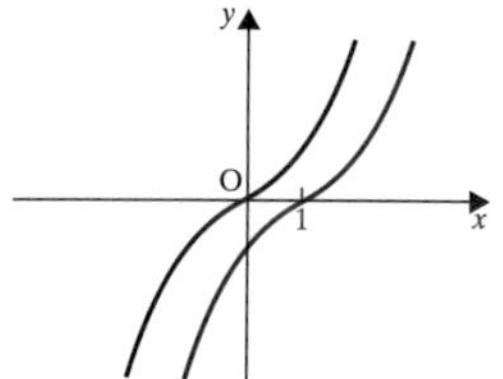

The shape and position of the curve $y = x^3$ is known. If $f(x) = x^3$, then $(x-1)^3 = f(x-1)$, so the curve $y = (x-1)^3$ is a translation of the first curve by one unit in the positive direction of the x-axis.

EXERCISE 11F

Sketch each of the following curves.

1 $y = -x^2$

2 $y = -\dfrac{1}{x}$

3 $y = -3^x$

4 $y = 1 + \dfrac{1}{x}$

5 $y = 2^x - 3$

6 $y = \dfrac{1}{x} - 2$

7 $y = (x-4)^4$

8 $y = x^2 - 9$

9 $y = \dfrac{1}{x-2}$

10 On the same set of axes sketch the graphs of $f(x) = x^3$, $f(x) = (x+1)^3$, $f(x) = -(x+1)^3$ and $f(x) = 2 - (x+1)^3$.

11 On the same set of axes sketch the lines $y = 2x - 1$ and $y = \frac{1}{2}(x+1)$. Describe a transformation which maps the first line to the second line.

12 Repeat Question 11 for the curves $y = 1 + \dfrac{1}{x}$ and $y = \dfrac{1}{x-1}$.

13 Find the coordinates of the reflection of the point $(2, 5)$ in the line $y = x$.

14 P' is the reflection of the point $P(a, b)$ in the line $y = x$. Find the coordinates of P' in terms of a and b.

Inverse Functions

Consider the function f where $f(x) = 2x$ for $x = 2, 3, 4$.

Under this function, the domain $\{2, 3, 4\}$ maps to the image-set $\{4, 6, 8\}$ and this is illustrated by the arrow diagram.

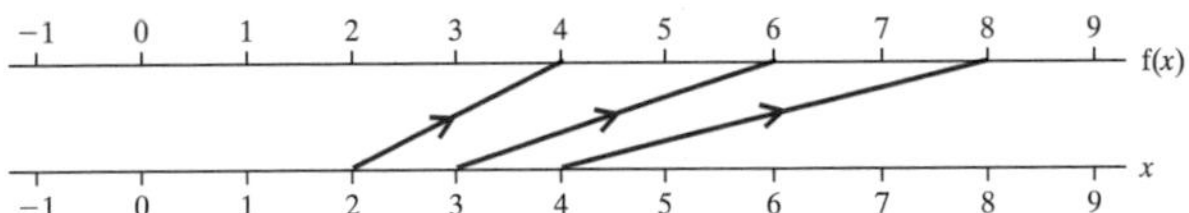

It is possible to reverse this mapping, i.e. we can map each member of the image-set back to the corresponding member of the domain by halving each member of the image-set.

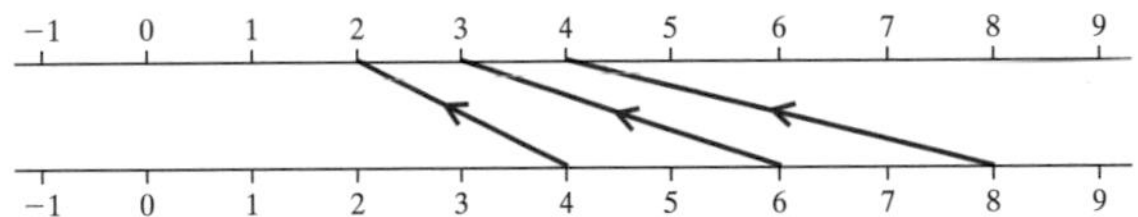

This procedure can be expressed algebraically, i.e.
for $x = 4, 6, 8$, $x \to \frac{1}{2}x$ maps 4 to 2, 6 to 3 and 8 to 4.

This reverse mapping is a function in its own right and it is called the *inverse* function of f where $f(x) = 2x$.

Denoting this inverse function by f^{-1} we can write $f^{-1}(x) = \frac{1}{2}x$. In fact, $f(x) = 2x$ can be reversed for all real values of x and the rule for doing this is a function.

Therefore, for $f(x) = 2x$, $f^{-1}(x) = \frac{1}{2}x$ is such that f^{-1} reverses f for all real values of x, i.e. f^{-1} maps the output of f to the input of f.

In general, for any function f,

if there exists a function, g, that maps the output of f back to its input, i.e. g : f(x) → x, then this function is called the inverse of f and it is denoted by f^{-1}.

The Graph of a Function and its Inverse

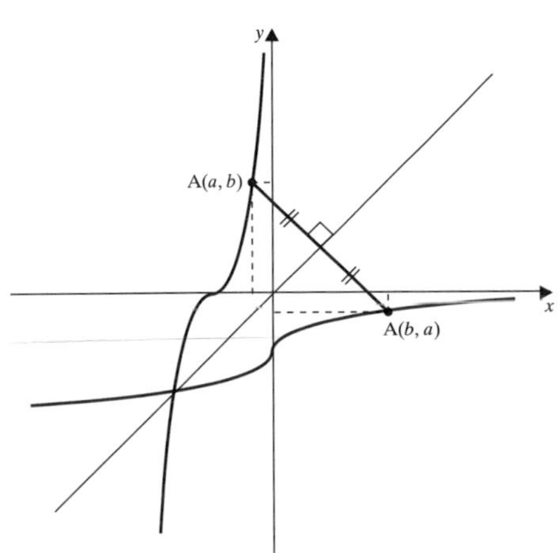

Consider the curve that is obtained by reflecting $y = f(x)$ in the line $y = x$. The reflection of a point $A(a, b)$ on the curve $y = f(x)$, is the point A′ whose coordinates are (b, a), i.e. interchanging the x- and y-coordinates of A gives the coordinates of A′.

We can therefore obtain the equation of the reflected curve by interchanging x and y in the equation $y = f(x)$.

Now the coordinates of A on $y = f(x)$ can be written as $[a, f(a)]$. Therefore the coordinates of A′ on the reflected curve are $[f(a), a]$, i.e. the equation of the reflected curve is such that the output of f is mapped to the input of f.

Hence if the equation of the reflected curve can be written in the form $y = g(x)$, then g is the inverse of f, i.e. $g = f^{-1}$.

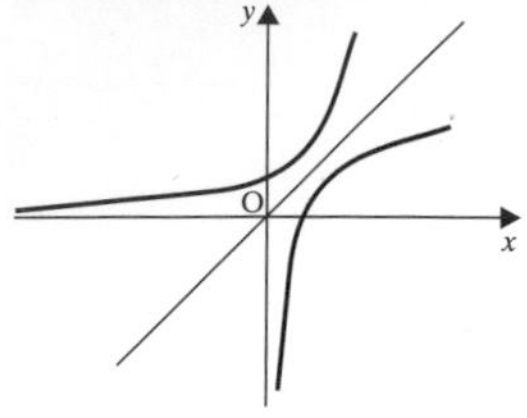

To illustrate these properties, consider the curve $y = 2^x$ and its reflection in the line $y = x$.

The equation of the reflected curve is given by $x = 2^y$.

Using the 'log' notation introduced in Chapter 2, we can write this equation in the form $y = \log_2 x$.

Therefore for the function $f(x) = 2^x$ the inverse function is given by $f^{-1}(x) = \log_2 x$.

Any curve whose equation can be written in the form $y = f(x)$ can be reflected in the line $y = x$. However this reflected curve may not have an equation that can be written in the form $y = f^{-1}(x)$.

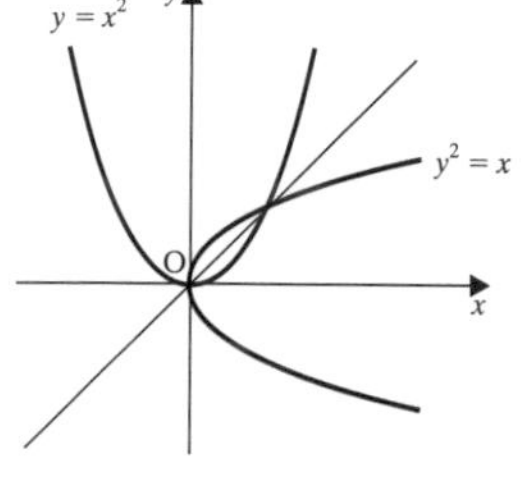

Consider the curve $y = x^2$ and its reflection in the line $y = x$.

The equation of the image curve is $x = y^2 \quad \Rightarrow \quad y = \pm\sqrt{x}$ and $x \to \pm\sqrt{x}$ is not a function.

(We can see this from the diagram as, on the reflected curve, one value of x maps to two values of y. So in this case y cannot be written as a function of x.)

Therefore the function $f : x \to x^2$ does not have an inverse, i.e.

not every function has an inverse.

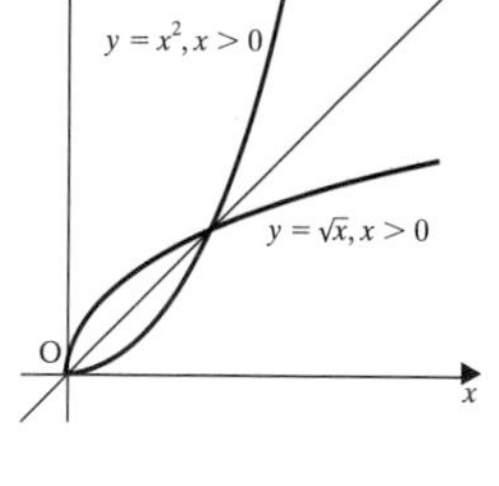

If we change the definition of f to $f : x \to x^2$ for $x \in \mathbb{R}^+$ then the inverse mapping is

$x \to \sqrt{x}$ for $x \in \mathbb{R}^+$ and this is a function, i.e.

$f^{-1}(x) = \sqrt{x}$ for $x \in \mathbb{R}^+$

To summarise:

The inverse of a function undoes the function, i.e. it maps the output of a function back to its input.

The inverse of the function f is written f^{-1}.

Not all functions have an inverse.

When the curve whose equation is $y = f(x)$ is reflected in the line $y = x$, the equation of the reflected curve is $x = f(y)$.

If this equation can be written in the form $y = g(x)$ then g is the inverse of f, i.e. $g(x) = f^{-1}(x)$, and the domain of g is the range of f.

EXAMPLES 11G

1 Determine whether there is an inverse of the function f given by $f(x) = 2 + \dfrac{1}{x}$.

If f^{-1} exists, express it as a function of x.

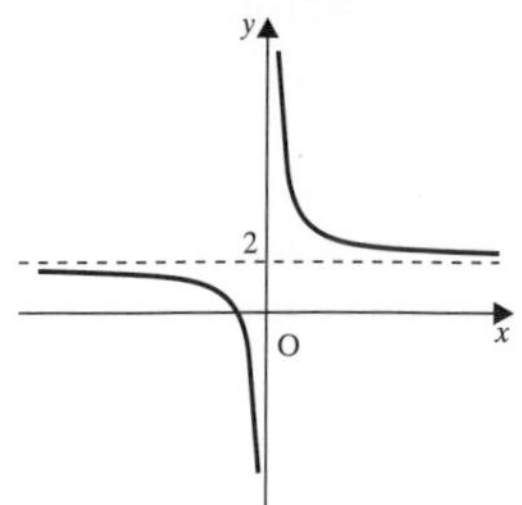

From the sketch of $f(x) = 2 + \dfrac{1}{x}$, we see that one value of $f(x)$ maps to one value of x, therefore the reverse mapping is a function. The equation of the reflection of $y = 2 + \dfrac{1}{x}$ can be written as

$$x = 2 + \frac{1}{y} \quad \Rightarrow \quad y = \frac{1}{x-2}$$

$\therefore$ when $f(x) = 2 + \dfrac{1}{x}$, $f^{-1}(x) = \dfrac{1}{x-2}$, provided that $x \neq 2$

2 Find $f^{-1}(4)$ when $f(x) = 5x - 1$.

If $y = f(x)$, i.e. $y = 5x - 1$

then for the reflected curve $x = 5y - 1 \quad \Rightarrow \quad y = \frac{1}{5}(x+1)$

i.e. $\qquad f^{-1}(x) = \frac{1}{5}(x+1)$

$\therefore \qquad f^{-1}(4) = \frac{1}{5}(4+1) = 1$

EXERCISE 11G

1 Sketch the graphs of $f(x)$ and $f^{-1}(x)$ on the same axes.

a $f(x) = 3x - 1$ **b** $f(x) = 2^{x}$ **c** $f(x) = (x-1)^3$

d $f(x) = 2 - x$ **e** $f(x) = \dfrac{1}{x-3}$ **f** $f(x) = \dfrac{1}{x}$

2 Which of the functions given in Question 1 are their own inverses?

3 Determine whether f has an inverse function, and if it does, find it when

a $f(x) = x + 1$ **b** $f(x) = x^2 + 1$ **c** $f(x) = x^3 + 1$

d $f(x) = x^2 - 4, x \geqslant 0$ **e** $f(x) = (x+1)^4, x \geqslant -1$

4 The function f is given by $f(x) = 1 - \dfrac{1}{x}$. Find

a $f^{-1}(4)$ **b** the value of x for which $f^{-1}(x) = 2$

c any values of x for which $f^{-1}(x) = x$

5 If $f(x) = 3^x$, find

a $f(2)$ **b** $f^{-1}(9)$ **c** $f^{-1}(\frac{1}{3})$

Compound Functions

Consider the two functions f and g given by

$$f(x) = x^2 \quad \text{and} \quad g(x) = \frac{1}{x}$$

These two functions can be combined in several ways.

1 They can be added or subtracted,

i.e. $f(x) + g(x) = x^2 + \dfrac{1}{x}$ and $f(x) - g(x) = x^2 - \dfrac{1}{x}$.

2 They can be multiplied or divided,

i.e. $f(x)\,g(x) = (x^2) \times \left(\dfrac{1}{x}\right) = x$ and $\dfrac{f(x)}{g(x)} = \dfrac{x^2}{1/x} = x^3$.

3 The output of f can be made the input of g,

i.e. $x \xrightarrow{f} x^2 \xrightarrow{g} \dfrac{1}{x^2}$ or $g[f(x)] = g(x^2) = \dfrac{1}{x^2}$

Therefore the function $x \rightarrow 1/x^2$ is obtained by taking the function g of the function f.

Function of a Function

A compound function formed in the way described in (3) above is known as a *function of a function* and it can be denoted by gf (or by g ∘ f).

For example, if $f(x) = 3^x$ and $g(x) = 1 - x$ then gf(x) means the function g of the function $f(x)$,

i.e. $\quad gf(x) = g(3^x) = 1 - 3^x$

Similarly $\quad fg(x) = f(1-x) = 3^{(1-x)}$

Note that $gf(x)$ is *not* the same as $fg(x)$.

EXERCISE 11H

1 If f, g and h are functions defined by $f(x) = x^2$, $g(x) = \dfrac{1}{x}$, $h(x) = 1 - x$ find as a function of x

a fg **b** fh **c** hg **d** hf **e** gf

2 If $f(x) = 2x - 1$ and $g(x) = x^3$ find the value of

a $g \circ f(3)$ **b** $f \circ g(2)$

c $f \circ g(0)$ **d** $g \circ f(0)$

3 Given that $f(x) = 2x$, $g(x) = 1 + x$ and $h(x) = x^2$, find as a function of x

a hg **b** fhg **c** ghf

4 The function $f(x) = (2-x)^2$ can be expressed as a function of a function. Find g and h as functions of x such that $gh(x) = f(x)$.

5 Repeat Question 4 when $f(x) = (x+1)^4$.

6 Express the function $f(x)$ as a combination of functions $g(x)$ and $h(x)$, and define $g(x)$ and $h(x)$, where $f(x)$ is

a $10^{(x+1)}$ **b** $\dfrac{1}{(3x-2)^2}$

c $2^x + x^2$ **d** $\dfrac{(2x+1)}{x}$

e $(5x-6)^4$ **f** $(x-1)(x^2-2)$

MIXED EXERCISE 11

1 A function f is defined by $f(x) = \dfrac{1}{(1-x)}$, $x \neq 1$.

a Why is 1 excluded from the domain of f?

b Find the value of $f(-3)$.

c Sketch the curve $y = f(x)$.

d Find $f^{-1}(x)$ in terms of x and give the domain of f^{-1}.

2 Find the greatest or least value of each of the following functions, stating the value of x at which they occur.

a $f(x) = x^2 - 3x + 5$

b $f(x) = 2x^2 - 7x + 1$

c $f(x) = (x-1)(x+5)$

3 If $f(x) = 10^x$, sketch the following curves on the same set of axes.

a $y = f(x)$ **b** $y = f(x+3)$

c $y = f^{-1}(x)$

4 Given that $f(x) = 10^x$, $g(x) = x^2$ and $h(x) = \frac{1}{x}$,

a find fg(2), hg(3) and gf(−1)

b find, in terms of x, hfg(x) and gfh(x)

c if they exist, find $f^{-1}(x)$, $g^{-1}(x)$ and $h^{-1}(x)$

d the value(s) of x for which gh$(x) = 9$

e does the function $(gh)^{-1}$ exist?

5 Draw sketches of the following curves, showing any asymptotes.

a $y = (x-2)(x-3)(x-4)$

b $y = \frac{1}{3-x}$

c $y = 2^{4-x}$

6 The function f is given by $f(x) = 2^{(3x-2)}$

a find $g(x)$ and $h(x)$ such that $f = gh$

b evaluate ff(2) and $f^{-1}(2)$.

7 The functions f, g and h are defined by $f(x) = 2x$, $g(x) = 3x^2$ and $h(x) = x - 1$

a Sketch the curves $y = g(x-3)$ and $y = gf^{-1}(x)$.

b Find the value(s) of x for which $f^{-1}(x) = g(x)$.

8 If $f(x) = 3x$, $g(x) = \frac{1}{x}$ and $h(x) = x^2 - 1$, find

a $f \circ g(x)$ **b** $g \circ f \circ h(x)$

c $g^{-1} \circ f^{-1}(x)$ **d** $(g \circ f)^{-1}(x)$

CHAPTER 12

Inequalities

Manipulating Inequalities

An inequality compares two unequal quantities.

Consider, for example, the two real numbers 3 and 8 for which

$$8 > 3$$

The inequality remains true, i.e. the inequality sign is unchanged, when the same term is added or subtracted on both sides, e.g.

$$8 + 2 > 3 + 2 \quad \Rightarrow \quad 10 > 5$$

and $$8 - 1 > 3 - 1 \quad \Rightarrow \quad 7 > 2$$

The inequality sign is unchanged also when both sides are multiplied or divided by a positive quantity, e.g.

$$8 \times 4 > 3 \times 4 \quad \Rightarrow \quad 32 > 12$$

and $$8 \div 2 > 3 \div 2 \quad \Rightarrow \quad 4 > 1\tfrac{1}{2}$$

If, however, both sides are multiplied or divided by a *negative* quantity the inequality is no longer true. For example, if we multiply by -1, the LHS becomes -8 and the RHS becomes -3 so the correct inequality is now LHS $<$ RHS, i.e.

$$8 \times -1 < 3 \times -1 \quad \Rightarrow \quad -8 < -3$$

Similarly, dividing by -2 gives $-4 < -1\tfrac{1}{2}$.

These examples are illustrations of the following general rules.

Adding or subtracting a term, or multiplying or dividing both sides by a positive number, does not alter the inequality sign.

Multiplying or dividing both sides by a *negative* number reverses the inequality sign.

i.e. if a, b and k are real numbers, and $a > b$ then,

$a + k > b + k$ for *all* values of k.

$ak > bk$ for *positive* values of k.

$ak < bk$ for *negative* values of k.

Solving Linear Inequalities

When an inequality contains an unknown quantity, the rules given above can be used to 'solve' it. Whereas the solution of an equation is a value, or values, of the variable, the solution of an inequality is a range, or ranges, of values of the variable.

If the unknown quantity appears only in linear form, we have a *linear inequality* and the solution range has only *one boundary*.

EXAMPLE 12A

Find the set of values of x that satisfy the inequality $x - 5 < 2x + 1$.

$x - 5 < 2x + 1 \Rightarrow x < 2x + 6$ (adding 5 to each side)

$\Rightarrow -x < 6$ (subtracting $2x$ from each side)

$\Rightarrow x > -6$ (multiplying both sides by -1)

So the set of values of x satisfying the given inequality is $x > -6$

EXERCISE 12A

Solve the following inequalities.

1 $x - 4 < 3 - x$

2 $x + 3 < 3x - 5$

3 $x < 4x + 9$

4 $7 - 3x < 13$

5 $x > 5x - 2$

6 $2x - 1 < x - 4$

7 $1 - 7x > x + 3$

8 $2(3x - 5) > 6$

9 $3(3 - 2x) < 2(3 + x)$

Solving Quadratic Inequalities

A quadratic inequality is one in which the variable appears to the power 2, e.g. $x^2 - 3 > 2x$.

The solution is a range or ranges of values of the variable with *two boundaries*.

If the terms in the inequality can be collected and factorised, a graphical solution is easy to find.

EXAMPLE 12B

Find the range(s) of values of x that satisfy the inequality $x^2 - 3 > 2x$.

$$x^2 - 3 > 2x \Rightarrow x^2 - 2x - 3 > 0$$

$$\Rightarrow (x - 3)(x + 1) > 0$$

or $\quad f(x) > 0$ where $f(x) = (x - 3)(x + 1)$

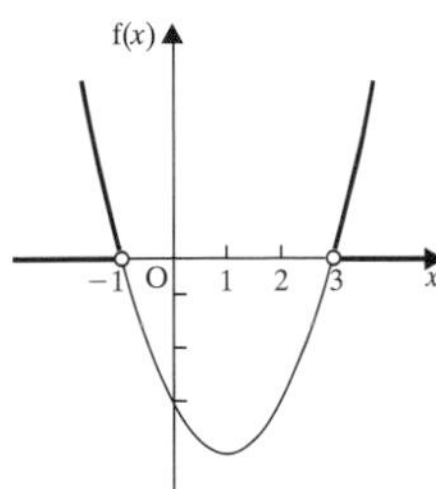

If we sketch the graph of $f(x)$ then $f(x) > 0$ where the graph is above the x-axis. The values of x corresponding to these portions of the graph satisfy $f(x) > 0$. The points where $f(x) = 0$, i.e. where $x = 3$ and -1 are not part of this solution and this is indicated on the sketch by open circles.

From the graph we see that the ranges of values of x which satisfy the given inequality are $x < -1$ and $x > 3$.

In the example above there are two boundary values and two separate ranges that lie *outside* these values. On the other hand, the solution of the inequality $(x - 3)(x + 1) < 0$ is the part of the graph below the x-axis *between* $x = -1$ and $x = 3$. There are still two boundaries but only one range, i.e. $-1 < x < 3$.

EXERCISE 12B

Find the ranges of values of x that satisfy the following inequalities.

1 $(x-2)(x-1)>0$

2 $(x+3)(x-5)\geqslant 0$

3 $(x-2)(x+4)<0$

4 $(2x-1)(x+1)\geqslant 0$

5 $x^2-4x>3$

6 $4x^2<1$

7 $(2-x)(x+4)\geqslant 0$

8 $5x^2>3x+2$

9 $(3-2x)(x+5)\leqslant 0$

10 $(x-1)^2>9$

11 $(x+1)(x+2)\leqslant 4$

12 $(1-x)(4-x)>x+11$

Rational Fractions in Inequalities

Consider the inequality $\dfrac{x-2}{x+5}>3$.

We do not know whether $x+5$ is positive or negative, and this prevents the apparently obvious step of multiplying both sides by $x+5$.

A number of different ways of solving inequalities of this type are demonstrated in the worked examples that follow.

No one method is ideal in all cases and you should consider a variety of approaches before deciding how to solve a particular example.

EXAMPLES 12C

1 Find the range of values of x for which $\dfrac{x-2}{x-5}>3$.

> Although we cannot multiply both sides by $x-5$ because its sign is not known, we *can* multiply both sides by $(x-5)^2$ which cannot be negative.

$$\frac{x-2}{x-5}>3$$

$$\therefore \quad (x-2)(x-5)>3(x-5)^2$$

$$\Rightarrow \quad (x-5)\{(x-2)-3(x-5)\}>0$$

> Note that $(x-5)$ must not be cancelled because we do not know its sign.

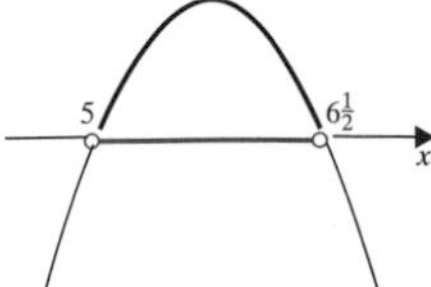

$$\Rightarrow \quad (x-5)(13-2x)>0$$

$\therefore$ the required range is $5<x<6\frac{1}{2}$.

2 Find the possible values of x for which $\dfrac{(x-2)(x+3)(x-4)}{x-1} < 0$.

The critical values of x in this inequality are −3, 1, 2 and 4 so we construct a table in which the columns are separated by these values, i.e.

	$x < -3$	$-3 < x < 1$	$1 < x < 2$	$2 < x < 4$	$x > 4$
$x+3$	−	+	+	+	+
$x-1$	−	−	+	+	+
$x-2$	−	−	−	+	+
$x-4$	−	−	−	−	+
$f(x)$	+	−	+	−	+

Therefore $\dfrac{(x-2)(x+3)(x-4)}{x-1} < 0$ when $-3 < x < 1$ or $2 < x < 4$

3 What values of x satisfy the inequality $\dfrac{(x-2)^2-8}{5-4x} > 1$?

$$\frac{(x-2)^2-8}{5-4x} > 1 \quad \Rightarrow \quad \frac{(x-2)^2-8}{5-4x} - 1 > 0$$

$$\Rightarrow \quad \frac{(x-2)^2-8-(5-4x)}{(5-4x)} > 0$$

$$\Rightarrow \quad \frac{x^2-9}{5-4x} > 0 \quad \Rightarrow \quad \frac{(x-3)(x+3)}{5-4x}$$

i.e. $\dfrac{f(x)}{g(x)} > 0$ where $f(x) = (x-3)(x+3)$ and $g(x) = 5-4x$

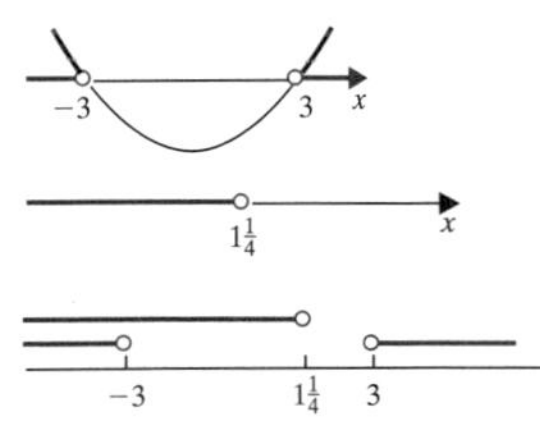

This fraction is positive if $f(x)$ and $g(x)$ have the same sign.

$f(x) > 0$ when $x < -3$ or $x > 3$ and $g(x) > 0$ when $x < 1\frac{1}{4}$

$\therefore$ $f(x)$ and $g(x)$ are *both* positive when $x < -3$.

Similarly,

$f(x) < 0$ if $-3 < x < 3$ and $g(x) < 0$ if $x > 1\frac{1}{4}$

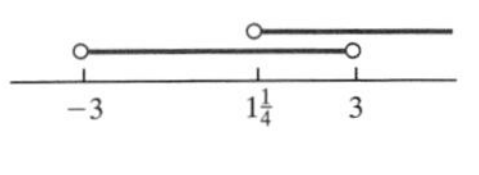

$\therefore$ $f(x)$ and $g(x)$ are *both* negative if $1\frac{1}{4} < x < 3$.

$\therefore$ $\dfrac{x^2-9}{5-4x}$ is positive, i.e. $\dfrac{(x-2)^2-8}{5-4x} > 1$

for values of x in *both* of the ranges found above,

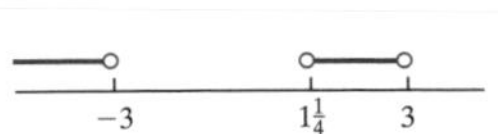

i.e. for $1\frac{1}{4} < x < 3$ *and* $x < -3$.

4 Find the range of values of x for which $3x + 4 < x^2 - 6 < 9 - 2x$

> $3x + 4 < x^2 - 6 < 9 - 2x$ is called a *double inequality* because it contains two inequalities, i.e. $3x + 4 < x^2 - 6$ and $x^2 - 6 < 9 - 2x$.

> We are looking for the set of values of x for which *both* inequalities are satisfied so we first solve each of them separately.

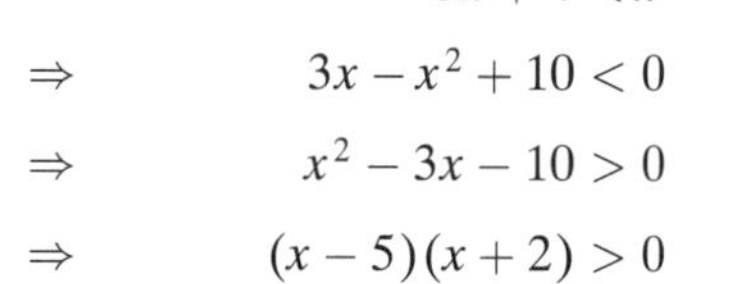

$$3x + 4 < x^2 - 6$$

$$\Rightarrow \quad 3x - x^2 + 10 < 0$$

$$\Rightarrow \quad x^2 - 3x - 10 > 0$$

$$\Rightarrow \quad (x - 5)(x + 2) > 0$$

$\therefore \; x < -2$ or $x > 5$

$$x^2 - 6 < 9 - 2x$$

$$\Rightarrow \quad x^2 + 2x - 15 < 0$$

$$\Rightarrow \quad (x + 5)(x - 3) < 0$$

$\therefore \; -5 < x < 3$

The required set of values of x must satisfy *both* of these conditions,

i.e.

$\therefore$ the range is $-5 < x < -2$.

EXERCISE 12C

For what range(s) of values of x are the following inequalities valid?

1 $\dfrac{x-1}{2+x} < 1$

2 $\dfrac{x-1}{2+x} > 1$

3 $\dfrac{3}{x-1} > 2$

4 $\dfrac{x}{2x-8} > 3$

5 $\dfrac{12}{x-3} < x + 1$

6 $\dfrac{x}{x-2} < \dfrac{x}{x-1}$

7 $\dfrac{(x-2)}{(x-1)(x-3)} > 0$

8 $\dfrac{2x}{(x-4)^2} > 1$

9 $\dfrac{(x-1)(x-2)}{(x+1)(x-3)} < 0$

10 $(3x-2)(2x+1) < 6x - 3$

11 $(x+1)(x+3)(x+5) > 0$

12 $2 - x > 2x + 4 > x$

13 $x - 1 < 3x + 1 < x + 5$

14 $2x - 1 < x^2 - 4 < 12$

15 $x - 4 < x(x-4) < 5$

16 $x - 3 > x^2 - 9 > -5$

Problems

The types of problem which involve inequalities are very varied. Their solutions depend not only on all the methods used so far in this chapter but also on other known facts, for example:

1 a perfect square can never be negative,

2 the nature of the roots of a quadratic equation $ax^2 + bx + c = 0$ depends upon whether $b^2 - 4ac = 0$ or $b^2 - 4ac > 0$ or $b^2 - 4ac < 0$. As two of the above conditions are inequalities, many problems about the roots of a quadratic equation require the solution, or interpretation, of inequalities.

The worked examples that follow are intended to give somc ideas to use in problem solving and make no claim to cover every situation.

EXAMPLES 12D

1 Find the range(s) of values of k for which the roots of the equation $kx^2 + kx - 2 = 0$ are real.

$kx^2 + kx - 2 = 0$

For real roots '$b^2 - 4ac$' $\geqslant 0$

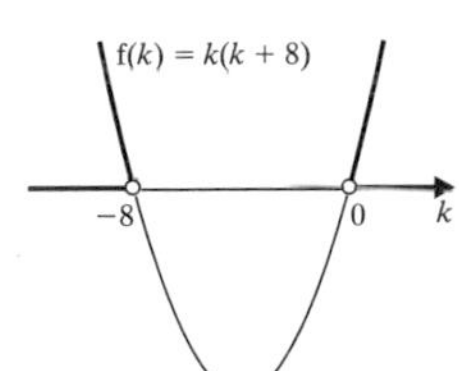

i.e. $\quad k^2 - 4(k)(-2) \geqslant 0$

$\Rightarrow \quad k(k+8) \geqslant 0$

Therefore the equation $kx^2 + kx - 2 = 0$ has real roots if the value of k lies in either of the ranges $k \leqslant -8$ or $k \geqslant 0$.

Note This type of question is sometimes expressed in another, less obvious, way, i.e. 'If x is real and $kx^2 + kx - 2 = 0$, find the values that k can take'. Once we realise that, because x is real the roots of the equation are real, the solution is identical to that above.

2 Prove that $x^2 + 2xy + 2y^2$ cannot be negative.

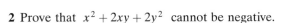

> Knowing that a perfect square cannot be negative, we rearrange the given expression in the form of perfect squares.

$$x^2 + 2xy + 2y^2 = x^2 + 2xy + y^2 + y^2$$

$$= (x+y)^2 + y^2$$

Each of the two terms on the RHS is a square and so cannot be negative.

Therefore $x^2 + 2xy + 2y^2$ cannot be negative.

3 Find the values of p for which $x^2 - 2px + p + 6$ is positive for all real values of x.

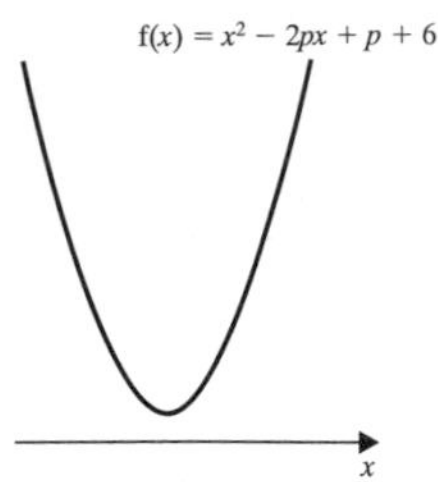

As $x^2 - 2px + p + 6$ is positive for all values of x, the graph of $f(x)$, where $f(x) = x^2 - 2px + p + 6$, is entirely above the x-axis, i.e.

the graph never crosses the x-axis so there are no values of x for which $f(x) = 0$, i.e. $x^2 - 2px + p + 6 = 0$ has no real roots.

$$\therefore \quad 'b^2 - 4ac' < 0 \quad \Rightarrow \quad (-2p)^2 - 4(1)(p+6) < 0$$
$$\Rightarrow \quad 4p^2 - 4p - 24 < 0$$
$$\Rightarrow \quad p^2 - p - 6 < 0$$
$$\Rightarrow \quad (p+2)(p-3) < 0$$

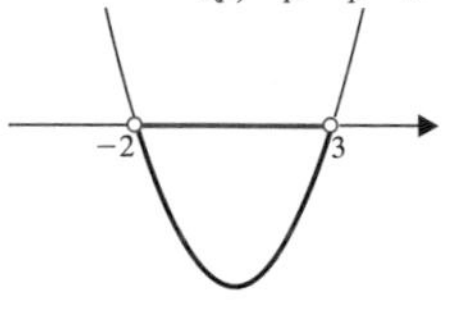

From the graph of $f(p) = p^2 - p - 6$ we see that $f(p) < 0$ for values of p between -2 and 3.

Therefore $x^2 - 2px + p + 6$ is positive for all real values of x provided that $-2 < p < 3$.

4 If x is real find the set of possible values of the function $\dfrac{x^2}{x+1}$.

> If we use $y = \dfrac{x^2}{x+1}$ we are looking for the range of values of y. To make use of the fact that x is real we need a quadratic equation in x.

$$y = \frac{x^2}{x+1} \quad \Rightarrow \quad x^2 - yx - y = 0$$

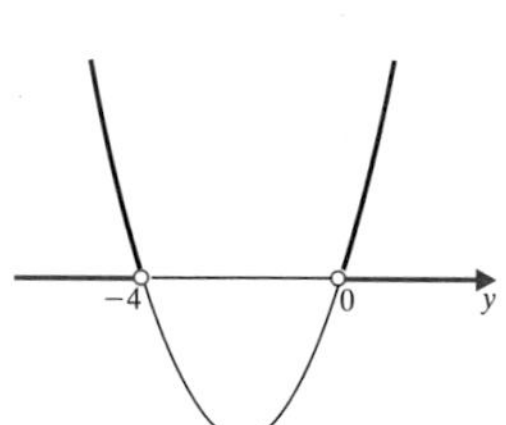

Since x is real, the roots of this equation are real, so $'b^2 - 4ac' \geqslant 0$

i.e. $\quad (-y)^2 - 4(1)(-y) \geqslant 0 \quad \Rightarrow \quad y(y+4) \geqslant 0$

$\therefore \; y \leqslant -4$ or $y \geqslant 0$

Therefore, for real values of x,

$$\frac{x^2}{x+1} \leqslant -4 \quad \text{or} \quad \frac{x^2}{x+1} \geqslant 0$$

EXERCISE 12D

1 Find the values of p for which the given equation has real roots.

a $x^2 + (p+3)x + 4p = 0$

b $x^2 + 3x + 1 = px$

2 Find the range of values of a for which the equation $x^2 - ax + (a+3) = 0$ has no real roots.

3 What is the set of values of p for which $p(x^2 + 2) < 2x^2 + 6x + 1$ for all real values of x?

In Questions 4 to 9 find the set of possible values of the given function.

4 $\dfrac{x+1}{2x^2+x+1}$

5 $\dfrac{1+x^2}{x}$

6 $\dfrac{x-2}{(x+2)(x-3)}$

7 $\dfrac{x}{1+x^2}$

8 $\dfrac{x-1}{x(x+1)}$

9 $x+1+\dfrac{1}{x+1}$

10 Find the set of values of k for which $x^2+3kx+k$ is positive for all real values of x.

11 Show that, if x is real, the function $\dfrac{2-x}{x^2-4x+1}$ can take any real value.

12 If x is real, find the range of the function $\dfrac{(2x+1)}{(x^2+2)}$.

13 Show that $x^2-4xy+5y^2 \geqslant 0$ for all real values of x and y.

14 Prove that $(a+b)^2 \geqslant 4ab$ for all real values of a and b.

MIXED EXERCISE 12

Solve each of the inequalities given in Questions 1 to 10.

1 $2x+1<4-x$

2 $x-5>1-3x$

3 $6x-5>1+2x$

4 $(x-3)(x+2)>0$

5 $(2x-3)(3x+2)<0$

6 $x^2-3<10$

7 $(x-3)^2>2$

8 $(3-x)(2-x)<20$

9 $x(4x+3)>2x-1$

10 $(x-6)(x+1)>2x-12$

In Questions 11 to 16 find the set of values of x for which:

11 $-3<5-2x<3$

12 $x^2+x+1<x+2<x^2-6x+12$

13 $\dfrac{2x-4}{x-1}<1$

14 $\dfrac{x-1}{x+1} \leqslant x$

15 $2 \geqslant \dfrac{x-1}{x+1} \geqslant 0$

16 $\dfrac{(x-1)(x+1)}{(x+2)(x-2)} \leqslant 0$

17 Prove that $x^2+y^2-10y+25 \geqslant 0$ for all real values of x and y.

18 For what values of k does the equation $4x^2+8x-8=k(4x-3)$ have real roots?

19 If x is real find the set of possible values of the function

$$\frac{x^2+1}{x^2+x+1}$$

20 Provided that x is real, prove that the function $\dfrac{2(3x+1)}{3(x^2-9)}$ can take all real values.

CHAPTER 13 Differentiation 1

Chords, Tangents, Normals and Gradients

Consider any two points, A and B, on any curve.

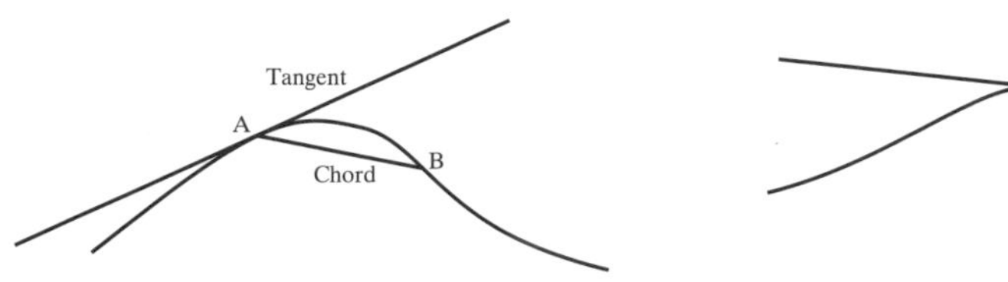

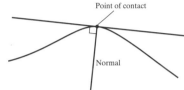

The line joining A and B is called a chord.

The line that touches the curve at A is called the tangent at A.

The word *touch* has a precise mathematical meaning. A line that meets a curve at a point and carries on without crossing to the other side of the curve at that point, *touches* the curve at the *point of contact*.

The line perpendicular to the tangent at A is called the normal at A.

Gradient, or slope, defines the direction of a line (lines can be straight or curved).

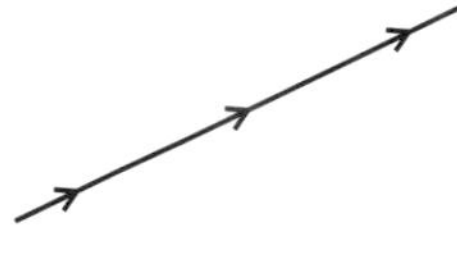

If we walk along a straight line, we move in the same direction all the time, i.e. the gradient of a straight line is constant.

Now if we walk along a curve, our direction is continually changing. It follows that the gradient of a curve is not constant but has different values at different points on the curve.

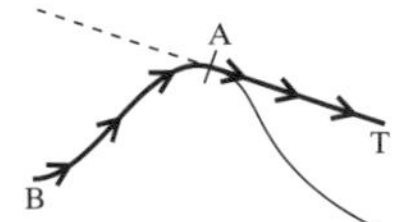

Moving from B to A along the curve in the diagram here, our direction is changing all the time. Now if at A we continue to move, but without any further change in direction, we go along the straight line AT. i.e. along the tangent to the curve at A, so

the gradient of the curve at A is the same as the gradient of the tangent to the curve at A.

Before a numerical value can be given to a gradient, the line or curve must be drawn on a pair of x- and y-axes. Then the gradient is the rate at which y increases with respect to x.

For a straight line this is found by taking the coordinates of two points and working out $\dfrac{\text{increase in } y}{\text{increase in } x}$.

We can use this to find the gradient of a tangent to a curve but, if the tangent is just drawn by eye, the value obtained can only be approximate and a more accurate method is needed.

Consider the problem of finding the gradient of the tangent at a point A on a curve.

If B is another point on the curve, fairly close to A, then the gradient of the chord AB gives an *approximate* value for the gradient of the tangent at A. As B gets nearer to A, the chord AB gets closer to the tangent at A, so the approximation becomes more accurate.

So we can say,

as B → A
the gradient of chord AB → the gradient of the tangent at A.

or

$$\lim_{\text{as B} \to \text{A}} \text{(gradient of chord AB)} = \text{gradient of tangent at A}$$

We can see how this works if we apply the definition to a particular point on a particular curve. Suppose, for instance, that we want the gradient of the curve $y = x^2$ at the point A where $x = 1$.

We need a succession of points $B_1, B_2 \ldots$ getting closer to $A(1, 1)$. Taking the x-coordinates of these points as 1.2, 1.1, 1.05, 1.01, 1.001, we can calculate the corresponding y-coordinates and hence the gradient of the chord joining A to each position of B.

x	1.2	1.1	1.05	1.01	1.001
$y \quad (= x^2)$	1.44	1.21	1.1025	1.0201	1.002 001
Increase in y	0.44	0.21	0.1025	0.0201	0.002 001
Increase in x	0.2	0.1	0.05	0.01	0.001
Gradient of chord AB	2.2	2.1	2.05	2.01	2.001

From the numbers in the last row of the table it is clear that, as B gets nearer to A, the gradient of the chord gets nearer to 2, i.e.

$$\lim_{\text{as B} \to \text{A}} \text{(gradient of chord AB)} = 2$$

It is much too tedious to go through this process each time we want the gradient at just one point on just one curve so we need a more general method.
For this we use a general point $A(x, y)$ and a variable small change in the value of x between A and B.

A new symbol, δ, is used to denote this small change.

When δ appears as a prefix to any letter representing a variable quantity, it denotes a small increase in that quantity.

e.g. δx means a small increase in x
δy means a small increase in y
δt means a small increase in t

Note that δ is only a prefix. It does not have an independent value and cannot be treated as a factor.

Now consider again the gradient of the curve with equation $y = x^2$.

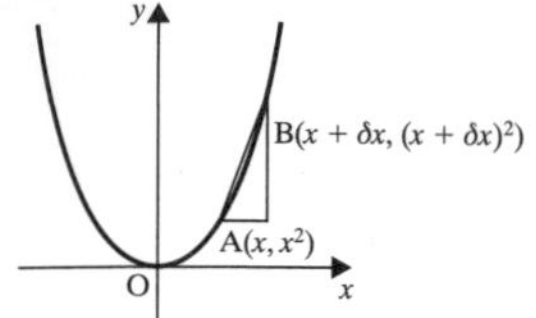

This time we will look for the gradient at *any* point $A(x, y)$ on the curve and use a point B where the x-coordinate of B is $x + \delta x$.

For any point on the curve, $y = x^2$.

So, at B, the y-coordinate is $(x + \delta x)^2 = x^2 + 2x\delta x + (\delta x)^2$.

The gradient of chord AB is given by $\dfrac{\text{increase in } y}{\text{increase in } x}$,

$$\text{which is } \frac{(x+\delta x)^2 - x^2}{(x+\delta x) - x} = \frac{2x\delta x + (\delta x)^2}{\delta x}$$

$$= 2x + \delta x$$

Now as $B \to A$, $\delta x \to 0$, therefore

$$\text{the gradient of the curve at A} = \lim_{\text{as B} \to \text{A}} (\text{gradient of chord AB})$$

$$= \lim_{\text{as } \delta x \to 0} (2x + \delta x)$$

$$= 2x$$

This result can now be used to find the gradient at any particular point on the curve with equation $y = x^2$,

e.g. at the point where $x = 3$, the gradient is $2(3) = 6$ and at the point $(4, 16)$, the gradient is $2(4) = 8$.

Similarly at the point where $x = 1$ the gradient is 2×1 i.e. 2. This confirms the value found by the longer method on p. 151.

Differentiation

The process of finding a general expression for the gradient of a curve at any point is known as differentiation.

The general expression for the gradient of a curve $y = f(x)$ is itself a function so it is called the *gradient function*. For the curve $y = x^2$, for example, the gradient function is $2x$.

Because the gradient function is derived from the given function, it is more often called the *derived function* or the *derivative*.

The method used above, in which the limit of the gradient of a chord was used to find the derived function, is known as *differentiating from first principles*. It is the fundamental way in which the gradient of each new type of function is found and, although many short cuts can be developed, it is important to understand, and be able to use, this basic method.

EXAMPLE 13A

By differentiating from first principles, find the gradient function of the expression x^3+5.
Find also the gradient at the point $(2, 13)$ on the curve $y = x^3+5$.

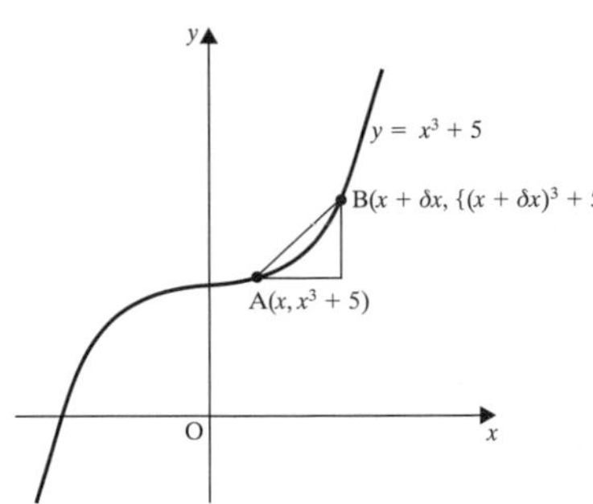

Let A and B be two neighbouring points on the curve, where A is the point (x,y) and the x-coordinate of B is $x+\delta x$.

Therefore at A $y = x^3+5$
and at B $y = (x+\delta x)^3+5$

The gradient of chord AB is $\dfrac{\{(x+\delta x)^3+5\}-\{x^3+5\}}{\{x+\delta x\}-x}$

which simplifies to $3x^2+3x\delta x+(\delta x)^2$

The gradient of the curve at A $= \underset{\text{as } \delta x \to 0}{\text{limit}} \{3x^2+3x\delta x+(\delta x)^2\}$

$= 3x^2$

Therefore the gradient function of x^3+5 is $3x^2$.

At the point $(2, 13)$, $x = 2$, so the gradient is $3(2)^2 = 12$.

EXERCISE 13A

By differentiating from first principles, find the derivative (i.e. the gradient function) of each of the following expressions. Hence find the gradient of the curve $y = f(x)$ at the given point.

1 $4x$; $(1, 4)$
2 $x-1$; $(3, 2)$
3 x^3; $(1, 1)$
4 x^2+2; $(2, 6)$
5 x^2-x; $(1, 0)$

Notation

We now need a concise way to write the fact that the derivative of x^2 is $2x$ (and all similar results of differentiating with respect to x).

One notation, based on the equation of the curve, is:

If $y = x^2$, $\dfrac{dy}{dx} = 2x$ (we say 'dy by dx')

The result of Example 13A can also be written this way, i.e.

for $y = x^3+5$, $\dfrac{dy}{dx} = 3x^2$

Note that d has no independent meaning and must never be regarded as a factor.

The complete symbol $\dfrac{d}{dx}$ means 'the derivative with respect to x of'.

So, $\dfrac{dy}{dx}$ means 'the derivative with respect to x of y'

and $\dfrac{d}{dx}(x^2-x)$ means 'the derivative with respect to x of (x^2-x)'.

An alternative notation concentrates on the function of x rather than the equation of the curve.

An example is $\quad \mathrm{f}(x) = x^2, \; \mathrm{f}'(x) = 2x$

In this form, f′ means 'the gradient function' or 'the derived function'.

Again we can illustrate this notation using the results of Example 13A, e.g. for $\mathrm{f}(x) = x^3, \; \mathrm{f}'(x) = 3x^2$.

Either of these notations can be used for variables other than x, e.g. if $y = z^3$ then we differentiate y with respect to z and write $\dfrac{\mathrm{d}y}{\mathrm{d}z} = 3z^2$.

Similarly, if $s = t^2 - t$, we differentiate s with respect to t and write $\dfrac{\mathrm{d}s}{\mathrm{d}t} = 2t - 1$.

(Because the phrase 'with respect to' is used very frequently, it is often abbreviated to w.r.t.)

The General Gradient Function

Taking two points, $\mathrm{A}(x,y)$ and $\mathrm{B}(x + \delta x, y + \delta y)$, on the curve with equation $y = \mathrm{f}(x)$, the gradient of AB is $\dfrac{\delta y}{\delta x}$, which is equal to $\dfrac{\mathrm{f}(x+\delta x) - \mathrm{f}(x)}{\delta x}$.

Then $$\frac{\mathrm{d}y}{\mathrm{d}x} = \mathrm{f}'(x) = \lim_{\text{as } \delta x \to 0} \frac{\mathrm{f}(x+\delta x) - \mathrm{f}(x)}{\delta x}$$

EXAMPLE 13B Find the derivative of the function $\dfrac{1}{x}$.

$$\mathrm{f}(x) = \frac{1}{x} \quad \Rightarrow \quad \mathrm{f}(x+\delta x) = \frac{1}{x+\delta x}$$

$$\mathrm{f}(x+\delta x) - \mathrm{f}(x) = \frac{1}{x+\delta x} - \frac{1}{x} = \frac{x - (x+\delta x)}{x(x+\delta x)}$$

$$= \frac{-\delta x}{x(x+\delta x)}$$

$$\frac{\mathrm{f}(x+\delta x) - \mathrm{f}(x)}{\delta x} = \frac{-\delta x}{x(x+\delta x)(\delta x)} = \frac{-1}{x(x+\delta x)}$$

Then $$\mathrm{f}'(x) = \lim_{\text{as } \delta x \to 0} \frac{\mathrm{f}(x+\delta x) - \mathrm{f}(x)}{\delta x} = \lim_{\text{as } \delta x \to 0} \frac{-1}{x(x+\delta x)} = \frac{-1}{x^2}$$

i.e. the derivative of $\dfrac{1}{x}$ is $-\dfrac{1}{x^2}$.

EXERCISE 13B Use the general formula for the derivative of $\mathrm{f}(x)$ to differentiate

1 $1/x^2$ **2** $2/x$

Differentiating x^n with Respect to x

Some of the results that have been produced so far can now be collected and tabulated.

y	x^2	x^3	x^4	x^{-1}
$\frac{dy}{dx}$	$2x$	$3x^2$	$4x^3$	$-x^{-2}\left(\text{or } -\frac{1}{x^2}\right)$

From this table it *appears* that when we differentiate a power of x we multiply by that power and then reduce the power by 1, i.e. it looks as though

$$\text{if } y = x^n, \text{ then } \frac{dy}{dx} = nx^{n-1}$$

This result, although deduced from just a few examples, is in fact valid for all powers, including those that are fractional or negative. It is not possible to give a proof at this stage and this is one example of a 'rule' which, for the moment, we must just take on trust. It is easy to apply and makes the task of differentiating a power of x very much simpler, e.g.

$$\frac{d}{dx}(x^7) = 7x^6 \qquad \frac{d}{dx}(x^3) = 3x^2$$

$$\frac{d}{dx}(x^{-2}) = -2x^{-3} \qquad \frac{d}{dx}(x^{3/2}) = \tfrac{3}{2}x^{1/2}$$

EXAMPLE 13C Differentiate with respect to x

a $x^{-1/3}$ **b** $\sqrt[4]{(x^3)}$

a Use $\frac{d}{dx}(x^n) = nx^{n-1}$, where $n = -\frac{1}{3}$

$$\frac{dy}{dx} = -\frac{1}{3}x^{-\frac{1}{3}-1} = -\frac{1}{3}x^{-\frac{4}{3}} = -\frac{1}{3x^{4/3}}$$

b $\sqrt[4]{(x^3)}$ can be written $x^{\frac{3}{4}}$, i.e. $n = \frac{3}{4}$

$$\frac{d}{dx}(x^{\frac{3}{4}}) = \frac{3}{4}x^{\frac{3}{4}-1} = \frac{3}{4}x^{-\frac{1}{4}} \text{ or } \frac{3}{4\sqrt[4]{x}}$$

EXERCISE 13C Differentiate with respect to x.

1 x^5

2 x^{-3}

3 $x^{4/3}$

4 $\frac{1}{x}$

5 x^{10}

6 $\frac{1}{x^2}$

7 $\sqrt{x^3}$

8 $x^{-1/2}$

9 $\frac{1}{x^4}$

10 $x^{1/3}$

11 $x^{-1/4}$

12 x

13 $\sqrt{x^7}$

14 $\frac{1}{x^7}$

15 $x^{1/7}$

16 $\sqrt{(x^2)^3}$

Differentiating Constants and Multiples of x

Any line with equation $y = c$ is a horizontal straight line whose gradient is zero,

$\therefore$ if $y = c$ then $\frac{dy}{dx} = 0$

Any line with equation $y = kx$ is a sloping line with gradient k,

$\therefore$ if $y = kx$ then $\frac{dy}{dx} = k$

If y is a constant multiple of a function of x, i.e. $y = af(x)$ then $\frac{dy}{dx} = af'(x)$,

e.g. if $y = 3x^5$, $\frac{dy}{dx} = 3 \times 5x^4 = 15x^4$

and if $y = 4x^{-2}$, $\frac{dy}{dx} = 4 \times -2x^{-3} = -8x^{-3}$

In general, if a is a constant

$$\frac{d}{dx}ax^n = anx^{n-1}$$

A function of x which contains the sum or difference of a number of separate terms can be differentiated term by term, applying the basic rule to each in turn.

For example, if $y = x^4 + \frac{1}{x} - 6x$

then $\frac{dy}{dx} = \frac{d}{dx}(x^4) + \frac{d}{dx}(x^{-1}) - \frac{d}{dx}(6x) = 4x^3 - \frac{1}{x^2} - 6$

(This property is demonstrated, though not proved, by Question 5 in Exercise 13A.)

EXERCISE 13D

Differentiate each of the following functions w.r.t. x.

1 $x^3 - x^2 + 5x - 6$

2 $3x^2 + 7 - \frac{4}{x}$

3 $\sqrt{x} + \frac{1}{\sqrt{x}}$

4 $2x^4 - 4x^2$

5 $x^3 - 2x^2 - 8x$

6 $x^2 + 5\sqrt{x}$

7 $x^{-3/4} - x^{3/4} + x$

8 $3x^3 - 4x^2 + 9x - 10$

9 $x^{3/2} - x^{1/2} + x^{-1/2}$

10 $\sqrt{x} + \sqrt{x^3}$

11 $\frac{1}{x^2} - \frac{1}{x^3}$

12 $\frac{1}{\sqrt{x}} - \frac{2}{x}$

13 $x^{-1/2} + 3x^{3/2}$

14 $x^{1/4} - x^{1/5}$

15 $\frac{4}{x^3} + \frac{x^3}{4}$

16 $\frac{4}{x} + \frac{5}{x^2} - \frac{6}{x^3}$

17 $3\sqrt{x} - 3x$

18 $x - 2x^{-1} - 3x^{-3}$

19 $x\sqrt{x} - x^2\sqrt{x}$

20 $\frac{\sqrt{x}}{x^2} + \frac{x^2}{\sqrt{x}}$

Differentiating Products and Fractions

All the rules given above can be applied to the differentiation of expressions containing products or quotients if they can be multiplied out or divided into separate terms.

Further on in the book you will find better methods for differentiating products and quotients, but these use techniques that you have not yet met.

EXAMPLES 13E

1 If $y = (x-3)(x^2+7x-1)$, find $\frac{dy}{dx}$.

$$y = (x-3)(x^2+7x-1) = x^3+4x^2-22x+3$$

$$\Rightarrow \quad \frac{dy}{dx} = 3x^2+8x-22$$

2 Find $\frac{dt}{dz}$ given that $t = \frac{6z^2+z-4}{2z}$.

$$t = \frac{6z^2+z-4}{2z} = \frac{6z^2}{2z}+\frac{z}{2z}-\frac{4}{2z}$$

$$= 3z+\tfrac{1}{2}-\frac{2}{z}$$

$$\Rightarrow \quad \frac{dt}{dz} = 3+0-2(-z^{-2})$$

$$= 3+\frac{2}{z^2}$$

EXERCISE 13E

Differentiate each of the following equations with respect to the variable concerned.

1 $y = (x+1)^2$

2 $z = x^{-2}(2-x)$

3 $y = (3x-4)(x+5)$

4 $y = (4-z)^2$

5 $s = \frac{t^{-1}+3t^2}{2t^2}$

6 $s = \frac{t^2+t}{2t}$

7 $y = \left(\frac{1}{x}\right)(x^2+1)$

8 $y = \frac{z^3-z}{\sqrt{z}}$

9 $y = 2x(3x^2-4)$

10 $s = (t+2)(t-2)$

11 $s = \frac{t^3-2t^2+7t}{t^2}$

12 $y = \frac{\sqrt{x}+7}{x^2}$

Gradients of Tangents and Normals

If the equation of a curve is known, and the gradient function can be found, then the gradient, m say, at a particular point A on that curve can be calculated. This is also the gradient of the tangent to the curve at A.

The normal at A is perpendicular to the tangent at A, therefore its gradient is $-\frac{1}{m}$.

EXAMPLES 13F

1 The equation of a curve is $s = 6 - 3t - 4t^2 - t^3$. Find the gradient of the tangent and of the normal to the curve at the point $(-2, 4)$.

$$s = 6 - 3t - 4t^2 - t^3 \quad \Rightarrow \quad \frac{ds}{dt} = 0 - 3 - 8t - 3t^2$$

At the point $(-2, 4)$, $\dfrac{ds}{dt} = -3 - 8(-2) - 3(-2)^2 = 1$

Therefore the gradient of the tangent at $(-2, 4)$ is 1 and the gradient of the normal is $-1/1$, i.e. -1.

2 Find the coordinates of the points on the curve $y = 2x^3 - 3x^2 - 8x + 7$ where the gradient is 4.

$$y = 2x^3 - 3x^2 - 8x + 7 \quad \Rightarrow \quad \frac{dy}{dx} = 6x^2 - 6x - 8$$

If the gradient is 4 then $\dfrac{dy}{dx} = 4$

i.e. $\quad 6x^2 - 6x - 8 = 4 \quad \Rightarrow \quad 6x^2 - 6x - 12 = 0 \quad \Rightarrow \quad x^2 - x - 2 = 0$

$\therefore \quad (x - 2)(x + 1) = 0 \quad \Rightarrow \quad x = 2$ or -1

When $x = 2$, $y = 16 - 12 - 16 + 7 = -5$

when $x = -1$, $y = -2 - 3 + 8 + 7 = 10$

Therefore the gradient is 4 at the points $(2, -5)$ and $(-1, 10)$.

EXERCISE 13F

Find the gradient of the tangent and the gradient of the normal at the given point on the given curve.

1 $y = x^2 + 4$ where $x = 1$

2 $y = \dfrac{3}{x}$ where $x = -3$

3 $y = \sqrt{z}$ where $z = 4$

4 $s = 2t^3$ where $t = -1$

5 $v = 2 - \dfrac{1}{u}$ where $u = 1$

6 $y = (x + 3)(x - 4)$ where $x = 3$

7 $y = z^3 - z$ where $z = 2$

8 $s = t + 3t^2$ where $t = -2$

9 $z = x^2 - \dfrac{2}{x}$ where $x = 1$

10 $y = \sqrt{x} + \dfrac{1}{\sqrt{x}}$ where $x = 9$

11 $s = \sqrt{t}(1 + \sqrt{t})$ where $t = 4$

12 $y = \dfrac{x^2 - 4}{x}$ where $x = -2$

Find the coordinates of the point(s) on the given curve where the gradient has the value specified.

13 $y = 3 - \dfrac{2}{x}$; $\frac{1}{2}$

14 $z = x^2 - x^3$; -1

15 $s = t^3 - 12t + 9$; 15

16 $v = u + \dfrac{1}{u}$; 0

17 $s = (t + 3)(t - 5)$; 0

18 $y = \dfrac{1}{x^2}$; $\frac{1}{4}$

19 $y = (2x - 5)(x + 1)$; -3

20 $y = z^3 - 3z$; 0

MIXED EXERCISE 13

1 Differentiate $3x^2+x$ with respect to x from first principles.

2 Find the derivative of

a $x^{-3}-x^3+7$ **b** $x^{1/2}-x^{-1/2}$ **c** $\frac{1}{x^2}+\frac{2}{x^3}$

3 Differentiate w.r.t. x.

a $y=x^{3/2}-x^{2/3}+x^{-1/3}$ **b** $y=\sqrt{x}-\frac{1}{x}+\frac{1}{x^3}$ **c** $\frac{1}{x^{3/4}}-\frac{1}{x^{1/4}}$

4 Find the gradient of the curve $y=2x^3-3x^2+5x-1$ at the point

a $(0,-1)$ **b** $(1,3)$ **c** $(-1,-11)$

5 Find the gradient of the given curve at the given point.

a $y=x^2+x-9;\ x=2$ **b** $y=x(x-4);\ x=5$

6 The equation of a curve is $y=(x-3)(x+4)$. Find the gradient of the curve

a at the point where the curve crosses the y-axis

b at each of the points where the curve crosses the x-axis.

7 If the equation of a curve is $y=2x^2-3x-2$ find

a the gradient at the point where $x=0$

b the coordinates of the points where the curve crosses the x-axis

c the gradient at each of the points found in part **b**.

8 Find the coordinates of the point(s) on the curve $y=3x^3-x+8$ at which the gradient is **a** 8 **b** 0

9 Find $\frac{dy}{dx}$ if **a** $y=x^4-x^2$ **b** $y=(3x+4)^2$ **c** $y=\frac{x-3}{\sqrt{x}}$

10 Find the gradient of the tangent at the point where $x=2$ on the curve $y=(2-\sqrt{x})^2$.

11 Find the coordinates of the point on the curve $y=x^2$ where the gradient of the normal is $\frac{1}{4}$.

12 The equation of a curve is $s=4t^2+5t$. Find the gradient of the normal at each of the points where the curve crosses the t-axis.

13 Find the coordinates of the points on the curve $y=x^3-6x^2+12x+2$ at which the tangent is parallel to the line $y=3x$.

14 The curve $y=(x-2)(x-3)(x-4)$ cuts the x-axis at the points P(2,0), Q(3,0) and R(4,0). Prove that the tangents at P and R are parallel and find the gradient of the normal at Q.

15 For a certain equation, $\frac{dy}{dx}=2x+1$.
Which of the following could be the given equation?

a $y=2x^2+x$ **b** $y=x^2+x-1$ **c** $y=x^2+1$ **d** $y=x^2+x$

CHAPTER 14

Tangents, normals and stationary values

The Equations of Tangents and Normals

We have seen how to find the gradient of a tangent at a particular point, A, on a curve. We also know that the tangent passes through the point A. Therefore the tangent is a line passing through a known point and having a known gradient and its equation can be found using $y - y_1 = m(x - x_1)$.

The equation of a normal can be found in the same way.

EXAMPLES 14A

1 Find the equation of the normal to the curve $y = \frac{4}{x}$ at the point where $x = 1$.

$$y = \frac{4}{x} \quad \Rightarrow \quad \frac{dy}{dx} = -\frac{4}{x^2}$$

When $x = 1, y = 4$ and $\frac{dy}{dx} = -4$

The gradient of the tangent at (1, 4) is -4, therefore the gradient of the normal at (1, 4) is $-\frac{1}{-4}$, i.e. $\frac{1}{4}$.

The equation of the normal is given by $y - y_1 = m(x - x_1)$

i.e. $\quad y - 4 = \frac{1}{4}(x - 1) \quad \Rightarrow \quad 4y = x + 15$

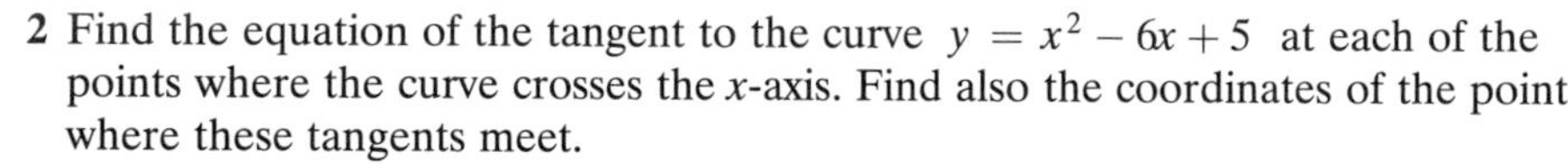

2 Find the equation of the tangent to the curve $y = x^2 - 6x + 5$ at each of the points where the curve crosses the x-axis. Find also the coordinates of the point where these tangents meet.

The curve crosses the x-axis where $y = 0$,

i.e. where $x^2 - 6x + 5 = 0 \quad \Rightarrow \quad (x - 5)(x - 1) = 0$

$\Rightarrow \qquad x = 5$ and $x = 1$

Therefore the curve crosses the x-axis at $(5, 0)$ and $(1, 0)$

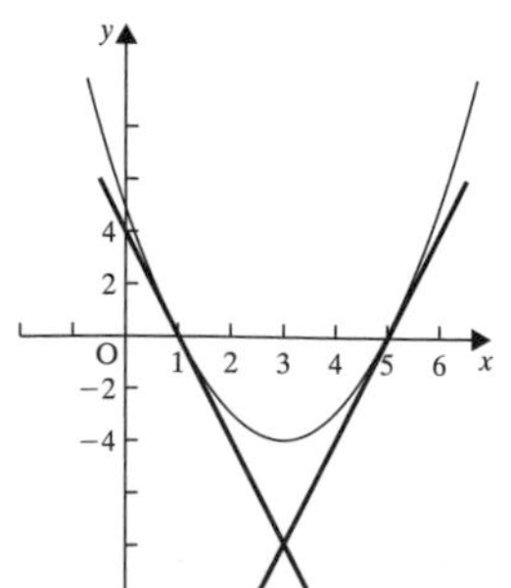

$$y = x^2 - 6x + 5 \quad \Rightarrow \quad \frac{dy}{dx} = 2x - 6$$

At $(5, 0)$, the gradient of the tangent is given by $\frac{dy}{dx} = 10 - 6 = 4$

therefore the equation of this tangent is

$y - 0 = 4(x - 5) \quad \Rightarrow \quad y = 4x - 20$

At $(1, 0)$ the gradient of the tangent is given by $\frac{dy}{dx} = 2 - 6 = -4$

Therefore the equation of the tangent is $y - 0 = -4(x - 1) \quad \Rightarrow \quad y + 4x = 4$

If the two tangents meet at P then, at P,

$$y + 4x = 4 \text{ and } y - 4x = -20$$

Solving these equations gives $2y = -16 \Rightarrow y = -8$

Using $y = -8$ in $y + 4x = 4$ gives $-8 + 4x = 4 \Rightarrow x = 3$

Therefore the tangents meet at $(3, -8)$.

EXERCISE 14A

In each question from 1 to 6 find, at the given point,

a the equation of the tangent **b** the equation of the normal.

1 $y = x^2 - 4$ where $x = 1$

2 $y = x^2 + 4x - 2$ where $x = 0$

3 $y = 1/x$ where $x = -1$

4 $y = x^2 + 5$ where $x = 0$

5 $y = x^2 - 5x + 7$ where $x = 2$

6 $y = (x - 2)(x^2 - 1)$ where $x = -2$

7 Find the equation of the normal to the curve $y = x^2 + 4x - 3$ at the point where the curve cuts the y-axis.

8 Find the equation of the tangent to the curve $y = x^2 - 3x - 4$ at the point where this curve cuts the line $x = 5$.

9 Find the equation of the tangent to the curve $y = (2x - 3)(x - 1)$ at each of the points where this curve cuts the x-axis. Find the point of intersection of these tangents.

10 Find the equation of the normal to the curve $y = x^2 - 6x + 5$ at each of the points where the curve cuts the x-axis.

11 Find the equation of the tangent to the curve $y = 3x^2 + 5x - 1$ at each of the points of intersection of the curve and the line $y = x - 1$.

12 Find the equations of the tangent to the curve $y = x^2 + 5x - 3$ at the points where the line $y = x + 2$ crosses the curve.

13 Find the coordinates of the point on the curve $y = 2x^2$ where the gradient is 8. Hence find the equation of the tangent to $y = 2x^2$ whose gradient is 8.

14 Find the coordinates of the point on the curve $y = 3x^2 - 1$ where the gradient is 3.

15 Find the equation of the tangent to the curve $y = 4x^2 + 3x$ whose gradient is -1.

16 Find the equation of the normal to the curve $y = 2x^2 - 2x + 1$ whose gradient is $\frac{1}{2}$.

17 Find the value of k for which $y = 2x + k$ is a tangent to the curve $y = 2x^2 - 3$.

18 Find the equation of the tangent to the curve $y = (x - 5)(2x + 1)$ that is parallel to the x-axis.

19 Find the coordinates of the point(s) on the curve $y = x^2 - 5x + 3$ where the gradient of the normal is $\frac{1}{3}$.

20 A curve has the equation $y = x^3 - px + q$. The tangent to this curve at the point $(2, -8)$ is parallel to the x-axis. Find the values of p and q.
Find also the coordinates of the other point where the tangent is parallel to the x-axis.

Stationary Values

Consider a function $f(x)$. The derived function, $f'(x)$, expresses the rate at which $f(x)$ increases with respect to x.

At a particular point,

if f′(x) is positive then f(x) is increasing as x increases,
whereas if f′(x) is negative then f(x) is decreasing as x increases.

There may be points where $f'(x)$ is zero, i.e. $f(x)$ is momentarily neither increasing nor decreasing with respect to x.

The value of $f(x)$ at such a point is called a *stationary value*

i.e. $\quad f'(x) = 0 \quad \Rightarrow \quad f(x)$ has a stationary value.

To look at this situation graphically, consider the curve with equation $y = f(x)$.

At A and B, $f(x)$, and therefore y, is neither increasing nor decreasing with respect to x. So the values of y at A and B are stationary values,

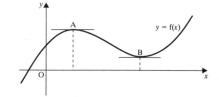

i.e. $\quad \dfrac{dy}{dx} = 0 \quad \Rightarrow \quad y$ has a stationary value.

The point on a curve where y has a stationary value is called a *stationary point* and we see that, at any stationary point, the gradient of the tangent to the curve is zero, i.e. the tangent is parallel to the x-axis.

To sum up:

at a stationary point
- **y, or f(x) has a stationary value**
- **$\dfrac{dy}{dx}$, or f′(x), is zero**
- **the tangent is parallel to the x-axis.**

EXAMPLE 14B Find the stationary values of the function $x^3 - 4x^2 + 7$.

If $\quad f(x) = x^3 - 4x^2 + 7$

then $\quad f'(x) = 3x^2 - 8x$

At stationary points, $f'(x) = 0$ i.e. $3x^2 - 8x = 0$

$\Rightarrow \quad x(3x - 8) = 0 \quad \Rightarrow \quad x = 0$ and $x = \frac{8}{3}$

Therefore there are stationary points where $x = 0$ and $x = \frac{8}{3}$

When $x = 0$, $f(x) = 0 - 0 + 7 = 7$

When $x = \frac{8}{3}$, $f(x) = (\frac{8}{3})^3 - 4(\frac{8}{3})^2 + 7 = -2\frac{13}{27}$

Therefore the stationary values of $x^3 - 4x^2 - 5$ are 7 and $-2\frac{13}{27}$.

EXERCISE 14B Find the value(s) of x at which the following functions have stationary values.

1 x^2+7 **3** x^3-4x^2+6 **5** x^3-2x^2+11

2 $2x^2-3x-2$ **4** $4x^3-3x-9$ **6** x^3-3x-5

Find the value(s) of x for which y has a stationary value.

7 $y=x^2-8x+1$ **9** $y=2x^3+x^2-8x+1$ **11** $y=2x^3+9x^2-24x+7$

8 $y=x+\dfrac{9}{x}$ **10** $y=9x^3-25x$ **12** $y=3x^3-12x+19$

Find the coordinates of the stationary points on the following curves.

13 $y=\dfrac{x^2+9}{2x}$ **15** $y=(x-3)(x+2)$ **17** $y=\sqrt{x}+\dfrac{1}{\sqrt{x}}$

14 $y=x^3-2x^2+x-7$ **16** $y=x^{3/2}-x^{1/2}$ **18** $y=8+\dfrac{x}{4}+\dfrac{4}{x}$

Turning Points

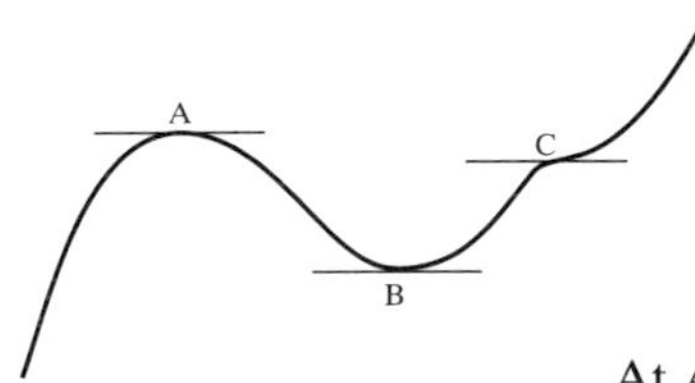

In the immediate neighbourhood of a stationary point a curve can have any one of the shapes shown in the diagram.

Moving through A from left to right we see that the curve is rising, then turns at A and begins to fall, i.e. the gradient changes from positive, to zero at A, and then becomes negative.

At A there is a *turning point.*

The value of y at A is called a *maximum value* and A is called a *maximum point.*

Moving through B from left to right the curve is falling, then turns at B and begins to rise, i.e. the gradient changes from negative, to zero at B, and then becomes positive.

At B there is a *turning point.*

The value of y at B is called a *minimum value* and B is called a *minimum point.*

The tangent is always horizontal at a turning point.

Note that a maximum value of y is *not necessarily the greatest value of* y *overall.* The terms maximum and minimum apply only to the behaviour of the curve in the neighbourhood of a stationary point; the point can be called a local maximum or a local minimum.

At C the curve does not turn. The gradient goes from positive, to zero at C and then becomes positive again, i.e. the gradient does not change sign at C.

C is not a turning point but, because there is a change in the sense in which the curve is turning (from clockwise to anti-clockwise), C is called a *point of inflexion.*

Any point on a curve where the sense of turning changes is a point of inflexion. In the diagram there are two points of inflexion other than C; one is between A and B and the other is between B and C. We see from these points that

the tangent is not necessarily horizontal at a point of inflexion.

Investigating the Nature of Stationary Points

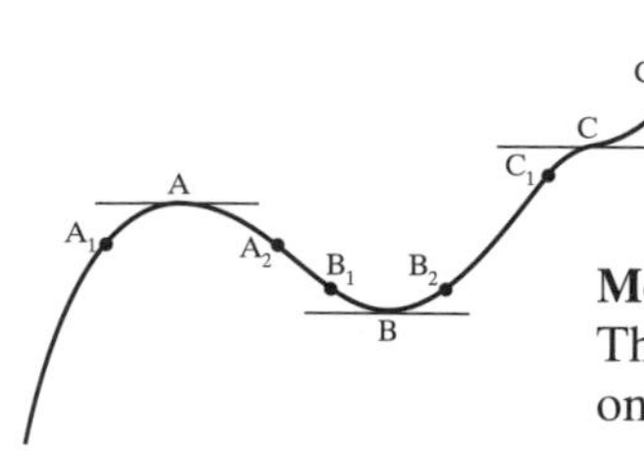

We already know how to locate stationary points on a curve and now look at several ways of distinguishing between the different types of stationary point.

Method 1

This method compares the value of y at the stationary point with values of y at points on either side of, and near to, the stationary point.

For a maximum value, e.g. at A

$$y \text{ at } A_1 < y \text{ at } A$$
$$y \text{ at } A_2 < y \text{ at } A$$

For a minimum point, e.g. at B

$$y \text{ at } B_1 > y \text{ at } B$$
$$y \text{ at } B_2 > y \text{ at } B$$

For a point of inflexion, e.g. at C

$$y \text{ at } C_1 < y \text{ at } C$$
$$y \text{ at } C_2 > y \text{ at } C$$

Collecting these conclusions we have:

	Maximum	Minimum	Inflexion
y values on each side of the stationary point	both smaller	both larger	one larger and one smaller

Note that the points chosen on either side of the stationary point must be such that no *other* stationary point, nor any break in the graph, lies between them.

Method 2

This method examines the sign of the gradient at points close to, and on either side of, the stationary point.

For a maximum point, A

$$\frac{dy}{dx} \text{ at } A_1 \text{ is +ve,} \quad \frac{dy}{dx} \text{ at } A_2 \text{ is −ve}$$

For a minimum point, B

$$\frac{dy}{dx} \text{ at } B_1 \text{ is −ve,} \quad \frac{dy}{dx} \text{ at } B_2 \text{ is +ve}$$

For a point of inflexion, C

$$\frac{dy}{dx} \text{ at } C_1 \text{ is +ve,} \quad \frac{dy}{dx} \text{ at } C_2 \text{ is +ve}$$

Collecting these conclusions we have:

Sign of $\frac{dy}{dx}$	Passing through maximum $+\ 0\ -$	Passing through minimum $-\ 0\ +$	Passing through point of inflexion $+\ 0\ +$ or $-\ 0\ -$
Gradient of tangent	/ — \	\ — /	/ — / or \ — \

Method 3

In this method we observe how $\frac{dy}{dx}$ changes with respect to x as we pass through a stationary point. Now the rate at which $\frac{dy}{dx}$ increases with respect to x could be written $\frac{d}{dx}\left(\frac{dy}{dx}\right)$ but this notation is clumsy and it is condensed to $\frac{d^2y}{dx^2}$ (we say d 2 y by dx squared). $\frac{d^2y}{dx^2}$ is the *second derivative* of y with respect to x.

If $\frac{dy}{dx}$ is increasing as x increases we can write $\frac{d^2y}{dx^2}$ is +ve.

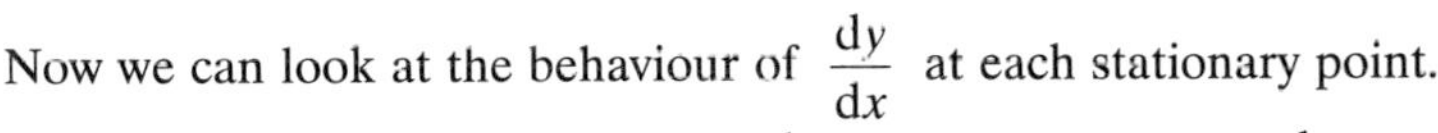

Now we can look at the behaviour of $\frac{dy}{dx}$ at each stationary point.

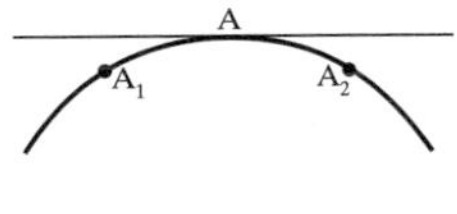

For the maximum point A; at A_1 $\frac{dy}{dx}$ is +ve and at A_2 $\frac{dy}{dx}$ is −ve

so, passing through A, $\frac{dy}{dx}$ goes from + to −, i.e. $\frac{dy}{dx}$ decreases

$\Rightarrow$ at A, $\frac{d^2y}{dx^2}$ is negative.

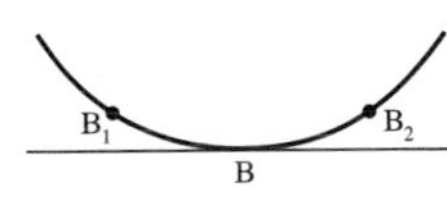

For the minimum point B; at B_1 $\frac{dy}{dx}$ is −ve and at B_2 $\frac{dy}{dx}$ is +ve

so, passing through B, $\frac{dy}{dx}$ goes from − to +, i.e. $\frac{dy}{dx}$ increases

$\Rightarrow$ at B, $\frac{d^2y}{dx^2}$ is positive.

Summing up method 3 we have:

	Maximum	Minimum
Sign of $\frac{d^2y}{dx^2}$	negative (or zero)	positive (or zero)

Points of inflexion are not so easily dealt with by this method because, although $\frac{d^2y}{dx^2}$ is zero at a point of inflexion, it is also possible for $\frac{d^2y}{dx^2}$ to be zero at a turning point. So for any stationary point where $\frac{d^2y}{dx^2} = 0$, this method fails and one of the other two approaches must be used.

Non-stationary Points of Inflexion

So far we have looked at points of inflexion that are also stationary points. However there are other points of inflexion where this is not the case. At these points the sense in which the curve is turning changes but the gradient is not zero,

i.e. $\dfrac{d^2y}{dx^2} = 0$ but $\dfrac{dy}{dx} \neq 0$

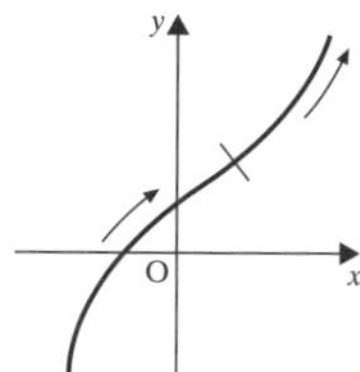

EXAMPLES 14C

1 Locate the stationary points on the curve $y = 4x^3 + 3x^2 - 6x - 1$ and determine the nature of each one.

$y = 4x^3 + 3x^2 - 6x - 1$

$\Rightarrow \quad \dfrac{dy}{dx} = 12x^2 + 6x - 6$

At stationary points, $\dfrac{dy}{dx} = 0$

i.e. $\qquad 12x^2 + 6x - 6 = 0$

$\Rightarrow \quad 6(2x-1)(x+1) = 0$

$\therefore$ there are stationary points where $x = \frac{1}{2}$ and $x = -1$.

When $x = \frac{1}{2}$, $y = -2\frac{3}{4}$ and when $x = -1$, $y = 4$

i.e. the stationary points are $(\frac{1}{2}, -2\frac{3}{4})$ and $(-1, 4)$.

Differentiating $\dfrac{dy}{dx}$ w.r.t x gives

$\dfrac{d^2y}{dx^2} = 24x + 6$

When $x = \frac{1}{2}$,

$\dfrac{d^2y}{dx^2} = 12 + 6$ which is positive

$\Rightarrow \quad (\frac{1}{2}, -2\frac{3}{4})$ is a minimum point.

When $x = -1$,

$\dfrac{d^2y}{dx^2} = -24 + 6$ which is negative

$\Rightarrow \quad (-1, 4)$ is a maximum point.

2 Find the stationary values of $3x^4 - 8x^3 + 6x^2 - 3$ and investigate their nature.

$f(x) = 3x^4 - 8x^3 + 6x^2 - 3 \quad \Rightarrow \quad f'(x) = 12x^3 - 24x^2 + 12x$

At stationary values $f'(x) = 0$

i.e. $12x^3 - 24x^2 + 12x = 0 \quad \Rightarrow \quad 12x(x^2 - 2x + 1) = 0$

$\Rightarrow \quad 12x(x-1)(x-1) = 0$

so there are stationary values when $x = 0$ and $x = 1$

$x = 0 \quad \Rightarrow \quad f(x) = -3$

$x = 1 \quad \Rightarrow \quad f(x) = 3 - 8 + 6 - 3 = -2$

i.e. the stationary values of $f(x)$ are -2 and -3.

Differentiating $f'(x)$ w.r.t. x gives

$f''(x) = 36x^2 - 48x + 12 = 12(3x^2 - 4x + 1)$

> $f''(x)$ is the notation for $f'\{f'(x)\}$

When $x = 0$, $f''(x) = 12$ which is positive

$\Rightarrow \quad f(x) = -3$ is a minimum value.

When $x = 1$, $f''(x) = 12(3 - 4 + 1)$ which is zero.

> This is inconclusive so we will look at the signs of $f'(x)$ on either side of $x = 1$.

x	$\frac{1}{2}$	1	$1\frac{1}{2}$
$f'(x)$	+	0	+
Gradient	/	–	/

From this table we see that the stationary value at $x = 1$, i.e. -2, is an inflexion.

3 A is a point of inflexion on the curve whose equation is $3y = x^3 - 6x^2 + 9x + 1$

a Find the coordinates of A and show that it is not a stationary point.

b Find the equation of the tangent at A.

c Sketch the curve and the tangent in the region of A.

a $3y = x^3 - 6x^2 + 9x + 1$

$$\Rightarrow \quad 3\frac{dy}{dx} = 3x^2 - 12x + 9 = 3(x^2 - 4x + 3)$$

$$\Rightarrow \quad \frac{dy}{dx} = x^2 - 4x + 3 \quad \Rightarrow \quad \frac{d^2y}{dx^2} = 2x - 4$$

At the point of inflexion, A, $\frac{d^2y}{dx^2} = 0 \quad \Rightarrow \quad x = 2$

When $x = 2$, $y = 1 \quad \Rightarrow \quad$ A is the point $(2, 1)$

At A, $\frac{dy}{dx} = -1$ i.e. the gradient is -1 so A is not a stationary point.

b The equation of the tangent at A is $y - 1 = -1(x - 2) \quad \Rightarrow \quad y + x = 3$

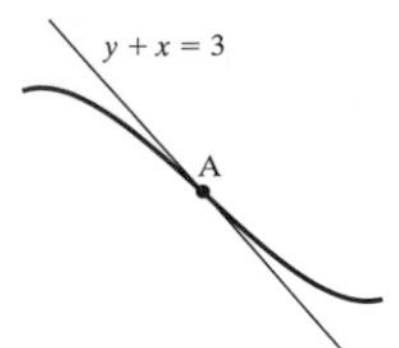

c When $x = 1.5$, $\frac{dy}{dx} < 0$ and $\frac{d^2y}{dx^2} < 0$, i.e. $\frac{dy}{dx}$ is decreasing so, as it is negative, it is becoming more negative.

When $x = 2.5$, $\frac{dy}{dx} < 0$ and $\frac{d^2y}{dx^2} > 0$, i.e. $\frac{dy}{dx}$, which is still negative, is increasing, i.e. becoming less negative.

EXERCISE 14C

Find the stationary points on the following curves and distinguish between them.

1 $y = 2x - x^2$

2 $y = 3x - x^3$

3 $y = \frac{9}{x} + x$

4 $y = x^2(x - 5)$

5 $y = x^2$

6 $y = x + \frac{1}{2x^2}$

7 $y = 2x^2 - x^4$

8 $y = x^4$

9 $y = (2x + 1)(x - 3)$

10 $y = x^5 - 5x$

11 $y = x^2(x^2 - 8)$

12 $y = x^2 + \frac{16}{x^2}$

Find the stationary value(s) of each of the following functions and determine their character.

13 $x + \frac{1}{x}$

14 $3 - x + x^2$

15 $4x^3 - x^4$

16 $8 - x^3$

17 $x^3 + 7$

18 $x^2(3x^2 - 2x - 3)$

19 Show that the curve with equation $y = x^5 + x^3 + 4x - 3$ has no stationary points but does have a point of inflexion. Find the coordinates of this point and the equation of the tangent there.

Applications

EXAMPLES 14D

1 An open box is made from a square sheet of cardboard, with sides half a metre long, by cutting out a square from each corner, folding up the sides and joining the cut edges. Find the maximum capacity of the box.

> The capacity of the box depends on the unknown length of the side of the square cut from each corner so we denote this by x metres. The side of the cardboard sheet is $\frac{1}{2}$m, so we know that $0 < x < \frac{1}{4}$.

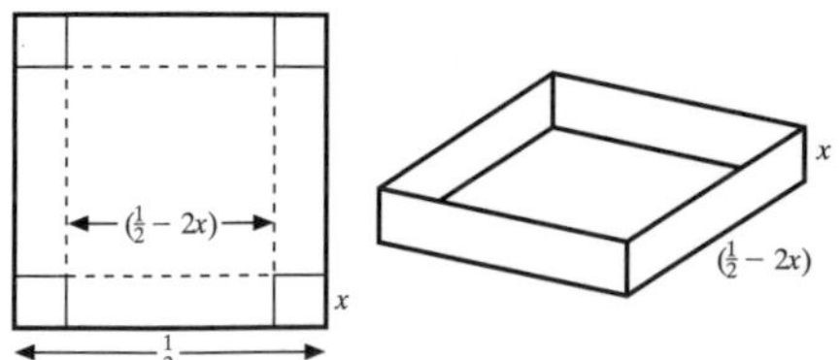

Using metres throughout,

the base of the box is a square of side $(\frac{1}{2} - 2x)$ and the height of the box is x,

$\therefore$ the capacity, C, of the box is given by

$C = x(\frac{1}{2} - 2x)^2 = \frac{1}{4}x - 2x^2 + 4x^3$ for $0 < x < \frac{1}{4}$

$$\Rightarrow \qquad \frac{dC}{dx} = \frac{1}{4} - 4x + 12x^2$$

At a stationary value of C, $\dfrac{dC}{dx} = 0$,

i.e. $\quad 12x^2 - 4x + \frac{1}{4} = 0 \quad \Rightarrow \quad 48x^2 - 16x + 1 = 0$

$\Rightarrow \quad (4x - 1)(12x - 1) = 0 \quad \Rightarrow \quad x = \frac{1}{4}$ or $x = \frac{1}{12}$

There are stationary values of C when $x = \frac{1}{4}$ and when $x = \frac{1}{12}$.

It is obvious that it is not possible to make a box if $x = \frac{1}{4}$ so we need only check that $x = \frac{1}{12}$ gives a maximum capacity.

$\dfrac{d^2C}{dx^2} = -4 + 24x$ which is negative when $x = \frac{1}{12}$.

Therefore C has a maximum value of $\frac{1}{12}(\frac{1}{2} - \frac{1}{6})^2$, i.e. $\frac{1}{108}$,

so the maximum capacity of the box is $\frac{1}{108}\,m^3$

or, correct to 3 s.f., $9260\,cm^3$.

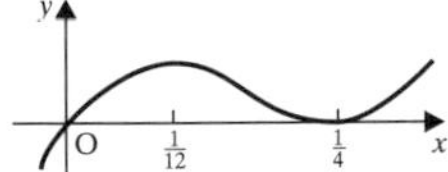

Alternatively the nature of the stationary point where $x = \frac{1}{12}$ can be investigated by using the sketch given by a graphics calculator for the curve $C = \frac{1}{4}x - 2x^2 + 4x^3$ and looking at the section for which $0 < x < \frac{1}{4}$.

The sketch shows that there is a maximum point between $x = 0$ and $x = \frac{1}{4}$ so $x = \frac{1}{12}$ must give the maximum value of C.

The sketch also shows that there is a minimum point *on the curve* where $x = \frac{1}{4}$ but this is *not* a minimum value of the *capacity*, as a box cannot be made if $x = \frac{1}{4}$.

2 The function $ax^2 + bx + c$ has a gradient function $4x + 2$ and a stationary value of 1. Find the values of a, b and c.

$f(x) = ax^2 + bx + c \quad \Rightarrow \quad f'(x) = 2ax + b$

But we know that $f'(x) = 4x + 2$

$\therefore \quad 2ax + b$ is identical to $4x + 2$

i.e. $\quad a = 2$ and $b = 2$

The stationary value of $f(x)$ occurs when $f'(x) = 0$

i.e. when $4x + 2 = 0 \quad \Rightarrow \quad x = -\frac{1}{2}$

the stationary value of $f(x)$ is $2(-\frac{1}{2})^2 + 2(-\frac{1}{2}) + c = -\frac{1}{2} + c$

But the stationary value of $f(x)$ is also 1,

$\therefore \quad -\frac{1}{2} + c = 1 \quad \Rightarrow \quad c = \frac{3}{2}$

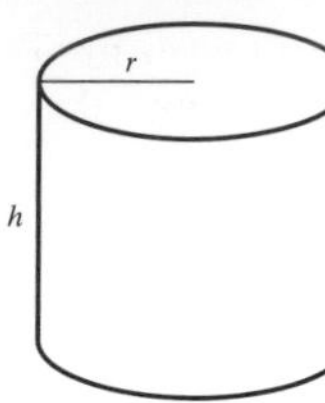

3 A cylinder has a radius r metres and a height h metres. The sum of the radius and height is 2 m. Find an expression for the volume, V cubic metres, of the cylinder in terms of r only. Hence find the maximum volume.

$V = \pi r^2 h$ and $r + h = 2$

$\therefore \quad V = \pi r^2(2 - r) = \pi(2r^2 - r^3)$

Now for maximum volume, $\dfrac{\mathrm{d}V}{\mathrm{d}r} = 0,$

i.e. $\quad \pi(4r - 3r^2) = 0 \quad \Rightarrow \quad \pi r(4 - 3r) = 0$

Therefore there are stationary values of V when $r = 0$ and $r = \frac{4}{3}$.

It is obvious that, when $r = 0$, $V = 0$ and no cylinder exists, so we check the sign of $\dfrac{\mathrm{d}^2V}{\mathrm{d}r^2}$ only for $r = \frac{4}{3}$.

$\dfrac{\mathrm{d}^2V}{\mathrm{d}r^2} = \pi(4 - 6r)$ which is negative when $r = \frac{4}{3}$

Therefore the maximum value of V occurs when $r = \frac{4}{3}$ and is $\pi(\frac{4}{3})^2(2 - \frac{4}{3})$,

i.e. the maximum volume is $\dfrac{32\pi}{27}\,\mathrm{m}^3$.

EXERCISE 14D

1 A farmer has an 80 m length of fencing. He wants to use it to form three sides of a rectangular enclosure against an existing fence which provides the fourth side. Find the maximum area that he can enclose and give its dimensions.

2 A large number of open cardboard boxes are to be made and each box must have a square base and a capacity of 4000 cm^3. Find the dimensions of the box which contains the minimum area of cardboard.

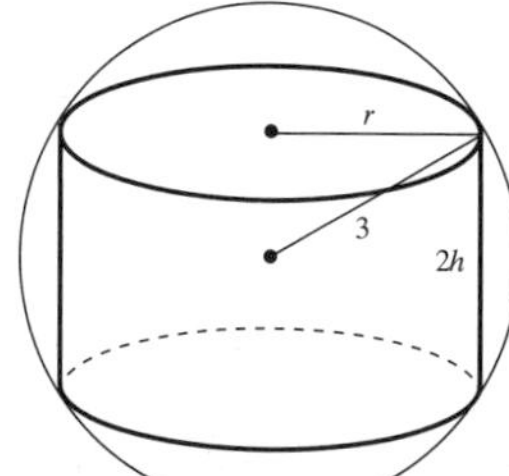

3 The diagram shows a cylinder cut from a solid sphere of radius 3 cm. Given that the cylinder has a height of $2h$, find its radius in terms of h. Hence show that the volume, V cubic metres, of the cylinder is given by

$$V = 2\pi h(9 - h^2)$$

Find the maximum volume of the cylinder as h varies.

4 A variable rectangle has a constant perimeter of 20 cm. Find the lengths of the sides when the area is maximum.

5 A variable rectangle has a constant area of 35 cm^2. Find the lengths of the sides when the perimeter is minimum.

6 The curve $y = ax^2 + bx + c$ crosses the y-axis at the point $(0, 3)$ and has a stationary point at $(1, 2)$. Find the values of a, b and c.

7 The gradient of the tangent to the curve $y = px^2 - qx - r$ at the point $(1, -2)$ is 1. If the curve crosses the x-axis where $x = 2$, find the values of p, q and r. Find the other point of intersection with the x-axis and sketch the curve.

8 y is a quadratic function of x. The line $y = 2x$ is a tangent to the curve at the point $(3, 6)$. The turning point on the curve occurs where $x = -2$. Find the equation of the curve.

MIXED EXERCISE 14

1 Find the gradient of the curve with equation $y = 6x^2 - x$ at the point where $x = 1$. Find the equation of the tangent at this point. Where does this tangent meet the line $y = 2x$?

2 Find the equation of the normal to the curve $y = 1 - x^2$ at the point where the curve crosses the positive x-axis. Find also the coordinates of the point where the normal meets the curve again.

3 Find the coordinates of the points on the curve $y = x^3 + 3x$ where the gradient is 15.

4 Find the equations of the tangents to the curve $y = x^3 - 6x^2 + 12x + 2$ that are parallel to the line $y = 3x$.

5 Find the equation of the normal to the curve $y = x^2 - 6$ which is parallel to the line $x + 2y - 1 = 0$.

6 Locate the turning points on the curve $y = x(x^2 - 12)$, determine their nature and draw a rough sketch of the curve.

7 Find the stationary values of the function $x + \frac{1}{x}$ and sketch the function.

8 If the perimeter of a rectangle is fixed in length, show that the area of the rectangle is greatest when it is square.

9 A door is in the shape of a rectangle surmounted by a semicircle whose diameter is equal to the width of the rectangle. If the perimeter of the door is 7 m, and the radius of the semicircle is r metres, express the height of the rectangle in terms of r. Show that the area of the door has a maximum value when the radius is $\frac{7}{4 + \pi}$.

10 An open tank is constructed with a square base and vertical sides, to hold 32 cubic metres of water. Find the dimensions of the tank if the area of sheet metal used to make it is to have a minimum value.

11 Triangle ABC has a right angle at C. The shape of the triangle can vary but the sides BC and CA have a fixed total length of 10 cm. Find the maximum area of the triangle.

CHAPTER 15 Trigonometric functions

The Trigonometric Functions

The General Definition of an Angle

The angles that were used in Chapter 7 were limited to the range 0 to 180°, i.e. only those angles that can be found in a triangle. In order to be able to work with angles of unlimited size we need a broader definition and we now define an angle as a measure of rotation.

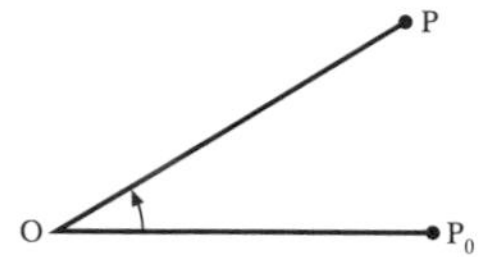

Consider a line which can rotate from its initial position OP_0 about the point O to any other position OP.

The amount of rotation is indicated by the angle between OP_0 and OP, i.e.

an angle is a measure of the rotation of a line about a fixed point.

The anticlockwise sense of rotation is taken as positive and clockwise rotation is negative. It follows that an angle formed by the anticlockwise rotation of OP is a positive angle.

The rotation of OP is not limited to one revolution, so an angle can be as big as we choose to make it.

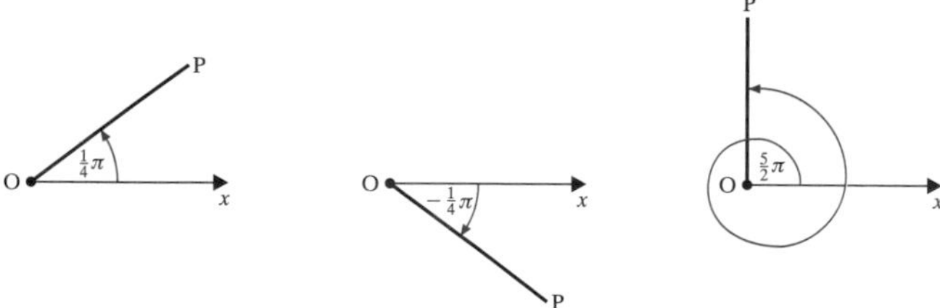

If θ is any angle, then θ can be measured either in degrees or in radians and in either case θ can take all real values.

The Trig Functions

Since angles, and therefore the trig ratios of angles, are no longer restricted to triangles we also need general definitions for the sine, cosine and the tangent of an angle that are valid for angles of all values.

If OP is drawn on x- and y-axes as shown
and if, for all values of θ, the length of OP is r and the coordinates of P are (x,y), then the sine, cosine and tangent functions are defined as follows.

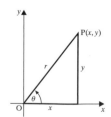

$$\sin\theta = \frac{y}{r}$$

$$\cos\theta = \frac{x}{r}$$

$$\tan\theta = \frac{y}{x}$$

The Trig Ratios of 30°, 45°, 60°

The sine, cosine and tangent of 30°, 45°, and 60°, can be expressed exactly in surd form and are worth remembering.

This triangle shows that

$$\sin 45^\circ = \frac{1}{\sqrt{2}}$$

$$\cos 45^\circ = \frac{1}{\sqrt{2}}$$

$$\tan 45^\circ = 1$$

And this triangle gives

$$\sin 60^\circ = \frac{\sqrt{3}}{2}, \quad \sin 30^\circ = \frac{1}{2}$$

$$\cos 60^\circ = \frac{1}{2}, \quad \cos 30^\circ = \frac{\sqrt{3}}{2}$$

$$\tan 60^\circ = \sqrt{3}, \quad \tan 30^\circ = \frac{1}{\sqrt{3}}$$

The Sine Function

From the definition $f(\theta) = \sin\theta$, and measuring θ in radians, we can see that:

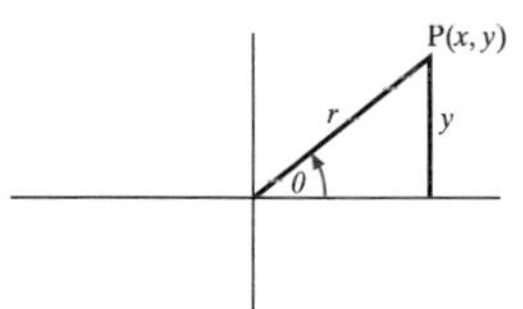

For $0 \leqslant \theta \leqslant \frac{1}{2}\pi$, OP is in the first quadrant;
y is positive and increases in value from 0 to r as θ increases from 0 to $\frac{1}{2}\pi$.
Now r is always positive, so $\sin\theta$ increases from 0 to 1.

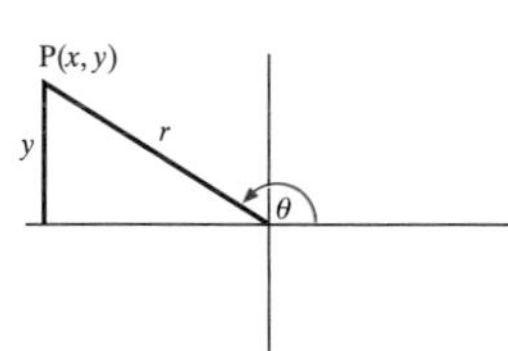

For $\frac{1}{2}\pi \leqslant \theta \leqslant \pi$, OP is in the second quadrant;
again y is positive but decreases in value from r to 0,
so $\sin\theta$ decreases from 1 to 0.

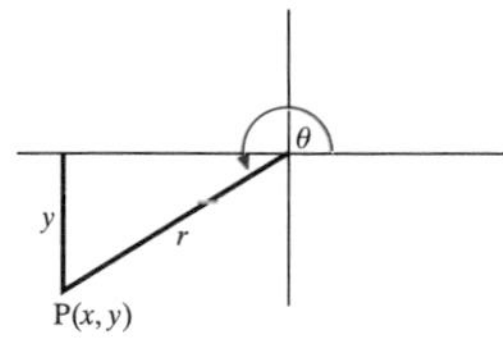

For $\pi \leqslant \theta \leqslant \frac{3}{2}\pi$, OP is in the third quadrant;
y is negative and decreases from 0 to $-r$,
so $\sin\theta$ decreases from 0 to -1.

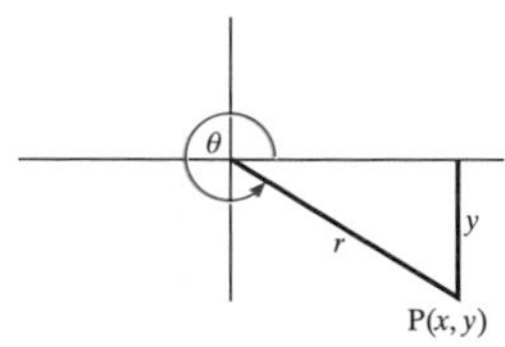

For $\frac{3}{2}\pi \leqslant \theta \leqslant 2\pi$, OP is in the fourth quadrant;
y is still negative but increases from $-r$ to 0,
so $\sin\theta$ increases from -1 to 0.

For $\theta \geqslant 2\pi$, the cycle repeats itself as OP travels round the quadrants again. For negative values of θ, OP rotates clockwise round the quadrants in the order 4th, 3rd, 2nd, 1st, etc. So $\sin\theta$ decreases from 0 to -1, then increases to 0 and on to 1 before decreasing to zero and repeating the pattern.

This shows that $\sin\theta$ is positive for $0 < \theta < \pi$ and negative when $\pi < \theta < 2\pi$.

Further, $\sin\theta$ varies in value between -1 and 1 and the pattern repeats itself every revolution.

A plot of the graph of $f(\theta) = \sin\theta$ confirms these observations.

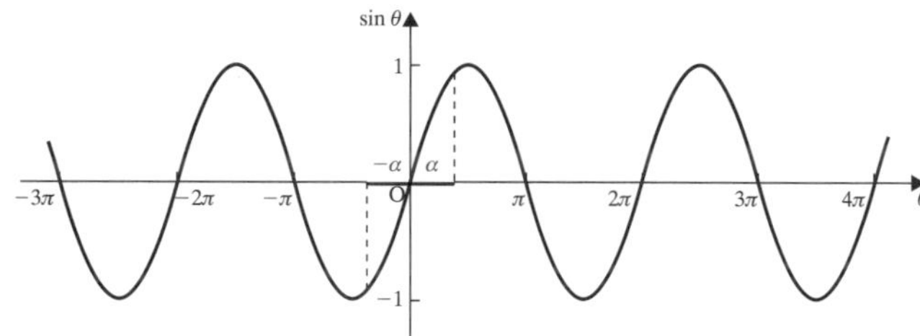

A graph of this shape is called, for obvious reasons, a *sine wave* and shows clearly the following characteristics of the sine function.

The curve is continuous (i.e. it has no breaks).

$-1 \leqslant \sin\theta \leqslant 1$

The shape of the curve from $\theta = 0$ to $\theta = 2\pi$ is repeated for each complete revolution. Any function with a repetitive pattern is called *periodic* or *cyclic*. The width of the repeating pattern, as measured on the horizontal scale, is called the *period*.

The period of the sine function is 2π.

Other properties of the sine function shown by the graph are as follows.

$$\sin\theta = 0 \text{ when } \theta = n\pi \text{ where } n \text{ is an integer.}$$

The curve has rotational symmetry about the origin so, for any angle α

$$\sin(-\alpha) = -\sin\alpha, \text{ e.g. } \sin(-30^\circ) = -\sin 30^\circ = -\tfrac{1}{2}$$

An enlarged section of the graph for $0 \leqslant \theta \leqslant 2\pi$, shows further relationships.

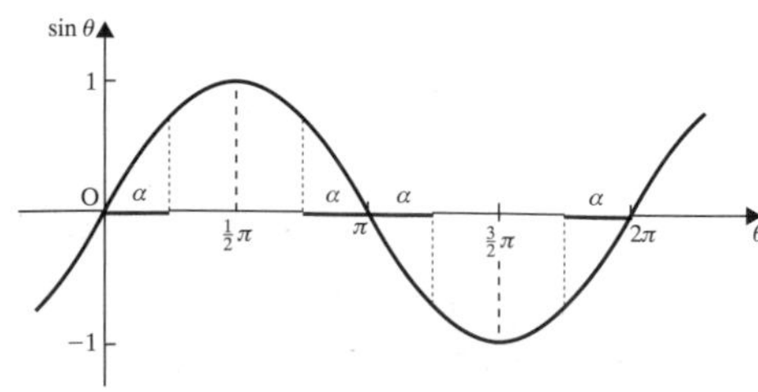

The curve is symmetrical about the line $\theta = \frac{1}{2}\pi$, so

$\sin(\pi - \alpha) = \sin\alpha$,

e.g. $\sin 130^\circ = \sin(180^\circ - 130^\circ) = \sin 50^\circ$

The curve has rotational symmetry about $\theta = \pi$,

so $\sin(\pi + \alpha) = -\sin\alpha$

and $\sin(2\pi - \alpha) = -\sin\alpha$

EXAMPLES 15A

1 Find the exact value of $\sin\frac{4}{3}\pi$.

$$\sin\tfrac{4}{3}\pi = \sin(\pi + \tfrac{1}{3}\pi) = -\sin\tfrac{1}{3}\pi = -\frac{\sqrt{3}}{2}$$

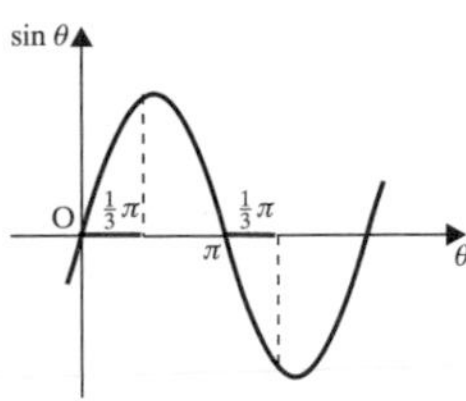

2 Sketch the graph of $y = \sin(\theta - \frac{1}{4}\pi)$ for values of θ between 0 and 2π.

> Remember that the curve $y = f(x - a)$ is a translation of the curve $y = f(x)$ by a units in the positive direction of the x-axis.

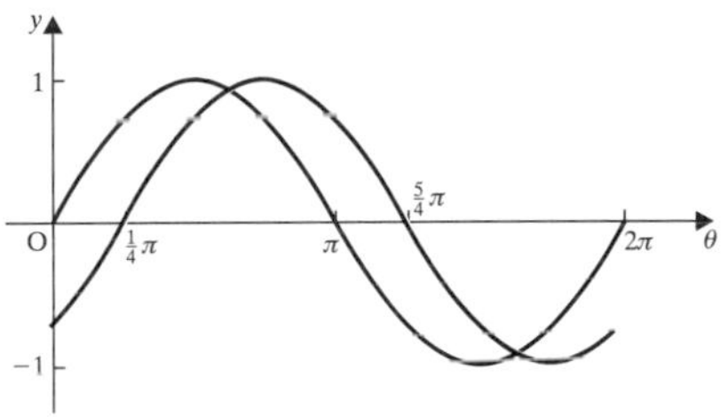

EXERCISE 15A

Find the exact value of

1 $\sin 120°$ **2** $\sin -2\pi$ **3** $\sin 300°$ **4** $\sin -210°$

5 Write down all the values of θ between 0 and 6π for which $\sin\theta = 1$.

6 Write down all the values of θ between 0 and -4π for which $\sin\theta = -1$.

Express in terms of the sine of an acute angle

7 $\sin 125°$ **8** $\sin 290°$ **9** $\sin -120°$ **10** $\sin\frac{7}{6}\pi$

Sketch each of the following curves for values of θ in the range $0 \leqslant \theta \leqslant 3\pi$.

11 $y = \sin(\theta + \frac{1}{3}\pi)$

12 $y = -\sin\theta$

13 $y = \sin(-\theta)$

14 $y = 1 - \sin\theta$

15 $y = \sin(\pi - \theta)$

16 $y = \sin(\frac{1}{2}\pi - \theta)$

Use a graphics calculator or computer for Questions 17 to 19 and set the range for θ as -2π to 4π.

17 On the same set of axes draw the graphs of $y = \sin\theta, y = 2\sin\theta,$ and $y = 3\sin\theta$. What can you deduce about the relationship between the curves $y = \sin\theta$ and $y = a\sin\theta$?

18 On the same set of axes draw the curves $y = \sin\theta$ and $y = \sin 2\theta$.

19 On the same set of axes draw the curves $y = \sin\theta$ and $y = \sin 3\theta$. What can you deduce about the relationship between the two curves?

20 *Sketch* the curves

a $y = \sin 4\theta$ **b** $y = 4\sin\theta$

One-way Stretches

Questions 17 to 20 in the last exercise show examples of one-way stretches. For example, the curve $y = 2\sin\theta$ is a one-way stretch of the curve $y = \sin\theta$ by a factor 2 parallel to the y-axis.

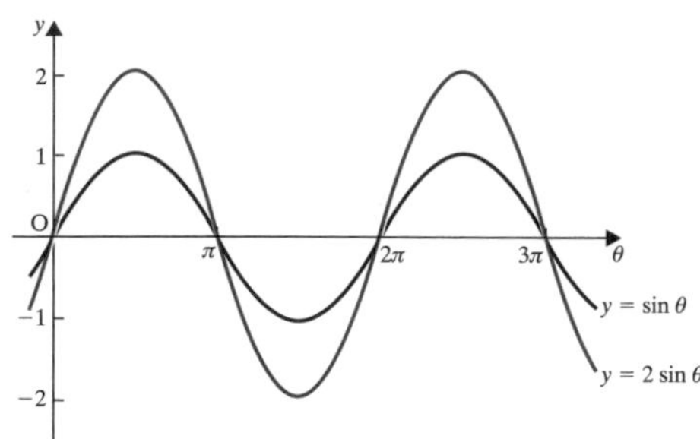

In general, if we compare points with the same x-coordinate on the curves $y = \mathrm{f}(x)$ and $y = a\mathrm{f}(x)$, then the y-coordinate of the point on $y = a\mathrm{f}(x)$ is a times the y-coordinate of the point on $y = \mathrm{f}(x)$. Therefore

the curve $y = a\mathrm{f}(x)$ is a one-way stretch of the curve $y = \mathrm{f}(x)$ by a factor a parallel to the y-axis.

Also, the curve $y = \sin 2\theta$ is a one-way stretch of the curve $y = \sin\theta$ by a factor $\frac{1}{2}$ parallel to the x-axis (or a one-way reduction by a factor 2).

Now consider points with the same y-coordinate on the curves $y = \mathrm{f}(x)$ and $y = \mathrm{f}(ax)$. The x-coordinate on $y = \mathrm{f}(ax)$ must be $\dfrac{1}{a}$ times the x-coordinate on $y = \mathrm{f}(x)$. Therefore, in general,

the curve $y = \mathrm{f}(ax)$ is a one-way stretch of the curve $y = \mathrm{f}(x)$ by a factor $\dfrac{1}{a}$ parallel to the x-axis.

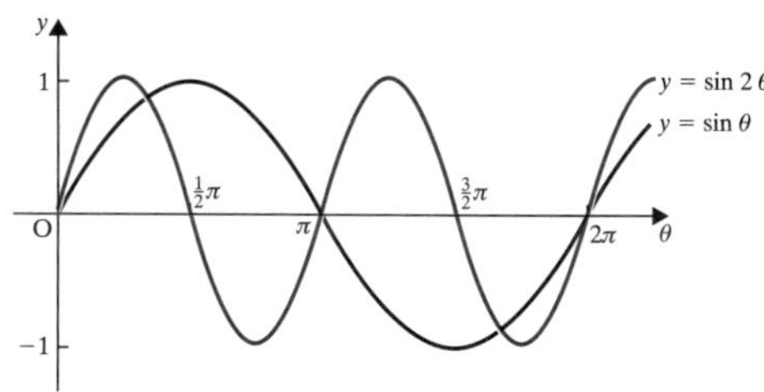

The Cosine Function

For any position of P, $\cos\theta = \dfrac{x}{r}$

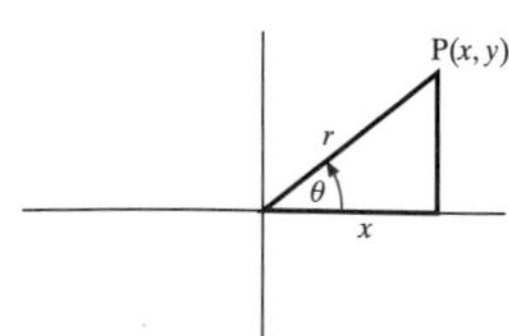

When P is in the first quadrant, x decreases from r to 0 as θ increases, so $\cos\theta$ decreases from 1 to 0.

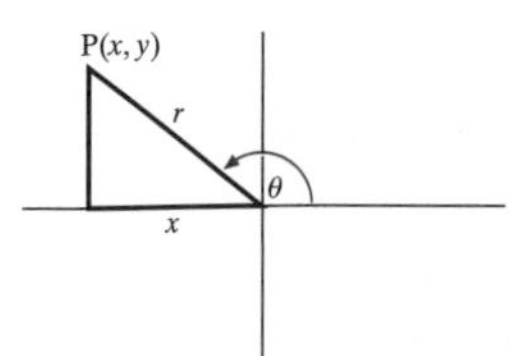

When P is in the second quadrant, x decreases from 0 to $-r$, so $\cos\theta$ decreases from 0 to -1.

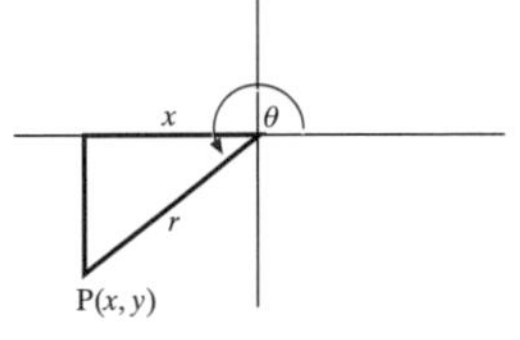

Similarly, when P is in the third quadrant, $\cos\theta$ increases from -1 to 0,

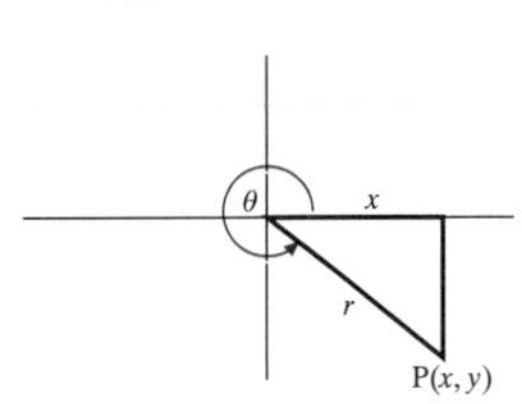

and when P is in the fourth quadrant, $\cos\theta$ increases from 0 to 1.

The cycle then repeats itself, and we get this graph of $f(\theta) = \cos\theta$.

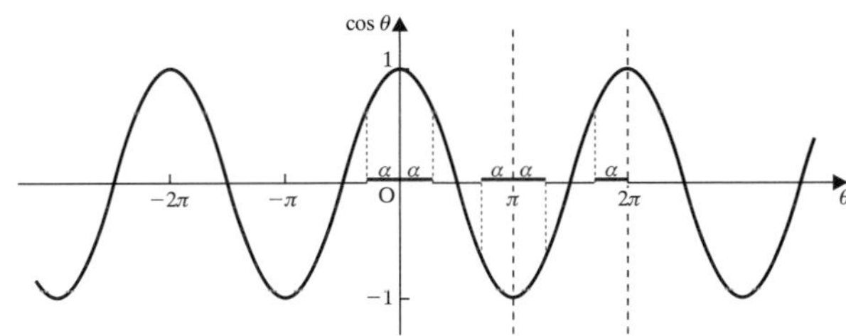

The characteristics of this graph are as follows.

The curve is continuous.
$-1 \leqslant \cos\theta \leqslant 1$
It is periodic with a period of 2π.

It is the same shape as the sine wave but is translated a distance $\frac{1}{2}\pi$ to the left. Such a translation of a sine wave is called a *phase shift*.

$$\cos\theta = 0 \text{ when } \theta = \ldots -\tfrac{1}{2}\pi, \tfrac{1}{2}\pi, \tfrac{3}{2}\pi, \tfrac{5}{2}\pi, \ldots$$

The curve is symmetric about $\theta = 0$, so $\cos -\alpha = \cos\alpha$

The curve has rotational symmetry about $\theta = \frac{1}{2}\pi$, so

$$\cos(\pi - \alpha) = -\cos\alpha$$

Further considerations of symmetry show that

$$\cos(\pi + \alpha) = -\cos\alpha \text{ and } \cos(2\pi - \alpha) = \cos\alpha$$

EXERCISE 15B

1 Write in terms of the cosine of an acute angle

a $\cos 123°$ **b** $\cos 250°$ **c** $\cos(-20°)$ **d** $\cos(-154°)$

2 Find the exact value of

a $\cos 150°$ **b** $\cos\frac{3}{2}\pi$ **c** $\cos\frac{5}{4}\pi$ **d** $\cos 6\pi$

3 *Sketch* each of the following curves.

a $y = \cos(\theta + \pi)$ **b** $y = \cos(\theta - \frac{1}{3}\pi)$ **c** $y = \cos(-\theta)$

4 Sketch the graph of $y = \cos(\theta - \frac{1}{2}\pi)$. What relationship does this suggest between $\sin\theta$ and $\cos(\theta - \frac{1}{2}\pi)$? Is there a similar relationship between $\cos\theta$ and $\sin(\theta - \frac{1}{2}\pi)$?

5 Sketch the curve $y = \cos(\theta - \frac{1}{4}\pi)$ for values of θ between $-\pi$ and π. Use the sketch to find the values of θ in this range for which

a $\cos(\theta - \frac{1}{4}\pi) = 1$ **b** $\cos(\theta - \frac{1}{4}\pi) = -1$ **c** $\cos(\theta - \frac{1}{4}\pi) = 0$

6 On the same set of axes, sketch the graphs $y = \cos\theta$ and $y = 3\cos\theta$.

7 On the same set of axes, sketch the graphs $y = \cos\theta$ and $y = \cos 3\theta$.

8 Sketch the graphs of $f(\theta) = \cos 4\theta$ for $0 \leqslant \theta \leqslant \pi$. Hence find the values of θ in this range for which $f(\theta) = 0$.

The Tangent Function

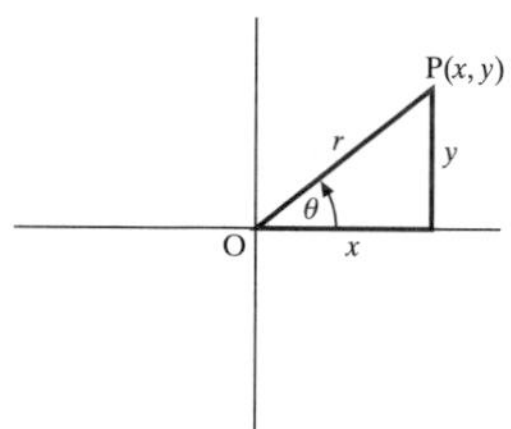

For any position of P, $\tan\theta = \dfrac{y}{x}$.

As OP rotates through the first quadrant, x decreases from r to 0, while y increases from 0 to r. This means that the fraction y/x increases from 0 to very large values indeed. In fact, as $\theta \to \frac{1}{2}\pi$, $\tan\theta \to \infty$.

Similar analysis of the behaviour of y/x in the other quadrants shows that in the second quadrant, $\tan\theta$ is negative and increases from $-\infty$ to 0, in the third quadrant, $\tan\theta$ is positive and increases from 0 to ∞, and in the fourth quadrant, $\tan\theta$ is negative and increases from $-\infty$ to 0. The cycle then repeats itself and we can draw the graph of $f(\theta) = \tan\theta$.

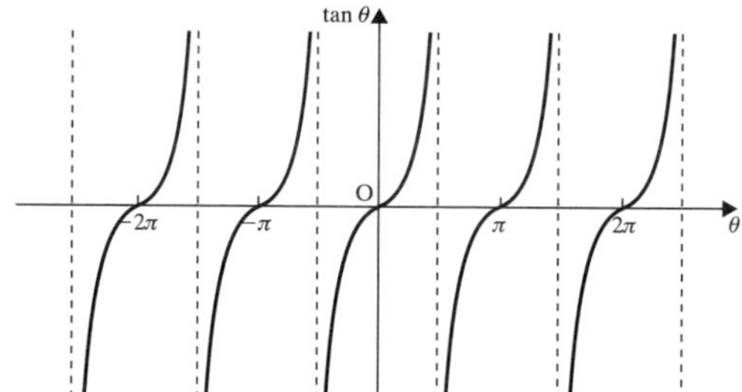

From the graph we can see that the characteristics of the tangent function are different from those of the sine and cosine functions in several respects.

It is not continuous, being *undefined* when $\theta = \ldots -\frac{1}{2}\pi, \frac{1}{2}\pi, \frac{3}{2}\pi, \ldots$
The range of values of $\tan\theta$ is unlimited.
It is periodic with a period of π (not 2π as in the other cases).

The graph has rotational symmetry about $\theta = 0$, so

$$\tan(-\alpha) = -\tan\alpha$$

The graph has rotational symmetry about $\theta = \frac{1}{2}\pi$, giving

$$\tan(\pi - \alpha) = -\tan\alpha$$

As the cycle repeats itself from $\theta = \pi$ to 2π, we have

$$\tan(\pi + \alpha) = \tan\alpha \quad \text{and} \quad \tan(2\pi - \alpha) = -\tan\alpha$$

EXAMPLE 15C Express $\tan \frac{11}{4}\pi$ as the tangent of an acute angle.

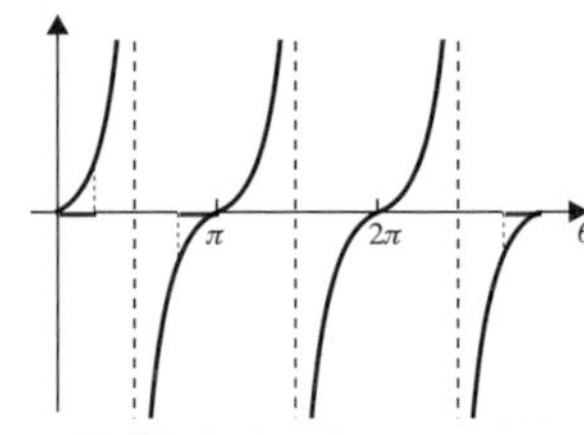

$$\tan\left(\tfrac{11}{4}\pi\right) = \tan\left(2\pi + \tfrac{3}{4}\pi\right) = \tan\left(\tfrac{3}{4}\pi\right)$$
$$= \tan\left(\pi - \tfrac{1}{4}\pi\right)$$
$$= -\tan \tfrac{1}{4}\pi$$

EXERCISE 15C

1 Find the exact value of

a $\tan \frac{9}{4}\pi$ **b** $\tan 120°$ **c** $\tan -\frac{2}{3}\pi$ **d** $\tan \frac{7}{4}\pi$

2 Write in terms of the tangent of an acute angle

a $\tan 220°$ **b** $\tan \frac{12}{7}\pi$ **c** $\tan 310°$ **d** $\tan -\frac{7}{5}\pi$

3 Sketch the graph of $y = \tan\theta$ for values of θ in the range 0 to 2π. From this sketch find the values of θ in this range for which

a $\tan\theta = 1$ **b** $\tan\theta = -1$ **c** $\tan\theta = 0$ **d** $\tan\theta = \infty$

4 Using the basic definitions of $\sin\theta$, $\cos\theta$ and $\tan\theta$, show that, for all values of θ,

$$\tan\theta = \frac{\sin\theta}{\cos\theta}$$

Relationships Between sin θ, cos θ and tan θ

Because each trig function is a ratio of two of the three quantities x, y and r, we would expect to find several relationships between $\sin\theta$, $\cos\theta$ and $\tan\theta$. Most of these relationships will be investigated in later chapters, but here is a summary of the results from the various exercises so far.

If the graph of $\cos\theta$ is shifted by $\frac{1}{2}\pi$ to the right we get the graph of $\sin\theta$.
So $\cos\left(\theta - \frac{1}{2}\pi\right) = \sin\theta$

But $\cos\left(\theta - \frac{1}{2}\pi\right) = \cos\left(\frac{1}{2}\pi - \theta\right)$

Therefore $\cos\left(\frac{1}{2}\pi - \theta\right) = \sin\theta$

Two angles which add up to $\frac{1}{2}\pi$ (90°) are called *complementary* angles.

i.e. **the sine of an angle is equal to the cosine of the complementary angle and vice-versa.**

Now $\sin\theta = \dfrac{y}{r}$, $\cos\theta = \dfrac{x}{r}$ and $\tan\theta = \dfrac{y}{x}$

$$\therefore \quad \frac{\sin\theta}{\cos\theta} = \frac{y/r}{x/r} = \frac{y}{x} = \tan\theta$$

i.e. for all values of θ. $\boldsymbol{\tan\theta \equiv \dfrac{\sin\theta}{\cos\theta}}$

We have also seen that the sign of each trig ratio depends on the size of the angle, i.e. the quadrant in which P is.
We can summarise the sign of each ratio in a quadrant diagram.

sin + ve	All + ve
tan + ve	cos + ve

EXAMPLES 15D

1 Give all the values of x between 0 and 360° for which $\sin x = -0.3$.

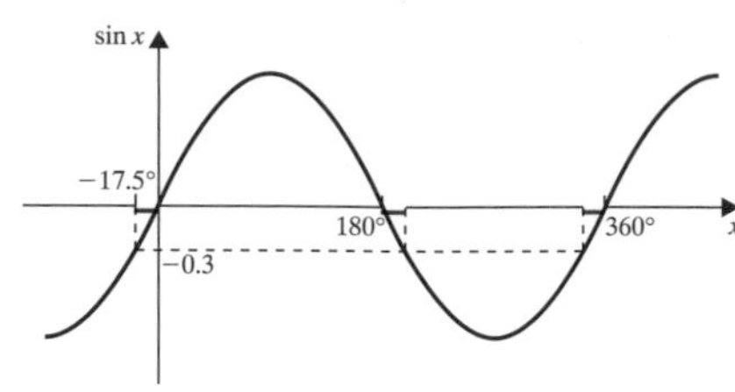

> The value given for x by a calculator is $-17.5°$.
> From the graph, we see that, when $\sin x = -0.3$, the values of x in the specified range are $180° + 17.5°$ and $360° - 17.5°$.

When $\sin x = -0.3$, $x = 197.5°$ and $342.5°$.

> Note that when the range of values is given in degrees, the answer should also be given in degrees and the same applies for radians.

2 Find the smallest positive value of θ for which $\cos\theta = 0.7$ and $\tan\theta$ is negative.

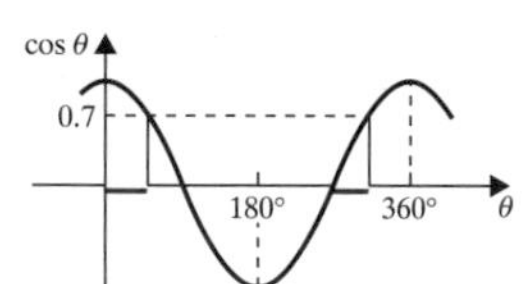

If $\cos\theta = 0.7$, the possible values of θ are 45.6°, 314.4°, ...

Now $\tan\theta$ is positive if θ is in the first quadrant and negative if θ is in the fourth quadrant.

cos +, tan + (first quadrant); cos +, tan − (fourth quadrant)

Therefore the required value of θ is 314.4°.

EXERCISE 15D

1 Within the range $-2\pi \leqslant \theta \leqslant 2\pi$, give all the values of θ for which

a $\sin\theta = 0.4$ **b** $\cos\theta = -0.5$

c $\tan\theta = 1.2$

2 Within the range $0 \leqslant \theta \leqslant 720°$, give all the values of θ for which

a $\tan\theta = -0.8$ **b** $\sin\theta = -0.2$

c $\cos\theta = 0.1$

3 Find the smallest angle (positive or negative) for which

a $\cos\theta = 0.8$ and $\sin\theta \geqslant 0$

b $\sin\theta = -0.6$ and $\tan\theta \leqslant 0$

c $\tan\theta = \sin\frac{1}{6}\pi$.

4 Using $\tan\theta \equiv \dfrac{\sin\theta}{\cos\theta}$, show that the equation $\tan\theta = \sin\theta$ can be written as $\sin\theta(\cos\theta - 1) = 0$, provided that $\cos\theta \neq 0$.
Hence find the values of θ between 0 and 2π for which $\tan\theta = \sin\theta$.

5 Sketch the graph of $y = \sin 2\theta$. Use your sketch to help find the values of θ in the range $0 \leqslant \theta \leqslant 360°$ for which $\sin 2\theta = 0.4$.

6 Sketch the graph of $y = \cos 3\theta$. Hence find the values of θ in the range $0 \leqslant \theta \leqslant 2\pi$ for which $\cos 3\theta = -1$.

The Reciprocal Trigonometric Functions

The reciprocals of the three main trig functions have their own names and are sometimes referred to as the *minor* trig ratios.

$$\frac{1}{\sin\theta} \equiv \textbf{cosec}\,\theta, \quad \frac{1}{\cos\theta} \equiv \textbf{sec}\,\theta, \quad \frac{1}{\tan\theta} \equiv \textbf{cot}\,\theta$$

The names given above are abbreviations of cosecant, secant and cotangent respectively.

The graph of $f(\theta) = \operatorname{cosec}\theta$ is given below.

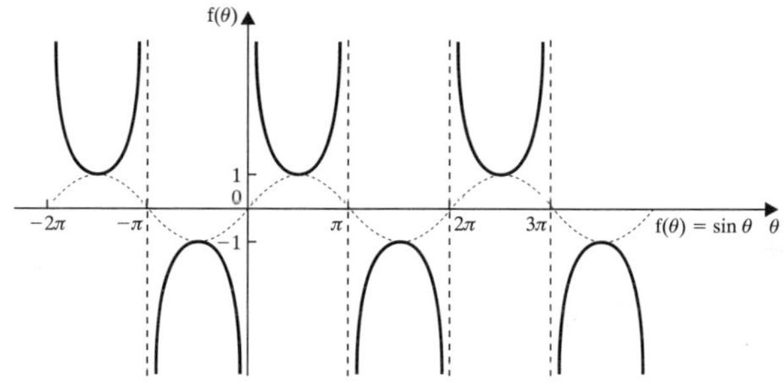

From this graph we can see that

> the cosec function is not continuous, being undefined when θ is any integral multiple of π (we would expect this because these are values of θ where $\sin\theta = 0$ and $\frac{1}{0}$ is undefined).

The pattern of the graph $f(\theta) = \sec\theta$ is similar to that of the cosec graph, as we would expect.

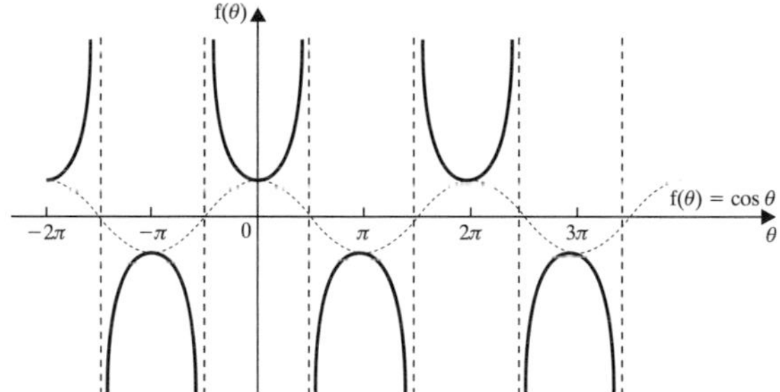

The graph of $f(\theta) = \cot\theta$ is given below.

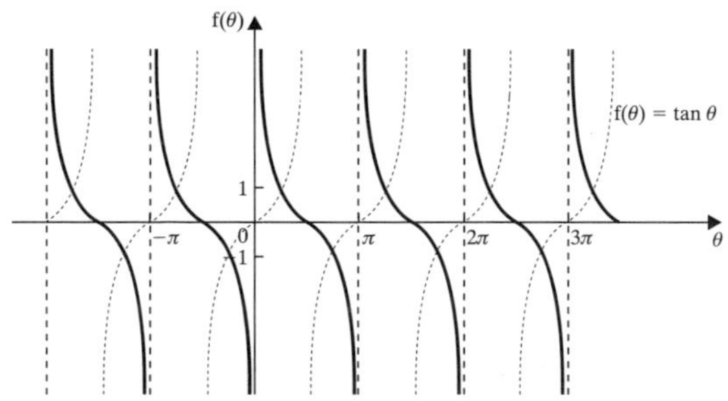

In this right-angled triangle,

$$\tan\alpha = \frac{a}{b} \quad \text{and} \quad \cot\beta = \frac{a}{b} \quad \left(\cot\beta = \frac{1}{\tan\beta}\right)$$

Now $\alpha + \beta = 90°$, i.e. α and β are complementary angles. Hence the cotangent of an angle is equal to the tangent of its complement.

In fact, for *any* angle θ, $\mathbf{\cot\theta \equiv \tan(\tfrac{1}{2}\pi - \theta)}$

This can be seen from the graph.

Reflecting the curve $y = \tan\theta$ in the vertical axis gives $y = \tan(-\theta)$.

Then translating this curve $\frac{1}{2}\pi$ to the left gives $y = \tan(\frac{1}{2}\pi - \theta)$, which is the curve $y = \cot\theta$.

EXAMPLE 15E

For $0 \leqslant \theta \leqslant 360°$, find the values of θ for which $\operatorname{cosec}\theta = -8$.

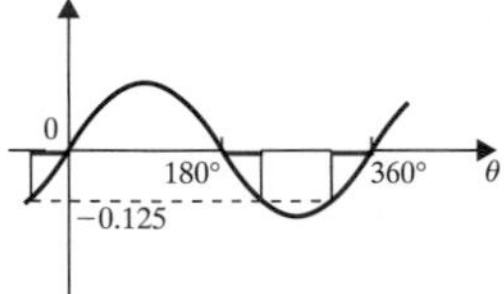

$$\sin\theta = \frac{1}{\operatorname{cosec}\theta} = -\frac{1}{8} = -0.125$$

$\therefore$ from a calculator $\theta = -7.2°$

From the sketch, the required values of θ are $187.2°$ and $352.8°$.

EXERCISE 15E

1 Find, in the range $0 \leqslant \theta \leqslant 360°$, the values of θ for which

a $\sec\theta = 2$ **b** $\cot\theta = 0.6$ **c** $\operatorname{cosec}\theta = 1.5$

2 In the range $-180° \leqslant \theta \leqslant 180°$ find the values of θ for which

a $\cot\theta = 1.2$ **b** $\sec\theta = -1.5$ **c** $\operatorname{cosec}\theta = -2$

3 Given that $\tan\theta \equiv \dfrac{\sin\theta}{\cos\theta}$, write $\cot\theta$ in terms of $\sin\theta$ and $\cos\theta$.
Hence show that $\cot\theta - \cos\theta = 0$ can be written in the form $\cos\theta(1 - \sin\theta) = 0$, provided that $\sin\theta \neq 0$.
Thus find the values in the range $-\pi \leqslant \theta \leqslant \pi$ for which $\cot\theta - \cos\theta = 0$.

4 Find, in surd form, the values of

a $\cot\frac{1}{4}\pi$ **b** $\sec\frac{5}{4}\pi$ **c** $\operatorname{cosec}\frac{11}{6}\pi$

5 Sketch the graph of $f(\theta) = \sec(\theta - \frac{1}{4}\pi)$ for $0 \leqslant \theta \leqslant 2\pi$ and give the values of θ for which $f(\theta) = 1$.

6 Sketch the graph of $f(\theta) = \cot(\theta + \frac{1}{3}\pi)$ for $-\pi \leqslant \theta \leqslant \pi$.
Hence give the values of θ in this range for which $f(\theta) = 1$.

Graphical Solutions of Trig Equations

If you have worked through all the exercises in this chapter, you will already have solved several trig equations with the help of sketch graphs. In this section we are going to look at more complicated equations which require accurate plots of the graphs to solve them.

Consider the equation $\theta = 3\sin\theta$ where θ is measured in radians.

The values of θ for which $\theta = 3\sin\theta$ can be found by plotting the graphs of $y = \theta$ and $y = 3\sin\theta$ on the same axes and hence finding the values of θ at points of intersection.

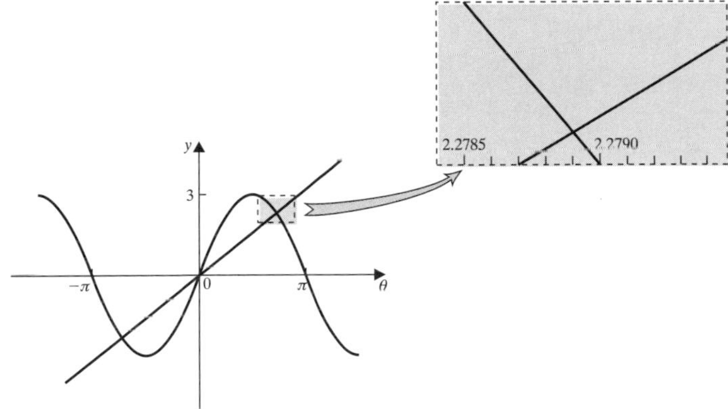

From the enlarged section of the graph $\theta = 2.2789$ rad

Therefore the three points of intersection occur where

$$\theta = -2.2789\text{ rad}, 0, 2.2789\text{ rad} \quad \text{correct to 4 d.p.}$$

If these graphs are produced on a graphics calculator or on a computer using suitable software, then it is possible to zoom in on the points of intersection and get very accurate values for θ. If the graphs are hand drawn, the accuracy of the results will depend on the patience and accuracy of the drawer! Plotting accurate graphs manually is tedious, so if you do not have either of the tools mentioned above, try just Question 1 in the following exercise and do not attempt to get answers correct to more than 2 s.f.

EXERCISE 15F

1 Plot the graphs of $y = \theta$ and $y = 2\cos\theta$ for values of θ in the range $-\pi \leqslant \theta \leqslant \pi$. Hence find the values of θ for which $\theta = 2\cos\theta$.

Repeat this question using *sketch* graphs and measuring θ in degrees. If this was plotted accurately, would it give the same solutions as when the angle is measured in radians?

2 Measuring the angle in radians throughout, find graphically the values of θ for which

a $2\theta = 4\sin\theta$ **b** $\sin\theta = \theta^2$ **c** $\cos\theta = \theta - 1$

CHAPTER 16 Trigonometry 1

Identities

In this chapter we concentrate on some trigonometric identities and some of their uses. Remember that an identity is the equivalence between two different forms of the same expression. One such identity was introduced in Chapter 15, namely

$$\tan\theta \equiv \frac{\sin\theta}{\cos\theta}$$

The Pythagorean Identities

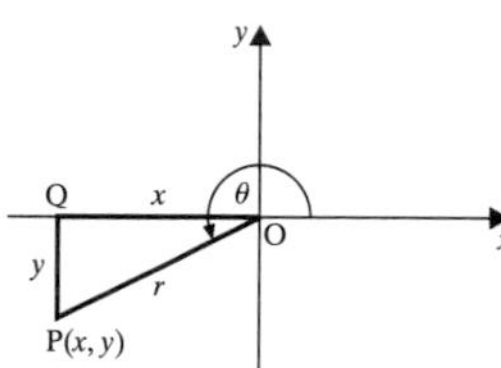

For any angle θ,

$\sin\theta = \dfrac{y}{r}$, $\cos\theta = \dfrac{x}{r}$ and $\tan\theta = \dfrac{y}{x}$.

Also, in right-angled triangle OPQ, $x^2 + y^2 = r^2$ (Pythagoras)

Therefore, $(\cos\theta)^2 + (\sin\theta)^2 = \left(\dfrac{x}{r}\right)^2 + \left(\dfrac{y}{r}\right)^2 = \dfrac{x^2+y^2}{r^2} = 1$

Using the notation $\cos^2\theta$ to mean $(\cos\theta)^2$, etc., we have

$$\cos^2\theta + \sin^2\theta \equiv 1 \qquad [1]$$

Using the identity $\tan\theta \equiv \dfrac{\sin\theta}{\cos\theta}$ we can write [1] in two other forms.

$[1] \div \cos^2\theta \quad \Rightarrow \quad 1 + \dfrac{\sin^2\theta}{\cos^2\theta} \equiv \dfrac{1}{\cos^2\theta}$

$\Rightarrow$ $\mathbf{1 + \tan^2\theta \equiv \sec^2\theta}$

$[1] \div \sin^2\theta \quad \Rightarrow \quad \dfrac{\cos^2\theta}{\sin^2\theta} + 1 \equiv \dfrac{1}{\sin^2\theta}$

$\Rightarrow$ $\mathbf{\cot^2\theta + 1 \equiv \operatorname{cosec}^2\theta}$

These identities can be used to

simplify trig expressions,
eliminate trig terms from pairs of equations,
derive a variety of further trig relationships,
calculate other trig ratios of any angle for which one trig ratio is known.

These identities are also very useful in the solution of certain types of trig equation and we will look at this application later in this chapter.

EXAMPLES 16A

1 Simplify $\dfrac{\sin\theta}{1+\cot^2\theta}$.

$$\frac{\sin\theta}{1+\cot^2\theta} \equiv \frac{\sin\theta}{\operatorname{cosec}^2\theta}$$

$$\equiv \sin^3\theta$$

Using $1+\cot^2\theta \equiv \operatorname{cosec}^2\theta$ and $\operatorname{cosec}\theta \equiv \dfrac{1}{\sin\theta}$

2 Eliminate θ from the equations $x = 2\cos\theta$ and $y = 3\sin\theta$.

$\cos\theta = \dfrac{x}{2}$ and $\sin\theta = \dfrac{y}{3}$

Using $\cos^2\theta + \sin^2\theta \equiv 1$ gives

$$\left(\frac{x}{2}\right)^2 + \left(\frac{y}{3}\right)^2 = 1$$

$$\Rightarrow \quad 9x^2 + 4y^2 = 36$$

In Example 2, both x and y initially depend on θ, a variable angle. Used in this way, θ is called a *parameter*, and is a type of variable that plays an important part in the analysis of curves and functions.

3 If $\sin A = -\frac{1}{3}$ and A is in the third quadrant, find $\cos A$ without using a calculator.

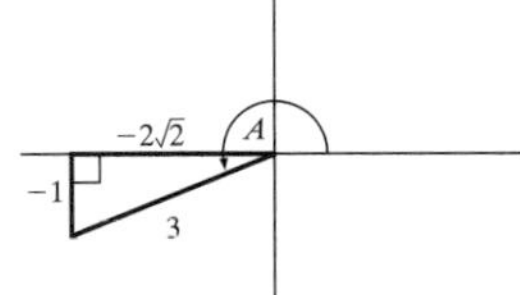

There are two ways of doing this problem. The first method involves drawing a quadrant diagram and working out the remaining side of the triangle, using Pythagoras theorem.

From the diagram, $x = -2\sqrt{2}$

$$\therefore \quad \cos A = \frac{x}{r} = -\frac{2\sqrt{2}}{3}$$

The second method uses the identity $\cos^2 A + \sin^2 A \equiv 1$ giving

$$\cos^2 A + \frac{1}{9} = 1$$

$$\Rightarrow \quad \cos A = \pm\sqrt{\tfrac{8}{9}} = \pm\tfrac{2\sqrt{2}}{3}$$

As A is between π and $\frac{3}{2}\pi$, $\cos A$ is negative, i.e.

$$\cos A = -\frac{2\sqrt{2}}{3}$$

4 Prove that $(1 - \cos A)(1 + \sec A) \equiv \sin A \tan A$

> Because the relationship has yet to be proved, we must not assume its truth by using the complete identity in our working. The left and right hand sides must be isolated throughout the proof, preferably by working on only one of these sides. In general, start with the more complicated side. It often helps to express all ratios in terms of sine and/or cosine as, in general, these are easier to work with.

Consider the LHS:

$$(1 - \cos A)(1 + \sec A) \equiv 1 + \sec A - \cos A - \cos A \sec A$$

$$\equiv 1 + \sec A - \cos A - \cos A\left(\frac{1}{\cos A}\right)$$

$$\equiv \sec A - \cos A \equiv \frac{1}{\cos A} - \cos A$$

$$\equiv \frac{1 - \cos^2 A}{\cos A} \equiv \frac{\sin^2 A}{\cos A}$$

> $\cos^2 A + \sin^2 A \equiv 1$

$$\equiv \sin A\left[\frac{\sin A}{\cos A}\right] \equiv \sin A \tan A \equiv \text{RHS}$$

EXERCISE 16A

1 Without using a calculator, complete the following table.

	$\sin\theta$	$\cos\theta$	$\tan\theta$	type of angle
a		$-\frac{5}{13}$		reflex
b	$\frac{3}{5}$			obtuse
c			$\frac{7}{24}$	acute
d				straight line

Simplify the following expressions.

2 $\dfrac{1 - \sec^2 A}{1 - \operatorname{cosec}^2 A}$

3 $\dfrac{\sin\theta}{\sqrt{(1 - \cos^2\theta)}}$

4 $\dfrac{\sin\theta}{\cos\theta} + \dfrac{\cos\theta}{\sin\theta}$

5 $\dfrac{\sqrt{(1 + \tan^2\theta)}}{\sqrt{(1 - \sin^2\theta)}}$

6 $\dfrac{1}{\cos\theta\sqrt{(1 + \cot^2\theta)}}$

7 $\dfrac{\sin\theta}{1 + \cot^2\theta}$

Eliminate θ from the following pairs of equations.

8 $x = 4\sec\theta$
$y = 4\tan\theta$

9 $x = a\operatorname{cosec}\theta$
$y = b\cot\theta$

10 $x = 2\tan\theta$
$y = 3\cos\theta$

11 $x = 1 - \sin\theta$
$y = 1 + \cos\theta$

12 $x = 2 + \tan\theta$
$y = 2\cos\theta$

13 $x = a\sec\theta$
$y = b\sin\theta$

Prove the following identities.

14 $\cot\theta + \tan\theta \equiv \sec\theta \operatorname{cosec}\theta$

15 $\dfrac{\cos A}{1-\tan A} + \dfrac{\sin A}{1-\cot A} \equiv \sin A + \cos A$

16 $\tan^2\theta + \cot^2\theta \equiv \sec^2\theta + \operatorname{cosec}^2\theta - 2$

17 $\dfrac{\sin A}{1+\cos A} \equiv \dfrac{1-\cos A}{\sin A}$

(*Hint* Multiply top and bottom of LHS by $(1-\cos A)$.)

18 $\dfrac{\sin A}{1+\cos A} + \dfrac{1+\cos A}{\sin A} \equiv \dfrac{2}{\sin A}$

Solving Equations

We have already solved some simple trig equations in Chapter 15. We can now solve more varied equations by using the Pythagorean identities.

EXAMPLES 16B

1 Solve the equation $2\cos^2\theta - \sin\theta = 1$ for values of θ in the range 0 to 2π.

> The given equation is quadratic, but it involves the sine and the cosine of θ, so we use $\cos^2\theta + \sin^2\theta \equiv 1$ to express the equation in terms of $\sin\theta$ only.

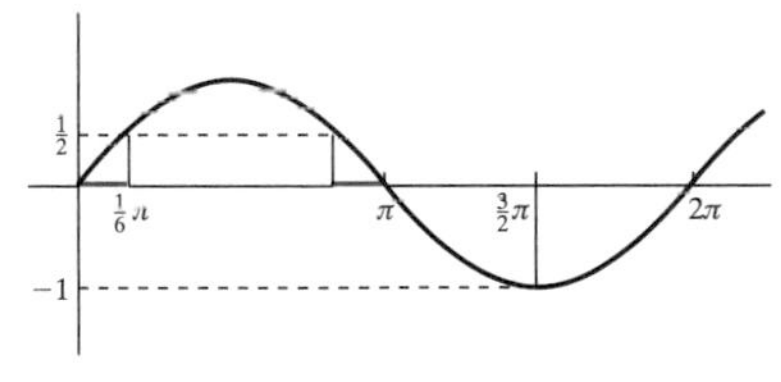

$$2\cos^2\theta - \sin\theta = 1$$
$$\Rightarrow \quad 2(1-\sin^2\theta) - \sin\theta = 1$$
$$\Rightarrow \quad 2\sin^2\theta + \sin\theta - 1 = 0$$
$$\Rightarrow \quad (2\sin\theta - 1)(\sin\theta + 1) = 0 \quad \Rightarrow \quad \sin\theta = \tfrac{1}{2} \text{ or } -1$$

If $\sin\theta = \frac{1}{2}$, $\theta = \frac{1}{6}\pi, \frac{5}{6}\pi$

If $\sin\theta = -1$, $\theta = \frac{3}{2}\pi$

Therefore the solution of the equation is $\theta = \frac{1}{6}\pi, \frac{5}{6}\pi, \frac{3}{2}\pi$.

2 a Explain why $\sin x = 0$ is not a possible solution of the equation $\cot x = \sin x$.

b Solve the equation $\cot x = \sin x$ for values of x from 0 to 360°.

a Using $\cot x \equiv \dfrac{\cos x}{\sin x}$ gives

$$\frac{\cos x}{\sin x} = \sin x$$

Both sides of an equation can be multiplied by any number *except* zero. We can simplify the equation by multiplying by $\sin x$ provided that $\sin x \neq 0$. So we must exclude any values of x for which $\sin x = 0$ from the solutions.

b $\dfrac{\cos x}{\sin x} = \sin x \quad \Rightarrow \quad \cos x = \sin^2 x$

$\Rightarrow \quad \cos^2 x + \cos x - 1 = 0$

This equation does not factorise, so we use the formula, giving

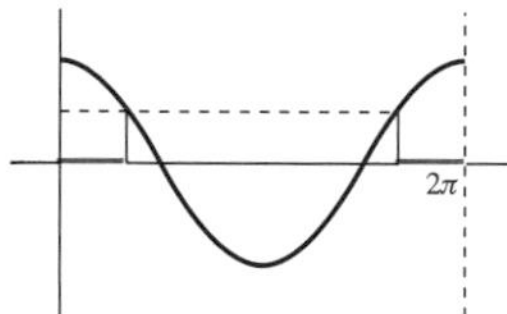

$\cos x = \frac{1}{2}(-1 \pm \sqrt{5})$

$\therefore \quad \cos x = 0.618$

$\Rightarrow \quad x = 51.8^\circ$ or 308.2° $\quad$ $\sin x \neq 0$ for either value of x

or $\quad \cos x = -1.618$ and there is no value of x for which this is true.

EXERCISE 16B Solve the following equations for angles in the range $0 \leqslant \theta \leqslant 360^\circ$.

1 $\sin\theta = \dfrac{\sqrt{3}}{2}$

2 $\cos\theta = 0$

3 $\tan\theta = -\sqrt{3}$

4 $\sin\theta = -\frac{1}{4}$

5 $\cos\theta = -\frac{1}{2}$

6 $\tan\theta = 1$

7 $\sin^2\theta = \frac{1}{4}$

8 $\sec^2\theta + \tan^2\theta = 6$

9 $4\cos^2\theta + 5\sin\theta = 3$

10 $\cot^2\theta = \operatorname{cosec}\theta$

11 $\tan\theta + \cot\theta = 2\sec\theta$

12 $\tan\theta + 3\cot\theta = 5\sec\theta$

Solve the following equations for angles in the range $-\pi \leqslant \theta \leqslant \pi$.

13 $5\cos\theta - 4\sin^2\theta = 2$

14 $4\cot^2\theta + 12\operatorname{cosec}\theta + 1 = 0$

15 $4\sec^2\theta - 3\tan\theta = 5$

16 $2\cos\theta - 4\sin^2\theta + 2 = 0$

17 $2\sin\theta\cos\theta + \sin\theta = 0$

18 $\sqrt{3}\tan\theta = 2\sin\theta$

Equations involving Multiple Angles

Many trig equations involve ratios of a multiple of θ, for example

$$\cos 2\theta = \tfrac{1}{2}, \quad \tan 3\theta = -2$$

(Note that $\cos 2\theta = \frac{1}{2} \quad \not\Rightarrow \quad \cos\theta = \frac{1}{4}$, etc.)

Simple equations of this type can be solved by finding first the values of the multiple angle and then, by division, the corresponding values of θ.

However, if values of θ are required in the range $\alpha \leqslant \theta \leqslant \beta$, then values of the multiple angle, $k\theta$ say, must be found in a range that is multiplied by the same factor, i.e. $k\alpha \leqslant k\theta \leqslant k\beta$,

e.g. if values of θ are required for $0 \leqslant \theta \leqslant 360^\circ$,

then 2θ must be found in the range $0 \leqslant 2\theta \leqslant 720^\circ$,
$\frac{1}{2}\theta$ must be found in the range $0 \leqslant \frac{1}{2}\theta \leqslant 180^\circ$, and so on.

EXAMPLES 16C

1 Find the values of θ in the range $-\pi \leqslant \theta \leqslant \pi$, for which $\cos 2\theta = \frac{1}{2}$.

Using $2\theta = \phi$ gives $\cos\phi = \frac{1}{2}$

As values of θ are required in the range $-\pi \leqslant \theta \leqslant \pi$, we want values of ϕ (i.e. 2θ) in the range $-2\pi \leqslant \phi \leqslant 2\pi$.

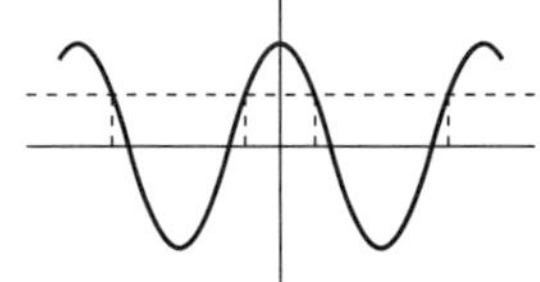

In the range $-2\pi \leqslant \phi \leqslant 2\pi$ the solutions of $\cos\phi = \frac{1}{2}$ are $\pm\frac{1}{3}\pi, \pm\frac{5}{3}\pi$.

But $\phi = 2\theta$, therefore $2\theta = \pm\frac{1}{3}\pi, \pm\frac{5}{3}\pi$

$\Rightarrow \quad \theta = \pm\frac{1}{6}\pi, \pm\frac{5}{6}\pi$

2 Find the solution of the equation $\cot(\frac{1}{3}\theta - 90°) = 1$, for which $0 \leqslant \theta \leqslant 540°$.

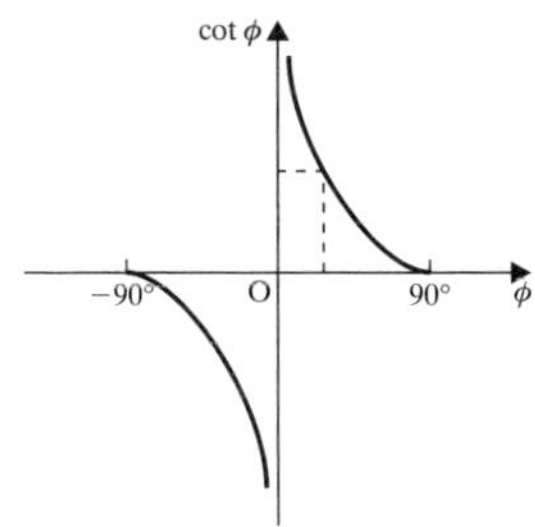

Using $\frac{1}{3}\theta - 90° = \phi$ gives

$\cot(\frac{1}{3}\theta - 90°) = \cot\phi$

As θ is required in the range $0 \leqslant \theta \leqslant 540°$ we must find ϕ in the range $\frac{1}{3}(0) - 90° \leqslant \phi \leqslant \frac{1}{3}(540°) - 90°$ i.e. $-90° \leqslant \phi \leqslant 90°$.

The solution of the equation $\cot\phi = 1$ is

$$\phi = 45°$$

But $\quad \phi = \frac{1}{3}\theta - 90°$,

so $\quad \frac{1}{3}\theta - 90° = 45°$

$\Rightarrow \quad \frac{1}{3}\theta = 135°$

i.e. $\quad \theta = 405°$

EXERCISE 16C

Find the solutions of the following equations, for values of θ in the range $0 \leqslant \theta \leqslant 180°$.

1 $\tan 2\theta = 1$

2 $\cos 3\theta = -0.5$

3 $\sin\frac{1}{2}\theta = -\frac{\sqrt{2}}{2}$

4 $\cos(2\theta - 45°) = 0$

5 $\sin(\frac{1}{4}\theta + 30°) = -1$

6 $\tan(\theta - 60°) = 0$

Solve the equations for values of θ in the range $-180° \leqslant \theta \leqslant 180°$.

7 $\tan 2\theta = 1.8$

8 $\sin 3\theta = 0.7$

9 $\cos\frac{1}{2}\theta = 0.85$

Solve the equations for values of θ in the range $0 \leqslant \theta \leqslant \pi$.

10 $\tan 4\theta = -\sqrt{3}$

11 $\sec 5\theta = 2$

12 $\cot\frac{1}{2}\theta = -1$

13 $\cos(\theta + \frac{1}{4}\pi) = \frac{1}{2}$

14 $\tan(2\theta - \frac{1}{3}\pi) = -1$

MIXED EXERCISE 16

1 Eliminate α from the equations $x = \cos\alpha$, $y = \operatorname{cosec}\alpha$.

2 If $\cos\beta = 0.5$, find possible values for $\sin\beta$ and $\tan\beta$, giving your answers in exact form.

3 Simplify the expression $\dfrac{1}{1+\cos\theta} + \dfrac{1}{1-\cos\theta}$. Hence solve the equation

$\dfrac{1}{1+\cos\theta} + \dfrac{1}{1-\cos\theta} = 4$ for values of θ in the range $0 \leqslant \theta \leqslant 2\pi$.

4 Find the solution of the equation $\sec\theta + \tan^2\theta = 5$ for values of θ in the range $0 \leqslant \theta \leqslant 360°$.

5 Prove that $(\cot\theta + \operatorname{cosec}\theta)^2 \equiv \dfrac{1+\cos\theta}{1-\cos\theta}$.

6 Find the values of θ for which $\tan(3\theta - \frac{1}{3}\pi) = 1$ in the interval $[-\pi, \pi]$.

7 Eliminate θ from the equations

a $x - 2 = \sin\theta,\ y + 1 = \cos\theta$ **b** $x = \sec\theta - 3,\ y = 2 - \tan\theta$

8 Solve the equation $\tan 2\alpha = \cot 2\alpha$ given that $0 \leqslant \alpha \leqslant 180°$.

9 Prove that $(\cos A + \sin A)^2 + (\cos A - \sin A)^2 \equiv 2$.

10 Simplify $(1 + \cos A)(1 - \cos A)$.

11 Find the solution of the equation $\tan\theta = 3\sin\theta$ for values of θ in the range $-180° \leqslant \theta \leqslant 180°$.

12 Simplify $\sec^4\theta - \sec^2\theta$.

13 Solve the equation $\cos 3\theta = \frac{1}{2}\sqrt{3}$ giving values of θ from 0 to 180°.

14 Find, in the range $-180° < \theta < 180°$, the values of θ that satisfy the equation $2\cos^2\theta - \sin\theta = 1$.

15 Find the solutions, in the range from 0 to π, of the equation $\tan(2\theta - \frac{1}{2}\pi) = \frac{1}{3}\sqrt{3}$.

SUMMARY C

INEQUALITIES

If $a > b$ then $a + k > b + k$ for all values of k

$ak > bk$ for all positive values of k

$ak < bk$ for all negative values of k

Functions

A function f is a rule that maps a number x to another single number $\mathrm{f}(x)$. The domain of a function is the set of input numbers, i.e. the set of values of x. When $x \in \mathbb{R}$, x can have any value.

The range of a function is the set of output values, i.e. the set of values of $\mathrm{f}(x)$.

The general form of a quadratic function is $\mathrm{f}(x) = ax^2 + bx + c$ where $a \neq 0$.

If $a > 0$, $\mathrm{f}(x)$ has a minimum value where $x = -b/2a$.

If $a < 0$, $\mathrm{f}(x)$ has a maximum value where $x = -b/2a$.

The general form of a polynomial function is
$$\mathrm{f}(x) = a_n x^n + a_{x-1}x^{n-1} + \ldots + a_0$$
where n is a positive integer and $a_n, a_{n-1}, \ldots$ are constants.

The general form of a rational function is $\dfrac{\mathrm{f}(x)}{\mathrm{g}(x)}$ where $\mathrm{f}(x)$ and $\mathrm{g}(x)$ are polynomials.

The function that maps the output of f to its input is called the inverse function of f, and is denoted by f^{-1}, i.e. $\mathrm{f}^{-1} : \mathrm{f}(x) \to x$. The range of f is the domain of f^{-1}.

Note that while it is always possible to reverse a mapping, the rule that does this may not be a function, so not all functions have an inverse.

When a function f operates on a function g we have a function of a function, or a composite function, which is denoted by fg, or $\mathrm{f} \circ \mathrm{g}$.

TRANSFORMATIONS OF CURVES

$y = \mathrm{f}(x) + c$ is a translation of $y = \mathrm{f}(x)$ by c units in the direction Oy.

$y = \mathrm{f}(x + c)$ is a translation of $y = \mathrm{f}(x)$ by c units in the direction xO (i.e. the $-$ve x-axis).

$y = -\mathrm{f}(x)$ is the reflection of $y = \mathrm{f}(x)$ in the x-axis.

$y = \mathrm{f}(-x)$ is the reflection of $y = \mathrm{f}(x)$ in the y-axis.

$y = a\mathrm{f}(x)$ is a one-way stretch of $y = \mathrm{f}(x)$ by a factor a parallel to Oy.

$y = \mathrm{f}(ax)$ is a one-way stretch of $y = \mathrm{f}(x)$ by a factor $1/a$ parallel to Ox.

CURVES

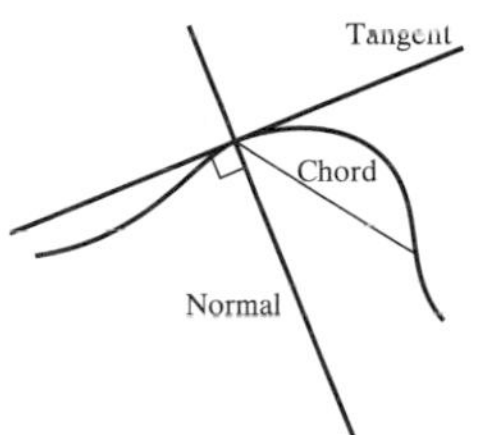

A chord is a straight line joining two points on a curve.

A tangent to a curve is a line that touches the curve at one point, called the point of contact.

A normal to a curve is the line perpendicular to a tangent and through its point of contact.

The gradient of a curve at a point on the curve is the gradient of the tangent at that point.

DIFFERENTIATION

Differentiation is the process of finding a general expression for the gradient of a curve at any point on the curve.
This general expression is called the gradient function, or the derived function or the derivative.

The derivative is denoted by $\dfrac{\mathrm{d}y}{\mathrm{d}x}$ or by $\mathrm{f}'(x)$, where

$$\frac{\mathrm{d}y}{\mathrm{d}x} = \lim_{\delta x \to 0}\left[\frac{\mathrm{f}(x + \delta x) - \mathrm{f}(x)}{\delta x}\right]$$

When $y = x^n$, $\dfrac{\mathrm{d}y}{\mathrm{d}x} = nx^{n-1}$

When $y = ax^n$, $\dfrac{\mathrm{d}y}{\mathrm{d}x} = anx^{n-1}$

When $y = c$, $\dfrac{\mathrm{d}y}{\mathrm{d}x} = 0$

Stationary Values

A stationary value of $f(x)$ is its value where $f'(x) = 0$.

The point on the curve $y = f(x)$ where $f(x)$ has a stationary value is called a stationary point.

At all stationary points, the tangents to the curve $y = f(x)$ are parallel to the x-axis.

Turning Points

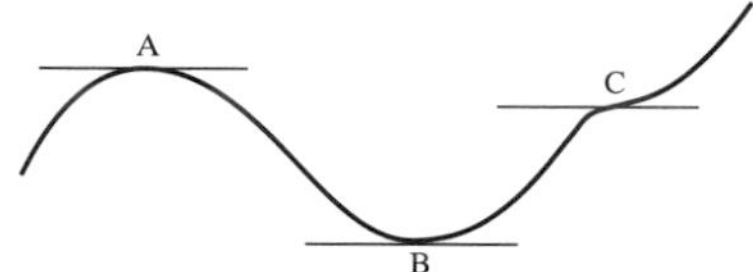

A, B and C are stationary points on the curve $y = f(x)$. The points A and B are called turning points. The point C is called a point of inflexion.

At A, $f(x)$ has a maximum value and A is called a maximum point.

At B, $f(x)$ has a minimum value and B is called a minimum point.

There are three methods for distinguishing stationary points:

	Max	Min	Inflexion
1 Find value of y on each side of stationary value	Both smaller	Both larger	One smaller One larger
2 Find sign of $\frac{dy}{dx}$ on each side of stationary value	+ 0 −	− 0 +	+ 0 + or − 0 −
Gradient			
3 Find sign of $\frac{d^2y}{dx^2}$ at stationary value	−ve (or 0)	+ve (or 0)	

Method 3 is often the easiest to apply but it fails if $\frac{d^2y}{dx^2}$ is zero. In this case use one of the other methods.

TRIG RATIOS OF 30°, 60°, 45°

	30°	60°	45°
sin	$\frac{1}{2}$	$\frac{\sqrt{3}}{2}$	$\frac{1}{\sqrt{2}}$
cos	$\frac{\sqrt{3}}{2}$	$\frac{1}{2}$	$\frac{1}{\sqrt{2}}$
tan	$\frac{1}{\sqrt{3}}$	$\sqrt{3}$	1

TRIGONOMETRIC FUNCTIONS

$f(x) = \sin x$

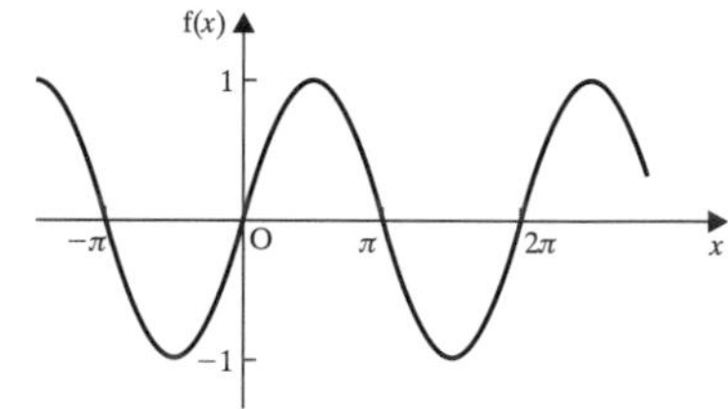

The sine function, $f(x) = \sin x$,
is defined for all values of x
is periodic with a period 2π
has a maximum value of 1 when $x = (2n + \frac{1}{2})\pi$
and a minimum value of -1 when $x = (2n + \frac{3}{2})\pi$
is zero when $x = n\pi$.

$f(x) = \cos x$

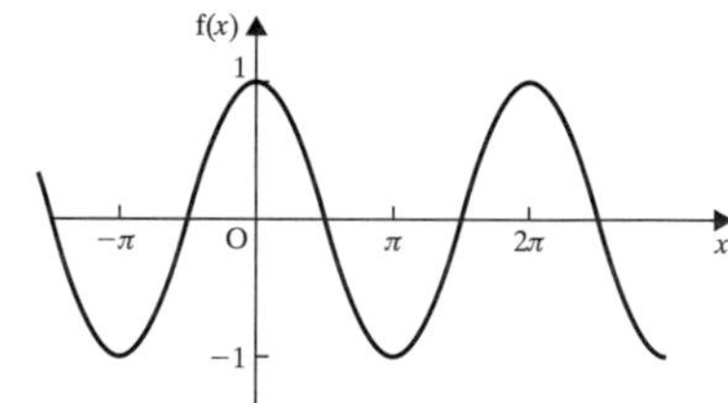

The cosine function, $f(x) = \cos x$,
is defined for all values of x
is periodic with a period 2π
has a maximum value of 1 when $x = 2n\pi$
and a minimum value of -1 when $x = (2n + 1)\pi$
is zero when $x = \frac{1}{2}(2n + 1)\pi$.

$f(x) = \tan x$

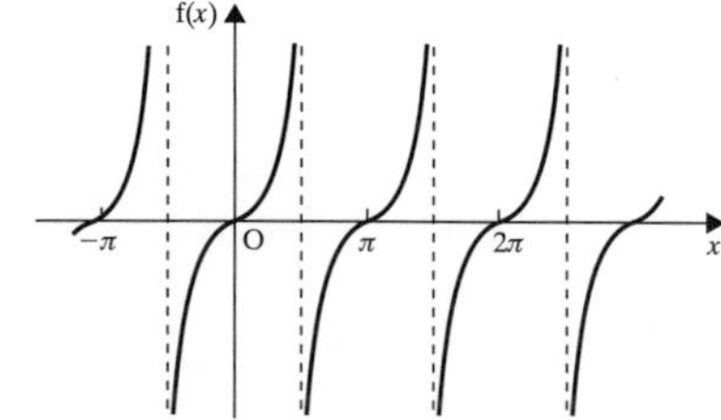

The tangent function, $y = \tan x$,
is undefined for some values of x,
these values being all odd multiples of $\frac{1}{2}\pi$,
is periodic with period π.

The reciprocal trig functions are

$\sec x \left(= \dfrac{1}{\cos x}\right)$ $\text{cosec}\, x \left(= \dfrac{1}{\sin x}\right)$ $\cot x \left(= \dfrac{1}{\tan x}\right)$

TRIGONOMETRIC IDENTITIES

$\sin\theta \equiv \cos\left(\tfrac{1}{2}\pi - \theta\right)$

$\cos\theta \equiv \sin\left(\tfrac{1}{2}\pi - \theta\right)$

$\tan\theta \equiv \dfrac{\sin\theta}{\cos\theta}$

Pythagorean Identities

$\cos^2\theta + \sin^2\theta \equiv 1$

$1 + \tan^2\theta \equiv \sec^2\theta$

$\cot^2\theta + 1 \equiv \text{cosec}^2\theta$

MULTIPLE CHOICE EXERCISE C

In Questions 1 to 26 write down the letter or letters corresponding to a correct answer.

1 The minimum value of $(x-1)(x-3)$ is when x equals

A 1 **B** 3 **C** 2 **D** 0 **E** -2

2 If $f(x) = 2x - 1$ then $f^{-1}(x)$ is

A $1 - 2x$

B $\tfrac{1}{2}(x+1)$

C $2y - 1$

D $\tfrac{1}{2}x - 1$

E $2x + 1$

3 The values of x for which $(x-1)(x-5) < 0$ are

A $x < 3$

B $x < 1,\ x > 5$

C $x < 0$

D $1 < x < 5$

E $1 \leqslant x \leqslant 5$

4 The curve $y = f(x)$ is

The curve $y = f(-x)$ could be

A

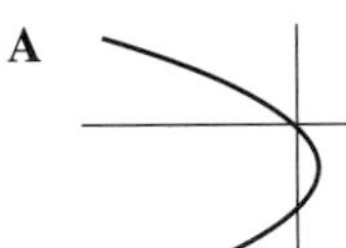

B

C

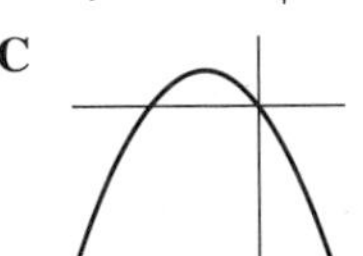

D

5 The curve $y = \sin(x - 30°)$ could be

A

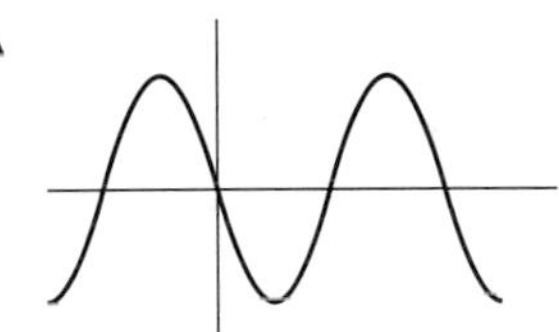

B

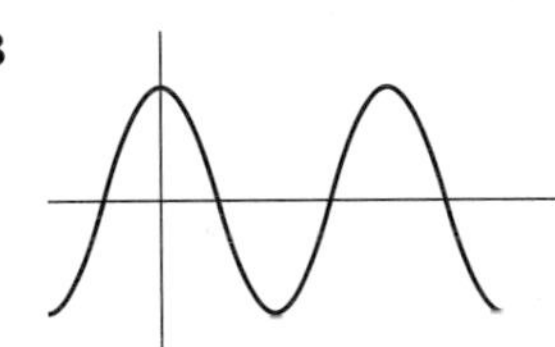

C

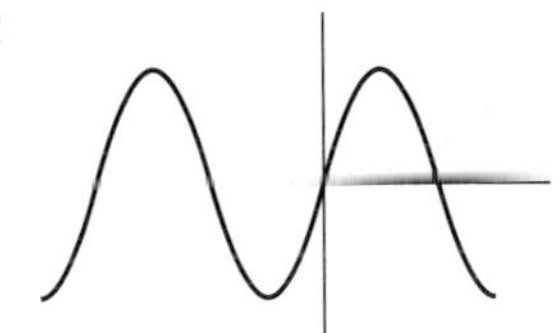

D

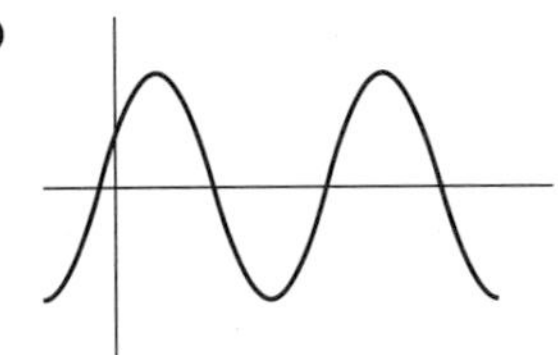

E

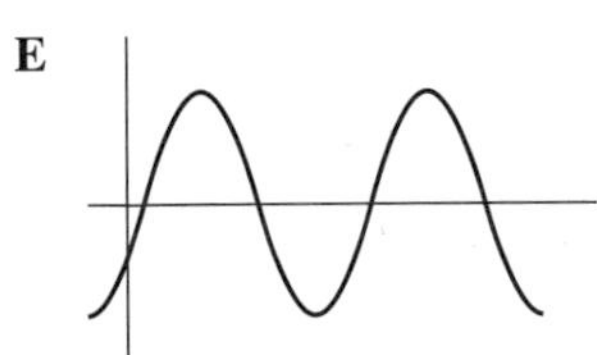

6 The equation of a curve C is $y = ax^2 + 1$, where a is a constant. Which of these curves cannot be C?

A

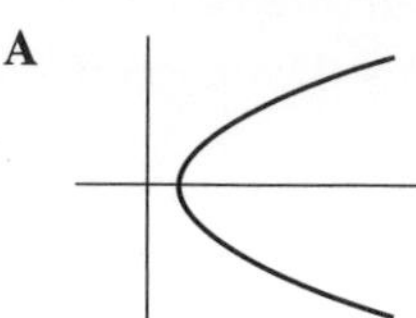

B

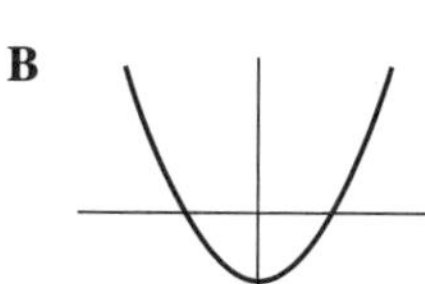

C

D

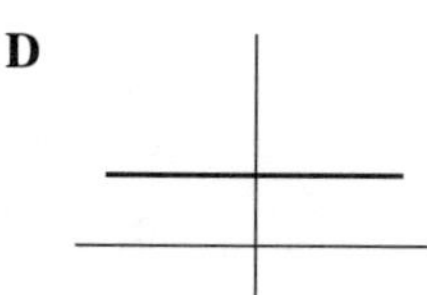

E

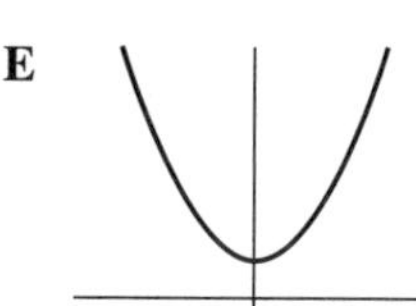

F

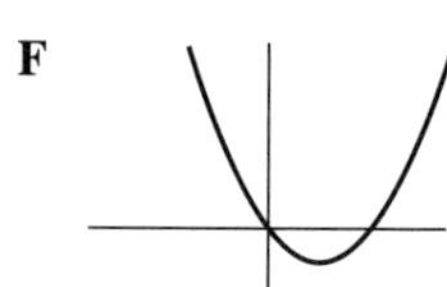

7 One of these is not an identity. Which one is it?

A $\cos^2\theta = 1 - \sin^2\theta$

B $\cos^2\theta = 1 + \sin^2\theta$

C $1 + \tan^2\theta = \sec^2\theta$

8 The function $x^3 - 12x + 5$ has a stationary value when

A $x = \sqrt{6}$ **D** $x = 4$

B $x = -2$ **E** $x = 1$

C $x = 2$

9 When $x = 1$ the function $x^3 - 3x^2 + 7$ is

A stationary **D** decreasing

B increasing **E** minimum

C maximum

10 The rate of increase w.r.t. x of the function $x^2 - \dfrac{1}{x^2}$ is

A $2x + \dfrac{2}{x^3}$ **D** $2x + \dfrac{3}{x^3}$

B $2x - \dfrac{2}{2x}$ **E** $2x + \dfrac{1}{2x}$

C $2x - \dfrac{2}{x^3}$ **F** $\dfrac{2x^4 + 2}{x^3}$

11 The gradient function of $y = (x - 3)(x^2 + 2)$ is

A $2x$ **D** $-3(2x + 2)$

B $2x - 3$ **E** $x^3 - 3x^2 + 2x - 6$

C $3x^2 - 6x + 2$

12 If $\cos\theta = \frac{1}{2}$, the solution in the interval $[-\frac{1}{2}\pi, \frac{1}{2}\pi]$ is

A $\theta = \pm\frac{1}{6}\pi$

B $\theta = \frac{1}{2}\pi$

C $\theta = +\frac{1}{3}\pi$

D $\theta = \pm\frac{1}{3}\pi$

E there is no solution

13 The graph of the function $f(\theta) \equiv \cos(2\theta - \frac{1}{2}\pi)$ has a period

A 2π **D** $-\frac{1}{2}\pi$

B π **E** none of these

C $\frac{1}{2}\pi$

14 $\sin\theta = \frac{1}{2} \Leftarrow$

A $\theta = \dfrac{\pi}{6}$ **D** $\theta = 390°$

B $\theta = 30°$ **E** $\theta = \dfrac{\pi}{3}$

C $\theta = \dfrac{5\pi}{6}$ **F** $\sin 2\theta = 1$

15 If $\frac{x-2}{x-3} < 1$ then

A $x - 2 < x - 3$

B $(x-2)(x-3) < (x-3)^2$

C $-2 < -3$

16 If $f(x) = x^2$ then

A $f(x+a)$ has a minimum value when $x = 0$

B $f^{-1}(x) = \pm\sqrt{x}$

C $f(ax)$ goes through the origin.

17 When $\tan\theta = 1$ and $0 \leqslant \theta \leqslant 2\pi$

A $\cos\theta = \sqrt{2}$

B $\theta = \frac{1}{4}\pi$

C θ lies in quadrants 1 and 3

18 $f(x) = x + \frac{1}{x}$

A $f(x)$ is stationary when $x = -1$

B $\frac{d}{dx}f(x) = 1 - \frac{1}{x^2}$

C $y = f(x)$ has no turning points.

19 $y = x^3 - 4x + 5$

A $\frac{d^2y}{dx^2} = 9x$

B The curve has two turning points.

C y is increasing when $x = 2$

20 $y = x^4$

A y is decreasing when $x = 1$

B x^4 has only one stationary value

C $\frac{dy}{dx} = 4x^3$

21 An angle θ is such that $\tan\theta = 1$ and $\cos\theta$ is negative.

A $\sin\theta$ is positive

B $\cos\theta = -\frac{1}{2}\sqrt{2}$

C $\cot\theta = -1$

22 $f(\theta) = \cos\theta$

A For $-\frac{1}{2}\pi < \theta < \frac{1}{2}\pi$, $f(\theta) > 0$

B $f(\theta)$ is undefined when $\theta = \frac{1}{2}\pi$

C $-1 \leqslant f(\theta) \leqslant 1$

23 The graph of $f(\theta) = \cos\theta$ compared with the graph of $f(\theta) = \sin\theta$ is

A inverted

B 90° to the left

C 90° to the right

24 The solution of the equation $\cos 2\theta = \frac{1}{2}$ is the same as the solution of the equation

A $\tan 2\theta = \sqrt{3}$

B $\cos(-2\theta) = \frac{1}{2}$

C $4\cos^2\theta = 3$

25 If $x = 1 - \tan\theta$ and $y = \sec\theta$ the Cartesian equation given by eliminating θ is

A $x^2 - y^2 = 2x$

B $x^2 - y^2 + 2 = 2x$

C $(1-x)^2 = (y-1)(y+1)$

26 $\sin(\theta - \frac{1}{2}\pi)$ is identical to

A $\cos(\theta + \frac{1}{2}\pi)$

B $\sin(\frac{1}{2}\pi - \theta)$

C $\sin(\theta + \frac{3}{2}\pi)$

In Questions 27 to 32 a single statement is made. Write T if it is true, F if it is false.

27 The function $f(\theta) = \cos\theta$ is such that $|\theta| \leqslant 1$.

28 $\sin\theta = 0$ when $\theta = n\pi$

29 $\frac{d^2y}{dx^2} = 0$ when $x = a \iff y$ has an inflexion where $x = a$.

30 $\cos\theta = 0 \quad \Leftarrow \quad \theta = \frac{\pi}{2}$

31 $y = \frac{1}{x^2} \quad \Rightarrow \quad \frac{dy}{dx} = -\frac{1}{2x}$

32 If $\frac{1}{x} < 2$ then $\frac{1}{2} < x$

Examination Questions C

1 Use calculus to find the values of x for which $y = 3x^4 + 8x^3 + 6x^2 - 1$ has stationary points.

(AQA)

2

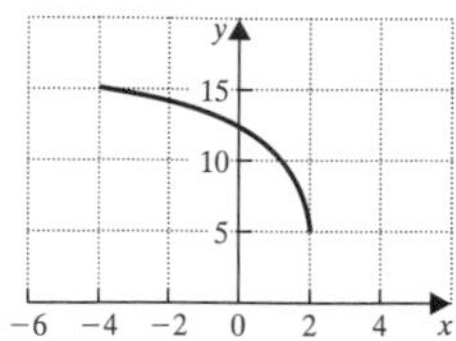

The entire graph of a function $y = \mathrm{f}(x)$ is illustrated above.

a Write down the domain of the function $\mathrm{f}(x)$.

b Sketch the graph of the inverse function $y = \mathrm{f}^{-1}(x)$ marking appropriate values on the axes.

c Write down the range of $\mathrm{f}^{-1}(x)$.

(AQA)

3 Find the equation of the tangent to the curve $y = x^3 + 2x^2 + 3x + 6$ at the point where $x = -1$.

(AQA)

4 Functions f and g are defined by

$$\mathrm{f} : x \mapsto \frac{3}{x+3}, \quad x \in \mathbb{R}, \ x \geqslant 0,$$

$$\mathrm{g} : x \mapsto x + 1, \quad x \in \mathbb{R}, \ x \geqslant 0,$$

Show that $\quad \mathrm{gf} : x \mapsto \dfrac{x+6}{x+3}, \quad x \in \mathbb{R}, \ x \geqslant 0$

Express fg in similar form. Find $(\mathrm{gf})^{-1}(x)$.

(OCR)

5 **i**

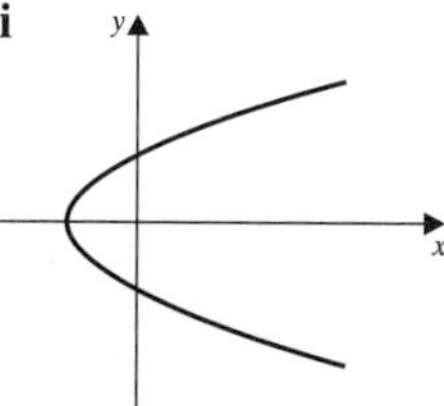

ii

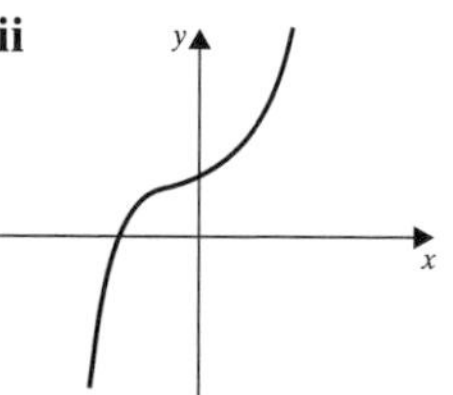

iii

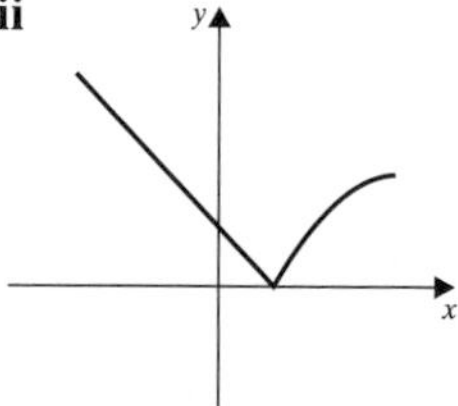

a For each of the graphs **i**, **ii** and **iii**, state whether or not it represents a function.

b One of the graphs above is such that the function f represented by the graph has an inverse, f^{-1}.
Assuming equal scales on the axes of the graphs drawn, sketch the graph of f^{-1}.

(AQA)

6 Find, in degrees to 1 decimal place, the values of x which lie in the interval $-180° \leqslant x \leqslant 180°$ and satisfy the equation $\sin 2x = -0.57$.

(Edexcel)

7 The diagram shows the graph of the function f defined for $x \geqslant 0$ by $f : x \mapsto 1 + \sqrt{x}$.
Copy the sketch, and show on the same diagram the graph of f^{-1}, making clear the relation between the two graphs.

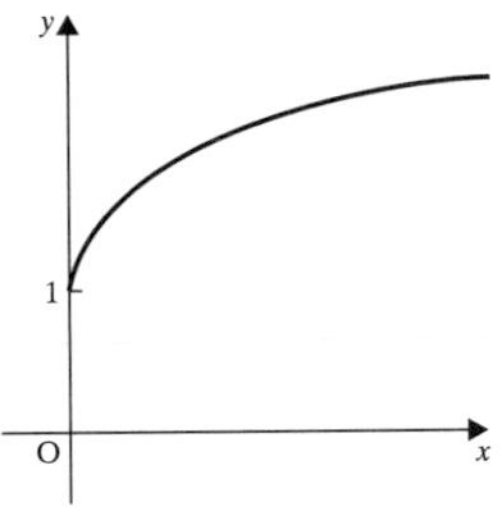

Give an expression in terms of x for $f^{-1}(x)$, and state the domain of f^{-1}.
There is one value of x for which $f(x) = f^{-1}(x)$. By considering your diagram, explain why this value of x satisfies the equation $1 + \sqrt{x} = x$.

By treating the equation $1 + \sqrt{x} = x$ as a quadratic equation for $\sqrt{x}$, or otherwise, show that the value of x satisfying $f(x) = f^{-1}(x)$ is $\frac{1}{2}(3 + \sqrt{5})$. (OCR)

8 Given that $\cos^2 x = \frac{2}{9}$, where $\pi \leqslant x \leqslant 2\pi$, find the exact value, or values, of

a $\sin x$ **b** $\tan x$ (Edexcel)

9 Let $f(A) = \dfrac{\cos A}{1 + \sin A} + \dfrac{1 + \sin A}{\cos A}$.

a Prove that $f(A) = 2\sec A$.

b Solve the equation $f(A) = 4$,
giving your answers for A, in degrees, in the interval $0° < A < 360°$. (OCR)

10 *No credit will be given for numerical answers without supporting working.*

Solve the equation

$$4\cot^2\theta + 12\operatorname{cosec}\theta + 1 = 0$$

giving all answers of θ to the nearest degree in the interval $0° \leqslant \theta \leqslant 360°$. (AQA)

11 *No credit will be given for a numerical approximation or for a numerical answer without supporting working.*

a Find all solutions of the equation

$$\tan 2x = 2$$

in the range $0 \leqslant x \leqslant 360°$. Give your answers correct to one decimal place.

b Find all solutions of the equation

$$2\sin^2 x + 3\cos x = 0$$

in the interval $0 \leqslant x \leqslant 360°$. (WJEC)

12 A piece of wire of total length 12 m is cut into two pieces. One piece of wire is bent into a rectangle of sides x m and $3x$ m and the other is bent to form the boundary of a square. Show that the total area enclosed by the rectangle and the square is given by

$$A = 7x^2 - 12x + 9 \text{ m}^2.$$

Find the value of x if the total area enclosed is a minimum. Verify that your stationary value is a minimum. (WJEC)

13 The functions f and g are defined by

$$f(x) = 2\sqrt{x-5}, \quad x \in \mathbb{R}, x \geqslant 5,$$

$$g(x) = x^2 + 3, \quad x \in \mathbb{R}.$$

a Find the range of f and the range of g.

b Explain *briefly* why the function fg does not exist.

c Find
i an expression for $gf(x)$
ii the domain and the range of the function gf.

d Explain briefly why the inverse function of g does not exist.

e Find an expression for $f^{-1}(x)$, giving the domain and range of f^{-1}. (WJEC)

14 A curve has equation $y = x^3 + 3x^2 - 9x + 4$.

a Find $\frac{dy}{dx}$.

b Find the coordinates of the two stationary points on the curve.

c Use calculus to determine the nature of each of the stationary points and show these on a sketch of the curve.

d Deduce the range of k for which $x^3 + 3x^2 - 9x + 4 = k$ has three real roots. (WJEC)

15

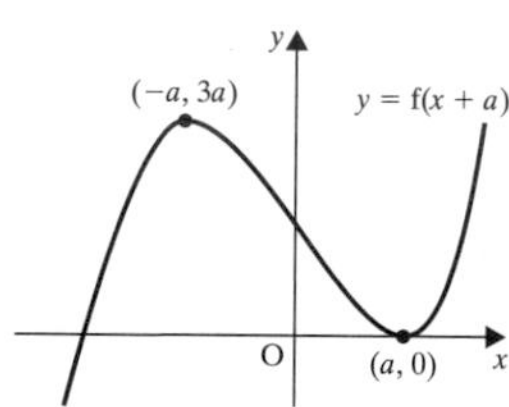

The diagram shows the curve $y = f(x+a)$, where a is a positive constant. The maximum and minimum points on the curve are $(-a, 3a)$ and $(a, 0)$ respectively. Sketch the following curves, on separate diagrams, in each case stating the coordinates of the maximum and minimum points:

i $y = f(x)$, **ii** $y = -2f(x+a)$. (OCR)

16 A large tank in the shape of a cuboid is to be made from $54\,\text{m}^2$ of sheet metal. The tank has a horizontal rectangular base and no top. The height of the tank is x metres. Two of the opposite vertical faces are squares.

a Show that the volume, $V\,\text{m}^3$, of the tank is given by

$$V = 18x - \tfrac{2}{3}x^3.$$

b Given that x can vary, use differentiation to find the maximum value of V.

c Justify that the value of V you have found is a maximum. (Edexcel)

17 The functions f and g are defined as follows:

$$\mathrm{f}(x) = \frac{1}{x}\ (x \neq 0); \quad \mathrm{g}(x) = \frac{1}{x+1}\ (x \neq -1)$$

a Find an expression, in terms of x, for $\mathrm{g}^{-1}(x)$.

b Describe geometrically the transformation which maps
i the graph of $y = \mathrm{f}(x)$ into the graph of $y = \mathrm{g}(x)$
ii the graph of $y = \mathrm{f}(x)$ into the graph of $y = \mathrm{g}^{-1}(x)$.

c Find an expression, in terms of x, for the composite function $\mathrm{gf}(x)$. Simplify your answer.

d Given that there is a function h such that $\mathrm{gh}(x) = \mathrm{f}(x)$, find $\mathrm{h}(x)$. (OCR)

18 The specification for a new rectangular car park states that the length x m is to be 5 m more than the breadth. The perimeter of the car park is to be greater than 32 m.

a Form a linear inequality in x.

The area of the car park is to be less than $104\ \mathrm{m}^2$.

b Form a quadratic inequality in x.

c By solving your inequalities, determine the set of possible values of x. (Edexcel)

19 A curve has equation $y = \frac{1}{x} - \frac{1}{x^2}$. Use differentiation to find the coordinates of the stationary point and determine, showing your working, whether the stationary point is a maximum point or a minimum point. Deduce, or obtain otherwise, the coordinates of the stationary points of each of the following curves:

i $y = \frac{1}{x} - \frac{1}{x^2} + 5$, **ii** $y = \frac{2}{x-1} - \frac{2}{(x-1)^2}$ (OCR)

20 Solve the equation $4\tan^2 x + 12\sec x + 1 = 0$, giving all solutions in degrees, to the nearest degree, in the interval $-180° < x < 180°$. (AQA)

21 Find, in degrees, the values of θ in the interval [0, 360°] for which

$$4\sin^2\theta - 2\sin\theta = 4\cos^2\theta - 1.$$

In your answers, distinguish clearly between those that are exact and those which are given to a degree of accuracy of your choice, which you should state. (Edexcel)

22 A student models the evening lighting-up time by the equation

$$L = 6.125 - 2.25\cos\left(\frac{\pi t}{6}\right)$$

where the time, L hours pm, is always in GMT (Greenwich Mean Time) and t is in months, starting in mid-December. The model assumes that all months are equally long.

a Calculate the value of L for mid-January and for mid-May.

b Find, by solving an appropriate equation, the two months in the year when the lighting-up time will be 5 pm (GMT)

c Write down an equation for L if t were to be in months starting in mid-March. (AQA)

23 i Draw sketch graphs of $y = \sin x$ and $y = -\cos x$ for $0° \leqslant x \leqslant 360°$.

ii Mo needs to draw the graph of $y = 2\sin^2 x$, also for $0° \leqslant x \leqslant 360°$, and makes two attempts, P and Q shown in the figure below.

Realising that both cannot be right, Mo tests each of P and Q by finding the value of y when $x = 270°$.
Carry out the test for each of P and Q. In each case state which one of the statements A, B, C below is a correct conclusion from the test.

A The curve is correctly drawn.

B The curve is incorrectly drawn.

C There is no evidence that the curve is incorrectly drawn.

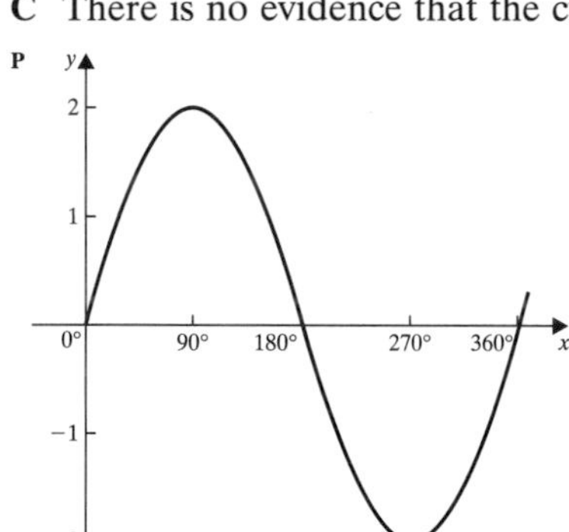

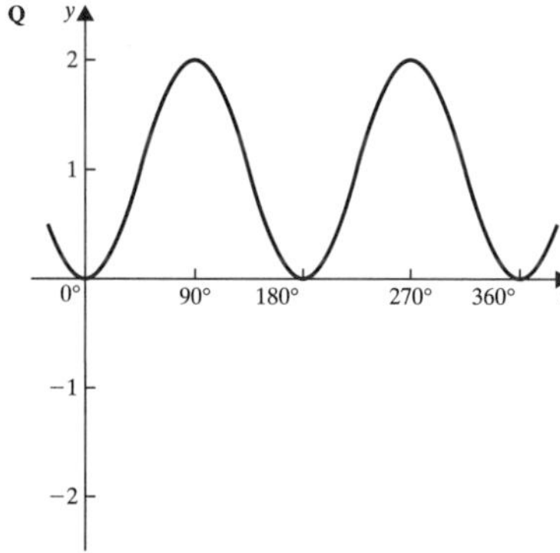

iii a Solve the equation $2\sin^2 x + 3\cos x = 0$ for $0° \leqslant x \leqslant 360°$.

b Given that one of the curves, P and Q, in part **ii** is in fact correct, illustrate your answers to part **iii a** by drawing two suitable curves. (OCR)

24 a Given that $\tan 75° = 2 + \sqrt{3}$, find in the form $m + n\sqrt{3}$, where m and n are integers, the value of **i** $\tan 15°$ **ii** $\tan 105°$

b Find, in radians to two decimal places, the values of x in the interval $0 \leqslant x \leqslant 2\pi$, for which $3\sin^2 x + \sin x - 2 = 0$. (Edexcel)

25

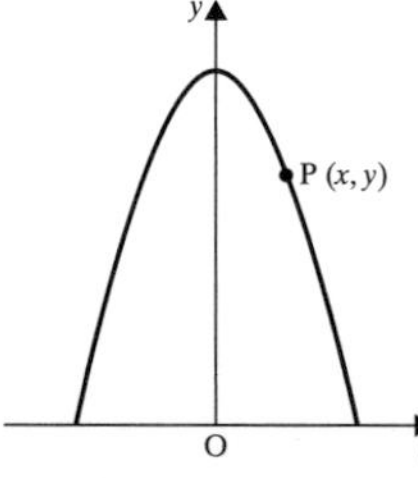

The figure shows the part of the curve with equation $y = 5 - \frac{1}{2}x^2$ for which $y \geqslant 0$.
The point P(x,y) lies on the curve and O is the origin.

a Show that $\text{OP}^2 = \frac{1}{4}x^4 - 4x^2 + 25$.

Taking $f(x) \equiv \frac{1}{4}x^4 - 4x^2 + 25$,

b find the values of x for which $f'(x) = 0$.

c Hence, or otherwise, find the minimum distance from O to the curve, showing that your answer is a minimum. (Edexcel)

26 The figure shows a minor sector OMN of a circle centre O and radius r cm. The perimeter of the sector is 100 cm and the area of the sector is A cm^2.

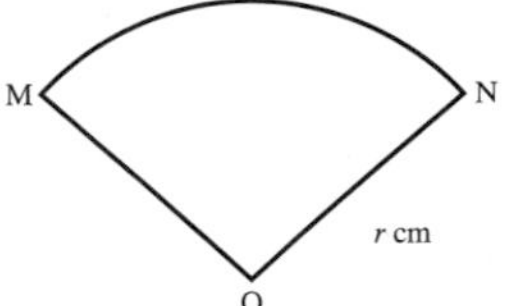

a Show that $A = 50r - r^2$.

Given that r varies, find

b the value of r for which A is a maximum and show that A is a maximum

c the value of $\angle$MON for this maximum area

d the maximum area of the sector OMN. (Edexcel)

27 Given that $y = x^3 - x + 6$,

a find $\dfrac{dy}{dx}$.

On the curve representing y, P is the point where $x = -1$.

b Calculate the y-coordinate of the point P.

c Calculate the value of $\dfrac{dy}{dx}$ at P.

d Find the equation of the tangent at P.

The tangent at the point Q is parallel to the tangent at P.

e Find the coordinates of Q.

f Find the equation of the normal to the curve at Q. (OCR)

28 Find the set of values of x for which $2(x^2 - 5) < x^2 + 6$ (AQA)

29 The curve with equation $y = 2 + k\sin x$ passes through the point with coordinates $(\frac{1}{2}\pi, -2)$.
Find **a** the value of k **b** the greatest value of y,
c the values of x in the interval $0 \leqslant x \leqslant 2\pi$ for which $y = 2 + \sqrt{2}$. (Edexcel)

30 The diagram shows the graph of $y = x^2(3 - x)$.
The coordinates of the points A and B on the graph are $(2, 4)$ and $(3, 0)$ respectively.

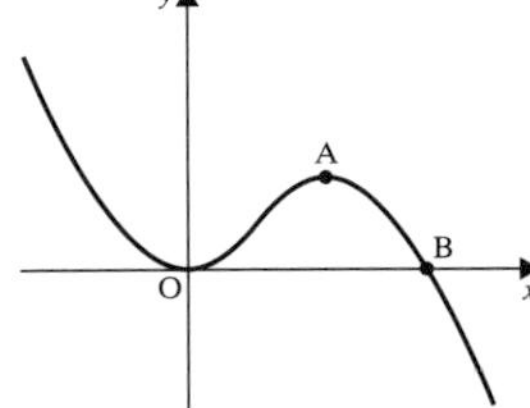

a Write down the solution set of the inequality $x^2(3 - x) \leqslant 0$.

b The equation $3x^2 - x^3 = k$ has three real solutions for x.
Write down the set of possible values for k.

c Functions f and g are defined as follows:

$$f : x \rightarrow x^2(3 - x), \quad 0 \leqslant x \leqslant 2,$$
$$g : x \rightarrow x^2(3 - x), \quad 0 \leqslant x \leqslant 3.$$

Explain why f has an inverse while g does not.

d State the domain and range of f^{-1}, and sketch the graph of f^{-1}. (OCR)

31 Find the set of values of x for which

$$x^2 - x - 12 > 0.$$

(Edexcel)

32 Find, correct to the nearest degree, all the values of θ between $0°$ and $360°$ satisfying the equation

$$8\cos^2\theta + 2\sin\theta = 7.$$

(WJEC)

33 a Write down an example of a polynomial in x of order 4.

b In an experiment, Ama measures the value of s at different times, t. Her results are shown as the curve on the graph below

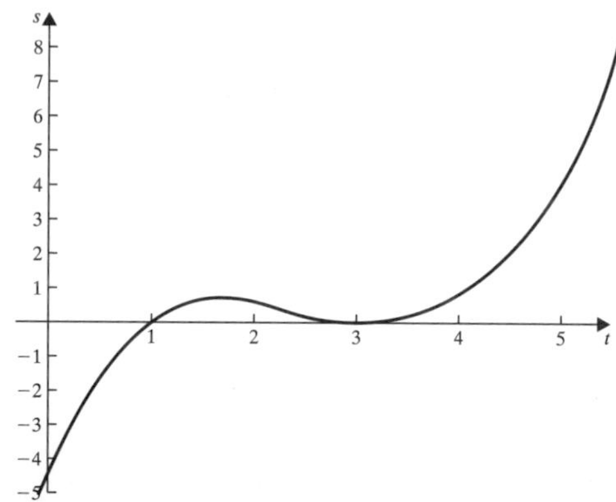

Ama believes that it is possible to model s as a polynomial in t.

i Explain why it is reasonable to think that the order of such a polynomial might be 3.

Ama proposes a model of the form

$$s = a(t-p)(t-q)^2$$

ii Write down the points where the curve meets the coordinate axes and use them to find values for p, q and a.

iii Compare the values obtained from the model with those on the graph when $t = 2$, 4 and 5, and comment on the quality of the model.

Ama proposes a refinement to the model making it into

$$s = a(t-p)(t-q)^2(1-ht)$$

where a, p and q have the same values as before and h is a small positive constant. Ama chooses the value of h so that the model and the graph are in agreement when $t = 5$.

iv Find the value of h.

(OCR)

34 The diagram shows part of the graph of $y = \sin x$, where x is measured in radians, and the values of α on the x-axis and k on the y-axis are such that $\sin\alpha = k$. Write down, in terms of α,

a a value of x between $\frac{1}{2}\pi$ and π such that $\sin x = k$

b two values of x between 3π and 4π such that $\sin x = -k$.

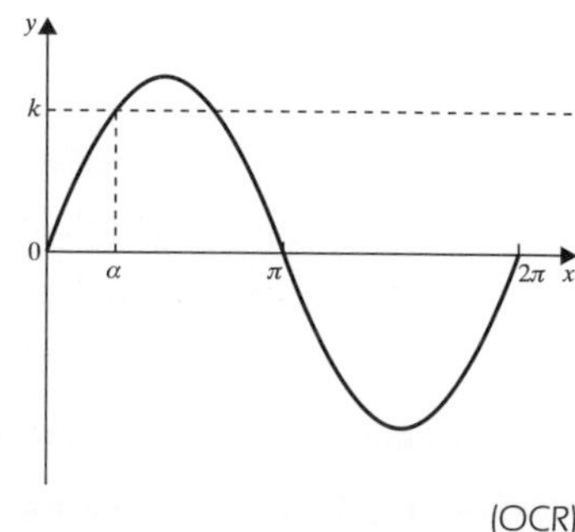

(OCR)

35 A piece of wire of length 4 metres is bent into the shape of a sector of a circle of radius r metres and angle θ radian.

a State, in terms of θ and r.

i the length of the arc **ii** the area A of the sector.

b Hence show that $A = 2r - r^2$.

c Find the value of r which will make the area a maximum. Deduce the corresponding value of θ.

d The figures labelled A–F below show, all to the same scale, six possible sectors which can be made from the piece of wire. Which of them has the largest area?

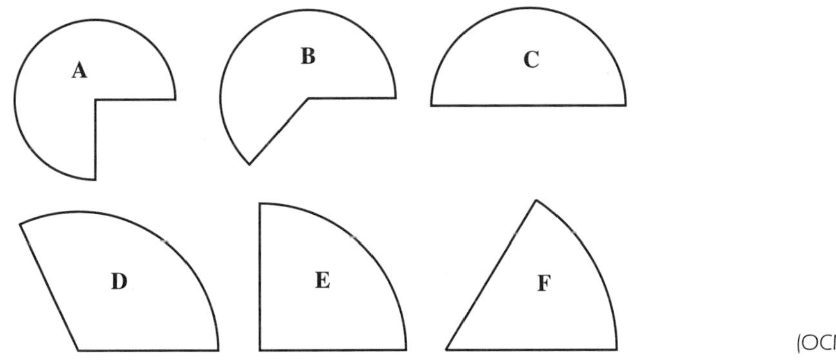

(OCR)

36 The function f with domain $\{x : x \geqslant 0\}$ is defined by $f(x) = \dfrac{8}{x+2}$.

a Sketch the graph of f and state the range of f.

b Find $f^{-1}(x)$, where f^{-1} denotes the inverse of f.

c Calculate the value of x for which $f(x) = f^{-1}(x)$. (AQA)

37 Use differentiation to find the coordinates of the stationary points on the curve

$$y = x + \frac{4}{x},$$

and determine whether each stationary point is a maximum point or a minimum point. Find the set of values of x for which y increases as x increases. (OCR)

38 a Find the values of $\cos x$ for which

$$6\sin^2 x = 5 + \cos x.$$

b Find all the values of x in the interval $180° < x < 540°$ for which

$$6\sin^2 x = 5 + \cos x.$$

(Edexcel)

39 Given that $y = x^3 - 4x^2 + 5x - 2$, find $\dfrac{dy}{dx}$.

P is the point on the curve where $x = 3$.

a Calculate the y-coordinate of P.

b Calculate the gradient at P.

c Find the equation of the tangent at P.

d Find the equation of the normal at P.

Find the values of x for which the curve has a gradient of 5. (OCR)

40 A landscape gardener is given the following instructions about laying a rectangular lawn. The length x m is to be 2 m longer than the width. The width must be greater than 6.4 m and the area is to be less than 63 m^2.
By forming an inequality in x, find the set of possible values of x. (Edexcel)

41 Solve the equation

$$9\cos^2 x - 6\cos x - 0.21 = 0, \quad 0° \leqslant x \leqslant 360°,$$

giving each answer in degrees to 1 decimal place. (Edexcel)

42 Express $2x^2 + 5x + 4$ in the form $a(x+b)^2 + c$, stating the numerical values of a, b and c.
Hence, or otherwise, write down the coordinates of the minimum point on the graph of $y = 2x^2 + 5x + 4$. (OCR)

43 The function f is given by

$$\text{f} : x \mapsto x^2 - 8x, \quad x \in \mathbb{R}, x \leqslant 4.$$

a Determine the range of f.

b Find the value of x for which $\text{f}(x) = 20$.

c Find $\text{f}^{-1}(x)$ in terms of x. (Edexcel)

44 The diagram shows triangle PQR, in which PQ $=$ 1 unit, QR $=$ 2 units and RP $= k$ units. Express $\cos R$ in terms of k.

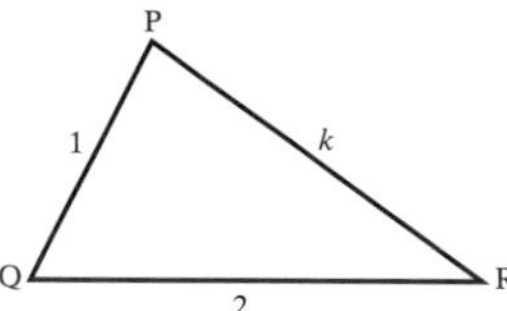

Given that $\cos R < \frac{7}{8}$, show that $2k^2 - 7k + 6 < 0$. Find the set of values of k satisfying this inequality. (OCR)

45 The diagram shows a rectangular cake-box, with no top, which is made from thin card. The volume of the box is 500 cm^3. The base of the box is a square with sides of length x cm.

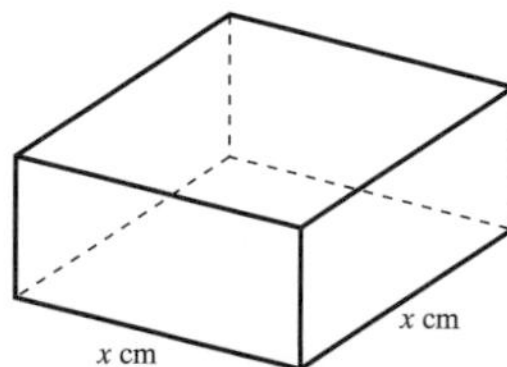

a Show that the area, A cm^2, of card used to make such an open box is given by

$$A = x^2 + \frac{2000}{x}.$$

b Given that x varies, find the value of x for which $\frac{dA}{dx} = 0$.

c Find the height of the box when x has this value.

d Show that when x has this value, the area of the card used is least. (Edexcel)

46 The function f is defined by

$$\mathrm{f}: x \to \frac{3x+1}{x-2}, \quad x \in \mathbb{R}, x \neq 2$$

Find, in a similar form, the functions

a ff **b** f^{-1} (Edexcel)

47 Given that $\mathrm{f}(x) = x^4$ and $\mathrm{g}(x) = x + 2$, simplify

$\mathrm{fg}(x) - \mathrm{gf}(x)$ (AQA)

48 Given that $\mathrm{f}(x) \equiv x^3 + 2x^2 - 5x - 6$, find

a $\mathrm{f}(2)$

b the complete set of values of x for which $\mathrm{f}(x) < 0$. (Edexcel)

49 Sketch the graph of the curve with equation $y = x(1 - x)$. Determine the greatest and least values of y when $-1 \leqslant x \leqslant 1$. (OCR)

CHAPTER 17

Exponential and logarithmic functions

Exponential Growth and Decay

There are many situations where a quantity grows or decays by a constant factor over equal time intervals.

For example, if a debt of £100 has 2% interest added each month then, if no repayment is made,

the debt grows to £100 $\times$ 1.02 after one month

to £100 $\times$ 1.02 $\times$ 1.02, i.e. £100 $\times$ $(1.02)^2$ after two months,

and so on, to £100 $\times$ $(1.02)^n$ after n months.

1.02 is called the *growth factor* per month.

So if £y is the debt after x months, $y = 100(1.02)^x$.

This is an exponential function that *increases* in value as the exponent, x, increases; we say that y grows exponentially.

Some business assets (e.g. vehicles) depreciate over time. If each year the value of a lorry is 'written off' by half its value at the start of the year each year, then starting with a value of £A, it is worth £$A \times 0.5$ after one year, £$A \times (0.5)^2$ after two years and so on.

If £y is the value after x years, then
$y = A \times (0.5)^x = A(\frac{1}{2})^x = A \times 2^{-x}$.

Now 2^{-x} is an exponential function that *decreases* in value as x increases, so we say that y decays exponentially.

In each case considered above, the mathematical expression for the relationship is obtained by making certain assumptions and it is not necessarily valid at all times. The first example assumes that interest rates remain constant, and we know that this is not the case. So this relationship is valid only for the time during which the interest rate is 2%.

The assumption in the second case is that the rule for writing down the value of assets never changes and again this is not true in practice; after a few years the vehicle will have a value small enough to be written off completely even if it is not lost through other causes in the mean time.

There are many other situations which, while not giving exact relationships like those above, approximate very closely to exponential, or natural, growth and decay.

So exponential functions are important because they provide mathematical models for these situations. At this stage exponential functions are defined and in Chapter 33 we see how they are used in problems about rates of growth and decay.

The Exponential Function

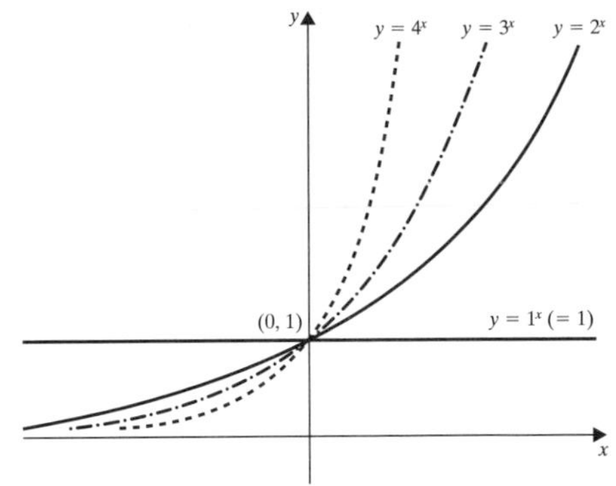

The general shape of an exponential curve was seen in Chapter 11. The diagram shows a few more members of the exponential family.

Note that these curves and, in fact, *all* exponential curves, pass through the point $(0, 1)$.
This is because, if $y = a^x$ for any positive base a,

$$\text{when } x \text{ is } 0, \ y = a^0 = 1.$$

Each exponential curve has a unique property which you can discover experimentally by using an accurate plot of the curve $y = 2^x$. Choose three or four points on the curve and, at each one, draw the tangent as accurately as possible and find its gradient. Then complete the following table.

Point	Gradient of tangent, i.e. $\frac{dy}{dx}$	y-coordinate	$\frac{dy}{dx} \div y$
1			
2			
3			
4			

An accurate drawing should result in numbers in the last column that are all reasonably close to 0.7.
When this experiment is carried out for 3^x and 4^x we find again that $\frac{dy}{dx} \div y$ has a constant value; for 3^x the constant is about 1.1 and for 4^x it is about 1.4. So we have

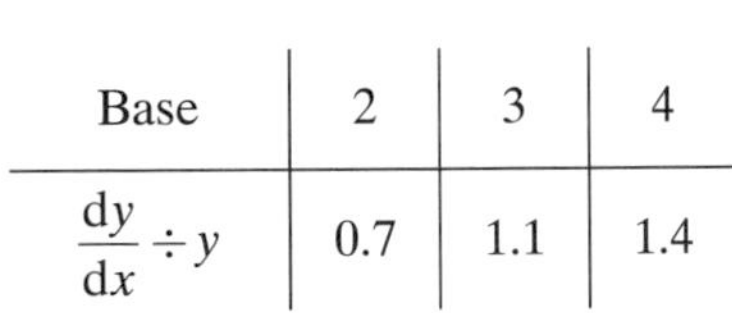

Base	2	3	4
$\frac{dy}{dx} \div y$	0.7	1.1	1.4

$\Rightarrow$

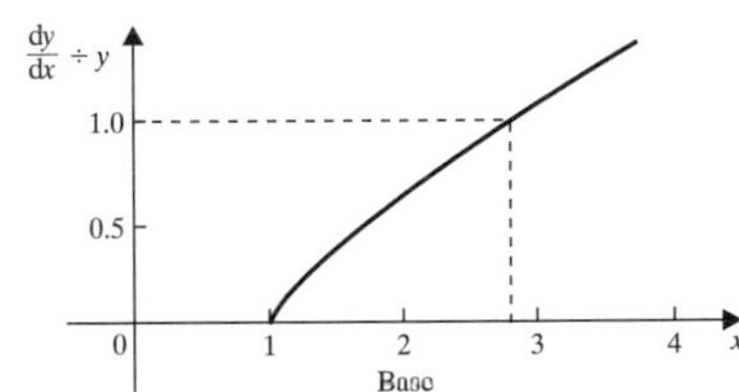

From this graph we can see that there is a base, somewhere between 2 and 3, for which $\frac{dy}{dx} \div y = 1$, i.e. $\frac{dy}{dx} = y$.

Calling this base e we have **if $y = e^x$ then $\frac{dy}{dx} = e^x$**

The function e^x is the only function which is unchanged when differentiated.

We first met e in Chapter 2; e is irrational, i.e. like π, $\sqrt{2}$, etc., it cannot be given an exact decimal value but, to 4 significant figures, $e = 2.718$.

Summing up:

for any value of a ($a > 0$), a^x is *an* exponential function,
for the base e ($e \approx 2.718$), e^x is *the* exponential function

$$\frac{d}{dx}(e^x) = e^x$$

The following diagrams show sketches of $y = e^x$ and of some simple variations.

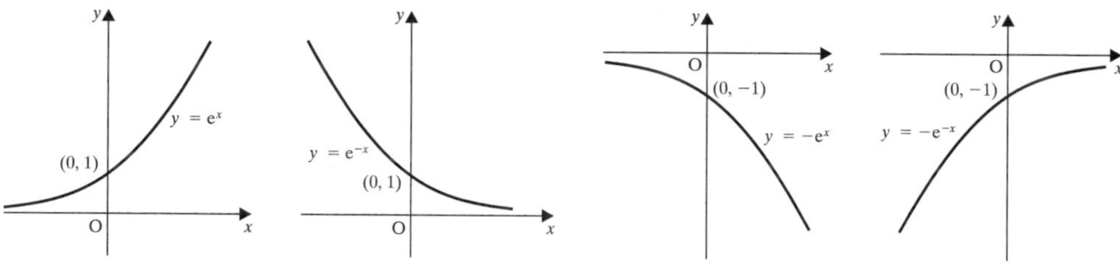

EXAMPLE 17A Find the coordinates of the stationary point on the curve $y = e^x - x$, and determine its type. Sketch the curve showing the stationary point clearly.

$$y = e^x - x \quad \Rightarrow \quad \frac{dy}{dx} = e^x - 1$$

At a stationary point, $\frac{dy}{dx} = 0$ therefore $e^x - 1 = 0$

i.e. $e^x = 1 \quad \Rightarrow \quad x = 0$

When $x = 0$, $y = e^0 - 0 = 1$.

Therefore (0, 1) is a stationary point.

$\frac{d^2y}{dx^2} = e^x$ and this is positive when $x = 0$

Therefore (0, 1) is a minimum point.

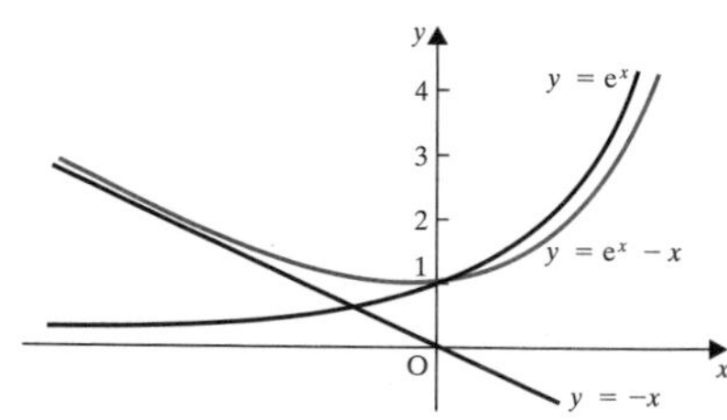

This curve is made up from separate sketches of $y = e^x$ and $y = -x$ by adding their ordinates.

EXERCISE 17A

1 Evaluate, correct to 3 s.f.

a e^2 **b** e^{-1} **c** $e^{1.5}$ **d** $e^{-0.3}$

2 Write down the derivative of

a $2e^x$ **b** $x^2 - e^x$ **c** e^x

In questions 3 to 5 find the gradient of each curve at the specified value of x.

3 $y = e^x - 2x$ where $x = 2$

4 $y = x^2 + 2e^x$ where $x = 1$

5 $y = e^x - 3x^3$ where $x = 0$

6 Find the value of x at which the function $e^x - x$ has a stationary value.

7 Sketch the given curve.

a $y = 1 - e^x$ **b** $y = e^x + 1$ **c** $y = x - e^x$

d $y = 1 - e^{-x}$ **e** $y = 1 + e^{-x}$ **f** $y = x^2 + e^x$

8 A culture dish was seeded with $4\,\text{mm}^2$ of mould. The table shows the area of the mould at six-hourly intervals.

Time, x hours	0	6	12	18	24	30	36
Area, $y\,\text{mm}^2$	4	8.1	15.9	33	68	118	190

a Show that, for a time, the growth factor in the area for each interval of 6 hours is roughly 2 (i.e. the area approximately doubles every 6 hours). Hence show that it is reasonable to use an exponential function as a model for the area and suggest a possible function.

b On the same set of axes, plot the points in the table and draw the curve given by the equation you chose as the model.

c Give reasons why

i the model is approximate

ii when and why the model ceases to be reasonable.

Natural (Naperian) Logarithms

From Chapter 2 we know that logarithms to the base e are called *natural* or *Naperian* logarithms, and that $\log_e a$ is written as $\ln a$, i.e.

$$\ln a = b \quad \Leftrightarrow \quad a = e^b$$

Logarithms to the base 10 are called *common logarithms* and are denoted by log or lg. They used to be an important tool for calculations but calculators have eliminated their usefulness in that respect.

The laws used for working with logarithms to a general base, given in Chapter 2, apply equally well to natural logarithms, i.e.

$$\ln a + \ln b = \ln ab$$
$$\ln a - \ln b = \ln a/b$$
$$\ln a^n = n \ln a$$

One further rule can be added to this list. It is needed when we want to change the base of a logarithm.

Changing the Base of a Logarithm

Suppose that $x = \log_a c$ and that we wish to express x as a logarithm to the base b.

$$\log_a c = x \quad \Rightarrow \quad c = a^x$$

Now taking logs to the base b gives

$$\log_b c = x \log_b a \quad \Rightarrow \quad x = \frac{\log_b c}{\log_b a}$$

i.e. $$\log_a c = \frac{\log_b c}{\log_b a}$$

In the special case when $c = b$, i.e. when $\log_b c = 1$, this relationship becomes

$$\log_a b = \frac{1}{\log_b a}$$

The change of base rule can be used to convert a logarithm into a natural logarithm,

e.g. $$\log_a c = \frac{\ln c}{\ln a}$$

The base of an exponential function can be changed in a similar way. Suppose that we wish to express 3^x as a power of e.

Using $\quad 3^x = e^p$ gives $x \ln 3 = p$

$\therefore \quad 3^x = e^{x \ln 3}$

In general $\quad a^x = e^{x \ln a}$

EXAMPLES 17B

1 Separate $\ln(\tan x)$ into two terms.

$$\ln(\tan x) = \ln\left(\frac{\sin x}{\cos x}\right)$$

$$= \ln\sin x - \ln\cos x$$

2 Express $4\ln(x+1) - \frac{1}{2}\ln x$ as a single logarithm.

$$4\ln(x+1) - \tfrac{1}{2}\ln x = \ln(x+1)^4 - \ln\sqrt{x}$$

$$= \ln\frac{(x+1)^4}{\sqrt{x}}$$

EXERCISE 17B

1 Evaluate

a $\ln 48$ **b** $\ln e$ **c** $\ln 1$

2 Express as a sum or difference of logarithms or as a product

a $\ln\dfrac{x^2}{x+1}$ **b** $\ln(a^2 - b^2)$ **c** $\ln\cot x$ **d** $\ln\sin^2 x$

3 Express as a single logarithm

a $\ln\cos x - \ln\sin x$ **b** $1 + \ln x$ **c** $\frac{2}{3}\ln(x-1)$

4 Given that $\ln a = 3$

a express $\log_a x^2$ as a simple natural logarithm

b express as a single logarithm $\ln x^3 + 6\log_a x$.

5 Solve the following equations for x

a $e^x = 8.2$ **b** $e^{2x} + e^x - 2 = 0$ (Hint. Use $e^{2x} = (e^x)^2$)

c $e^{2x-1} = 3$ **d** $e^{4x} - e^x = 0$

6 Given that $\ln a = 2$ solve the following equations for x

a $a^x = e^2$ **b** $a^x = e^6$ **c** $a^x = 1$

7 The equation $y = A(1.3)^{-t}$, where A is constant, is used to predict the value £y of a company vehicle when it is t years old.

a What is the meaning of the constant A?

b A new car is valued at £15 000. Use the model to predict the value of this car when it is two years old. Give one reason why this value should only be considered as approximate.

c By changing the base of the exponent (i.e. index) to e, express y in terms of A, e and t.

The Logarithmic Function

Consider the curve with equation $y = f(x)$ where $f(x) = \ln x$.

If $y = \ln x$ then $x = e^y$, i.e.

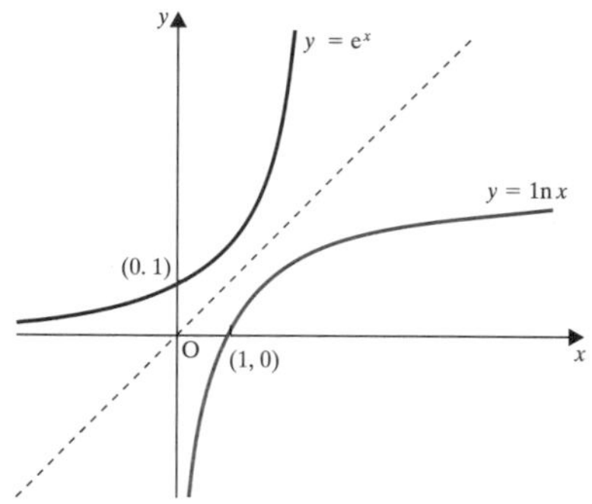

the logarithmic function is the inverse of the exponential function.

It follows that the curve $y = \ln x$ is the reflection of the curve $y = e^x$ in the line $y = x$.

There is no part of the curve $y = \ln x$ in the second and third quadrants.

This is because, if $x = e^y$ (i.e. if $y = \ln x$), x is positive for all real values of y.

Therefore

$\ln x$ does not exist for negative values of x.

The Derivative of ln x

We know that $y = \ln x \iff x = e^y$ and we also know how to differentiate the exponential function. So a relationship between $\frac{d}{dx}(y)$ and $\frac{d}{dy}(x)$ would help in finding the derivative of $\ln x$.

Consider the equation $y = f(x)$ where $f(x)$ is any function of x,

$$\frac{dy}{dx} = \lim_{\delta x \to 0} \frac{\delta y}{\delta x} = \lim_{\delta x \to 0} \left(1 \bigg/ \frac{\delta x}{\delta y}\right)$$

Now $\delta y \to 0$ as $\delta x \to 0$, $\quad \therefore \quad \frac{dy}{dx} = \lim_{\delta y \to 0} \left(1 \bigg/ \frac{\delta x}{\delta y}\right)$

i.e. $\quad \mathbf{\frac{dy}{dx} = 1 \bigg/ \frac{dx}{dy}}$

This relationship can be used to find the derivative of *any* function if the derivative of its inverse is known. We will now apply it to differentiate $\ln x$.

$$y = \ln x \iff x = e^y$$

Differentiating e^y w.r.t. y gives $\frac{dx}{dy} = e^y = x$

Therefore $\frac{dy}{dx} = 1 \bigg/ \frac{dx}{dy} = \frac{1}{x}$,

i.e. $\quad \mathbf{\frac{d}{dx} \ln x = \frac{1}{x}}$

This result can be used to differentiate many log functions if they are first simplified by applying the laws given on page 210.

EXAMPLES 17C

1 Find the derivative of $\ln(2x)$.

$$f(x) = \ln(2x) = \ln 2 + \ln x$$

$$\frac{d}{dx}\{f(x)\} = 0 + \frac{1}{x} \quad (\ln 2 \text{ is a number})$$

The derivative of $\ln(2x)$ is $\frac{1}{x}$.

2 Find the derivative of **a** $\ln(1/x^3)$ **b** $\ln(4\sqrt{x})$

a $$f(x) = \ln(1/x^3) = \ln(x^{-3}) = -3\ln x$$

$$\frac{d}{dx}\{f(x)\} = \frac{d}{dx}\{-3\ln x\} = \frac{-3}{x}$$

b $$f(x) = \ln(4\sqrt{x}) = \ln 4 + \ln(\sqrt{x}) = \ln 4 + \tfrac{1}{2}\ln x$$

$$\frac{d}{dx}\{f(x)\} = \frac{d}{dx}(\ln 4) + \frac{d}{dx}(\frac{1}{2}\ln x)$$

$$= 0 + \frac{\frac{1}{2}}{x} = \frac{1}{2x}$$

3 Find $\frac{dy}{dx}$ if $y = \log_a x^2$

We only know how to differentiate natural logs, so first we must change the base from *a* to e.

$$y = \log_a x^2 = \frac{\ln x^2}{\ln a} = \frac{2\ln x}{\ln a} = \frac{2}{\ln a}(\ln x)$$

$$\therefore \quad \frac{dy}{dx} = \frac{2}{x\ln a}$$

EXERCISE 17C

1 Write down the derivative of each of the following functions.

a $\ln x^3$ **b** $\ln(3x)$ **c** $\ln(x^{-2})$ **d** $\ln(3/\sqrt{x})$

e $\ln(1/x^5)$ **f** $\ln(2x^{1/2})$ **g** $\ln(x^{-3/2})$ **h** $\ln(x^3/\sqrt{x})$

2 Locate the stationary points on each curve.

a $y = \ln x - x$ **b** $y = x^3 - 2\ln x^3$ **c** $y = \ln x - \sqrt{x}$

3 Sketch each of the following curves.

a $y = -\ln x$ **b** $y = \ln(-x)$ **c** $y = 2 + \ln x$ **d** $y = \ln x^2$

4 You will need a graphics calculator for this question.
The equation of a curve is $y = x^2 - 2\ln 4x$.
Use your calculator to

a draw the curve

b find the range of values of x for which $x^2 - 2\ln 4x < 0$, giving answers correct to 1 decimal place, and showing how you justify that accuracy.

c find the coordinates of the minimum point on the curve.

Use calculus to find exact values for the coordinates of the minimum point on the curve and comment on the accuracy of your answer to part **c**.

CHAPTER 18 Functions 2

In this chapter we start by looking at some methods for sketching curves without using a graphics calculator.

Transformations

The graphs of many functions can be obtained from transformations of the curves representing basic functions. These transformations are covered in Chapters 11 and 15 and you need to be familiar with them.

Consider the curve $y = \dfrac{x}{x+1}$ which looks as if $\dfrac{x}{x+1}$ is related to $f(x) = \dfrac{1}{x}$.

We see that we can write $y = \dfrac{x}{x+1} = 1 - \dfrac{1}{x+1}$, i.e. $y = 1 - f(x+1)$

We can now build up a picture of the curve $y = \dfrac{x}{x+1}$ in stages, with rough sketches.

1 Sketch $f(x) = \dfrac{1}{x}$.

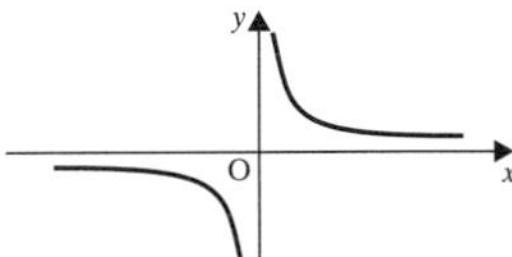

2 Sketch $g(x) = \dfrac{1}{x+1} = f(x+1)$

($f(x)$ moved one unit to the left.)

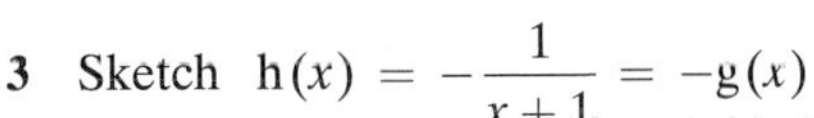

3 Sketch $h(x) = -\dfrac{1}{x+1} = -g(x)$

($g(x)$ reflected in the x-axis.)

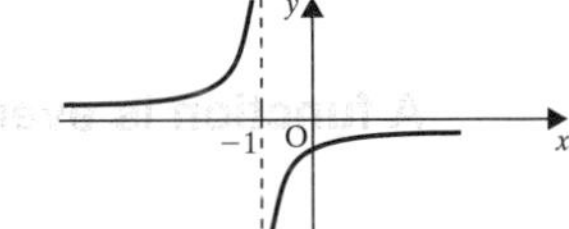

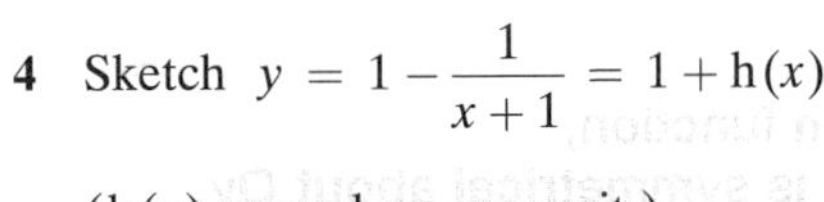

4 Sketch $y = 1 - \dfrac{1}{x+1} = 1 + h(x)$

($h(x)$ moved up one unit.)

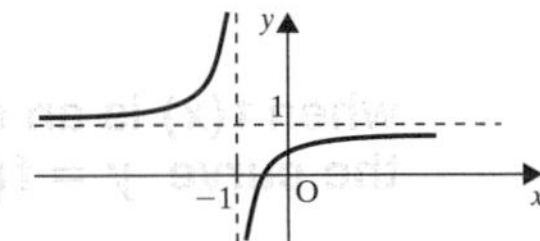

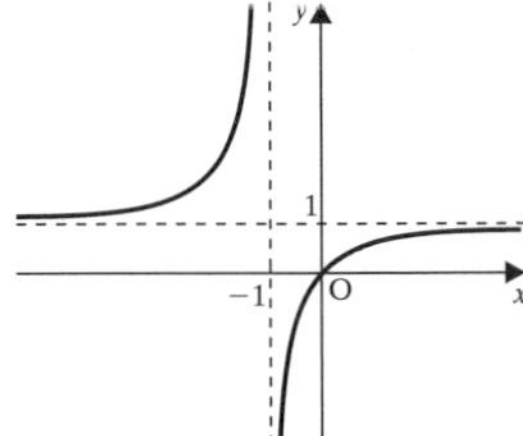

From the last sketch, we see that the asymptotes to the curve are $y = 1$ and $x = -1$.

From the equation of the curve, $y = 0$ when $x = 0$, so the curve goes through the origin and there are no other intercepts on the axes.

We can now draw a more accurate sketch.

In general, a sketch graph should clearly show the following features of a curve: asymptotes, intercepts on the axes and turning points when they exist.

EXERCISE 18A

1 Find the values of a and b such that $x^2 - 4x + 1 \equiv (x-a)^2 + b$. On the same set of axes sketch the curves $y = x^2$ and $y = x^2 - 4x + 1$.

2 For $-2\pi < x < 2\pi$ and on the same set of axes sketch the graphs of

a $y = \cos x$ and $y = 3\cos x$

b $y = \sin x$ and $y = \sin 2x$

c $y = \cos x$ and $y = \cos(x - \frac{1}{6}\pi)$

d $y = \sin x$ and $y = 2\sin(\frac{1}{6}\pi - x)$

3 On the same set of axes sketch the graphs of $y = \dfrac{1}{x}$, $y = \dfrac{3}{x}$ and $y = \dfrac{1}{3x}$.

4 Show that $\dfrac{x-2}{x-3} \equiv 1 + \dfrac{1}{x-3}$.

Sketch the curve $y = \dfrac{x-2}{x-3}$, clearly showing the asymptotes and the intercepts on the axes.

Sketch the following curves clearly showing any asymptotes, turning points and intercepts on the axes.

5 $y = \dfrac{1}{x-1}$

6 $y = 1 - 2\sin x$

7 $y = 1 - (x-2)^2$

8 $y = \dfrac{1}{1-x}$

9 $y = \dfrac{1-x}{x}$

10 $y = 1 - x^3$

11 $y = \dfrac{2x+1}{x}$

12 $y = \dfrac{x+1}{x-1}$

13 $y = 3 - (x-2)^2$

14 $y = 2\sin(x - \frac{1}{3}\pi)$

15 $y = 3\cos(x + \frac{1}{6}\pi)$

16 $y = \dfrac{1+2x}{1-x}$

17 $y = 3x^3 - 4$

18 $y = 3 - 2x^4$

19 $y = 3 - (x+2)^3$

20 $y = (x-1)^4$

Next we look at some functions with interesting properties.

Even Functions

A function is even if $f(x) = f(-x)$.

Since the curve $y = f(-x)$ is the reflection of the curve $y = f(x)$ in the y-axis, it follows that

when $f(x)$ is an even function,
the curve $y = f(x)$ is symmetrical about Oy.

Some familiar even functions and their graphs are shown below.

$f(x) = \cos x$

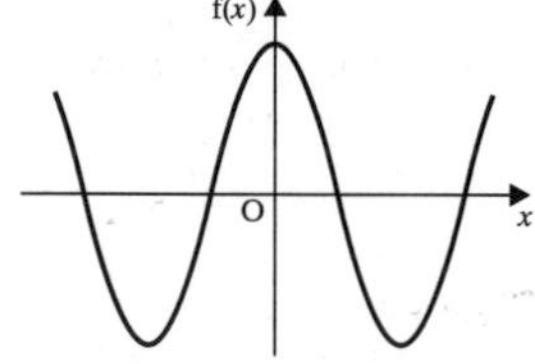

$f(-x) = \cos(-x) = \cos x = f(x)$

$f(x) = x^2$

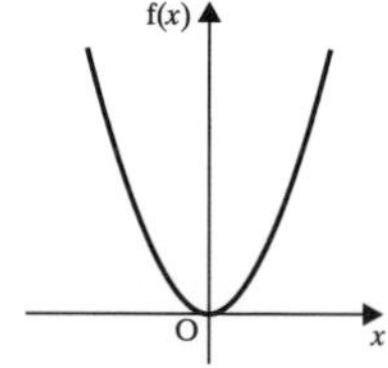

$f(-x) = (-x)^2 = x^2 = f(x)$

Odd Functions

A function is odd if $f(x) = -f(-x)$.

As the curve $y = -f(-x)$ is a reflection of the curve $y = f(x)$ in O*y* followed by a reflection in O*x*, it follows that

when $f(x) = -f(-x)$ the curve $y = f(x)$ has rotational symmetry of order 2 about the origin.

Some familiar odd functions and their graphs are shown below.

$f(x) = \sin x$ 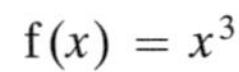 $f(x) = x^3$ $f(x) = \frac{1}{x}$

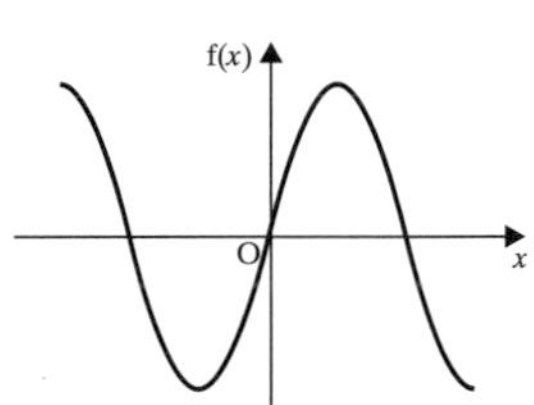

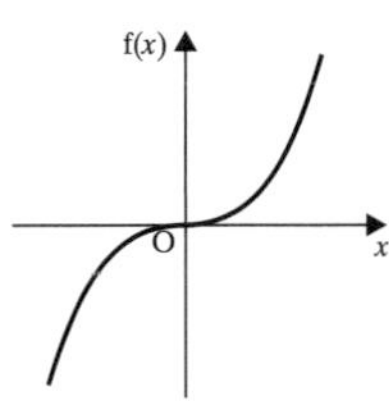

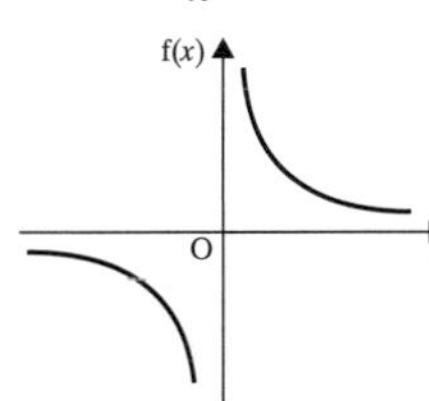

$$-f(-x) = -\sin(-x) = -(-\sin x) = \sin x = f(x)$$

$$-f(-x) = -(-x)^3 = -(-x^3) = x^3 = f(x)$$

$$-f(-x) = -\left(\frac{1}{-x}\right) = \frac{1}{x} = f(x)$$

Periodic Functions

A function whose graph consists of a basic pattern which repeats at regular intervals is called a *periodic* function. The width of the basic pattern is the *period* of the function.

$f(x) = \sin x$, for example, is periodic and its period is 2π.

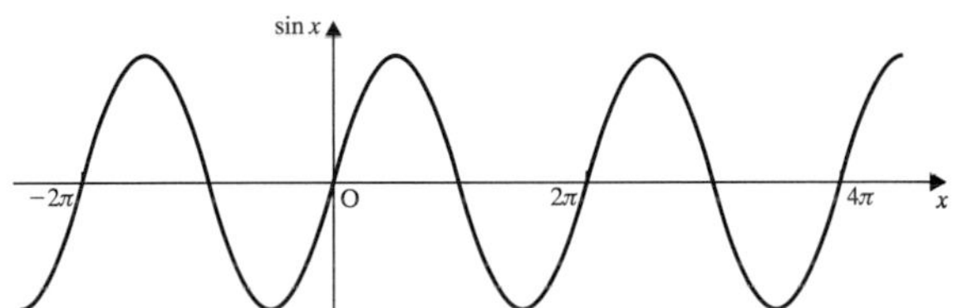

$f(x)$ is periodic with period a, so it follows that $f(x + a) = f(x)$ for all x.

Therefore the definition of $f(x)$ within one period (e.g. for $0 < x \leqslant a$) together with the definition $f(x+a) = f(x)$ for all x, defines a periodic function.

For example, if $\begin{cases} f(x) = 2x - 1 \text{ for } 0 < x \leqslant 1 \\ f(x+1) = f(x) \text{ for all values of } x \end{cases}$

then we know that the function is periodic with a period of 1.

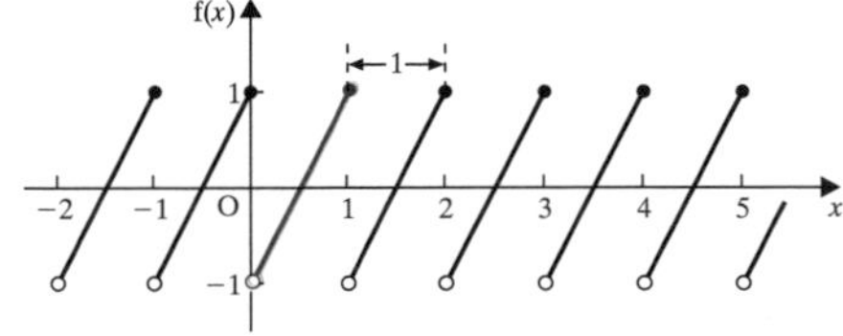

The graph of this function can be sketched by drawing

$f(x) = 2x - 1$ for $0 < x \leqslant 1$

and repeating the pattern at unit intervals in both directions.

The basic pattern in the graph of a periodic function can be made up of two or more different definitions. The next worked example illustrates such a compound periodic function.

EXAMPLE 18B Sketch the graph of the function f defined by

$$\begin{aligned} f(x) &= x && \text{for } 0 \leqslant x < 1 \\ f(x) &= 4 - x^2 && \text{for } 1 \leqslant x < 2 \\ f(x+2) &= f(x) && \text{for all real values of } x \end{aligned}$$

From the last line of the definition, f is periodic with period 2.

The graph of this function is built up by first drawing $f(x) = x$ in the interval $0 \leqslant x < 1$, then drawing $f(x) = 4 - x^2$ in the interval $1 \leqslant x < 2$. This pattern is then repeated every 2 units along the x-axis.

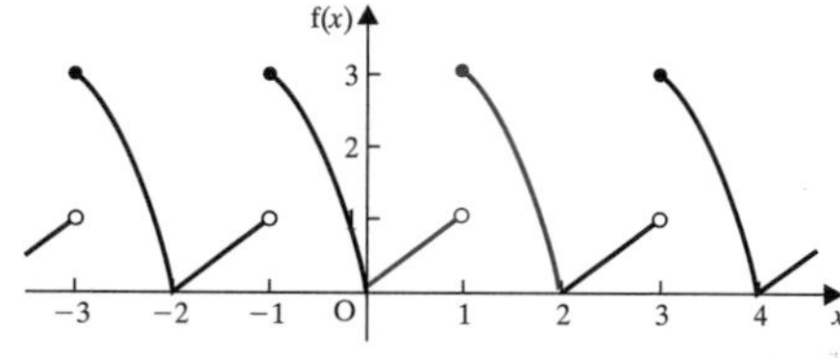

EXERCISE 18B

1 Sketch each function and state whether it is even, odd and/or periodic.

a $\cos x$ **b** $\tan x$ **c** e^{-x} **d** $\ln(1+x)$

e $\cot x$ **f** $(x-1)^2$ **g** $x - 1$ **h** $\dfrac{1}{x}$

2 Sketch the graph of f(x) within the interval $-4 < x \leqslant 6$ if

$$f(x) = 4 - x^2 \quad \text{for } 0 < x \leqslant 2$$

and $$f(x) = f(x-2) \quad \text{for all values of } x.$$

3 If

$$\begin{aligned} f(\theta) &= \sin\theta && \text{for } 0 < x \leqslant \tfrac{1}{2}\pi \\ f(\theta) &= \cos\theta && \text{for } \tfrac{1}{2}\pi < x \leqslant \pi \end{aligned}$$

and $f(\theta + \pi) = f(\theta)$ for all values of θ,

sketch the function $f(\theta)$ for the range $-2\pi < \theta \leqslant 2\pi$.

4 A function f(x) is periodic with a period of 4. Sketch the graph of the function for $-6 \leqslant x \leqslant 6$, given that

$$\begin{aligned} f(x) &= -x && \text{for } 0 < x \leqslant 3 \\ f(x) &= 3x - 12 && \text{for } 3 < x \leqslant 4 \end{aligned}$$

The Modulus of x

The modulus of x is written as $|x|$ and it means the positive value of x whether x itself is positive or negative, e.g. $|2| = 2$ and $|-2| = 2$.

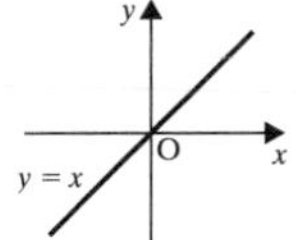

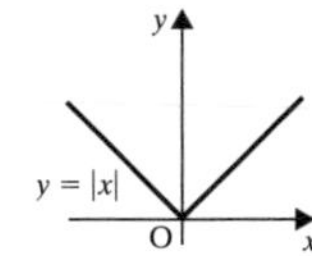

Therefore the graph of $y = |x|$ can be obtained from the graph of $y = x$ by changing the part of the graph for which y is negative to the equivalent positive values, i.e. by reflecting the part of the graph where y is negative in the x-axis.

The Modulus of a Function

In general, the curve C_1 whose equation is $y = |f(x)|$ is obtained from the curve C_2 with equation $y = f(x)$, by reflecting in the x-axis the parts of C_2 for which $f(x)$ is negative. The remaining sections of C_1 are not changed.

For example, to sketch $y = |(x-1)(x-2)|$ we start by sketching the curve $y = (x-1)(x-2)$. We then reflect in the x-axis the part of this curve which is below the x-axis.

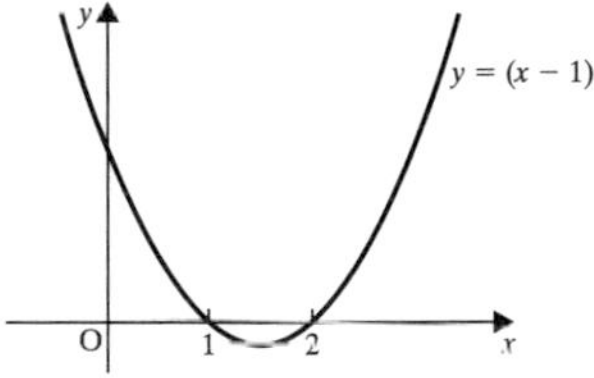

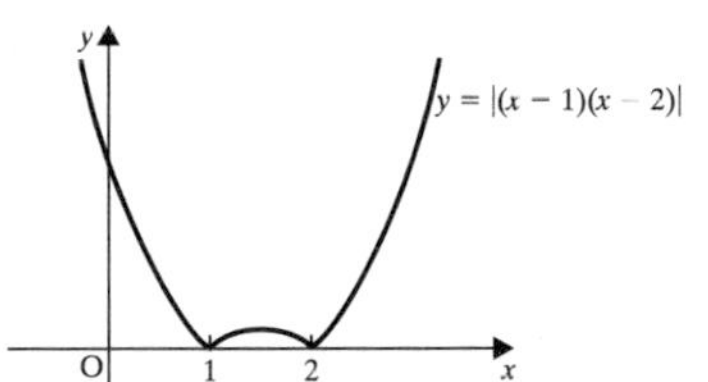

Note that for any function f, the mapping $x \to |f(x)|$ is also a function.

EXAMPLE 18C Sketch the graph of $y = 3 - |1 - 2x|$.

We can use transformations to build up the picture in stages.

1. Draw $f(x) = 1 - 2x$

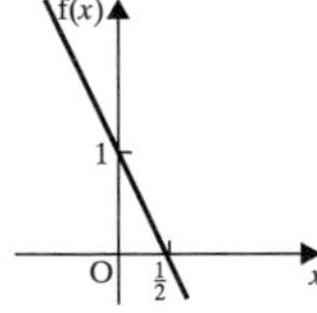

2. Draw $g(x) = |1 - 2x|$

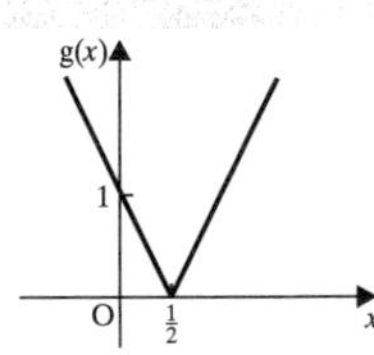

3. Draw $-|1 - 2x|$ ($-g(x)$ is the reflection of $g(x)$ in Ox.)

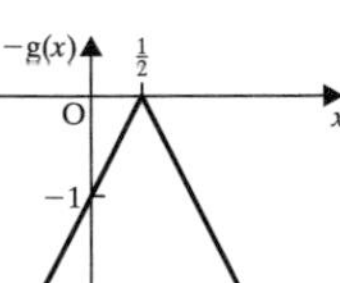

4. $y = 3 - |1 - 2x|$ is the graph of $-g(x)$ translated 3 units upwards.

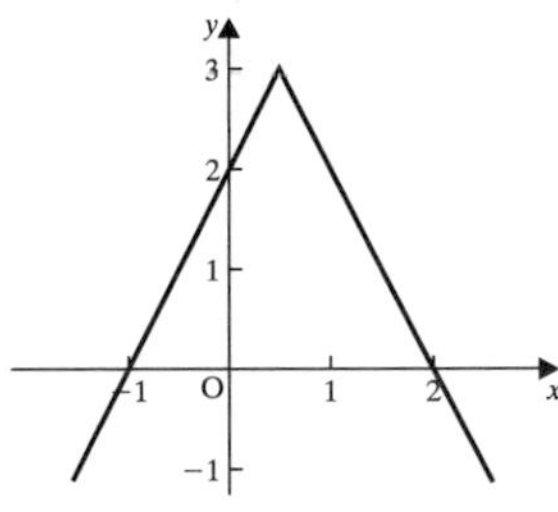

EXERCISE 18C

Sketch the following graphs.

1 $y = \|2x-1\|$	**5** $y = \|\sin x\|$	**9** $y = \|2x+5\|-4$
2 $y = \|x(x-1)(x-2)\|$	**6** $y = \|\ln x\|$	**10** $y = \|x^2-x-20\|$
3 $y = \|x^2-1\|$	**7** $y = \|\cos x\|$	**11** $y = 1+\|2-x^2\|$
4 $y = \|x^2+1\|$	**8** $y = 3+\|x+1\|$	**12** $y = \|\tan x\|$

A Function of |x|

When $y = f(|x|)$, then for positive values of x, $y = f(x)$ and for negative values of x, $y = f(-x)$.

For example, when $y = \ln|x|$, for $x = 2$, $y = \ln 2$
and for $x = -2$, $y = \ln 2$.

Therefore the curve whose equation is $y = f(|x|)$ is symmetrical about the y-axis.

To draw the curve whose equation is $y = f(|x|)$, we draw the part of the curve for which $x \geqslant 0$ and then add the reflection of this in the y-axis to get the rest of the curve.

The diagram shows the curve $y = \ln|x|$.

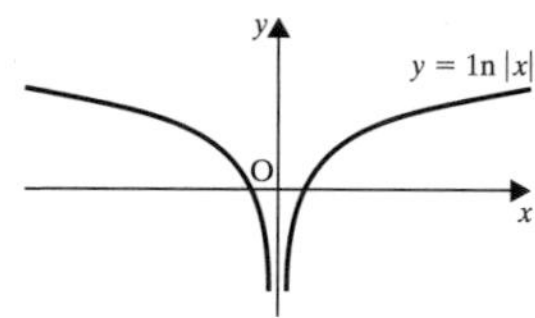

There is a fundamental difference between $f(|x|)$ (a function of a modulus) and $|f(x)|$ (a modulus of a function); $f(|x|)$ can be negative (e.g. $\ln|-0.5| = \ln 0.5 = -0.69\ldots$) but $|f(x)|$ is always positive.

EXERCISE 18D

Sketch the following graphs.

1 $y = \sin\|x\|$	**3** $y = e^{\|x\|}$
2 $y = (\|x\|)^3$	**4** $y = (\|x\|-1)(\|x\|-2)$

Cartesian Equations

When a section of the curve $y = f(x)$ is reflected in Oy, the equation of that part of the curve becomes $y = -f(x)$.

e.g., if $y = |x|$ for $x \in \mathbb{R}$ we can write this equation as

$$\begin{cases} y = x & \text{for } x \geqslant 0 \\ y = -x & \text{for } x < 0 \end{cases}$$

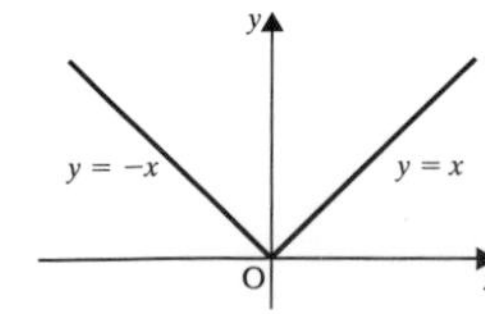

Intersection

To find the points of intersection between two graphs whose equations involve a modulus, we first sketch the graphs to locate the points roughly. Then we identify the equations in non-modulus form for each part of the graph. If these equations are written on the sketch then the correct pair of equations for solving simultaneously can be identified.

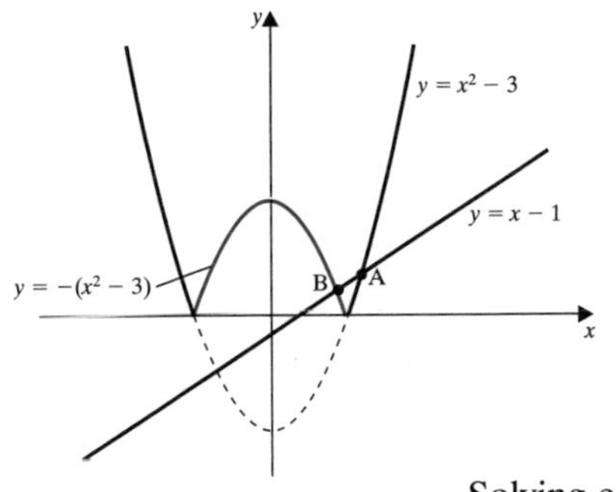

For example, the points common to $y = x - 1$ and $y = |x^2 - 3|$ can be seen from the sketch.

We can also see from the sketch that the coordinates of A satisfy the equations

$$y = x - 1 \text{ and } y = x^2 - 3 \qquad [1]$$

and the coordinates of B satisfy the equations

$$y = x - 1 \text{ and } y = 3 - x^2 \qquad [2]$$

Solving equations [1] gives $x^2 - x - 2 = 0 \quad \Rightarrow \quad x = -1$ or 2

It is clear from the diagram that $x \neq 1$, so A is the point $(2, 1)$.

Similarly, solving equations [2] gives $x^2 + x - 4 = 0$

$$\Rightarrow \qquad x = \tfrac{1}{2}(-1 \pm \sqrt{17})$$

Again from the diagram, it is clear that the x-coordinate of B is positive, so at B, $x = \tfrac{1}{2}(-1 + \sqrt{17})$

Then using $y = x - 1$ gives $y = \tfrac{1}{2}(-3 + \sqrt{17})$

This example illustrates the importance of checking solutions to see if they are relevant to the given problem.

EXAMPLE 18E Find the coordinates of the points of intersection of the graphs

$$y = 2 - |x - 1| \text{ and } y = |x + 2| - 3$$

> The sequence of diagrams show how the graphs and the non-modulus forms of the equations are built up from transformations.

For $y = 2 - |x - 1|$

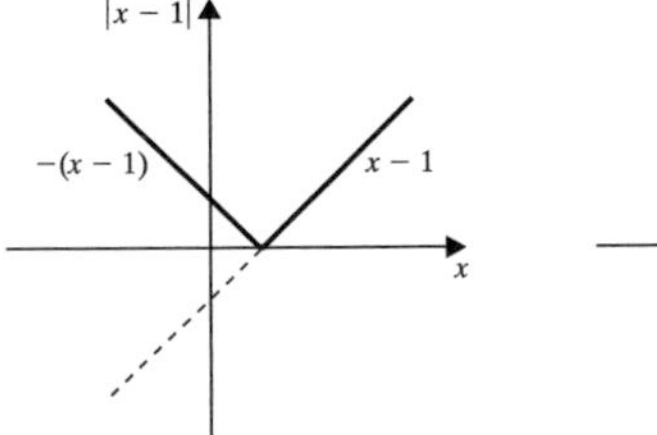

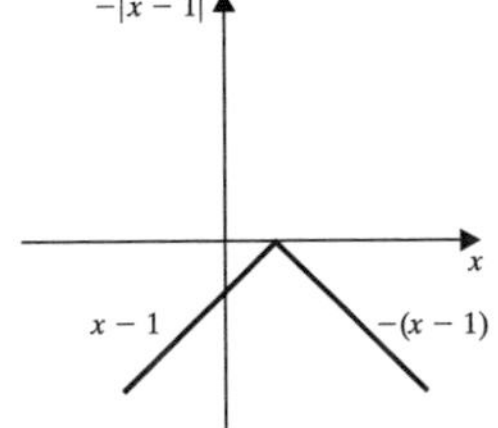

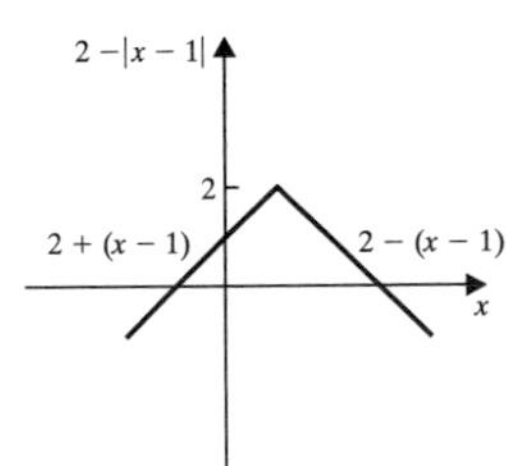

For $y = |x+2| - 3$

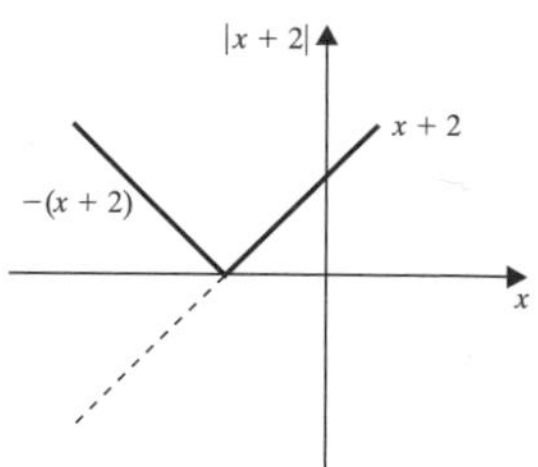

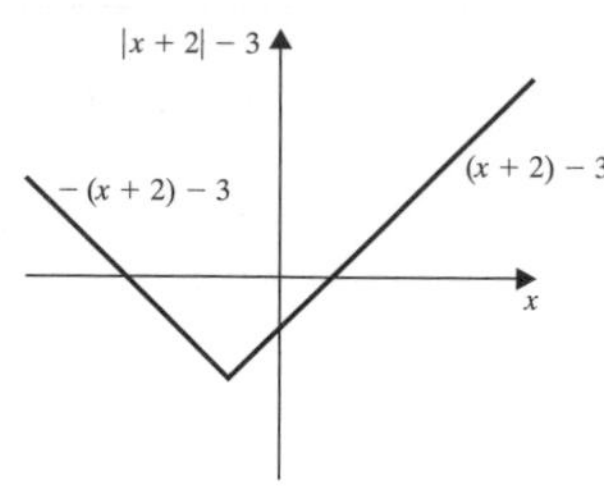

Both graphs are then drawn on the same set of axes.

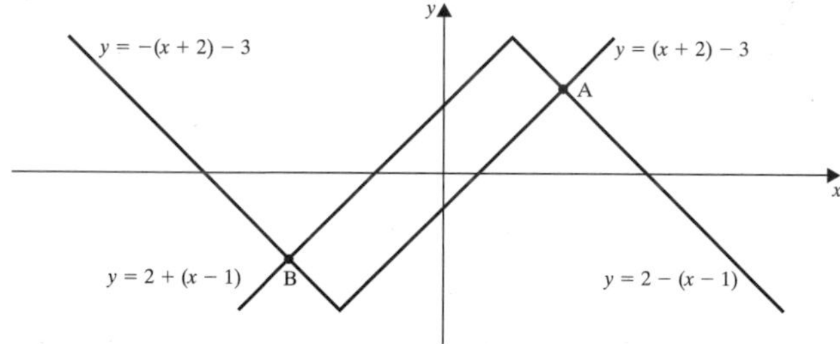

From this graph we can see that the coordinates of A satisfy the equations

$y = (x+2) - 3$ and $y = 2 - (x-1) \quad \Rightarrow \quad x = 2$ and $y = 1$

and the coordinates of B satisfy the equations

$y = 2 + (x-1)$ and $y = -(x+2) - 3 \quad \Rightarrow \quad x = -3$ and $y = -2$

Therefore the points of intersection are $(2, 1)$ and $(-3, -2)$.

Solving Equations Involving Modulus Signs

We can solve an equation such as $|2x-1| = 3x$ by sketching the graphs of $y = |2x-1|$ and $y = 3x$ and finding their points of intersection as illustrated above.

Alternatively we can use the fact that if $|a| = b$ then $a^2 = b^2$ to produce an equation in non-modulus form.

e.g.
$$
\begin{aligned}
|2x-1| = 3x \quad \Rightarrow \quad (2x-1)^2 &= 9x^2 \\
\Rightarrow \quad 4x^2 - 4x + 1 &= 9x^2 \\
\Rightarrow \quad 5x^2 + 4x - 1 &= 0 \\
\Rightarrow \quad (5x-1)(x+1) &= 0 \text{ giving } x = \tfrac{1}{5} \text{ or } x = -1.
\end{aligned}
$$

Remember that when the solution of an equation involves squaring, it is vital that the roots are checked in the original equation.

Checking in the original equation shows that when $x = \frac{1}{5}$, $|2x-1| = \frac{3}{5} = 3x$ and when $x = -1$, $|2x-1| = 3$, but $3x = -3$, so $x = \frac{1}{5}$ is the only solution.

Note that, although $|a| = b \Rightarrow a^2 = b^2$, the converse is *not* true, i.e. $a^2 = b^2 \not\Rightarrow |a| = b$.

EXERCISE 18E

Find the points of intersection of the graphs.

1 $y = |x|$ and $y = 1 - |x|$

2 $y = x$ and $y = |x^2 - 2x|$

3 $y = 2|x|$ and $y = 3 + 2x - x^2$

4 $y = \left|\frac{1}{x}\right|$ and $y = |x|$

5 $y = |x^2 - 4|$ and $y = 2x + 1$

Solve the following equations.

6 $|x^2 - 1| - 1 = 3x - 2$

7 $2 - |x + 1| = |4x - 3|$

8 $2|1 - x| = x$

9 $|2 - x^2| + 2x + 1 = 0$

Inequalities

Many inequalities can be solved easily with the aid of sketch graphs. The following worked examples illustrate how to use graphs to solve a variety of inequality problems. In some cases they provide an alternative, though not always as direct, method for solving similar problems discussed in Chapter 12.

EXAMPLES 18F

1 Find the set of values of x for which $|x - 3| < 5 - |x|$.

We start by drawing on the same set of axes the graphs $y = |x - 3|$ and $y = 5 - |x|$

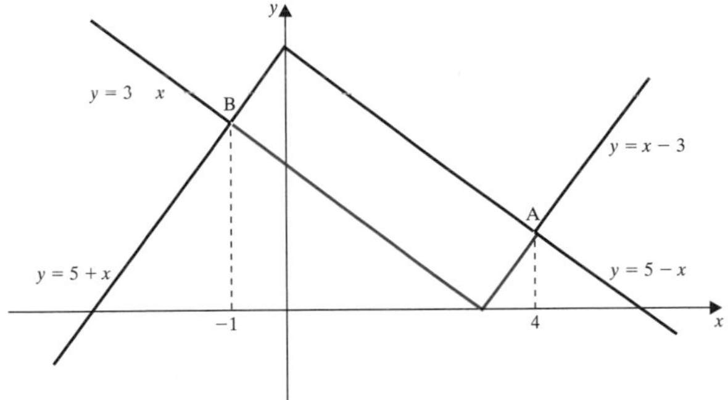

At A $\quad x - 3 = 5 - x \quad \Rightarrow \quad x = 4$

and at B $\quad 3 - x = 5 + x \quad \Rightarrow \quad x = -1$

From the graph we see that

$$|3 - x| < 5 - |x| \text{ for } -1 < x < 4.$$

2 Find the set of values of x for which $\dfrac{(x-1)^2}{x+5} < 1$.

> Divide out $\dfrac{(x-1)^2}{x+5}$ so that it contains only proper fractions.

$$\frac{(x-1)^2}{x+5} = x - 7 + \frac{36}{x+5}$$

The inequality then becomes $x - 7 + \dfrac{36}{x+5} < 1 \quad \Rightarrow \quad \dfrac{36}{x+5} < 8 - x$

> On the same set of axes we now draw sketches of $y = 8 - x$ and $y = \dfrac{36}{x+5}$.

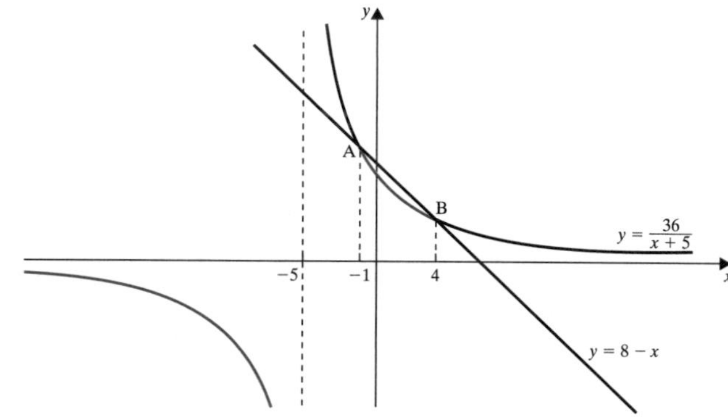

The x coordinates of A and B are the solutions of the equation

$\dfrac{36}{x+5} = 8 - x \quad \Rightarrow \quad x^2 - 3x - 4 = 0 \quad \Rightarrow \quad x = -1$ and 4

$\dfrac{36}{x+5} < 8 - x$ when the curve is below the line.

From the graph we see that this is when $x < -5$ and when $-1 < x < 4$

> Notice that we adjusted the inequality so that each side was an expression whose graph was easily recognised.

EXERCISE 18F

Solve the following inequalities.

1 $\dfrac{1}{x-1} < x$

2 $\dfrac{1}{x} > x^2$

3 $\dfrac{x}{x-1} < 0$

4 $\dfrac{x+1}{x} < 0$

5 $\dfrac{x+1}{x+2} < 0$

6 $\dfrac{1+x^2}{x} > 0$

7 $|x| < 1 - |x|$

8 $|x-1| < |x+2|$

9 $2|x| < |1-x|$

10 $|x+1| < 2x$

11 $3x - 1 < 1 + |x|$

12 $|3x+2| > 2 - |x+1|$

13 $1 + x^2 > 2x + 1$

14 $1 + x^2 > |2x+1|$

15 $|1 - x^2| < 2x + 1$

16 $1 - |x| > x^2 - 1$

17 Find the set of values of x between 0 and 2π for which $|\sin x| < |\cos x|$.

18 On the same set of axes sketch the curves whose equations are

$$y = x \quad \text{and} \quad y = \frac{1}{x}$$

Deduce the shape of the curve whose equation is $y = x + \frac{1}{x}$ and the range of $f : x \to x + \frac{1}{x}$.

19 Use any method to find the range of the function, f, where

a $f(x) = \dfrac{1}{1-x}$ **b** $f(x) = \dfrac{1}{(x-2)(x-6)}$

c $f(x) = \dfrac{x}{(1-x)^2}$ **d** $f(x) = \dfrac{4}{(x-1)(x-3)}$

MIXED EXERCISE 18

1 Sketch the curve whose equation is $y = 1 - \dfrac{1}{x-3}$.

Give the equations of the asymptotes and the coordinates of the intercepts on the axes.

2 The function f is defined by

$$f(x) = \sin x \quad \text{for} \quad 0 \leqslant x < \tfrac{1}{2}\pi$$
$$f(x) = \pi - x \quad \text{for} \quad \tfrac{1}{2}\pi \leqslant x < \pi$$
$$f(x) = f(x + \pi) \quad \text{for} \quad x \in \mathbb{R}$$

Sketch the graph of $f(x)$ for $0 \leqslant x < 4\pi$.

3 Sketch the graph of $g(x) = \dfrac{x+2}{x+1}$.

On the same set of axes sketch the graph of the function g^{-1} and hence state the range of g^{-1}.

4 Find the set of values of x for which $|x-1| > 1 + |1+x|$.

5 Find the range of values of x for which $\dfrac{x^2-2}{x} > 1$.

6 The function f is periodic with period 2 and

$$f(x) = x^2 \quad \text{for} \quad 0 \leqslant x < 1$$
$$f(x) = 3 - 2x \quad \text{for} \quad 1 \leqslant x < 2$$

Sketch the graph of f for $-2 \leqslant x \leqslant 4$.

7 State whether $f(x)$ is odd, even, periodic (in which case give the period) or none of these.

a $f(x) = \tan x$

b $f(x) = (x+1)(x)(x-1)$

c $f(x) = x^4$

d $f(x) = \sin(x - \frac{1}{2}\pi)$

8 Sketch the graph of $y = |\sin x|$.
Which of the following statements apply to $f(x) = |\sin x|$?

a $f(x)$ is even

b $f(x)$ is odd

c $f(x)$ is periodic.

9 Sketch the graph of $y = \cos|x|$
Is it true that $f(x) = \cos|x| \Rightarrow$

a $f(x)$ is even

b $f(x)$ is odd

c $f(x)$ is periodic.

10 Solve the equations

a $|x+2| = 1 - x$

b $|x| + 1 = x - |x|$

c $|3x - 1| = 1 + |x|$

CHAPTER 19 Sequences

Sequences

Consider the following sets of numbers,

2, 4, 6, 8, 10, ...
1, 2, 4, 8, 16, ...
4, 9, 16, 25, 36, ...

Each set of numbers, in the order given, has a pattern and there is an obvious rule for obtaining the next number and as many subsequent numbers as we wish to find.
Sets like these are called *sequences* and each member of the set is a term of the sequence.

The terms in a sequence are denoted by $u_1, u_2, \ldots, u_r, \ldots$ where u_r is the rth term.

Defining a Sequence

Consider the sequence 1, 2, 4, 8, 16, ...

Each term is a power of 2 so we can write the sequence as

$$2^0, 2^1, 2^2, 2^3, 2^4, \ldots$$

All the terms are of the form 2^r, so 2^r is a general term.
However, 2^r is *not* the rth term, as $u_1 = 2^0, u_2 = 2^1, u_3 = 2^2, \ldots$

Now we can see that $u_r = 2^{r-1}$ and we can use this to find any term of the sequence.

e.g. the ninth term, u_9, is given by $u_9 = 2^{9-1} = 256$

Hence the rule $u_r = 2^{r-1}$ for $r = 1, 2, 3, \ldots$ enables the whole sequence to be generated and so defines it completely.

However it is not always possible, or desirable, to define a sequence by giving u_r in terms of r.

Consider the sequence 2, 4, 6, 8, 10, ...

The obvious way to describe this sequence is 'starting with 2, add 2 to each term to get the next term',
or, using symbolic notation, $u_1 = 2,\ u_{r+1} = u_r + 2$

This is also a definition of the sequence since it can be generated as follows.

$$u_1 = 2,\ u_2 = u_1 + 2 = 4,\ u_3 = u_2 + 2 = 6, \text{ and so on.}$$

When a sequence is generated from one known term together with a relationship between consecutive terms, the process is called *iteration* and the sequence is said to be defined iteratively. The relationship between u_r and u_{r+1} is called a *recurrence* relation.

EXAMPLE 19A Write down the first four terms of the sequence defined by

a $u_r = \dfrac{r}{r+1}$ **b** $u_1 = 2, u_{r+1} = \dfrac{u_r}{u_r + 1}$

a $u_r = \dfrac{r}{r+1} \quad \Rightarrow \quad u_1 = \dfrac{1}{1+1} = \dfrac{1}{2}$

$$u_2 = \frac{2}{2+1} = \frac{2}{3}$$

$$u_3 = \frac{3}{3+1} = \frac{3}{4}$$

$$u_4 = \frac{4}{4+1} = \frac{4}{5}$$

b $u_1 = 2$, and $u_{r+1} = \dfrac{u_r}{u_r + 1} \quad \Rightarrow \quad u_2 = \dfrac{2}{2+1} = \dfrac{2}{3}$

$$u_3 = \frac{\frac{2}{3}}{\frac{2}{3}+1} = \frac{2}{5}$$

$$u_4 = \frac{\frac{2}{5}}{\frac{2}{5}+1} = \frac{2}{7}$$

The Behaviour of u_r as $r \to \infty$

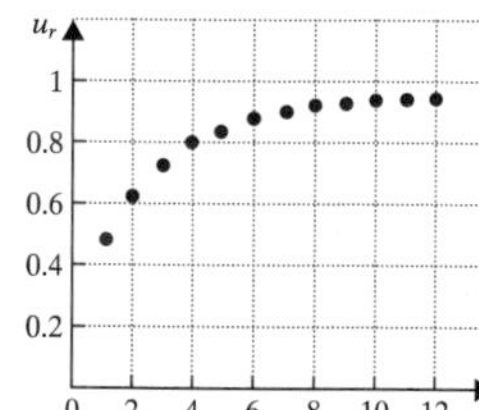

Consider the sequence $\frac{1}{2}, \frac{2}{3}, \frac{3}{4}, \frac{4}{5} \ldots$ All the terms are less than 1, and the values of the terms are increasing as r increases, i.e. as the sequence progresses, the value of the terms is getting closer to 1. Expressing this in symbols we have

$$u_r \to 1 \text{ as } r \to \infty \quad \text{or} \quad \lim_{r \to \infty} u_r = 1$$

and we say that the sequence *converges*.
We can illustrate this on a graph by plotting values of u_r against values of r.

Any sequence whose terms approach one finite value is said to be convergent.

A sequence that is not convergent is called divergent.

Some examples of divergent sequences are

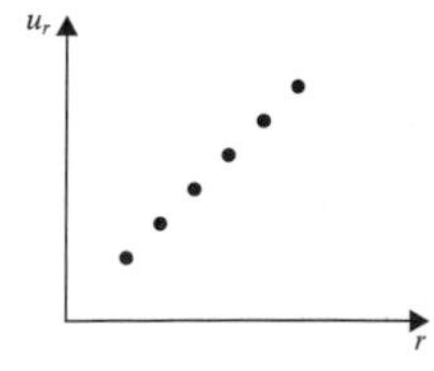

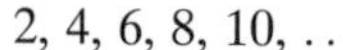

2, 4, 6, 8, 10, ...
The terms are increasing without limit, so this sequence is clearly divergent.

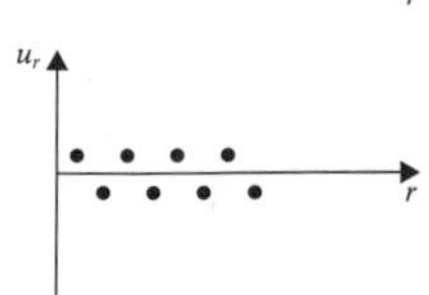

1, –1, 1, –1, 1, –1, ...
The terms form a repeating pattern, i.e. 1, –1, so they do not approach *one* finite value and the sequence is divergent.

A divergent sequence whose terms form a repeating pattern is called a periodic sequence.

For example, 1, 2, 3, 1, 2, 3, 1, 2, 3, ... is a periodic sequence.

As the sequence progresses, we can describe the behaviour of the terms as periodic, or say that it cycles or oscillates in a regular pattern.

It is usually obvious from a few terms of a sequence what the nature of that sequence is. If this is not the case, a sketch showing values of u_r plotted against r will usually make it clear.

The examples below show such plots for convergent, periodic and simple divergent sequences.

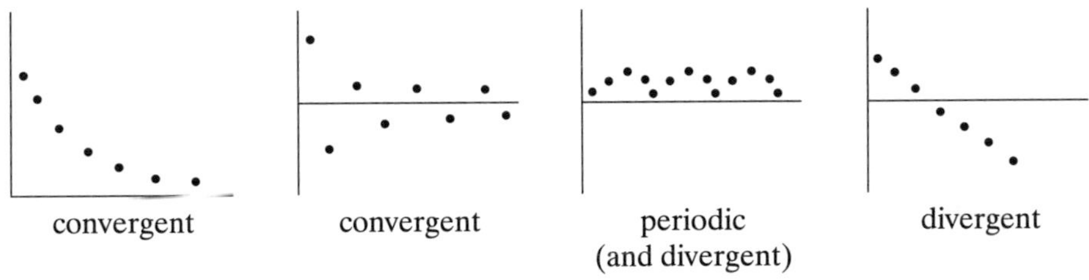

convergent convergent periodic (and divergent) divergent

EXERCISE 19A

1 Write down the first six terms of each sequence and state whether it is convergent, divergent, or divergent and periodic.

a $u_r = \dfrac{1}{r^r}$ b $u_r = (-1)^r 2^r$ c $u_r = \dfrac{(-1)^r}{2^r}$ d $u_r = \sin\dfrac{\pi r}{6}$

2 Write down the first six terms of the sequence given by the recurrence relation.

a $u_{r+1} = \dfrac{1+u_r}{1+2u_r}$; $u_1 = 3$ b $u_{r+1} = 2 - (u_r)^2 - u_r$; $u_1 = -4$

c $u_r = \sqrt{2 - u_{r-1}}$; $u_1 = 0.5$ d $u_r = \frac{1}{5}(5 - \{u_{r-1}\}^2)$; $u_1 = 1$

3 Describe what happens to the terms of the sequence generated by the iteration formula $u_{r+1} = \dfrac{2}{2-u_r}$ as r increases if

a $u_1 = 1$ b $u_1 = 0.5$ c $u_1 = -2$

4 Repeat Question 3 for the sequence defined by $u_{r+1} = \frac{1}{4}(2 - u_r)$ when

a $u_1 = 1$ b $u_1 = -1$ c $u_1 = 2$

5 A sequence is generated by the recurrence relation $u_{r+1} = \dfrac{1}{u_r - 3}$

Given that $u_2 = u_1$ find the possible values of u_1.

Series

When the terms of a sequence are added, a series is formed,

e.g. $1+2+4+8+16+\dots$ is a series.

If the series stops after a finite number of terms it is called a finite series,

e.g. $1+2+4+8+16+32+64$ is a finite series of seven terms.

If the series continues indefinitely it is called an infinite series,

e.g. $1+\frac{1}{2}+\frac{1}{4}+\frac{1}{8}+\frac{1}{16}+\frac{1}{32}+\dots+\frac{1}{1024}+\dots$ is an infinite series.

Consider again the series $1+2+4+8+16+32+64$
As each term is a power of 2 we can write this series in the form

$$2^0+2^1+2^2+2^3+2^4+2^5+2^6$$

All the terms of this series are of the form 2^r, so 2^r is a general term. We can then define the series as the sum of terms of the form 2^r, where r takes all integral values in order from 0 to 6 inclusive.

Using $\sum$ as a symbol for 'the sum of terms such as' we can redefine our series more concisely as $\sum 2^r$, r taking all integral values from 0 to 6 inclusive, or, even more briefly as $\sum_{r=0}^{6} 2^r$.

Placing the lowest and highest value that r takes below and above the sigma symbol respectively, indicates that r takes all integral values between these extreme values.

Thus $\sum_{r=2}^{10} r^3$ means 'the sum of all terms of the form r^3, where r takes all integral values from 2 to 10 inclusive',

i.e. $$\sum_{r=2}^{10} r^3 = 2^3+3^3+4^3+5^3+6^3+7^3+8^3+9^3+10^3$$

Note that a finite series, when written out, should always end with the last term even if intermediate terms are omitted, e.g. $3+6+9+\dots+99$.

The infinite series $1+\frac{1}{2}+\frac{1}{4}+\frac{1}{8}+\frac{1}{16}+\dots$ may also be written in the sigma notation. The continuing dots after the last written term indicate that the series is infinite, i.e. there is *no* last term. Each term of this series is a power of $\frac{1}{2}$ so a general term can be written $(\frac{1}{2})^r$. The first term is 1 or $(\frac{1}{2})^0$, so the first value that r takes is zero. There is no last term of this series, so there is no upper limit for the value of r.

Therefore $1+\frac{1}{2}+\frac{1}{4}+\frac{1}{8}+\frac{1}{16}+\dots$ may be written as $\sum_{r=0}^{\infty} (\frac{1}{2})^r$

Note that when a given series is rewritten in sigma notation it is as well to check that the first few values of r give the correct first few terms of the series.

Writing a series in sigma notation, apart from the obvious advantage of being brief, allows us to select a particular term of a series without having to write down all the earlier terms.

For example, in the series $\sum_{r=3}^{10}(2r+5)$,

the first term is the value of $2r+5$ when $r = 3$, i.e. $2 \times 3 + 5 = 11$,
the last term is the value of $2r+5$ when $r = 10$, i.e. 25,
the fourth term is the value of $2r+5$ when r takes its fourth value in order from $r = 3$, i.e. when $r = 6$.

Thus the fourth term of $\sum_{r=3}^{10}(2r+5)$ is $2 \times 6 + 5 = 17$.

EXAMPLE 19B

Write the following series in the sigma notation,

a $1 - x + x^2 - x^3 + \dots$ **b** $2 - 4 + 8 - 16 + \dots + 128$

a A general term of this series is $\pm x^r$, having a positive sign when r is even and a negative sign when r is odd.
Because $(-1)^r$ is positive when r is even and negative when r is odd, the general term can be written $(-1)^r x^r$.

The first term of this series is 1, or x^0.

Hence $1 - x + x^2 - x^3 + \dots = \sum_{r=0}^{\infty}(-1)^r x^r$

b $2 - 4 + 8 - 16 + \dots + 128 = 2 - (2)^2 + (2)^3 - (2)^4 + \dots + (2)^7$

So a general term is of the form $\pm 2^r$, being positive when r is odd and negative when r is even,
i.e. the general term is $(-1)^{r+1}2^r$

Hence $2 - 4 + 8 - 16 + \dots + 128 = \sum_{r=1}^{7}(-1)^{r+1}2^r$

EXERCISE 19B

1 Write the following series in the sigma notation:

a $1 + 8 + 27 + 64 + 125$ **b** $2 + 4 + 6 + 8 + \dots + 20$

c $\frac{1}{2} + \frac{1}{3} + \frac{1}{4} + \frac{1}{5} + \dots + \frac{1}{50}$ **d** $1 + \frac{1}{3} + \frac{1}{9} + \frac{1}{27} + \dots$

e $-4 - 1 + 2 + 5 \dots + 17$ **f** $8 + 4 + 2 + 1 + \frac{1}{2} + \dots$

2 Write down the first three terms and, where there is one, the last term of each of the following series:

a $\sum_{r=1}^{\infty}\frac{1}{r}$ **b** $\sum_{r=0}^{5}r(r+1)$

c $\sum_{r=0}^{20}\frac{r+2}{(r+1)(2r+1)}$ **d** $\sum_{r=0}^{\infty}\frac{1}{(r^2+1)}$

e $\sum_{r=-1}^{8}r(r+1)(r+2)$ **f** $\sum_{r=0}^{\infty}a^r(-1)^{r+1}$

3 For the following series, write down the term indicated, and the number of terms in the series.

a $\sum_{r=1}^{9} 2^r$, 3rd term

b $\sum_{r=-1}^{8} (2r+3)$, 5th term

c $\sum_{r=-6}^{-1} \frac{1}{(2r+1)}$, last term

d $\sum_{r=0}^{\infty} \frac{1}{(r+1)(r+2)}$, 20th term

e $\sum_{r=1}^{\infty} \left(\frac{1}{2}\right)^r$, nth term

f $8+4+0-4-8-12\ldots-80$, 15th term

g $\frac{1}{16}+\frac{1}{8}+\frac{1}{4}+\frac{1}{2}+\ldots+32$, 7th term

Arithmetic Progression

Consider the sequence $5, 8, 11, 14, 17, \ldots, 29$

Each term of this sequence exceeds the previous term by 3, so the sequence can be written in the form

$$5, (5+3), (5+2\times 3), (5+3\times 3), (5+4\times 3), \ldots, (5+8\times 3)$$

This sequence is an example of an arithmetic progression (AP) which is a sequence where any term differs from the preceding term by a constant, called the *common difference*. The common difference may be positive or negative. For example, the first six terms of an AP whose first term is 8 and whose common difference is -3, are $8, 5, 2, -1, -4, -7$.

In general, if an AP has a first term a, and a common difference d, the first four terms are $a, (a+d), (a+2d), (a+3d)$, and the nth term, u_n, is $a+(n-1)d$.

So **an AP with n terms can be written as**
$a, (a+d), (a+2d), \ldots, [a+(n-1)d]$

EXAMPLES 19C

1 The 8th term of an AP is 11 and the 15th term is 21. Find the common difference, the first term of the series, and the nth term.

If the first term of the series is a and the common difference is d, then the 8th term is $a+7d$,

$\therefore \quad a+7d = 11 \quad [1]$

and the 15th term is $a+14d$,

$\therefore \quad a+14d = 21 \quad [2]$

[2] – [1] gives $7d = 10 \Rightarrow d = \frac{10}{7}$ and $a = 1$

so the first term is 1 and the common difference is $\frac{10}{7}$.

Hence the nth term is $a+(n-1)d = 1+(n-1)\frac{10}{7} = \frac{1}{7}(10n-3)$

2 The nth term of an AP is $12-4n$. Find the first term and the common difference.

If the nth term is $12-4n$, the first term $(n=1)$ is 8.
The second term $(n=2)$ is 4.

Therefore the common difference is -4.

The Sum of an Arithmetic Progression

Consider the sum of the first ten even numbers, which is an AP.

Writing it first in normal, then in reverse, order we have

$$S = 2 + 4 + 6 + 8 + \ldots + 18 + 20$$
$$S = 20 + 18 + 16 + 14 + \ldots + 4 + 2$$

Adding gives $\quad 2S = 22 + 22 + 22 + 22 + \ldots + 22 + 22$

As there are ten terms in this series, we have

$$2S = 10 \times 22 \quad \Rightarrow \quad S = 110$$

Applying this method to a general AP gives formulae for the sum, which may be quoted and used.

If S_n is the sum of the first n terms of an AP with last term l,

then $\quad S_n = a + (a+d) + (a+2d) + \ldots + (l-d) + l$

reversing $\quad S_n = l + (l-d) + (l-2d) + \ldots + (a+d) + a$

adding $\quad 2S_n = (a+l) + (a+l) + (a+l) + \ldots + (a+l) + (a+l)$

as there are n terms we have $2S_n = n(a+l)$

$$\Rightarrow \quad \mathbf{S_n = \tfrac{1}{2}n(a+l)}$$

i.e. $\quad S_n$ = (number of terms) × (average term)

Also, because the nth term, l, is equal to $a + (n-1)d$, we have

$$S_n = \tfrac{1}{2}n[a + a + (n-1)d]$$

i.e. $\quad \mathbf{S_n = \tfrac{1}{2}n[2a + (n-1)d]}$

Either of these formulae can now be used to find the sum of the first n terms of an AP.

EXAMPLES 19C (continued)

3 Find the sum of the following series

a an AP of eleven terms whose first term is 1 and whose last term is 6

b $\displaystyle\sum_{r=1}^{8}\left(2 - \frac{2r}{3}\right)$

a We know the first and last terms, and the number of terms so we use $S_n = \frac{1}{2}n(a+l)$.

$$\Rightarrow \quad S_{11} = \tfrac{11}{2}(1+6) = \tfrac{77}{2}$$

b $\displaystyle\sum_{r=1}^{8}\left(2 - \frac{2r}{3}\right) = \frac{4}{3} + \frac{2}{3} + 0 - \frac{2}{3} - \ldots - \frac{10}{3}$

This is an AP with 8 terms where $a = \frac{4}{3}$, $d = -\frac{2}{3}$.

Using $S_n = \frac{1}{2}n[2a + (n-1)d]$ gives

$$S = 4[\tfrac{8}{3} + 7(-\tfrac{2}{3})] = -8$$

4 In an AP the sum of the first ten terms is 50 and the 5th term is three times the 2nd term. Find the first term and the sum of the first 20 terms.

> If a is the first term and d is the common difference, and there are n terms, using $S = \frac{1}{2}n[2a + (n - 1)d]$ gives

$S_{10} = 50 = 5(2a + 9d)$ [1]

Now using $u_n = a + (n - 1)d$ gives

$u_5 = a + 4d$ and $u_2 = a + d$

Therefore $a + 4d = 3(a + d)$ [2]

From [1] and [2] we get $d = 1$ and $a = \frac{1}{2}$.
So the first term is $\frac{1}{2}$ and the sum of the first 20 terms is S_{20} where

$S_{20} = 10(1 + 19 \times 1) = 200$

5 Show that the terms of $\sum_{r=1}^{n} \ln 2^r$ are in arithmetic progression.

Find the sum of the first 10 terms of this series.

By taking $r = 1, 2, 3 \ldots$ we have

$$\begin{aligned}\sum_{r=1}^{n} \ln 2^r &= \ln 2 + \ln 2^2 + \ln 2^3 + \ldots + \ln 2^n \\ &= \ln 2 + 2\ln 2 + 3\ln 2 + \ldots + n\ln 2\end{aligned}$$

We now see that there is a common difference of $\ln 2$ between successive terms, so the terms of this series are in arithmetic progression.

$$\begin{aligned}\text{Hence} \quad \sum_{r=1}^{10} \ln 2^r &= \ln 2 + 2\ln 2 + 3\ln 2 + \ldots + 10\ln 2 \\ &= (1 + 2 + 3 + \ldots + 10)\ln 2 \\ &= \tfrac{10}{2}(1 + 10)\ln 2 = 55\ln 2\end{aligned}$$

Note that the sum of the first n natural numbers,

i.e. $1 + 2 + 3 + \ldots + n$

is an AP in which $a = 1$ and $d = 1$ so

$$\sum_{r=1}^{n} r = \tfrac{1}{2}n(n + 1)$$

This is a result that may be quoted, unless a proof is specifically asked for.

6 The sum of the first n terms of a series is given by $S_n = n(n + 3)$
Find the fourth term of the series and show that the terms are in arithmetic progression.

If the terms of the series are $a_1, a_2, a_3, \ldots, a_n$

then $S_n = a_1 + a_2 + \ldots + a_n = n(n + 3)$

So $\quad S_4 = a_1 + a_2 + a_3 + a_4 = 28$

and $\quad S_3 = a_1 + a_2 + a_3 = 18$

Hence the fourth term of the series, a_4, is 10.

Now $\quad S_n = a_1 + a_2 + \ldots + a_{n-1} + a_n = n(n+3)$

and $\quad S_{n-1} = a_1 + a_2 + \ldots + a_{n-1} = (n-1)(n+2)$

Hence the nth term of the series, a_n, is given by

$a_n = n(n+3) - (n-1)(n+2) = 2n + 2$

Replacing n by $n-1$ gives the $(n-1)$th term

i.e. $\quad a_{n-1} = 2(n-1) + 2 = 2n$

Then $\quad a_n - a_{n-1} = (2n+2) - 2n = 2$

i.e. there is a common difference of 2 between successive terms, showing that the series is an AP.

EXERCISE 19C

1 Write down the fifth term and the nth term of the following APs.

a $\sum_{r=1}^{n}(2r-1)$ **b** $\sum_{r=1}^{n}4(r-1)$

c $\sum_{r=0}^{n}(3r+3)$

d first term 5, common difference 3

e first term 6, common difference -2

f first term p, common difference q

g first term 10, last term 30, 11 terms

h $1, 5, \ldots$

i $2, 1\frac{1}{2}, \ldots$

j $-4, -1, \ldots$

2 Find the sum of the first ten terms of each AP given in Question 1.

3 The 9th term of an AP is 8 and the 4th term is 20. Find the first term and the common difference.

4 The 6th term of an AP is twice the 3rd term and the first term is 3. Find the common difference and the 10th term.

5 The nth term of an AP is $\frac{1}{2}(3-n)$. Write down the first three terms and the 20th term.

6 Find the sum, to the number of terms indicated, of each of the following APs.

a $1 + 2\frac{1}{2} + \ldots$, 6 terms

b $3 + 5 + \ldots$, 8 terms

c the first twenty odd integers

d $a_1 + a_2 + a_3 + \ldots + a_8$ where $a_n = 2n + 1$

e $4 + 6 + 8 + \ldots + 20$

f $\sum_{r=1}^{3n}(3-4r)$

g $S_n = n^2 - 3n$, 8 terms

h $S_n = 2n(n+3)$, m terms

7 The sum of the first n terms of an AP is S_n where $S_n = n^2 - 3n$. Write down the fourth term and the nth term.

8 The sum of the first n terms of a series is given by S_n where $S_n = n(3n-4)$. Show that the terms of the series are in arithmetic progression.

9 In an arithmetic progression, the 8th term is twice the 4th term and the 20th term is 40. Find the common difference and the sum of the terms from the 8th to the 20th inclusive.

10 How many terms of the AP, $1+3+5+\dots$ are required to make a sum of 1521?

11 Find the least number of terms of the AP, $1+3+5+\dots$, that are required to make a sum exceeding 4000.

12 If the sum of the first n terms of a series is S_n where $S_n = 2n^2 - n$,

a prove that the series is an AP, stating the first term and the common difference

b find the sum of the terms from the 3rd to the 12th inclusive.

13 In an AP the 6th term is half the 4th term and the 3rd term is 15.

a Find the first term and the common difference.

b How many terms are needed to give a sum that is less than 65?

Geometric Progressions

Consider the sequence

$$12,\ 6,\ 3,\ 1.5,\ 0.75,\ 0.375,\ \dots$$

Each term of this sequence is half the preceding term so the sequence may be written

$$12,\ 12(\tfrac{1}{2}),\ 12(\tfrac{1}{2})^2,\ 12(\tfrac{1}{2})^3,\ 12(\tfrac{1}{2})^4,\ 12(\tfrac{1}{2})^5,\ \dots$$

Such a sequence is called a geometric progression (GP) which is a sequence where each term is a constant multiple of the preceding term. This constant multiplying factor is called the *common ratio*, and it can have any value.

Hence, if a GP has a first term of 3 and a common ratio of -2, the first four terms are

$$3,\ 3(-2),\ 3(-2)^2,\ 3(-2)^3$$

i.e. $3,\ -6,\ 12,\ -24$

In general if a GP has a first term a, and a common ratio r, the first four terms are a, ar, ar^2, ar^3 and the nth term, u_n, is ar^{n-1},

i.e. **a GP with n terms can be written $a,\ ar,\ ar^2,\ \dots,\ ar^{n-1}$**

The Sum of a Geometric Progression

Consider the sum of the first eight terms, S_8, of the GP with first term 1 and common ratio 3,

i.e. $$S_8 = 1 + 1(3) + 1(3)^2 + 1(3)^3 + \dots + 1(3)^7$$

$\Rightarrow$ $$3S_8 = 3 + 3^2 + 3^3 + \dots + 3^7 + 3^8$$

Hence $$S_8 - 3S_8 = 1 + 0 + 0 + 0 + \dots + 0 - 3^8$$

So $$S_8(1-3) = 1 - 3^8$$

$\Rightarrow$ $$S_8 = \frac{1-3^8}{1-3} = \frac{3^8-1}{2}$$

This process can be applied to a general GP.

Consider the sum, S_n, of the first n terms of a GP with first term a and common ratio r,

i.e. $S_n = a + ar + \dots + ar^{n-2} + ar^{n-1}$

Multiplying by r gives $rS_n = ar + ar^2 + \dots + ar^{n-1} + ar^n$

Hence $S_n - rS_n = a - ar^n$

$\Rightarrow$ $S_n(1-r) = a(1-r^n)$

$$\Rightarrow \quad \mathbf{S_n = \frac{a(1-r^n)}{1-r}}$$

If $r > 1$ the formula may be written $\dfrac{a(r^n-1)}{r-1}$.

EXAMPLES 19D

1 The 5th term of a GP is 8, the third term is 4, and the sum of the first ten terms is positive. Find the first term, the common ratio, and the sum of the first ten terms.

> For a first term a and common ratio r, the nth term is ar^{n-1}.

Therefore $ar^4 = 8 \quad (n = 5)$

and $ar^2 = 4 \quad (n = 3)$

dividing gives $r^2 = 2$

$\Rightarrow$ $r = \pm\sqrt{2}$ and $a = 2$

Using the formula $S_n = \dfrac{a(r^n-1)}{r-1}$ gives,

when $r = \sqrt{2}$, $S_{10} = \dfrac{2[(\sqrt{2})^{10} - 1]}{\sqrt{2}-1} = \dfrac{62}{\sqrt{2}-1}$

when $r = -\sqrt{2}$, $S_{10} = \dfrac{2[(-\sqrt{2})^{10} - 1]}{-\sqrt{2}-1} = \dfrac{-62}{\sqrt{2}+1}$

But we are told that $S_{10} > 0$, so we deduce that

$r = \sqrt{2}$ and $S_{10} = \dfrac{62}{\sqrt{2}-1} = 62(\sqrt{2}+1)$

2 A prize fund is set up with a single investment of £2000 to provide an annual prize of £150. The fund accrues interest at 5% p.a. paid yearly. If the first prize is awarded one year after the investment, find the number of years for which the full prize can be awarded.

> After one year the value of the fund is the initial investment of £2000, plus 5% interest, less one £150 prize, i.e. £{(1.05) (2000) – 150}.

If £P_n is the value of the fund after n years then

$P_1 = 2000(1.05) - 150$

$P_2 = 1.05P_1 - 150 = 2000(1.05)^2 - 150(1.05) - 150$

$P_3 = 1.05P_2 - 150 = 2000(1.05)^3 - 150^2(1.05)^2 - 150(1.05) - 150$

$$P_n = 2000(1.05)^n - 150(1.05)^{n-1} - 150(1.05)^{n-2} - \ldots - 150$$
$$= 2000(1.05)^n - 150[1 + 1.05 + \ldots + (1.05)^{n-1}]$$

The expression in square brackets is a GP of n terms with $a = 1$ and $r = 1.05$ and hence

$$P_n = 2000(1.05)^n - 150\left[\frac{(1.05)^n - 1}{1.05 - 1}\right]$$
$$= 3000 - 1000(1.05)^n$$

The fund can award the full prize as long as there is money left in the fund at the end of a year, i.e. as long as $P_n \geqslant 0$

$\Rightarrow \quad 3000 - 1000(1.05)^n \geqslant 0$

$\Rightarrow \quad 1.05^n \leqslant 3$

$\Rightarrow \quad n \ln 1.05 \leqslant \ln 3$

$\therefore \quad n \leqslant \ln 3 \div \ln 1.05 = 22.5\ldots$

Dividing by ln 1.05 does not alter the inequality as ln 1.05 is positive.

Therefore the prize fund contains some money after 22 years but would not after 23 years, so the full prize can be awarded for 22 years.

3 The sum of the first n terms of a series is $3^n - 1$. Show that the terms of this series are in geometric progression and find the first term, the common ratio and the sum of the second n terms of this series.

If the series is $\quad a_1 + a_2 + \ldots + a_n$

then $\quad S_n = a_1 + a_2 + \ldots + a_{n-1} + a_n = 3^n - 1$

and $\quad S_{n-1} = a_1 + a_2 + \ldots + a_{n-1} = 3^{n-1} - 1$

therefore $\quad a_n = 3^n - 1 - (3^{n-1} - 1)$

i.e. the nth term is $3^n - 3^{n-1} = 3^{n-1}(3 - 1) = (2)3^{n-1}$

Similarly $a_{n-1} = (2)3^{n-2}$ so $a_n \div a_{n-1} = 3$

showing that successive terms in the series have a constant ratio of 3. Hence this series is a GP with first term 2 and common ratio 3.

The sum of the second n terms is

(the sum of the first $2n$ terms) – (the sum of the first n terms)

$= S_{2n} - S_n$

$= (3^{2n} - 1) - (3^n - 1)$

$= 3^{2n} - 3^n$

EXERCISE 19D

1 Write down the fifth term and the nth term of the following GPs.

a 2, 4, 8, ... **b** $2, 1, \frac{1}{2}, \ldots$ **c** $3, -6, 12, \ldots$

d first term 8, common ratio $-\frac{1}{2}$ **e** first term 3, last term $\frac{1}{81}$, 6 terms

2 Find the sum, to the number of terms given, of the following GPs.

a $3 + 6 + \ldots$, 6 terms **b** $3 - 6 + \ldots$, 8 terms

c $1 + \frac{1}{2} + \frac{1}{4} + \ldots$, 20 terms **d** first term 5, common ratio $\frac{1}{5}$, 5 terms

e first term $\frac{1}{2}$, common ratio $-\frac{1}{2}$, 10 terms

f first term 1, common ratio -1, 2001 terms.

3 The 6th term of a GP is 16 and the 3rd term is 2. Find the first term and the common ratio.

4 Find the common ratio, given that it is negative, of a GP whose first term is 8 and whose 5th term is $\frac{1}{2}$.

5 The nth term of a GP is $(-\frac{1}{2})^n$. Write down the first term and the 10th term.

6 Evaluate $\sum_{r=1}^{10} (1.05)^r$

7 Find the sum to n terms of the following series.

a $x + x^2 + x^3 + \ldots$ **b** $x + 1 + \frac{1}{x} + \ldots$

c $1 - y + y^2 - \ldots$ **d** $x + \frac{x^2}{2} + \frac{x^3}{4} + \frac{x^4}{8} + \ldots$

e $1 - 2x + 4x^2 - 8x^3 + \ldots$

8 Find the sum of the first n terms of the GP $2 + \frac{1}{2} + \frac{1}{8} + \ldots$ and find the least value of n for which this sum exceeds 2.65.

9 The sum of the first 3 terms of a GP is 14. If the first term is 2, find the possible values of the sum of the first 5 terms.

10 Evaluate $\sum_{r=1}^{10} 3(\frac{3}{4})^r$

11 A mortgage is taken out for £10 000 and is repaid by annual instalments of £2000. Interest is charged on the outstanding debt at 10%, calculated annually. If the first repayment is made one year after the mortgage is taken out find the number of years it takes for the mortgage to be repaid.

12 A bank loan of £500 is arranged to be repaid in two years by equal monthly instalments. Interest, *calculated monthly*, is charged at 11% p.a. on the remaining debt. Calculate the monthly repayment if the first repayment is to be made one month after the loan is granted.

Convergence of Series

If a piece of string, of length l, is cut up by first cutting it in half and keeping one piece, then cutting the remainder in half and keeping one piece, and so on, the sum of the lengths retained is

$$\frac{l}{2}+\frac{l}{4}+\frac{l}{8}+\frac{l}{16}+\dots$$

As this process can (in theory) be carried on indefinitely, the series formed above is infinite.

After several cuts have been made the remaining part of the string will be very small indeed, so the sum of the cut lengths will be very nearly equal to the total length, l, of the original piece of string. The more cuts that are made the closer to l this sum becomes,
i.e. if after n cuts, the sum of the cut lengths is

$$\frac{l}{2}+\frac{l}{2^2}+\frac{l}{2^3}+\dots+\frac{l}{2^n}$$

then, as $n \to \infty$, $\frac{l}{2}+\frac{l}{2^2}+\dots+\frac{l}{2^n} \to l$

or $$\lim_{n\to\infty}\left[\frac{l}{2}+\frac{l}{2^2}+\dots+\frac{l}{2^n}\right]=l$$

l is called the sum to infinity of this series.

In general, if S_n is the sum of the first n terms of any series and if $\lim_{n\to\infty}[S_n]$ exists and is finite, the series is said to be *convergent*.

In this case the sum to infinity, S_∞, is given by

$$\mathbf{S_\infty = \lim_{n\to\infty}[S_n]}$$

The series $l/2+l/2^2+l/2^3+\dots$ for example, is convergent as its sum to infinity is l.

However, for the series $1+2+3+\dots+n$, we have $S_n = \frac{1}{2}n(n+1)$.
As $n \to \infty$, $S_n \to \infty$ so this series does not converge and is said to be divergent.

For any AP, $S_n = \frac{1}{2}n[2a+(n-1)d]$, which always approaches infinity as $n \to \infty$. Therefore any AP is divergent.

The Sum to Infinity of a GP

Consider the general GP, $a+ar+ar^2+\dots$

Now $$S_n = \frac{a(1-r^n)}{1-r}$$

and if $|r|<1$, then $\lim_{n\to\infty} r^n = 0$

So $$\lim_{n\to\infty} S_n = \lim_{n\to\infty}\left[\frac{a(1-r^n)}{1-r}\right] = \frac{a}{1-r}$$

If $|r| > 1$, $\lim_{n \to \infty} r^n = \infty$ and the series does not converge.

Therefore, provided that $|r| < 1$, a GP converges to a sum of $\dfrac{a}{1-r}$.

i.e. **for a GP, $S_\infty = \dfrac{a}{1-r}$ provided that $|r| < 1$.**

EXAMPLES 19E

1 Determine whether each series converges. If it does, give its sum to infinity.

a $3 + 5 + 7 + \ldots$ **b** $1 - \frac{1}{4} + \frac{1}{16} - \frac{1}{64} + \ldots$ **c** $3 + \frac{9}{2} + \frac{27}{4} + \ldots$

a $3 + 5 + 7 + \ldots$ is an AP $(d = 2)$ and so does not converge.

b $1 - \frac{1}{4} + \frac{1}{16} - \frac{1}{64} + \ldots = 1 + (-\frac{1}{4}) + (-\frac{1}{4})^2 + (-\frac{1}{4})^3 + \ldots$

which is a GP where $r = -\frac{1}{4}$, i.e. $|r| < 1$

So this series converges and $S_\infty = \dfrac{a}{1-r} = \dfrac{1}{1-(-\frac{1}{4})} = \dfrac{4}{5}$

c $3 + \frac{9}{2} + \frac{27}{4} + \ldots = 3 + 3(\frac{3}{2}) + 3(\frac{9}{4}) + \ldots = 3 + 3(\frac{3}{2}) + 3(\frac{3}{2})^2 + \ldots$

This series is a GP where $r = \frac{3}{2}$ and, as $|r| > 1$, the series does not converge.

2 Find the condition satisfied by x so that $\displaystyle\sum_{r=0}^{\infty} \frac{(x-1)^r}{2^r}$ converges.

Evaluate this expression when $x = 1.5$.

$$\sum_{r=0}^{\infty} \frac{(x-1)^r}{2^r} = 1 + \frac{x-1}{2} + \left(\frac{x-1}{2}\right)^2 + \ldots$$

This series is a GP with common ratio $\dfrac{x-1}{2}$ and so converges if $\left|\dfrac{x-1}{2}\right| < 1$,

i.e. if $-1 < \dfrac{x-1}{2} < 1$

$\Rightarrow \quad -1 < x < 3$

When $x = 1.5$, the series converges

and $\displaystyle\sum_{r=0}^{\infty} \frac{(x-1)^r}{2^r} = \sum_{r=0}^{\infty} (\tfrac{1}{4})^r = 1 + \tfrac{1}{4} + (\tfrac{1}{4})^2 + \ldots$

using $S_\infty = \dfrac{a}{1-r}$ where $r = \frac{1}{4}$ and $a = 1$ gives

$$S_\infty = \frac{1}{1 - \frac{1}{4}} = \frac{4}{3}$$

3 The 3rd term of a convergent GP is the mean of the 1st and 2nd terms. Find the common ratio and, if the first term is 1, find the sum to infinity.

If the series is $a + ar + ar^2 + ar^3 + \dots$

then $\quad ar^2 = \frac{1}{2}(a + ar)$

$a \neq 0$, so $\quad 2r^2 - r - 1 = 0$

$\Rightarrow \quad (2r+1)(r-1) = 0$

i.e. $\quad r = -\frac{1}{2}$ or 1

As the series is convergent, the common ratio is $-\frac{1}{2}$.

When $r = -\frac{1}{2}$ and $a = 1$,

$$S_\infty = \frac{1}{1+\frac{1}{2}} = \frac{2}{3}$$

EXERCISE 19E

1 Determine whether each of the series given below converges.

a $4 + \frac{4}{3} + \frac{4}{3^2} + \dots$ **b** $9 + 7 + 5 + 3 + \dots$

c $20 - 10 + 5 - 2.5 + \dots$ **d** $\frac{5}{10} + \frac{5}{100} + \frac{5}{1000} + \dots$

e $p + 2p + 3p + \dots$ **f** $3 - 1 + \frac{1}{3} - \frac{1}{9} + \dots$

2 Find the range of values of x for which the following series converge,

a $1 + x + x^2 + x^3 + \dots$ **b** $x + 1 + \frac{1}{x} + \frac{1}{x^2} + \dots$

c $1 + 2x + 4x^2 + 8x^3 + \dots$ **d** $1 - (1-x) + (1-x)^2 - (1-x)^3 + \dots$

e $(a+x) + (a+x)^2 + (a+x)^3 + \dots$ **f** $(a+x) + 1 + \frac{1}{a+x} + \frac{1}{(a+x)^2} + \dots$

3 Find the sum to infinity of those series in Question 1 that are convergent.

4 The sum to infinity of a GP is twice the first term. Find the common ratio.

5 The sum to infinity of a GP is 16 and the sum of the first 4 terms is 15. Find the first four terms.

6 If a, b and c are the first three terms of a GP, prove that $\sqrt{a}$, $\sqrt{b}$ and $\sqrt{c}$ form another GP.

MIXED EXERCISE 19

Write down the first six terms of each sequence and describe the behaviour of the terms as the sequence progresses.

1 $u_r = \dfrac{r}{r^2 + 1}$

2 $u_r = \cos \pi r$

3 $u_{r+1} = (u_r)^2 - u_r$

a when $u_1 = 1$

b when $u_1 = 0.5$

4 $u_{r+1} = \dfrac{1}{1 - u_r}$ when

a $u_1 = 2$

b $u_1 = 1$

Find the sum of each of the following series.

5 $1 - \frac{1}{2} + \frac{1}{4} - \frac{1}{8} + \ldots$

6 $2 - (2)(3) + (2)(3)^2 - (2)(3)^3 + \ldots + (2)(3)^{10}$

7 $\displaystyle\sum_{r=2}^{n} ab^{2r}$

8 $\displaystyle\sum_{r=5}^{n} 4r$

9 $e + e^2 + e^3 + \ldots + e^n$

10 $\displaystyle\sum_{r=1}^{\infty} \frac{1}{2^r}$

11 $u_1 = u_2$ and $u_{r+1} = \frac{1}{3}(2u_r{}^2 - 5)$. Find the possible values of u_1.

12 The sum of the first n terms of a series is n^3. Write down the first four terms and the nth term of the series.

13 The fourth term of an AP is 8 and the sum of the first ten terms is 40. Find the first term and the tenth term.

14 The second, fourth and eighth terms of an AP are the first three terms of a GP. Find the common ratio of the GP.

15 Find the value of x for which the numbers $x + 1, x + 3, x + 7$, are in geometric progression.

16 The second term of a GP is $\frac{1}{2}$ and the sum to infinity of the series is 4. Find the first term and the common ratio of the series.

CHAPTER 20 Integration 1

Differentiation Reversed

When x^2 is differentiated with respect to x the derivative is $2x$.

Conversely, if the derivative of an unknown function is $2x$ then it is clear that the unknown function could be x^2.

This process of finding a function from its derivative, which reverses the operation of differentiating, is called *integration*.

The Constant of Integration

As seen above, $2x$ is the derivative of x^2, but it is also the derivative of x^2+3, x^2-9, and, in fact, the derivative of $x^2 +$ any constant.

Therefore the result of integrating $2x$, which is called *the integral of* $2x$, is not a unique function but is of the form

$$x^2 + K \quad \text{where } K \text{ is any constant}$$

K is called *the constant of integration*.

This is written $\int 2x\,\mathrm{d}x = x^2 + K$

where $\int \ldots \mathrm{d}x$ means 'the integral of ... w.r.t. x'.

Integrating *any* function reverses the process of differentiating so, for any function $\mathrm{f}(x)$ we have

$$\boldsymbol{\int \frac{\mathrm{d}}{\mathrm{d}x}\mathrm{f}(x)\,\mathrm{d}x = \mathrm{f}(x) + K}$$

e.g. because differentiating x^3 w.r.t. x gives $3x^2$ we have $\int 3x^2\,\mathrm{d}x = x^3 + K$

and it follows that $\int x^2\,\mathrm{d}x = \frac{1}{3}x^3 + K$

Note that it is not necessary to write $\frac{1}{3}K$ in the second form, as K represents *any* constant in either expression.

In general, the derivative of x^{n+1} is $(n+1)x^n$ so $\int x^n\,\mathrm{d}x = \frac{1}{n+1}x^{n+1} + K$

i.e. **to integrate a power of x, *increase* that power by 1 and *divide* by the new power.**

This rule can be used to integrate any power of x *except* -1, which is considered later.

Integrating e^x

We know that $\frac{d}{dx}e^x = e^x$, $\quad\therefore\quad$ $\int \mathbf{e^x\,dx = e^x + K}$

Integrating a Sum or Difference of Functions

We saw in Chapter 13 that a function can be differentiated term by term. Therefore, as integration reverses differentiation, integration also can be done term by term.

To integrate products or quotients of functions, first express them as sums or differences of functions

e.g. $$\int \frac{2x-1}{\sqrt{x}}\,dx = \int \frac{2x}{\sqrt{x}} - \frac{1}{\sqrt{x}}\,dx = \int 2x^{\frac{1}{2}}\,dx - \int x^{-\frac{1}{2}}\,dx$$

EXAMPLE 20A Find the integral of $1 + x^7 + \frac{1}{x^2} - \sqrt{x}$

$$\int\left(1 + x^7 + \frac{1}{x^2} - \sqrt{x}\right)dx = \int (1 + x^7 + x^{-2} - x^{1/2})\,dx$$

$$= \int 1\,dx + \int x^7\,dx + \int x^{-2}\,dx - \int x^{1/2}\,dx$$

$$= x + \frac{1}{8}x^8 + \frac{1}{-1}x^{-1} - \frac{1}{\frac{3}{2}}x^{3/2} + K$$

$$= x + \frac{1}{8}x^8 - \frac{1}{x} - \frac{2}{3}x^{3/2} + K$$

EXERCISE 20A Integrate with respect to x

1 x^5

2 $\frac{1}{x^5}$

3 $\sqrt[4]{x}$

4 x^{-3}

5 $\frac{1}{x^{5/2}}$

6 $x^{-1/2}$

7 x^1

8 $\frac{1}{\sqrt[3]{x}}$

9 $1 + x^2$

10 $x + e^x$

11 $2x - \sqrt{x}$

12 $1 + \frac{1}{x^2}$

13 $2x - 3e^x$

14 $e^x + 5x - 2x^{-2}$

15 $x(1 + x)$

16 $(2 - 3x)(1 + 5x)$

17 $\frac{1 + x}{\sqrt{x}}$

18 $\frac{1 - 2x}{x^3}$

19 $\frac{1}{2}(x - e^x)$

20 $\frac{1 + x + x^3}{\sqrt{x}}$

21 $(1 - x)^2$

22 $x(1 + x)(1 - x)$

23 $(1 + e^{\frac{1}{2}x})(1 - e^{\frac{1}{2}x})$

24 $\frac{1 - \sqrt{x}}{x^2}$

Using Integration to Find an Area

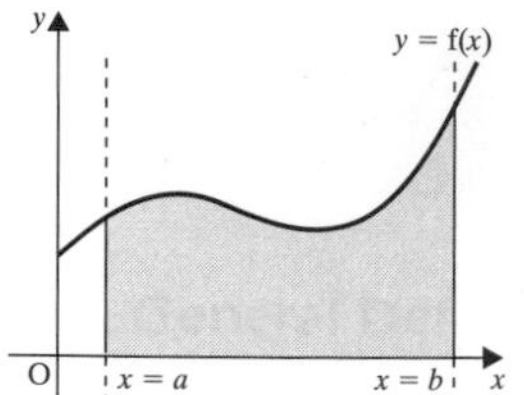

The area shown in the diagram is bounded by the curve $y = f(x)$, the x-axis and the lines $x = a$ and $x = b$.

There are several ways in which this area can be estimated, e.g. by counting squares on graph paper. A better method is to divide the area into thin vertical strips and treat each strip, or *element*, as being approximately rectangular.

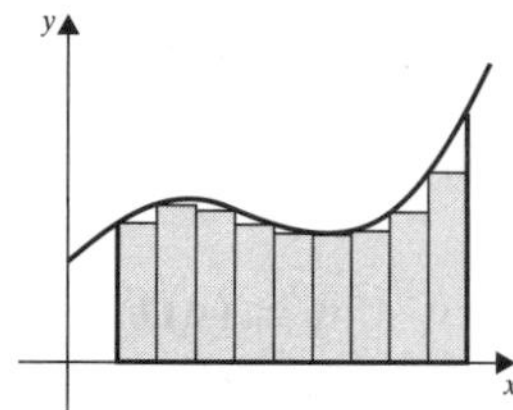

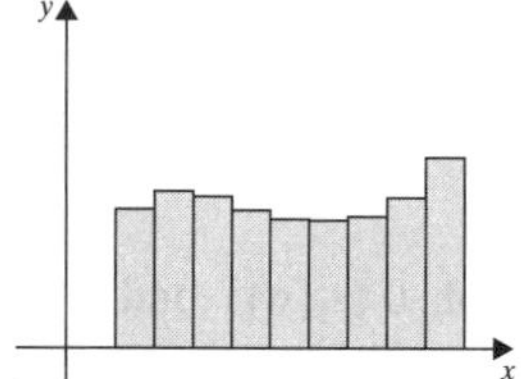

The sum of the areas of the rectangular strips then gives an approximate value for the required area. The thinner the strips are, the better is the approximation.

Note that every strip has one end on the x-axis, one end on the curve and two vertical sides, i.e. all the strips have the same type of boundaries.

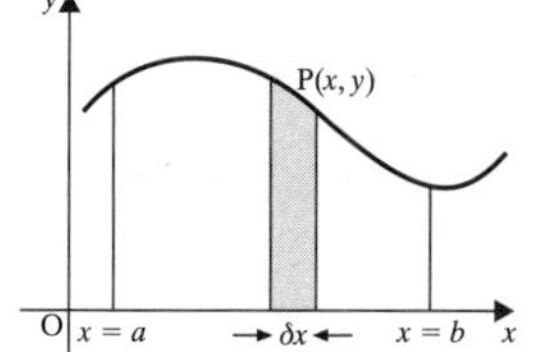

Now, considering a typical element bounded on the left by the ordinate through a general point $P(x,y)$, we see that

the width of the element represents a small increase in the value of x and so can be called δx.

Also, if A represents the part of the area up to the ordinate through P, then

the area of the element represents a small increase in the value of A and so can be called δA.

A typical strip is approximately a rectangle of height y and width δx.

Therefore, for any element $\quad \delta A \approx y\,\delta x$ [1]

The required area can now be found by adding the areas of all the strips from $x = a$ to $x = b$.

The notation for this is $\displaystyle\sum_{x=a}^{x=b} \delta A$

so, $\quad$ total area $= \displaystyle\sum_{x=a}^{x=b} \delta A$

$\Rightarrow \quad$ total area $\approx \displaystyle\sum_{x=a}^{x=b} y\,\delta x$

As δx gets smaller the accuracy of the results increases until, in the limiting case,

$$\textbf{total area} = \lim_{\delta x \to 0} \sum_{x=a}^{x=b} y\,\delta x$$

The equation $\delta A \approx y\delta x$ can also be written in the alternative form $\dfrac{\delta A}{\delta x} \approx y$

This form too becomes more accurate as δx gets smaller, giving $\lim\limits_{\delta x \to 0} \dfrac{\delta A}{\delta x} = y$

But $\quad \lim\limits_{\delta x \to 0} \dfrac{\delta A}{\delta x}$ is $\dfrac{dA}{dx}$ $\quad$ so $\quad \dfrac{dA}{dx} = y$

Hence $\quad A = \int y \, dx$

The boundary values of x defining the total area are $x = a$ and $x = b$ and we indicate this by writing

$$\textbf{total area} = \int_a^b y\,dx$$

The total area can therefore be found in two ways, either as the limit of a sum or by integration,

i.e. $\quad \lim\limits_{\delta x \to 0} \sum\limits_{x=a}^{x=b} y\,\delta x = \int_a^b y\,dx$

and we conclude that **integration is a process of summation.**

In this chapter we will find areas bounded by straight lines and a curve by using integration. This means that we will be using $\int_a^b y\,dx$, so now we must find how to calculate the value of this expression.

Definite Integration

Suppose that we wish to find the area bounded by the x-axis, the lines $x = a$ and $x = b$ and the curve $y = 3x^2$.

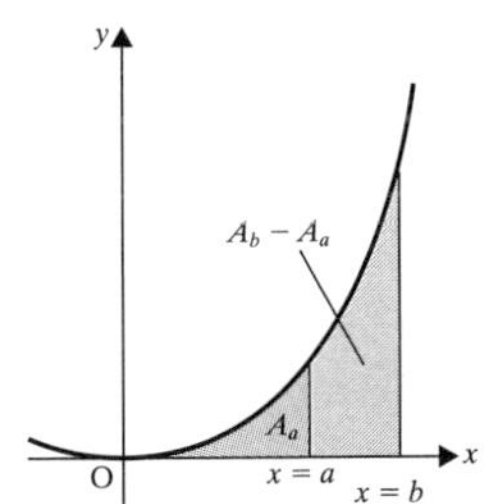

Using the method above gives $A = \int 3x^2\,dx$, i.e. $A = x^3 + K$

From this area function we can find the value of A corresponding to a particular value of x.

Hence using $x = a$ gives $\quad A_a = a^3 + K$

and $\quad$ using $x = b$ gives $\quad A_b = b^3 + K$

Then the area between $x = a$ and $x = b$ is given by $A_b - A_a$ where $A_b - A_a = (b^3 + K) - (a^3 + K) = b^3 - a^3$

Now $A_b - A_a$ is referred to as the definite integral from a to b of $3x^2$ and is denoted by $\int_a^b 3x\,dx$, i.e. $\int_a^b 3x^2\,dx = (x^3)_{x=b} - (x^3)_{x=a}$

The RHS of this equation is usually written in the form $\left[x^3\right]_a^b$ where a and b are called the *boundary values* or *limits of integration*; b is the *upper limit* and a is the *lower limit*.

$\int_a^b y\,dx$ is called the definite integral from a to b of y w.r.t. x.

Whenever a definite integral is calculated, the constant of integration disappears.

Note. A definite integral can be found in this way only if the function to be integrated is defined for every value of x from a to b,

e.g. $\int_{-1}^{1} \frac{1}{x^2}\,dx$ cannot be found directly as $\frac{1}{x^2}$ is undefined when $x = 0$,

(there is a break in the graph of $y = \frac{1}{x^2}$ where $x = 0$).

EXAMPLE 20B

Evaluate $\int_1^4 \frac{1}{x^2}\,dx$

$$\int_1^4 \frac{1}{x^2}\,dx \equiv \int_1^4 x^{-2}\,dx$$

$$= \left[-x^{-1}\right]_1^4 = \{-4^{-1}\} - \{-1^{-1}\} = -\tfrac{1}{4} + 1 = \tfrac{3}{4}$$

EXERCISE 20B

Evaluate each of the following definite integrals.

1 $\int_0^2 x^3\,dx$

2 $\int_1^2 \sqrt{x^5}\,dx$

3 $\int_2^4 (x^2 + 4)\,dx$

4 $\int_4^9 \sqrt{x}\,dx$

5 $\int_0^3 (x^2 + 2x - 1)\,dx$

6 $\int_0^2 (x^3 - 3x)\,dx$

7 $\int_{-1}^0 (1 - x)^2\,dx$

8 $\int_1^2 \frac{3x + 1}{\sqrt{x}}\,dx$

9 $\int_{-1}^0 (2 + 3x)^2\,dx$

10 $\int_{1/2}^7 (2e^x + 1)\,dx$

Finding Area by Definite Integration

As we have seen, the area bounded by a curve $y = f(x)$, the lines $x = a$, $x = b$, and the x-axis, can be found from the definite integral

$$\int_a^b f(x)\,dx$$

It is best, though, not to regard this as a *formula* but to consider the summation of the areas of elements, with a typical element being shown in a diagram.

EXAMPLE 20C

Find the area in the first quadrant bounded by the x and y axes and the curve $y = 1 - x^2$.

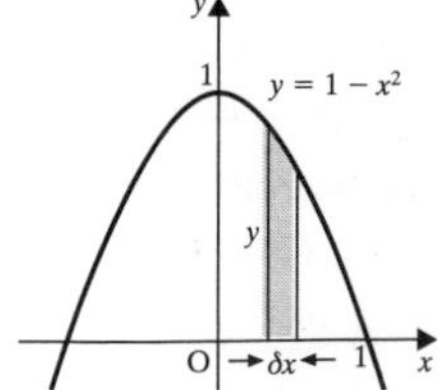

> The required area starts at the y-axis, i.e. at $x = 0$ and ends where the curve crosses the x-axis, i.e. where $x = 1$.

$$\text{Area} = \lim_{\delta x \to 0} \sum_{x=0}^{x=1} y\,\delta x = \int_0^1 (1 - x^2)\,dx = \left[x - \frac{x^3}{3}\right]_0^1 = \left(1 - \frac{1}{3}\right) - (0 - 0) = \frac{2}{3}$$

The required area is $\frac{2}{3}$ of a square unit.

EXERCISE 20C

In each question find the area with the given boundaries.

1 The x-axis, the curve $y = x^2 + 3$ and the lines $x = 1, x = 2$.

2 The curve $y = \sqrt{x}$, the x-axis and the lines $x = 4, x = 9$.

3 The x-axis, the lines $x = -1$, $x = 1$, and the curve $x^2 + 1$.

4 The curve $y = x^2 + x$, the x-axis and the line $x = 3$.

5 The positive x and y axes and the curve $y = 4 - x^2$.

6 The lines $x = 2$, $x = 4$, the x-axis and the curve $y = x^3$.

7 The curve $y = 4 - x^2$, the positive y-axis and the negative x-axis.

8 The x-axis, the lines $x = 1$ and $x = 2$, and the curve $y = \frac{1}{2}x^3 + 2x$.

9 The x-axis and the lines $x = 1$, $x = 5$, and $y = 2x$. Check the result by sketching the required area and finding it by mensuration.

The Meaning of a Negative Result

Consider the area bounded by $y = 4x^3$, the x-axis and the lines

a $x = 1$ and $x = 2$

b $x = -2$ and $x = -1$

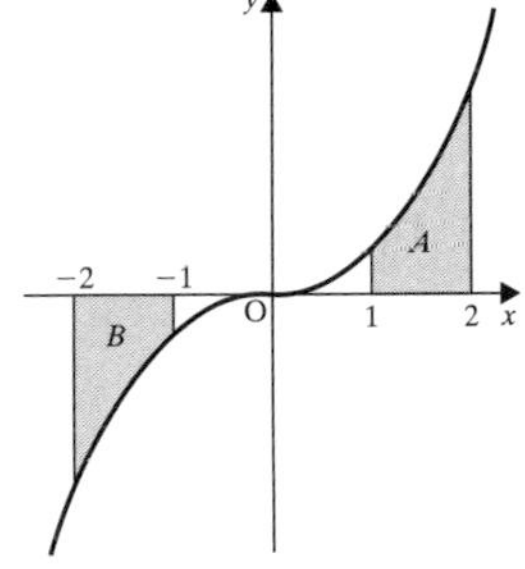

This curve is symmetrical about the origin so the two shaded areas are equal.

a Considering A

$$\lim_{\delta x \to 0} \sum_{x=1}^{x=2} y\,\delta x = \int_1^2 y\,dx$$

$$= \int_1^2 4x^3\,dx$$

$$= \left[x^4\right]_1^2 = 16 - 1 = 15$$

b Considering B

$$\lim_{\delta x \to 0} \sum_{x=-2}^{x=-1} y\,\delta x = \int_{-2}^{-1} y\,dx$$

$$= \int_{-2}^{-1} 4x^3\,dx$$

$$= \left[x^4\right]_{-2}^{-1} = 1 - 16 = -15$$

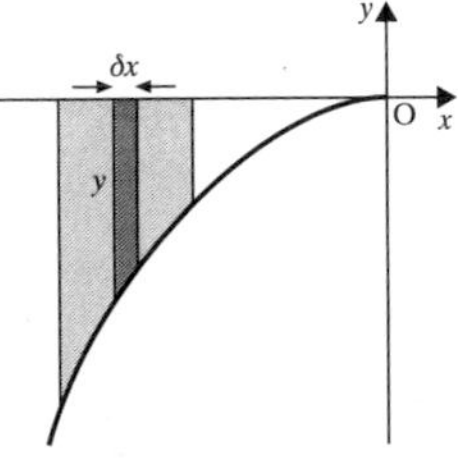

This integral has a negative value because, from -2 to -1, the value of y which gives the length of the strip, is negative. Area cannot be negative; the minus sign simply means that area A is below the x-axis. The actual area is 15 square units.

Take care with problems involving a curve that crosses the x-axis between the boundary values.

EXAMPLE 20D Find the area enclosed between the curve $y = x(x-1)(x-2)$ and the x-axis.

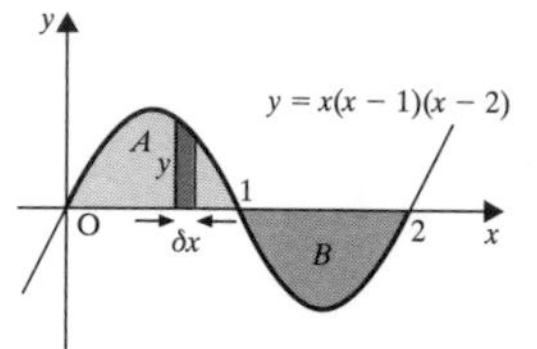

The area enclosed between the curve and the x-axis is the sum of the areas A and B.

For A we use
$$\int_0^1 y\,dx = \int_0^1 (x^3 - 3x^2 + 2x)\,dx$$
$$= \left[\frac{x^4}{4} - x^3 + x^2\right]_0^1$$
$$= \tfrac{1}{4}$$

For B we use
$$\int_1^2 (x^3 - 3x^2 + 2x)\,dx = \left[\frac{x^4}{4} - x^3 + x^2\right]_1^2$$
$$= (4 - 8 + 4) - (\tfrac{1}{4} - 1 + 1)$$
$$= -\tfrac{1}{4}$$

The minus sign refers only to the *position* of area B relative to the x-axis. The actual area is $\frac{1}{4}$ of a square unit.

So the total shaded area is $(\frac{1}{4} + \frac{1}{4})$ sq unit $= \frac{1}{2}$ sq unit.

EXERCISE 20D In each Question from 1 to 5 find the specified area.

1 The area below the x-axis and above the curve $y = x^2 - 1$.

2 The area bounded by the curve $y = 1 - x^3$, the x-axis and the lines $x = 2, x = 3$.

3 The area between the x and y axes and the curve $y = (x-1)^2$.

4 Sketch the curve $y = x(x^2 - 1)$, showing where it crosses the x-axis. Find

a the area enclosed above the x-axis and below the curve

b the area enclosed below the x-axis and above the curve

c the total area between the curve and the x-axis.

5 Repeat Question 4 for the curve $y = x(4 - x^2)$.

6 Evaluate

a $\displaystyle\int_0^2 (x-2)\,dx$

b $\displaystyle\int_2^4 (x-2)\,dx$

c $\displaystyle\int_0^4 (x-2)\,dx$

Interpret your results by means of a sketch.

Using Horizontal Elements

Suppose that area between the curve $x = y(4-y)$ and the y-axis is required.

The curve crosses the y-axis where $y = 0$ and $y = 4$ as shown.

A vertical element is not suitable in this case because it has *both ends on the curve* and its length cannot be easily found.

However it is easy to find the approximate area of a *horizontal* strip, by treating it as a rectangle with length x and width δy.

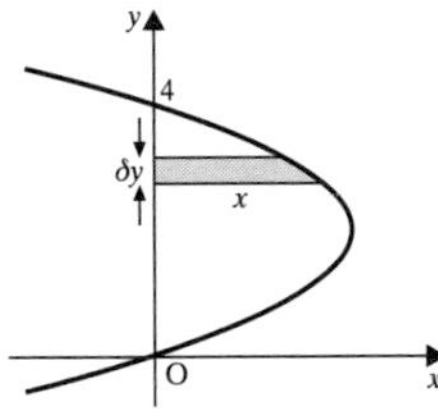

i.e. area of element $\approx x\,\delta y$ and the required area is therefore given by

$$\lim_{\delta y \to 0} \sum_{y=0}^{y=4} x\,\delta y = \int_0^4 x\,\mathrm{d}y = \int_0^4 y(y-4)\,\mathrm{d}y$$

EXERCISE 20E

1 Evaluate

a $\displaystyle\int_3^6 (y^2 - y)\,\mathrm{d}y$ **b** $\displaystyle\int_1^8 (5 + \sqrt{2y})\,\mathrm{d}y$ **c** $\displaystyle\int_{-1}^8 (4 - y^{-2/3})\,\mathrm{d}y$

In Questions 2 to 5 find the area specified by the given boundaries.

2 The y-axis and the curve $x = 9 - y^2$.

3 The curve $x = y^2$, the y-axis and the lines $y = 1, y = 2$.

4 The y-axis, the curve $x = \sqrt{y}$ and the line $y = 4$.

5 The curve $x = (y-2)(y+1)$ and the y-axis.

6 Find the area in the first quadrant bounded by the x and y axes, the curve $y = x^2$ and the line $y = 16$

a by using vertical elements

b by using horizontal elements and the equation of the curve in the form $x = \sqrt{y}$.

7 If $y = x^2$, show by means of sketch graphs and *not* by evaluating the integrals, that $\displaystyle\int_0^1 y\,\mathrm{d}x = 1 - \int_0^1 x\,\mathrm{d}y$.

Finding Compound Areas

EXAMPLES 20F

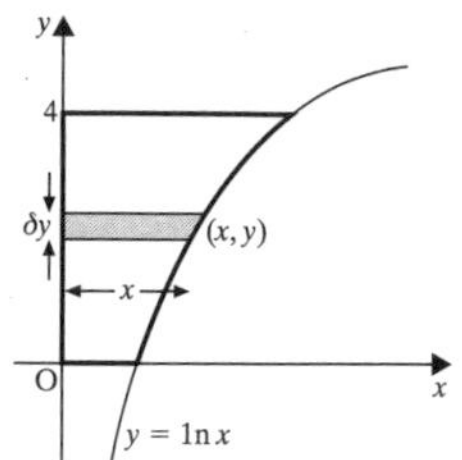

1 A plane region is defined by the line $y = 4$, the x- and y-axes and part of the curve $y = \ln x$. Find the area of the region.

> A vertical element is unsuitable in this case as the top and bottom are not always on the same boundaries, but a horizontal element is satisfactory.

The area, δA, of a typical horizontal element is given by $\delta A \approx x\,\delta y$.

> Because the width of our element is δy we will have to integrate w.r.t. y, so we need the equation of the curve in the form $x = f(y)$

$y = \ln x \quad \Rightarrow \quad x = e^y$

$\therefore \quad \delta A \approx e^y\,\delta y$

$$\Rightarrow \quad A = \lim_{\delta y \to 0} \sum_{y=0}^{y=4} e^y\,\delta y = \int_0^4 e^y\,dy$$

$$= \Big[e^y\Big]_0^4$$

$$= e^4 - e^0$$

The defined area is $(e^4 - 1)$ square units.

A similar approach can be made in a variety of circumstances provided that an element can be found

- which has the same format throughout, i.e. the ends of all the elements are on the same boundaries
- whose length and width are measured parallel to the x- and y-axes.

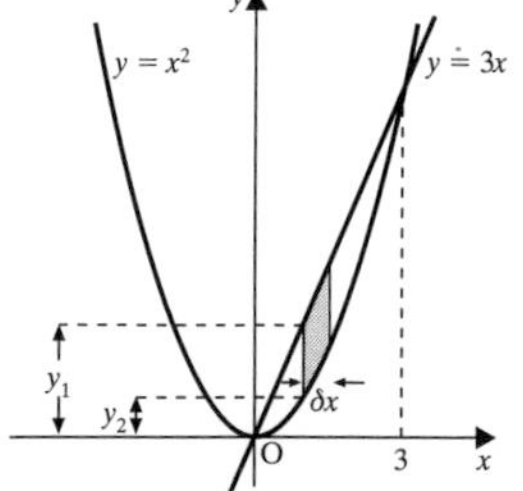

2 Find the area between the curve $y = x^2$ and the line $y = 3x$.

The line and curve meet where $x^2 = 3x$, i.e. where $x = 0$ and $x = 3$.

> A vertical strip always has its top on the line and its foot on the curve so it is a suitable element. It is approximately a rectangle whose width is δx and whose height is the vertical distance between the line and the curve. The area of the element, δA, is given by

$\delta A \approx (y_1 - y_2)\,\delta x = (3x - x^2)\,\delta x$

$$\therefore \quad A = \lim_{\delta x \to 0} \sum_{x=0}^{x=3} (3x - x^2)\,\delta x = \int_0^3 (3x - x^2)\,dx$$

$$= \left[\tfrac{3}{2}x^2 - \tfrac{1}{3}x^3\right]_0^3$$

$$= 4\tfrac{1}{2}$$

The required area is $4\frac{1}{2}$ square units.

EXERCISE 20F

1 Calculate the area bounded by the curve $y = \sqrt{x}$, the y-axis and the line $y = 3$.

2 Find, by integration, the area bounded by

a the x-axis, the line $x = 2$ and the curve $y = x^2$

b the y-axis, the line $y = 4$ and the curve $y = x^2$

Sketch these two areas on the same diagram and hence check the sum of the answers to **a** and **b**.

3 A region in the xy plane is bounded by the lines $y = 1$ and $x = 1$, and the curve $y = e^x$. Find its area.

4 Find the area defined by the inequalities $y \leqslant 1 - x^2$ and $y \geqslant 1 - x$.

5 Find the area of the region of the xy plane defined by $y \geqslant e^x, x \geqslant 0, y \leqslant e$.

6 Calculate the area in the first quadrant between the curve $y^2 = x$ and the line $x = 9$.

7 Find the area between the y-axis and the curve $y^2 = 1 - x$.

The next two questions are a little harder.

8 Evaluate the area between the line $y = x - 1$ and the curve

a $y = x(1 - x)$

b $y = (2x + 1)(x - 1)$

9 Calculate the area of the region of the xy plane defined by the inequalities $y \geqslant (x + 1)(x - 2)$ and $y \leqslant x$.

To Integrate $\frac{1}{x}$

At first sight it looks as though we can write $\frac{1}{x} = x^{-1}$ and integrate by using the rule $\int x^n \, dx = \frac{1}{n+1} x^{(n+1)} + K$.

However, this method fails when $n = -1$ because the resulting integral is meaningless.

Taking a second look at $\frac{1}{x}$ it can be *recognised* as the derivative of $\ln x$.

It must be remembered, however, that $\ln x$ is defined only when $x > 0$. Hence, provided that $x > 0$ we have

$$\frac{d}{dx}(\ln x) = \frac{1}{x} \quad \Longleftrightarrow \quad \int \frac{1}{x} \, dx = \ln x + K$$

Now if $x < 0$ the statement $\int \frac{1}{x} \, dx = \ln x$ is not valid because the log of a negative number does not exist.

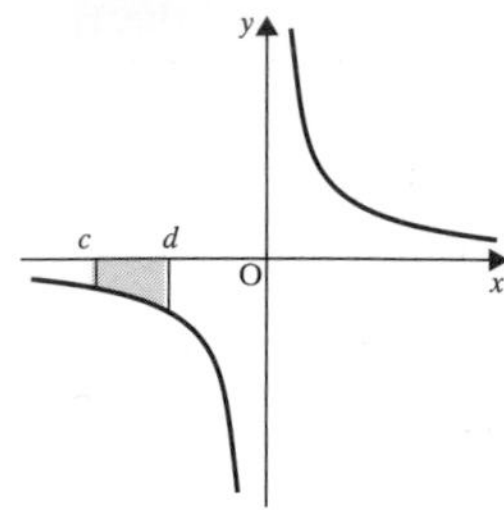

However, the function $\dfrac{1}{x}$ exists for negative values of x, as the graph of $y = \dfrac{1}{x}$ shows.

Also, the definite integral $\displaystyle\int_c^d \frac{1}{x}\,\mathrm{d}x$, which is represented by the shaded area, clearly exists. It must, therefore, be possible to integrate $\dfrac{1}{x}$ when x is negative.

If $x < 0$ then $-x > 0$

i.e. $$\int \frac{1}{x}\,\mathrm{d}x = \int \frac{-1}{(-x)}\,\mathrm{d}x = \ln(-x) + K$$

Thus, when $x < 0$, $\displaystyle\int \frac{1}{x}\,\mathrm{d}x = \ln(-x) + K$

and when $x > 0$, $\displaystyle\int \frac{1}{x}\,\mathrm{d}x = \ln x + K$

These two results can be combined using $|x|$ so that, for both positive and negative values of x, we have

$$\boldsymbol{\int \frac{1}{x}\,\mathrm{d}x = \ln|x| + K}$$

The expression $\ln|x| + K$ can be simplified if K is replaced by $\ln A$, where A is a positive constant, giving

$$\int \frac{1}{x}\,\mathrm{d}x = \ln|x| + \ln A = \ln A|x|$$

Further $\dfrac{\mathrm{d}}{\mathrm{d}x}(\ln x^a) = \dfrac{\mathrm{d}}{\mathrm{d}x}(a\ln x) = \dfrac{a}{x}$

$$\therefore \quad \boldsymbol{\int \frac{a}{x}\,\mathrm{d}x = a\ln|x| + K \quad \text{or} \quad a\ln A|x|}$$

e.g. $$\int \frac{4}{x}\,\mathrm{d}x = 4\ln|x| + K \quad \text{or} \quad 4\ln A|x|$$

EXERCISE 20G

Integrate with respect to x.

1 $\dfrac{2}{x}$

2 $\dfrac{1}{4x}$

3 $\dfrac{3}{2x}$

4 $\dfrac{x+1}{x}$

5 $\dfrac{x^2+x-1}{x}$

Evaluate

6 $\displaystyle\int_1^2 \frac{1}{3x}\,\mathrm{d}x$

7 $\displaystyle\int_1^3 \left(1 - \frac{1}{x}\right)\mathrm{d}x$

8 $\displaystyle\int_1^2 \left(\frac{1-x}{x}\right)\mathrm{d}x$

9 $\displaystyle\int_2^3 \left(\mathrm{e}^x - \frac{1}{x}\right)\mathrm{d}x$

10 $\displaystyle\int_4^5 \frac{2-x}{3x}\,\mathrm{d}x$

The Approximate Value of a Definite Integral

We know that the definite integral $\int_a^b f(x)\,dx$ can be used to evaluate the area between the curve $y = f(x)$, the x-axis and the ordinates at $x = a$ and $x = b$. It is not always possible, however, to find a function whose derivative is $f(x)$. In such cases the definite integral, and hence the exact value of the specified area, cannot be found.

If, on the other hand, we divide the area into a *finite* number of strips then the sum of their areas gives an approximate value for the required area and hence an approximate value of the definite integral. When using this method, choose strips whose widths are all the same.

The Trapezium Rule

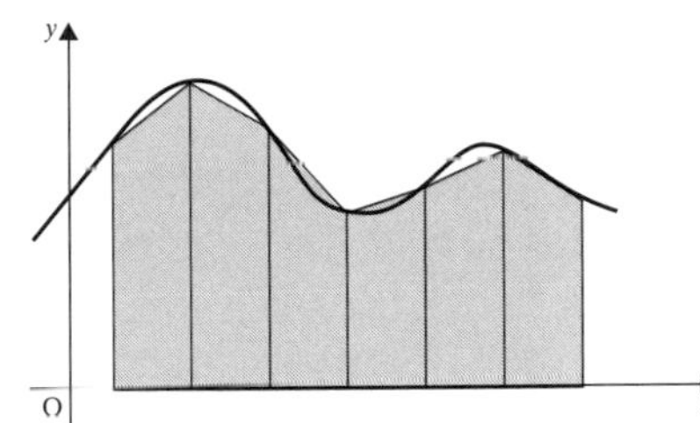

When the area shown in the diagram is divided into vertical strips, each strip is approximately a trapezium.

If the width of the strip and its two vertical sides are known, the area of the strip can be found using the formula

$$\text{area} = \tfrac{1}{2}(\text{sum of parallel sides}) \times \text{width}$$

The sum of the areas of all the strips then gives an approximate value for the area under the curve.

Now suppose that there are n strips, *all with the same width*, d say, and that the vertical edges of the strips (i.e. the ordinates) are labelled $y_0, y_1, y_2, \ldots, y_{n-1}, y_n$.

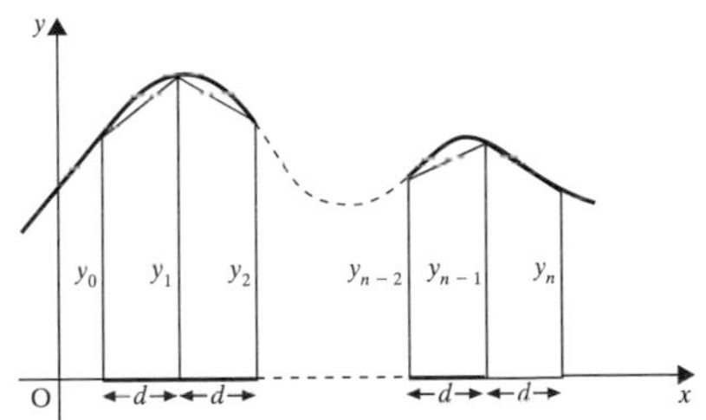

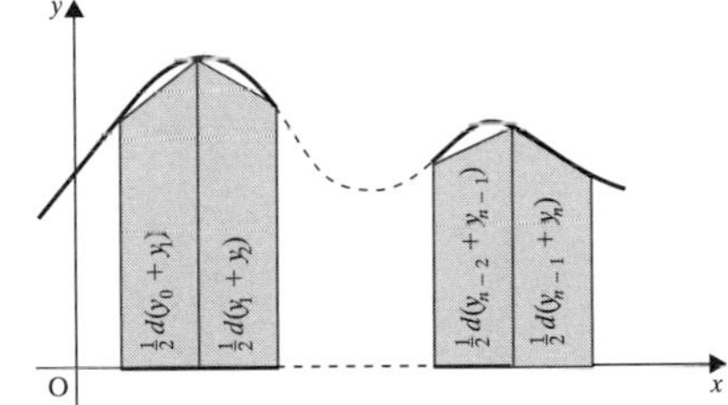

The sum of the areas of all the strips is

$$\tfrac{1}{2}(y_0 + y_1)(d) + \tfrac{1}{2}(y_1 + y_2)(d) + \tfrac{1}{2}(y_2 + y_3)(d) + \ldots$$
$$\ldots + \tfrac{1}{2}(y_{n-2} + y_{n-1})(d) + \tfrac{1}{2}(y_{n-1} + y_n)(d)$$

Therefore the area, A, under the curve is given approximately by

$$\boldsymbol{A \approx \tfrac{1}{2}(d)\,[y_0 + 2y_1 + 2y_2 + \ldots + 2y_{n-1} + y_n]}$$

This formula is known as the *Trapezium Rule*.

An easy way to remember the formula in terms of ordinates is

half width of strip × (first + last + twice all the others)

Be careful not to confuse the number of strips and the number of ordinates – they are not the same.

The Mid-ordinate Rule

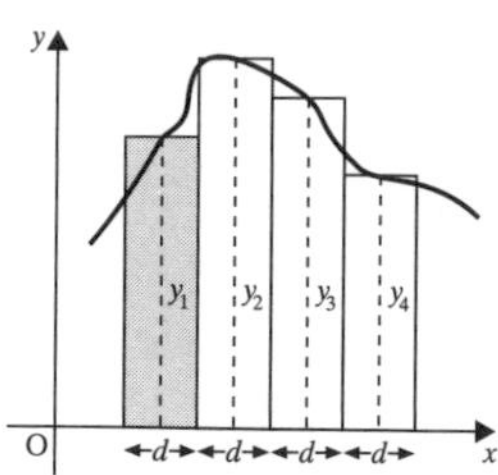

This rule is similar to the trapezium rule but it uses rectangular elements instead of trapeziums. The required area is divided into equal width strips, the top boundary of each strip being a horizontal line through the point on the curve at the centre of the strip.

The sum of the areas of these rectangles gives an approximate value for the area under the curve,

i.e. $$A \approx d\left[y_1 + y_2 + \ldots + y_n\right]$$

Simpson's Rule

A formula which gives a better approximation than that obtained from the trapezium rule is known as *Simpson's Rule*.
Using the same notation as before it states that

$$A \approx \tfrac{1}{3}(d)[y_0 + 4y_1 + 2y_2 + 4y_3 + 2y_4 + \ldots + 2y_{n-2} + 4y_{n-1} + y_n]$$

i.e. $$A \approx \tfrac{1}{3}d[\{y_0 + y_n\} + 4\{y_1 + y_3 + \ldots\} + 2\{y_2 + y_4 + \ldots\}]$$

This formula, given here without proof, is based on dividing the required area into equal width strips and, for each *pair* of strips, finding a parabola which passes through the top of the three ordinates bounding the two strips.

Because this formula is based on pairs of strips it follows that it can be used only when the number of strips is even, i.e. when the number of ordinates is odd.

The degree of accuracy of an answer given by any of these rules depends upon the number of strips into which the required area is divided, because the narrower the strip, the nearer its shape at the top becomes to the shape of the curve.

EXAMPLE 20H

Use four strips to find an approximate value for the definite integral $\int_1^5 x^3$ using

a the trapezium rule **b** the mid-ordinate rule **c** Simpson's rule

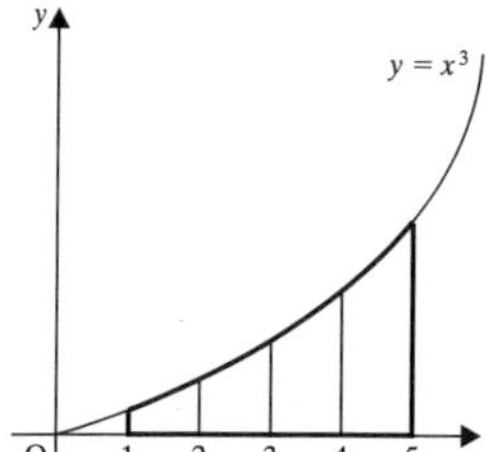

The given definite integral represents the area bounded by the x-axis, the lines $x = 1$ and $x = 5$, and the curve $y = x^3$.

> Five ordinates are used when there are four strips whose widths must all be the same. From $x = 1$ to $x = 5$ there are four units so the width of each strip must be 1 unit. Hence the five ordinates are where $x = 1, x = 2, x = 3, x = 4$ and $x = 5$.

a Using the trapezium rule,

$y_0 = 1^3 = 1, \quad y_1 = 2^3 = 8, \quad y_2 = 3^3 = 27, \quad y_3 = 64, \quad y_4 = 125$

The required area, A, is given by

$A \approx \tfrac{1}{2}(1)[1 + 125 + 2\{8 + 27 + 64\}] = 162$

The required area is approximately 162 square units.

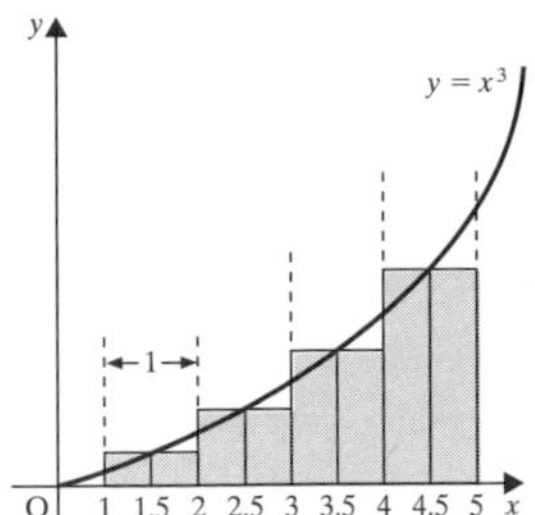

b The values of the mid-ordinates are given by:

$$y_1 = 1.5^3, \quad y_2 = 2.5^3, \quad y_3 = 3.5^3, \quad y_4 = 4.5^3$$

The width of the strip is 1 unit so using the mid-ordinate rule gives

$$A = (1)[3.375 + 15.63 + 42.88 + 91.13] = 153.0$$

The required area is approximately 153 square units.

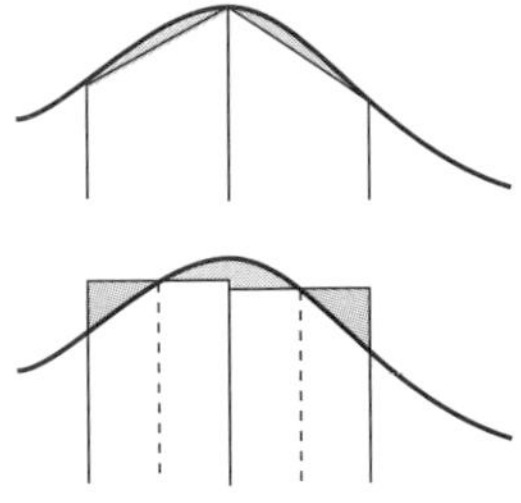

The values given by these two rules for the approximate area under the curve are slightly different and, in fact, the mid-ordinate value is usually more accurate. This is because the top of the trapezium is often entirely on one side of the curve, so that part of the area under the curve is not allowed for. The top line of the rectangular strip, on the other hand, usually cuts across the curve so that part of the area included in the rectangle but which is not under the curve, tends to balance the area under the curve that is not included in the rectangle. This is seen in the diagrams showing the tops of two enlarged strips.

c There is an odd number of ordinates so Simpson's rule can be used.

$$A \approx \tfrac{1}{3}(1)[\{1 + 125\} + 4\{8 + 64\} + 2\{27\}] = 156$$

The required area is approximately 156 square units.

Simpson's rule gives an even more accurate approximation than the other two rules because, being based on a parabola, the tops of the strips are even nearer to the shape of the curve.

EXERCISE 20H

In Questions 1 to 4 estimate the value of each definite integral, using the trapezium rule with 5 ordinates.

1 $\displaystyle\int_0^4 x^2\,\mathrm{d}x$ **2** $\displaystyle\int_1^3 \frac{1}{x^2}\,\mathrm{d}x$ **3** $\displaystyle\int_0^{2\pi/3} \sqrt{\sin x}\,\mathrm{d}x$ **4** $\displaystyle\int_1^3 \ln x\,\mathrm{d}x$

5 Find the true value of the definite integrals given in Questions 1 and 2 and estimate the value of each integral using 5 ordinates and

a the mid-ordinate rule **b** Simpson's rule.

Complete the table below and note the comparative accuracy of the results given by the three rules.

Value using	$\int_0^4 x^2\,\mathrm{d}x$	$\int_1^3 \frac{1}{x^2}\,\mathrm{d}x$
Trapezium rule		
Mid-ordinate rule		
Simpson's rule		
Definite integration		

MIXED EXERCISE 20

Integrate with respect to x

1 $x^2 - \dfrac{1}{x^2}$

2 $\sqrt[3]{x}$

3 $\sqrt{x} + \dfrac{1}{\sqrt{x}}$

4 $\dfrac{x\,e^x - 1}{x}$

5 $\dfrac{x^3 - 1}{x}$

6 $\dfrac{x^2 - 1}{\sqrt{x}}$

Evaluate

7 $\displaystyle\int_3^6 (6-x)^2\,dx$

8 $\displaystyle\int_{-1}^8 \frac{3y}{\sqrt[3]{8y}}\,dy$

9 $\displaystyle\int_1^{32} \left(\sqrt[5]{x} - \frac{1}{\sqrt[5]{x}}\right) dx$

Find the areas specified in Questions 10 to 12.

10 Bounded by the x and y axes and the curve $y = 1 - x^3$.

11 Bounded by the curve $x = y^2 - 4$ and the y-axis.

12 The *total* area between the curve $y = (x-1)(x-2)(x-3)$ and the x-axis.

13 **a** Find an approximate value for the area between the x-axis and the curve $y = (x-1)(x-4)$, using

i the trapezium rule with 4 ordinates

ii the mid-ordinate rule with 3 strips

iii Simpson's rule using 3 ordinates.

b Evaluate $\displaystyle\int_1^4 (x-1)(x-4)\,dx$

14 **a** Use the trapezium rule with 3 ordinates to estimate the value of $\displaystyle\int_0^5 (3+x)\,dx$.

b Find the value of $\displaystyle\int_0^5 (3+x)\,dx$.

c Explain the connection between the results of **a** and **b**.

15 A region of the xy plane is defined by the inequalities $0 \leqslant x \leqslant 4$ and $0 \leqslant y \leqslant e^x$. Find the area of the region.

16 Find the area of the region in the first quadrant bounded by the y-axis, the line $y = 6$ and the curve $y = x^2 + 2$.

17 A plane region is bounded by the curve $y = 6 - x^2$ and the line $y = 2$. Find the area of the region.

SUMMARY D

LOGARITHMS

The formula for changing the base of a logarithm from a to b is

$$\log_a x = \frac{\log_b x}{\log_b a}$$

where a and b are both positive numbers.

FUNCTIONS

Exponential Functions

a^x is an exponential function.
e^x is *the* exponential function where

$e = 2.718\,28\ldots$ and $\frac{d}{dx}(e^x) = e^x$

Logarithmic Function

$\log_a x$ is a logarithmic function.
$\log_e x = \ln x$ is the natural logarithmic function

and $\frac{d}{dx}(\ln x) = \frac{1}{x}$

The natural logarithmic function is the inverse of the exponential function, i.e.
$f(x) = e^x \quad \Rightarrow \quad f^{-1}(x) = \ln x$

Modulus Functions

$|x|$ is the modulus function where $|x|$ is the positive numerical value of x, i.e. when $x = -3$, $|x| = 3$.
The curve $y = |f(x)|$ is obtained from the curve $y = f(x)$ by reflecting in the x-axis the parts of the curve for which $f(x)$ is negative. The section(s) for which $f(x)$ is positive remain unchanged.

Even Functions

A function is even if $f(x) = f(-x)$.
Even functions are symmetrical about the y-axis.

Odd Functions

A function is odd if $f(x) = -f(-x)$. Odd functions have rotational symmetry about the origin.

Periodic Functions

A periodic function has a basic pattern that repeats at regular intervals. The width of the interval is called the period.

INTEGRATION

Standard Integrals

Function	Integral
x^n	$\frac{1}{n+1}x^{n+1} (n \neq -1)$
e^x	e^x
$\frac{1}{x}$	$\ln \lvert x \rvert$

Integration as a Process of Summation

$$\lim_{\delta x \to 0} \sum_{x=a}^{x=b} f(x)\delta x = \int_a^b f(x)\,dx$$

Area

The area bounded by the x-axis, the lines $x = a$ and $x = b$ and the curve $y = f(x)$ can be found by summing the areas of vertical strips of width δx and using

$$\text{Area} = \lim_{\delta x \to 0} \sum_{x=a}^{x=b} y\,\delta x = \int_a^b y\,dx$$

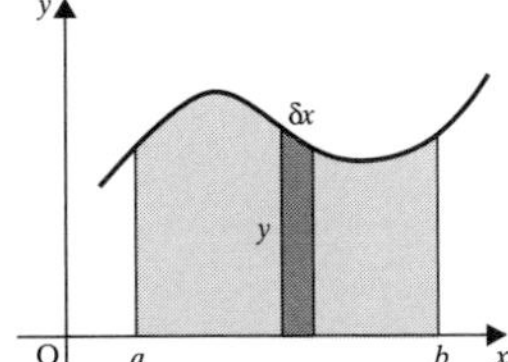

Similarly, for horizontal stripes,

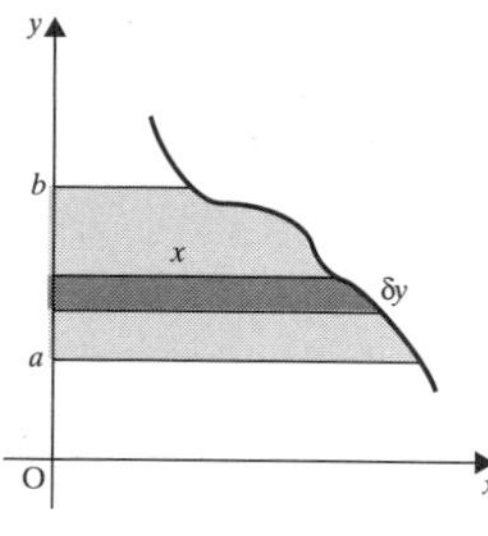

$$\text{Area} = \lim_{\delta y \to 0} \sum_{y=a}^{y=b} x\,\delta y = \int_a^b x\,dy$$

For compound areas the length of a strip is usually a difference of two quantities.

The area shown is given by

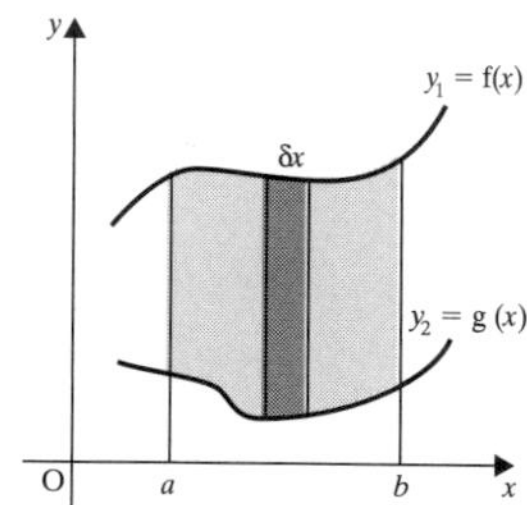

$$\text{Area} = \lim_{\delta x \to 0} \sum_{x=a}^{x=b} (y_1 - y_2)\,\delta x = \int_a^b (y_1 - y_2)\,dx$$

An approximate value for the area A under a curve can be found by taking strips of equal width d and using

The Trapezium Rule

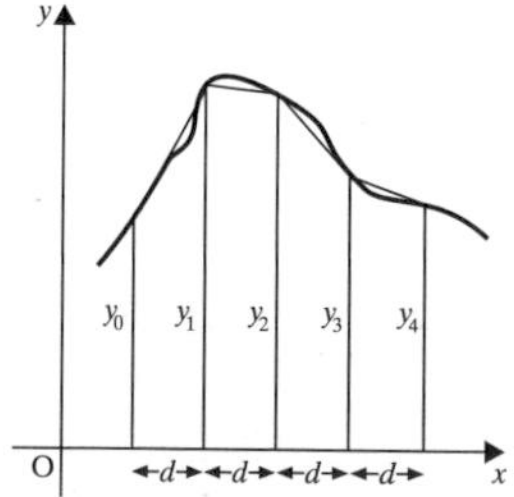

$$A = \int_a^b f(x)\,dx \approx \tfrac{1}{2}d\,[y_0 + 2y_1 + \ldots + 2y_{n-1} + y_n]$$

where the values of y are the lengths of the parallel sides of the trapeziums, i.e. the y-coordinates of the points on the curve at the edge of each strip.

The Mid-ordinate Rule

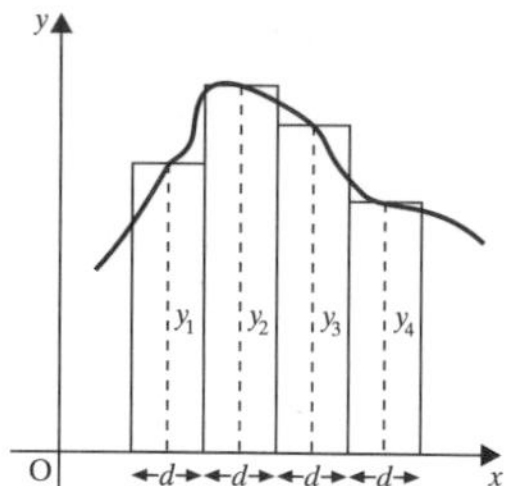

$$A = \int_a^b f(x)\,dx \approx d[y_1 + y_2 + \ldots + y_n]$$

where the values of y are heights of the rectangles, i.e. the y-coordinates of the points on the curve at the centre of each strip.

Simpson's Rule

In this case there must be an even number of strips.

$$\int_a^b f(x)\,dx \approx \tfrac{1}{3}d\,[(1\text{st} + \text{last}) + 4(2\text{nd} + 4\text{th} + \ldots) + 2(3\text{rd} + 5\text{th} + \ldots)]$$

SEQUENCES

A sequence is an ordered progression of numbers which can be generated from a rule. The rule may give the rth term as a function of r, e.g. $u_r = 2r$, or it may give a relationship between successive terms together with the first term, e.g. $u_r = 2 + u_{r-1}$ and $u_1 = 1$, in which case it is called a recurrence relationship.

A sequence converges if u_r approaches a single finite value as r increases. If a sequence does not converge it is called divergent.

NUMBER SERIES

Each term in a number series has a fixed numerical value.

A finite series has a finite number of terms.

e.g. $a_1 + a_2, + a_3 + \ldots + a_{10}$ is a finite series with ten terms.

The sum of the first n terms of a series is denoted by S_n, i.e. $S_n = a_1 + a_2 + a_3 + \ldots + a_n$

An infinite series has no last term.

If, as $n \to \infty$, S_n tends to a finite value, S, then the series converges and S is called its sum to infinity.

ARITHMETIC PROGRESSIONS

In an arithmetic progression, each term differs from the preceding term by a constant (called the common difference).

An AP with first term a, common difference d and n terms is

$$a,\ a+d,\ a+2d,\ \dots,\ \{a+(n-1)d\}$$

The sum of the first n terms is given by

$$S_n = \tfrac{1}{2}n(a+l) \text{ where } l \text{ is the last term}$$
$$= \tfrac{1}{2}n\{2a+(n-1)d\}$$

GEOMETRIC PROGRESSIONS

In a geometric progression each term is a constant multiple of the preceding term. This multiple is called the common ratio.

A GP with first term a, common ratio r and n terms is

$$a,\ ar,\ ar^2,\ \dots,\ ar^{n-1}$$

The sum of the first n terms is given by

$$S_n = \frac{a(1-r^n)}{1-r}$$

The sum to infinity is given by $S = \dfrac{a}{1-r}$ provided that $|r| < 1$.

MULTIPLE CHOICE EXERCISE D

In Questions 1 to 13 write down the letters of the correct responses. There may be more than one.

1 $\dfrac{d}{dx}e^x$ is

A xe^{x-1} **C** $\ln x$

B $\dfrac{1}{e^x}$ **D** e^x

2

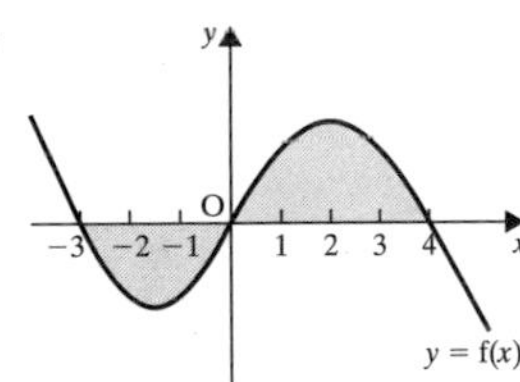

The shaded area in the diagram is given by

A $\displaystyle\int_{-3}^{4} f(x)\,dx$

B $\displaystyle\int_{-3}^{0} f(x)\,dx + \int_{0}^{4} f(x)\,dx$

C $\displaystyle\int_{0}^{4} f(x)\,dx - \int_{-3}^{0} f(x)\,dx$

D $\displaystyle\int_{0}^{4} f(x)\,dx - \int_{0}^{-3} f(x)\,dx$

3

x	1	2	3
y	0	3	6

Using the information in the table, the trapezium rule with 3 ordinates gives $\displaystyle\int_1^3 y\,dx \approx$

A 12 **C** 9 **E** $4\frac{1}{2}$

B $\frac{9}{2}$ **D** 6

4 As the sequence $u_{r+1} = \dfrac{1}{u_r}$, $u_1 = 2$, progresses, the terms

A get close to zero

B cycle between 2 and $\frac{1}{2}$

C are all positive after the first two

5 The series $\sum_{r=2}^{12} \frac{2^r}{r}$ has

A eleven terms

B is a GP

C a first term of 2

D a second term of 2

6 When $|x| = |2-x|$, $x =$

A 2 **C** 1 and -1

B 1 **D** -1

7 The graph of $y = |x+1|$ could be

A 

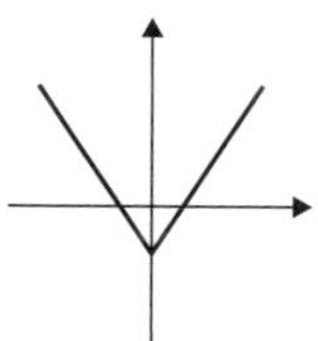

B

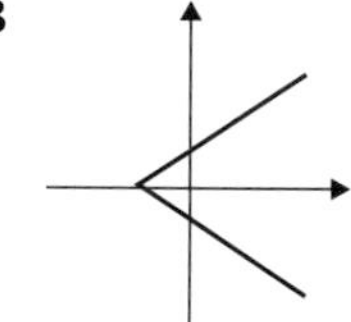

C

D

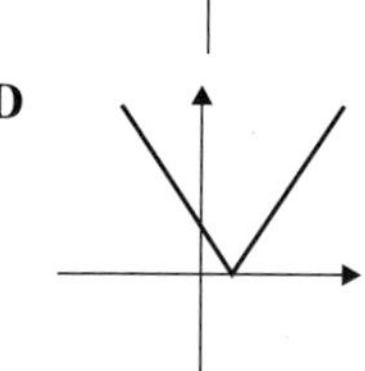

8 In the interval $0 < x < 1$

A $|x+1| > 0$

B $|x-1| < 0$

C $|x+1| = |x|+1$

9 $\int \frac{2}{x}\,dx =$

A $2\ln|x| + c$ **C** $\ln|2x| + c$

B $\ln x^2 + c$ **D** $-\frac{2}{x^2}$

10 The series
$1 + 2 + 4 + 8 + 16 + \ldots + 64$

A is an AP

B has a sum to infinity

C has 7 terms

D is equal to 127

11 $y = x(x+1)$, $\frac{dy}{dx} =$

A 2 **C** x **E** $2x$

B 0 **D** $2x+1$

12 $y = \frac{x+1}{x}$, $\frac{dy}{dx} =$

A $\frac{-1}{x^2}$ **C** 1

B $x + \ln|x|$ **D** 2

13

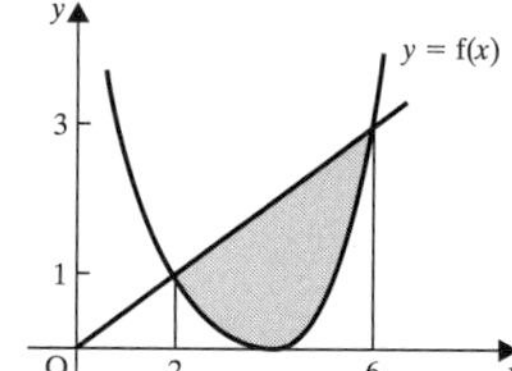

The shaded area in the diagram can be found from

A $\int_2^6 f(x)\,dx$

B $\int_2^6 (f(x) - \frac{1}{2}x)\,dx$

C $\int_2^6 (\frac{1}{2}x - f(x))\,dx$

D $8 - \int_2^6 f(x)\,dx$

For Questions 14 to 21 write down whether the statement is true or false.

14 $\left[f(x)\right]_0^a = f(a) - 0$

15 The area between the curve $y = 1 - x^2$ and the x-axis is given by $\int_{-1}^{1} y\,dx$.

16 If $2^n - 1$ is the sum of the first n terms of a series, the sum of the first $2n$ terms is $4^n - 1$.

17 $\int y\,dx = \frac{1}{2}y^2 + c$

18 The sum to infinity of the GP $1, \frac{1}{2}, \frac{1}{4}, \ldots$ is 2.

19 $|x| - 1 > 0$ for all values of x.

20 The function f, given by $f(x) = (|x| - 2)^2$, is an even function.

21 If $S = 1 - 2 + 4 - 8 + 16 - \ldots$ then $S = \frac{1}{3}$.

EXAMINATION QUESTIONS D

1 The equation of a curve is $y = 2x^2 - \ln x$, where $x > 0$. Find by differentiation the x-coordinate of the stationary point on the curve, and determine whether this point is a maximum point or a minimum point. (OCR)

2 **a** Sketch the graph of $y = e^x$ for all real values of x.

b On the same axes sketch the graph of $y = e^{-x}$. Describe a simple transformation which maps the graph of $y = e^x$ onto the graph of $y = e^{-x}$. (AQA)

3 **a** Sketch the graph of $y = e^x$.

b Given that $f(x) = e^x - x - 1$, show that f is an increasing function for $x > 0$. (WJEC)

4 The fourth term of an arithmetic series is 20 and the ninth term is 40. Find

a the common difference **b** the first term **c** the sum of the first 20 terms. (WJEC)

5 It is given that $f(x) \equiv (x - \alpha)(x - \beta), x \in \mathbb{R}$, where α and β are positive constants. Sketch, on separate diagrams, the curves with the following equations, giving in each case the coordinates of the points at which the curve meets the x-axis.

i $y = f(x)$, **ii** $y = |f(x)|$, **iii** $y = f(x + 2\alpha)$ (OCR)

6 An arithmetic series has first term 82 and common difference -10.

i Show that the sum of the first n terms is

$$n(87 - 5n).$$

ii Find the value of n for which this sum is equal to 370. (AQA)

7 Find $\int_1^4 \left(x + \frac{6}{\sqrt{x}}\right)^2 dx$

No credit will be given for a numerical approximation or for a numerical answer without supporting working. (AQA)

8 The function f is defined by $f : x \mapsto 5 + 8x^2 - 36 \ln x$ and has domain $x \geqslant 1$.

a Using calculus, determine the stationary value of f, giving your answer to 2 significant figures.

b Find the range of the function f″. (Edexcel)

9 Given that a is a positive constant, sketch the graph of $y = |3x - a|$, indicating clearly in terms of a where the graph crosses or touches the coordinate axes.
Solve the inequality $|3x - a| < x$. (AQA)

10 The graph of $y = (x+a)^2 + b$ is sketched opposite.

a Write down the values of a and b.

b Use algebraic integration to find the area of the shaded region shown.

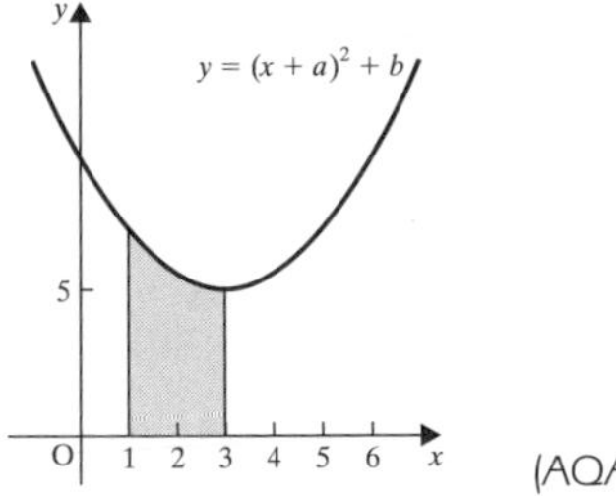

(AQA)

11 The functions f and g are defined over the set of real numbers by

$$\mathrm{f} : x \mapsto 3x - 5,$$
$$\mathrm{g} : x \mapsto \mathrm{e}^{-2x}.$$

a State the range of g.

b Sketch the graphs of the inverse functions f^{-1} and g^{-1} and write on your sketches the coordinates of any points at which a graph meets the coordinates axes.

c State, giving a reason, the number of roots of the equation $\mathrm{f}^{-1}(x) = \mathrm{g}^{-1}(x)$.

d Evaluate $\mathrm{fg}(-\frac{1}{3})$, giving your answer to 2 decimal places. (Edexcel)

12 An arithmetic progression has 241 terms and a common difference of 0.1. Given that the sum of all the terms is 964, find the first term. (OCR)

13 The diagram shows the curve $y = 3x - x^2$. The curve meets the x-axis at the origin O and at the point A. The tangent to the curve at the point B(2, 2) intersects the x-axis at C.

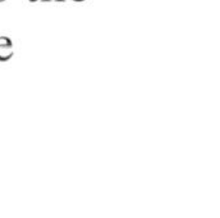

a Find the equation of the tangent to the curve at B.

b Find the shaded area.

(WJEC)

14 Sketch the graph of $y = |x - 2a|$, where a is a positive constant. (You should indicate the coordinates of the points where the graph meets the axes.)
Find, in terms of a, the two values of x satisfying $|x - 2a| = \frac{1}{2}a$. (OCR)

15 The functions f and g are given by

$$\mathrm{f} : x \mapsto 3x - 1,\ x \in \mathbb{R},$$
$$\mathrm{g} : x \mapsto \mathrm{e}^{\frac{x}{2}},\ x \in \mathbb{R}.$$

a Find the value of $\mathrm{fg}(4)$, giving your answer to 2 decimal places.

b Express the inverse function f^{-1} in the form $\mathrm{f}^{-1} : x \mapsto \ldots$

c Using the same axes, sketch the graphs of the functions f and gf. Write on your sketch the value of each function at $x = 0$.

d Find the values of x for which $\mathrm{f}^{-1}(x) = \dfrac{5}{\mathrm{f}(x)}$. (Edexcel)

16 a Given that the first and second terms of an arithmetic progression are 12 and 6 respectively, find the sum of the first hundred terms.

b Given that the first and second terms of a geometric progression are 12 and 6 respectively, show that the sum of the first ten terms is $\frac{3069}{128}$. (OCR)

17 An equation of a curve C is $y = \ln 3 + \ln x$.

a Find the coordinates of the point where C crosses the x-axis.

b Sketch, in a single diagram, both C and the curve with equation $y = \ln x$.

c Express $\ln 3 + \ln x$ as a single logarithm, and hence show that the inverse function of the function $(\ln 3 + \ln x)$ is $\frac{1}{3}e^x$

d Sketch, in a single diagram, the graphs of $y = e^x$ and $y = \frac{1}{3}e^x$, and describe briefly the relationship between these graphs and the graphs you sketched in answer to part (b). (Edexcel)

18

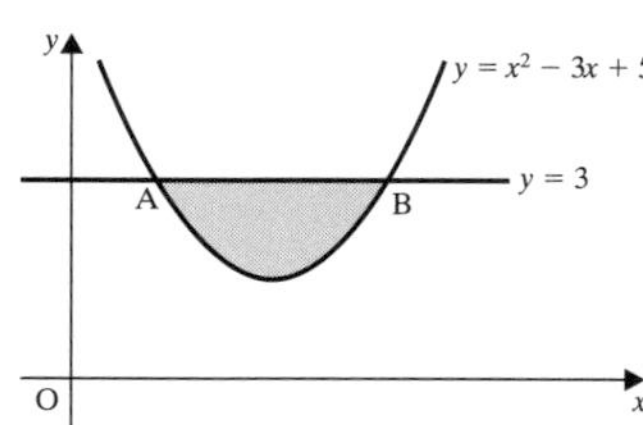

The graph shows sketches of the line $y = 3$ and the curve $y = x^2 - 3x + 5$ (not drawn to scale); they intersect at the points A and B. The shaded region is bounded by the arc AB and the chord AB.

a Find the coordinates of A and B.

b Find the area of the shaded region.

c Show that the equation of the tangent to the curve at A is

$$y + x - 4 = 0$$

and find the equation of the tangent to the curve at B.

d The tangents to the curve at A and B meet at the point C. Show that the coordinates of C are $\left(\frac{3}{2}, \frac{5}{2}\right)$. (AQA)

19 An arithmetic series has common difference 1.6 and first term 6. The sum of the first n terms is denoted by S_n.

a Show that S_n may be expressed in the form $pn^2 + qn$ for some constants p and q. State the value of p and the value of q.

b Find the value of the positive integer n for which $S_n = n^2$. (OCR)

20 The function f is defined by $f : x \mapsto e^x + k,\ x \in \mathbb{R}$ and k is a positive constant.

a State the range of f.

b Find $f(\ln k)$, simplifying your answer.

c Find f^{-1}, the inverse function of f, in the form $f^{-1} : x \mapsto \ldots$, stating its domain.

d On the same axes, sketch the curves with equations $y = f(x)$, and $y = f^{-1}(x)$, giving the coordinates of all points where the graphs cut the axes. (Edexcel)

21 A young person decides to save £50 at the start of each month to supplement her pension when she retires. Interest is calculated at the end of each month and is added to her account. The total in her account after n months can be modelled by the expression

$$\sum_{i=1}^{n} 50 \times 1.004^i.$$

a Find the total amount in her account after 3 months. Give your answer to the nearest 10p.

b Calculate the total amount in her account if she continues this method of saving without a break for 35 years. Give your answer to the nearest £100.

c Find the annual rate of interest assumed in this model. Give your answer to 1 decimal place. (AQA)

22 A curve has equation $y = e^x - kx$ where k is a constant ($k > 0$).

a The curve has a single stationary point at M. Calculate the x-coordinate of M and hence show that the y-coordinate of M can be written as $k(1 - \ln k)$.

b Use the result from part (a) to determine the exact value of k for which the curve touches the x-axis.

c In the case when $k = 2$,

i find the value of $\frac{d^2y}{dx^2}$ at M and hence determine the nature of M.
Deduce that the curve lies entirely above the x-axis.

ii calculate the area of the finite region bounded by the curve, the coordinate axes and the line with equation $x = 3$, leaving your answer in terms of e. (AQA)

23 The first four terms of three series A, B and C are given below. One series is an arithmetic series, one is a geometric series and one is neither.

A: $8 + 4 + 2 + 1 + \ldots$
B: $1 + 4 + 8 + 13 + \ldots$
C: $23 + 20 + 17 + 14 + \ldots$

a State the value of the common difference and calculate the 21st term of the arithmetic series.

b Find the sum to infinity of the geometric series. (AQA)

24 $f(x) \equiv \dfrac{(2\sqrt{x}+3)^2}{x}, \ x > 0$

a Show that $f(x)$ can be expressed as $A + Bx^{-\frac{1}{2}} + Cx^{-1}$, giving the values of the constants A, B and C.

b Find $\int f(x)\,dx$.

c Find the area of the finite region bounded by the curve with equation $y = f(x)$ and the lines with equations $x = 4$, $x = 9$ and $y = 0$, giving your answer in terms of natural logarithms.

(Edexcel)

25 The diagram below shows sketches of the line with equation $x + y = 4$ and the curve with equation $y = x^2 - 2x + 2$ intersecting at points P and Q. The minimum point of the curve is M. The shaded region R is bounded by the line and the curve.

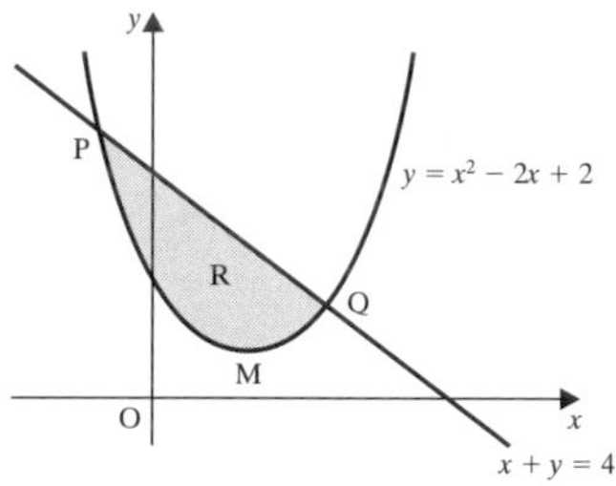

a Show that the coordinates of M are $(1, 1)$

b Find the coordinates of the points P and Q.

c Prove that the triangle PMQ is right-angled and hence show that the area of the triangle PMQ is 3 square units.

d Show that the area of the region R is $1\frac{1}{2}$ times that of the triangle PMQ. (AQA)

26

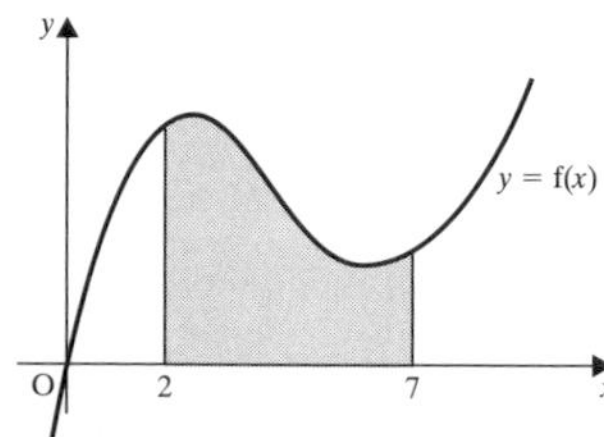

For some function f, part of the graph of $y = f(x)$ is illustrated above. It is given that the shaded region has an area equal to 20 square units.

a Sketch the graph of $y = f(x) + 3$ and calculate

$$\int_2^7 (f(x) + 3)\,dx$$

b Find the value of the constant k for which

$$\int_2^7 (f(x) + k)\,dx = 0$$

(AQA)

27 The nth term of a sequence is ar^{n-1}, where a and r are constants. The first term is 3 and the second term is $-\frac{3}{4}$. Find the values of a and r.

Hence find the value of $\sum_{n=1}^{\infty} ar^{n-1}$. (OCR)

28

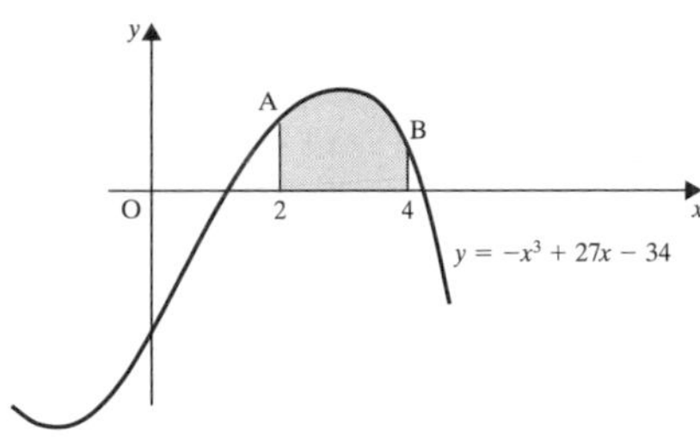

The figure shows a sketch of part of the curve with equation $y = \mathrm{f}(x)$ where $\mathrm{f}(x) = -x^3 + 27x - 34$.

a Find $\int \mathrm{f}(x)\,\mathrm{d}x$.

The lines $x = 2$ and $x = 4$ meet the curve at points A and B as shown.

b Find the area of the finite region bounded by the curve and the lines $x = 2$, $x = 4$ and $y = 0$.

c Find the area of the finite region bounded by the curve and the straight line AB. (Edexcel)

29 The figure below shows the graph of $y = \ln x$ for $x > 0$.

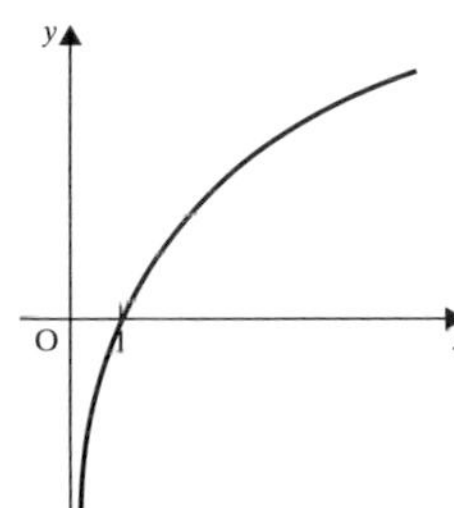

Use this diagram to sketch the graphs of

a $y = |\ln x|$ for $x > 0$; **b** $y = \ln|x|$ for all real x, $x \neq 0$. (OCR)

30 Solve the equation $|x| = |2x + 1|$ (OCR)

31 At the beginning of 1990, an investor decided to invest £6000 in a Personal Equity Plan (PEP), believing that the value of the investment should increase, on average, by 6% each year. Show that, if this percentage rate of increase is in fact maintained for 10 years, the value of the original investment will be about £10 745.

The investor added a further £6000 to the PEP at the beginning of each year between 1991 and 1995 inclusive. Assuming that the 6% annual rate of increase continues to apply, show that the total value, in £, of the PEP at the beginning of the year 2000 may be written as

$$6000 \sum_{r=5}^{10} (1.06)^r$$

and evaluate this, correct to the nearest £. (OCR)

32 The sequence $u_1, u_2, u_3, \ldots$ is defined by $u_n = 2n^2$

i Write down the value of u_3.

ii Express $u_{n+1} - u_n$ in terms of n, simplifying your answer.

iii The differences between successive terms of the sequence form an arithmetic progression. For this arithmetic progression, state its first term and its common difference, and find the sum of its first 1000 terms. (OCR)

33

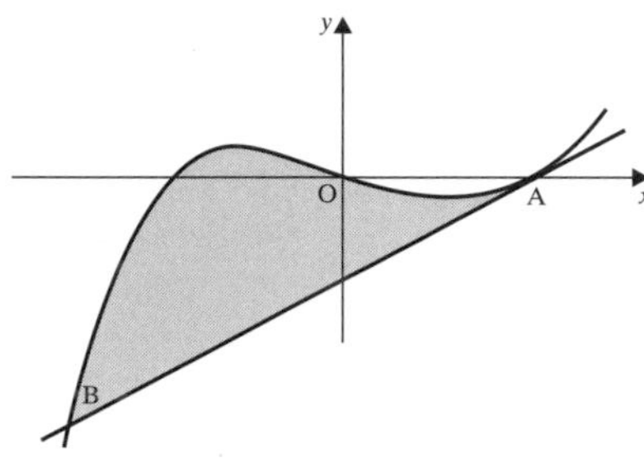

The diagram shows a sketch of the graph of the curve $y = x^3 - x$ together with the tangent to the curve at the point A$(1, 0)$.

i Use differentiation to find the equation of the tangent to the curve at A, and verify that the point B where the tangent cuts the curve again has coordinates $(-2, -6)$.

ii Use integration to find the area of the region bounded by the curve and the tangent (shaded in the diagram), giving your answer as a fraction in its lowest terms. (OCR)

34 At the start of a particular year, Mrs Brown made a single investment of £2000. At the end of that year and at the end of each subsequent year the value of her investment was 10% greater than its value at the start of the year.
Find, to the nearest £, the value of Mrs Brown's investment at the end of

a the fifth year **b** the tenth year.

Mrs Chan decided to invest £2000 at the start of each year with the same broker and at a fixed rate of interest of 10% per annum.

c Write down the first three terms of a series whose sum is the total value of Mrs Chan's annual investment at the end of the 12 years.

d Hence determine the value, to the nearest £, of Mrs Chan's investment at the end of 12 years. (Edexcel)

35 All the integers which are exactly divisible by 3 and lie between 1 and 100 form a series. Find

a the number of terms in the series

b the sum of the terms in the series. (Edexcel)

36 It is given that $y = x^{\frac{3}{2}} + \frac{48}{x}, \quad x > 0.$

a Find the value of x and the value of y when $\frac{dy}{dx} = 0$.

b Show that the value of y which you found in part (a) is a minimum.

The finite region R is bounded by the curve with equation $y = x^{\frac{3}{2}} + \frac{48}{x}$, the lines $x = 1, x = 4$ and the x-axis.

c Find, by integration, the area of R giving your answer in the form $p + q \ln r$, where the numbers p, q and r are to be found. (Edexcel)

37 A pump is used to extract air from a bottle. The first operation of the pump extracts $56\,\text{cm}^3$ of air and subsequent extractions follow a geometric progression. The third operation of the pump extracts $31.5\,\text{cm}^3$ of air.

a Determine the common ratio of the geometric progression and calculate the total amount of air that could be extracted from the bottle, if the pump were to extract air indefinitely.

b After how many operations of the pump does the total amount of air extracted from the bottle first exceed $220\,\text{cm}^3$? (AQA)

38

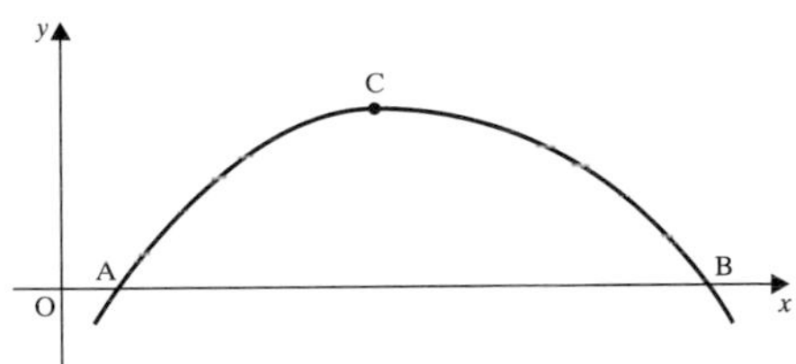

The function f is defined for positive real values of x by

$$f(x) = 12 \ln x - x^{\frac{3}{2}}.$$

The figure shows a sketch of the curve with equation $y = f(x)$. The curve crosses the x-axis at the points A and B. The gradient of the curve is zero at the point C.

a By calculation, show that the value of x at the point A lies between 1.1 and 1.2.

The value of x at the point B lies in the interval $(n, n + 1)$, where n is an integer.

b Determine the value of n.

c Show that $x = 4$ at the point C and hence find the greatest positive value of $f(x)$, giving your answer to 2 decimal places.

d Write down the set of values of x for which $f(x)$ is an increasing function of x. (Edexcel)

39 The ninth term of an arithmetic progression is 52 and the sum of the first twelve terms is 414. Find the first term and the common difference. (AQA)

40

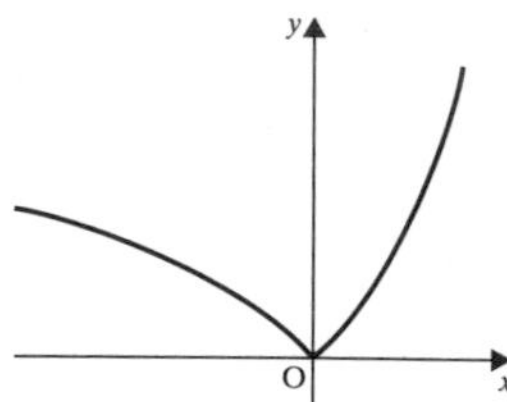

The diagram shows the graph of $y = |f(x)|$, for a certain function f with domain IR. Sketch, on separate diagrams, two possibilities for the graph of $y = f(x)$. (OCR)

41 An employer offers the following schemes of salary payments over a five-year period:

Scheme X: 60 monthly payments, starting with £1000 and increasing by £6 each month [£1000, £1006, £1012, ...];

Scheme Y: 5 annual payments, starting with £12 000 and increasing by £d each year [£12 000, £$(12\,000 + d)$, ...].

a Over the complete five-year period, find the total salary payable under Scheme X.

b Find the value of d which gives the same total salary for both schemes over the complete five-year period. (Edexcel)

42 An athlete plans a training schedule which involves running 20 km in the first week of training; in each subsequent week the distance is to be increased by 10% over the previous week. Write down an expression for the distance to be covered in the nth week according to the schedule, and find in which week the athlete would first cover more than 100 km. (OCR)

43 i The tenth term of an arithmetic progression is 36, and the sum of the first ten terms is 180. Find thc first term and the common difference.

ii Evaluate $\sum_{r=1}^{1000}(3r - 1)$. (OCR)

44 The functions f and g are defined by

$$f : x \mapsto 3x - 1, \quad x \in \mathbb{R},$$
$$g : x \mapsto x^2 + 1, \quad x \in \mathbb{R},$$

a Find the range of g.

b Calculate the value of gf(2).

c Determine the values of x for which $gf(x) = fg(x)$,

d Determine the values of x for which $|f(x)| = 8$. (Edexcel)

45 The functions f and g are defined by

$$f : x \mapsto x^2 - 10, \quad x \in \mathbb{R},$$
$$g : x \mapsto |x - 2|, \quad x \in \mathbb{R}$$

a Show that $f \circ f : x \mapsto x^4 - 20x^2 + 90, \ x \in \mathbb{R}$.
Find all the values of x for which $f \circ f(x) = 26$.

b Show that $g \circ f(x) = |x^2 - 12|$. Sketch a graph of $g \circ f$.
Hence, or otherwise, solve the equation $g \circ f(x) = x$. (AQA)

46 A small ball is dropped from a height of 1 m onto a horizontal floor. Each time the ball strikes the floor it rebounds to $\frac{3}{5}$ of the height from which it has just fallen.

a Show that, when the ball strikes the floor for the third time, it has travelled a distance 2.92 m.

b Show that the total distance travelled by the ball cannot exceed 4 m. (Edexcel)

47

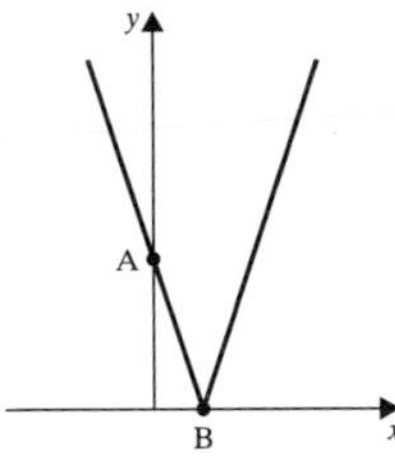

The figure shows a sketch of the graph $y = |3x - 2|$.

i State the coordinates of the points labelled A and B.

ii Make a copy of the figure, and shade in the area represented by

$$\int_0^2 |3x - 2|\,dx.$$

iii Evaluate this area. (OCR)

48 Solve the inequality $f|x + 1| < |x - 2|$. (OCR)

CHAPTER 21

Differentiation 2

Differentiating a Function of a Function

Suppose that we want to differentiate $(2x-1)^3$. We could expand the bracket and differentiate term by term, but this is tedious and, for powers higher than three, very long and not easy. We obviously need a more direct method for differentiating an expression of this kind.

Now $(2x-1)^3$ is a cubic function of the linear function $(2x-1)$, i.e. it is a *function of a function*.

A function of this type is of the form $\mathrm{gf}(x)$, i.e. $\mathrm{g}\{\mathrm{f}(x)\}$.

For example

$(x^2-3)^3$ is a cubic function g, of a quadratic function, f.

$\sqrt{1+x^4}$ is a square root function g, of a quartic function, f.

Consider any equation of the form $y = \mathrm{gf}(x)$

If we make the substitution $u = \mathrm{f}(x)$, then $y = \mathrm{gf}(x)$ can be expressed in two simple parts, i.e.

$$u = \mathrm{f}(x) \quad \text{and} \quad y = \mathrm{g}(u)$$

A small increase of δx in the value of x causes a corresponding small increase of δu in the value of u.
Then if $\delta x \to 0$, it follows that $\delta u \to 0$

Hence $$\frac{dy}{dx} = \lim_{\delta x \to 0}\left(\frac{\delta y}{\delta x}\right) = \lim_{\delta x \to 0}\left(\frac{\delta y}{\delta u} \times \frac{\delta u}{\delta x}\right)$$

$$\Rightarrow \quad \frac{dy}{dx} = \left(\lim_{\delta u \to 0}\frac{\delta y}{\delta u}\right) \times \left(\lim_{\delta x \to 0}\frac{\delta u}{\delta x}\right)$$

i.e. $$\mathbf{\frac{dy}{dx} = \frac{dy}{du} \times \frac{du}{dx}}$$

This is known as *the chain rule*.

EXAMPLES 21A

1 Find $\dfrac{dy}{dx}$ if $y = (2x-4)^4$.

If $u = 2x-4$ then $y = u^4$

Then $\dfrac{dy}{dx} = \dfrac{dy}{du} \times \dfrac{du}{dx}$ gives

$$\frac{dy}{dx} = (4u^3)(2) = 8u^3$$

But $u = 2x-4$

$$\therefore \quad \frac{dy}{dx} = 8(2x-4)^3$$

Example 1 is a particular case of the equation $y = (ax + b)^n$. Similar working shows that, in general,

if $y = (ax+b)^n$ then $\dfrac{dy}{dx} = an(ax+b)^{n-1}$

This fact is needed very often and is quotable, e.g. $\dfrac{d}{dx}(3x-2)^5 = 3(5)(3x-2)^4$.

2 Given $y = (x^3+1)^4$ find $\dfrac{dy}{dx}$.

If $u = x^3 + 1$ then $y = u^4$

Using $\dfrac{dy}{dx} = \dfrac{dy}{du} \times \dfrac{du}{dx}$ gives

$$\frac{dy}{dx} = (4u^3)(3x^2) = 12x^2u^3$$

Replacing u by $x^3 + 1$ we have

$$\frac{dy}{dx} = 12x^2(x^3+1)^3$$

3 Differentiate w.r.t. x the function $\dfrac{1}{(1-x^2)^5}$.

$y = (1-x^2)^{-5} \quad \Rightarrow \quad y = u^{-5}$ where $u = 1 - x^2$

Now $\quad \dfrac{dy}{dx} = \dfrac{dy}{du} \times \dfrac{du}{dx}$

$$\therefore \quad \frac{d}{dx}\left[\frac{1}{(1-x^2)^5}\right] = (-5u^{-6})(-2x) = 10x(u^{-6})$$

$$= \frac{10x}{(1-x^2)^6}$$

After some time you will find that in most cases the necessary substitution can be done mentally and the answer written down directly, e.g. to differentiate $(x^3-x)^{3/2}$ we mentally use the substitutions $u = x^3 - x$ and $y = u^{3/2}$ giving

$$\frac{d}{dx}(x^3-x)^{3/2} = [\tfrac{3}{2}(x^3-x)^{1/2}](3x^2-1)$$

This skill is important and well worth the practice required for achieving it.

EXERCISE 21A

Use a substitution to differentiate each function with respect to x.

1 $(3x+1)^2$

2 $(3-x)^4$

3 $(4x-5)^5$

4 $(x^2+1)^3$

5 $(2+3x)^7$

6 $(2-6x)^3$

7 $(2x^4-5)^{1/2}$

8 $(x^2+3)^{-1}$

9 $\sqrt{3x^3-4}$

10 $\dfrac{1}{\sqrt{x}+3x}$

11 $\dfrac{3}{\sqrt{4-x^2}}$

12 $\dfrac{7}{(x^3+3x)^{1/3}}$

Differentiate each function directly.

13 $(4-2x)^5$

14 $(x^2+3)^2$

15 $(3x-4)^7$

16 $(x^2+4)^2$

17 $(1-2x^2)^3$

18 $(2-x^3)^4$

19 $(2+x^2)^{3/4}$

20 $\sqrt[3]{x^2-x}$

21 $(2-3x^2)^{-1}$

22 $(4-x^2)^{-2}$

23 $(x^5-3)^{-1/2}$

24 $\sqrt[4]{6-\sqrt{x}}$

Differentiating a Product

Suppose that $y = uv$ where u and v are both functions of x, e.g. $y = x^2(x^4-1)$

It is dangerously tempting to think that $\frac{dy}{dx}$ is given by $\left(\frac{du}{dx}\right)\left(\frac{dv}{dx}\right)$

But this is *not so* as is clearly shown by a simple example such as $y = (x^2)(x^3)$ where, because $y = x^5$, we know that $\frac{dy}{dx} = 5x^4$ which is *not* equal to $(2x)(3x^2)$.

Returning to $y = uv$ where $u = f(x)$ and $v = g(x)$, we see that if x increases by a small amount δx then there are corresponding small increases of δu, δv and δy in the values of u, v and y.

$$\therefore \quad y+\delta y = (u+\delta u)(v+\delta v)$$

$$= uv + u\delta v + v\delta u + \delta u\delta v$$

But $y = uv$

$$\therefore \quad \delta y = u\delta v + v\delta u + \delta u\delta v$$

$$\Rightarrow \quad \frac{\delta y}{\delta x} = u\frac{\delta v}{\delta x} + v\frac{\delta u}{\delta x} + \delta u\frac{\delta v}{\delta x}$$

Now as $\delta x \to 0$, $\frac{\delta v}{\delta x} \to \frac{dv}{dx}$, $\frac{\delta u}{\delta x} \to \frac{du}{dx}$ and $\delta u \to 0$

$$\text{Therefore} \quad \frac{dy}{dx} = \lim_{\delta x \to 0}\frac{\delta y}{\delta x}$$

$$= u\frac{dv}{dx} + v\frac{du}{dx} + 0$$

i.e. $$\mathbf{\frac{d}{dx}(uv) = v\frac{du}{dx} + u\frac{dv}{dx}}$$

This formula is demonstrated by the simple example we considered above, i.e. $y = (x^2)(x^3)$.

Using $u = x^2$ and $v = x^3$ gives

$$\frac{dy}{dx} = (x^3)(2x) + (x^2)(3x^2) = 5x^4, \text{ which is correct.}$$

EXAMPLE 21B Differentiate with respect to x

a $(x+1)^3(2x-5)^2$ **b** $\dfrac{(x-1)^2}{(x+2)}$

a If $u = (x+1)^3$, $\dfrac{du}{dx} = 3(x+1)^2$

and if $v = (2x-5)^2$, $\dfrac{dv}{dx} = 2(2)(2x-5)$

$\dfrac{d}{dx}(uv) = v\dfrac{du}{dx} + u\dfrac{dv}{dx}$ gives

$$\frac{d}{dx}(x+1)^3(2x-5)^2 = \{(2x-5)^2\}\{3(x+1)^2\} + \{(x+1)^3\}\{2(2)(2x-5)\}$$
$$= (2x-5)(x+1)^2\{3(2x-5)+4(x+1)\}$$
$$= (2x-5)(x+1)^2(10x-11)$$

b If we write $\dfrac{(x-1)^2}{(x+2)}$ as $(x-1)^2(x+2)^{-1}$

then $u = (x-1)^2$ gives $\dfrac{du}{dx} = 2(x-1)$

and $v = (x+2)^{-1}$ gives $\dfrac{dv}{dx} = -(x+2)^{-2}$

Using $\dfrac{d}{dx}(uv) = v\dfrac{du}{dx} + u\dfrac{dv}{dx}$ we have

$$\frac{d}{dx}\left[\frac{(x-1)^2}{(x+2)}\right] = (x+2)^{-1}\{2(x-1)\} + (x-1)^2\{-(x+2)^{-2}\}$$
$$= \frac{(x-1)}{(x+2)^2}\{2(x+2)-(x-1)\}$$
$$= \frac{(x-1)(x+5)}{(x+2)^2}$$

EXERCISE 21B Differentiate each function with respect to x.

1 $x^2(x-3)^2$

2 $x\sqrt{x-6}$

3 $(x+2)(x-2)^5$

4 $x(2x+3)^3$

5 $(x+1)^2(x-1)^4$

6 $\sqrt{x}(x-3)^3$

7 $\dfrac{(x+5)^4}{(x-3)}$

8 $\dfrac{x}{(3x+2)^2}$

9 $\dfrac{(2x-7)^2}{\sqrt{x}}$

10 $x^3\sqrt{x-1}$

11 $x(x+3)^{-1}$

12 $x^2(2x-3)^2$

Differentiating a Quotient

To differentiate a function of the form $\frac{u}{v}$, where u and v are both functions of x, it is sometimes helpful to rewrite the function as uv^{-1} and differentiate it as a product. This method was used in part (b) of the previous worked example but it is not always the neatest way to differentiate a quotient. The alternative is to apply the formula derived below.

When a function is of the form $\frac{u}{v}$, where u and v are both functions of x, a small increase of δx in the value of x causes corresponding small increases of δu and δv in the values of u and v. Then, as $\delta x \to 0$, δu and δv also tend to zero.

If $y = \frac{u}{v}$ then $y + \delta y = \frac{(u + \delta u)}{(v + \delta v)}$

$$\therefore \quad \delta y = \frac{u + \delta u}{v + \delta v} - \frac{u}{v} = \frac{v\delta u - u\delta v}{v(v + \delta v)}$$

$$\therefore \quad \frac{\delta y}{\delta x} = \left(v\frac{\delta u}{\delta x} - u\frac{\delta v}{\delta x}\right) \bigg/ v(v + \delta v)$$

$$\Rightarrow \quad \frac{dy}{dx} = \lim_{\delta x \to 0} \frac{\delta y}{\delta x} = \left(v\frac{du}{dx} - u\frac{dv}{dx}\right) \bigg/ v^2$$

i.e. $$\frac{dy}{dx} = \frac{v\frac{du}{dx} - u\frac{dv}{dx}}{v^2}$$

EXAMPLE 21C

If $y = \frac{(4x - 3)^6}{(x + 2)}$ find $\frac{dy}{dx}$.

Using $u = (4x - 3)^6$ gives $\frac{du}{dx} = 24(4x - 3)^5$

and $v = x + 2$ gives $\frac{dv}{dx} = 1$

Then $$\frac{dy}{dx} = \left(v\frac{du}{dx} - u\frac{dv}{dx}\right) \bigg/ v^2$$

$$= \frac{(x + 2)\{24(4x - 3)^5\} - (4x - 3)^6}{(x + 2)^2} = \frac{(4x - 3)^5(20x + 51)}{(x + 2)^2}$$

EXERCISE 21C

Use the quotient formula to differentiate each of the following functions with respect to x.

1 $\frac{(x - 3)^2}{x}$

2 $\frac{x^2}{(x + 3)}$

3 $\frac{(4 - x)}{x^2}$

4 $\frac{(x + 1)^2}{x^3}$

5 $\frac{4x}{(1 - x)^3}$

6 $\frac{2x^2}{(x - 2)}$

7 $\frac{x^{5/3}}{(3x - 2)}$

8 $\frac{(1 - 2x)^3}{x^3}$

9 $\frac{\sqrt{(x + 1)^5}}{x}$

To Differentiate e^u where $u = f(x)$

If $y = e^u$ then $\dfrac{dy}{dx} = \dfrac{dy}{du} \times \dfrac{du}{dx}$ gives

$$\frac{dy}{dx} = e^u \times \frac{du}{dx}$$

This can also be expressed in the form $\dfrac{d}{dx} e^{f(x)} f'(x)$

i.e. $\quad \mathbf{\dfrac{d}{dx} e^{f(x)} = f'(x) e^{f(x)}}$

e.g. $\quad$ if $y = e^{(x^2+1)}$ then $\dfrac{dy}{dx} = 2xe^{(x^2+1)}$

The case when u is a linear function of x is particularly useful,

i.e. $\quad y = e^{(ax+b)} \quad \Rightarrow \quad \dfrac{dy}{dx} = ae^{(ax+b)}$

To Differentiate $\ln u$ where $u = f(x)$

If $y = \ln u$ then $\dfrac{dy}{dx} = \dfrac{dy}{du} \times \dfrac{du}{dx}$ gives $\dfrac{dy}{dx} = \dfrac{1}{u} \times \dfrac{du}{dx}$

This can also be expressed in the form

$$\mathbf{\frac{d}{dx}\{\ln f(x)\} = \frac{1}{f(x)} \times f'(x) = \frac{f'(x)}{f(x)}}$$

e.g. if $y = \ln(2 + x^3)$ then $\dfrac{dy}{dx} = \dfrac{3x^2}{2 + x^3}$

Again the case when u is $ax + b$ occurs frequently and is worth noting,

i.e. $\quad y = \ln(ax + b) \quad \Rightarrow \quad \dfrac{dy}{dx} = \dfrac{a}{ax + b}$

EXERCISE 21D Differentiate w.r.t. x

1 $\ln 2x$

2 e^{4x-1}

3 $\ln(5x + 2)$

4 e^{x^2-3}

5 $\ln(2x^2 - 3)$

6 $e^{(3x^3-2x)}$

7 $\ln(1 + x + x^2)$

8 $\ln(1 + 2x)^2$

9 $e^{x(x+1)}$

Compound Exponential and Logarithmic Functions

For any given function the first step is to identify its category, e.g. x^2e^x and $(1+x)\ln x$ are both products.
Whereas e^{x^2} and $\ln(1-x^2)$ are both functions of a function.

EXERCISE 21E

In Questions 1 to 9

a identify the type of function

b express the function in terms of u and/or v, stating clearly the substitutions that have been made.

1 $e^x(x^2+1)$ **4** $\sqrt{e^{(x+1)}}$ **7** $(\ln x)^2$

2 $e^{(x^2+1)}$ **5** $e^x \ln x$ **8** e^{-2x}

3 $x\ln x$ **6** $\ln(3-x^2)$ **9** $1/\ln x$

10 If f and g are the functions defined by $f: x \to x^2$ and $g: x \to e^x$ write down the functions $fg(x)$ and $gf(x)$.

11 The functions f, g and h are defined as follows

$$f: x \to x^2 \qquad g: x \to \frac{1}{x} \qquad h: x \to \ln x$$

Write down the functions

a $fg(x)$ **b** $hf(x)$ **c** $hg(x)$ **d** $fh(x)$ **e** $hfg(x)$ **f** $fg^{-1}(x)$

EXAMPLES 21F

1 Find the derivative of x^3e^x.

$y = x^3e^x$ becomes $y = uv$ if $u = x^3$ and $v = e^x$

$$\Rightarrow \quad \frac{du}{dx} = 3x^2 \text{ and } \frac{dv}{dx} = e^x$$

$$\therefore \quad \frac{dy}{dx} = v\frac{du}{dx} + u\frac{dv}{dx} = e^x(3x^2) + x^3(e^x)$$

i.e.
$$\frac{dy}{dx} = x^2(3+x)e^x$$

2 Differentiate $\ln(x\sqrt{x^2-4})$ w.r.t. x.

First we simplify the log expression by changing it into a sum.

$$\ln(x\sqrt{x^2-4}) = \ln x + \ln\sqrt{x^2-4} = \ln x + \tfrac{1}{2}\ln(x^2-4)$$

$$\therefore \quad \frac{d}{dx}\ln\{x\sqrt{x^2-4}\} = \frac{d}{dx}\ln x + \frac{d}{dx}\{\tfrac{1}{2}\ln(x^2-4)\}$$

$$= \frac{1}{x} + \frac{1}{2}\left(\frac{2x}{x^2-4}\right) = \frac{1}{x} + \frac{x}{x^2-4}$$

Simplifying the given function at the start, made the differentiation in this problem much easier. *Before differentiating any function, all possible simplification should be done,* particularly when complicated log expressions are involved.

EXERCISE 21F Differentiate the following functions with respect to x

1 xe^x

2 $x^2\ln x$

3 $e^x(x^3-2)$

4 $x^2\ln(x-2)^6$

5 $(x-1)e^x$

6 $(x^2+4)\ln\sqrt{x}$

7 $x\sqrt{2+x}$

8 $x\ln\sqrt{x-5}$

9 $(x^2-2)e^x$

10 $\dfrac{x}{e^x}$

11 $\dfrac{e^x}{x^2}$

12 $\dfrac{(\ln x)}{x^3}$

13 $\dfrac{\sqrt{x+1}}{\ln x}$

14 $\dfrac{e^x}{x^2-1}$

15 $\dfrac{e^x}{e^x-e^{-x}}$

16 e^{4x}

17 $\ln(x^2-1)$

18 e^{x^2}

19 $6e^{(1-x)}$

20 $e^{(x^2+1)}$

21 $\ln\sqrt{x+2}$

22 $(\ln x)^2$

23 $1/(\ln x)$

24 $\sqrt{e^x}$

Identifying the Category of a Function

Before any of the techniques explained earlier can be used to differentiate a given function, it is important to recognise which category the function belongs to, e.g. is it a product or a fraction or a function of a function. It somtimes seems difficult to distinguish between a product and a function of a function. It should not be a problem if you remember that

a a product has two separate parts, each being an independent function of x; you can actually put brackets round the parts and each is complete in itself, e.g. $(e^x)(\sin x)$.

b whereas if *one operation* is carried out *on another function of* x we have a function of a function e.g. $e^{\sin x}$ is an exponential function of a trig function and it cannot be separated into independent parts.

Remember also that sometimes a fraction can be expressed more simply as a product, e.g. $\dfrac{\sin x}{e^x}$ can be written as $e^{-x}\sin x$.

MIXED EXERCISE 21 This exercise contains a mixture of compound functions. In each case first identify the type of function and then use the appropriate method to find its derivative. Some of the functions can be differentiated by using a basic rule so do not assume that special techniques are always needed.

1 $x\sqrt{x+1}$

2 $(x^2-8)^3$

3 $\dfrac{x}{x^2+1}$

4 $\sqrt[3]{2-x^4}$

5 $\dfrac{x^2+1}{x^2+2}$

6 $x^2(\sqrt{x}-2)$

7 $(x^2-2)^3$

8 $\sqrt{x-x^2}$

9 $\dfrac{x}{\sqrt{x}+1}$

10 $x^2\sqrt{x-2}$

11 $\dfrac{\sqrt{x+1}}{x^2}$

12 $(x^4+x^2)^3$

13 $\sqrt{x^2-8}$

14 $x^3(x^2-6)$

15 $(x^2-6)^3$

16 $\dfrac{x}{x^2-6}$

17 $(x^4+3)^{-2}$

18 $\sqrt{x}(2-x)^3$

19 $\dfrac{\sqrt{x}}{(2-x)^3}$

20 $(x-1)(x-2)^2$

21 $(2x^3+4)^5$

22 $x\ln x$

23 $(4x-1)^{2/3}$

24 $\dfrac{e^x}{x-1}$

25 $\dfrac{\sqrt{1+x^3}}{x^2}$

26 $\dfrac{\ln x}{\ln(x-1)}$

27 10^{3x}

28 $\dfrac{(1+2x^2)}{1+x^2}$

29 $e^{-2/x}$

30 $\ln(1-e^x)$

31 $e^{3x}x^3$

32 $\dfrac{2x}{(2x-1)(x-3)}$

33 $\dfrac{e^{x/2}}{x^5}$

34 $\ln\left[\dfrac{x^2}{(x+3)(x^2-1)}\right]$

35 $\ln 4x^3(x+3)^2$

36 $(\ln x)^4$

37 $\dfrac{(x+3)^3}{x^2+2}$

38 $\sqrt{e^x-x}$

39 $4\ln(x^2+1)$

Find and simplify $\dfrac{dy}{dx}$ and hence find $\dfrac{d^2y}{dx^2}$

40 $y=\dfrac{1+2x}{1-2x}$

41 $y=\ln\dfrac{x}{x+1}$

42 $y=\dfrac{e^x}{e^x-4}$

43 The graph shows the number of stick insects in a vivarium at various times.

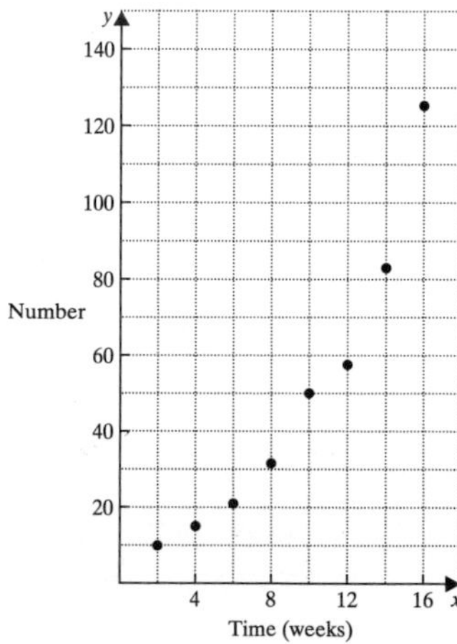

a Find the growth factor between the numbers of stick insects in consecutive time intervals of two weeks. Hence find a relationship between x (time in days) and y (number of insects) that can be used as an approximate model for the data given in the graph.

b Use the model to estimate
  i the number of insects 45 days after the first recorded entry on the graph
  ii how many days after the first recorded entry the number of insects had grown to 100.

Explain why these results can only be estimates.

c Find the rate at which the numbers are growing 10 weeks after the first entry and comment on accuracy of the result.

CHAPTER 22 **Trigonometry 2**

Compound Angles

It is often useful to be able to express a trig ratio of an angle $A+B$ in terms of trig ratios of A and of B.

It is dangerously easy to think, for instance, that $\sin(A+B)$ is $\sin A + \sin B$. However, this is *false* as can be seen by considering

$$\sin(45^\circ + 45^\circ) = \sin 90^\circ = 1$$

whereas $\quad \sin 45^\circ + \sin 45^\circ = \frac{1}{2}\sqrt{2} + \frac{1}{2}\sqrt{2} \neq 1$

Similarly $\quad \cos(A+B)$ is NOT $\cos A + \cos B$

and $\quad \tan(A+B)$ is NOT $\tan A + \tan B$

The correct identity for $\sin(A+B)$ is $\sin(A+B) \equiv \sin A\cos B + \cos A\sin B$.

This is proved geometrically, when A and B are both acute, from the diagram.

The right-angled triangles OPQ and OQR contain angles A and B as shown. From the diagram, $\angle\text{URQ} = A$

$$\sin(A+B) = \frac{\text{TR}}{\text{OR}} = \frac{\text{TS}+\text{SR}}{\text{OR}} = \frac{\text{PQ}+\text{SR}}{\text{OR}}$$

$$= \frac{\text{PQ}}{\text{OQ}} \times \frac{\text{OQ}}{\text{OR}} + \frac{\text{SR}}{\text{QR}} \times \frac{\text{QR}}{\text{OR}}$$

$\therefore \quad \sin(A+B) \equiv \sin A\cos B + \cos A\sin B$

This identity is in fact valid for all angles and it can be adapted to give the full set of compound angle formulae. You are left to do this in the following exercise.

EXERCISE 22A

1 In the identity $\sin(A+B) \equiv \sin A\cos B + \cos A\sin B$, replace B by $-B$ to show that $\sin(A-B) \equiv \sin A\cos B - \cos A\sin B$.

2 In the identity $\sin(A-B) \equiv \sin A\cos B - \cos A\sin B$, replace A by $(\frac{1}{2}\pi - A)$ to show that $\cos(A+B) \equiv \cos A\cos B - \sin A\sin B$.

3 In the identity $\cos(A+B) \equiv \cos A\cos B - \sin A\sin B$, replace B by $-B$ to show that $\cos(A-B) \equiv \cos A\cos B + \sin A\sin B$.

4 Use $\dfrac{\sin(A+B)}{\cos(A+B)}$ to show that $\tan(A+B) \equiv \dfrac{\tan A + \tan B}{1 - \tan A\tan B}$.

5 Replace B by $-B$ in the formula for $\tan(A+B)$ to show that

$$\tan(A-B) \equiv \frac{\tan A - \tan B}{1 + \tan A\tan B}$$

Collecting these results we have:

$$\sin(A+B) \equiv \sin A\cos B + \cos A\sin B$$
$$\sin(A-B) \equiv \sin A\cos B - \cos A\sin B$$
$$\cos(A+B) \equiv \cos A\cos B - \sin A\sin B$$
$$\cos(A-B) \equiv \cos A\cos B + \sin A\sin B$$
$$\tan(A+B) \equiv \frac{\tan A + \tan B}{1 - \tan A\ \tan B}$$
$$\tan(A-B) \equiv \frac{\tan A - \tan B}{1 + \tan A\ \tan B}$$

The Factor Formulae

The formulae above can be combined to give another group that allow us to convert a sum or difference of the sines (or cosines) of two different angles, into a product.

We will carry out the rearrangement on just one pair of the formulae above, to show how it is done.

Add $\quad \sin(A+B) \equiv \sin A\cos B + \cos A\sin B$

and $\quad \sin(A-B) \equiv \sin A\cos B - \cos A\sin B$

$\Rightarrow \quad \sin(A+B) + \sin(A-B) \equiv 2\sin A\cos B$

Subtracting the same pair of formulae gives a similar result, while adding and subtracting the formulae for $\cos(A+B)$ and $\cos(A-B)$ completes the following set of four that convert a sum or difference of two sines or cosines into a product.

$$\sin(A+B) + \sin(A-B) \equiv 2\sin A\cos B$$
$$\sin(A+B) - \sin(A-B) \equiv 2\cos A\sin B$$
$$\cos(A+B) + \cos(A-B) \equiv 2\cos A\cos B$$
$$\cos(A+B) - \cos(A-B) \equiv -2\sin A\sin B$$

Note the minus sign.

If you need to use any of these you can either

a learn how to convert the original formulae or

b memorise the extra four above. If you choose this option it often helps to remember them in words, e.g. $\sin + \sin = \text{twice } \sin \times \cos$

EXAMPLES 22B

1 Find exact values for **a** $\sin 75°$ **b** $\cos 105°$

To find exact values, we need to express the given angle in terms of angles whose trig ratios are known as exact values, e.g. 30°, 60°, 45°, 90°, 120°, ...
Now $75° = 45° + 30°$ (or $120° - 45°$ or other alternative compound angles).

a $\sin 75° = \sin(45° + 30°) = \sin 45°\cos 30° + \cos 45°\sin 30°$

$$= \left(\frac{\sqrt{2}}{2}\right)\left(\frac{\sqrt{3}}{2}\right) + \left(\frac{\sqrt{2}}{2}\right)\left(\frac{1}{2}\right) = \frac{\sqrt{2}}{4}(\sqrt{3}+1)$$

b $\cos 105° = \cos(60° + 45°) = \cos 60° \cos 45° - \sin 60° \sin 45°$

$$= \left(\frac{1}{2}\right)\left(\frac{\sqrt{2}}{2}\right) - \left(\frac{\sqrt{3}}{2}\right)\left(\frac{\sqrt{2}}{2}\right)$$

$$= \frac{\sqrt{2}}{4}(1 - \sqrt{3})$$

2 A is obtuse and $\sin A = \frac{3}{5}$, B is acute and $\sin B = \frac{12}{13}$. Find the exact value of $\cos(A + B)$

$\cos(A + B) \equiv \cos A \cos B - \sin A \sin B$

> In order to use this formula, we need values for $\cos A$ and $\cos B$. These can be found using Pythagoras' theorem in the appropriate right-angled triangle.

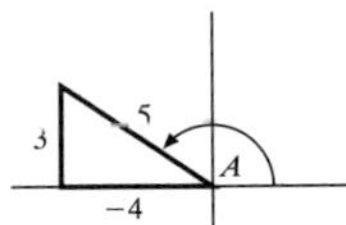

$\cos A = -\frac{4}{5}$

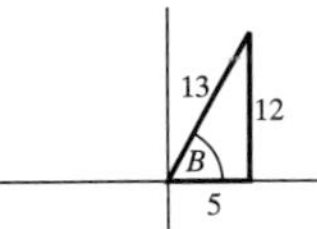

$\cos B = \frac{5}{13}$

$$\therefore \quad \cos(A + B) = \left(-\frac{4}{5}\right)\left(\frac{5}{13}\right) - \left(\frac{3}{5}\right)\left(\frac{12}{13}\right) = -\frac{56}{65}$$

3 Simplify $\sin\theta \cos\frac{1}{3}\pi - \cos\theta \sin\frac{1}{3}\pi$ and hence find the smallest positive value of θ for which the expression has a minimum value.

> $\sin\theta \cos\frac{1}{3}\pi - \cos\theta \sin\frac{1}{3}\pi$ is the expansion of $\sin(A - B)$ with $A = \theta$ and $B = \frac{1}{3}\pi$.

$\mathrm{f}(\theta) = \sin\theta \cos\frac{1}{3}\pi - \cos\theta \sin\frac{1}{3}\pi = \sin(\theta - \frac{1}{3}\pi)$

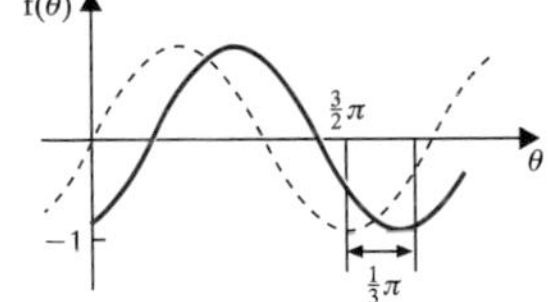

The graph of $\mathrm{f}(\theta) = \sin(\theta - \frac{1}{3}\pi)$ is a sine wave, but translated $\frac{1}{3}\pi$ in the direction of the positive θ-axis.

Therefore $\mathrm{f}(\theta)$ has a minimum value of -1 and the smallest +ve value of θ at which this occurs is $\frac{3}{2}\pi + \frac{1}{3}\pi = \frac{11}{6}\pi$.

4 Prove that $\dfrac{\sin(A - B)}{\cos A \cos B} \equiv \tan A - \tan B$.

Expanding the numerator, the LHS becomes

$$\frac{\sin A \cos B - \cos A \sin B}{\cos A \cos B}$$

$$\equiv \frac{\sin A \cos B}{\cos A \cos B} - \frac{\cos A \sin B}{\cos A \cos B}$$

$$\equiv \tan A - \tan B \equiv \text{RHS}$$

5 Find, in the range $0 \leqslant \theta \leqslant 2\pi$, the solution of the equation

$$2\cos\theta = \sin(\theta + \tfrac{1}{6}\pi).$$

$$2\cos\theta = \sin(\theta + \tfrac{1}{6}\pi)$$
$$= \sin\theta\cos\tfrac{1}{6}\pi + \cos\theta\sin\tfrac{1}{6}\pi = \tfrac{\sqrt{3}}{2}\sin\theta + \tfrac{1}{2}\cos\theta$$
$$\therefore \quad \tfrac{3}{2}\cos\theta = \tfrac{\sqrt{3}}{2}\sin\theta$$
$$\Rightarrow \quad \frac{3}{\sqrt{3}} = \frac{\sin\theta}{\cos\theta} \quad \Rightarrow \quad \tan\theta = \sqrt{3}$$

$\sqrt{3}$ 2π

Now $\tan\tfrac{1}{3}\pi = \sqrt{3}$, so the solution is $\theta = \tfrac{1}{3}\pi, \tfrac{4}{3}\pi$.

6 Find the values of θ, in the range $0 \leqslant \theta \leqslant \pi$, that satisfy the equation

$$\sin 5\theta + \sin 3\theta = 0.$$

Comparing with $\sin(A+B) + \sin(A-B) \equiv 2\sin A\cos B$ gives

$A + B = 5\theta$ and $A - B = 3\theta \quad \Rightarrow \quad A = 4\theta$ and $B = \theta$

$$\therefore \quad 2\sin 4\theta\cos\theta = 0 \quad \Rightarrow \quad \sin 4\theta = 0 \text{ or } \cos\theta = 0$$
$$\sin 4\theta = 0 \quad \Rightarrow \quad 4\theta = 0, \pi, 2\pi, 3\pi, 4\pi$$
$$\text{and} \quad \cos\theta = 0 \quad \Rightarrow \quad \theta = \tfrac{1}{2}\pi$$

$\therefore$ the solution of the equation is $\theta = 0, \tfrac{1}{4}\pi, \tfrac{1}{2}\pi, \tfrac{3}{4}\pi, \pi$

EXERCISE 22B

Find the exact value of each expression, leaving your answer in surd form where necessary.

1 $\cos 40^\circ\cos 50^\circ - \sin 40^\circ\sin 50^\circ$

2 $\sin 37^\circ\cos 7^\circ - \cos 37^\circ\sin 7^\circ$

3 $\cos 75^\circ$

4 $\tan 105^\circ$

5 $\sin 165^\circ$

6 $\cos 15^\circ$

Simplify each of the following expressions.

7 $\sin\theta\cos 2\theta + \cos\theta\sin 2\theta$

8 $\cos\alpha\cos(90^\circ - \alpha) - \sin\alpha\sin(90^\circ - \alpha)$

9 $\dfrac{\tan A + \tan 2A}{1 - \tan A\tan 2A}$

10 $\dfrac{\tan 3\beta - \tan 2\beta}{1 + \tan 3\beta\tan 2\beta}$

11 A is acute and $\sin A = \tfrac{7}{25}$, B is obtuse and $\sin B = \tfrac{4}{5}$. Find an exact expression for

a $\sin(A+B)$ **b** $\cos(A+B)$ **c** $\tan(A+B)$

12 Find the greatest value of each expression and the value of θ between 0 and 360° at which it occurs.

a $\sin\theta\cos 25^\circ - \cos\theta\sin 25^\circ$

b $\sin\theta\sin 30^\circ + \cos\theta\cos 30^\circ$

c $\cos\theta\cos 50^\circ - \sin\theta\sin 50^\circ$

d $\sin 60^\circ\cos\theta - \cos 60^\circ\sin\theta$

Prove the following identities.

13 $\cot(A+B) \equiv \dfrac{\cot A \cot B - 1}{\cot A + \cot B}$

14 $(\sin A + \cos A)(\sin B + \cos B) \equiv \sin(A+B) + \cos(A-B)$

15 $\sin(A+B) + \sin(A-B) \equiv 2\sin A \cos B$

16 $\cos(A+B) + \cos(A-B) \equiv 2\cos A \cos B$

17 $\dfrac{\sin(A+B)}{\cos A \cos B} \equiv \tan A + \tan B$

Solve the following equations for values of θ in the range $0 \leqslant \theta \leqslant 360°$.

18 $\cos(45° - \theta) = \sin\theta$

19 $3\sin\theta = \cos(\theta + 60°)$

20 $\tan(A - \theta) = \frac{2}{3}$ and $\tan A = 3$

21 $\sin(\theta + 60°) = \cos\theta$

22 $\sin 4\theta + \sin 2\theta = 0$

The Double Angle Identities

The compound angle formulae deal with any two angles A and B and can therefore be used for two equal angles, i.e. when $B = A$.

Replacing B by A in the trig identities for $(A+B)$ gives the following set of double angle identities.

$$\sin 2A \equiv 2\sin A\cos A$$

$$\cos 2A \equiv \cos^2 A - \sin^2 A$$

$$\tan 2A \equiv \frac{2\tan A}{1 - \tan^2 A}$$

The second of these identities can be expressed in several forms because

$$\cos^2 A - \sin^2 A \equiv \begin{cases} (1 - \sin^2 A) - \sin^2 A = 1 - 2\sin^2 A \\ \cos^2 A - (1 - \cos^2 A) \equiv 2\cos^2 A - 1 \end{cases}$$

i.e.

$$\cos 2A \equiv \begin{cases} \cos^2 A - \sin^2 A \\ 1 - 2\sin^2 A \\ 2\cos^2 A - 1 \end{cases}$$

There is one other group of identities that can be useful, and these are derived from two of the forms of the cosine double angle formulae above.

Starting with $\cos 2A \equiv 2\cos^2 A - 1$ we have

$$\cos^2 A \equiv \tfrac{1}{2}(1 + \cos 2A)$$

Similarly starting with $\cos 2A \equiv 1 - 2\sin^2 A$, we get

$$\sin^2 A \equiv \tfrac{1}{2}(1 - \cos 2A)$$

EXAMPLES 22C

1 If $\tan\theta = \frac{3}{4}$, find the values of $\tan 2\theta$ and $\tan 4\theta$.

Using $\tan 2A \equiv \dfrac{2\tan A}{1-\tan^2 A}$ with $A = \theta$ and $\tan\theta = \frac{3}{4}$ gives

$$\tan 2\theta = \frac{2(\frac{3}{4})}{1-(\frac{3}{4})^2} = \frac{24}{7}$$

Using the identity for $\tan 2A$ again, but this time with $A = 2\theta$, gives

$$\tan 4\theta = \frac{2\tan 2\theta}{1-\tan^2 2\theta} = \frac{2(\frac{24}{7})}{1-(\frac{24}{7})^2} = -\frac{336}{527}$$

2 Eliminate θ from the equations $x = \cos 2\theta$, $y = \sec\theta$.

Using $\cos 2\theta \equiv 2\cos^2\theta - 1$ gives

$$x = 2\cos^2\theta - 1 \text{ and } y = \frac{1}{\cos\theta}$$

$$\therefore \quad x = 2\left(\frac{1}{y}\right)^2 - 1 \quad \Rightarrow \quad (x+1)y^2 = 2$$

Note that this is a Cartesian equation which has been obtained by *eliminating the parameter* θ from a *pair of parametric equations.*

3 Prove that $\sin 3A \equiv 3\sin A - 4\sin^3 A$

$$\begin{aligned}
\sin 3A &\equiv \sin(2A + A)\\
&\equiv \sin 2A\cos A + \cos 2A\sin A\\
&\equiv (2\sin A\cos A)\cos A + (1-2\sin^2 A)\sin A\\
&\equiv 2\sin A\cos^2 A + \sin A - 2\sin^3 A\\
&\equiv 2\sin A(1-\sin^2 A) + \sin A - 2\sin^3 A\\
&\equiv 3\sin A - 4\sin^3 A
\end{aligned}$$

4 Find the solution of the equation $\cos 2x + 3\sin x = 2$ giving values of θ in the interval $[-\pi, \pi]$.

When a trig equation involves different multiples of an angle, it is usually sensible to express the equation in a form where the trig ratios are all of the same angle and, when possible, involving only one trig ratio.

Using $\cos 2x \equiv 1 - 2\sin^2 x$ gives

$$1 - 2\sin^2 x + 3\sin x = 2$$

$$\Rightarrow \quad 2\sin^2 x - 3\sin x + 1 = 0$$

$$\Rightarrow \quad (2\sin x - 1)(\sin x - 1) = 0$$

$$\therefore \quad \sin x = \tfrac{1}{2} \text{ or } \sin x = 1$$

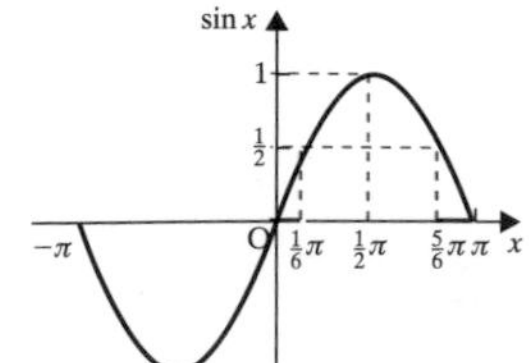

When $\sin x = \frac{1}{2}$, $x = \frac{1}{6}\pi, \frac{5}{6}\pi$ and when $\sin x = 1$, $x = \frac{1}{2}\pi$

Therefore the solution is $x = \frac{1}{6}\pi, \frac{5}{6}\pi, \frac{1}{2}\pi$.

5 Express $4\cos^2 x + 1$ in terms of the angle $2x$.

Using $\cos^2 x = \frac{1}{2}(1 + \cos 2x)$ gives

$$\begin{aligned} 4\cos^2 x + 1 &= 4 \times \tfrac{1}{2}(1 + \cos 2x) + 1 \\ &= 2(1 + \cos 2x) + 1 \\ &= 3 + 2\cos 2x \end{aligned}$$

6 a Use the formula $\tan 2\theta = \dfrac{2\tan\theta}{1 - \tan^2\theta}$ to mark two sides of a right-angled triangle in terms of $\tan\theta$ where one of the angles in the triangle is 2θ. Use Pythagoras' theorem to find the length of the third side in terms of $\tan\theta$.

b Hence write down a formula for $\sin 2\theta$ in terms of $\tan\theta$.

c Use the formula from part **b** to find the smallest positive value of θ for which $2\tan\theta = \operatorname{cosec} 2\theta$.

a Using Pythagoras' theorem,

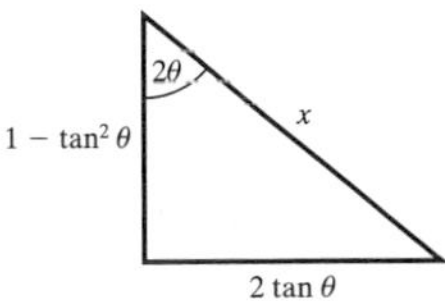

$$\begin{aligned} x^2 &= (2\tan\theta)^2 + (1 - \tan^2\theta)^2 \\ &= 4\tan^2\theta + 1 - 2\tan^2\theta + \tan^4\theta \\ &= 1 + 2\tan^2\theta + \tan^4\theta \\ &= (1 + \tan^2\theta)^2 \end{aligned}$$

$\Rightarrow \quad x = 1 + \tan^2\theta$

b $\sin 2\theta = \dfrac{\text{opp}}{\text{hyp}} = \dfrac{2\tan\theta}{1 + \tan^2\theta}$

c $2\tan\theta = \operatorname{cosec} 2\theta = \dfrac{1}{\sin 2\theta}$

Using the formula from **b** gives $2\tan\theta = \dfrac{1 + \tan^2\theta}{2\tan\theta}$

$\Rightarrow \quad 4\tan^2\theta = 1 + \tan^2\theta$

i.e. $\quad 3\tan^2\theta = 1 \quad \Rightarrow \quad \tan\theta = \dfrac{1}{\sqrt{3}}$

The smallest positive value of θ for which $\tan\theta = \dfrac{1}{\sqrt{3}}$ is $\dfrac{\pi}{6}$.

EXERCISE 22C

Simplify, giving an exact value where this is possible.

1 $2\sin 15° \cos 15°$

2 $\cos^2 \frac{1}{8}\pi - \sin^2 \frac{1}{8}\pi$

3 $\sin\theta\cos\theta$

4 $1 - 2\sin^2 4\theta$

5 $\dfrac{2\tan 75°}{1-\tan^2 75°}$

6 $\dfrac{2\tan 3\theta}{1-\tan^2 3\theta}$

7 $2\cos^2 \frac{3}{8}\pi - 1$

8 $1 - 2\sin^2 \frac{1}{8}\pi$

9 Find the value of $\cos 2\theta$ and $\sin 2\theta$ when θ is acute and when

a $\cos\theta = \frac{3}{5}$ **b** $\sin\theta = \frac{7}{25}$ **c** $\tan\theta = \frac{12}{5}$

10 If $\tan\theta = -\frac{7}{24}$ and θ is obtuse, find

a $\tan 2\theta$ **b** $\cos 2\theta$ **c** $\sin 2\theta$ **d** $\cos 4\theta$

11 Eliminate θ from the following pairs of equations.

a $x = \tan 2\theta,\ y = \tan\theta$ **b** $x = \cos 2\theta,\ y = \cos\theta$

c $x = \cos 2\theta,\ y = \operatorname{cosec}\theta$ **d** $x = \sin 2\theta,\ y = \sec 4\theta$

12 Express in terms of $\cos 2x$

a $2\sin^2 x - 1$ **b** $4 - 2\cos^2 x$

c $2\cos^2 x + \sin^2 x$ **d** $2\cos^2 x(1+\cos^2 x)$

e $\cos^4 x$ (Hint: $\cos^4 x \equiv (\cos^2 x)^2$) **f** $\sin^4 x$

13 Prove the following identities.

a $\dfrac{1-\cos 2A}{\sin 2A} \equiv \tan A$

b $\sec 2A + \tan 2A \equiv \dfrac{\cos A + \sin A}{\cos A - \sin A}$

c $\cos 4A \equiv 8\cos^4 A - 8\cos^2 A + 1$

d $\sin 2\theta \equiv \dfrac{2\tan\theta}{1+\tan^2\theta}$

e $\cos 2\theta \equiv \dfrac{1-\tan^2\theta}{1+\tan^2\theta}$

14 Find the solutions of the following equations in the interval $[0, 2\pi]$.

a $\cos 2x = \sin x$ **b** $\sin 2x + \cos x = 0$

c $\cos 2x = \cos x$ **d** $\sin 2x = \cos x$

e $4 - 5\cos\theta = 2\sin^2\theta$ **f** $\sin 2\theta - 1 = \cos 2\theta$

15 a Use the identities in Question 13 parts **d** and **e** to express in terms of $\tan\theta$

i $\cos 2\theta - \sin 2\theta$ **ii** $\dfrac{\sin 2\theta}{1-\cos 2\theta}$

b Hence find, in degrees, the smallest value of θ greater than 0 for which

i $\cos 2\theta - \sin 2\theta = 1$ **ii** $\dfrac{\sin 2\theta}{1-\cos 2\theta} = 1$

General Solutions

Consider the equation $\sin\theta = \frac{1}{2}$.

So far we have solved such trig equations for angles within a range of values but, as you can see from the graph, there are an infinite number of angles whose sines are $\frac{1}{2}$. Because the graph is periodic, with period 2π, these angles occur at regular intervals of 2π.

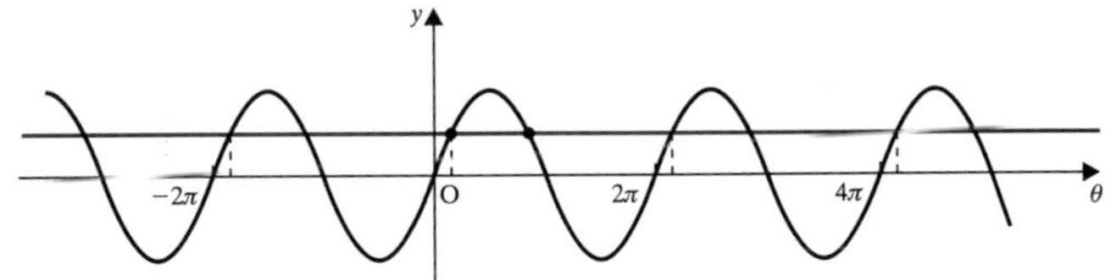

In the range $0 \leqslant \theta \leqslant 2\pi$ there are two values of θ for which $\sin\theta = \frac{1}{2}$, i.e. $\dfrac{\pi}{6}$ and $\dfrac{5\pi}{6}$.

We can see from the graph that $\sin\theta = \frac{1}{2}$ for angles that differ from $\dfrac{\pi}{6}$ and from $\dfrac{5\pi}{6}$ by intervals of 2π in both directions,

i.e. $\sin\theta = \frac{1}{2}$ when $\theta = \dfrac{\pi}{6} + 2n\pi$ and when $\theta = \dfrac{5\pi}{6} + 2n\pi$

where n is an integer (+ve or −ve).

This is called the *general solution* of the equation.

We can find the general solution of any trig equation that simplifies to $\sin\theta = a$ in the same way, i.e. we find the values of θ in the interval 0 to 2π, then add $2n\pi$ to each of these.

Note that in general there are two solutions for $\sin\theta = a$ in the interval 0 to 2π but, when $\sin\theta = 0$, there are three, namely $\theta = 0,\ \pi,\ 2\pi$.

Similarly, when $\sin\theta = 1$, there is only one solution: $\theta = \frac{1}{2}\pi$, and when $\sin\theta = -1$ the only solution is $\theta = \frac{3}{2}\pi$.

The cosine curve also has a period of 2π, so we can find the general solution of any trig equation that simplifies to $\cos\theta = b$ in exactly the same way.

The general solution of $\tan\theta = c$ can be found in a similar way but, because the tangent curve has a period of π, we can find the value of θ in the interval 0 to π, and then add $n\pi$ to this to give the general solution, e.g. the general solution of $\tan\theta = \sqrt{3}$ is $\theta = \frac{1}{3}\pi + n\pi$.

When a multiple angle is involved, e.g. $\sin 2x = 1$, find the general solution for the multiple angle first and then use it to find the general solution for x,

i.e. $\sin 2x = 1 \quad\Rightarrow\quad 2x = \frac{1}{2}\pi + 2n\pi \quad\Rightarrow\quad x = \frac{1}{4}\pi + n\pi$

EXAMPLE 22D Find the general solution of the equation $\cos 3x(1 - \tan x) = 0$

$\cos 3x(1 - \tan x) = 0 \quad \Rightarrow \quad \cos 3x = 0$ or $\tan x = 1$

When $\cos 3x = 0$, $3x = \frac{\pi}{2}$ or $\frac{3\pi}{2}$ for $0 \leqslant 3x \leqslant 2\pi$

$$\therefore\ 3x = \frac{\pi}{2} + 2n\pi \text{ or } 3x = \frac{3\pi}{2} + 2n\pi \quad \Rightarrow \quad x = \frac{\pi}{6} + \frac{2n\pi}{3} \text{ or } x = \frac{\pi}{2} + \frac{2n\pi}{3}$$

When $\tan x = 1$, $x = \frac{\pi}{4}$ for $0 \leqslant x \leqslant \pi \quad \Rightarrow \quad x = \frac{\pi}{4} + n\pi$

$\therefore$ the general solution is $x = \frac{\pi}{6} + \frac{2n\pi}{3}, \frac{\pi}{2} + \frac{2n\pi}{3}$ or $\frac{\pi}{4} + n\pi$.

EXERCISE 22D Find the general solution of each equation

1 $\sin x = -1$

2 $\cos 2x = \frac{1}{2}$

3 $\tan 2x = 1$

4 $\sin x + 2\sin x \cos 2x = 0$

MIXED EXERCISE 22

1 Eliminate θ from the equations $x = \sin\theta$ and $y = \cos 2\theta$.

2 Prove the identity $\dfrac{\sin 2\theta}{1 + \cos 2\theta} \equiv \tan\theta$.

3 Prove that $\tan(\theta + \frac{1}{4}\pi)\tan(\frac{1}{4}\pi - \theta) \equiv 1$.

4 If $\cos A = \frac{4}{5}$ and $\cos B = \frac{5}{13}$ find the possible values of $\cos(A + B)$.

5 Eliminate θ from the equations $x = \cos 2\theta$ and $y = \cos^2\theta$.

6 Solve the equation $8\sin\theta\cos\theta = 3$ for values of θ from $-180°$ to $180°$.

7 Find in the interval $[-\pi, \pi]$ the solution of the equation $\cos^2\theta - \sin^2\theta = 1$.

8 Prove the identity $\cos^4\theta - \sin^4\theta \equiv \cos 2\theta$.

9 Simplify the expression $\dfrac{1 + \cos 2x}{1 - \cos 2x}$.

10 Find the values of A between 0 and 360° for which $\sin(60° - A) + \sin(120° - A) = 0$.

11 a Express $2\sin^2\theta + 1$ in terms of $\cos 2\theta$.

b Express $4\cos^2 2A$ in terms of $\cos 4A$ (Hint: Use $2A = x$).

CHAPTER 23 Trigonometry 3

$f(\theta) = a\cos\theta + b\sin\theta$

The diagrams below show the graphs of $f(\theta) = a\cos\theta + b\sin\theta$ for a variety of values of a and b.

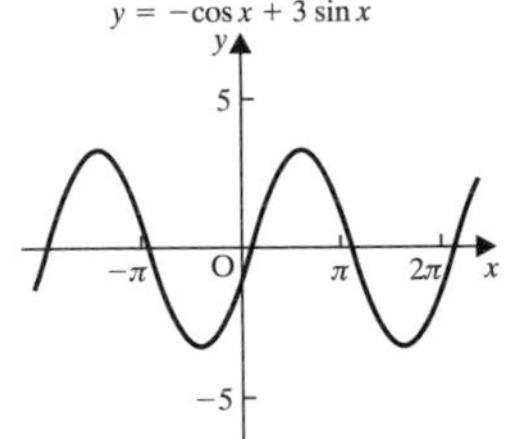

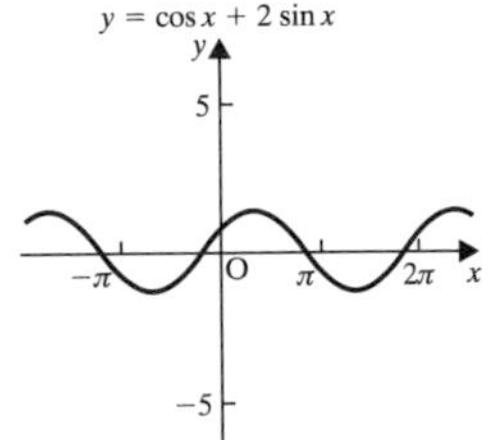

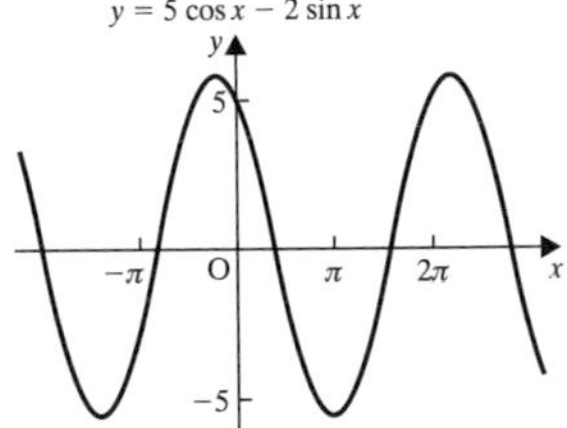

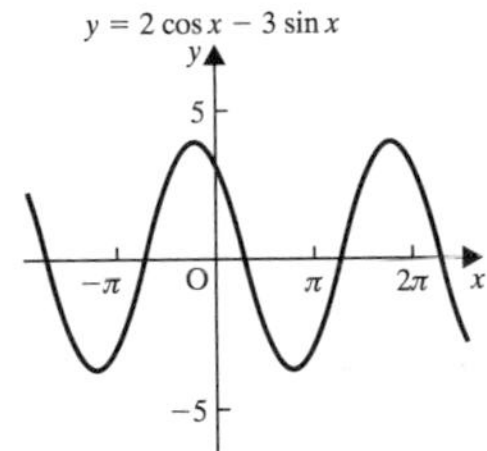

If you have access to the appropriate technology, we suggest that you draw the graph of $f(\theta)$ for some values of a and b. Each of these graphs is a sine wave, although with differing amplitude and phase shift.

These diagrams suggest that it is possible to express $a\cos\theta + b\sin\theta$ as $r\sin(\theta + \alpha)$ where the values of r and α depend on the values of a and b. This is possible provided that we can find values of r and α such that

$$r\sin(\theta + \alpha) \equiv a\cos\theta + b\sin\theta$$

i.e. $$r\underline{\sin\theta}\cos\alpha + r\underline{\underline{\cos\theta}}\sin\alpha \equiv a\underline{\underline{\cos\theta}} + b\underline{\sin\theta}$$

Since this is an identity we can compare coefficients of $\cos\theta$ and of $\sin\theta$

$\Rightarrow$ $$r\sin\alpha = a \quad [1]$$

and $$r\cos\alpha = b \quad [2]$$

Equations [1] and [2] can now be solved to give r and α in terms of a and b.

Squaring and adding equations [1] and [2] gives

$$r^2(\sin^2\alpha + \cos^2\alpha) = a^2 + b^2 \quad \Rightarrow \quad r = \sqrt{a^2 + b^2}$$

Dividing equation [1] by equation [2] gives

$$\frac{r\sin\alpha}{r\cos\alpha} = \frac{a}{b} \quad \Rightarrow \quad \tan\alpha = \frac{a}{b}$$

Therefore $$\boldsymbol{r\sin(\theta + \alpha) \equiv a\cos\theta + b\sin\theta}$$

$$\textbf{where } \boldsymbol{r = \sqrt{a^2 + b^2}} \textbf{ and } \boldsymbol{\tan\alpha = \frac{a}{b}}$$

It is also possible to express $a\cos\theta + b\sin\theta$ as $r\sin(\theta - \alpha)$ or as $r\cos(\theta \pm \alpha)$, using a similar method.

EXAMPLES 23A

1 Express $3\sin\theta - 2\cos\theta$ as $r\sin(\theta - \alpha)$.

$3\sin\theta - 2\cos\theta \equiv r\sin(\theta - \alpha)$

$\Rightarrow \qquad 3\sin\theta - 2\cos\theta \equiv r\sin\theta\cos\alpha - r\cos\theta\sin\alpha$

Comparing coefficients of $\sin\theta$ and of $\cos\theta$ gives

$$\left.\begin{aligned} 3 &= r\cos\alpha \\ 2 &= r\sin\alpha \end{aligned}\right\} \quad \Rightarrow \quad \left\{\begin{aligned} 13 &= r^2 &\Rightarrow\quad r &= \sqrt{13} \\ \tan\alpha &= \tfrac{2}{3} &\Rightarrow\quad \alpha &= 33.7^\circ \end{aligned}\right.$$

$\therefore \qquad 3\sin\theta - 2\cos\theta = \sqrt{13}\sin(\theta - 33.7^\circ)$

2 Find the maximum value of $f(x) = 3\cos x + 4\sin x$ and the smallest positive value of x at which it occurs.

> Expressing f(x) in the form $r\sin(x + \alpha)$ enables us to 'read' its maximum value, and the values of x at which it occurs, from the resulting sine wave. Note also that in this question you can choose the form in which to express f(x): in this case it is sensible to choose $r\cos(x - \alpha)$ as this fits f(x) better than $r\sin(x + \alpha)$ does.

$3\cos x + 4\sin x \equiv r\cos(x - \alpha) \equiv r\cos x\cos\alpha + r\sin x\sin\alpha$

Hence $$\left.\begin{aligned} r\cos\alpha &= 3 \\ r\sin\alpha &= 4 \end{aligned}\right\} \quad \Rightarrow \quad \left\{\begin{aligned} r^2 &= 25 &\Rightarrow\quad r &= 5 \\ \tan\alpha &= \tfrac{4}{3} &\Rightarrow\quad \alpha &= 53.1^\circ \end{aligned}\right.$$

$\therefore \qquad f(x) \equiv 5\cos(x - 53.1^\circ)$

The graph of $f(x)$ is a cosine wave with amplitude 5 and phase shift 53.1°.

$\therefore$ $f(x)$ has a maximum value of 5 and, from the sketch, the smallest positive value of x at which it occurs is 53.1°.

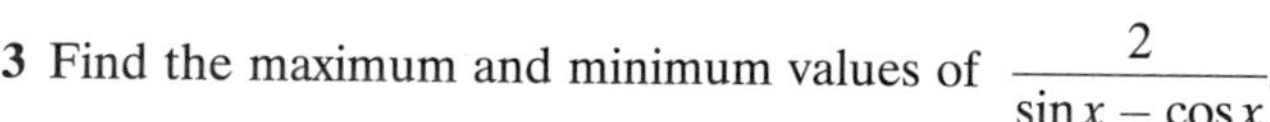

3 Find the maximum and minimum values of $\dfrac{2}{\sin x - \cos x}$.

> We first express $\sin x - \cos x$ in the form $r\sin(x - \alpha)$ then the given function can be expressed as a cosec function and we can sketch its graph. Note that values of x are not required so we do not need the value of α.

If $f(x) \equiv \sin x - \cos x \equiv r\sin(x - \alpha) \equiv r\sin x\cos\alpha - r\cos x\sin\alpha$

then $$\left.\begin{aligned} r\cos\alpha &= 1 \\ r\sin\alpha &= 1 \end{aligned}\right\} \quad \Rightarrow \quad r^2 = 2, \text{ i.e. } r = \sqrt{2}$$

$\therefore \qquad \sin x - \cos x \equiv \sqrt{2}\sin(x - \alpha)$

Hence $$\frac{2}{f(x)} \equiv \frac{2}{\sqrt{2}\sin(x - \alpha)} \equiv \sqrt{2}\operatorname{cosec}(x - \alpha)$$

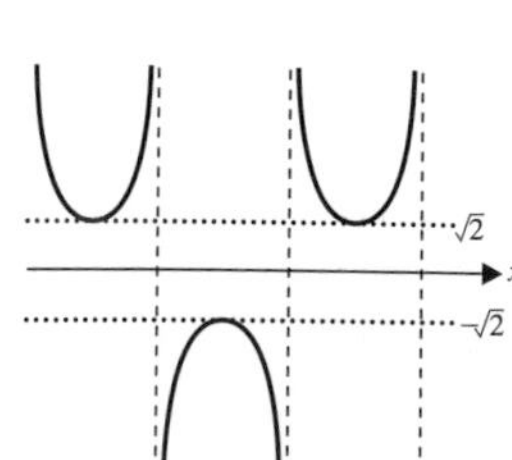

From the sketch, the maximum value of $\dfrac{2}{\sin x - \cos x}$ is $-\sqrt{2}$ and the minimum value is $\sqrt{2}$.

> Note that the greatest and least values of $\dfrac{2}{f(x)}$ are $+\infty$ and $-\infty$.

The Equation $a\cos x + b\sin x = c$

An equation of this type can be solved by expressing the LHS of the equation in the form $r\cos(x+\alpha)$ or an equivalent form. This method is illustrated in the next example.

EXAMPLES 23A (continued)

4 Find, for $-\pi \leqslant x \leqslant \pi$, the solution of the equation
$$\sqrt{3}\cos x + \sin x = 1$$

If $\sqrt{3}\cos x + \sin x \equiv r\cos(x-\alpha) \equiv r\cos x\cos\alpha + r\sin x\sin\alpha$

then $\begin{cases} r\cos\alpha = \sqrt{3} \\ r\sin\alpha = 1 \end{cases} \Rightarrow \begin{cases} r^2 = 4 \Rightarrow r = 2 \\ \tan\alpha = \frac{1}{\sqrt{3}} \Rightarrow \alpha = \frac{1}{6}\pi \end{cases}$

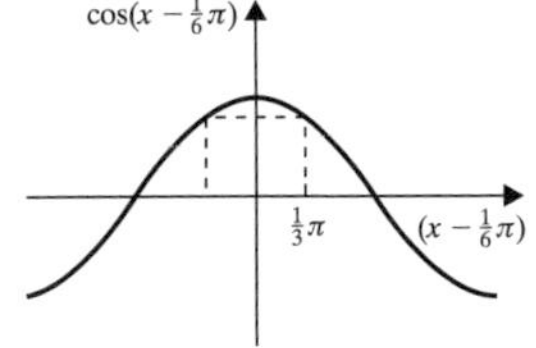

i.e. $\sqrt{3}\cos x + \sin x \equiv 2\cos(x - \frac{1}{6}\pi)$

$\therefore$ the equation becomes

$$2\cos(x-\tfrac{1}{6}\pi) = 1$$
$$\Rightarrow \quad \cos(x-\tfrac{1}{6}\pi) = \tfrac{1}{2}$$
$$\Rightarrow \quad x - \tfrac{1}{6}\pi = \pm\tfrac{1}{3}\pi$$
$$\therefore \quad x = \tfrac{1}{2}\pi, \ -\tfrac{1}{6}\pi$$

EXERCISE 23A

1 Find the values of r and α for which

a $\sqrt{3}\cos\theta - \sin\theta \equiv r\cos(\theta+\alpha)$ **b** $\cos\theta + 3\sin\theta \equiv r\cos(\theta-\alpha)$

c $4\sin\theta - 3\cos\theta \equiv r\sin(\theta-\alpha)$

2 Express $\cos 2\theta - \sin 2\theta$ in the form $r\cos(2\theta+\alpha)$.

3 Express $2\cos 3\theta + 5\sin 3\theta$ in the form $r\sin(3\theta+\alpha)$.

4 Express $\cos\theta - \sqrt{3}\sin\theta$ in the form $r\sin(\theta-\alpha)$. Hence sketch the graph of $f(\theta) = \cos\theta - \sqrt{3}\sin\theta$. Give the maximum and minimum values of $f(\theta)$ and the values of θ between 0 and 360° at which they occur.

5 Express $7\cos\theta - 24\sin\theta$ in the form $r\cos(\theta+\alpha)$. Hence sketch the graph of $f(\theta) = 7\cos\theta - 24\sin\theta + 3$ and give the maximum and minimum values of $f(\theta)$ and the values of θ between 0 and 360° at which they occur.

6 Find the greatest and least values of $\cos x + \sin x$. Hence find the maximum and minimum values of $\dfrac{1}{\cos x + \sin x}$.

7 Find the maximum and minimum values of $\dfrac{\sqrt{2}}{\cos\theta - \sqrt{2}\sin\theta}$.

8 Find the solution of the following equations in the interval $0 \leqslant x \leqslant 360°$.

a $\cos x + \sin x = \sqrt{2}$ **b** $7\cos x + 6\sin x = 2$

c $\cos x - 3\sin x = 1$ **d** $2\cos x - \sin x = 2$

The Inverse Trigonometric Functions

Consider the function given by $f : x \to \sin x$ for $x \in \mathbb{R}$.

The inverse mapping is given by $\sin x \to x$ but this is not a function because one value of $\sin x$ maps to many values of x, i.e. $f(x) = \sin x$ does not have an inverse function for the domain $x \in \mathbb{R}$.

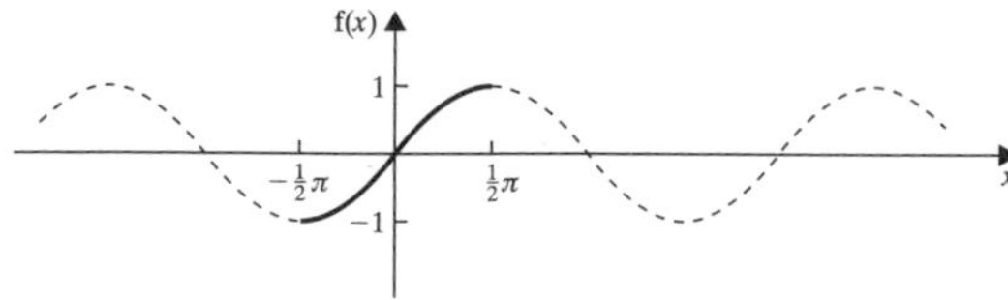

However, if we now consider the function $f : x \to \sin x$ for $-\frac{1}{2}\pi \leqslant x \leqslant \frac{1}{2}\pi$ then the inverse mapping, $\sin x \to x$, is such that one value of $\sin x$ maps to only one value of x. Therefore $f : x \to \sin x$ for $-\frac{1}{2}\pi \leqslant x \leqslant \frac{1}{2}\pi$ does have an inverse, i.e. f^{-1}exists.

Now the equation of the graph of f is $y = \sin x$ for $-\frac{1}{2}\pi \leqslant x \leqslant \frac{1}{2}\pi$ and the curve $y = f^{-1}(x)$ is obtained by reflecting $y = \sin x$ in the line $y = x$.

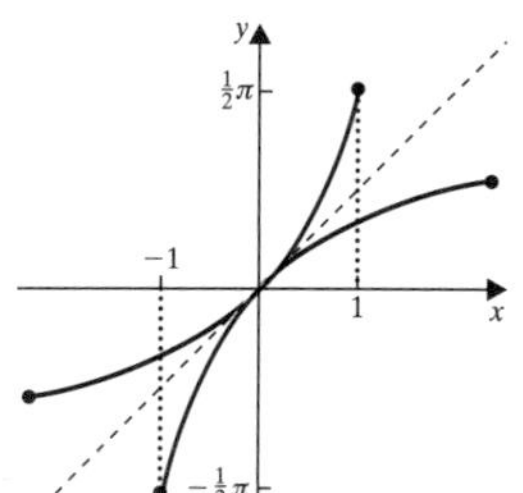

Therefore interchanging x and y gives the equation of this curve,
i.e. $\sin y = x$, so $y =$ the angle between $-\frac{1}{2}\pi$ and $\frac{1}{2}\pi$ whose sine is x.
Using $\sin^{-1}$ to mean 'the angle between $-\frac{1}{2}\pi$ and $\frac{1}{2}\pi$ whose sine is',
we have $y = \sin^{-1} x$.

Thus **if** $\quad \mathbf{f : x \to \sin x} \quad -\frac{1}{2}\pi \leqslant x \leqslant \frac{1}{2}\pi$

then $\quad \mathbf{f^{-1} : x \to \sin^{-1} x} \quad -1 \leqslant x \leqslant 1$

It is important to realise that $\sin^{-1} x$ is an angle, and that this angle is in the interval $[-\frac{1}{2}\pi, \frac{1}{2}\pi]$.

Thus, for example, $\sin^{-1} 0.5$ is the angle between $-\frac{1}{2}\pi$ and $\frac{1}{2}\pi$ whose sine is 0.5, i.e. $\sin^{-1} 0.5 = \frac{1}{6}\pi$.

Note that the alternative notation arcsin is sometimes used to mean 'the angle whose sine is'.

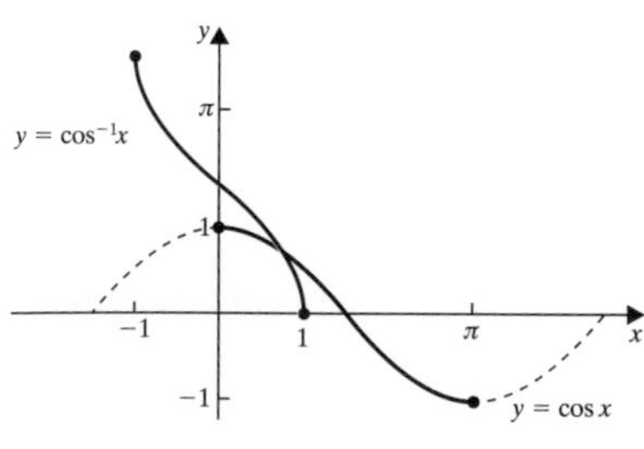

Now consider the function given by $f : x \to \cos x$, $0 \leqslant x \leqslant \pi$.

From the diagram, we see that f^{-1} exists and it is denoted by $\cos^{-1}$ where $\cos^{-1} x$ means 'the angle between 0 and π whose cosine is x'.

Thus $\quad$ if $f : x \to \cos x \quad 0 \leqslant x \leqslant \pi$

$\quad$ then $f^{-1} : x \to \cos^{-1} x \quad -1 \leqslant x \leqslant 1$

Note that $\cos^{-1} x$ is an angle in the range $0 \leqslant x \leqslant \pi$.

Thus, for example, $\cos^{-1}(-0.5) = x \quad \Rightarrow \quad x = \frac{2}{3}\pi$

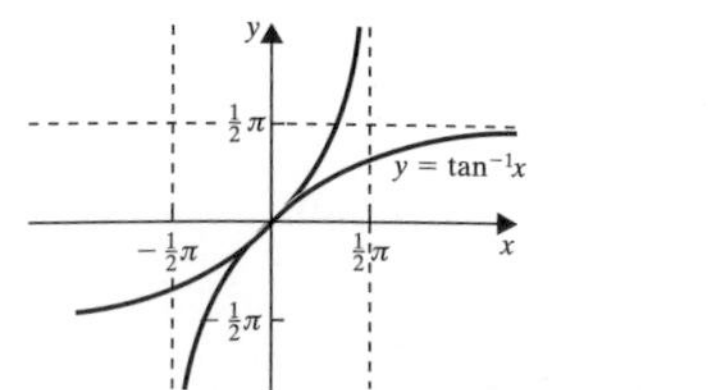

Similarly, if $f : x \to \tan x$ for $-\frac{1}{2}\pi \leqslant x \leqslant \frac{1}{2}\pi$, then f^{-1} exists and is written $\tan^{-1}$ where $\tan^{-1} x$ means 'the angle between $-\frac{1}{2}\pi$ and $\frac{1}{2}\pi$ whose tangent is x'.

Note that the domain of $\tan^{-1} x$ is all values of x.

EXAMPLES 23B

1 Express $\sin(2\cos^{-1}x)$ in terms of x only.

Using $\cos^{-1}x = \alpha$ gives $\sin(2\cos^{-1}x) = \sin 2\alpha$

$$= 2\sin\alpha\cos\alpha$$

Also $\cos^{-1}x = \alpha$ gives $\cos\alpha = x$

and $\sin\alpha = \sqrt{1-\cos^2\alpha} = \sqrt{1-x^2}$

$\therefore \quad 2\sin\alpha\cos\alpha = 2\{\sqrt{1-x^2}\}(x) = 2x\sqrt{1-x^2}$

$\therefore \quad \sin(2\cos^{-1}x) = 2x\sqrt{1-x^2}$

2 Find the exact value of $\tan^{-1}\frac{3}{4} + \tan^{-1}\frac{5}{12}$.

If $\alpha = \tan^{-1}\frac{3}{4}$ and $\beta = \tan^{-1}\frac{5}{12}$ then $\tan\alpha = \frac{3}{4}$ and $\tan\beta = \frac{5}{12}$

Now $\tan^{-1}\frac{3}{4} + \tan^{-1}\frac{5}{12} = \alpha + \beta$

and $\tan(\alpha+\beta) = \dfrac{\tan\alpha + \tan\beta}{1 - \tan\alpha\tan\beta} = \dfrac{(\frac{3}{4}) + (\frac{5}{12})}{1 - (\frac{3}{4})(\frac{5}{12})}$

$$= \tfrac{56}{33}$$

$\therefore \quad (\alpha + \beta) = \tan^{-1}\frac{56}{33}$

i.e. $\tan^{-1}\frac{3}{4} + \tan^{-1}\frac{5}{12} = \alpha + \beta = \tan^{-1}\frac{56}{33}$

EXERCISE 23B

1 Find the value of the following in terms of π.

a $\tan^{-1}\sqrt{3}$ **b** $\sin^{-1}(-1)$ **c** $\cos^{-1}0$

d $\sin^{-1}\left(-\dfrac{\sqrt{3}}{2}\right)$ **e** $\cos^{-1}(-\frac{1}{2})$ **f** $\tan^{-1}(-1)$

2 Find the value of the following in terms of π.

a $\tan^{-1}\frac{1}{3} + \tan^{-1}\frac{1}{2}$ **b** $\sin^{-1}\frac{1}{3} + \cos^{-1}\frac{1}{2}$

3 Simplify

a $\sin^{-1}x + \cos^{-1}x$ **b** $\sin(2\tan^{-1}x)$

c $\tan^{-1}x + \tan^{-1}\dfrac{1}{x}$ **d** $\tan(\tan^{-1}\frac{1}{3} + \tan^{-1}\frac{1}{4})$

4 Prove that

a $\cos(2\sin^{-1}x) \equiv 1 - 2x^2$ **b** $\sin(\cos^{-1}x) \equiv \sqrt{1-x^2}$

MIXED EXERCISE 23

1 Express $4\sin\theta - 3\cos\theta$ in the form $r\sin(\theta - \alpha)$. Hence find the maximum and minimum values of $\dfrac{7}{4\sin\theta - 3\cos\theta + 2}$.

State the greatest and least values.

2 Find, in degrees correct to 1 decimal place, the value of

a $\sin^{-1} 0.8$ **b** $\cos^{-1}(-0.3)$ **c** $\tan^{-1}(0.5)$

3 Express $\sin 2\theta - \cos 2\theta$ in the form $r\sin(2\theta - \alpha)$. Hence find the smallest positive value of θ for which $\sin 2\theta - \cos 2\theta$ has a maximum value.

4 Express $\cos x + \sin x$ in the form $r\cos(x - \alpha)$. Hence find the smallest positive value of x for which $\dfrac{1}{(\cos x + \sin x)}$ has a minimum value.

5 Express $4\sin x + 3\cos x$ in the form $r\sin(x + \alpha)$. Hence find all the values of x in the range $0 \leqslant x \leqslant 360°$ for which $\cos 3x = \cos 2x$.

6 Find all the values of x between 0 and 180° for which $\cos x - 2\sin x = 1$.

7 Solve the equation $3\cos x - 2\sin x = 1$ for values of x in the range $0 \leqslant x \leqslant 180°$.

8 If $\arctan x + \arctan y = \frac{1}{4}\pi$, show that $x + y = 1 - xy$.

9 Express $3\cos x - 4\sin x$ in the form $r\cos(x + \alpha)$.

Hence express $4 + \dfrac{10}{3\cos x - 4\sin x}$ in the form $4 + k\sec(x + \alpha)$

and sketch the graph of $y = 4 + \dfrac{10}{3\cos x - 4\sin x}$.

CHAPTER 24

Differentiation 3

Investigating the Gradient Function when $y = \sin x$

The diagram shows the graph of $y = \sin x$ for values of x between $-\pi$ and $\frac{5}{2}\pi$. It is drawn with equal scales on the axes.

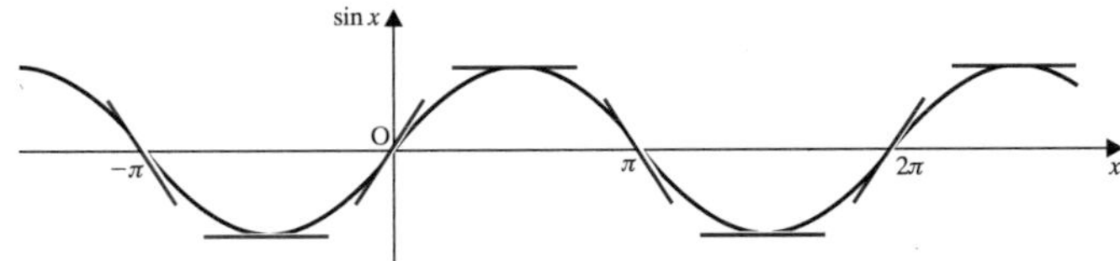

The tangents to the curve are drawn by eye and their gradients used to complete the following table.

x	$-\pi$	$-\frac{\pi}{2}$	0	$\frac{\pi}{2}$	π	$\frac{3\pi}{2}$	2π	$\frac{5\pi}{2}$
approximate gradient of $\sin x$ i.e. $\frac{d(\sin x)}{dx}$	−1	0	1	0	−1	0	1	0

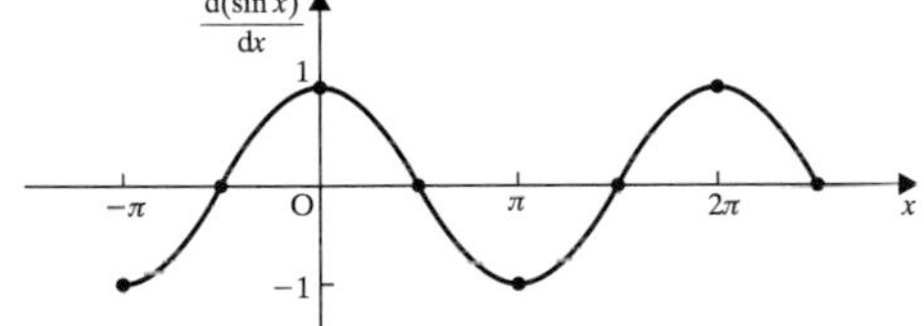

Plotting these approximate values of $\frac{d(\sin x)}{dx}$ against x, and drawing a smooth curve through them, gives a curve that look remarkably like the cosine wave.

The next exercise explores this relationship further.

EXERCISE 24A

For this exercise you will need a large plot of the graph of $y = \sin x$ for values of x in the range $0 \leqslant x \leqslant 2\pi$ drawn with equal scales on the axes.

There is a copy of this on p. 530, but if you have access to the appropriate hardware and software, a landscape print of this plot on A4 paper is easier to use.

1 Draw tangents to the curve $y = \sin x$ at each value of x given in the table and then find the gradient of each tangent to complete the second row of the table.

x	0	1	2	3	4	5	6
approximate gradient							
$\cos x$							

Now use your calculator to complete the third row of the table. Comment on the values in the last two rows of the completed table.

2 Repeat Question 1 using a plot of $y = \sin x°$ for values of x from 0 to 360 at intervals of 60 in the table. Is $\frac{d}{dx}(\sin x°)$ approximately equal to $\cos x°$?

3 a Use a *sketch* graph of $y = \cos x$ for $-\pi \leqslant x \leqslant 3\pi$ to fill in the second row of the table.

x	$-\pi$	$-\frac{\pi}{2}$	0	$\frac{\pi}{2}$	π	$\frac{3\pi}{2}$	2π	$\frac{5\pi}{2}$	3π
gradient of $\cos x$									

b Plot these values of the gradient of $\cos x$ against x and draw a curve through the points.

c Deduce the equation of the curve.

d State, with a reason, whether the result for part **c** is valid when x is measured in degrees.

The Derivatives of sin *x* and cos *x*

The results from the last exercise demonstrate that, when x is measured in radians the gradient function of $\sin x$ is $\cos x$, and the gradient function of $\cos x$ is $-\sin x$.

i.e. **if** $y = \sin x$ **then** $\frac{dy}{dx} = \cos x$

and **if** $y = \cos x$ **then** $\frac{dy}{dx} = -\sin x$

These two results can be quoted whenever they are needed.

It is important to realise that they are valid only when x is measured in radians and, throughout all subsequent work on the calculus of trig functions, the angle is measured in radians unless it is stated otherwise.

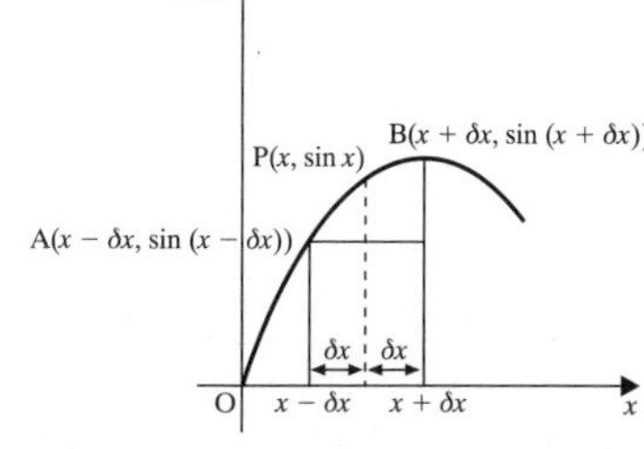

These results can be obtained by differentiating from first principles, e.g. for $y = \sin x$; we take three points on the curve, a general point $P(x, y)$, and points A and B on either side of P.
The x-coordinates of A and B are $x - \delta x$ and $x + \delta x$ respectively.

The gradient of AB is $\dfrac{\sin(x + \delta x) - \sin(x - \delta x)}{2\delta x}$

which, using the compound angle formulae, simplifies to

$$\frac{2\cos x \sin \delta x}{2\delta x}$$

When δx is small, $\sin \delta x \approx \delta x$ (this is shown on page 434),

$\therefore$ as $\delta x \to 0$, this fraction simplifies to $\cos x$,

and the gradient of AB $\to$ the gradient of the tangent at P,

i.e. $$\frac{dy}{dx} = \cos x$$

EXAMPLES 24B

1 Find the smallest positive value of x for which there is a stationary value of the function $x + 2\cos x$.

$f(x) = x + 2\cos x \quad \Rightarrow \quad f'(x) = 1 - 2\sin x$

For stationary values $f'(x) = 0$

i.e. $\quad 1 - 2\sin x = 0 \quad \Rightarrow \quad \sin x = \frac{1}{2}$

The smallest positive angle with a sine of $\frac{1}{2}$ is $\frac{1}{6}\pi$.

> NOTE that the answer *must* be given in radians because the rule used to differentiate $\cos x$ is valid only for an angle in radians.

2 Find the smallest positive value of θ for which the curve $y = 2\theta - 3\sin\theta$ has a gradient of $\frac{1}{2}$.

$y = 2\theta - 3\sin\theta$ gives $\dfrac{dy}{d\theta} = 2 - 3\cos\theta$

When $\dfrac{dy}{d\theta} = \frac{1}{2}, \quad 2 - 3\cos\theta = \frac{1}{2}$

$$3\cos\theta = \tfrac{3}{2}$$

$$\cos\theta = \tfrac{1}{2}$$

The smallest positive value of θ for which $\cos\theta = \frac{1}{2}$, is $\frac{1}{3}\pi$.

EXERCISE 24B

1 Write down the derivative of each of the following expressions.

a $\sin x - \cos x$

b $\sin\theta + 4$

c $3\cos\theta$

d $5\sin\theta - 6$

e $2\cos\theta + 3\sin\theta$

f $4\sin x - 5 - 6\cos x$

2 Find the gradient of each curve at the point whose x-coordinate is given.

a $y = \cos x;\ \frac{1}{2}\pi$

b $y = \sin x;\ 0$

c $y = \cos x + \sin x;\ \pi$

d $y = x - \sin x;\ \frac{1}{2}\pi$

e $y = 2\sin x - x^2;\ -\pi$

f $y = -4\cos x;\ \frac{1}{2}\pi$

3 For each of the following curves find the smallest positive value of θ at which the gradient of the curve has the given value.

a $y = 2\cos\theta;\ -1$

b $y = \theta + \cos\theta;\ \frac{1}{2}$

c $y = \sin\theta + \cos\theta;\ 0$

d $y = \sin\theta + 2\theta;\ 1$

4 Considering only positive values of x, locate the first two turning points on each of the following curves and determine whether they are maximum or minimum points.

a $2\sin x - x$ b $x + 2\cos x$

In each case illustrate your solution by a sketch.

5 Find the equation of the tangent to the curve $y = \cos\theta + 3\sin\theta$ at the point where $\theta = \frac{1}{2}\pi$.

6 Find the equation of the normal to the curve $y = x^2 + \cos x$ at the point where $x = \pi$.

7 Find the coordinates of a point on the curve $y = \sin x + \cos x$ at which the tangent is parallel to the line $y = x$.

Compound Functions

The various functions that can be handled when they occur in products, quotients and functions of a function, now include the sine and cosine ratios.

Differentiation of sin f(x)

If $y = \sin f(x)$ then using $u = f(x)$ gives $y = \sin u$.

Then $\quad \dfrac{dy}{dx} = \dfrac{dy}{du} \times \dfrac{du}{dx} \quad \Rightarrow \quad \dfrac{dy}{dx} = (\cos u)\dfrac{du}{dx}$

i.e. $\quad \mathbf{\dfrac{d}{dx}\{\sin f(x)\} = f'(x)\cos f(x)}$

Similarly $\quad \mathbf{\dfrac{d}{dx}\{\cos f(x)\} = -f'(x)\sin f(x)}$

e.g. $\quad \dfrac{d}{dx}\sin e^x = e^x \cos e^x$ and $\dfrac{d}{dx}\cos(\ln x) = -\dfrac{1}{x}\sin(\ln x)$

In particular $\quad \mathbf{\dfrac{d}{dx}(\sin ax) = \boldsymbol{a}\cos ax}$

and $\quad \mathbf{\dfrac{d}{dx}(\cos ax) = -\boldsymbol{a}\sin ax}$

These results are quotable.

EXAMPLES 24C

1 Differentiate $\cos(\frac{1}{6}\pi - 3x)$ with respect to x.

$$\frac{d}{dx}\{\cos(\tfrac{1}{6}\pi - 3x)\} = -(-3)\sin(\tfrac{1}{6}\pi - 3x)$$
$$= 3\sin(\tfrac{1}{6}\pi - 3x)$$

2 Find the derivative of $\dfrac{e^x}{\sin x}$.

$y = \dfrac{e^x}{\sin x} = \dfrac{u}{v}$ where $u = e^x$ and $v = \sin x$

$\Rightarrow \quad \dfrac{du}{dx} = e^x$ and $\dfrac{dv}{dx} = \cos x$

$$\frac{dy}{dx} = \left(v\frac{du}{dx} - u\frac{dv}{dx}\right) \div v^2 = \frac{e^x \sin x - e^x \cos x}{\sin^2 x}$$

$$\frac{d}{dx}\left(\frac{e^x}{\sin x}\right) = \frac{e^x}{\sin^2 x}(\sin x - \cos x)$$

3 Find $\dfrac{dy}{d\theta}$ if $y = \cos^3\theta$.

$$y = \cos^3\theta = [\cos\theta]^3$$

$$y = u^3 \text{ where } u = \cos\theta$$

$$\frac{dy}{d\theta} = \frac{dy}{du} \times \frac{du}{d\theta} = (3u^2)(-\sin\theta) = 3(\cos\theta)^2(-\sin\theta)$$

$$\therefore \quad y = \cos^3\theta \quad \Rightarrow \quad \frac{dy}{d\theta} = -3\cos^2\theta\sin\theta$$

This is one example of a general rule, i.e.

if $y = \cos^n x$ **then** $\dfrac{dy}{dx} = -n\cos^{n-1}x\sin x$

and

if $y = \sin^n x$ **then** $\dfrac{dy}{dx} = n\sin^{n-1}x\cos x$

EXERCISE 24C

Differentiate each of the following functions with respect to x.

1 $\sin 4x$

2 $\cos(\pi - 2x)$

3 $\sin(\frac{1}{2}x + \pi)$

4 $\dfrac{\sin x}{x}$

5 $\dfrac{\cos x}{e^x}$

6 $\sqrt{\sin x}$

7 $\sin^2 x$

8 $\sin x \cos x$

9 $e^{\sin x}$

10 $\ln(\cos x)$

11 $e^x \cos x$

12 $x^2 \sin x$

13 $\sin x^2$

14 $e^{\cos x}$

15 $\ln \sin^3 x$

16 $\sec x$, i.e. $\dfrac{1}{\cos x}$

17 $\tan x$, i.e. $\dfrac{\sin x}{\cos x}$

18 $\operatorname{cosec} x$

19 $\cot x$

Using the answers to Questions 16 to 19, we can now make a complete list of the derivatives of the basic trig functions:

function	derivative
$\sin x$	$\cos x$
$\cos x$	$-\sin x$
$\tan x$	$\sec^2 x$
$\cot x$	$-\operatorname{cosec}^2 x$
$\sec x$	$\sec x \tan x$
$\operatorname{cosec} x$	$-\operatorname{cosec} x \cot x$

Extending the Chain Rule

We have already seen that the chain rule can be used to write down directly the derivative of $y = \mathrm{fg}(x)$ where $u = \mathrm{g}(x)$

i.e. $$\frac{\mathrm{d}y}{\mathrm{d}x} = \frac{\mathrm{d}y}{\mathrm{d}u} \times \frac{\mathrm{d}u}{\mathrm{d}x} = \mathrm{f}'(u)\mathrm{g}'(x)$$

e.g. $$\frac{\mathrm{d}}{\mathrm{d}x}\ln(\tan x) = \frac{1}{\tan x}(\sec^2 x)$$

This direct differentiation of a succession of functions of x can be extended to deal quickly and easily with $y = \mathrm{fgh}(x)$ where $y = \mathrm{f}(u), u = \mathrm{g}(v)$ and $v = \mathrm{h}(x)$,

i.e. $$\frac{\mathrm{d}}{\mathrm{d}x}\{\mathrm{fgh}(x)\} = \left(\frac{\mathrm{d}y}{\mathrm{d}u}\right)\left(\frac{\mathrm{d}u}{\mathrm{d}v}\right)\left(\frac{\mathrm{d}v}{\mathrm{d}x}\right) = \mathrm{f}'(u)\mathrm{g}'(v)\mathrm{h}'(x)$$

This extended use of the chain rule allows quite complex functions to be differentiated without *writing down* detailed substitutions. You will find it helps to express the given function in words before applying the chain rule as this automatically places the operators in the correct order for differentiation.

Consider, for example, the function $\ln(\sin \mathrm{e}^x)$, which is

a log function of a sine function of an exponential function.

Using the chain rule to write down the derivative, we apply in succession,

the rule for differentiating a log function

the rule for differentiating a sine function

the rule for differentiating an exponential function

i.e. $$\frac{\mathrm{d}}{\mathrm{d}x}\{\ln(\sin \mathrm{e}^x)\} = \left(\frac{1}{\sin \mathrm{e}^x}\right)(\cos \mathrm{e}^x)(\mathrm{e}^x)$$

The real benefit of the chain rule is that the derivatives of apparently complicated functions, $\mathrm{fgh}(x)$ or even $\mathrm{fghj}(x)$, can be written down just as easily as for simpler ones.

e.g. $$\frac{\mathrm{d}}{\mathrm{d}x}(\mathrm{e}^{\sin x^2}) = (\mathrm{e}^{\sin x^2})(\cos x^2)(2x)$$

and $$\frac{\mathrm{d}}{\mathrm{d}x}\{\ln(\cos x)\}^3 = 3\{\ln(\cos x)\}^2\left(\frac{1}{\cos x}\right)(-\sin x)$$

EXERCISE 24D

Questions 1 to 4 are simple expressions. Use these to revise the chain rule. Some of the remaining questions look more complicated than would be found in most A-level examination papers but they are fun to try and surprisingly easy.

Using the chain rule to differentiate each function

1 $\sqrt{1 + \ln x}$

2 $\cos(x^2 + 3)$

3 $\mathrm{e}^{(x^3 - x)}$

4 $\sin(\ln x)$

5 $\ln(\cos x^2)$

6 $(1 + \mathrm{e}^{x^2})^2$

7 $\sqrt{3 - \sin^2 x}$

8 $\{\ln(\tan x)\}^2$

9 $\mathrm{e}^{\sqrt{(2-x^2)}}$

10 $\cos^2(x^2 + 1)$

Compound Functions of More than One Type

EXAMPLE 24E Differentiate $e^{x^2}\sin x$ with respect to x.

If $y = e^{x^2}\sin x$ then y is a product of e^{x^2} and $\sin x$ and we will use the product formula. But first, as e^{x^2} itself is a function of a function, we differentiate it mentally giving $2x e^{x^2}$.
Now use $y = uv$ where $u = e^{x^2}$ and $v = \sin x$.

$$\frac{dy}{dx} = v\frac{du}{dx} + u\frac{dv}{dx} = \{\sin x\}\{2xe^{x^2}\} + e^{x^2}\cos x$$

EXERCISE 24E Differentiate each function with respect to x.

1 $e^{x^2}\cos x$

2 $\dfrac{\cos^2 x}{x}$

3 $x(\ln \sin x)$

4 $\dfrac{x^2 - 1}{\ln x}$

5 $e^x\sqrt{x^2+2}$

6 $\dfrac{\ln x}{\sqrt{\cos x}}$

7 $(x+1)\,e^{\sin x}$

8 $\dfrac{x^2}{\sin^2 x}$

9 $e^{x^2}\ln x$

MIXED EXERCISE 24 This exercise contains a variety of functions. Consider carefully what method to use in each case and do not forget to check first whether a given function has a standard derivative.

Find the derivative of each function in Questions 1 to 19.

1 a $-\sin 4\theta$ **b** $\theta - \cos\theta$ **c** $\sin^3\theta + \sin 3\theta$

2 a $x^3 + e^x$ **b** $e^{(2x+3)}$ **c** $e^x\sin x$

3 a $\ln \frac{1}{3}x^{-3}$ **b** $\ln\dfrac{2}{x^2}$ **c** $\ln\dfrac{\sqrt{x}}{4}$

4 a $3\sin x - e^{-x}$ **b** $\ln x^{1/2} - \frac{1}{2}\cos x$

c $x^4 + 4e^x - \ln 4x$ **d** $\frac{1}{2}e^{-x} + x^{-1/2} - \ln\frac{1}{2}x$

5 $(x+1)\ln x$

6 $\sin^2 3x$

7 $(4x-1)^{2/3}$

8 $(3\sqrt{x} - 2x)^2$

9 $\dfrac{(x^4-1)}{(x+1)^3}$

10 $\dfrac{\ln x}{\ln(x-1)}$

11 $\ln\cot x$

12 $x^2\sin x$

13 $\dfrac{e^x}{x-1}$

14 $\dfrac{1+\sin x}{1-\sin x}$

15 $x^2\sqrt{x-1}$

16 $(1-x^2)(1-x)^2$

17 $\ln\sqrt{\dfrac{(x+3)^3}{x^2+2}}$

18 $\sin x\cos^3 x$

19 $e^{\cos^2 x}$

20 Find the value(s) of x for which the following functions have stationary values.

a $3x - e^x$ **b** $x^2 - 2\ln x$ **c** $\ln\frac{1}{x} + 4x$

In each Question from 21 to 24, find

a the gradient of the curve at the given point,

b the equation of the tangent to the curve at that point,

c the equation of the normal to the curve at that point.

21 $y = \sin x - \cos x$; $x = \frac{1}{2}\pi$

22 $y = x + e^x$; $x = 1$

23 $y = 1 + x + \sin x$; $x = 0$

24 $y = 3 - x^2 + \ln x$; $x = 1$

25 Considering only positive values of x, locate the first two turning points, if there are two, on each of the following curves and determine whether they are maximum or minimum points.

a $y = 1 - \sin x$ **b** $y = \frac{1}{2}x + \cos x$ **c** $y = e^x - 3x$

26 Find the coordinates of a point on the curve where the tangent is parallel to the given line.

a $y = 3x - 2\cos x$; $y = 4x$ **b** $y = 2\ln x - x$; $y = x$

CHAPTER 25 **Differentiation 4**

Implicit Functions

All the differentiation carried out so far has involved equations that could be expressed in the form $y = \mathrm{f}(x)$.
Some curves have equations that cannot easily be written in this way. For example it is difficult to isolate y in the equation $x^2 - y^2 + y = 1$.
A relationship of this type, where y is not given explicitly as a function of x, is called an *implicit function*, because it is *implied* in the equation that $y = \mathrm{f}(x)$.

To Differentiate an Implicit Function

The method we use is to differentiate, term by term, with respect to x, but first we need to know how to differentiate terms like y^2 with respect to x.

If $\quad \mathrm{g}(y) = y^2$ and $y = \mathrm{f}(x)$

then $\quad \mathrm{g}(y) = \{\mathrm{f}(x)\}^2$ which is a function of a function.

Using the mental substitution $u = \mathrm{f}(x)$ we have

$$\frac{\mathrm{d}}{\mathrm{d}x}\{\mathrm{f}(x)\}^2 = 2\{\mathrm{f}(x)\}\left(\frac{\mathrm{d}}{\mathrm{d}x}\mathrm{f}(x)\right) = 2y\left(\frac{\mathrm{d}y}{\mathrm{d}x}\right) = \left(\frac{\mathrm{d}}{\mathrm{d}y}\mathrm{g}(y)\right)\left(\frac{\mathrm{d}y}{\mathrm{d}x}\right)$$

In general, $$\frac{\mathrm{d}}{\mathrm{d}x}\mathrm{g}(y) = \left(\frac{\mathrm{d}}{\mathrm{d}y}\mathrm{g}(y)\right)\left(\frac{\mathrm{d}y}{\mathrm{d}x}\right)$$

e.g. $\frac{\mathrm{d}}{\mathrm{d}x}y^3 = 3y^2\frac{\mathrm{d}y}{\mathrm{d}x}$ and $\frac{\mathrm{d}}{\mathrm{d}x}\mathrm{e}^y = \mathrm{e}^y\frac{\mathrm{d}y}{\mathrm{d}x}$

Now suppose that we have to differentiate a term containing both x and y, with respect to x, e.g. x^2y^3.

This term is a product so we differentiate it using the product rule.

$$\frac{\mathrm{d}}{\mathrm{d}x}(x^2y^3) = y^3\frac{\mathrm{d}}{\mathrm{d}x}(x^2) + x^2\frac{\mathrm{d}}{\mathrm{d}x}(y^3) = 2xy^3 + x^2 3y^2\frac{\mathrm{d}y}{\mathrm{d}x}$$

We can now differentiate any expression, term by term, with respect to x,

e.g. if $$x^2 - y^2 + x^2y = 1$$

then $$\frac{\mathrm{d}}{\mathrm{d}x}(x^2) - \frac{\mathrm{d}}{\mathrm{d}x}(y^2) + \frac{\mathrm{d}}{\mathrm{d}x}(x^2y) = \frac{\mathrm{d}}{\mathrm{d}x}(1)$$

$\Rightarrow$ $$2x - 2y\frac{\mathrm{d}y}{\mathrm{d}x} + 2xy + x^2\frac{\mathrm{d}y}{\mathrm{d}x} = 0$$

Hence $$2x(1+y) = \frac{\mathrm{d}y}{\mathrm{d}x}(2y - x^2) \quad \Rightarrow \quad \frac{\mathrm{d}y}{\mathrm{d}x} = \frac{2x(1+y)}{2y - x^2}$$

EXAMPLES 25A

1 Differentiate each equation with respect to x and hence find $\frac{dy}{dx}$ in terms of x and y.

a $x^3 + xy^2 - y^3 = 5$ **b** $y = xe^y$

a If $x^3 + xy^2 - y^3 = 5$ then differentiating term by term gives

$$\frac{d}{dx}(x^3) + \frac{d}{dx}(xy^2) - \frac{d}{dx}(y^3) = \frac{d}{dx}(5)$$

$$\therefore \quad 3x^2 + y^2 + 2xy\frac{dy}{dx} - 3y^2\frac{dy}{dx} = 0$$

Hence $\frac{dy}{dx} = \frac{(3x^2 + y^2)}{y(3y - 2x)}$

b If $y = xe^y$ then $\frac{dy}{dx} = \frac{d}{dx}(xe^y)$

$$= e^y\frac{d}{dx}(x) + x\frac{d}{dx}(e^y)$$

$$\Rightarrow \quad \frac{dy}{dx} = e^y + xe^y\frac{dy}{dx}$$

Hence $\frac{dy}{dx} = \frac{e^y}{1 - xe^y}$

2 If $e^x y = \sin x$ show that $\frac{d^2y}{dx^2} + 2\frac{dy}{dx} + 2y = 0$

> In a problem of this type it is tempting to express $e^x y = \sin x$ in the form $y = e^{-x}\sin x$, find $\frac{dy}{dx}$ and $\frac{d^2y}{dx^2}$ and show that they satisfy the given equation, which is called a *differential equation*. However it is much more direct to differentiate the implicit equation as given.

Differentiating $e^x y = \sin x$ w.r.t. x gives

$$e^x y + e^x\frac{dy}{dx} = \cos x$$

Differentiating again w.r.t. x gives

$$\left(e^x y + e^x\frac{dy}{dx}\right) + \left(e^x\frac{dy}{dx} + e^x\frac{d^2y}{d^2x}\right) = -\sin x = -e^x y$$

Hence $e^x\frac{d^2y}{d^2x} + 2e^x\frac{dy}{dx} + 2e^x y = 0$

> There is no finite value of x for which $e^x = 0$ so we can divide the equation by e^x

i.e. $\frac{d^2y}{dx^2} + 2\frac{dy}{dx} + 2y = 0$

3 Given that $y = \sin^{-1}x$, show that $\dfrac{dy}{dx} = \dfrac{1}{\sqrt{1-x^2}}$.

$y = \sin^{-1}x \quad \Rightarrow \quad \sin y = x \quad \Rightarrow$

Differentiating w.r.t. x gives

$$\cos y\frac{dy}{dx} = 1 \quad \Rightarrow \quad \frac{dy}{dx} = \frac{1}{\cos y}$$

But $\cos y = \sqrt{1-x^2}$ so $\dfrac{dy}{dx} = \dfrac{1}{\sqrt{1-x^2}}$

4 Show that the equation of the tangent at the point (x_1, y_1) to the curve with equation $x^2 - 2y^2 - 6y = 0$ is $xx_1 - 2yy_1 - 3(y + y_1) = 0$.

> To find the equation of the tangent we need the gradient of the curve and in this case it must be found by implicit differentiation.

$$x^2 - 2y^2 - 6y = 0 \quad \Rightarrow \quad 2x - 2\left(2y\frac{dy}{dx}\right) - 6\frac{dy}{dx} = 0$$

$$\Rightarrow \quad \frac{dy}{dx} = \frac{x}{(3+2y)}$$

$\therefore$ the gradient of the tangent at the point (x_1, y_1) is $\dfrac{x_1}{(3+2y_1)}$

and the equation of the tangent is $y - y_1 = \dfrac{x_1}{(3+2y_1)}(x - x_1)$

$\Rightarrow \quad (y - y_1)(3 + 2y_1) = x_1(x - x_1)$

which simplifies to $xx_1 - 2yy_1 - 3(y + y_1) = x_1^2 - 2y_1^2 - 6y_1$

> The terms collected on the LHS were chosen to match the required equation.

The RHS is zero because (x_1, y_1) is on the given curve, i.e. $x_1^2 - 2y_1^2 - 6y_1 = 0$.

So the equation of the tangent becomes $xx_1 - 2yy_1 - 3(y + y_1) = 0$

EXERCISE 25A

Differentiate the following equations with respect to x.

1 $x^2 + y^2 = 4$

2 $x^2 + xy + y^2 = 0$

3 $x(x + y) = y^2$

4 $\dfrac{1}{x} + \dfrac{1}{y} = e^y$

5 $\frac{1}{x^2}+\frac{1}{y^2}=\frac{1}{4}$

6 $\frac{x^2}{4}-\frac{y^2}{9}=1$

7 $\sin x+\sin y=1$

8 $\sin x\cos y=2$

9 $xe^y=x+1$

10 $\sqrt{(1+y)(1+x)}=x$

11 Find $\frac{dy}{dx}$ as a function of x if $y^2=2x+1$.

12 Find $\frac{d^2y}{dx^2}$ as a function of x if $\sin y+\cos y=x$.

13 Find the gradient of $x^2+y^2=9$ at the points where $x=1$.

14 If $y\cos x=e^x$ show that $\frac{d^2y}{dx^2}-2\tan x\frac{dy}{dx}-2y=0$.

15 Find $\frac{dy}{dx}$ given that $y=\tan^{-1}x$.

16 Write down the equation of the tangent to

a $x^2-3y^2=4y$ **b** $x^2+xy+y^2=3$

at the point (x_1, y_1).

17 Show that the equation of the tangent to $x^2+xy+y=0$ at the point (x_1, y_1) is

$$x(2x_1+y_1)+y(x_1+1)+y_1=0$$

18 Find the equation of the tangent at $(1, \frac{1}{3})$ to the curve whose equation is

$$2x^2+3y^2-3x+2y=0$$

19 Show that the equation of the tangent at (x_1, y_1) to the curve $ax^2+by^2+cxy+dx=0$ is

$$axx_1+byy_1+\tfrac{1}{2}c(xy_1+yx_1)+\tfrac{1}{2}d(x+x_1)=0$$

20 Given that $\sin y=2\sin x$ show that $\left(\frac{dy}{dx}\right)^2=1+3\sec^2 y$. By differentiating this equation with respect to x show that

$$\frac{d^2y}{dx^2}=3\sec^2 y\tan y$$

and hence that $\cot y\frac{d^2y}{dx^2}-\left(\frac{dy}{dx}\right)^2+1=0$.

Logarithmic Differentiation

We have already seen the advantage of simplifying a logarithmic expression before attempting to differentiate.

We are now going to examine some equations that are awkward to differentiate as they stand, but which are much easier to deal with if we first take logs of both sides of the equation. One type of equation where this method is useful is one in which the variable is contained in an index.

EXAMPLES 25B

1 Differentiate x^x with respect to x.

$$y = x^x$$

$$\ln y = x \ln x$$

thus $$\frac{1}{y}\frac{dy}{dx} = x\frac{1}{x} + \ln x$$

$\Rightarrow$ $$\frac{dy}{dx} = y(1 + \ln x)$$

Therefore $$\frac{d}{dx}(x^x) = x^x(1 + \ln x)$$

2 Differentiate the equation $x = y^x$ with respect to x.

$x = y^x \quad \Rightarrow \quad \ln x = x \ln y$

$\therefore$ $$\frac{1}{x} = \ln y + (x)\left(\frac{1}{y}\frac{dy}{dx}\right)$$

i.e. $$y = x^2\frac{dy}{dx} + xy\ln y$$

Note that in the second example it is not easy to express $\frac{dy}{dx}$ as a function of x, because it is difficult in the first place to find y in terms of x. So although we would usually give a derived function in terms of x it is not always possible, or sensible, to do so.

Differentiation of a^x where a is a Constant

The basic rule for differentiating an exponential function applies when the base is e but not for any other base. So for a^x, where we need another approach, we use logarithmic differentiation.

Using $y = a^x$ gives $\ln y = x \ln a$

Differentiating w.r.t. x gives $\frac{1}{y}\frac{dy}{dx} = \ln a$

Hence $$\frac{dy}{dx} = y \ln a = a^x \ln a$$

i.e. $$\mathbf{\frac{d}{dx}a^x = a^x \ln a}$$

This result is quotable.

Differentiating Inverse Trig Functions

We can use a technique similar to logarithmic differentiation to find the derivatives of inverse trig functions.
For example, when $y = \sin^{-1}x$, we can write $\sin y = x$, then differentiating implicitly gives

$$\cos y\frac{dy}{dx} = 1 \quad \Rightarrow \quad \frac{dy}{dx} = \frac{1}{\cos y} = \frac{1}{\sqrt{1-\sin^2 y}} = \frac{1}{\sqrt{1-x^2}},$$

i.e. $$\frac{d}{dx}(\sin^{-1}x) = \frac{1}{\sqrt{1-x^2}}$$

Notice that to give $\cos y$ in terms of $\sin y$, and hence in terms of x, we use the identity $\cos^2 y \equiv 1 - \sin^2 y$ but take only the positive square root; this is because the graph of $f(x) = \sin^{-1}x$ has positive gradient for all values of x in its domain (see Chapter 23).

EXERCISE 25B In each Question find $\frac{dy}{dx}$.

1 $y = 3^x$ **3** $y = 3^{2x}$ **5** $y = 3^{-x}$

2 $y = 2(1.5)^x$ **4** $y = x3^x$ **6** $y = 5a^x$

Differentiate each equation with respect to x.

7 $x^y = e^x$ **8** $x^y = y + 1$ **9** $y = x^{2x}$

10 Find, in terms of x, the derivative of

a $\cos^{-1}x$ **b** $\sin^{-1}2x$ **c** $\tan^{-1}x$

Parametric Equations

Sometimes a direct relationship between x and y is awkward to analyse and in such cases it may be easier to express x and y each in terms of a third variable, called a *parameter*.

Consider, for example, the equations

$$x = t^2, \quad y = t - 1$$

A point $P(x, y)$ is on the curve representing the relationship if and only if the coordinates of P are $(t^2, t-1)$.

By giving t any value we choose, we get a pair of corresponding values of x and y. For example, when $t = 3$, $x = 9$ and $y = 2$, therefore (9, 2) is a point on the curve.

The direct relationship between x and y can be found by eliminating t from these two *parametric equations*. In this case it is $(y+1)^2 = x$.

While the Cartesian equation can be used to find the gradient and general shape of a curve, as well as the equations of tangents and normals, it is often easier to derive them from the parametric equations.

Sketching a Curve from Parametric Equations

Consider the curve whose parametric equations are

$$x = t^2 \quad \text{and} \quad y = t - 1$$

To get an idea of the shape of this curve, we can plot some points by finding the values of x and y that correspond to some chosen values of t,

t	-2	-1	0	1	2
x	4	1	0	1	4
y	-3	-2	-1	0	1

Also we can see what happens to x and to y as t varies:

$x \geqslant 0$ for all values of t,

as $t \to \infty$, $x \to \infty$ and $y \to \infty$,

as $t \to -\infty$, $x \to \infty$ and $y \to -\infty$,

there are no values of t for which either x or y is undefined so it is reasonable to assume that the curve is continuous.

Based on this information, a sketch of the curve can now be made.

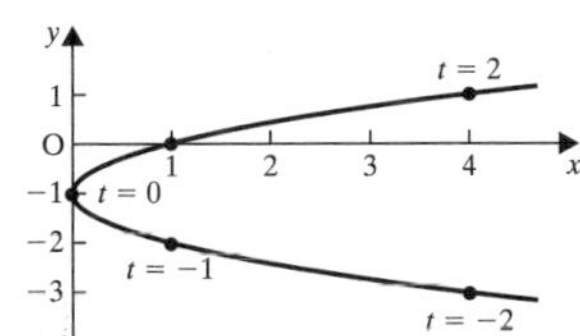

Finding the Gradient Function using Parametric Equations

If both x and y are given as functions of t then a small increase of δt in the value of t results in corresponding small increases of δx and δy in the values of x and y.

As $\delta t \to 0$, δx and δy also approach zero, therefore $\dfrac{dy}{dx} = \dfrac{dy}{dt} \times \dfrac{dt}{dx}$

But $\dfrac{dt}{dx} = 1 \Big/ \dfrac{dx}{dt}$

Therefore $$\frac{dy}{dx} = \frac{dy}{dt} \Big/ \frac{dx}{dt}$$

Hence, for the parametric equations given above, i.e. $x = t^2$ and $y = t - 1$, we have

$$\frac{dy}{dt} = 1, \quad \frac{dx}{dt} = 2t \quad \Rightarrow \quad \frac{dy}{dx} = \frac{1}{2t}$$

Each point on the curve is defined by a value of t which also gives the value of $\frac{dy}{dx}$ at that point. For example, when $t = 1$, $\frac{dy}{dx} = \frac{1}{2}$.

Conversely, the value(s) of t where $\frac{dy}{dx}$ has a particular value lead to the coordinates of the relevant point(s) on the curve.

For example, when $\frac{dy}{dx} = 2$, $\frac{1}{2t} = 2 \quad \Rightarrow \quad t = \frac{1}{4}$

and when $t = \frac{1}{4}$, $x = \frac{1}{16}$ and $y = -\frac{3}{4}$;

i.e. the gradient of the curve is 2 at the point $(\frac{1}{16}, -\frac{3}{4})$.

Note that there are no values of t for which $\frac{dy}{dx} = 0$, so there are no stationary points on this curve.

EXAMPLES 25C

1 Find the cartesian equation of the curve whose parametric equations are

a $x = t^2$, $y = 2t$ **b** $x = \cos\theta$, $y = \sin\theta$ **c** $x = 2t$, $y = 2/t$

a $y = 2t \quad \Rightarrow \quad t = \frac{1}{2}y$

$\therefore\ x = t^2 \quad \Rightarrow \quad x = (\frac{1}{2}y)^2 = \frac{1}{4}y^2 \quad \Rightarrow \quad y^2 = 4x$

b Using $\cos^2\theta + \sin^2\theta = 1$ where $\cos\theta = x$ and $\sin\theta = y$ gives

$x^2 + y^2 = 1$

c $y = 2/t \quad \Rightarrow \quad t = 2/y$

$\therefore\ x = 2t \quad \Rightarrow \quad x = 4/y \quad \Rightarrow \quad xy = 4$

2 Find the stationary point on the curve whose parametric equations are $x = t^3$, $y = (t+1)^2$ and determine its nature.

$$\frac{dy}{dx} = \frac{dy}{dt}\bigg/\frac{dx}{dt} = \frac{2(t+1)}{3t^2}$$

At stationary points $\frac{dy}{dx} = 0$ i.e. $t = -1$

When $t = -1$, $x = -1$ and $y = 0$.

Therefore the stationary point is $(-1, 0)$

> To determine the nature of the stationary point we examine the sign of $\frac{dy}{dx}$ in the neighbourhood of the point by first choosing appropriate values for x and then finding the corresponding values of t.

Value of x	-2	-1	0
Value of t	$-\sqrt[3]{2}$	-1	0
Sign of $\frac{dy}{dx}$	$-$ \	0 —	$+$ /

The equations $x = t^3$ and $y = (t+1)^2$ show that there is no finite value of t for which either x or y is not defined, so there are no breaks in the curve and there are no other stationary points.

Hence $(-1, 0)$ is a minimum point.

3 **Find the equation of the normal to the curve $x = t^2, y = t + 2/t$, at the point where $t = 1$. Show, without sketching the curve, that this normal does not cross the curve again.**

$x = t^2$ and $y = t + \frac{2}{t}$ give $\frac{dy}{dt} = 1 \quad \frac{2}{t^2}$ and $\frac{dx}{dt} = 2t$

$$\therefore \qquad \frac{dy}{dx} = \frac{dy}{dt} \div \frac{dx}{dt} = \frac{1 - 2/t^2}{2t} = \frac{t^2 - 2}{2t^3}$$

When $t = 1$; $x = 1$, $y = 3$ and $\frac{dy}{dx} = -\frac{1}{2}$.

Therefore the gradient of the normal at P(1, 3) is $\frac{-1}{-\frac{1}{2}} = 2$

The equation of this normal is $y - 3 = 2(x - 1)$ i.e. $y = 2x + 1$

> All points for which $x = t^2$ and $y = t + 2/t$ are on the given curve. For any point that is on both the curve and the normal, these coordinates also satisfy the equation of the normal.

At points common to the curve and the normal,

$$t + \frac{2}{t} = 2t^2 + 1 \quad \Rightarrow \quad 2t^3 - t^2 + t - 2 = 0 \qquad [1]$$

> If a cubic equation can be factorised, each factor equated to zero gives a root of the equation, just as in the case of a quadratic equation.

One point where the curve and normal meet is the point where $t = 1$, so $t = 1$ is a root of [1] and $(t - 1)$ is a factor of the LHS.

i.e. $\qquad (t - 1)(2t^2 + t + 2) = 0$

Therefore, at any other point where the normal meets the curve, the value of t is a root of the equation $2t^2 + t + 2 = 0$

Checking the value of $b^2 - 4ac$ shows that this equation has no real roots so there are no more points where the normal meets the curve.

EXERCISE 25C

1 Find the gradient function of each of the following curves in terms of the parameter.

a $x = 2t^2,\ y = t$ **b** $x = \cos\theta,\ y = \sin\theta$ **c** $x = t,\ y = \dfrac{4}{t}$

2 Sketch each curve given in question 1.

3 If $x = \dfrac{t}{1-t}$ and $y = \dfrac{t^2}{1-t}$, find $\dfrac{dy}{dx}$ in terms of t. What is the value of $\dfrac{dy}{dx}$ at the point where $x = 1$?

4 a If $x = t^2$ and $y = t^3$, find $\dfrac{dy}{dx}$ in terms of t.

b If $y = x^{3/2}$, find $\dfrac{dy}{dx}$.

c Explain the connection between these two results.

5 Find the cartesian equation of each of the curves given in question 1 and hence find $\dfrac{dy}{dx}$. Show in each case that $\dfrac{dy}{dx}$ agrees with the gradient function found in question 1.

6 Find the turning points of the curve whose parametric equations are $x = t$, $y = t^3 - t$, and distinguish between them.

7 A curve has parametric equations $x = \theta - \cos\theta$, $y = \sin\theta$. Find the smallest positive value of θ at which the gradient of this curve is zero.

8 Find the equation of the tangent to the curve $x = t^2$, $y = 4t$ at the point where $t = -1$.

9 Find the equation of the normal to the curve $x = 2\cos\theta$, $y = 3\sin\theta$ at the point where $\theta = \frac{1}{4}\pi$. Find the coordinates of the point where this normal cuts the curve again.

MIXED EXERCISE 25

1 Differentiate with respect to x.

a y^4 **b** xy^2 **c** $1/y$ **d** $x\ln y$

e $\sin y$ **f** e^y **g** $y\cos x$ **h** $y\cos y$

In each question from 2 to 7, find $\dfrac{dy}{dx}$ in terms of x and y.

2 $x^2 - 2y^2 = 4$

3 $1/x + 1/y = 2$

4 $x^2y^3 = 9$

5 $y = 3(1.1)^x$

6 $y = \dfrac{(x-1)^4}{(x+3)}$

7 $x^2y^2 = \dfrac{(y+1)}{(x+1)}$

In Questions 8 to 13 find $\frac{dy}{dx}$ in terms of the parameter.

8 $x = t^2,\ y = t^3$

9 $x = (t+1)^2,\ y = t^2 - 1$

10 $x = \sin^2\theta,\ y = \cos^3\theta$

11 $x = 4t,\ y = 4/t$

12 $x = e^t,\ y = 1 - t$

13 $x = \frac{t}{1-t},\ y = \frac{t^2}{1-t}$

14 If $x = \sin t$ and $y = \cos 2t$, find $\frac{dy}{dx}$ in terms of x and prove that $\frac{d^2y}{dx^2} + 4 = 0$.

15 If $x = e^t - t$ and $y = e^{2t} - 2t$, show that $\frac{dy}{dx} = 2(e^t + 1)$.

16 Differentiate $y^2 - 2xy + 3y = 7x$ w.r.t. x. Hence show that

$$\frac{d^2y}{dx^2}(2y - 2x + 3) = \frac{dy}{dx}\left(4 - 2\frac{dy}{dx}\right).$$

17 Find the equation of the tangent to the curve $x = \cos\theta,\ y = 2\sin\theta$ at the point where $\theta = \frac{3\pi}{4}$.

Differentiation 5

Rate of Increase

When the variation of y depends upon another variable x, then

$\mathbf{\frac{dy}{dx}}$ **gives the rate at which y increases compared with x.**

This fact forms the basis of methods which can be used to analyse practical situations in which two variables are related.

Small Increments

Consider two variables, x and y, related by the equation $y = f(x)$.

If x increases by a small increment δx

then y increases by a corresponding small amount δy.

Now $\displaystyle\lim_{\delta x \to 0} \frac{\delta y}{\delta x} = \frac{dy}{dx}$

so,

provided that δx is small, $\dfrac{\delta y}{\delta x} \approx \dfrac{dy}{dx} \quad \Rightarrow \quad \delta y \approx \dfrac{dy}{dx}(\delta x)$

or, alternatively, $\delta\{f(x)\} \approx f'(x)\,\delta x$

This approximation can be used to estimate the value of a function close to a known value, i.e. $y + \delta y$ can be estimated if y is known at a particular value of x.

For example, knowing that $\ln 1 = 0$, we can find an approximate value for ln 1.1 from $y = \ln x$ as follows.

$$y = \ln x \quad \Rightarrow \quad \frac{dy}{dx} = \frac{1}{x}$$

so $\quad \delta y \approx \dfrac{dy}{dx}(\delta x) = \dfrac{1}{x}(\delta x)$

Now x increases from 1 to 1.1, i.e. $\delta x = 0.1$

$\therefore \quad \delta y \approx \frac{1}{1}(0.1)$

Hence $\quad \ln 1.1 = y + \delta y \approx (\ln 1) + 0.1$

but $\ln 1 = 0$

$\therefore \quad \ln 1.1 \approx 0.1$

EXAMPLES 26A

1 Using $y = \sqrt{x}$, estimate the value of $\sqrt{101}$.

$$y = \sqrt{x} \quad \Rightarrow \quad \frac{dy}{dx} = \tfrac{1}{2}x^{-1/2}$$

$$\delta y \approx \frac{dy}{dx}\delta x \text{ gives } \delta y \approx \frac{1}{2\sqrt{x}}\delta x$$

> So that the value of y can be written down, the value we take for x must be a number with a known square root.

Taking $x = 100$, $y = \sqrt{100}$ and $\delta x = 1$ gives

$$\delta y \approx \frac{1}{2\sqrt{100}}(1) = \tfrac{1}{20}$$

Then $\quad \sqrt{101} = y + \delta y \approx \sqrt{100} + \tfrac{1}{20}$

i.e. $\quad \sqrt{101} \approx 10.05$

2 Given that $1° = 0.0175$ rad and that $\cos 30° = 0.8660$, use $f(\theta) = \sin\theta$ to find an approximate value for

a $\sin 31°$ **b** $\sin 29°$

Using $\delta y = \dfrac{dy}{dx}\delta\theta$ gives $\delta y \approx (\cos\theta)\delta\theta$ where $y = \sin\theta$.

a Taking $\theta = \frac{1}{6}\pi$, $\sin\theta = \frac{1}{2}$ and $\delta\theta = 0.0175$ gives

$$\delta y \approx (\cos\tfrac{1}{6}\pi)(0.0175) \text{ and } \sin 31° \approx \sin 30° + \delta y$$

Hence $\sin 31° \approx 0.5 + (0.8660)(0.0175)$

i.e. $\quad \sin 31° \approx 0.515$

b Again $\theta = \frac{1}{6}\pi$ and $\sin\theta = \frac{1}{2}$, but this time, because the angle *decreases* from 30° to 29°, $\delta\theta = -0.0175$

$$\delta\theta \approx (\cos\tfrac{1}{6}\pi)(-0.0175)$$

Hence $\sin 29° \approx \sin 30° + \delta y = 0.5 + (0.8660)(-0.0175)$

i.e. $\quad \sin 29° \approx 0.485$

Small Percentage Increases

In order to adapt the method used above to estimate the percentage change in a dependent variable caused by a small change in the independent variable we use the additional fact that

$$\text{if } x \text{ increases by } r\% \text{ then } \delta x = \frac{r}{100}(x)$$

and the corresponding percentage increase in y is $\dfrac{\delta y}{y} \times 100$

EXAMPLES 26A (continued)

3 The period, T, of a simple pendulum is calculated from the formula $T = 2\pi\sqrt{l/g}$, where l is the length of the pendulum and g is the constant gravitational acceleration. Find the percentage change in the period caused by lengthening the pendulum by 2 per cent.

$$T = 2\pi\sqrt{\frac{l}{g}} \quad \Rightarrow \quad \frac{dT}{dl} = \left(\frac{2\pi}{\sqrt{g}}\right)(\tfrac{1}{2}l^{-1/2}) = \frac{\pi}{\sqrt{lg}}$$

The length *increases* so the small increment in the length is positive and is given by

$\delta l = (\frac{2}{100})(l) = \frac{1}{50}l$

Using $\delta T \approx \dfrac{dT}{dl}\delta l$ gives

$$\delta T \approx (\tfrac{1}{50}l)\{\pi/\sqrt{lg}\} = (\tfrac{1}{50})\pi\sqrt{l/g}$$

The percentage change in the period is given by $\dfrac{\delta T}{T} \times 100$

i.e. $\frac{1}{50}\{\pi\sqrt{l/g}\} \div \{2\pi\sqrt{l/g}\} \times 100\% = 1\%$

This is a positive change, so we see that the period *increases* by 1%.

EXERCISE 26A

1 Using $y = \sqrt[3]{x}$, find, *without using a calculator*, an approximate value for

a $\sqrt[3]{1001}$ **b** $\sqrt[3]{9}$ **c** $\sqrt[3]{63}$

Work to 6 d.p.
Now use a calculator to find the accuracy of each approximation.

2 Given that $1° = 0.0175$ rad, $\sin 60° = 0.8660$ and $\sin 45° = 0.7071$, use $f(\theta) = \cos\theta$ to find an approximate value for

a $\cos 31°$ **b** $\cos 59°$ **c** $\cos 44°$

3 If $f(x) = x\ln(1+x)$ find an approximation for the increase in $f(x)$ when x increases by δx.
Hence estimate the value of $\ln(2.1)$ given that $\ln 2 = 0.6931$.

4 If $y = \tan x$ find an approximation for δy when x is increased by δx and use it to estimate, in terms of π, the value of $\tan\frac{9}{32}\pi$.

5 Use $f(x) = \sqrt[5]{x}$ to find the approximate value of $\sqrt[5]{33}$.

6 Given that $y = \sqrt{\dfrac{x-2}{x-1}}$ determine the value of $\dfrac{dy}{dx}$ when $x = 3$. Deduce the approximate increase in the value of y when x increases from 3 to $3+a$ where a is small.

Comparative Rates of Change

Some problems involving the rate of change of one variable compared with another do not provide a direct relationship between these two variables. Instead, each of them is related to a third variable.

The identity $\frac{dy}{dx} = \frac{dy}{dt} \times \frac{dt}{dx}$ is useful in solving problems of this type.

Suppose, for instance, that the radius, r, of a circle is increasing at a rate of 1 mm per second. This means that $\frac{dr}{dt} = 1$. The rate at which the area, A, of the circle is increasing is $\frac{dA}{dt}$.

We do not know A as a function of t but we do know that $A = \pi r^2$, and, from the chain rule, that

$$\frac{dA}{dt} = \frac{dA}{dr} \times \frac{dr}{dt}$$

Then $\frac{dA}{dt}$ can be calculated, as $\frac{dr}{dt}$ is given and $\frac{dA}{dr}$ can be found from $A = \pi r^2$.

In some cases, more than three variables may be involved but the same approach is used with a relationship of the form

$$\frac{dy}{dx} = \frac{dy}{dp} \times \frac{dp}{dq} \times \frac{dq}{dx}$$

EXAMPLES 26B

1 A spherical balloon is being blown up so that its volume increases at a constant rate of 1.5 $cm^3\,s^{-1}$. Find the rate of increase of the radius when the volume of the balloon is 56 cm^3.

If, at time t, the radius of the balloon is r and the volume is V then

$$V = \tfrac{4}{3}\pi r^3 \quad \Rightarrow \quad \frac{dV}{dr} = 4\pi r^2$$

> We are looking for $\frac{dr}{dt}$ and we are given $\frac{dV}{dt} = 1.5$ so we use

$$\frac{dr}{dt} = \frac{dr}{dV} \times \frac{dV}{dt} = \frac{dV}{dt} \div \frac{dV}{dr} \quad \Rightarrow \quad \frac{dr}{dt} = \frac{1.5}{4\pi r^2} = \frac{3}{8\pi r^2}$$

Now substituting $V = 56$ in $V = \frac{4}{3}\pi r^3$ gives $r = 2.373\ldots$

Therefore, when $V = 56$, $\frac{dr}{dt} = \frac{3}{8\pi(2.373\ldots)^2} = 0.021\,19\ldots$

i.e. the radius is increasing at a rate of 0.0212 $cm\,s^{-1}$ (correct to 3 s.f.)

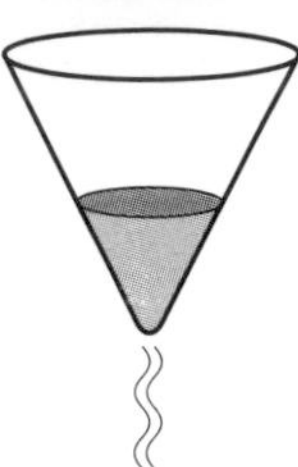

2 A funnel holding liquid has the shape of an inverted cone with a semi-vertical angle of 30°. The liquid is running out of a small hole at the vertex.

a If it is assumed that the hole is small enough to be ignored when finding the volume, show that the volume, $V\,\text{cm}^3$, of liquid left in the funnel when the depth of the liquid is h cm, is given by $V = \frac{1}{9}\pi h^3$

b If it is further assumed that the liquid is running out at a constant rate of $3\,\text{cm}^3\,\text{s}^{-1}$, find the rate of change of h when $V = 81\pi\,\text{cm}^3$.

c If the rate of flow of the liquid is not constant, but starts off at $3\,\text{cm}^3/\text{s}$ when the funnel is full and then decreases as h decreases, state with a reason the nature of the error in the answer to part **b**.

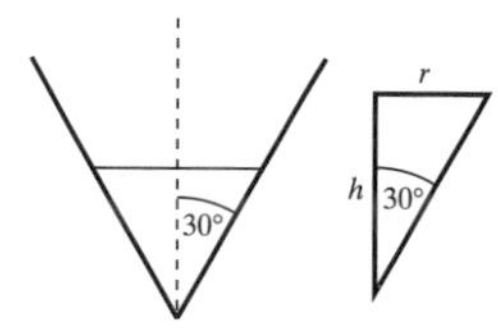

a At any time t, the volume V of liquid is given by $V = \frac{1}{3}\pi r^2 h$.

From the diagram, $r = h\tan 30° = \dfrac{1}{\sqrt{3}}h \quad \Rightarrow \quad r^2 = \frac{1}{3}h^2$

$\therefore \quad V = \frac{1}{3}\pi(\frac{1}{3}h^2)h = \frac{1}{9}\pi h^3 \qquad [1]$

b As liquid is running out at a constant rate of $3\,\text{cm}^3\,\text{s}^{-1}$, V is decreasing by $3\,\text{cm}^3\,\text{s}^{-1}$,

i.e. $\dfrac{dV}{dt} = -3$

We want $\dfrac{dh}{dt}$, so we use $\dfrac{dh}{dt} = \dfrac{dV}{dt} \times \dfrac{dh}{dV} = \dfrac{dV}{dt} \div \dfrac{dV}{dh}$,

and from [1], $\dfrac{dV}{dh} = \frac{1}{3}\pi h^2$

$\therefore \quad \dfrac{dh}{dt} = (-3) \div (\frac{1}{3}\pi h^2) = -\dfrac{9}{\pi h^2}$

When $V = 81\pi$, using [1] gives $81\pi = \frac{1}{9}\pi h^3 \quad \Rightarrow \quad h = 9$

$\therefore$ at this instant, $\dfrac{dh}{dt} = -\dfrac{1}{9\pi}$

i.e. the depth is decreasing at the rate of $\dfrac{1}{9\pi}\,\text{cm}\,\text{s}^{-1}$.

c When $V = 81\pi$, the rate of decrease in volume is less than $3\,\text{cm}^3\,\text{s}^{-1}$,

i.e. $\left|\dfrac{dV}{dt}\right| < 3 \quad \Rightarrow \quad \left|\dfrac{dh}{dt}\right|\left(= \left|\dfrac{dV}{dt} \times \dfrac{dh}{dV}\right|\right)$ is less than $\dfrac{1}{9\pi}$

$\therefore$ the rate of decrease of h is less than $\dfrac{1}{9\pi}$.

So the answer for **b** contains a positive error, i.e. it is larger than the true value.

If we also consider the effect of neglecting the hole, i.e. the missing part of the funnel at the vertex, when calculating V, then we are using values of V, and hence h, that are larger than their true values. Since $dV/dh = (1/3)\pi h^2$, this increases the value of dV/dh, but as we *divide* by it to find dh/dt, the error in the numerical value of dh/dt is reduced. To some extent these errors cancel each other out.

EXERCISE 26B

1 Ink is dropped on to blotting paper forming a circular stain which increases in area at a rate of $2.5\,\text{cm}^2/\text{s}$. Find the rate at which the radius is changing when the area of the stain is $16\pi\,\text{cm}^2$.

2 The surface area of a cube is increasing at a rate of $10\,\text{cm}^2/\text{s}$. Show that, when the edge is r cm, the surface area, $A\,\text{cm}^2$, can be modelled by the equation

$$A = 6r^2$$

Find the rate of increase of the volume of the cube when the edge is of length 12 cm.

3 The circumference of a circular patch of oil on the surface of a pond is assumed to be increasing at the constant rate of 2 m/s.

a When the radius is 4 m, at what rate is the area of the oil changing?

b If the circumference is actually increasing more quickly than 2 m/s when the radius is 4 m, is your answers for part **a** too large or too small?

4 A container in the form of a right circular cone of height 16 cm and base radius 4 cm is held vertex downward and filled with liquid. If the liquid leaks out from the vertex at a rate of $4\,\text{cm}^3/\text{s}$, find the rate of change of the depth of the liquid in the cone when half of the liquid has leaked out.

5 A right circular cone has a constant volume. The height h and the base radius r can both vary. Find the rate at which h is changing with respect to r at the instant when r and h are equal.

6 The radius of a hemispherical bowl is a cm. The bowl is being filled with water at a steady rate of $3\pi a^3\,\text{cm}^3$ per minute. Find, in terms of a, the rate at which the water is rising when the depth of water in the bowl is $\frac{1}{2}a$ cm.

The volume of the shaded part of this hemisphere is $\frac{1}{3}\pi h^2(3a - h)$.

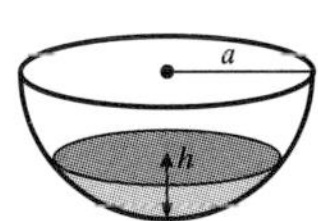

SUMMARY E

Throughout this summary, a and b represent constant quantities.

TRIGONOMETRY

Compound Angle Identities

$\sin(A \pm B) \equiv \sin A \cos B \pm \cos A \sin B$

$\cos(A \pm B) \equiv \cos A \cos B \mp \sin A \sin B$

$\tan(A \pm B) \equiv \dfrac{\tan A \pm \tan B}{1 \mp \tan A \tan B}$

Double Angle Identities

$\sin 2A \equiv 2 \sin A \cos A$

$$\cos 2A \equiv \begin{cases} \cos^2 A - \sin^2 A \\ 2\cos^2 A - 1 \\ 1 - 2\sin^2 A \end{cases}$$

and $\begin{cases} \cos^2 A \equiv \frac{1}{2}(1 + \cos 2A) \\ \sin^2 A \equiv \frac{1}{2}(1 - \cos 2A) \end{cases}$

$\tan 2A \equiv \dfrac{2\tan A}{1 - \tan^2 A}$

Expressing $a\cos\theta \pm b\sin\theta$ as a Single Term

For various values of a and b, $a\cos\theta \pm b\sin\theta$ can be expressed as

$$r\cos(\theta \pm \alpha) \quad \text{or} \quad r\sin(\theta \pm \alpha)$$

where $r = \sqrt{a^2 + b^2}$ and $\tan\alpha$ is either a/b or b/a.

LOGARITHMS

For any positive numbers a and b the formula for changing the base of a logarithm from b to a is

$$\log_a x = \frac{\log_b x}{\log_b a}$$

INVERSE TRIGONOMETRIC FUNCTIONS:

The inverse trig functions are $\sin^{-1}x$, $\cos^{-1}x$, $\tan^{-1}x$.

$\sin^{-1}x$ means 'the angle in the range $-\frac{1}{2}\pi \leqslant \theta \leqslant \frac{1}{2}\pi$ whose sine is x'.

$\cos^{-1}x$ means 'the angle in the range $0 \leqslant \theta \leqslant \pi$ whose cosine is x'.

$\tan^{-1}x$ means 'the angle in the range $0 \leqslant \theta \leqslant \frac{1}{2}\pi$ whose tangent is x'.

DIFFERENTIATION

Standard Results

$f(x)$	$\frac{d}{dx}f(x)$
x^n	nx^{n-1}
$\sin x$	$\cos x$
$\cos x$	$-\sin x$
$\tan x$	$\sec^2 x$
$\sec x$	$\sec x \tan x$
$\operatorname{cosec} x$	$-\operatorname{cosec} x \cot x$
$\cot x$	$-\operatorname{cosec}^2 x$
a^x	$a^x \ln a$

Further Quotable Results

$f(x)$	$\frac{d}{dx}f(x)$
$\sin ax$	$a\cos ax$
e^{ax}	$a e^{ax}$
$\ln(ax)$	$1/x$ (*not* a/x)

COMPOUND FUNCTIONS

If u and v are both functions of x then

$$y = uv \quad \Rightarrow \quad \frac{dy}{dx} = v\frac{du}{dx} + u\frac{dv}{dx}$$

$$y = \frac{u}{v} \quad \Rightarrow \quad \frac{dy}{dx} = \left(v\frac{du}{dx} - u\frac{dv}{dx}\right)\Big/ v^2$$

If $y = f(u)$ and $u = g(x)$ then

$$\frac{dy}{dx} = \frac{dy}{du} \times \frac{du}{dx}$$

This is known as the Chain Rule, and can be extended,
e.g. if $y = f(u), u = g(v), v = h(x)$ then

$$\frac{dy}{dx} = \frac{dy}{du} \times \frac{du}{dv} \times \frac{dv}{dx}$$

IMPLICIT DIFFERENTIATION

When y cannot be isolated, each term can be differentiated with respect to x. Remember that $\frac{d}{dx}(y) = \frac{dy}{dx}$ and that $\frac{d}{dx}[f(y)] = f'(y)\frac{dy}{dx}$,

e.g. $\frac{d}{dx}(y^2) = (2y)\left(\frac{dy}{dx}\right)$

and $\frac{d}{dx}(xy) = y + (x)\left(\frac{dy}{dx}\right)$ (by product rule)

LOGARITHMIC DIFFERENTIATION

Sometimes it is easier to differentiate $y = f(x)$ with respect to x if we first take logs of both sides.

When doing this remember that $\frac{d}{dx}(\ln y) = \frac{1}{y}\frac{dy}{dx}$.

This process is called logarithmic differentiation. It is *essential* when differentiating functions such as x^x.

Parametric Differentiation

If $y = f(t)$ and $x = g(t)$ then $\frac{dy}{dx} = \frac{dy}{dt} \div \frac{dx}{dt}$

Small Increments

If $y = f(x)$ and x increases by a small amount, δx, then

$$\delta y \approx \left(\frac{dy}{dx}\right)(\delta x)$$

Comparative Rates of Change

If a quantity p depends on a quantity q and the rate at which q increases with time t is known, then

$$\frac{dp}{dt} = \frac{dp}{dq} \times \frac{dq}{dt}$$

MULTIPLE CHOICE EXERCISE E

In Questions 1 to 11 write down the letter or letters corresponding to a correct answer.

1 If $x^2 + y^2 = 4$ then $\frac{dy}{dx}$ is

A $2x + 2y$ **B** $4 - x^2$

C $-\frac{x}{y}$ **D** $\frac{y}{x}$

2 $\frac{d}{dx}\left(\frac{1}{1+x}\right)$ **is**

A $\frac{-1}{(1+x)^2}$ **B** $\frac{1}{1-x}$

C $\ln(1+x)$ **D** $\frac{-1}{1+x^2}$

E $-(1+x)^{-2}$

3 $\frac{d}{dx}\ln\left(\frac{x+1}{2x}\right)$ is

A $\frac{1}{2}$ B $\frac{1}{x+1}-\frac{1}{2x}$

C $\frac{2x}{x+1}$ D $\frac{1}{x+1}-\frac{1}{x}$

E $-\frac{1}{x(x+1)}$

4 $\frac{d}{dx}a^x$ is

A xa^{x-1} B a^x

C $x\ln a$ D $a^x\ln a$

5 If $x = \cos\theta$ and $y = \cos\theta + \sin\theta$, $\frac{dy}{dx}$ is

A $1-\cot\theta$ B $1-\tan\theta$

C $\cot\theta - 1$ D $\cot\theta + 1$

6 The greatest value of $5\cos\theta - 4\sin\theta$ is

A 3 B 1 C $\sqrt{41}$ D ± 5

7 $3\cos\theta - 4\sin\theta =$

A $5\cos(\theta+\alpha)$ where $\tan\alpha = \frac{3}{4}$

B $5\sin(\alpha-\theta)$ where $\tan\alpha = \frac{3}{4}$

C $5\cos(\theta+\alpha)$ where $\tan\alpha = \frac{4}{3}$

D $-5\cos(\theta-\alpha)$ where $\tan\alpha = \frac{4}{3}$

8 If $y = \ln(\ln x)$ and $x > 1$ then

A $\frac{dy}{dx} = \frac{1}{\ln x}$ C $\frac{dy}{dx} = \frac{1}{x\ln x}$

B $e^y = \ln x$ D $y = \ln x^2$

9 Given that $x = \cos^2\theta$ and $y = \sin^2\theta$,

A $x^2+y^2 = 1$ C $0 \leqslant y \leqslant 1$

B $\frac{dy}{dx} = \tan\theta$ D $y = x - \frac{1}{2}\pi$

10 $2\cos(2\theta - 60°) \equiv$

A $\cos 2\theta - \sqrt{3}\sin 2\theta$

B $\sin 2\theta - \sqrt{3}\cos 2\theta$

C $\cos 2\theta + 2\sqrt{3}\cos\theta\sin\theta$

D $\cos 2\theta + \sqrt{3}\sin 2\theta$

11 $\tan\theta = 0.8 \Rightarrow \tan 2\theta =$

A 0.4 C 1.6

B 40/9 D $4.\dot{4}$

In Questions 12 to 19 a single statement is made. Write T if it is true and F if it is false.

12 $\frac{d}{dx}(uv) = \frac{du}{dx} \times \frac{dv}{dx}$

13 $\frac{d}{dx}(x^2y^2) = 2xy^2 + 2x^2y$

14 Given that $y = \ln x^2$ and x increases by δx then $\delta y \approx \left(\frac{1}{x^2}\right)(\delta x)$.

15 When $y = \cos 2\theta$ and $x = \sin\theta$, $\frac{dy}{dx} = -4x\sqrt{1-x^2}$.

16 If $y = f(t)$ and $x = g(t)$ then $\frac{dy}{dx} = \frac{dy}{dt} \div \frac{dx}{dt}$.

17 $y = e^{3x} \iff \frac{dy}{dx} = 3e^{3x}$

18 $y = 5\sin\theta + 12\cos\theta \iff y = 13\sin(\theta+\alpha)$ where $\tan\alpha = \frac{12}{5}$.

19 $\sin 2\theta = 0.6 \Rightarrow \sin\theta = 0.3$.

Examination Questions E

1 A student is asked to express $3\sin\theta + 4\cos\theta$ in the form $R\sin(\theta+\alpha)$. She writes

$$3\sin\theta + 4\cos\theta \equiv 5\sin\left(\theta + \frac{\pi}{3}\right)$$

Determine, with a reason, whether she is correct. (AQA)

2 Find the equation of the tangent to the curve $y = (4x+3)^5$ at the point $(-\frac{1}{2}, 1)$, giving your answer in the form $y = mx + c$. (OCR)

3 Differentiate each of the following functions with respect to x.

a $\dfrac{1}{\sqrt{x}}$ **b** e^{-x} **c** $x^2\cos(2x)$ (OCR)

4 The parametric equations of a curve are $x = \ln t,\ y = t + t^2$, where $t > 0$.

Express $\dfrac{dy}{dx}$ in terms of t, simplifying your answer. (OCR)

5 The function f is defined by $f : x \mapsto 5 + 3x^2 - 36\ln(x-1)$ and has domain $x \geqslant 2$.

a Determine the value of x for which f is stationary.

b The second derivative of $f(x)$ is $f''(x)$. Find the range of f''. (AQA)

6 Find the four solutions of the equation $\sin 2\theta = \cos^2\theta$ in the interval $0° < \theta < 360°$. Give each of these solutions correct to the nearest degree. (AQA)

7 Express $\sin 4\theta$ in terms of $\sin 2\theta$ and $\cos 2\theta$, and hence express $\dfrac{\sin 4\theta}{\sin\theta}$ in terms of $\cos\theta$ only. (OCR)

8 Find the x-coordinates of the stationary points of $y = x^3e^{-kx}$, where k is a positive constant. (OCR)

9 Find all values of θ, for $0 \leqslant \theta \leqslant 180°$, which satisfy the equation $\cos 2\theta = \cos\theta$. (OCR)

10 Given that

$$3\cos x - 4\sin x \equiv R\cos(x+\alpha),$$

where $R > 0$ and $0° < \alpha < 90°$, find the values of R and α, giving the value of α correct to two decimal places.

Hence solve the equation

$$3\cos 2\theta - 4\sin 2\theta = 2,$$

for $0° < \theta < 360°$, giving your answers correct to one decimal place. (OCR)

11 The curve C has parametric equations

$$x = 4\cos 2t,\ y = 3\sin t,\ -\frac{\pi}{2} < t < \frac{\pi}{2}.$$

A is the point $(2, 1\frac{1}{2})$ and lies on C.

a Find the value of t at the point A.

b Find $\frac{dy}{dx}$ in terms of t.

c Show that an equation of the normal to C at A is $6y - 16x + 23 = 0$.

The normal at A cuts C again at the point B.

d Find the y-coordinate of the point B. (Edexcel)

12

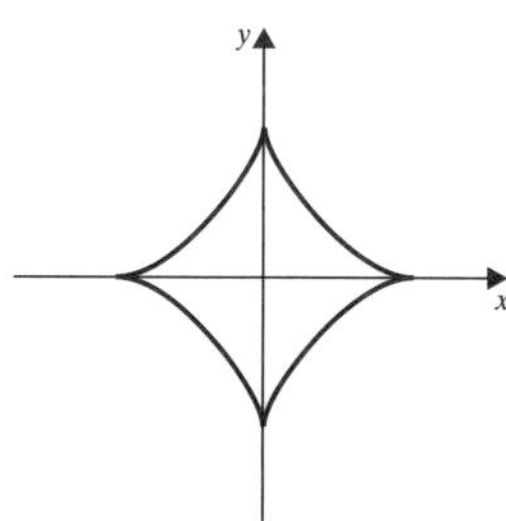

The curve shown above is called an astroid and is defined by the parametric equations

$x = 8\cos^3\theta,\ y = 8\sin^3\theta$ for $0 \leqslant \theta \leqslant 2\pi$.

a By obtaining expressions for $\frac{dx}{d\theta}$ and $\frac{dy}{d\theta}$, show clearly that the gradient of the curve is given by $\frac{dy}{dx} = -\tan\theta$.

The astroid above can also be defined by the Cartesian equation

$$x^{\frac{2}{3}} + y^{\frac{2}{3}} = k.$$

b Find the value of the constant k by first finding values of x and y for a particular value of θ.

c Use implicit differentiation to find an expression for $\frac{dy}{dx}$ in terms of x and y. (AQA)

13 Find all values of θ in the range $0°$ to $360°$ satisfying

a $2\sin 2\theta = \sin\theta$ **b** $3\sec^2\theta + 5\tan\theta - 5 = 0$ (WJEC)

14 Differentiate the following with respect to x, simplifying your answers where possible.

a $x^2 \ln x$ **b** $\frac{3x^2 - 5}{2x^2 + 7}$ **c** $(x^3 + 5)^{10}$ (WJEC)

15 Given that

$$a\cos\theta + b\sin\theta = r\cos(\theta - \alpha) \quad \text{where } r > 0,$$

show that

$$r = \sqrt{a^2 + b^2}$$

and find an expression for $\tan\alpha$ in terms of a and b.

Hence find all values of θ between $0°$ and $360°$ satisfying the equation

$$2\cos\theta + 3\sin\theta = 1.$$

Give your answers correct to the nearest degree. (AQA)

16 a Write down the exact value for $\cos 30°$.

b Use an appropriate double angle formula to write $\cos 15°$ as $\sqrt{a + \sqrt{b}}$, where a and b are rational. (OCR)

17 Write down the derivative of $\tan 2x$ with respect to x. (OCR)

18 Differentiate the following functions with respect to x, simplifying your answers as far as possible.

a $x^2 \sin 2x$ **b** $\dfrac{x}{\sqrt{x^2 - 1}}$ (AQA)

19 The parametric equations of a curve are $x = e^{2t} - 5t,\ y = e^{2t} - 2t$. Find $\dfrac{dy}{dx}$ in terms of t.

Find the exact value of t at the point on the curve where the gradient is 2. (OCR)

20 a Let $u = \sin v$, write down $\dfrac{du}{dv}$ as a function of v.

b Hence, or otherwise, obtain $\dfrac{dv}{du}$ in terms of u. (OCR)

21 a Express the function $2\sin x° + \cos x°$ in the form $R\sin(x + \alpha)°$, stating the values of R and α. Using these values, write down the coordinates of the maximum turning point on the graph of

$$2\sin x° + \cos x° \text{ for } 0 \leqslant x \leqslant 90.$$

b Express $3\cos 2x + \sin x$ in terms of $\sin x$. Hence calculate all of the values of x between 0 and 360 which satisfy the equation

$$3\cos 2x° + \sin x° = 1.$$

(AQA)

22 Express $2\cos\theta + 2\sin\theta$ in the form $R\cos(\theta - \alpha)$, where $R > 0$ and $0 < \alpha < \frac{1}{2}\pi$, giving the values of R and α in an exact form.
Hence, or otherwise, show that one of the acute angles θ satisfying the equation $2\cos\theta + 2\sin\theta = \sqrt{6}$ is $\frac{5}{12}\pi$, and find the other acute angle. (OCR)

23 Differentiate $x^2 \ln x$ with respect to x. (OCR)

24 Given that θ lies in the interval $0° < \theta < 45°$,

a show that $\cot\theta + \tan\theta = \dfrac{2}{\sin 2\theta}$

b find the values of θ for which $\cot\theta + \tan\theta > 4$. (AQA)

25 A curve C is defined by the parametric equations

$$x = e^t + t, \quad y = e^{-t} + t.$$

Express $\frac{dy}{dx}$ in terms of t. (AQA)

26 The curve C has parametric equations

$$x = t^3, \quad y = t^2, \quad t > 0.$$

a Find an equation of the tangent to C at A(1, 1).

Given that the line l with equation $3y - 2x + 4 = 0$ cuts the curve C at point B,

b find the coordinates of B,

c prove that the line l only cuts C at the point B. (Edexcel)

27 $$8\cos x^\circ - 15\sin x^\circ \equiv R\cos(x + A)^\circ, \quad 0 \leqslant A \leqslant 90, \quad R > 0$$

Find

a the value of R and, to one decimal place, the value of A,

b the maximum value of $8\cos x^\circ - 15\sin x^\circ$, and the smallest positive value of x for which this occurs,

c the two smallest values of x for which $8\cos x^\circ - 15\sin x^\circ = 6$. (Edexcel)

28 A curve C is defined by the equation

$$x^2 - 2xy + 3y^2 = 9$$

i Show that the point P(3, 2) lies on the curve.

ii Show that at P

$$\frac{dy}{dx} = -\frac{1}{3}.$$

iii The point $Q(3 + h, 2 + k)$ also lies on the curve C and is close to P. Using the result of part **ii**, write down an approximate expression for k in terms of h. (AQA)

29 Differentiate with respect to x, **a** $\frac{\sin x}{x}$, $x > 0$ **b** $\ln\left(\frac{1}{x^2 + 9}\right)$

Given that $y = x^x$, $x > 0$, $y > 0$, by taking logarithms

c show that $\frac{dy}{dx} = x^x(1 + \ln x)$. (Edexcel)

30 Given that $y = \cos 2x + \sin x$, $0 < x < 2\pi$, and x is in radians,

a find, in terms of π, the values of x for which $y = 0$.

b Find, to 2 decimal places, the values of x for which $\frac{dy}{dx} = 0$. (Edexcel)

31 Starting from the identity for $\cos(A+B)$, prove that

$$\cos 2x = 1 - 2\sin^2 x.$$

Find, in radians to 2 decimal places, the values of x in the interval $0 \leqslant x < 2\pi$ for which

a $2\cos 2x + 1 = \sin x$ **b** $2\cos x + 1 = \sin \frac{1}{2}x$ (Edexcel)

32 a Express $5\sin x + 12\cos x$ in the form $R\sin(x+\theta)$ where $R > 0$, and $0° < \theta < 90°$.

b Hence, or otherwise, find the maximum and minimum values of $f(x)$ where

$$f(x) = \frac{30}{5\sin x + 12\cos x + 17}$$

State also the values of x, in the range $0° < x < 360°$, at which they occur. (OCR)

33 a Find the equation of the tangent to the curve $y = e^{-x}$ at the point $P(t, e^{-t})$.

b The tangent at P meets the x- and y-axes at the points Q and R respectively. Show that A, the area of triangle OQR, where O is the origin, is given by

$$A = \tfrac{1}{2}(t+1)^2 e^{-t}.$$

c Find, showing your working, the stationary values of A and determine their nature. (WJEC)

34 A curve is defined parametrically by $x = (2t - 1)$, $y = t^3$ and P is the point on the curve where $t = 2$.

a Obtain an expression for $\frac{dy}{dx}$ in terms of t and calculate the gradient of the curve at P.

b Determine a cartesian equation of the curve, expressing your answer in the form $y = f(x)$.

c Sketch the curve, showing clearly the values of the intercepts on the axes.

d Write down the equation of the tangent to the curve at P. This tangent intersects the curve again at Q, with parameter q. Show that $q^3 = 12q - 16$. Hence determine the coordinates of the point Q.

e Prove that the normal to the curve at P does not intersect the curve at any other point. (AQA)

35 i Differentiate with respect to x **a** $e^{\sqrt{x}}$, **b** $\dfrac{1+x^2}{\ln x}$

ii A curve is given by the parametric equations

$$x = \cos 2t,\ y = 4\sin^3 t,\ 0 \leqslant t \leqslant \frac{\pi}{4}.$$

a Show that $\frac{dy}{dx} = -3\sin t$.

b Find an equation of the normal to the curve at the point where $t = \frac{\pi}{6}$. (Edexcel)

36 The volume, V, of a sphere of radius r is given by $V = \frac{4}{3}\pi r^3$.

a Obtain an expression for $\frac{dV}{dr}$.

b A balloon when almost fully inflated can be modelled by a sphere. When the radius of the balloon is 15 cm, it is observed that the rate of increase of the radius is $0.1\ \text{cm s}^{-1}$. Find, to two significant figures, the rate of increase of the volume at this time. (AQA)

37 The volume of liquid, $V\,\text{cm}^3$, in a container when the depth is x cm $(x > 0)$ is given by

$$V = \frac{30\sqrt{x}}{(x+9)}$$

The container has height h cm.

a Given that $x = h$ when $\dfrac{dV}{dx} = 0$, find the value of h, and hence determine the maximum capacity of the container.

b Calculate the rate of change of volume when the depth is 1 cm and increasing at a rate of $0.02\,\text{cm s}^{-1}$, giving your answer in $\text{cm}^3\,\text{s}^{-1}$. (AQA)

38 In a simple model of the tides in a harbour, the depth, y m, is given by

$$y = 20 + 5\cos t,$$

where t is the time measured in suitable units. For $0 \leqslant t \leqslant 4\pi$, write down the maximum and minimum values of y and the values of t when they occur.

In a more refined model, y is given by

$$y = 20 + \{5 + \cos(\tfrac{1}{2}t)\}\cos t.$$

Using this model, show that the values of t when y is stationary satisfy the equation

$$\sin(\tfrac{1}{2}t)\{6\cos^2(\tfrac{1}{2}t) + 20\cos(\tfrac{1}{2}t) - 1\} = 0$$

(OCR)

39 A packaging designer places a right circular cone inside a hollow sphere of fixed radius R and centre O, as shown in the figure, where the vertical angle of the cone is θ.

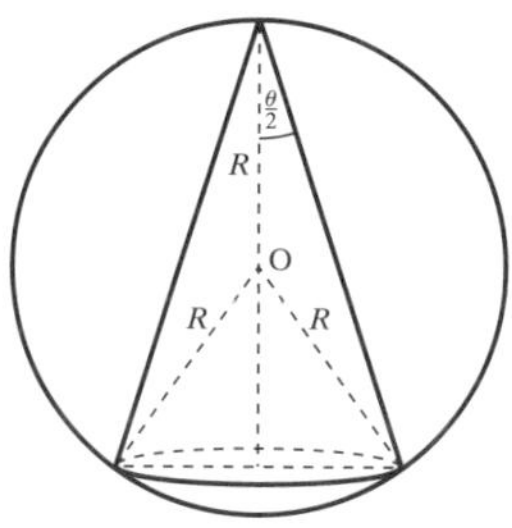

a Prove that the volume V of the cone is given by $V = \dfrac{\pi}{3}R^3(1 + \cos\theta)\sin^2\theta$.

The designer needs to find the maximum volume of the cone when R is fixed and θ varies.

b Show, by differentiation, that $\arccos\frac{1}{3}$ is the value of θ for which V is a maximum.

c Find the maximum value of V, in terms of R. (Edexcel)

40 A population P is growing at the rate of 9% each year and at time t years may be approximated by the formula

$$P = P_0(1.09)^t, \quad t \geqslant 0,$$

where P is regarded as a continuous function of t and P_0 is the starting population at time $t = 0$.

a Find an expression for t in terms of P and P_0.

b Find the time T years when the population has doubled from its value at $t = 0$, giving your answer to 3 significant figures.

c Find, as a multiple of P_0, the rate of change of population $\dfrac{dP}{dt}$ at time $t = T$. (Edexcel)

41 Find all the values of θ, such that $0 \leqslant \theta \leqslant 180°$, which satisfy the equation $2\sin 2\theta = \tan\theta$. (OCR)

42 The curve with equation $ky = a^x$ passes through the points with coordinates (7, 12) and (12, 7). Find, to 2 significant figures, the values of the constants k and a.

Using your values of k and a, find the value of $\frac{dy}{dx}$ at $x = 20$, giving your answer to 1 decimal place. (Edexcel)

43 A speaker uses an amplifier to carry her words to members of the audience x metres away. The power output, P watts, is given by the formula $P = 0.0004x^2$.

i To increase the distance by a small amount δx metre, the output must be increased by δP watt. Find an approximate expression for δP in terms of x and δx.

ii Show that $\frac{\delta P}{P} \approx 2\frac{\delta x}{x}$.

iii If the power output of the amplifier is increased by 2%, by what percentage approximately is the distance her voice will carry increased? (OCR)

44 The equation of a closed curve is $(x+y)^2 + 2(x-y)^2 = 24$.

i Show, by using differentiation, that the gradient at the point (x, y) on the curve may be expressed in the form $\frac{3x-y}{x-3y}$.

ii Find the coordinates of all the points on the curve at which the tangent is parallel to either the x-axis or the y-axis.

iii Find the exact coordinates of all the points at which the curve crosses the axes, and the gradient of the curve at each of these points. (OCR)

45 A forest fire spreads so that the number of hectares burnt after t hours is given by

$$h = 30(1.65)^t$$

i By what constant factor is the burnt area multiplied from time $t = N$ to time $t = N+1$? Express this as a percentage increase.

ii 1.65 can be written as e^K. Find the value of K.

iii Hence show that $\frac{dh}{dt} = 15e^{Kt}$.

iv This shows that $\frac{dh}{dt}$ is proportional to h. Find the constant of proportionality. (OCR)

CHAPTER 27 Methods of proof

We saw in Chapter 5 how a demonstration can illustrate that a result holds in some given cases but a particular case, as yet unfound, may exist for which the result is not true. Because mathematics is used as a tool for modelling real-life situations and for predicting the behaviour of systems in untried conditions, it is vital that the mathematics is reliable. This means that the mathematics is based on general proofs that rule out the possibility of exceptions. (This does not mean that the predicted behaviour of a system is reliable, because the modelling process also involves assumptions about the nature of a system and assumptions are not certainties.)

We also saw in Chapter 5 that the most familiar form of proof is direct, i.e. we start with a known result and then use correct implications to arrive at the required result, e.g. to prove that $\cos 2x = 2\cos^2 x - 1$, start with

$\cos(A+B) = \cos A \cos B - \sin A \sin B.$

Then, replacing both A and B with x, we have

$$\cos(x+x) = \cos x \cos x - \sin x \sin x$$
$$\Rightarrow \quad \cos 2x = \cos^2 x - \sin^2 x$$
$$\Rightarrow \quad \cos 2x = \cos^2 x - (1 - \cos^2 x) = 2\cos^2 x - 1$$

There are some results that are difficult to prove directly but that can be proved using different forms of reasoning. We will look at one such method here.

Proof by Contradiction

This method of proof relies on the following reasoning.

To prove that a statement is true, start with the assumption that it is not true. Then use correct implications to arrive at a result that contradicts an earlier statement. This shows the assumption to be impossible, therefore the assumption is false so the original statement must be true.

The worked examples illustrate two classic uses of proof by contradiction.

EXAMPLES 27A

1 Prove that a tangent to a circle is perpendicular to the radius through the point of contact.

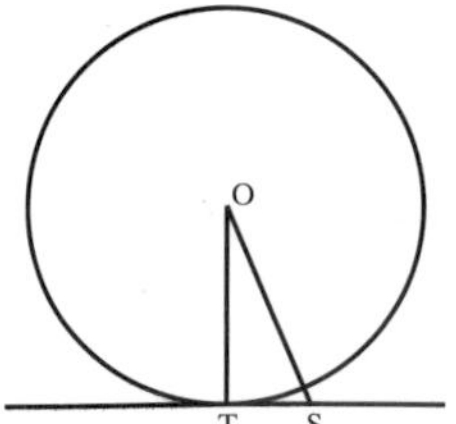

Assume that the tangent is not perpendicular to the radius at the point of contact, i.e. in the diagram, assume that the line through T is a tangent to the circle, centre O, where the angle at T is not 90°.

$\Rightarrow$ It is possible to draw a line OS such that $\angle$OST = 90°.
$\Rightarrow$ $\triangle$OST is a right-angled triangle in which OT is the hypotenuse.
$\Rightarrow$ OT > OS
$\Rightarrow$ S is inside the circle since OT is a radius.
$\Rightarrow$ The line through T and S cuts the circle in two distinct points.
$\Rightarrow$ The line through T and S is not a tangent to the circle.

The last statement contradicts the first, so the assumption is wrong, i.e. a tangent to a circle is perpendicular to the radius through the point of contact.

2 Prove that $\sqrt{2}$ is an irrational number.

Assume that $\sqrt{2}$ is not an irrational number $\Rightarrow$ $\sqrt{2}$ is a rational number

$\Rightarrow$ $\sqrt{2} = \dfrac{a}{b}$ where a and b are integers with no common factors

i.e. $\dfrac{a}{b}$ is a fraction in its lowest terms,

$\Rightarrow$ $2 = \dfrac{a^2}{b^2}$ $\Rightarrow$ $2b^2 = a^2$

$\Rightarrow$ a^2 is an even number so a is an even number

$\Rightarrow$ $a = 2c$ where c is an integer.

$\therefore$ $2b^2 = 4c^2$ $\Rightarrow$ $b^2 = 2c^2$

$\Rightarrow$ b^2 is an even number so b is an even number.

Now 'a is an even number' and 'b is an even number' contradict the statement '$\sqrt{2} = \dfrac{a}{b}$ where a and b are integers with no common factors'.

Therefore $\sqrt{2}$ is not a rational number, i.e. $\sqrt{2}$ is irrational.

Use of a Counter Example

We have seen that a mathematical result is accepted as correct only if it has been proved to be true in all cases; so to show that a statement is false, all we have to do is to demonstrate that it is not true in just *one* particular case,

e.g. to show that 'the sum of two irrational numbers is itself irrational' is not true, all we have to do is to demonstrate that $(5+\sqrt{2})+(5-\sqrt{2}) = 10$ which is rational.

This demonstration is called a *counter example*.

EXERCISE 27A

1 Prove by contradiction that if n is an integer such that n^2 is even then n is even. (Start with the assumption that 'n^2 is even and n is odd, i.e. $n = 2k+1$'.)

2 Prove by contradiction that $\sqrt{3}$ is an irrational number.

3 Find a counter example to show that it is not true that the product of two irrational numbers is irrational.

4 Prove by contradiction that, if n is an integer and n^2 is odd then n is odd.

5 Find a counter example to show that $\sin(A+B) = \sin A + \sin B$ is untrue.

6 Give a counter example to show that the converse of the statement

ABCD is a square $\Rightarrow$ Angles A, B, C and D are right angles is untrue.

7 Show that $a > b \not\Rightarrow a^2 > b^2$ where a and b are any two real numbers.

8 Prove by contradiction that 2 is the only even prime number. (Start with the assumption that 2 is not the only even prime number $\Rightarrow$ there is an even prime number $n > 2$.)

9 Prove by contradiction that if a is not zero then a^{-1} is not zero. (Start with $a \neq 0$ and $a^{-1} = 0$.)

10 Prove by contradiction that if p is a prime number then $\sqrt{p}$ is irrational and give a counter example to show that the converse is not true.

CHAPTER 28

Coordinate geometry 3

Loci

In general, within a plane a point P can be anywhere and, further, if (x, y) are the coordinates of P, then x and y can take any values independently of each other.

However, when the possible positions of P are restricted by some condition to a line (curved or straight), the set of points satisfying this condition is called the *locus* of P.

Further, the relationship between x and y which applies only to the locus of P defines that locus and is called the *Cartesian equation* of P.

EXAMPLES 28A

1 A point, P, is restricted so that it is equidistant from the points A(1, 2) and B(−2, −1). Find the Cartesian equation of P.

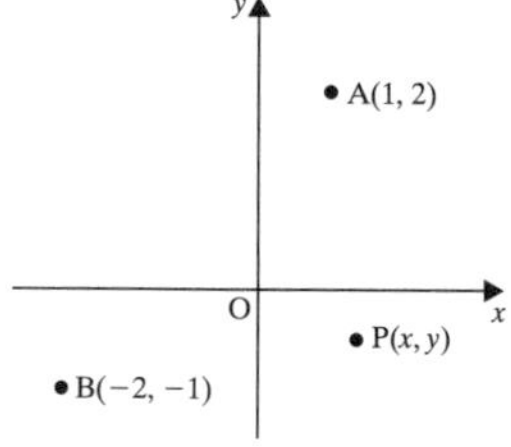

P is restricted to those positions where PA = PB.

> Translating this condition into a relationship between x and y gives the equation of the locus of P.

$PA^2 = (x-1)^2 + (y-2)^2$ and $PB^2 = (x+2)^2 + (y+1)^2$

If PA = PB then $PA^2 = PB^2$

> Using the given condition in this form avoids introducing square roots.

$\therefore$ PA = PB $\Rightarrow$ $(x-1)^2 + (y-2)^2 = (x+2)^2 + (y+1)^2$

$\Rightarrow$ $0 = 6x + 6y$

Therefore $x + y = 0$ is the equation of the locus of P.

> Note that the line $x + y = 0$ is the perpendicular bisector of AB.

2 A point P(x, y) is twice as far from the point A(3, 0) as it is from the line $x = 5$. Find the equation of the locus of P.

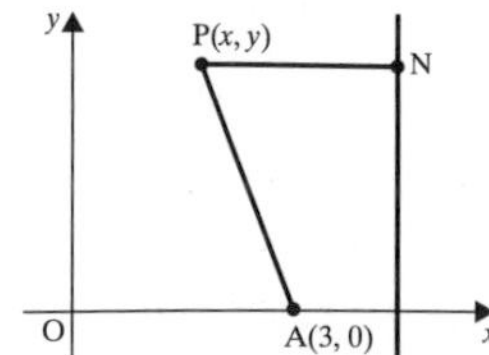

The restriction on P is that PA = 2PN

Now PA = 2PN $\Rightarrow$ $PA^2 = 4PN^2$

But $PA^2 = y^2 + (x-3)^2$ and $PN^2 = (5-x)^2$

$\therefore$ P satisfies the given condition $\iff$ $y^2 + (x-3)^2 = 4(5-x)^2$

i.e. the equation of the locus of P is $y^2 - 3x^2 + 34x = 91$

EXERCISE 28A Find the Cartesian equation of the locus of the set of points P in each case.

1 P is equidistant from the point (4, 1) and the line $x = -2$.

2 P is equidistant from (3, 5) and (−1, 1).

3 P is three times as far from the line $x = 8$ as from the point (2, 0).

4 P is equidistant from the lines $3x + 4y + 5 = 0$ and $12x - 5y + 13 = 0$.

5 P is at a constant distance of two units from the point (3, 5).

6 P is at a constant distance of five units from the line $4x - 3y = 1$.

7 A is the point (−1, 0), B is the point (1, 0) and angle APB is a right angle.

Circles

If a point P is at a constant distance r from a fixed point C then the locus of P is a circle whose centre is C and whose radius is r. In this section we look at a variety of methods for dealing with coordinate geometry problems involving circles.

The Equation of a Circle

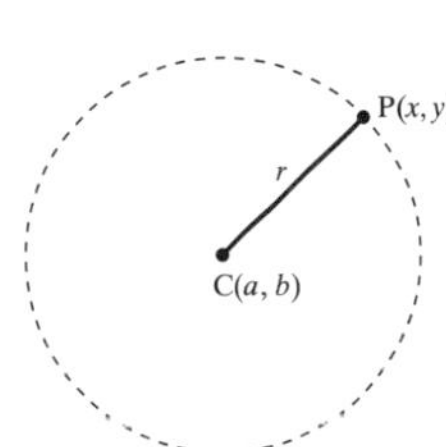

A point $P(x, y)$ is at a constant distance, r, from the point $C(a, b)$.

P is on the circle if and only if $CP = r$, i.e. $CP^2 = r^2$.

Now $\qquad CP^2 = (x - a)^2 + (y - b)^2$

$\therefore$ **P(x, y) is on the circle** $\Leftrightarrow$ $\mathbf{(x - a)^2 + (y - b)^2 = r^2}$.

i.e. $\mathbf{(x - a)^2 + (y - b)^2 = r^2}$

is the equation of a circle with centre (a, b) and radius r.

For example, the equation of a circle with centre (−2, 3) and radius 1 is

$$[x - (-2)]^2 + [y - 3]^2 = 1 \quad \Rightarrow \quad x^2 + y^2 + 4x - 6y + 12 = 0$$

As well as being able to write down the equation of a circle given its centre and radius, it is equally important to be able to recognise an equation as that of a circle.
Expanding and simplifying the equation of a circle with centre (a, b) and radius r gives

$$x^2 + y^2 - 2ax - 2ay + (a^2 + b^2 - r^2) = 0$$

which can be expressed as $x^2 + y^2 + 2gx + 2fy + c = 0$ where g, f and c are constants.

Comparing coefficients gives

$$g = -a, \ f = -b, \ c = a^2 + b^2 - r^2 \quad \Rightarrow \quad r^2 = f^2 + g^2 - c$$

So $\quad$ $\mathbf{x^2 + y^2 + 2gx + 2fy + c = 0}$ **is the general equation of a circle provided that** $\mathbf{g^2 + f^2 - c > 0}$

The centre of the circle is $\mathbf{(-g, -f)}$ **and the radius is** $\mathbf{\sqrt{g^2 + f^2 - c}}$.

Note that the coefficients of x^2 and y^2 are equal and that there is no xy term.

EXAMPLES 28B

1 Find the centre and radius of the circle whose equation is

$$x^2 + y^2 + 8x - 2y + 13 = 0$$

> There are two ways of finding the centre and radius of this circle. The first method involves forming perfect squares so that we can compare the given equation with $(x-a)^2 + (y-b)^2 = r^2$.

$$x^2 + 8x + 16 + y^2 - 2y + 1 = 16 + 1 - 13$$

$$\Rightarrow \quad (x+4)^2 + (y-1)^2 = 4$$

$\therefore$ the centre is $(-4, 1)$ and the radius is 2.

> Alternatively we can compare the given equation with the general equation of a circle.

$2g = 8 \quad \Rightarrow \quad g = 4, \quad 2f = -2 \quad \Rightarrow \quad f = -1$ and $c = 13$

The centre, $(-g, -f)$, is $(-4, 1)$ and the radius, $\sqrt{g^2 + f^2 - c}$ is 2.

2 Show that $2x^2 + 2y^2 - 6x + 10y = 1$ is the equation of a circle and find its centre and radius.

> Before we can compare this equation with the general form for the equation of a circle, we must divide the given equation by 2.

$2x^2 + 2y^2 - 6x + 10y - 1 = 0 \quad \Rightarrow \quad x^2 + y^2 - 3x + 5y - \frac{1}{2} = 0$

Comparing with $x^2 + y^2 + 2gx + 2fy + c = 0$ gives $2g = -3$, $2f = 5$, $c = -\frac{1}{2}$

$\Rightarrow (g^2 + f^2 - c) = 9$ which is greater than 0.

Therefore the equation does represent a circle.

The centre is $(\frac{3}{2}, -\frac{5}{2})$ and the radius is 3.

EXERCISE 28B

1 Write down the equation of the circle with

a centre (1, 2), radius 3

b centre (0, 4), radius 1

c centre (−3, −7), radius 2

d centre (4, 5), radius 3

2 Find the centre and radius of the circle whose equation is

a $x^2 + y^2 + 8x - 2y - 8 = 0$

b $x^2 + y^2 + x + 3y - 2 = 0$

c $x^2 + y^2 + 6x - 5 = 0$

d $2x^2 + 2y^2 - 3x + 2y + 1 = 0$

e $x^2 + y^2 = 4$

f $(x-2)^2 + (y+3)^2 = 9$

g $2x + 6y - x^2 - y^2 = 1$

h $3x^2 + 3y^2 + 6x - 3y - 2 = 0$

3 Determine which of the following equations represent circles.

a $x^2+y^2 = 8$

b $2x^2+y^2+3x-4 = 0$

c $x^2-y^2 = 8$

d $x^2+y^2+4x-2y+20 = 0$

e $x^2+y^2+8 = 0$

f $x^2+y^2+4x-2y-20 = 0$

Tangents to Circles and Other Problems

The following worked examples illustrate how a variety of problems concerning circles can be solved easily with the aid of a diagram and the use of geometric properties of a circle.

It is unnecessary to use calculus to find the equation of a tangent to a circle.

EXAMPLES 28C

1 Find the equation of the tangent at the point $(3, 1)$ on the circle with equation $x^2+y^2-4x+10y-8 = 0$. What is the angle between this tangent and the positive direction of the x-axis?

The centre of the circle is $C(2, -5)$.

> The tangent at A is perpendicular to the radius CA.

The gradient of CA is $\dfrac{1-(-5)}{3-2} = 6.$

Therefore the gradient of the tangent at A is $-\frac{1}{6}$ and the tangent goes through $A(3, 1)$.

So its equation is $y-1 = -\frac{1}{6}(x-3)$, i.e. $6y+x = 9$

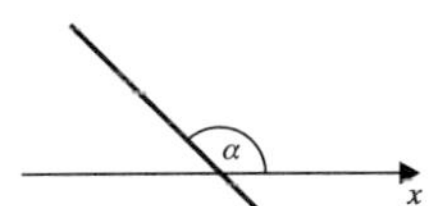

If α is the angle between the tangent and the positive direction of the x-axis,

then $\tan\alpha = -\frac{1}{6} \quad\Rightarrow\quad \alpha = 170.5°$

2 Find the equation of the circle whose diameter is the line joining the points $A(1, 5)$ and $B(-2, 3)$.

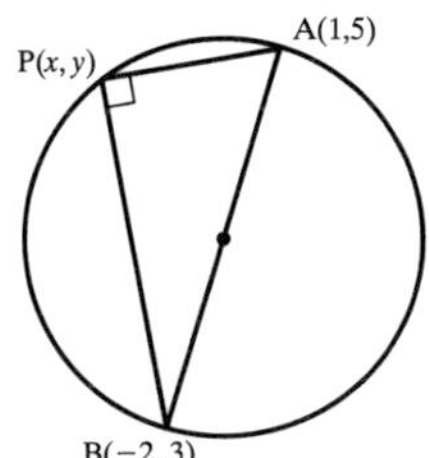

> We can use the fact that the angle in a semicircle is 90°.

$P(x, y)$ is a point on the circle if and only if

$$(\text{gradient AP}) \times (\text{gradient BP}) = -1$$

The gradient of AP is $\dfrac{y-5}{x-1}$ and the gradient of PB is $\dfrac{y-3}{x+2}$

$$\therefore \quad P(x, y) \text{ is on the circle} \quad \Longleftrightarrow \quad \left(\frac{y-5}{x-1}\right)\left(\frac{y-3}{x+2}\right) = -1$$

$$\Rightarrow \quad (y-5)(y-3) = -(x-1)(x+2)$$

$\therefore$ the equation of the circle is $x^2+y^2+x-8y+13 = 0$

3 Find the equation of the circle that goes through the points A(0, 1), B(4, 7) and C(4, −1).

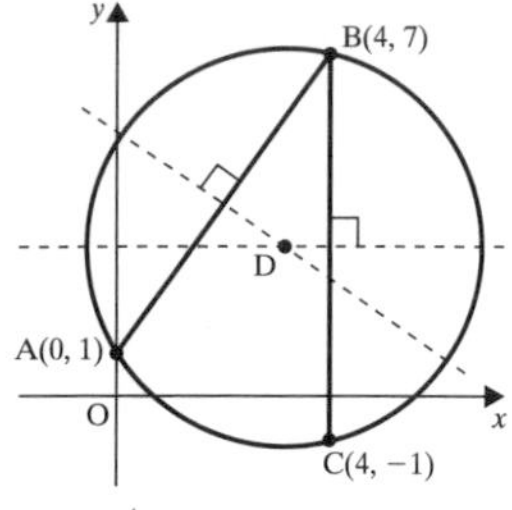

We can use the fact that the centre of a circle lies on the perpendicular bisector of a chord.

The midpoint of AB is the point (2, 4) and the gradient of AB is $\frac{3}{2}$.

$\therefore$ the perpendicular bisector of AB is the line

$$2x + 3y = 16 \qquad [1]$$

The midpoint of BC is the point (4, 3) and BC is vertical,

$\therefore$ the perpendicular bisector of BC is horizontal and its equation is

$$y = 3 \qquad [2]$$

Solving equations [1] and [2] gives $x = \frac{7}{2}$ and $y = 3$.

$\therefore$ D is the point $(\frac{7}{2}, 3)$

The radius, r, is the length of DA or DC or DB.

$$r^2 = (3-1)^2 + (\tfrac{7}{2} - 0)^2 = \tfrac{65}{4}$$

Therefore the equation of the circle is

$$(x - \tfrac{7}{2})^2 + (y-3)^2 = \tfrac{65}{4} \quad \Rightarrow \quad x^2 + y^2 - 7x - 6y + 5 = 0$$

EXERCISE 28C

1 Write down the equation of the tangent to the given circle at the given point.

a $x^2 + y^2 - 2x + 4y - 20 = 0$; (5, 1)

b $x^2 + y^2 - 10x - 22y + 129 = 0$; (6, 7)

c $x^2 + y^2 - 8y + 3 = 0$; (−2, 7)

Find the equations of the following circles (in some cases more than one circle is possible).

2 A circle passes through the points (1, 4), (7, 5) and (1, 8).

3 A circle has its centre on the line $x + y = 1$ and passes through the origin and the point (4, 2).

4 The line joining (2, 1) to (6, 5) is a diameter of a circle.

5 A circle with centre (2, 7) passes through the point (−3, −5).

6 A circle intersects the y-axis at the origin and at the point (0, 6) and also touches the x-axis.

7 Find the equation of the tangent at the origin to the circle

$$x^2 + y^2 + 2x + 4y = 0$$

Parameters

We have seen that the relationship between the x- and y-coordinates of a point on a curve can often be expressed more simply in the form of two equations, i.e.

$$\left.\begin{array}{l} x = f(t) \\ y = g(t) \end{array}\right\} \text{ where } t \text{ is a parameter.}$$

The use of parametric equations to plot curves, find gradients and hence tangents and normals to curves at particular points is covered in Chapter 25.

We now look at some particular curves that you should recognise from their parametric equations.

Parametric Equations of a Circle

Consider the curve whose parametric equations are

$$x = a\cos\theta \quad \text{and} \quad t = a\sin\theta$$

Squaring and adding to eliminate θ gives $x^2 + y^2 = a^2(\cos^2\theta + \sin^2\theta) = a^2$

Hence the Cartesian equation of this curve is $x^2 + y^2 = a^2$ which we recognise as a circle, centre the origin, and radius a.

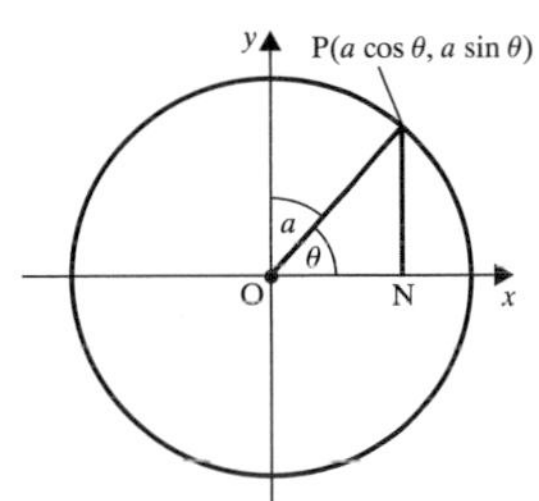

In general, a parameter has no geometric significance, but in the case of the circle, it does. Since P($a\cos\theta, a\sin\theta$) is any point on this circle, we can see from the diagram that

$$\text{ON} = a\cos\theta \quad \text{and} \quad \text{PN} = a\sin\theta,$$

i.e. θ is the angle between the radius and the positive direction of the x-axis.

The curve whose parametric equations are $x = a\cos\theta$ and $y = b\sin\theta$ is a circle, centre the origin and radius a.

The Ellipse

Consider the curve whose parametric equations are $x = a\cos\theta$ and $y = b\sin\theta$.

Eliminating θ gives $\dfrac{x^2}{a^2} + \dfrac{y^2}{b^2} = \cos^2\theta + \sin^2\theta = 1$

so the Cartesian equation of this curve is $\dfrac{x^2}{a^2} + \dfrac{y^2}{b^2} = 1.$

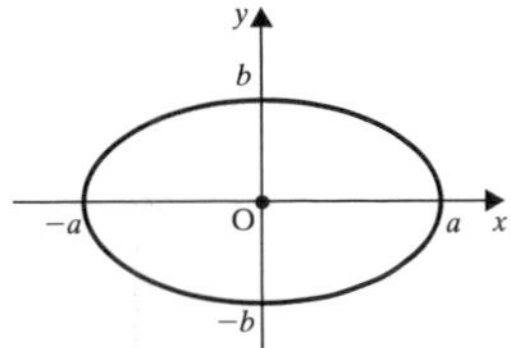

Sketching this curve shows that it is an ellipse which cuts the x-axis at $(a, 0)$ and $(-a, 0)$ and cuts the y-axis at $(b, 0)$ and $(-b, 0)$.

The parametric equations of this ellipse are $x = a\cos\theta$ and $y = b\sin\theta$ and the Cartesian equation is $\dfrac{x^2}{a^2} + \dfrac{y^2}{b^2} = 1$.

The Parabola

Consider the curve whose parametric equations are $x = at^2$ and $y = 2at$.

Eliminating t gives the Cartesian equation of the curve as $y^2 = 4ax$.

Sketching this curve shows that it is a parabola that goes through the origin and whose axis of symmetry is the x-axis.

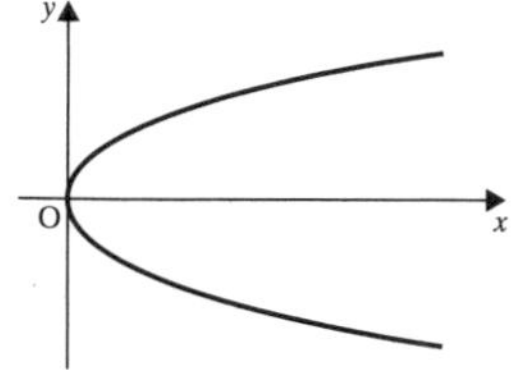

We would expect the curve $y^2 = 4ax$ to be this parabola because if we exchange x and y in the equation we get $y = \frac{1}{4a}x^2$ which we know is a parabola through O whose line of symmetry is the y-axis.

A parabola through O, with the x-axis as its line of symmetry, has parametric equations $x = at^2$, $y = 2at$ and its Cartesian equation is $y^2 = 4ax$.

EXAMPLES 28D

1 The parametric equations of a curve are $x = 8\cos t$, $y = 8\sin t$.
Find the equation of the tangent to the curve at the point where $t = \frac{\pi}{3}$.

The equations $x = 8\cos t$ and $y = 8\sin t$ are those of a circle, centre O and radius 8. Knowing what t represents, we can find the equation of the tangent using the geometric properties of the circle.

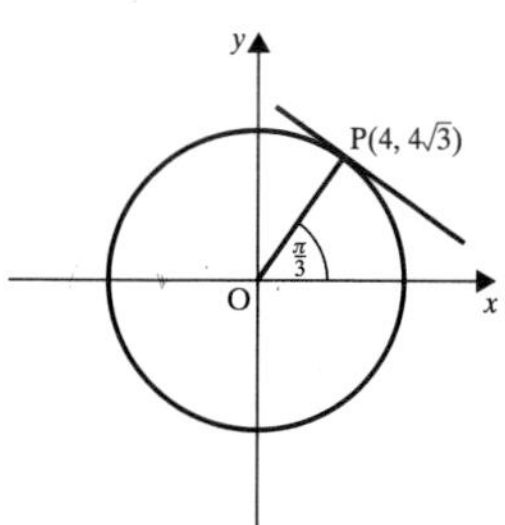

The curve is a circle, centre O and radius 8.

The coordinates of P, where $t = \frac{\pi}{3}$, are $\left(8\cos\frac{\pi}{3}, 8\sin\frac{\pi}{3}\right)$, i.e. $(4, 4\sqrt{3})$

The tangent at P is perpendicular to the radius OP.

The gradient of $\text{OP} = \tan\frac{\pi}{3} = \sqrt{3}$

$\therefore$ the gradient of the tangent at P is $-\frac{1}{\sqrt{3}}$

and its equation is $y - 4\sqrt{3} = -\frac{1}{\sqrt{3}}(x - 4) \quad \Rightarrow \quad x + y\sqrt{3} - 16 = 0$

2 Sketch the curve whose parametric equations are $x = 2k\cos\theta$ and $y = 5k\sin\theta$ where k is a positive constant.

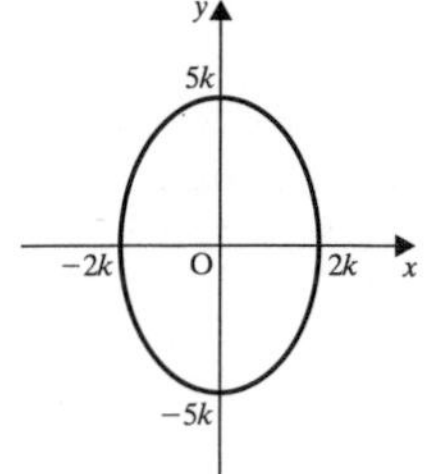

Comparing $x = 2k\cos\theta$ and $y = 5k\sin\theta$ with $x = a\cos\theta$ and $y = b\sin\theta$, we can see that the curve is an ellipse that cuts the x-axis at $2k$ and at $-2k$ and cuts the y-axis at $5k$ and at $-5k$.

Note that, for an ellipse or a parabola, the calculus methods given in Chapter 25 are needed for any problem involving the gradient.

EXERCISE 28D

In Questions 1 to 5, sketch the curve showing clearly where the curve cuts the axes.

1 $x = 3\cos\theta$
$y = 3\sin\theta$

2 $x = 4\cos t$
$y = 2\sin t$

3 $x = 2a\cos\theta$
$y = 2a\sin\theta$

4 $x = 3a\cos\theta$
$y = a\sin\theta$

5 $x = 2t^2$
$y = 4t$

6 The parametric equations of a curve are $x = \cos\theta$ and $y = 1 + \sin\theta$.

a Find the Cartesian equation of the curve and hence sketch the curve.

b Find the equation of the tangent at the point on the curve where $\theta = \dfrac{\pi}{4}$.

7 a Sketch the curve whose parametric equations are $x = 5\cos\theta$ and $y = 3\sin\theta$.

b Write down the equation of the tangent to the curve at the point where $\theta = 0$.

8 The parametric equations of a curve are $x = 2 + 2\cos\theta$ and $y = 1 + 2\sin\theta$.

a Find the Cartesian equation of the curve and sketch the curve.

b Find the values of θ at the points where the line $x = 2$ cuts the curve.

9 The parametric equations of a curve are $x = 2a\cos\theta$ and $y = 3a\sin\theta$.

a Sketch the curve.

b Find the equation of the line through the points where $\theta = 0$ and $\theta = \dfrac{\pi}{3}$.

Further Work on Parametric Equations

We now look at other ways in which parameters can be applied.

EXAMPLES 28E

1 Find the coordinates of the points where the line $y = 3x - 1$ cuts the curve whose parametric equations are $x = t$, $y = 2t^2$.

> The coordinates of any point on the curve are $(t, 2t^2)$.
>
> The line cuts the curve where the coordinates of a point on the curve satisfy the equation of the line, i.e. where $x = t$ and $y = 2t^2$ satisfies $y = 3x - 1$.

The line cuts the curve where

$$2t^2 = 3t - 1$$

$$\Rightarrow \quad 2t^2 - 3t + 1 = 0 \quad \Rightarrow \quad (t-1)(2t-1) = 0$$

> This is a quadratic equation in t, giving two values of t and therefore giving two distinct points on the curve.

So $t = 1$ or $\frac{1}{2}$

Therefore the points of intersection are $(1, 2)$ and $(\frac{1}{2}, \frac{1}{2})$.

2 The parametric equations of a curve are $x = \cos\theta$ and $y = \sin\theta$. Find the equation of the tangent to the curve at the point P $(\cos\alpha, \sin\alpha)$ on the curve.

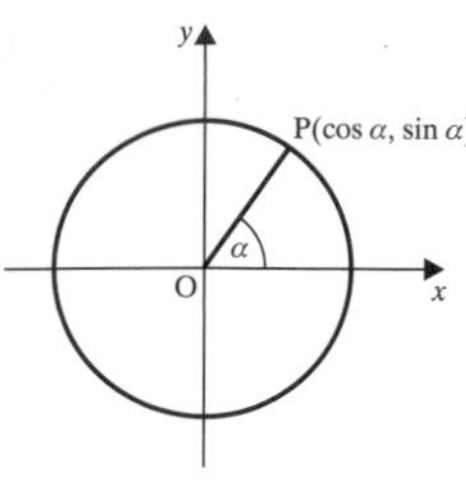

> Because $\left.\begin{array}{r} x = \cos\theta \\ y = \sin\theta \end{array}\right\}$ are the parametric equations of a circle, centre O and radius 1, we can find the gradient of the tangent without using calculus.

The gradient of OP is $\tan\alpha$ so the gradient of the tangent at P is $-\cot\alpha$.

> Use $y - y_1 = m(x - x_1)$ to find the equation of the tangent, with $y_1 = \sin\alpha$, $x_1 = \cos\alpha$ and $m = \cot\alpha$.

The equation of the tangent at P is $y - \sin\alpha = -\cot\alpha(x - \cos\alpha)$

$\Rightarrow \qquad y = -x\cot\alpha + \operatorname{cosec}\alpha$

In the example above α can be *any* of the possible values of θ, so we can replace α by θ in the equation of the tangent giving

$$y = -x\cot\theta + \operatorname{cosec}\theta$$

This is the equation of the tangent at *any* point $(\cos\theta, \sin\theta)$ on the curve. It is a *general equation* for a tangent to the curve because, by taking a particular value of θ, we can use it to find the equation of the tangent at the point on the curve corresponding to that value of θ. It can be found directly without using α as the parameter first.

3 The parametric equations of a curve are $x = 2t^2$ and $y = t$. The point A on the curve has parameter p and another point B on the curve has parameter q.

a Find, in terms of p and q, the gradient of the chord AB.

b Given that the gradient of AB is 0.5, show that $p + q = 1$.

c Hence find, in terms of p only, the coordinates of the midpoint of AB when the gradient of AB is 0.5.

a Since A is on the given curve, the coordinates of A are $(2p^2, p)$.
Similarly, the coordinates of B are $(2q^2, q)$.

$\therefore$ gradient of AB is $\dfrac{p - q}{2(p^2 - q^2)} = \dfrac{1}{2(p + q)}$

b As the gradient of AB is 0.5, $\dfrac{1}{2(p + q)} = 0.5$

$\Rightarrow \qquad 5p + q = 1$

c The midpoint of AB is the point where

$$x = \tfrac{1}{2}(2p^2 + 2q^2) = p^2 + q^2 \text{ and } y = \tfrac{1}{2}(p + q)$$

When the gradient of AB is 0.5, $p + q = 1$,

$\Rightarrow \qquad x = p^2 + (1 - p)^2$ and $y = \tfrac{1}{2}$

i.e. the midpoint is the point $(2p^2 - 2p + 1, \tfrac{1}{2})$

It is interesting to note that AB is any of the chords of the given curve that have a gradient 0.5. Therefore as the position of AB varies, the midpoint also varies but is always subject to the condition that $x = p^2 + (1-p)^2$ and $y = \frac{1}{2}$.

Hence these two equations are the parametric equations of the locus of the midpoint of AB.

In this case, y is constant, so this midpoint always lies on the straight line $y = \frac{1}{2}$.

EXERCISE 28E

1 Find the equation of the normal to the curve $x = t,\ y = \dfrac{1}{t}$ at the point on the curve with parameter T.

2 Find the equation of the tangent to the curve $x = t^2,\ y = 4t$ at the point on the curve where the parameter is p.

3 Find the equation of the tangent to the curve $x = \cos\theta, y = 2\sin\theta$ at the point where $\theta = \alpha$.

4 The coordinates of any point on a curve are $\left(2s^2, \dfrac{1}{s}\right)$ where s can take any value except zero. Find the gradient of the chord joining the points on the curve with parameters s_1 and s_2.

5 Find the coordinates of the points where the line $y = x - 1$ cuts the curve whose parametric equations are $x = p,\ y = p^2 - 1$.

6 **a** Find the equation of the normal at the point $\left(2s, \dfrac{2}{s}\right)$ to the curve whose parametric equations are $x = 2s,\ y = \dfrac{2}{s}$.

b Find, in terms of s, the coordinates of the point where this normal cuts the curve again.

7 The parametric equations of a curve are $x = t$ and $y = 1/t$. Find the general equation of the tangent to this curve (i.e. the equation of the tangent at the point $(t, 1/t)$). Find, in terms of t, the coordinates of the points at which the tangent cuts the x and y axes. Hence show that the area enclosed by this tangent and the coordinate axes is constant.

8 A curve has parametric equations $x = t^2,\ y = 4t$. Find the equation of the normal to this curve at the point $(t^2, 4t)$. Find the coordinates of the points where this normal cuts the coordinate axes. Hence find, in terms of t, the area of the triangle enclosed by the normal and the axes.

9 A curve is given by the parametric equations $x = t,\ y = \dfrac{1}{t}$.

a Find the equation of the normal to the curve at the point $\left(a, \dfrac{1}{a}\right)$.

b If this normal cuts the curve again at the point with parameter b, find a relationship between a and b.

10 A curve has parametric equations $x = 2t^2$, $y = 4t$. Find

a the Cartesian equation of the curve

b the equation of the tangent at the point where $y = 8$

c the equation of the chord joining the points on the curve where $t = p$ and $t = q$

d the coordinates of the points where $y = x - 6$ cuts the curve

e the value of k for which $y = x + k$ is a tangent to the curve

f the coordinates of the point(s) of intersection of the curve and the circle $x^2 + y^2 - 2x = 16$

g the coordinates of the point(s) of intersection of the curve and the curve given parametrically by $x = 8s$, $y = \dfrac{8}{s}$.

11 The parametric equations of a curve are $x = \cos\theta$ and $y = 1 + \sin\theta$.

a Find the gradient of the chord joining the points on the curve with parameters θ_1 and θ_2.

b Find the coordinates of the midpoint of the chord.

c If the chord is parallel to the x-axis, find a relationship between θ_1 and θ_2.

12 The parametric equations of a curve are

$$x = a\cos\theta, \; y = b\sin\theta.$$

a Show that the Cartesian equation of the curve is

$$\frac{x^2}{a^2} + \frac{y^2}{b^2} = 1$$

b Find in the range $-\pi \leqslant \theta \leqslant \pi$

i the values of θ for which $x = 0$

ii the values of θ for which $y = 0$

c State the range of possible values of **i** x **ii** y

d Sketch the curve when $a = 3$ and $b = 2$.

e What is the shape of the curve when $b = a$?

CHAPTER 29

Integration 2

Standard Integrals

Whenever a function $f'(x)$ is *recognised* as the derivative of a function $f(x)$ then

$$\frac{d}{dx}f(x) = f'(x) \quad \Rightarrow \quad \int f'(x)dx = f(x) + K$$

Thus any function whose derivative is known can be established as a standard integral.

Integrating $(ax + b)^n$

First consider the function $f(x) = (2x+3)^4$.

To differentiate $f(x)$ we make a substitution

i.e. $\quad u = 2x+3 \quad \Rightarrow \quad f(x) = u^4$

giving $\quad \dfrac{d}{dx}(2x+3)^4 = (4)(2)(2x+3)^3$

Hence $\quad \displaystyle\int (4)(2)(2x+3)^3 dx = (2x+3)^4 + K$

or $\quad \displaystyle\int (2x+3)^3 dx = \frac{1}{(2)(4)}(2x+3)^4 + K$

Considering $f(x) = (ax+b)^{n+1}$ in a similar way gives the general result

$$\boldsymbol{\int (ax+b)^n dx = \frac{1}{(a)(n+1)}(ax+b)^{n+1} + K}$$

Integrating Exponential Functions

It is already known that $\dfrac{d}{dx}e^x = e^x$

hence $\displaystyle\boldsymbol{\int e^x dx = e^x + K}$

Further, we have $\dfrac{d}{dx}(ce^x) = ce^x$

and $\quad \dfrac{d}{dx}e^{(ax+b)} = ae^{(ax+b)}$

Hence $\displaystyle\boldsymbol{\int ce^x dx = ce^x + K}$ and $\displaystyle\boldsymbol{\int e^{(ax+b)} = \frac{1}{a}e^{(ax+b)} + K}$

e.g. $\displaystyle\int 2e^x dx = 2e^x + K$ and $\displaystyle\int 4e^{(1-3x)} dx = (4)(-\tfrac{1}{3})e^{(1-3x)} + K$

To integrate an exponential function where the given base is not e but is some other constant, a say, the base must first be changed to e as follows.

Using $a^x = e^z$ and taking logs to the base e we have

$$x \ln a = z$$

$$\text{Hence } a^x = e^{x\ln a} \quad \Rightarrow \quad \int a^x\,dx = \int e^{x\ln a}\,dx$$

$$\Rightarrow \quad \int a^x\,dx = \frac{1}{\ln a}e^{x\ln a} + K$$

i.e. $$\boldsymbol{\int a^x dx = \frac{1}{\ln a}a^x + K}$$

This result also follows directly from

$$\frac{d}{dx}(a^x) = (\ln a)a^x$$

EXAMPLE 29A Write down the integral of e^{3x} w.r.t. x and hence evaluate $\int_0^1 e^{3x}dx$.

$$\int e^{3x}\,dx = \tfrac{1}{3}e^{3x} + K$$

The constant of integration disappears when a definite integral is calculated, hence

$$\int_0^1 e^{3x}dx = \left[\tfrac{1}{3}e^{3x}\right]_0^1 = \tfrac{1}{3}e^3 - \tfrac{1}{3}e^0$$

i.e. $$\int_0^1 e^{3x}dx = \tfrac{1}{3}(e^3 - 1)$$

EXERCISE 29A Integrate each function w.r.t. x.

1 e^{4x}

2 $4e^{-x}$

3 $e^{(3x-2)}$

4 $2e^{(1-5x)}$

5 $6e^{-2x}$

6 $5e^{(x-3)}$

7 $e^{(2+x/2)}$

8 2^x

9 $4^{(2+x)}$

10 $e^{2x} + \dfrac{1}{e^{2x}}$

11 $a^{(1-2x)}$

12 $2^x + x^2$

13 $(x+3)^{-2}$

14 $(1+x)^{\frac{1}{2}}$

15 $(1+3x)^5$

16 $(2-5x)^4$

Evaluate the following definite integrals.

17 $\int_0^2 e^{2x}\,dx$

18 $\int_{-1}^1 2e^{(x+1)}\,dx$

19 $\int_2^3 e^{(2-x)}\,dx$

20 $\int_0^2 -e^x\,dx$

Functions whose Integrals are Logarithmic

We know that $\int \frac{1}{x}\,dx = \ln|x| + K$

We also know that $\frac{d}{dx}\ln(ax+b) = \frac{a}{ax+b}$

$$\therefore \quad \int \frac{1}{ax+b}\,dx = \frac{1}{a}\ln|ax+b| + K = \frac{1}{a}\ln A|ax+b|$$

e.g.
$$\int \frac{1}{2x+5}\,dx = \tfrac{1}{2}\ln|2x+5| + K$$
$$= \tfrac{1}{2}\ln A|2x+5|$$

and
$$\int \frac{1}{4-3x}\,dx = -\tfrac{1}{3}\ln|4-3x| + K$$
$$= -\tfrac{1}{3}\ln A|4-3x|$$
$$= \tfrac{1}{3}\ln\frac{A}{|4-3x|}$$

EXERCISE 29B

Integrate w.r.t. x giving each answer in a form which

a uses K

b uses $\ln A$ and is simplified.

1 $\frac{1}{2x}$

2 $\frac{4}{x-1}$

3 $\frac{1}{3x+1}$

4 $\frac{3}{1-2x}$

5 $\frac{6}{2+3x}$

6 $\frac{3}{4-2x}$

7 $\frac{4}{1-x}$

8 $\frac{5}{6-7x}$

Evaluate

9 $\int_1^2 \frac{3}{x+1}\,dx$

10 $\int_1^2 \frac{1}{2x-1}\,dx$

11 $\int_4^5 \frac{2}{x-3}\,dx$

12 $\int_0^1 \frac{1}{2-x}\,dx$

Integrating Trigonometric Functions

Knowing the derivatives of the six trig functions, we can recognise the following integrals.

$$\frac{d}{dx}(\sin x) = \cos x \iff \int \cos x \, dx = \sin x + K$$

$$\frac{d}{dx}(\cos x) = -\sin x \iff \int \sin x \, dx = -\cos x + K$$

$$\frac{d}{dx}(\tan x) = \sec^2 x \iff \int \sec^2 x \, dx = \tan x + K$$

$$\frac{d}{dx}(\sec x) = \sec x \tan x \iff \int \sec x \tan x \, dx = \sec x + K$$

$$\frac{d}{dx}(\operatorname{cosec} x) = -\operatorname{cosec} x \cot x \iff \int \operatorname{cosec} x \cot x \, dx = -\operatorname{cosec} x + K$$

$$\frac{d}{dx}(\cot x) = -\operatorname{cosec}^2 x \iff \int \operatorname{cosec}^2 x \, dx = -\cot x + K$$

Remembering the derivatives of some variations of the basic trig functions we also have

$$\int c \cos x \, dx = c \sin x + K$$

and $\displaystyle\int \cos(ax+b)\, dx = \frac{1}{a}\sin(ax+b) + K$

with similar results for the remaining trig integrals

e.g. $$\int 3\sec^2 x \, dx = 3\tan x + K$$

$$\int \sin 4\theta \, d\theta = -\tfrac{1}{4}\cos 4\theta + K$$

$$\int \operatorname{cosec}^2(2x + \tfrac{3}{4}\pi)\, dx = -\tfrac{1}{2}\cot(2x + \tfrac{3}{4}\pi) + K$$

$$\int \operatorname{cosec} 5\theta \cot 5\theta \, d\theta = -\tfrac{1}{5}\operatorname{cosec} 5\theta + K$$

$$\int \sec(\tfrac{1}{2}\pi - 6x)\tan(\tfrac{1}{2}\pi - 6x)\, dx = -\tfrac{1}{6}\sec(\tfrac{1}{2}\pi - 6x) + K$$

Note that there is no need to *learn* these standard integrals. Knowledge of the standard derivatives is sufficient.

EXERCISE 29C

Integrate each function w.r.t. x.

1 $\sin 2x$

2 $\cos 7x$

3 $\sec^2 4x$

4 $\sin(\frac{1}{4}\pi + x)$

5 $3\cos(4x - \frac{1}{2}\pi)$

6 $\sec^2(\frac{1}{3}\pi + 2x)$

7 $\operatorname{cosec}^2 4x$

8 $2\sin(3x - \alpha)$

9 $5\cos(\alpha - \frac{1}{2}x)$

10 $5\sec 4x \tan 4x$

11 $\cos 3x - \cos x$

12 $\sec^2 2x - \operatorname{cosec}^2 4x$

Evaluate

13 $\displaystyle\int_0^{\pi/6} \sin 3x\,dx$

14 $\displaystyle\int_{\pi/4}^{\pi/6} \cos(2x - \tfrac{1}{2}\pi)\,dx$

15 $\displaystyle\int_0^{\pi/2} 2\sin(2x - \tfrac{1}{2}\pi)\,dx$

16 $\displaystyle\int_0^{\pi/8} \sec^2 2x\,dx$

EXERCISE 29D

This exercise contains a variety of functions, including those dealt with in Chapter 20.

Check that an integral is correct by differentiating it mentally.

Integrate w.r.t. x

1 $\sin(\frac{1}{2}\pi - 2x)$

2 $e^{(4x-1)}$

3 $\sec^2 7x$

4 $\dfrac{1}{2x-3}$

5 $\dfrac{1}{\sqrt{2x-3}}$

6 $\dfrac{1}{(3x-2)^2}$

7 5^x

8 $\operatorname{cosec}\frac{1}{2}x \cot\frac{1}{2}x$

9 $(3x-5)^2$

10 $e^{(4x-5)}$

11 $\sqrt{4x-5}$

12 $\operatorname{cosec}^2 3x$

13 $\dfrac{3}{2(1-x)}$

14 $10^{(x+1)}$

15 $\cos(3x - \frac{1}{3}\pi)$

16 $(x-2)(2x+4)$

17 $x(x-3)^2$

18 $3 - 2x(1-3x)$

19 $\dfrac{2}{3x^3}$

20 $\dfrac{3}{2x}$

21 $\frac{2}{3}(x-3)^2$

Evaluate

22 $\displaystyle\int_{-1/2}^{1/2} \sqrt{1-2x}\,dx$

23 $\displaystyle\int_0^{2} e^{(x/2+1)}\,dx$

24 $\displaystyle\int_{\pi/4}^{\pi/2} \sin 4x\,dx$

The Recognition Aspect of Integration

We have already seen the importance of the recognition aspect of integration in compiling a set of standard integrals.
Recognition is equally important when it is used to avoid serious errors in integration.

Consider, for instance, the derivative of the product $x^2 \sin x$.

Using the product formula gives $\dfrac{d}{dx}(x^2 \sin x) = 2x\sin x + x^2\cos x$

Clearly the derivative is not a simple product, therefore

the integral of a product is not itself a product.

i.e. $\displaystyle\int uv\,dx$ is NOT $\left(\displaystyle\int u\,dx\right)\left(\displaystyle\int v\,dx\right)$.

On the other hand, differentiating the function of a function $(1+x^2)^3$ gives

$$\frac{d}{dx}(1+x^2)^3 = 6x(1+x^2)^2$$

This time the derivative *is* a product so clearly the integral of a product *may* be a function of a function.

Integrating Products

First consider the function e^u where u is a function of x.

Differentiating as a function of a function gives

$$\frac{d}{dx}(e^u) = \left(\frac{du}{dx}\right)(e^u)$$

Thus any product of the form $\left(\frac{du}{dx}\right)e^u$ can be integrated by recognition, since

$$\boldsymbol{\int\left(\frac{du}{dx}\right)e^u\,dx = e^u + K}$$

e.g.

$$\int 2xe^{x^2}\,dx = e^{x^2} + K \quad (u = x^2)$$

$$\int \cos x\, e^{\sin x}\,dx = e^{\sin x} + K \quad (u = \sin x)$$

$$\int x^2 e^{x^3}\,dx = \tfrac{1}{3}\int 3x^2 e^{x^3}\,dx = \tfrac{1}{3}e^{x^3} + K \quad (u = x^3)$$

In these simple cases the substitution of u for $f(x)$ can be done mentally. All the results can be checked by differentiating them mentally.

Similar, but slightly less simple functions, can also be integrated by changing the variable but for these the substitution is written down.

Changing the Variable

Consider a general function $g(u)$ where u is a function of x.

$$\frac{d}{dx}g(u) = \frac{du}{dx}g'(u) \quad \text{or} \quad g'(u)\frac{du}{dx}$$

Therefore $$\int g'(u)\frac{du}{dx}\,dx = g(u) + K \qquad [1]$$

We also know that $$\int g'(u)\,du = g(u) + K \qquad [2]$$

Comparing [1] and [2] gives

$$\int g'(u)\frac{du}{dx}\,dx = \int g'(u)\,du$$

Replacing $g'(u)$ by $f(u)$ gives

$$\int f(u)\frac{du}{dx}\,dx = \int f(u)\,du$$

i.e. $$\boldsymbol{\ldots\frac{du}{dx}\,dx \equiv \ldots du} \qquad [3]$$

Thus integrating (a function of u)$\frac{du}{dx}$ w.r.t. x, is *equivalent* to integrating (the same function of u) w.r.t. u

i.e. the relationship in [3] is neither an equation nor an identity but is a pair of equivalent operations.

Suppose, for example, that we want to find $\int 2x(x^2+1)^5\,dx$.

Writing the integral in the form $\int (x^2+1)^5 2x\,dx$ and making the substitution $u = x^2+1$ gives

$$\int (x^2+1)^5 2x\,dx = \int u^5(2x)\,dx$$

But $\quad \dfrac{du}{dx} = 2x$ and as $\ldots\dfrac{du}{dx}dx \equiv \ldots du$

we have $\quad \ldots 2x\,dx \equiv \ldots du$

i.e. $$\int (x^2+1)^5 2x\,dx = \int u^5\,du$$
$$= \tfrac{1}{6}u^6 + K = \tfrac{1}{6}(x^2+1)^6 + K$$

In practice we can go direct from $\dfrac{du}{dx} = 2x$ to the equivalent operators $\ldots 2x\,dx \equiv \ldots du$ by 'separating the variables'.

Products which can be integrated by this method are those in which one factor is basically the derivative of the function in the other factor, and we use u for this function.

EXAMPLES 29E

1 Integrate $x^2\sqrt{x^3+5}$ w.r.t. x.

> In this product x^2 is basically the derivative of x^3+5 so we choose the substitution $u = x^3+5$.

If $u = x^3+5$ then $\dfrac{du}{dx} = 3x^2$

$\Rightarrow \quad \ldots du \equiv \ldots 3x^2\,dx$

Hence $$\int x^2\sqrt{x^3+5}\,dx = \tfrac{1}{3}\int (x^3+5)^{1/2}(3x^2\,dx) = \tfrac{1}{3}\int u^{1/2}\,du$$
$$= (\tfrac{1}{3})(\tfrac{2}{3})u^{3/2} + K = \tfrac{2}{9}(x^3+5)^{3/2} + K$$

2 Find $\int \cos x \sin^3 x\,dx$.

> Writing the given integral in the form $\cos x(\sin x)^3$ suggests substituting $u = \sin x$.

If $u = \sin x$ then $\ldots du \equiv \ldots \cos x\,dx$

$\therefore$ $$\int \cos x \sin^3 x\,dx = \int (\sin x)^3 \cos x\,dx = \int u^3\,du$$
$$= \tfrac{1}{4}u^4 + K = \tfrac{1}{4}\sin^4 x + K$$

Applied generally, the method used above shows that

$$\int \cos x \sin^n x \, dx = \frac{1}{n+1}\sin^{(n+1)} x + K$$

and similarly that

$$\int \sin x \cos^n x \, dx = \frac{-1}{n+1}\cos^{(n+1)} x + K$$

3 Find $\int \frac{\ln x}{x} dx$.

> Initially this looks like a fraction but once it is recognised as the product of $\frac{1}{x}$ and $\ln x$, it is clear that $\frac{1}{x} = \frac{d}{dx}(\ln x)$ and that we can make the substitution $u = \ln x$.

If $u = \ln x$ then $\ldots du \equiv \ldots \frac{1}{x} dx$

Hence $\int \frac{1}{x} \ln x \, dx = \int u \, du = \frac{1}{2}u^2 + K$

i.e. $\int \frac{\ln x}{x} dx = \frac{1}{2}(\ln x)^2 + K$

Note that $(\ln x)^2$ is *not* the same as $\ln x^2$.

EXERCISE 29E

Integrate the following expressions w.r.t. x.

1 $4x^3 e^{x^4}$

2 $\sin x \, e^{\cos x}$

3 $\sec^2 x \, e^{\tan x}$

4 $(2x+1)e^{(x^2+x)}$

5 $\operatorname{cosec}^2 x \, e^{(1-\cot x)}$

6 $(1+\cos x)e^{(x+\sin x)}$

7 $2x e^{(1+x^2)}$

8 $(3x^2-2)e^{(x^3-2x)}$

Find the following integrals by making the substitution suggested.

9 $\int x(x^2-3)^4 dx$; $u = x^2 - 3$

10 $\int x\sqrt{1-x^2} dx$; $u = 1 - x^2$

11 $\int \cos 2x(\sin 2x + 3)^2 dx$; $u = \sin 2x + 3$

12 $\int x^2(1-x^3) dx$; $u = 1 - x^3$

13 $\int e^x\sqrt{1+e^x}\,dx$; $u = 1+e^x$

14 $\int \cos x \sin^4 x\,dx$; $u = \sin x$

15 $\int \sec^2 x \tan^3 x\,dx$; $u = \tan x$

16 $\int x^n(1+x^{n+1})^2\,dx$; $u = 1+x^{n+1}$

17 $\int \operatorname{cosec}^2 x \cot^2 x\,dx$; $u = \cot x$

18 $\int \sqrt{x}\sqrt{1+x^{3/2}}\,dx$; $u = 1+x^{3/2}$

By using a suitable substitution, or by integrating at sight, find

19 $\int x^3(x^4+4)^2\,dx$

20 $\int e^x(1-e^x)^3\,dx$

21 $\int \sin\theta\sqrt{1-\cos\theta}\,d\theta$

22 $\int (x+1)\sqrt{x^2+2x+3}\,dx$

23 $\int x e^{x^2+1}\,dx$

24 $\int \sec^2 x(1+\tan x)\,dx$

Definite Integration with a Change of Variable

When evaluating a definite integral we first carry out the necessary integration then substitute the limiting values. If the integration requires a change of variable, from x to u, say, it is usually best to change the limits of integration also from x values to u values.

EXAMPLE 29F By using the substitution $u = x^3+1$, evaluate $\int_0^1 x^2\sqrt{x^3+1}\,dx$

If $u = x^3+1$ then $\ldots du \equiv \ldots 3x^2\,dx$

and $\begin{cases} x = 0 & \Rightarrow \quad u = 1 \\ x = 1 & \Rightarrow \quad u = 2 \end{cases}$

Hence $$\int_0^1 x^2\sqrt{x^3+1}\,dx = \tfrac{1}{3}\int_1^2 \sqrt{u}\,du$$
$$= \tfrac{1}{3}\left[\tfrac{2}{3}u^{3/2}\right]_1^2 = \tfrac{2}{9}(2\sqrt{2}-1)$$

EXERCISE 29F Evaluate

1 $\int_0^1 x e^{x^2}\,dx$

2 $\int_0^{\pi/2} \cos x \sin^4 x\,dx$

3 $\int_1^2 \frac{1}{x}\ln x\,dx$

4 $\int_1^2 x^2(x^3-1)^4\,dx$

5 $\int_0^{\pi/4} (\sec^2 x)e^{\tan x}\,dx$

6 $\int_1^2 x(1+2x^2)\,dx$

7 $\int_2^3 (x-1)e^{(x^2-2x)}\,dx$

8 $\int_0^{\pi/6} \cos x(1+\sin^2 x)\,dx$

9 $\int_1^3 \frac{1}{x}(\ln x)^2\,dx$

10 $\int_0^{\sqrt{3}} x\sqrt{1+x^2}\,dx$

Improper Integrals

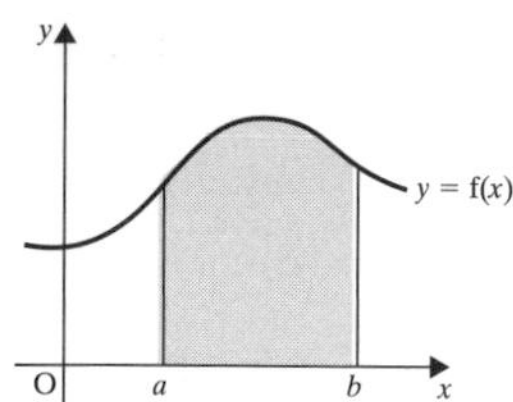

In this diagram the shaded area is bounded by the curve $y = f(x)$, the x-axis and the two lines $x = a$ and $x = b$. Both vertical lines intersect the curve, so the area is clearly defined and, as we know, can be evaluated using $\int_a^b f(x)\,dx$.

However, the area may not always be so clear.

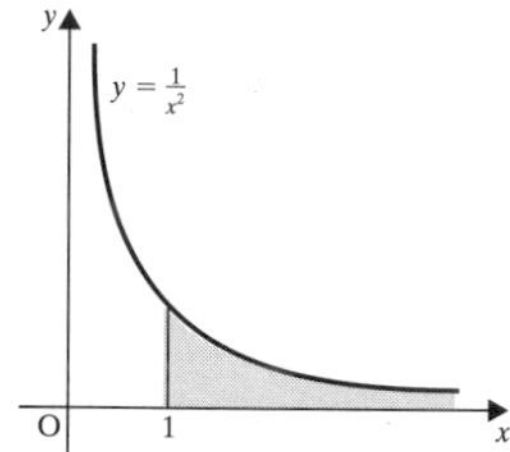

Consider, for example, the area bounded by the curve $y = \dfrac{1}{x^2}$, the x-axis and the line $x = 1$. In this case $a = 1$ and, as the x-axis is an asymptote to the curve, b is infinitely large and therefore indeterminate.

We cannot be sure what happens to the area as $x \to \infty$, so we will first take a large, *finite* value for b, $b = X$ say.

Finding this area by the usual method gives

$$\int_1^X \frac{1}{x^2}\,dx = \left[\frac{-1}{x}\right]_1^X \quad \Rightarrow \quad -\frac{1}{X}+1$$

Now as X approaches infinity, $\dfrac{1}{X}$ approaches 0.

Therefore the value of the area converges to 1, and we say that

$$\int_1^\infty \frac{1}{x^2}\,dx \text{ is convergent and equal to 1.}$$

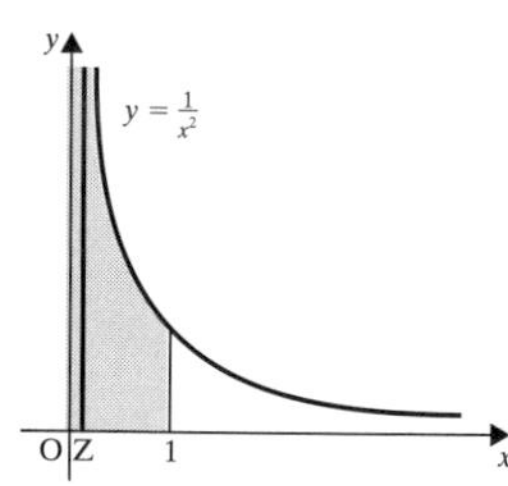

Now consider the area bounded by the same curve $y = \dfrac{1}{x^2}$, the y-axis and the line $x = 1$.

This time the y-axis is an asymptote, and we are not sure what happens to the area as $x \to 0$, so we take a very small value for a, $a = Z$ say.

Integrating as before gives

$$\int_Z^1 \frac{1}{x^2}\,dx = \left[\frac{-1}{x}\right]_Z^1 \quad \Rightarrow \quad -1+\frac{1}{Z}$$

As $Z \to 0$, $\dfrac{1}{Z} \to \infty$, so the value of the area becomes infinitely large.

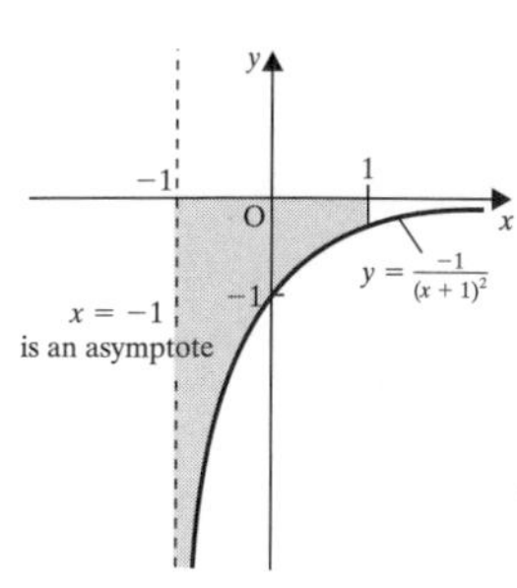

This area diverges so $\displaystyle\int_0^1 \frac{1}{x^2}\,dx$ is indeterminate.

Each of these cases is an example of an *improper integral* and they illustrate that some, but not all, improper integrals can be evaluated. Any definite integral which has one of its limits at a value of x where there is an asymptote, is an improper integral, whether or not the limit is zero or infinity.

An example of this type is $\displaystyle\int_{-1}^1 \frac{-1}{(x+1)^2}\,dx$.

EXERCISE 29G Find the value, if it exists, of each of the following integrals.

1 $\displaystyle\int_1^\infty \frac{1}{x^3}\,dx$

2 $\displaystyle\int_0^1 \frac{1}{x^3}\,dx$

3 $\displaystyle\int_0^2 \frac{1}{\sqrt{x}}\,dx$

4 $\displaystyle\int_2^\infty \frac{1}{\sqrt{x}}\,dx$

5 $\displaystyle\int_1^4 \frac{1}{(1-x)^2}\,dx$

Integration by Parts

It is not always possible to express a product in the form $f(u)\dfrac{du}{dx}$ so another approach is needed.

Looking again at the differentiation of a product uv where u and v are both functions of x we have

$$\frac{d}{dx}(uv) = v\frac{du}{dx} + u\frac{dv}{dx} \quad \Rightarrow \quad v\frac{du}{dx} = \frac{d}{dx}(uv) - u\frac{dv}{dx}$$

Now $v\dfrac{du}{dx}$ can be taken to represent a product which is to be integrated w.r.t. x.

Thus $$\int v\frac{du}{dx}\,dx = \int \frac{d}{dx}(uv)\,dx - \int u\frac{dv}{dx}\,dx$$

i.e $$\boldsymbol{\int v\frac{du}{dx}\,dx = uv - \int u\frac{dv}{dx}\,dx}$$

At this stage it may appear that the RHS is more complicated than the original product on the LHS.

However, by careful choice of the factor to be replaced by v we can ensure that $u\dfrac{dv}{dx}$ is easier to integrate than $v\dfrac{du}{dx}$.

The factor chosen to be replaced by v is usually the one whose derivative is a simpler function. It must also be remembered, however, that the other factor is replaced by $\dfrac{du}{dx}$ and therefore it must be possible to integrate it.

This method for integrating a product is called *integrating by parts*.

EXAMPLES 29H

1 Integrate xe^x w.r.t. x.

Taking $\quad v = x$ and $\dfrac{du}{dx} = e^x$

gives $\quad \dfrac{dv}{dx} = 1$ and $u = e^x$

Then $$\int v\frac{du}{dx}\,dx = uv - \int u\frac{dv}{dx}\,dx$$

gives $$\int xe^x\,dx = (e^x)(x) - \int (e^x)(1)\,dx$$
$$= xe^x - e^x + K$$

2 Find $\int x^2 \sin x \, dx$.

Taking $\quad v = x^2$ and $\dfrac{du}{dx} = \sin x$

gives $\quad \dfrac{dv}{dx} = 2x$ and $u = -\cos x$

Then $\quad \displaystyle\int v \frac{du}{dx} dx = uv - \int u \frac{dv}{dx} dx$

gives $\quad \displaystyle\int x^2 \sin x \, dx = (-\cos x)(x^2) - \int (-\cos x)(2x) \, dx$

$$= -x^2 \cos x + 2 \int x \cos x \, dx \qquad [1]$$

At this stage the integral on the RHS cannot be found without *repeating* the process of integrating by parts on the term $\int x \cos x \, dx$.

Taking $\quad v = x$ and $\dfrac{du}{dx} = \cos x$

gives $\quad \dfrac{dv}{dx} = 1$ and $u = \sin x$

Then $\quad \displaystyle\int x \cos x \, dx = (\sin x)(x) - \int (\sin x)(1) \, dx = x \sin x + \cos x + K$

Hence equation [1] becomes

$$\int x^2 \sin x \, dx = -x^2 \cos x + 2x \sin x + 2 \cos x + K$$

3 Find $\int x^4 \ln x \, dx$.

Because $\ln x$ can be differentiated but *not integrated*, we must use $v = \ln x$.

Taking $\quad v = \ln x$ and $\dfrac{du}{dx} = x^4$

gives $\quad \dfrac{dv}{dx} = \dfrac{1}{x}$ and $u = \frac{1}{5}x^5$

Integrating by parts then gives

$$\int x^4 \ln x \, dx = (\tfrac{1}{5}x^5)(\ln x) - \int (\tfrac{1}{5}x^5)\left(\frac{1}{x}\right) dx = \frac{1}{5}x^5 \ln x - \frac{1}{5}\int x^4 \, dx$$

$$\Rightarrow \quad \int x^4 \ln x \, dx = \tfrac{1}{5}x^5 \ln x - \tfrac{1}{25}x^5 + K$$

Special Cases of Integration by Parts

An interesting situation arises when an attempt is made to integrate $e^x \cos x$ or $e^x \sin x$.

4 Find $\int e^x \cos x \, dx$.

Taking $v = e^x$ and $\dfrac{du}{dx} = \cos x$

gives $\dfrac{dv}{dx} = e^x$ and $u = \sin x$

Hence $\int e^x \cos x \, dx = e^x \sin x - \int e^x \sin x \, dx$ [1]

> But $\int e^x \sin x \, dx$ is very similar to $\int e^x \cos x \, dx$ so it seems that we have made no progress.
>
> However, if we now apply integration by parts to $\int e^x \sin x \, dx$ an interesting situation emerges.

Taking $v = e^x$ and $\dfrac{du}{dx} = \sin x$

gives $\dfrac{dv}{dx} = e^x$ and $u = -\cos x$

so that $\int e^x \sin x \, dx = -e^x \cos x + \int e^x \cos x \, dx$

or $\int e^x \cos x \, dx = e^x \cos x + \int e^x \sin x \, dx$ [2]

Adding [1] and [2] gives

$$2\int e^x \cos x \, dx = e^x(\sin x + \cos x) + K$$

Clearly the same two equations can be used to give

$$2\int e^x \sin x \, dx = e^x(\sin x - \cos x) + K$$

Note that neither of the equations [1] and [2] contains a completed integration process, so the constant of integration is introduced only when these two equations have been combined.

Note also that the same choice of function for v must be made in both applications of integration by parts, i.e. we chose $v = e^x$ each time.

Integration of ln x

So far we have found no way to integrate $\ln x$. Now, however, if $\ln x$ is regarded as the product of 1 and $\ln x$ we can apply integration by parts as follows.

EXAMPLES 29H (continued)

5 Find $\int \ln x \, dx$.

Taking $\quad v = \ln x$ and $\dfrac{du}{dx} = 1$

gives $\quad \dfrac{dv}{dx} = \dfrac{1}{x}$ and $u = x$

Then $\quad \int v\dfrac{du}{dx}dx = uv - \int u\dfrac{dv}{dx}dx$

becomes $\quad \int \ln x \, dx = x\ln x - \int x\left(\dfrac{1}{x}\right)dx = x\ln x - x + K$

i.e. $\quad \int \ln x \, dx = x(\ln x - 1) + K$

EXERCISE 29H

Integrate the following functions w.r.t. x.

1 $x\cos x$

2 $x^2 e^x$

3 $x^3 \ln 3x$

4 xe^{-x}

5 $3x\sin x$

6 $e^x \sin 2x$

7 $e^{2x}\cos x$

8 $x^2 e^{4x}$

9 $e^{-x}\sin x$

10 $\ln 2x$

11 $e^x(x+1)$

12 $x(1+x)^7$

13 $x\sin\left(x + \frac{1}{6}\pi\right)$

14 $x\cos nx$

15 $x^n \ln x$

16 $3x\cos 2x$

17 $2e^x \sin x \cos x$

18 $x^2 \sin x$

19 $e^{ax}\sin bx$

20 By writing $\cos^3\theta$ as $(\cos^2\theta)(\cos\theta)$ use integration by parts to find $\int \cos^3\theta \, d\theta$.

Each of the following products can be integrated either:

a by immediate recognition, or

b by a suitable change of variable, or

c by parts.

Choose the best method in each case and hence integrate each function.

21 $(x-1)e^{x^2-2x+4}$

22 $(x+1)^2 e^x$

23 $\sin x(4+\cos x)^3$

24 $\cos x \, e^{\sin x}$

25 $x^4\sqrt{1+x^5}$

26 $e^x(e^x+2)^4$

27 $(x-1)e^{2x-1}$

28 $x(1-x^2)^9$

29 $\cos x \sin^5 x$

Definite Integration by Parts

When using the formula

$$\int v\frac{du}{dx}dx = uv - \int u\frac{dv}{dx}dx$$

remember that the term uv on the RHS is fully integrated. Consequently in a definite integration, uv must be *evaluated between the appropriate boundaries*

i.e. $$\int_a^b v\frac{du}{dx}dx = \Big[uv\Big]_a^b - \int_a^b u\frac{dv}{dx}dx$$

EXAMPLE 29I Evaluate $\int_0^1 xe^x\,dx$

$$\int xe^x\,dx = \int v\frac{du}{dx}dx$$

where $$v = x \text{ and } \frac{du}{dx} = e^x$$

Hence $$\int_0^1 xe^x\,dx = \Big[xe^x\Big]_0^1 - \int_0^1 e^x\,dx$$
$$= \Big[xe^x\Big]_0^1 - \Big[e^x\Big]_0^1$$
$$= (e^1 - 0) - (e^1 - e^0)$$
$$= e - e + 1$$

i.e. $$\int_0^1 xe^x\,dx = 1$$

EXERCISE 29I Evaluate

1 $\int_0^{\pi/2} x\sin x\,dx$

2 $\int_1^2 x^5\ln x\,dx$

3 $\int_0^1 (x+1)e^x\,dx$

4 $\int_0^{\pi} e^x\cos x\,dx$

5 $\int_1^2 x\sqrt{x-1}\,dx$

6 $\int_0^{\pi/2} x^2\cos x\,dx$

7 $\int_0^1 \ln(1+x)\,dx$

8 $\int_0^1 x^2e^x\,dx$

9 $\int_0^{\frac{\pi}{4}} x\sin 2x\,dx$

MIXED EXERCISE 29 Integrate the following functions, taking care to choose the best method in each case.

1 x^2e^{2x}

2 $2xe^{x^2}$

3 $\sec^2x(3\tan x - 4)$

4 $(x+1)\ln(x+1)$

5 $\sec^2x\tan^3x$

6 $x^2\cos x$

7 $(\sin x)e^{\cos x}$

8 $x(2x+3)^7$

9 $(1-x)e^{(1-x)^2}$

10 $xe^{(2-3x)}$

11 $\cos x \sin^4 x$

12 $\sin x(4+\cos x)^3$

13 $(x-1)e^{(x^2-2x+3)}$

14 $x^2(1-x^3)^9$

Evaluate each definite integral.

15 $\int_1^3 e^{3x}\,dx$

16 $\int_0^{\frac{\pi}{8}} \cos 4x\,dx$

17 $\int_0^1 \frac{1}{x-2}\,dx$

18 $\int_{-4}^{-2} \operatorname{cosec}^2 x\,dx$

19 $\int_0^1 x^2 e^{3x^3}\,dx$

20 $\int_0^{-4} x\cos 2x\,dx$

21 $\int_0^1 \frac{1}{2-x}\ln(2-x)\,dx$

22 $\int_1^2 x^2 \ln x\,dx$

Find the value, if it exists, of each of the following integrals.

23 $\int_0^{\pi/2} \tan x\,dx$

24 $\int_1^{\infty} e^{-x}\,dx$

25 $\int_0^1 x\ln x\,dx$

26 $\int_1^{\infty} \frac{1}{1+x^2}\,dx$

CHAPTER 30

Algebra 2

Partial Fractions

In Chapter 1 you were asked to express two separate fractions as a single fraction with a common denominator. Now we are going to reverse this process, i.e. we will take an expression such as $\dfrac{x-2}{(x+3)(x-4)}$ and express it as the sum of two separate fractions.

This process is called splitting up, or decomposing, into *partial fractions*.

Consider $\dfrac{x-2}{(x+3)(x-4)}$.

This fraction is a *proper fraction* because the highest power of x in the numerator (1 in this case) is less than the highest power of x in the denominator (2 in this case when the brackets are expanded).

Therefore its separate (or partial) fractions also will be proper,

i.e. $\dfrac{x-2}{(x+3)(x-4)}$ can be expressed as $\dfrac{A}{x+3}+\dfrac{B}{x-4}$

where A and B are numbers. The worked example which follows shows how the values of A and B can be found.

EXAMPLE 30A

Express $\dfrac{x-2}{(x+3)(x-4)}$ in partial fractions

$$\frac{x-2}{(x+3)(x-4)} \equiv \frac{A}{x+3}+\frac{B}{x-4}$$

Express the separate fractions on the RHS as a single fraction over a common denominator.

$$\frac{x-2}{(x+3)(x-4)} \equiv \frac{A(x-4)+B(x+3)}{(x+3)(x-4)}$$

This is not an equation because the RHS is just another way of expressing the LHS. It follows that, as the denominators are identical the numerators also are identical.

$\Rightarrow \qquad x-2 \equiv A(x-4)+B(x+3)$

Remembering that this is *not* an equation but two ways of writing the same expression, it follows that LHS = RHS for any value that we choose to give to x.

Choosing to substitute 4 for x (to eliminate A) gives

$2 = A(0)+B(7) \quad \Rightarrow \quad B=\frac{2}{7}$

Choosing to substitute -3 for x (to eliminate B) gives

$$-5 = A(-7) + B(0)$$

$$\Rightarrow \qquad A = \tfrac{5}{7}$$

Therefore $$\frac{x-2}{(x+3)(x-4)} \equiv \frac{5/7}{x+3} + \frac{2/7}{x-4}$$

$$\equiv \frac{5}{7(x+3)} + \frac{2}{7(x-4)}$$

EXERCISE 30A

Express the following fractions in partial fractions.

1 $\dfrac{x-2}{(x+1)(x-1)}$

2 $\dfrac{2x-1}{(x-1)(x-7)}$

3 $\dfrac{4}{(x+3)(x-2)}$

4 $\dfrac{7x}{(2x-1)(x+4)}$

5 $\dfrac{2}{x(x-2)}$

6 $\dfrac{2x-1}{x^2-3x+2}$

7 $\dfrac{3}{x^2-9}$

8 $\dfrac{6x+7}{3x(x+1)}$

9 $\dfrac{9}{2x^2+x}$

10 $\dfrac{x+1}{3x^2-x-2}$

The Cover-up Method

There is a shortcut to finding the values of A and B, called the *cover-up method*.

Consider $$f(x) = \frac{x}{(x-2)(x-3)} \equiv \frac{A}{x-2} + \frac{B}{x-3}$$

$$\equiv \frac{A(x-3)+B(x-2)}{(x-2)(x-3)}$$

$$\Rightarrow \qquad x \equiv A(x-3) + B(x-2)$$

When $x = 2$, $A = \dfrac{2}{2-3} = -2$,

which is the value of $\dfrac{x}{(x-3)}$ when $x = 2$,

i.e. $A = f(2)$ with the factor $(x-2)$ 'covered up'.

Similarly when $x = 3$, $B = \dfrac{3}{3-2} = 3$,

which is the value of $\dfrac{x}{(x-2)}$ when $x = 3$,

i.e. $B = f(3)$ with the factor $(x-3)$ covered up.

Hence $$\frac{x}{(x-2)(x-3)} \equiv \frac{-2}{x-2} + \frac{3}{x-3}$$

Note that this method can be used only for linear factors.

EXAMPLE 30B Express $f(x) = \dfrac{1}{(2x-1)(x+3)}$ in partial fractions.

$$\frac{1}{(2x-1)(x+3)} = \frac{f(\frac{1}{2}) \text{ with } (2x-1) \text{ covered up}}{(2x-1)} + \frac{f(-3) \text{ with } (x+3) \text{ covered up}}{(x+3)}$$

$$= \frac{2}{7(2x-1)} - \frac{1}{7(x+3)}$$

Note that the intermediate step does not need to be written down.

EXERCISE 30B Use the cover-up method to express in partial fractions,

1 $\dfrac{2}{(x+1)(x-1)}$

2 $\dfrac{3}{(x-2)(x+1)}$

3 $\dfrac{1}{x(x-3)}$

4 $\dfrac{4}{(x-1)(x+3)}$

5 $\dfrac{1}{(x^2-1)}$

6 $\dfrac{2}{(2x+1)(2x-1)}$

Quadratic Factors in the Denominator

It is also possible to decompose fractions with quadratic or higher degree factors in the denominator.

Consider $\dfrac{x^2+1}{(x^2+2)(x-1)}$.

This is a proper fraction, so its partial fractions are also proper, i.e.

$\dfrac{x^2+1}{(x^2+2)(x-1)}$ can be expressed in the form $\dfrac{Ax+B}{x^2+2} + \dfrac{C}{x-1}$.

Using the cover-up method gives $C = \frac{2}{3}$, but to find A and B the partial fraction form must be expressed as a single fraction, and the numerators compared, giving

$$x^2 + 1 \equiv (Ax+B)(x-1) + \tfrac{2}{3}(x^2+2)$$

The values of A and B can then be found by substituting any suitable values for x.

We will choose $x = 0$ and $x = -1$ as these are simple values to handle. (We do not choose $x = 1$ as it was used to find C.)

$x = 0$ gives $1 = B(-1) + \frac{2}{3}(2) \quad \Rightarrow \quad B = \frac{1}{3}$

$x = -1$ gives $2 = (-A + \frac{1}{3})(-2) + \frac{2}{3}(3) \quad \Rightarrow \quad A = \frac{1}{3}$

$$\therefore \quad \frac{x^2+1}{(x^2+2)(x-1)} \equiv \frac{x+1}{3(x^2+2)} + \frac{2}{3(x-1)}$$

A Repeated Factor in the Denominator

Consider the fraction $\dfrac{2x-1}{(x-2)^2}$.

This is a proper fraction, and can be expressed as two fractions with numerical numerators, as we can see if we adjust numerator,

i.e. $$\frac{2x-1}{(x-2)^2} \equiv \frac{2(x-2)-1+4}{(x-2)^2} \equiv \frac{2}{x-2} + \frac{3}{(x-2)^2}$$

Any fraction whose denominator is a repeated linear factor can be expressed as separate fractions with numerical numerators. For example,

$$\frac{2x^2-3x+4}{(x-1)^3} \text{ can be expressed as } \frac{A}{x-1} + \frac{B}{(x-1)^2} + \frac{C}{(x-1)^3}$$

When the numerator is not so easy to rearrange, the values of the numerators can be found using the method in the next worked example.

To summarise, a proper fraction can be decomposed into partial fractions and the form of the partial fractions depends on the form of the factors in the denominator where

a linear factor gives a partial fraction of the form $\dfrac{A}{ax+b}$

a quadratic factor gives a partial fraction of the form $\dfrac{Ax+B}{ax^2+bx+c}$

a repeated factor gives two partial fractions of the form $\dfrac{A}{ax+b} + \dfrac{B}{(ax+b)^2}$

EXAMPLE 30C

Express $\dfrac{x-1}{(x+1)(x-2)^2}$ in partial fractions.

$$\frac{x-1}{(x+1)(x-2)^2} \equiv \frac{-\frac{2}{9}}{x+1} + \frac{B}{(x-2)} + \frac{C}{(x-2)^2}$$

> Do not express the first fraction in the form $\dfrac{-2}{9(x+1)}$ until after all the other numerators have been found.

$\Rightarrow \qquad x-1 \equiv (-\frac{2}{9})(x-2)^2 + B(x+1)(x-2) + C(x+1)$

$x = 2$ gives $C = \frac{1}{3}$.

Comparing coefficients of x^2 gives $0 = -\frac{2}{9} + B \quad \Rightarrow \quad B = \frac{2}{9}$

$$\therefore \qquad \frac{x-1}{(x+1)(x-2)^2} \equiv -\frac{2}{9(x+1)} + \frac{2}{9(x-2)} + \frac{1}{3(x-2)^2}$$

> **Note** that C, i.e. the numerator of the highest power of $(x-2)$, can be found by the cover-up method, but B cannot.

EXERCISE 30C

Express in partial fractions

1 $\dfrac{2}{(x-1)(x+1)^2}$

2 $\dfrac{x^2+3}{x(x^2+2)}$

3 $\dfrac{2x^2+x+1}{(x-3)(x+1)^2}$

4 $\dfrac{x^2+1}{x(2x^2+1)}$

5 $\dfrac{x}{(x-1)(x-2)^2}$

6 $\dfrac{(x^2-1)}{x^2(2x+1)}$

7 $\dfrac{x^2-2}{(x+3)(x-1)^2}$

8 $\dfrac{(x-1)}{(x+1)(x+2)^2}$

9 Express $\dfrac{x}{(x^2-4)(x-1)}$ in partial fractions

a by first treating (x^2-4) as a quadratic factor

b by first factorising (x^2-4).

State which method you think is better and explain why.

Improper Fractions

When the highest power in the numerator of a fraction is greater than *or equal to* the highest power of x in the denominator, the fraction is *improper*.

For example, $\dfrac{x^3+4x^2-7}{x^2-3}$ and $\dfrac{x^2+8}{x^2-2}$ are both improper fractions.

Sometimes we need to express an improper fraction in a different form, where any remaining fractions are proper. Rearranging the numerator can often do this, by making it possible to cancel.

For example $\dfrac{x^3+4x^2-7}{x^2-3} \equiv \dfrac{(x^3-3x)+3x+(4x^2-12)+12-7}{x^2-3}$

$$\equiv \frac{x(x^2-3)+4(x^2-3)+3x+5}{x^2-3} \equiv x+4+\frac{3x+5}{x^2-3}$$

The improper fraction has been changed into a polynomial (in this case a linear function) plus a proper fraction.

If you cannot spot a suitable rearrangement, the alternative is long division which was covered in Chapter 1. Remember that if there are any 'missing' terms in either the numerator or the denominator you should include that term with a coefficient of zero (or at least leave a space for that term).

Consider, for example, dividing x^3+4x^2-7 by x^2-3.

There is no x term in either polynomial so the long division is set out with a term $0x$ in each one.

```
                 x   + 4          ← the quotient
x² + 0x − 3 ) x³ + 4x² + 0x − 7
              x³        − 3x
              ----------------
                   4x² + 3x − 7
                   4x²      − 12
                   -------------
                         3x + 5   ← the remainder
```

giving $\dfrac{x^3+4x^2-7}{x^2-3} \equiv x+4+\dfrac{3x+5}{x^2-3}$ as before.

EXERCISE 30D

1 Carry out each of the following divisions giving the quotient and the remainder.

a $(x^3 + x^2 - 3x + 6) \div (x^2 + 3)$

b $(x^4 - 5x^2 + 2) \div (x + 1)$

c $(3x^3 - 5) \div (x - 2)$

2 Express the following fractions as the sum of a polynomial and a proper fraction.

a $\dfrac{2x}{x-2}$ b $\dfrac{x^2+3}{x^2-1}$

c $\dfrac{x^2}{x-2}$

Expressing an Improper Fraction in Partial Fractions

Once an improper fraction has been converted into an equivalent polynomial function of x together with a proper fraction, the fractional part may be expressed in partial fractions.

EXAMPLE 30E Express $\dfrac{x^3}{(x+1)(x-3)}$ in partial fractions.

> This fraction is improper and it must be rearranged or divided out to obtain a mixed fraction before it can be expressed in partial fractions.

$$
\begin{array}{r}
x \quad +2 \\
x^2 - 2x - 3\,\overline{)\,x^3 + 0x^2 + 0x + 0} \\
\underline{x^3 - 2x^2 - 3x} \\
2x^2 + 3x \\
\underline{2x^2 - 4x - 6} \\
7x + 6
\end{array}
$$

$$
\begin{aligned}
\therefore \quad \frac{x^3}{(x+1)(x-3)} &\equiv x + 2 + \frac{7x+6}{(x+1)(x-3)} \\
&\equiv x + 2 + \frac{1}{4(x+1)} + \frac{27}{4(x-3)}
\end{aligned}
$$

EXERCISE 30E Express in partial fractions.

1 $\dfrac{x^2}{(x+1)(x-1)}$

2 $\dfrac{x^3+3}{(x-1)(x+1)}$

3 $\dfrac{x^2-2}{(x+3)(x-1)}$

4 $\dfrac{x^3}{(x+2)(x^2+1)}$

The Use of Partial Fractions in Differentiation

Rational functions with two or more factors in the denominator are often easier to differentiate if expressed in partial fractions. This method is an alternative to logarithmic differentiation which can also be used for functions of this type. Partial fractions are particularly useful when a second derivative is required.

EXAMPLE 30F Find the first and second derivatives of $\dfrac{x}{(x-1)(x+1)}$.

Taking $\quad y = \dfrac{x}{(x-1)(x+1)} = \dfrac{\frac{1}{2}}{(x-1)} + \dfrac{\frac{1}{2}}{(x+1)}$

$$= \tfrac{1}{2}(x-1)^{-1} + \tfrac{1}{2}(x+1)^{-1}$$

gives $\quad \dfrac{dy}{dx} = -\tfrac{1}{2}(x-1)^{-2} - \tfrac{1}{2}(x+1)^{-2}$

$$= \frac{-1}{2(x-1)^2} - \frac{1}{2(x+1)^2}$$

and $\quad \dfrac{d^2y}{dx^2} = (-2)(-\tfrac{1}{2})(x-1)^{-3} - (-2)(\tfrac{1}{2})(x+1)^{-3}$

$$= \frac{1}{(x-1)^3} + \frac{1}{(x+1)^3}$$

EXERCISE 30F In each question express the given function in partial fractions and hence find its first and second derivatives.

1 $\dfrac{2}{(x-2)(x-1)}$

2 $\dfrac{3x}{(2x-1)(x-3)}$

3 $\dfrac{x}{(x+2)(x-4)}$

4 $\dfrac{5}{(x+2)(x-3)}$

5 $\dfrac{x}{(2x+3)(x+1)}$

6 $\dfrac{3}{(3x-1)(x-1)}$

The Remainder Theorem

When $f(x) = x^3 - 7x^2 + 6x - 2$ is divided by $x - 2$, we get a quotient and a remainder. The relationship between these quantities can be written as

$$f(x) = x^3 - 7x^2 + 6x - 2 \equiv (\text{quotient})(x-2) + \text{remainder}$$

Now substituting 2 for x eliminates the term containing the quotient, giving

$$f(2) = \text{remainder}$$

This is a particular illustration of the more general case, namely if a polynomial $f(x)$, is divided by $(x - a)$ then

$$f(x) \equiv (\text{quotient})(x-a) + \text{remainder}$$

$$\Rightarrow \quad f(a) = \text{remainder}$$

This result is called the *remainder theorem* and can be summarised as

when a polynomial f(x) is divided by (x − a), the remainder is f(a).

If $(x - a)$ is a factor of $f(x)$, there is no remainder $\quad \Rightarrow \quad f(a) = 0$.

This is the factor theorem which we met in Chapter 1,

i.e. **if, for a polynomial f(x), f(a) = 0 then x − a is a factor of f(x),**

and this is very useful when solving cubic or higher degree equations or inequalities.

EXAMPLES 30G

1 Find the remainder when

a $x^3 - 2x^2 + 6$ is divided by $x + 3$

b $6x^2 - 7x + 2$ is divided by $2x - 1$

a When $f(x) = x^3 - 2x^2 + 6$ is divided by $x + 3$, the remainder is

$$f(-3) = (-3)^3 - 2(-3)^2 + 6 = -39$$

b If $\quad f(x) = 6x^2 - 7x + 2$, then

$$f(x) = (2x - 1)(\text{quotient}) + \text{remainder}$$

$$\Rightarrow \quad \text{remainder} = f(\tfrac{1}{2}) = 0$$

> **Note** that as the remainder is zero, $2x - 1$ is a factor of $f(x)$.

2 Find the values of x for which $x^3 - 2x^2 - x + 2 < 0$.

> Problems involving inequalities are often easier to solve if dealt with graphically.

Consider the curve $y = x^3 - 2x^2 - x + 2$.

> This is a cubic curve and it crosses the x-axis where $x^3 - 2x^2 - x + 2 = 0$ so we will try to factorise $x^3 - 2x^2 - x + 2$ using the factor theorem.

$f(x) = x^3 - 2x^2 - x + 2$

$f(1) = 0$, $\therefore$ $x - 1$ is a factor of $f(x)$

$\Rightarrow \quad x^3 - 2x^2 - x + 2 \equiv (x - 1)(x^2 - x - 2) \equiv (x - 1)(x + 1)(x - 2)$

$\therefore \quad y = f(x)$ cuts the x-axis at $x = -1, 1, 2 \quad \Rightarrow$

y
−1
O
1
2
x

From the sketch,

$x^3 - 2x^2 - x + 2 < 0$ when $x < -1$ and $1 < x < 2$.

3 The equation $f(x) = 0$ has a repeated root, where $f(x) = 4x^2 + px + q$. When $f(x)$ is divided by $x + 1$ the remainder is 1. Find the values of p and q.

$f(-1) = 4 - p + q = 1 \quad \Rightarrow \quad p = q + 3$ [1]

If $4x^2 + px + q = 0$ has a repeated root then '$b^2 - 4ac$' $= 0$

i.e. $\quad p^2 - 16q = 0$ [2]

Solving equations [1] and [2] simultaneously gives

$(q + 3)^2 - 16q = 0 \quad \Rightarrow \quad q^2 - 10q + 9 = 0 \quad \Rightarrow \quad (q - 9)(q - 1) = 0$

$\therefore$ either $q = 9$ and $p = 12$ or $q = 1$ and $p = 4$

EXERCISE 30G

1 Find the remainder when the following functions are divided by the linear factors indicated.

a $x^3 - 2x + 4,\ x - 1$

b $x^3 + 3x^2 - 6x + 2,\ x + 2$

c $2x^3 - x^2 + 2,\ x - 3$

d $x^4 - 3x^3 + 5x,\ 2x - 1$

e $9x^5 - 5x^2,\ 3x + 1$

f $x^3 - 2x^2 + 6,\ x - a$

g $x^2 + ax + b,\ x + c$

h $x^4 - 2x + 1,\ ax - 1$

2 If $x^2 - 7x + a$ has a remainder 1 when divided by $x + 1$, find a.

3 Factorise $2x^3 - x^2 - 2x + 1$.
Hence find the values of x for which $2x^3 - x^2 - 2x + 1 > 0$.

4 Given that $f(x) = x^3 - x^2 - x - 2$ show that $y = f(x)$ cuts the x-axis once only.
Find the values of x for which $x^3 - x^2 - x < 2$.

5 Show that the x-coordinates of the points of intersection of the curves $y = \dfrac{1}{x}$ and $x^2 + 4y^2 = 5$ satisfy the equation $x^4 - 5x^2 + 4 = 0$. Solve this equation.

6 Factorise $x^4 - 5x^3 + 5x^2 + 5x - 6$.
Hence sketch the curve $y = x^4 - 5x^3 + 5x^2 + 5x - 6$.

7 A function f is defined by

$$f(x) = 5x^3 - px^2 + x - q.$$

When $f(x)$ is divided by $x - 2$, the remainder is 3. Given that $(x - 1)$ is a factor of $f(x)$

a find p and q

b find the number of real roots of the equation

$$5x^3 - px^2 + x - q = 0.$$

Intersection of Curves

The points of intersection of any two curves can be found by solving their equations simultaneously. Each real root then gives a point of intersection. Some roots may be repeated and we will now look at the significance of this situation.

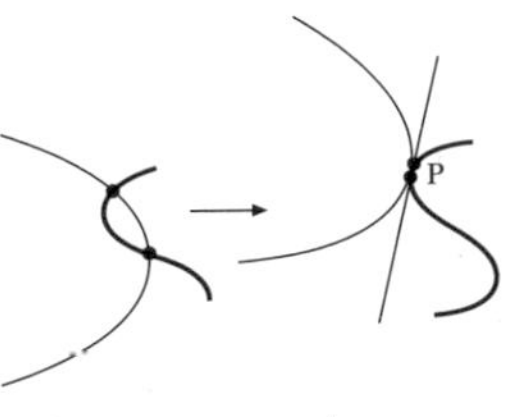

If there are two equal roots the curves meet twice at the same point P, i.e. they touch at P and have a common tangent at P.

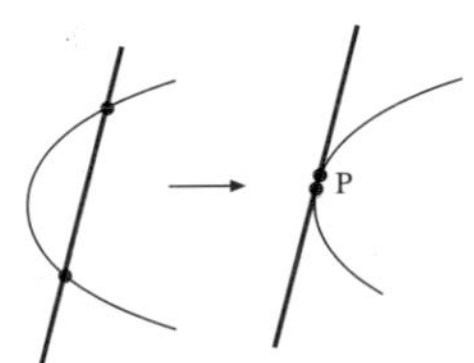

In particular, when a line and a curve meet twice at the same point P, the line is a tangent to the curve at P.

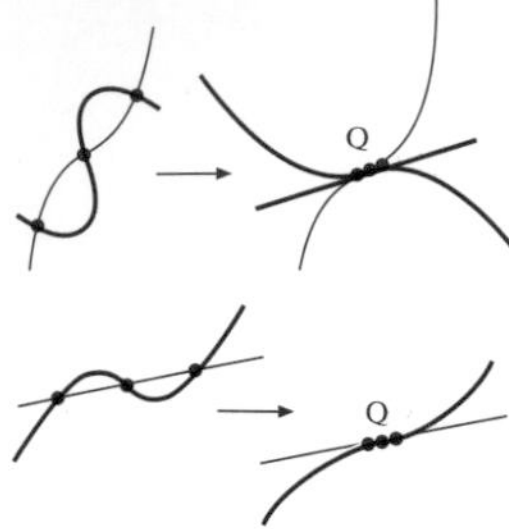

If there are three equal roots the curves meet three times at the same point Q. The curves have a common tangent at Q but this time each curve crosses, at Q, to the opposite side of the common tangent.

In particular, when a line and a curve meet at three coincident points Q, the line is a tangent to the curve at Q and the curve has a point of inflexion at Q.

Taking this argument further it becomes clear that,

a when the number of coincident points of intersection of a line and a curve is even, the curve touches the line and remains on the same side of the line;

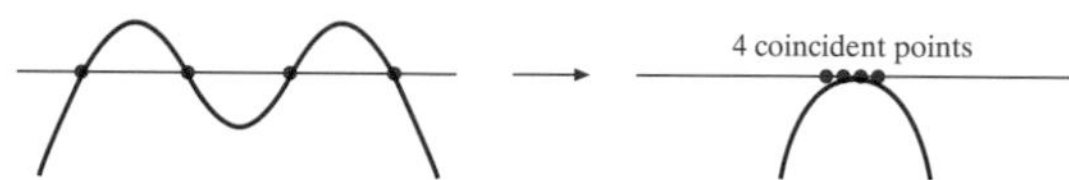

b when the number of coincident points of intersection is odd, the curve touches the line and crosses it. Thus the curve has a point of inflexion.

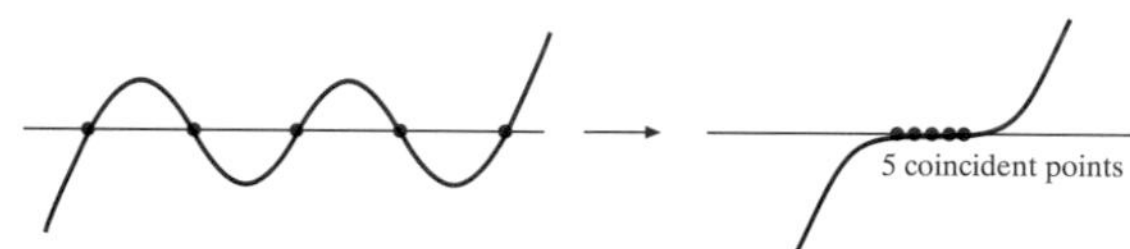

so if the solution of $\begin{cases} y = mx + c \\ y = \mathrm{f}(x) \end{cases}$

has $\begin{cases} \text{distinct roots, the curve crosses the line at distinct points.} \\ \text{repeated roots, the line touches the curve.} \end{cases}$

EXAMPLES 30H

1 Find the equations of the tangents with gradient 1 to the curve $x^2 + 2y^2 = 6$ and find the coordinates of their points of contact.

Any line with gradient 1 has equation $y = x + c$.

This line meets the curve $x^2 + 2y^2 = 6$ where $x^2 + 2(x + c)^2 = 6$

i.e. where $3x^2 + 4cx + (2c^2 - 6) = 0$ [1]

For the line to touch the curve, this equation must have equal roots.

So $b^2 - 4ac = 0$ gives $(4c)^2 - 12(2c^2 - 6) = 0 \quad \Rightarrow \quad c = \pm 3$

$\therefore$ the equations of the tangents are $y = x + 3$ and $y = x - 3$

> To find the coordinates of the points of contact we have to go back to equation [1].

When $c = 3$, [1] becomes $x^2 + 4x + 4 = 0 \quad \Rightarrow \quad x = -2$

When $c = -3$, [1] becomes $x^2 - 4x + 4 = 0 \quad \Rightarrow \quad x = 2$

> The corresponding values of y are found from the equations of the tangents.

The points of contact of the two tangents are $(2, -1)$ and $(-2, 1)$.

2 Find the points of intersection of the curves
$xy = 1$ and $x^2 + y^2 + 2x + 2y - 6 = 0$.

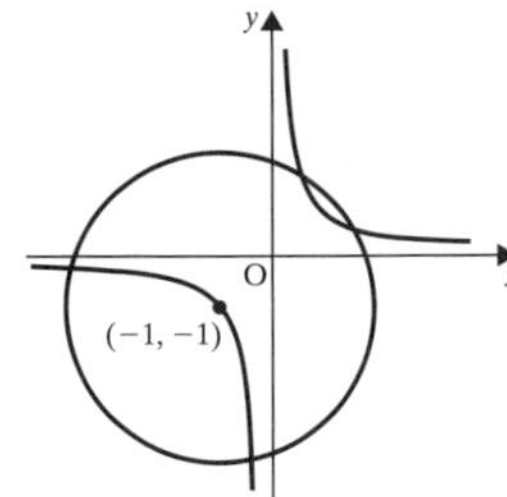

> We know the shape of the curve $xy = 1$ is a rectangular hyperbola and the other curve is a circle. Rearranging the equation of the circle as $(x+1)^2 + (y+1)^2 = 8$ tells us that its centre is $(-1, -1)$ and its radius is $2\sqrt{2}$.
>
> From the sketch, we expect 4 distinct points of intersection. However it is worth noting that the points in the first quadrant appear to be close together, and as this is only a sketch, it may be that these points are coincident or even do not exist.
>
> Now points of intersection satisfy both equations simultaneously so solve the equations by eliminating y.

$$x^2 + \left(\frac{1}{x}\right)^2 + 2x + 2\left(\frac{1}{x}\right) - 6 = 0 \quad \Rightarrow \quad x^4 + 2x^3 - 6x^2 + 2x + 1 = 0$$

Using the factor theorem gives $(x-1)(x-1)(x^2 + 4x + 1) = 0$

Hence $x = 1$ (twice) and $x = -2 \pm \sqrt{3}$

i.e. there are two equal roots and two distinct roots.

Therefore

> calculating the corresponding values of y from the equation $xy = 1$

the curves *touch* at $(1, 1)$

and *cut* at $\left(-2-\sqrt{3}, \frac{1}{-2-\sqrt{3}}\right)$ and $\left(-2+\sqrt{3}, \frac{1}{-2+\sqrt{3}}\right)$

3 Find the points of intersection of the line $y = x + 2$ and the curve
$y = x^4 - 2x^3 + 3x + 1$, showing that one of them is a point of inflexion on the curve.

The line and the curve meet at points whose x-coordinates are given by

$$x^4 - 2x^3 + 3x + 1 = x + 2 \quad \Rightarrow \quad x^4 - 2x^3 + 2x - 1 = 0$$

Using the factor theorem gives $(x-1)^3(x+1) = 0$

So there are three coincident points where $x = 1$ and one point at $x = -1$.

Therefore the line cuts the curve at a point of inflexion $(1, 3)$ and cuts it again at $(-1, 1)$.

EXERCISE 30H

Investigate the possible intersection of the following lines and curves giving the coordinates of all common points. State clearly those cases where the line touches the curve.

1 $y = x + 1;\ y^2 = 4x$

2 $2y + x = 3;\ x^2 - y^2 - 3y + 3 = 0$

3 $y = x - 5;\ x^2 + 2y^2 = 7$

4 $y = 0;\ y = x^2 - 3x + 2$

5 $y = 0;\ y = x^3 + 5x^2 + 6x$

6 $y = 0;\ y = (x-1)^2(x-2)^2$

7 $x = 0;\ x = y^4$

Find the value of k such that the given line touches the given curve.

8 $y = x + 2;\ y^2 = kx$

9 $y = kx + 3;\ xy + 9 = 0$

10 $y = 3x - k;\ x^2 + 2y^2 = 8$

Find the points of intersection or points of contact (if any) of the following pairs of curves. Illustrate your results by drawing diagrams.

11 $y^2 = 8x;\ xy = 1$

12 $x^2 + y^2 + 2x - 7 = 0;\ y^2 = 4x$

13 $xy = 2;\ 2x^2 + 2y^2 - 6x + 3y - 10 = 0$

14 $9x^2 = 2y;\ y^2 = 6x$

15 Find the value(s) or ranges of values of λ for which the line $y = 2x + \lambda$

a touches

b cuts in real points

c does not meet, the curve

$y^2 + 2x^2 = 4$

16 Sketch the curves $y = 3x^4$, $y = 4(2-x)^5, y = 2(x+3)^7$ $y = -5x^6$

17 Find the equation(s) of the tangent(s):

i from the point $(1, 0)$

ii with gradient $-\frac{1}{2}$

to each of the following curves,

a $y^2 + 4x = 0$

b $xy = 9$

c $x^2 = 6y$

MIXED EXERCISE 30

1 Use the cover-up method to express the given fraction in partial fractions.

a $\dfrac{4}{(2x+1)(x-3)}$ **b** $\dfrac{(3x-2)}{(x+1)(4x-3)}$ **c** $\dfrac{2t}{(t^2-1)}$

2 Express in partial fractions giving your calculation in full.

a $\dfrac{3}{x(2x+1)}$ **b** $\dfrac{x+4}{(x+3)(x-5)}$ **c** $\dfrac{(2x-3)}{(x-2)(4x-3)}$

d $\dfrac{4x}{4x^2-9}$ **e** $\dfrac{4}{x^2-7x-8}$ **f** $\dfrac{3x}{2x^2-2x-4}$

In Questions 3 to 8 use any suitable method to express the given fraction in partial fractions.

3 $\dfrac{3x-1}{x^2(x-3)}$

4 $\dfrac{1-4x}{(x^2+1)(x+4)}$

5 $\dfrac{8}{(x+3)(x-1)^2}$

6 $\dfrac{x^2}{(x+1)^2(x-1)}$

7 $\dfrac{x}{(x-1)(x^2+5)}$

8 $\dfrac{3-x}{(x^2+2)(x+2)}$

9 Express as the sum of a polynomial and partial fractions.

a $\dfrac{x^2}{(x+1)(x+2)}$ **b** $\dfrac{x^3+3}{x^2(x+1)}$

10 Express y in partial fractions and hence find $\frac{dy}{dx}$ and $\frac{d^2y}{dx^2}$.

a $y = \frac{2x}{(x-1)(x-2)}$ **b** $y = \frac{x}{(x+3)(x-2)}$ **c** $y = \frac{x^2+x+3}{(x+1)(x+2)(x+3)}$

11 Investigate the possible intersection of the given line and curve, giving the coordinates of any common points or points of contact.

a $2y - x = 4$ and $x^2 + y^2 - 4x = 4$

b $y = 0$ and $y = (x+3)^2(x+2)$

12 The line $y = mx$ and the curve $y = x^2 + a$ are drawn on the same axes.

a If $a = 2$, find the value(s) of m for which the line touches the curve.

b If $m = 3$ and $a = 1$, sketch the line and the curve, clearly marking any particular points where they meet or touch or cross the axes.

c If $a = 4$ find the range of values of m for which the line does not meet the curve.

13 If $f(x) \equiv 4x^3 - 7x + 3$, factorise $f(x)$ and hence sketch the graph of $y = f(x)$ marking the points where the curve crosses each axis. For what values of x is $f(x) < 0$?

14 Form an equation whose roots are the x-coordinates of the points of intersection of the line $y = 3 - x$ and the curve $y = x^3 - 4x^2 + 2x + 3$. Solve this equation and hence sketch the graphs of the line and the curve on the same axes.

CHAPTER 31 Integration 3

Integrating Fractions

Some expressions have an integral that is a function, but there are many others where an exact integral cannot be found. In this book we are concerned mainly with expressions that *can* be integrated. Even so you should be aware that, while the methods suggested usually work, they are not infallible.

There are several different methods for integrating fractions. It is important to recognise the best method for each type of fraction so as to avoid unnecessary working.

Method 1 Using Recognition

Consider the function $\ln u$ where $u = f(x)$.

Differentiating with respect to x gives

$$\frac{d}{dx}\ln u = \left(\frac{1}{u}\right)\left(\frac{du}{dx}\right) = \frac{du/dx}{u}$$

i.e. $$\frac{d}{dx}\ln f(x) = \frac{f'(x)}{f(x)}$$

Hence $$\int \frac{f'(x)}{f(x)}\,dx = \ln|f(x)| + K$$

so all fractions of the form $f'(x)/f(x)$ can be integrated *immediately* by recognition,

e.g. $$\int \frac{\cos x}{1+\sin x}\,dx = \ln|1+\sin x| + K \quad \text{as} \quad \frac{d}{dx}(1+\sin x) = \cos x$$

$$\int \frac{e^x}{e^x+4}\,dx = \ln|e^x+4| + K \quad \text{as} \quad \frac{d}{dx}(e^x+4) = e^x$$

Note, however, that $\int \frac{x}{\sqrt{1+x}}\,dx$ is *not* equal to $\ln|\sqrt{1+x}| + K$

because $\frac{d}{dx}\sqrt{1+x}$ is not equal to x

Method 1 applies only to an integral whose numerator is basically the derivative of *the complete denominator*.

An integral whose numerator is the derivative, not of the complete denominator but of a function *within* the denominator, belongs to the next type.

Method 2 Using Substitution

Consider the integral $\int \frac{2x}{\sqrt{x^2+1}}\,dx$

Noting that $2x$ is the derivative of x^2+1 we make the substitution $u = x^2+1$,

i.e. if $u = x^2+1$ then $\dots du \equiv \dots 2x\,dx$

This change of variable converts the given integral into the form $\int \frac{1}{\sqrt{u}}\,du$.

EXAMPLES 31A

1 Find $\int \frac{x^2}{1+x^3}\,dx$.

$$\int \frac{x^2}{1+x^3}\,dx = \tfrac{1}{3}\int \frac{3x^2}{1+x^3}\,dx$$

> This integral is of the form $\int \frac{f'(x)}{f(x)}\,dx$ so we use recognition.

$$= \tfrac{1}{3}\ln|1+x^3| + K$$

2 By writing $\tan x$ as $\frac{\sin x}{\cos x}$ find $\int \tan x\,dx$.

$$\int \tan x\,dx = \int \frac{\sin x}{\cos x}\,dx = -\int \frac{f'(x)}{f(x)}\,dx \quad \text{where } f(x) = \cos x$$

so $\int \frac{\sin x}{\cos x}\,dx = -\ln|\cos x| + K$

$\therefore \int \tan x\,dx = K - \ln|\cos x| \quad \text{or} \quad K + \ln|\sec x|$

Note that, similarly, $\int \cot x\,dx = \ln|\sin x| + K$

These results are quotable.

3 Find $\int \frac{e^x}{(1-e^x)^2}\,dx$.

> e^x is basically the derivative of $1-e^x$ but not of $(1-e^x)^2$ so we make the substitution $u = 1-e^x$.

If $u = 1-e^x$ then $\dots du \equiv \dots -e^x\,dx$

So $\int \frac{e^x}{(1-e^x)^2}\,dx = \int \frac{-1}{u^2}\,du = \frac{1}{u} + K$

$\therefore \int \frac{e^x}{(1-e^x)^2}\,dx = \frac{1}{1-e^x} + K$

4 Find $\displaystyle\int \frac{\sec^2 x}{\tan^3 x}\,dx$.

> $\sec^2 x$ is the derivative of $\tan x$ but not of $\tan^3 x$.

Taking $u = \tan x$ gives $\ldots du \equiv \ldots \sec^2 x\,dx$

Then $$\int \frac{\sec^2 x}{\tan^3 x}\,dx = \int \frac{1}{u^3}\,du = -\tfrac{1}{2}u^{-2} + K$$

i.e. $$\int \frac{\sec^2 x}{\tan^3 x}\,dx = \frac{-1}{2\tan^2 x} + K$$

EXERCISE 31A

In Questions 1 to 18 integrate each function w.r.t. x.

1 $\dfrac{\cos x}{4 + \sin x}$

2 $\dfrac{e^x}{3e^x - 1}$

3 $\dfrac{x}{(1 - x^2)^3}$

4 $\dfrac{\sin x}{\cos^3 x}$

5 $\dfrac{x^3}{1 + x^4}$

6 $\dfrac{2x + 3}{x^2 + 3x - 4}$

7 $\dfrac{x^2}{\sqrt{2 + x^3}}$

8 $\dfrac{\cos x}{(\sin x - 2)^2}$

9 $\dfrac{1}{x \ln x}$ i.e. $\dfrac{1/x}{\ln x}$

10 $\dfrac{\cos x}{\sin^6 x}$

11 $\dfrac{2x}{1 - x^2}$

12 $\dfrac{e^x}{\sqrt{1 - e^x}}$

13 $\dfrac{x - 1}{3x^2 - 6x + 1}$

14 $\dfrac{\cos x}{\sin^n x}$

15 $\dfrac{\sin x}{\cos^n x}$

16 $\dfrac{\sec x \tan x}{4 + \sec x}$

17 $\dfrac{x - 1}{x(x - 2)}$

18 $\dfrac{e^x - 1}{(e^x - x)^2}$

Evaluate

19 $\displaystyle\int_1^2 \frac{2x + 1}{x^2 + x}\,dx$

20 $\displaystyle\int_0^1 \frac{x}{x^2 + 1}\,dx$

21 $\displaystyle\int_2^3 \frac{2x}{(x^2 - 1)^3}\,dx$

22 $\displaystyle\int_0^1 \frac{e^x}{(1 + e^x)^2}\,dx$

23 $\displaystyle\int_{\pi/6}^{\pi/3} \frac{\sin 2x}{\cos(2x - \pi)}\,dx$

24 $\displaystyle\int_2^4 \frac{1}{x(\ln x)^2}\,dx$

Using Partial Fractions

If a fraction is not of either of the previous forms, it may be that it is easy to integrate when expressed in partial fractions. Remember, though, that only proper fractions can be converted directly into partial fractions. An improper fraction must first be rearranged so that it is made up of non-fractional terms and a proper fraction.

EXAMPLES 31B

1 Integrate $\dfrac{2x-3}{(x-1)(x-2)}$ w.r.t. x.

Use the cover-up method.

$$\frac{2x-3}{(x-1)(x-2)} = \frac{1}{x-1} + \frac{1}{x-2}$$

$$\therefore \quad \int \frac{2x-3}{(x-1)(x-2)}\,dx = \int \frac{1}{x-1}\,dx + \int \frac{1}{x-2}\,dx$$

$$= \ln|x-1| + \ln|x-2| + k$$

$$= \ln|(x-1)(x-2)| + k$$

Remember that $\ln A$ can be used instead of k, so $\ln|(x-1)(x-2)| + k$ becomes $\ln A|(x-1)(x-2)|$ which is neater.

2 Find $\displaystyle\int \frac{x^2+1}{x^2-1}\,dx$.

This fraction is improper so, before we can factorise the denominator and use partial fractions we must adjust the given fraction as follows.

$$\frac{x^2+1}{x^2-1} = \frac{(x^2-1)+2}{x^2-1} = 1 + \frac{2}{x^2-1} = 1 + \frac{2}{(x-1)(x+1)}$$

$$= 1 + \frac{1}{x-1} - \frac{1}{x+1}$$

Then $$\int \frac{x^2+1}{x^2-1}\,dx = \int 1\,dx + \int \frac{1}{x-1}\,dx - \int \frac{1}{x+1}\,dx$$

$$= x + \ln|x-1| - \ln|x+1| + \ln A$$

$$= x + \ln\frac{A|x-1|}{|x+1|}$$

Even when improper fractions do not need to be converted into partial fractions, it is still essential to reduce to proper form before attempting to integrate, i.e.

$$\int \frac{2x+4}{x+1}\,dx = \int \frac{2(x+1)+2}{x+1}\,dx = \int 2\,dx + \int \frac{2}{x+1}\,dx$$

$$= 2x + 2\ln|x+1| + k$$

Do not fall into the trap of thinking that, whenever the denominator of a fraction factorises, integration will involve partial fractions.

Look at each fraction carefully, as fractions requiring quite different integration techniques often *look* very similar. The following example shows this clearly.

EXAMPLES 31B (continued)

3 Integrate w.r.t. x

a $\dfrac{x+1}{x^2+2x-8}$ **b** $\dfrac{x+1}{(x^2+2x-8)^2}$ **c** $\dfrac{x+2}{x^2+2x-8}$

a This fraction is basically of the form f′(x)/f(x)

$$\int \frac{x+1}{x^2+2x-8}\,dx = \tfrac{1}{2}\int \frac{2x+2}{x^2+2x-8}\,dx = \frac{1}{2}\ln A|x^2+2x-8|$$

b This time the numerator is basically the derivative of the function *within* the denominator.

$$u = x^2+2x-8 \quad \Rightarrow \quad \ldots du \equiv \ldots(2x+2)\,dx \equiv \ldots 2(x+1)\,dx$$

$$\therefore \quad \int \frac{x+1}{(x^2+2x-8)^2}\,dx = \tfrac{1}{2}\int \frac{1}{u^2}\,du = -\frac{1}{2u}+K = K - \frac{1}{2(x^2+2x-8)}$$

c In this fraction the numerator is not related to the derivative of the denominator so, as the denominator factorises, we use partial fractions.

$$\int \frac{x+2}{x^2+2x-8}\,dx = \int \frac{\frac{1}{3}}{x+4}\,dx + \int \frac{\frac{2}{3}}{x-2}\,dx$$
$$= \tfrac{1}{3}\ln|x+4| + \tfrac{2}{3}\ln|x-2| + \ln A$$
$$= \ln A|(x+4)^{1/3}(x-2)^{2/3}|$$

EXERCISE 31B

Integrate each of the following functions w.r.t. x.

1 $\dfrac{2}{x(x+1)}$

2 $\dfrac{4}{(x-2)(x+2)}$

3 $\dfrac{x}{(x-1)(x+1)}$

4 $\dfrac{x-1}{x(x+2)}$

5 $\dfrac{x-1}{(x-2)(x-3)}$

6 $\dfrac{1}{x(x-1)(x+1)}$

7 $\dfrac{x}{x+1}$

8 $\dfrac{x+4}{x}$

9 $\dfrac{x}{x+4}$

10 $\dfrac{3x-4}{x(1-x)}$

11 $\dfrac{x^2-2}{x^2-1}$

12 $\dfrac{x^2}{(x+1)(x+2)}$

Choose the best method to integrate each function.

13 $\dfrac{x}{x^2-1}$

14 $\dfrac{2x}{(x^2-1)^2}$

15 $\dfrac{2}{x^2-1}$

16 $\dfrac{2x-5}{x^2-5x+6}$

17 $\dfrac{2x}{x^2-5x+6}$

18 $\dfrac{2x-3}{x^2-5x+6}$

Evaluate

19 $\displaystyle\int_0^4 \frac{x+2}{x+1}\,dx$

20 $\displaystyle\int_{-1}^1 \frac{5}{x^2+x-6}\,dx$

21 $\displaystyle\int_1^2 \frac{x+2}{x(x+4)}\,dx$

22 $\displaystyle\int_0^1 \frac{2}{3+2x}\,dx$

23 $\displaystyle\int_{1/2}^3 \frac{2}{(3+2x)^2}\,dx$

24 $\displaystyle\int_1^2 \frac{2x}{3+2x}\,dx$

Special Techniques for Integrating some Trigonometric Functions

To Integrate a Function Containing an Odd Power of sin *x* or cos *x*

When $\sin x$ or $\cos x$ appear to an *odd power* other than 1, the identity $\cos^2 x + \sin^2 x \equiv 1$ is often useful in converting the given function to an integrable form.

e.g. $\sin^3 x$ is converted to $(\sin^2 x)(\sin x) \Rightarrow (1-\cos^2 x)(\sin x)$

$\Rightarrow \sin x - \cos^2 x \sin x$

EXAMPLES 31C

1 Integrate w.r.t x, **a** $\cos^5 x$ **b** $\sin^3 x \cos^2 x$

a

$$\cos^5 x = (\cos^2 x)^2 \cos x$$
$$= (1-\sin^2 x)^2 \cos x$$
$$= (1 - 2\sin^2 x + \sin^4 x)\cos x$$

$$\therefore \quad \int \cos^5 x\,dx = \int \cos x\,dx - 2\int \sin^2 x \cos x\,dx + \int \sin^4 x \cos x\,dx$$

For any value of n we know that $\int \sin^n x \cos x\,dx = \frac{1}{n+1}\sin^{n+1} x + K.$

$$\therefore \quad \int \cos^5 x\,dx = \sin x - 2(\tfrac{1}{3})\sin^3 x + (\tfrac{1}{5})\sin^5 x + K$$
$$= \sin x - \tfrac{2}{3}\sin^3 x + \tfrac{1}{5}\sin^5 x + K$$

b $\sin^3 x \cos^2 x = \sin x(1-\cos^2 x)\cos^2 x$

$= \cos^2 x \sin x - \cos^4 x \sin x$

$$\therefore \quad \int \sin^3 x \cos^2 x\,dx = \int \cos^2 x \sin x\,dx - \int \cos^4 x \sin x\,dx$$
$$= -\tfrac{1}{3}\cos^3 x + \tfrac{1}{5}\cos^5 x + K$$

To Integrate a Function Containing only Even Powers of sin *x* or cos *x*

If $\sin x$ or $\cos x$ appear to an *even power*, the double angle identities are useful,

e.g. $\cos^4 x$ becomes $(\cos^2 x)^2 = \{\tfrac{1}{2}(1+\cos 2x)\}^2$

$$= \tfrac{1}{4}\{1 + 2\cos 2x + \cos^2 2x\}$$

then we can use a double angle identity again

$$= \tfrac{1}{4}(1+2\cos 2x) + \tfrac{1}{4}\{\tfrac{1}{2}(1+\cos 4x)\}$$
$$= \tfrac{3}{8} + \tfrac{1}{2}\cos 2x + \tfrac{1}{8}\cos 4x$$

Now each of these terms can be integrated.

EXAMPLES 31C (continued)

2 Integrate w.r.t. x, **a** $\sin^2 x$ **b** $16\sin^4 x\cos^2 x$

a $$\int \sin^2 \mathrm{d}x = \int \tfrac{1}{2}(1-\cos 2x)\,\mathrm{d}x$$
$$= \int \tfrac{1}{2}\,\mathrm{d}x - \tfrac{1}{2}\int \cos 2x\mathrm{d}x$$
$$= \tfrac{1}{2}x - \tfrac{1}{4}\sin 2x + K$$

b $$16\int \sin^4 x\cos^2 x\,\mathrm{d}x = 16\int \{\tfrac{1}{2}(1-\cos 2x)\}^2\{\tfrac{1}{2}(1+\cos 2x)\}\,\mathrm{d}x$$
$$= 2\int (1-\cos 2x - \cos^2 2x + \cos^3 2x)\mathrm{d}x$$
$$= 2x - \sin 2x - 2\int \cos^2 2x\,\mathrm{d}x + 2\int \cos^3 2x\,\mathrm{d}x$$

Now $$2\int \cos^2 2x\,\mathrm{d}x = \int (1+\cos 4x)\,\mathrm{d}x = x + \tfrac{1}{4}\sin 4x$$

and $$2\int \cos^3 2x\,\mathrm{d}x = 2\int \cos 2x(1-\sin^2 2x)\,\mathrm{d}x = \sin 2x - \tfrac{1}{3}\sin^3 2x$$

$$\therefore \quad 16\int \sin^4 x\cos^2 x\,\mathrm{d}x = x - \tfrac{1}{4}\sin 4x - \tfrac{1}{3}\sin^3 2x + K$$

Note that for a product with an odd power in one term and an even power in the other, the method for an odd power is usually best.

It is important to appreciate that the techniques used in these examples, although they are of most general use, are by no means exhaustive.

Because there are so many trig identities there is always the possibility that a particular integral can be dealt with in several different ways;
e.g. to integrate $\sin^2 x\cos^2 x$ w.r.t. x the quickest result is given by $2\sin x\cos x \equiv \sin 2x$, so that

$$\int \sin^2 x\cos^2 x\,\mathrm{d}x = \int (\tfrac{1}{2}\sin 2x)^2\,\mathrm{d}x = \tfrac{1}{4}\int \tfrac{1}{2}(1-\cos 4x)\,\mathrm{d}x$$

So it is always advisable to look for the identity which will make the given function integrable as quickly and simply as possible.

Remember that there are many expressions whose integrals cannot be found as a function at all. (In examination papers, of course, any integral asked for *can* be found!)

To Integrate any Power of tan x

The identity $\tan^2 x = \sec^2 x - 1$ is useful here,

e.g. $\tan^3 x$ becomes $\tan x(\sec^2 x - 1) \quad \Rightarrow \quad \sec^2 x\tan x - \tan x$

and we know that $\displaystyle\int \sec^2 x\tan^n x\,\mathrm{d}x = \frac{1}{n+1}\tan^{n+1}x + K$

EXAMPLES 31C (continued)

3 Integrate w.r.t. x, **a** $\tan^4 x$ **b** $\tan^5 x$

a $$\int \tan^4 x\,dx = \int \tan^2 x(\sec^2 x - 1)\,dx$$
$$= \int \sec^2 x \tan^2 x\,dx - \int \tan^2 x\,dx$$
$$= \tfrac{1}{3}\tan^3 x - \int (\sec^2 x - 1)\,dx$$
$$= \tfrac{1}{3}\tan^3 x - \tan x + x + K$$

b $$\int \tan^5 x\,dx = \int \tan^3 x(\sec^2 x - 1)\,dx$$
$$= \int \sec^2 x \tan^3 x - \int \tan x(\sec^2 x - 1)\,dx$$
$$= \int \sec^2 x \tan^3 x\,dx - \int \sec^2 x \tan x\,dx + \int \tan x\,dx$$
$$= \tfrac{1}{4}\tan^4 x - \tfrac{1}{2}\tan^2 x + \ln|\sec x| + K$$

Note that, to integrate *any* power of $\tan x$, the identity $\tan^2 x = \sec^2 x - 1$ is used to convert *$\tan^2 x$ only, one step at a time*, i.e. converting $\tan^4 x$ to $(1 - \sec^2 x)^2$ does not help.

To integrate $(\sin a\theta)(\cos b\theta)$

To integrate this type of product we use one of the factor formulae derived in Chapter 22. The one needed in this case is

$$\sin(A + B) + \sin(A - B) \equiv 2\sin A\cos B$$

but we use it in reverse, i.e.

$$\sin A\cos B \equiv \tfrac{1}{2}[\sin(A + B) + \sin(A - B)]$$

This formula changes a *product* of trig ratios into a *sum* of two ratios that we can integrate separately,

e.g. $$\int (\sin 3\theta)(\cos\theta)\,d\theta = \int \tfrac{1}{2}\{\sin(3\theta + \theta) + \sin(3\theta - \theta)\}\,d\theta$$
$$= \tfrac{1}{2}\int \{\sin 4\theta + \sin 2\theta\}\,d\theta$$
$$= -\tfrac{1}{8}\cos 4\theta - \tfrac{1}{4}\cos 2\theta + K$$

EXERCISE 31C

Integrate each function w.r.t. x.

1 $\cos^2 x$ **3** $\sin^5 x$ **5** $\sin^4 x$ **7** $\cos^4 x$

2 $\cos^3 x$ **4** $\tan^2 x$ **6** $\tan^3 x$ **8** $\sin^3 x$

Find

9 $\int \sin^2\theta\cos^3\theta\,\mathrm{d}\theta$

10 $\int \sin^{10}\theta\cos^3\theta\,\mathrm{d}\theta$

11 $\int \sin^n\theta\cos^3\theta\,\mathrm{d}\theta$

12 $\int \sin^2\theta\cos^2\theta\,\mathrm{d}\theta$

13 $\int (\sin 5\theta)(\cos 3\theta)\,\mathrm{d}\theta$

14 $\int (\sin\theta)(\cos 3\theta)\,\mathrm{d}\theta$

15 $\int_0^{\pi/2} \sin^3 x\cos^3 x\,\mathrm{d}x$

16 $\int_0^{\pi/4} \tan^6 x\,\mathrm{d}x$

17 $\int_{\pi/6}^{\pi/3} 2(\sin 4\theta)(\cos 2\theta)\,\mathrm{d}\theta$

18 $\int_0^{\pi/4} 2(\cos\theta)(\sin 5\theta)\,\mathrm{d}\theta$

CHAPTER 32 Integration 4

At this stage it is possible to classify most of the integrals that you are likely to meet.

Once correctly classified, a given expression can be integrated using the method best suited to its category.

The simplest category contains the quotable results listed below.

Standard Integrals

Function	Integral
x^n	$\frac{1}{n+1}x^{n+1} (n \neq -1)$
e^x	e^x
$\frac{1}{x}$	$\ln \lvert x \rvert$
$\cos x$	$\sin x$
$\sin x$	$-\cos x$
$\sec^2 x$	$\tan x$

Each of these should be recognised equally readily when x is replaced by ax or $(ax + b)$,

e.g. for e^{ax+b} the standard integral is $\frac{1}{a}e^{ax+b}$

and for $\cos ax$ the standard integral is $\frac{1}{a}\sin ax$.

EXERCISE 32A Integrate w.r.t. x

1 x^5

2 $x + e^x$

3 $x + \frac{1}{x}$

4 $2 + \cos x$

5 e^{2x-3}

6 $5 \sin 5x$

7 $\cos 8x$

8 $(1 - 4x)^3$

9 $\frac{1}{(2 - 3x)^2}$

10 $3 \sin (4 - 2x)$

Classification

When attempting to classify a particular function the following questions should be asked, *in order*, about the form of the integral.

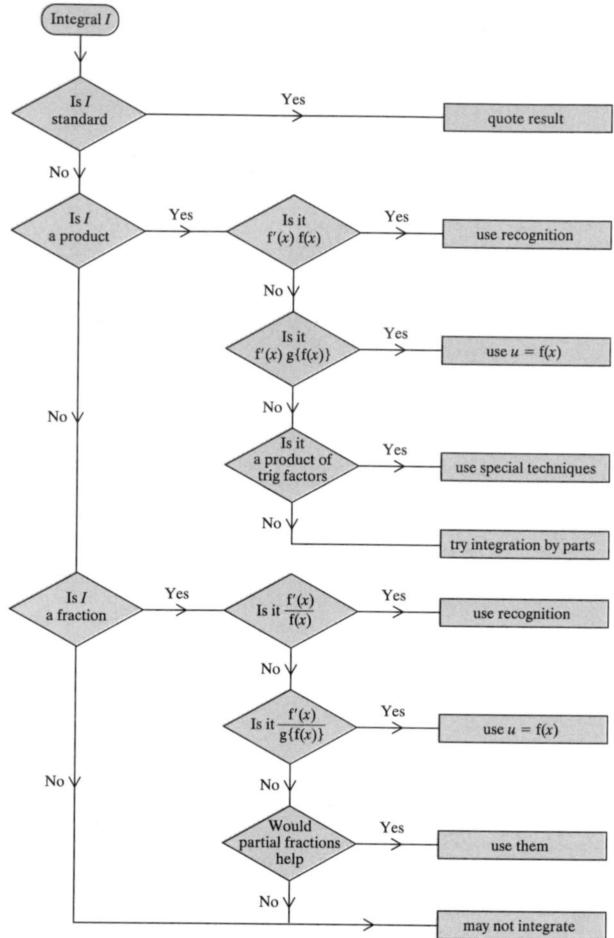

Integration by Substitution

Many expressions other than products and fractions can be integrated by making a suitable substitution. Because at this stage you cannot always be expected to 'spot' an appropriate change of variable, a substitution is suggested in all but the simplest of cases. The resulting integral must be converted so that it is expressed in terms of the original variable *except in the case of a definite integral* when it is usually much easier to change the limits (See Chapter 29).

The following examples illustrate some of the integrals that respond to a change of variable.

EXAMPLE 32B Use the substitution $u = 1 + 2x$ to find $\int x(1+2x)^{11}\,dx$.

$$u = 1 + 2x \quad \Rightarrow \quad \ldots du \equiv 2dx \quad \Rightarrow \quad \ldots \tfrac{1}{2}du \equiv \ldots dx$$

Hence
$$\int x(1+2x)^{11}\,dx = \int \tfrac{1}{2}(u-1)(u^{11})(\tfrac{1}{2}du)$$
$$= \tfrac{1}{4}\int (u^{12} - u^{11})du$$
$$= \tfrac{1}{4}(\tfrac{1}{13}u^{13} - \tfrac{1}{12}u^{12}) + K$$
$$= \tfrac{1}{624}u^{12}(12u - 13) + K$$

i.e.
$$\int x(1+2x)^{11}\,dx = \tfrac{1}{624}(1+2x)^{12}(24x-1) + K$$

EXERCISE 32B Find the following integrals using the suggested substitution.

1 $\int (x+1)(x+3)^5\,dx;\quad x+3 = u$

2 $\int \frac{x}{\sqrt{3-x}}\,dx;\quad 3-x = u^2$

3 $\int x\sqrt{x+1}\,dx;\quad x+1 = u^2$

4 $\int \frac{2x+1}{(x-3)^6}\,dx;\quad x-3 = u$

5 $\int 2x\sqrt{3x-4}\,dx;\quad 3x-4 = u^2$

Use a suitable substitution to find:

6 $\int 2x(1-x)^7\,dx$

7 $\int \frac{x+3}{(4-x)^5}\,dx$

EXERCISE 32C Use the flow chart to classify each of the following integrals. Then carry out the integration using an appropriate method.

1 $\int x(x-1)\,dx$

2 $\int (x-2)(x^2)\,dx$

3 $\int e^{2x+3}\,dx$

4 $\int x\sqrt{2x^2-5}\,dx$

5 $\int xe^x\,dx$

6 $\int \ln x\,dx$

7 $\int \sin^2 3x\,dx$

8 $\int xe^{-x^2}\,dx$

9 $\int \cos x \sin^2 x\,dx$

10 $\int u(u+7)^9\,du$

11 $\int \frac{x^2}{(x^3+9)^5}\,dx$

12 $\int \frac{\sin 2y}{1-\cos 2y}\,dy$

13 $\int \frac{1}{2x+7}\,dx$

14 $\int \sin 3x\sqrt{1+\cos 3x}\,dx$

15 $\int x \sin 4x \, dx$

16 $\int \frac{x+2}{x^2+4x-5} \, dx$

17 $\int \frac{x+1}{x^2+4x-5} \, dx$

18 $\int \frac{x+2}{(x^2+4x-5)^3} \, dx$

19 $\int 3y\sqrt{9-y^2} \, dy$

20 $\int e^{2x} \cos 3x \, dx$

21 $\int \ln 5x \, dx$

22 $\int \cos^3 2x \, dx$

23 $\int \cos x \, e^{\sin x} \, dx$

24 $\int \frac{\sin y}{\sqrt{7+\cos y}} \, dy$

25 $\int x^2 e^x \, dx$

26 $\int \frac{x}{x^2-4} \, dx$

27 $\int \frac{x^2}{x^2-4} \, dx$

28 $\int \frac{1}{x^2-4} \, dx$

29 $\int x \ln x \, dx$

30 $\int \cos^2 u \sin^3 du$

31 $\int \tan^2 \theta \, d\theta$

32 $\int \frac{2-x}{1-x} \, dx$

33 $\int \frac{\sec^2 x}{1-\tan x} \, dx$

34 $\int x\sqrt{7+x^2} \, dx$

35 $\int \sin(5\theta - \frac{1}{4}\pi) \, d\theta$

36 $\int \sec^2 u \, e^{\tan u} \, du$

Differential Equations

An equation in which at least one term contains $\frac{dy}{dx}, \frac{d^2y}{dx^2}$ etc., is called a *differential equation*. If it contains $\frac{dy}{dx}$ it is a first order differential equation whereas if it contains $\frac{d^2y}{dx^2}$ it is a second order differential equation, and so on.

For example, $x + 2\frac{dy}{dx} = 3y$ is a first order differential equation and

$\frac{d^2y}{dx^2} - 5\frac{dy}{dx} + 4y = 0$ is a second order differential equation.

Each of these examples is a *linear* differential equation because none of the differential coefficients $\left(\text{i.e. } \frac{dy}{dx}, \frac{d^2y}{dx^2}\right)$ is raised to a power higher than 1.

A differential equation represents a relationship between two variables. The same relationship can often be expressed in a form that does not contain a differential coefficient, e.g. $\frac{dy}{dx} = 2x$ and $y = x^2 + K$ express the same relationship between x and y,

but $\frac{dy}{dx} = 2x$ is a differential equation whereas $y = x^2 + K$ is not.

Converting a differential equation into a direct one is called *solving the differential equation*. This clearly involves some form of integration. There are many different types of differential equation, each requiring a specific technique for its solution. At this stage however we are going to deal with only one simple type, i.e. first order linear differential equations where we can separate the variables.

First Order Differential Equations with Separable Variables

Consider the differential equation $3y\frac{dy}{dx} = 5x^2$ [1]

Integrating both sides of the equation gives

$$\int 3y\frac{dy}{dx}\,dx = \int 5x^2\,dx$$

We saw in Chapter 29 that $\ldots\frac{dy}{dx}\,dx \equiv \ldots dy$

so $$\int 3y\,dy = \int 5x^2\,dx \qquad [2]$$

Temporarily removing the integral signs from this equation gives

$$3y\,dy = 5x^2\,dx \qquad [3]$$

This can be obtained direct from equation [1] by *separating the variables*, i.e. by *separating* dy *from* dx *and collecting on one side all the terms involving y together with* dy*, while all the x terms are collected, along with* dx *on the other side.*

It is vital to appreciate that what is shown in [3] above does not, in itself, have any meaning, and it *should not be written down as a step in the solution*. It simply provides a way of making a quick mental conversion from the differential equation [1] to the form [2] in which each side can be integrated separately.

Now returning to equation [2] and integrating each side we have

$$\tfrac{3}{2}y^2 = \tfrac{5}{3}x^3 + A$$

Note that it is unnecessary to introduce a constant of integration on both sides. A constant on one side only is sufficient.

When solving differential equations, the constant of integration is usually denoted by A, B, etc. and is called the *arbitrary constant*.

The solution of a differential equation including the arbitrary constant is called *the general solution*, or, very occasionally, *the complete primitive*. It represents a family of straight lines or curves, each member of the family corresponding to one value of A.

EXAMPLE 32D Find the general solution of the differential equation

$$\frac{1}{x}\frac{dy}{dx} = \frac{2y}{x^2+1}$$

$$\frac{1}{x}\frac{dy}{dx} = \frac{2y}{x^2+1} \quad \Rightarrow \quad \frac{1}{y}\frac{dy}{dx} = \frac{2x}{x^2+1}$$

So, after separating the variables we have,

$$\int \frac{1}{y}\,dy = \int \frac{2x}{x^2+1}\,dx$$

$$\Rightarrow \qquad \ln|y| = \ln|x^2+1| + A$$

Note that whenever we solve a differential equation some integration has to be done, so the systematic classification of each integral involved is an essential part of solving differential equations.

EXERCISE 32D Find the general solution of each differential equation.

1 $y\frac{dy}{dx} = \sin x$

2 $x^2\frac{dy}{dx} = y^2$

3 $\frac{1}{x}\frac{dy}{dx} = \frac{1}{y^2-2}$

4 $\tan y\frac{dy}{dx} = \frac{1}{x}$

5 $\frac{dy}{dx} = y^2$

6 $\frac{1}{x}\frac{dy}{dx} = \frac{1}{1-x^2}$

7 $(x-3)\frac{dy}{dx} = y$

8 $\tan y\frac{dx}{dy} = 4$

9 $u\frac{du}{dv} = v+2$

10 $\frac{y^2}{x^3}\frac{dy}{dx} = \ln x$

11 $e^x\frac{dy}{dx} = \frac{x}{y}$

12 $\sec x\frac{dy}{dx} = e^y$

13 $r\frac{dr}{d\theta} = \sin^2\theta$

14 $\frac{dv}{du} = \frac{v+1}{u+2}$

15 $xy\frac{dy}{dx} = \ln x$

16 $y(x+1) = (x^2+2x)\frac{dy}{dx}$

17 $v^2\frac{dv}{dt} = (2+t)^3$

18 $x\frac{dy}{dx} = \frac{1}{y} + y$

19 $r\frac{d\theta}{dr} = \cos^2\theta$

20 $y\sin^3 x\frac{dy}{dx} = \cos x$

21 $\frac{uv}{u-1} = \frac{du}{dv}$

22 $e^x\frac{dy}{dx} = e^{y-1}$

23 $\tan x\frac{dy}{dx} = 2y^2\sec^2 x$

24 $\frac{dy}{dx} = \frac{x(y^2-1)}{(x^2+1)}$

Calculation of the Arbitrary Constant

We saw on page 389 that

$$3y\frac{dy}{dx} = 5x^2 \quad \Longleftrightarrow \quad \tfrac{3}{2}y^2 = \tfrac{5}{3}x^3 + A$$

The equation $\tfrac{3}{2}y^2 = \tfrac{5}{3}x^3 + A$ represents a family of curves with similar characteristics. Each value of A gives one particular member of the family, i.e. a *particular solution*.
The value of A cannot be found from the differential equation alone; further information is needed.

Suppose that we require the equation of a curve that satisfies the differential equation $2\frac{dy}{dx} = \frac{\cos x}{y}$ and which passes through the point (0, 2).

We want one member of the family of curves represented by the differential equation, i.e. the particular value of the arbitrary constant must be found.

The general solution has to be found first so, separating the variables, we have

$$\int 2y\,dy = \int \cos x\,dx \quad \Rightarrow \quad y^2 = A + \sin x$$

In order to find the required curve we need the value of A such that the general solution is satisfied by $x = 0$ and $y = 2$

i.e. $\qquad 4 = A + 0 \quad \Rightarrow \quad A = 4$

Hence the equation of the specified curve is $y^2 = 4 + \sin x$.

EXAMPLES 32E

1 Describe the family of curves represented by the differential equation $y = x\frac{dy}{dx}$ and sketch any three members of this family.

Find the particular solution for which $y = 2$ when $x = 1$ and sketch this member of the family on the same axes as before.

By separating the variables.

$$y = x\frac{dy}{dx} \quad \Rightarrow \quad \int \frac{1}{y}\,dy = \int \frac{1}{x}\,dx$$

$$\Rightarrow \quad \ln|y| = \ln|x| + \ln|A| = \ln|Ax|$$

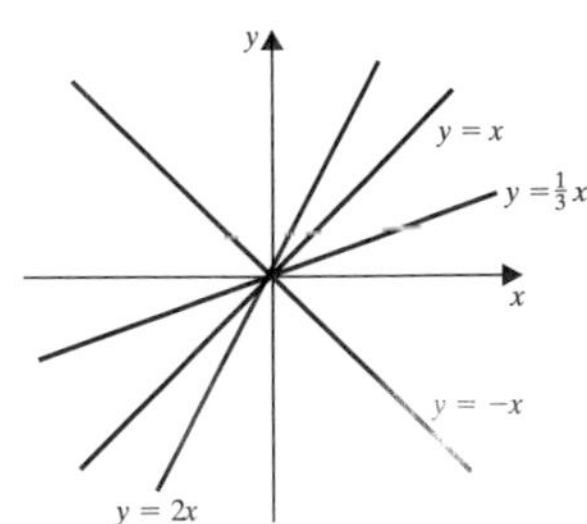

i.e. the general solution is $y = Ax$.

This equation represents a family of straight lines through the origin, each line having a gradient A, as shown in the following diagram.

If $y = 2$ when $x = 1$ then $A = 2$ and the corresponding member of the family is the line $y = 2x$.

2 A curve is such that the gradient is proportional to the product of the x- and y-coordinates. If the curve passes through the points (2, 1) and (4, e^2), find its equation.

First find the general solution.

$$\frac{dy}{dx} \propto xy \quad \Rightarrow \quad \frac{dy}{dx} = kxy \text{ where } k \text{ is a constant of proportion.}$$

$$\therefore \quad \int \frac{1}{y}\,dy = \int kx\,dx \quad \Rightarrow \quad \ln|y| = \tfrac{1}{2}kx^2 + A$$

There are *two* unknown constants this time so we need two extra pieces of information.

$$y = 1 \text{ when } x = 2 \quad \Rightarrow \quad \ln 1 = 2k + A$$

$$\ln 1 = 0 \text{ so } A + 2k = 0$$

and $\quad y = e^2$ when $x = 4 \quad \Rightarrow \quad \ln e^2 = 8k + A$

$$\ln e^2 = 2 \text{ so } A + 8k = 2$$

Solving these equations for A and k we get $k = \frac{1}{3}$ and $A = -\frac{2}{3}$.

$\therefore$ the equation of the specified curve is $\ln|y| = \frac{1}{6}x^2 - \frac{2}{3}$

$$= \tfrac{1}{6}(x^2 - 4)$$

$$\therefore \quad |y|^6 = e^{(x^2-4)} \quad \Rightarrow \quad y^6 = e^{(x^2-4)} \quad (|y|^6 \equiv y^6)$$

EXERCISE 32E

Find the particular solution of each of the following differential equations.

1 $y^2\dfrac{dy}{dx} = x^2 + 1$ and $y = 1$ when $x = 2$

2 $e^t\dfrac{ds}{dt} = \sqrt{s}$ and $s = 4$ when $t = 0$

3 $\dfrac{y}{x}\dfrac{dy}{dx} = \dfrac{y^2 - 1}{x^2 - 1}$ and $y = 3$ when $x = 2$

4 A curve passes through the origin and its gradient function is $2x - 1$. Find the equation of the curve and sketch it.

5 A curve for which $e^{-x}\dfrac{dy}{dx} = 1$, passes through the point (0, −1). Find the equation of the curve.

6 A curve passes through the points (1, 2) and ($\frac{1}{5}$, −10) and its gradient is inversely proportional to x^2. Find the equation of the curve.

7 If $y = 2$ when $x = 1$, find the coordinates of the point where the curve represented by $\dfrac{2y}{3}\dfrac{dy}{dx} = e^{-3x}$ crosses the y-axis.

8 Find the equation of the curve whose gradient function is $\dfrac{y+1}{x^2-1}$ and which passes through the point $(-3, 1)$.

9 The gradient function of a curve is proportional to $x+3$. If the curve passes through the origin and the point $(2, 8)$, find its equation.

10 Solve the differential equation $(1+x^2)\dfrac{dy}{dx} - y(y+1)x = 0$, given that $y = 1$ when $x = 0$.

MIXED EXERCISE 32

Integrate w.r.t. x

1 $x(1+x^2)^4$

2 xe^{-3x}

3 $\cos 2x \cos 3x$

4 $\dfrac{x+3}{x+2}$

5 $\dfrac{x^2}{(x^3+1)^2}$

6 $\dfrac{3}{(x-4)(x-1)}$

7 $\dfrac{(x+1)}{x(2x+1)}$

8 $\dfrac{x-1}{(x^2+1)}$

9 $\dfrac{\sin x}{\sqrt{\cos x}}$

Evaluate

10 $\displaystyle\int_{\pi/2}^{\pi} (\sin\tfrac{1}{2}x + \cos 2x)\,dx$

11 $\displaystyle\int_{2}^{5} x\sqrt{x-1}\,dx$

12 $\displaystyle\int_{0}^{\pi/4} \tan^3 x\,dx$

13 $\displaystyle\int_{1}^{2} x\sqrt{5-x^2}\,dx$

14 $\displaystyle\int_{4}^{6} \frac{5}{x^2-x-6}\,dx$

15 $\displaystyle\int_{2}^{3} \frac{1}{x\ln x}\,dx$

16 $\displaystyle\int_{-2}^{-1} \frac{2-x}{x(1-x)}\,dx$

17 $\displaystyle\int_{0}^{1} \frac{x(1-x)}{2+x}\,dx$

18 Solve the differential equation $\dfrac{dy}{dx} = 3x^2y^2$ given that $y = 1$ when $x = 0$.

19 If $\dfrac{dy}{dx} = x(y^2+1)$ and $y = 0$ when $x = 2$ find the particular solution of the differential equation.

20 Find the equation of the curve which passes through the point $(\frac{1}{2}, 1)$ and is defined by the differential equation $ye^{y^2}\dfrac{dy}{dx} = e^{2x}$. Show that the curve also passes through the point $(2, 2)$ and sketch the curve.

CHAPTER 33 Integration 5

General Rates of Increase

We have already seen that

$\dfrac{dy}{dx}$ **represents the rate at which *y* increases compared with x.**

Whenever the variation in one quantity, p say, depends upon the changing value of another quantity, q, then the rate of increase of p compared with q can be expressed as $\dfrac{dp}{dq}$.

There are many everyday situations where such relationships exist, e.g.

1) liquid expands when it is heated so, if V is the volume of a quantity of liquid and T is the temperature, then the rate at which the volume increases with temperature can be written $\dfrac{dV}{dT}$.

2) if the profit, P, made by a company selling radios depends upon the number, n, of radios sold, then $\dfrac{dP}{dn}$ represents the rate of increase of profit compared with the increase in sales.

Natural Occurrence of Differential Equations

Differential equations often arise when a physical situation is interpreted mathematically (i.e. when a mathematical model is made of the physical situation).

Consider the following examples.

1) Suppose that a body falls from rest in a medium which causes the velocity to decrease at a rate proportional to the velocity.

As the velocity is *decreasing* with time, its rate of increase is *negative*.

Using v for velocity and t for time, the rate of change of velocity can be written as $-\dfrac{dv}{dt}$, so

$$-\frac{dv}{dt} \propto v.$$

Then the motion of the body satisfies the differential equation

$$-\frac{dv}{dt} = kv$$

2) During the initial stages of the growth of yeast cells in a culture, the number of cells present increases in proportion to the number already formed.

Thus, n, the number of cells at a particular time t is such that $\frac{dn}{dt} \propto n$ and can be found from the differential equation

$$\frac{dn}{dt} = kn$$

Note. In forming (and subsequently solving) differential equations from naturally occurring data, it is not actually necessary to understand the background of the situation or experiment.

EXERCISE 33A

In Questions 1 to 4 form, but *do not solve*, the differential equations representing the given data.

1 A body moves with a velocity v which is inversely proportional to its displacement s from a fixed point.

2 The rate at which the height h of a certain plant increases is proportional to the natural logarithm of the difference between its present height and its final height H.

3 The manufacturers of a certain brand of soap powder are concerned that the number, n, of people buying their product at any time t has remained constant for some months. They launch a major advertising programme which results in the number of customers increasing at a rate proportional to the square root of n. Express as differential equations the progress of sales

a before advertising **b** after advertising.

4 In an isolated community, the number, n, of people suffering from an infectious disease is N_1 at a particular time. The disease then becomes epidemic and spreads so that the number of sick people increases at a rate proportional to n, until the total number of sufferers is N_2. The rate of increase then becomes inversely proportional to n until N_3 people have the disease. After this, the total number of sick people decreases at a constant rate. Write down the differential equation governing the incidence of the disease.

a for $N_1 \leqslant n \leqslant N_2$ **b** for $N_2 \leqslant n \leqslant N_3$ **c** for $n \geqslant N_3$.

Solving Naturally Occurring Differential Equations

We have seen that when one naturally occurring quantity varies with another, the relationship between them often involves a constant of proportion. Consequently, a differential equation that represents the relationship contains a constant of proportion whose value is not necessarily known. So the initial solution of the differential equation contains both this constant and the arbitrary constant. Extra given information may allow either or both constants to be evaluated.

EXAMPLES 33B

1 A particle moves in a straight line with an acceleration that is inversely proportional to its velocity. (Acceleration is the rate of increase of velocity.)

a Form a differential equation to represent this data.

b Given that the acceleration is $2\,\text{m}\,\text{s}^{-2}$ when the velocity is $5\,\text{m}\,\text{s}^{-1}$, solve the differential equation.

a Using $\frac{\mathrm{d}v}{\mathrm{d}t}$ for acceleration we have $\frac{\mathrm{d}v}{\mathrm{d}t} \propto \frac{1}{v} \quad \Rightarrow \quad \frac{\mathrm{d}v}{\mathrm{d}t} = \frac{k}{v}$

b If $v = 5$ when $\frac{\mathrm{d}v}{\mathrm{d}t} = 2$, then $2 = \frac{k}{5} \quad \Rightarrow \quad k = 10$

$$\therefore \quad \frac{\mathrm{d}v}{\mathrm{d}t} = \frac{10}{v}$$

Separating the variables gives $\int v\,\mathrm{d}v = \int 10\,\mathrm{d}t \quad \Rightarrow \quad \frac{1}{2}v^2 = 10t + A$

Natural Growth and Decay

There is a particular important type of naturally occurring relationship of which there are many examples in real life. These arise in situations where the rate of change of a quantity Q is proportional to the value of Q. Very often (though not invariably) the rate of change is taken with respect to time and then the relationship can be expressed as

$$\frac{\mathrm{d}Q}{\mathrm{d}t} \text{ varies with } Q.$$

If Q is increasing with time, this relationship can be expressed as the differential equation

$$\frac{\mathrm{d}Q}{\mathrm{d}t} = kQ \quad \text{where } k \text{ is a constant of proportion.}$$

Solving this differential equation by separating the variables gives

$$\int \frac{1}{Q}\,\mathrm{d}Q = \int k\,\mathrm{d}t \quad \Rightarrow \quad \ln AQ = kt$$

i.e $\quad AQ = \mathrm{e}^{kt}$ or $Q = B\mathrm{e}^{kt}$ (using B instead of $1/A$).

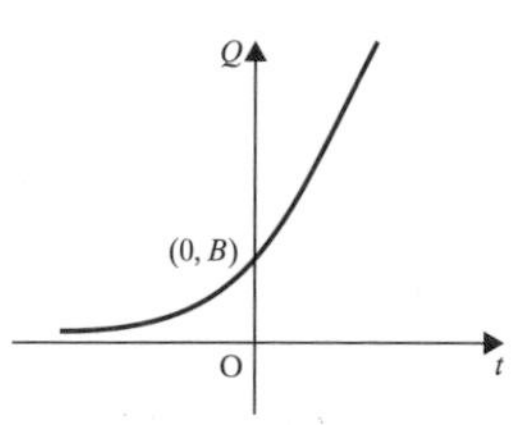

This equation shows that Q varies exponentially with time and the diagram shows a sketch of the corresponding graph.

Quantities that behave in this way are said to undergo exponential, or *natural*, growth. For example, a yeast undergoes natural growth when the rate of increase of the number of cells of yeast is proportional to the number of cells present.

Now if it is the rate of *decrease* of Q that is proportional to Q, then we have

$$\frac{-\mathrm{d}Q}{\mathrm{d}t} = kQ \quad \Rightarrow \quad Q = B\mathrm{e}^{-kt}$$

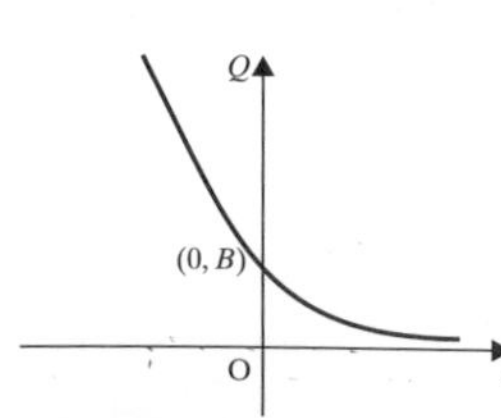

This graph is typical of a quantity undergoing exponential or *natural* decay. If, when $t = 0$, the value of Q is Q_0, the equations representing natural growth and decay become

$$Q = Q_0\mathrm{e}^{kt} \quad \text{and} \quad Q = Q_0\mathrm{e}^{-kt}$$

Half-life

When a substance is decaying naturally, the time taken for one-half of the original quantity to decay is called the *half-life* of the substance. So if the original amount is Q_0, the half-life is given by

$$\tfrac{1}{2}Q_0 = Q_0e^{-kt} \quad \Rightarrow \quad e^{kt} = 2 \quad \Rightarrow \quad t = \frac{1}{k}\ln 2$$

i.e. the value of the half-life is $\dfrac{1}{k}\ln 2$.

EXAMPLES 33B (continued)

2 When a uniform rod is heated it expands so that the rate of increase of its length, l, with respect to the temperature, θ°C, is proportional to the length. When the temperature is 0°C the length of the rod is L.

a Form and solve the differential equation that models this data; express l as a function of θ and illustrate this with a sketch.

b Given that the length of the rod has increased by 1% when the temperature is 20°C, find the value of θ at which the length of the rod has increased by 5%.

c Give a possible reason why the model may not be appropriate for very high temperatures.

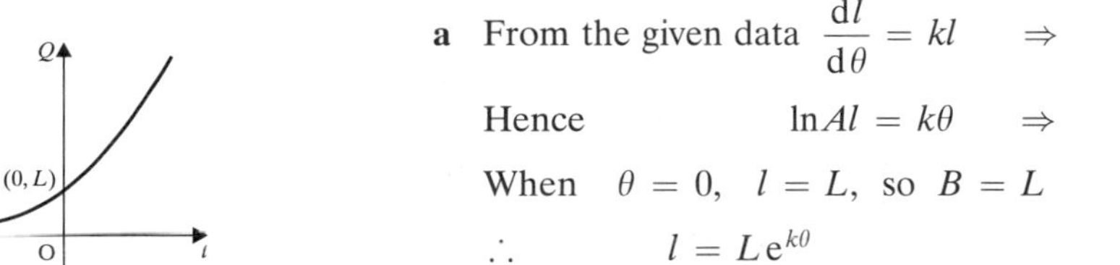

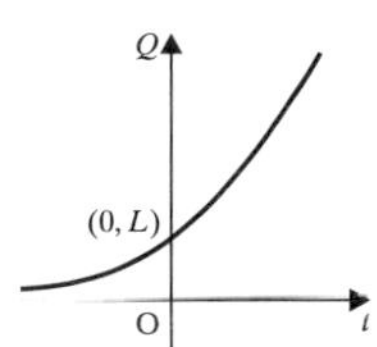

a From the given data $\dfrac{dl}{d\theta} = kl \quad \Rightarrow \quad \displaystyle\int \frac{1}{l}\,dl = \int k\,d\theta$

Hence $\ln Al = k\theta \quad \Rightarrow \quad l = Be^{k\theta}$

When $\theta = 0$, $l = L$, so $B = L$

$\therefore \quad l = Le^{k\theta}$

b When the length has increased by 1%, $l = L + 0.01L$

Then $1.01L = Le^{20k}$

$\therefore \quad e^{20k} = 1.01 \quad \Rightarrow \quad 20k = \ln 1.01$

i.e. $k = 0.000\,498$ (3 sf)

$\therefore \quad l = Le^{0.000\,498\,\theta} \quad \Rightarrow \quad 0.000\,498\,\theta = \ln(l/L)$

When $l = L + 0.05L = 1.05L, \quad 0.000\,498\,\theta = \ln 1.05$

$\Rightarrow \quad \theta = 98°$ (nearest degree)

c The rod might distort or melt at high temperatures.

3 The rate at which the atoms in a mass of radioactive material are disintegrating is proportional to N, the number of atoms present at any time. Initially the number of atoms is M.

a Form and solve the differential equation that represents this data.

b Given that half of the original mass disintegrates in 152 days, evaluate the constant of proportion in the differential equation.

c Sketch the graph of N against time.

a The rate at which the atoms are disintegrating is $-\mathrm{d}N/\mathrm{d}t$.

$\therefore \quad -\mathrm{d}N/\mathrm{d}t = kN$

Separating the variable gives $\displaystyle\int \frac{1}{N}\,\mathrm{d}N = -\int k\,\mathrm{d}t$

Hence $\ln AN = -kt \quad \Rightarrow \quad N = B\mathrm{e}^{-kt}$

When $t = 0,\ N = M \quad \Rightarrow \quad B = M$

$\therefore \quad N = M\mathrm{e}^{-kt}$

b When $N = \frac{1}{2}M,\ t = 152$

$\therefore \quad \frac{1}{2}M = M\mathrm{e}^{-152k} \quad \Rightarrow \quad \ln\left(\frac{1}{2}\right) = -152k$

$\therefore \quad 152k = \ln 2 \quad \Rightarrow \quad k = 0.004\,56 \ \text{(3 sf)}$

c

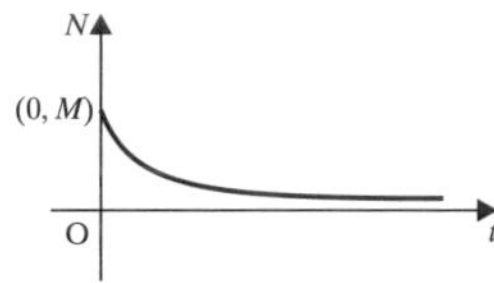

EXERCISE 33B

1 Grain is pouring from a hopper on to a barn floor where it forms a conical pile whose height h is increasing at a rate that is inversely proportional to h^3. The initial height of the pile is h_0 and the height doubles after a time T. Find, in terms of T, the time after which its height has grown to $3h_0$.

2 The gradient of any point of a curve is proportional to the square root of the x-coordinate. Given that the curve passes through the point $(1, 2)$ and at that point the gradient is 0.6, form and solve the differential equation representing the given relationship. Show that the curve passes through the point $(4, 4.8)$ and find the gradient at this point.

3 A colony of micro-organisms in a liquid is growing at a rate proportional to the number of organisms present at any time. Initially there are N organisms.

a Form a differential equation that models the growth in the size of the colony.

b Given that the colony increases by 50% in T hours, find the time that elapses from the start of the reaction before the size of the colony doubles.

c Under what conditions might the model be inappropriate?

4 If the half-life of a radioactive element that is decaying naturally is 500 years, find how many years it will be before the original mass of the element is reduced by 75%.

5 In a certain chemical reaction, a substance is transformed into a compound. The mass of the substance after any time t is m and the substance is being transformed at a rate that is proportional to the mass of the substance at that time. Given that the original mass is 50 g and that 20 g is transformed after 200 seconds,

a form and solve the differential equation relating m and t,

b find the mass of the substance transformed in 300 seconds.

6 The rate of decrease of the temperature of a liquid is proportional to the amount by which this temperature exceeds the temperature of its surroundings. This is known as Newton's Law of Cooling. Taking θ as the excess temperature at any time t, and θ_0 as the initial excess,

a show that $\theta = \theta_0 e^{-kt}$.

A pan of water at 65° is standing in a kitchen whose temperature is a steady 15°.

b Show that, after cooling for t minutes, the water temperature, ϕ, can be modelled by the equation

$$\phi = 15 + 50e^{-kt} \quad \text{where } k \text{ is a constant.}$$

c Given that after 10 minutes the temperature of the water has fallen to 50°, find the value of k.

d Find the temperature after 15 minutes and sketch the graph relating ϕ and t.

e Do you think that this model would be appropriate for a cooling time of 24 hours?

7 Use the knowledge gained from the last question to undertake the following piece of detective work.
You are a forensic doctor called to a murder scene. When the victim was discovered, the body temperature was measured and found to be 20°C. You arrive one hour later and find the body temperature at that time to be 18°C. Assuming that the ambient temperature remained constant in that intervening hour, give the police an estimate of the time of death. (Take 37°C as normal body temperature.)

Using Integration to Find Areas

The area bounded partly by a curve can be found by summing the areas of suitable elements as we saw in Chapter 20.
In this section we apply the method to curves whose equations involve trig functions and to curves whose equations are given parametrically.

EXAMPLES 33C

1 Find the area shaded in the diagram, which is bounded by the curves $y = \cos x$ and $y = \sin x$ and the y-axis.

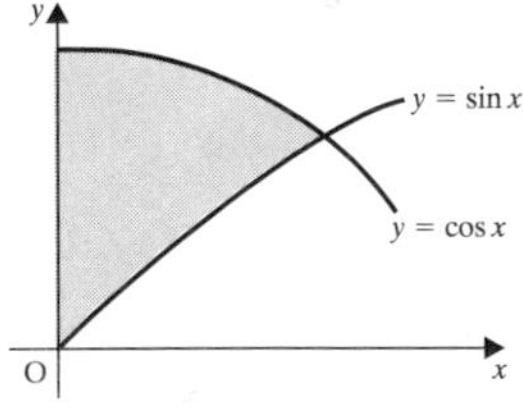

The curves intersect where $x = \frac{\pi}{4}$ so we need to sum vertical elements from $x = 0$ to $x = \frac{\pi}{4}$. Take a vertical strip as shown as the element.

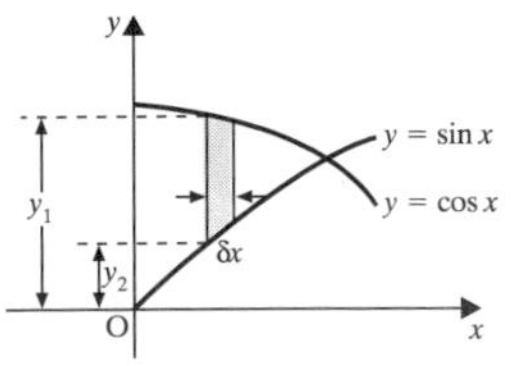

$$\delta A \approx (y_1 - y_2)\,\delta x = (\cos x - \sin x)\,\delta x$$

$$\therefore \quad A = \lim_{\delta x \to 0} \sum_{x=0}^{x=\frac{\pi}{4}} (\cos x - \sin x)\,\delta x = \int_0^{\frac{\pi}{4}} (\cos x - \sin x)\,dx$$

$$= \Big[\sin x + \cos x\Big]_0^{\frac{\pi}{4}} = \sqrt{2} - 1$$

2 Find the area above the x-axis bounded by the x-axis, the line $y = 2$ and the curve with parametric equations $x = 2t^2, y = 2t$.

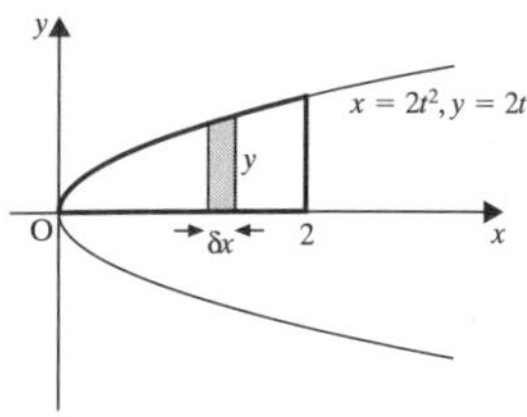

> The sketch of this curve need not be accurate but it is important to realise that it goes through the origin ($t = 0 \Rightarrow$ both $x = 0$ and $y = 0$).
>
> A suitable element is a vertical strip which is approximately a rectangle of height y, width δx and area δA.

$\delta A \approx y\,\delta x$

> Now $x = 2t^2$, so use $\delta y \approx \left(\frac{dy}{dx}\right)\delta x$ with x replacing y and t replacing x.

$$\delta x \approx \left(\frac{dx}{dt}\right)\delta t = 4t\,\delta t \quad \therefore \quad \delta A \approx (2t)(4t\,\delta t) \text{ so } A = \lim_{\delta x \to 0} \sum_{x=0}^{x=2} 8t^2\,\delta t$$

As $x = 2t^2$, $\delta t \to 0$ when $\delta x \to 0$.

Also when $x = 0, t = 0$ and when $x = 2, t = 1$
($t \neq -1$, as this gives $y = -2$ which is below the x-axis).

$$\therefore \qquad A = \lim_{\delta t \to 0} \sum_{t=0}^{t=1} 8t^2\delta t \quad \Rightarrow \quad A = \int_0^1 8t^2\,dt = \left[\tfrac{8}{3}t^3\right]_0^1 = \tfrac{8}{3}$$

The required area is $\frac{8}{3}$ square units.

EXERCISE 33C

1 The diagram shows a sketch of the curve $y = 1 + \cos x$.
Find the exact value of the shaded area.

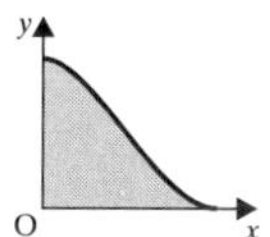

2 Calculate the area bounded by the curve $y = \sin x$ and the lines $y = \frac{1}{2}$ and $x = \frac{1}{2}\pi$.

3 The parametric equations of a curve are $x = 2t$ and $y = \dfrac{2}{t}$.

a Sketch the curve.

b Find the area bounded by this curve and the lines $x = 1$ and $x = 4$.

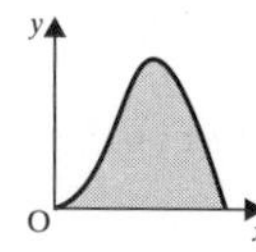

4 The sketch shows part of the graph $y = x\sin x$.
Calculate the value of the shaded area.

5 The parametric equations of a curve are $x = 4\cos\theta$ and $y = 2\sin\theta$.

a Sketch the curve for values of θ from 0 to $\pi/2$.

b Find the area in the first quadrant bounded by the curve and the axes.

6 The parametric equations of a curve are $x = t^3$ and $y = t^2$.
The sketch shows part of this curve and the tangent to the curve at the point where $t = 2$.

a Find the equation of the tangent. **b** Calculate the shaded area.

Volume of Revolution

If an area is rotated about a straight line, the three-dimensional object formed is called a *solid of revolution*, and its volume is a *volume of revolution*.

The line about which rotation takes place is always an axis of symmetry for the solid of revolution. Also, any cross-section of the solid which is perpendicular to the axis of rotation is circular.

Consider the solid of revolution formed when the shaded area is rotated about the x-axis.

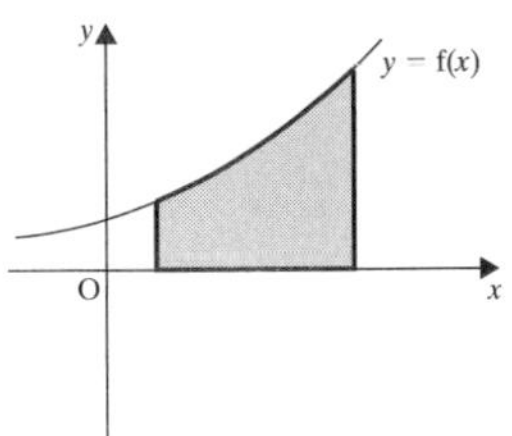

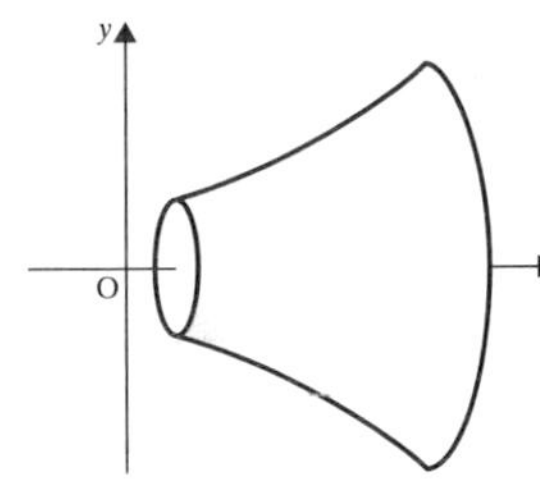

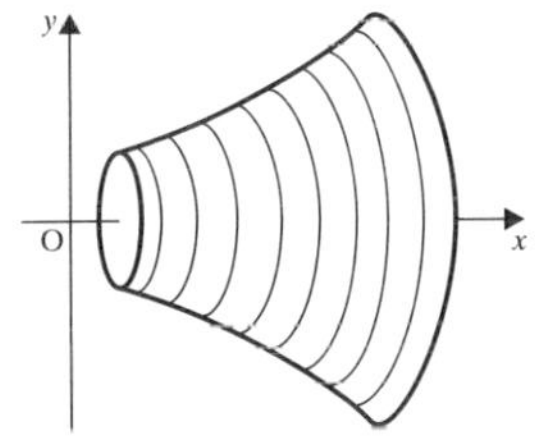

To calculate the volume of this solid we can divide it into 'slices' by making cuts perpendicular to the axis of rotation.

If the cuts are reasonably close together, each slice is approximately a cylinder and the approximate volume of the solid can be found by summing the volumes of these cylinders.

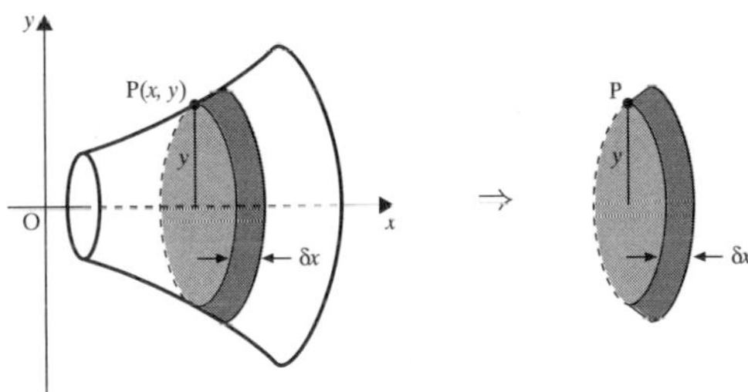

Consider an element formed by one cut through the point P(x, y) and the other cut distant δx from the first.
The volume, δV, of this element is approximately that of a cylinder of radius y and 'height' δx,

i.e. $\delta V \approx \pi y^2 \delta x$

Then the total volume of the solid is V, where $V \approx \sum \pi y^2 \delta x$

The smaller δx is, the closer is this approximation to V,

i.e. $\mathbf{V = \lim_{\delta x \to 0} \sum \pi y^2 \delta x = \int \pi y^2 dx}$

If the equation of the rotated curve is given, this integral can be evaluated and the volume of the solid of revolution found,
e.g. to find the volume generated when the area between part of curve $y = \mathrm{e}^x$ and the x-axis is rotated about the x-axis, we use $\int \pi(\mathrm{e}^x)^2\,\mathrm{d}x = \pi\int \mathrm{e}^{2x}\,\mathrm{d}x$

When an area rotates about the y-axis we can use a similar method based on slices perpendicular to the y-axis, giving $V = \int \pi x^2\,\mathrm{d}y$

EXAMPLES 33D

1 Find the volume generated when the area bounded by the x and y-axes, the line $x = 1$ and the curve $y = \mathrm{e}^x$ is rotated through one revolution about the x-axis.

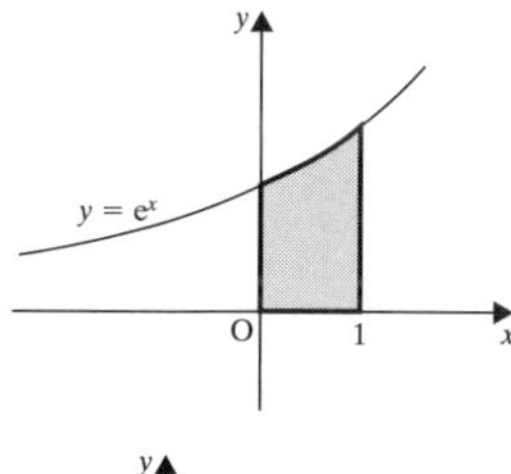

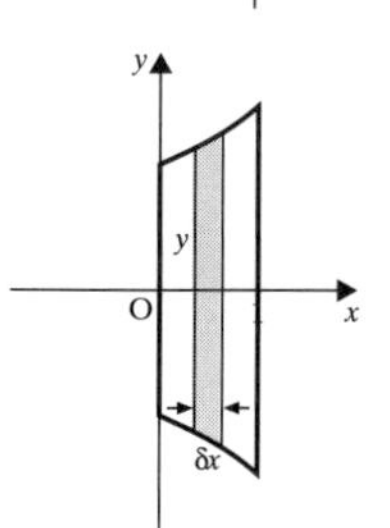

The volume, δV, of the element shown is approximately that of a cylinder of radius y and thickness δx, therefore $\delta V \approx \pi y^2 \delta x$

$\therefore$ the total volume is V, where $V \approx \sum_{x=0}^{x=1} \pi y^2\,\delta x$

$$\Rightarrow \quad V = \lim_{\delta x \to 0} \sum_{x=0}^{x=1} \pi y^2\,\delta x = \int_0^1 \pi y^2\,\mathrm{d}x = \pi\int_0^1 (\mathrm{e}^x)^2\,\mathrm{d}x$$

$$= \pi\int_0^1 \mathrm{e}^{2x}\,\mathrm{d}x = \pi\left[\tfrac{1}{2}\mathrm{e}^{2x}\right]_0^1 = \tfrac{1}{2}\pi(\mathrm{e}^2 - \mathrm{e}^0)$$

i.e. the specified volume of revolution is $\frac{1}{2}\pi(\mathrm{e}^2 - 1)$ cubic units.

2 The area defined by the inequalities $y \geqslant x^2 + 1,\ x \geqslant 0,\ y \leqslant 2,$ is rotated completely about the y-axis. Find the volume of the solid generated.

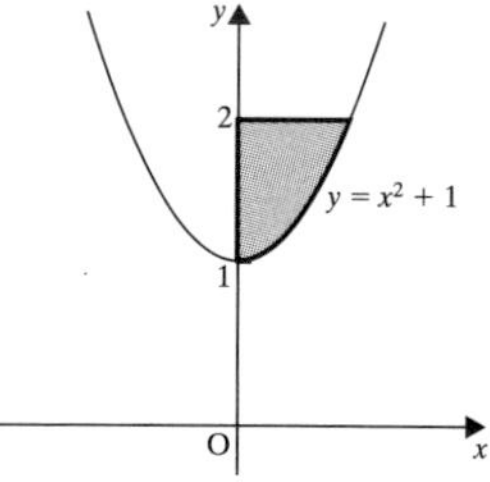

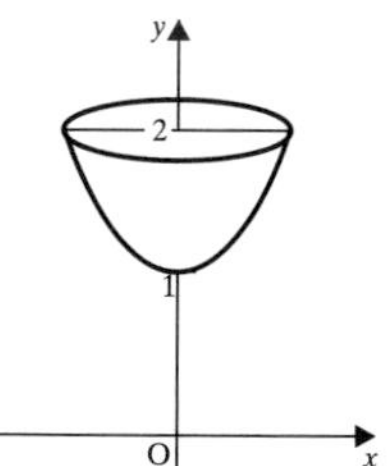

Rotating the shaded area about the y-axis gives the solid shown. This time we use horizontal cuts to form elements which are approximately cylinders with radius x and thickness δy.

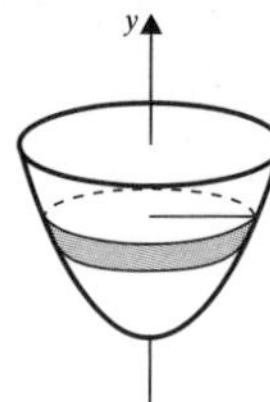

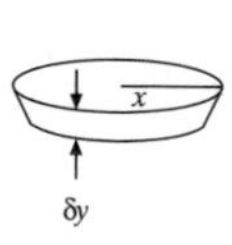

$$\delta V \approx \pi x^2 \delta y \quad \Rightarrow \quad V \approx \sum_{y=1}^{y=2} \pi x^2 \delta y$$

$$\therefore \quad V = \lim_{\delta y \to 0} \sum_{y=1}^{y=2} \pi x^2 \delta y = \int_1^2 \pi x^2 \, dy$$

Using the equation $y = x^2 + 1$ gives $x^2 = y - 1$

$$\therefore \quad V = \pi \int_1^2 (y-1)\, dy = \pi \left[\tfrac{1}{2}y^2 - y \right]_1^2$$

$$= \pi\{(2-2) - (\tfrac{1}{2} - 1)\}$$

i.e. the volume of the specified solid is $\frac{1}{2}\pi$ cubic units.

3 The area enclosed by the curve $y = 4x - x^2$ and the line $y = 3$ is rotated about the line $y = 3$. Find the volume of the solid generated.

The line $y = 3$ meets the curve $y = 4x - x^2$ at the points (1, 3) and (3, 3), therefore the volume generated has a cross-section as shown in the diagram.

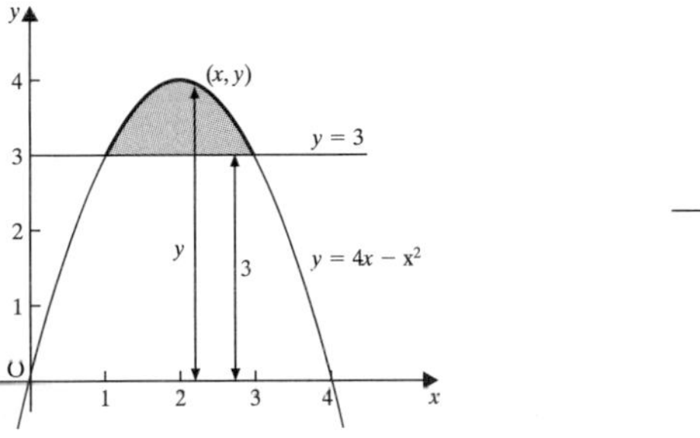

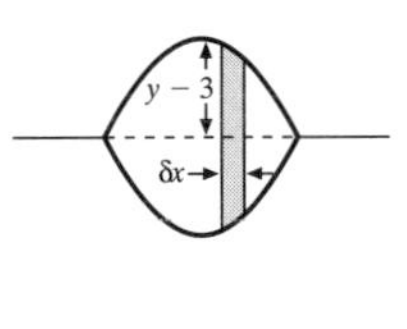

The element shown is approximately a cylinder with radius $(y - 3)$ and thickness δx, so its volume, δV, is given by $\delta V \approx \pi (y-3)^2 \delta x$

$$\text{i.e.} \quad V = \lim_{\delta x \to 0} \sum_{x=1}^{x=3} \pi (y-3)^2 \delta x = \pi \int_1^3 (y-3)^2 \, dx$$

$$\Rightarrow \quad V = \pi \int_1^3 (4x - x^2 - 3)^2 \, dx$$

$$= \pi \int_1^3 (9 - 24x + 22x^2 - 8x^3 + x^4)\, dx$$

$$= \pi \left[9x - 12x^2 + \tfrac{22}{3}x^3 - 2x^4 + \tfrac{1}{5}x^5 \right]_1^3$$

$$= \tfrac{16}{15}\pi$$

$\therefore$ the required volume is $\frac{16}{15}\pi$ cubic units.

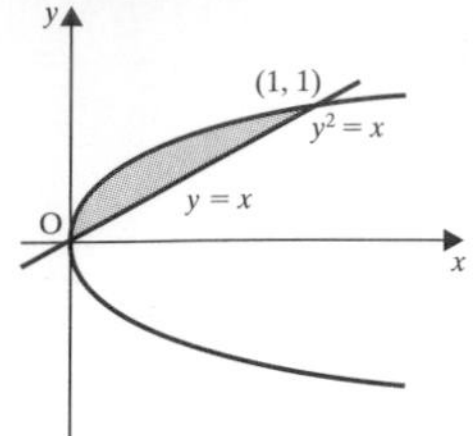

4 Find the volume generated when the area between the curve $y^2 = x$ and the line $y = x$ is rotated completely about the x-axis.

The defined area is shown in the diagram.

> When this area rotates about Ox, the solid generated is bowl-shaped on the outside, with a conical hole inside.
> The cross-section this time is not a simple circle but is an annulus, i.e. the area between two concentric circles.

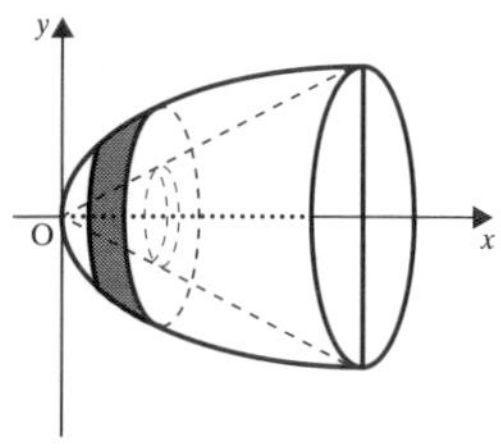

For a typical element the area of cross-section is $\pi {y_1}^2 - \pi {y_2}^2$

Therefore the volume of an element is given by $\delta V \approx \pi\{{y_1}^2 - {y_2}^2\}\delta x$

$$\therefore \qquad V \approx \sum_{x=0}^{x=1} \pi\{{y_1}^2 - {y_2}^2\}\delta x = \pi \int_0^1 ({y_1}^2 - {y_2}^2)\,dx$$

Now $y_1 = \sqrt{x}$ and $y_2 = x$

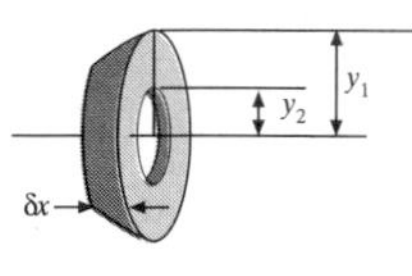

$$\therefore \qquad V = \pi \int_0^1 (x - x^2)\,dx = \pi\left[\tfrac{1}{2}x^2 - \tfrac{1}{3}x^3\right]_0^1 = \tfrac{1}{6}\pi$$

The volume generated is $\frac{1}{6}\pi$ cubic units.

Note that the volume specified in Example 4 could be found by calculating separately

1) the volume given when the curve $y^2 = x$ rotates about the x-axis;

2) the volume, by formula, of a cone with base radius 1 and height 1 and subtracting it from (**1**).

The method in which an annulus element is used, however, applies whatever the shape of the hollow interior.

EXERCISE 33D

In each of the following questions, find the volume generated when the area defined by the following sets of inequalities is rotated completely about the x-axis.

1 $0 \leqslant y \leqslant x(4 - x)$

2 $0 \leqslant y \leqslant e^x, \quad 0 \leqslant x \leqslant 3$

3 $0 \leqslant y \leqslant \dfrac{1}{x}, \quad 1 \leqslant x \leqslant 2$

4 $0 \leqslant y \leqslant x^2, \quad -2 \leqslant x \leqslant 2$

5 $y^2 \leqslant x, \quad x \leqslant 2$

In each of the following questions, the area bounded by the curve and line(s) given is rotated about the y-axis to form a solid. Find the volume generated.

6 $y = x^2,\ y = 4$

7 $y = 4 - x^2,\ y = 0$

8 $y = x^3,\ y = 1,\ y = 2,$ for $x \geqslant 0$

9 $y = \ln x,\ x = 0,\ y = 0,\ y = 1$

10 Find the volume generated when the area enclosed between $y^2 = x$ and $x = 1$ is rotated about the line $x = 1$.

11 The area defined by the inequalities

$$y \geqslant x^2 - 2x + 4, \;\; y \leqslant 4$$

is rotated about the line $y = 4$. Find the volume generated.

12 The area enclosed by $y = \sin x$ and the x-axis for $0 \leqslant x \leqslant \pi$ is rotated about the x-axis. Find the volume generated.

13 An area is bounded by the line $y = 1$, the x-axis and parts of the curve $y = 3 - x^2$. Find the volume generated when this area rotates completely about the y-axis.

14 The area enclosed between the curves $y = x^2$ and $y^2 = x$ is rotated about the x-axis. Find the volume generated.

MIXED EXERCISE 33

1 Find the area between the curves $y = x^2$ and $y^2 = x$.

2 The area in the first quadrant bounded by the curve $6 - x^2$ and the line $y = 2$ is rotated through one revolution about the line $y = 2$, find the volume generated.

3 Find the volume generated when the area between the curve $y = x^2 + 2$ and the line $y = 6$ is rotated about the x-axis.

4 The parametric equations of a curve are $x = t^3$, $y = \dfrac{1}{t^2 + 1}$.

The sketch shows the part of this curve given by $-2 \leqslant t \leqslant 2$.
Show that the area between this part of the curve and the x-axis is given by $\displaystyle\int_{-2}^{2} \frac{3t^2}{t^2 + 1}\,dt$.

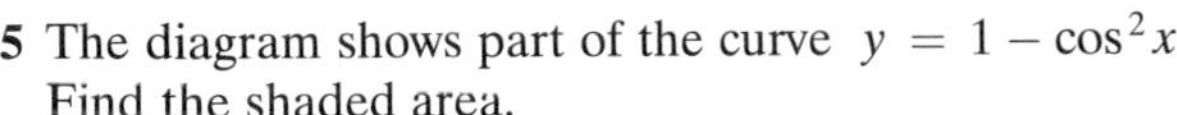

5 The diagram shows part of the curve $y = 1 - \cos^2 x$.
Find the shaded area.

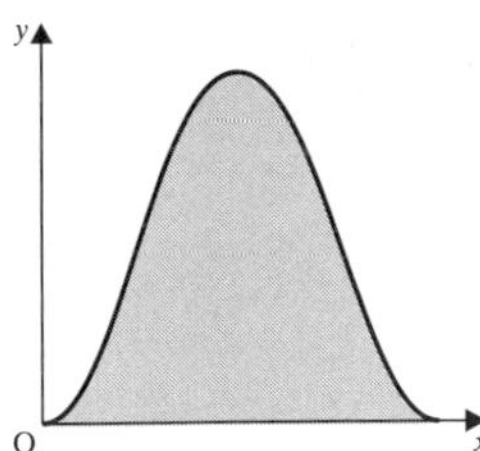

6 A virus has infected the population of rabbits on an isolated island and the evidence suggests that the growth in the number of rabbits infected is proportional to the number already infected. Initially 20 rabbits were recorded as being infected.

a Form a differential equation that models the growth in the number infected.

b Thirty days after the initial evidence was collected, 60 rabbits were infected. After how many further days does the model predict that 200 rabbits will be infected?

c In the event, only 100 rabbits were infected by that time. Give one reason why the model turned out to be unreliable.

7 A student needs to find a mathematical model for the spending power of a given sum of money over several years. For his first attempt at finding a model he assumes that the rate of decrease in spending power is proportional to the spending power at any given time.

a Using the data that £100 in January 1990 buys only £90 worth of goods in 1991, find the initial model connecting the value of goods, £y, that £100 in 1990 will buy x years later.

b Give one reason why this model is unlikely to be of any use in predicting the spending power that £100 in January 1990 would have in the year 2050.

SUMMARY F

COORDINATE GEOMETRY

Circles

The equation of a circle with centre (a, b) and radius r is $(x-a)^2+(y-b)^2=r^2$

The equation $x^2+y^2+2gx+2fy+c=0$ represents a circle with centre $(-g, -f)$ and radius $\sqrt{g^2+f^2-c}$ if $g^2+f^2-c>0$.

The parametric equations of a circle, centre O and radius r, are $x=r\cos\theta,\ y=r\sin\theta$

Ellipses

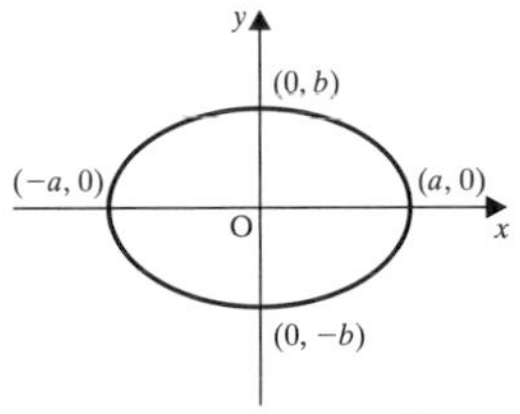

The equation of this ellipse is $\dfrac{x^2}{a^2}+\dfrac{y^2}{b^2}=1$

The parametric equations of this ellipse are

$$x=a\cos\theta,\ y=b\sin\theta$$

ALGEBRA

Partial Fractions

A proper fraction with a denominator that factorises can be expressed in partial fractions as follows:

The numerator which can be found by the cover-up method are screened.

$$\frac{f(x)}{(x-a)(x-b)}=\frac{A}{(x-a)}+\frac{B}{(x-b)}$$

$$\frac{f(x)}{(x-a)(x-b)^2}=\frac{A}{(x-a)}+\frac{B}{(x-b)}+\frac{C}{(x-b)^2}$$

$$\frac{f(x)}{(x-a)(x^2+b)}=\frac{A}{(x-a)}+\frac{Bx+C}{(x^2+b)}$$

The Remainder Theorem

When a polynomial, $f(x)$, is divided by $(x-a)$ the remainder is equal to $f(a)$.

It follows that if $f(a)=0$ then $(x-a)$ is a factor of $f(x)$.

INTEGRATION

Standard Integrals

Function	Integral
e^{ax}	$\frac{1}{a}e^{ax}$
$\frac{a}{x}$	$a\ln\lvert x\rvert$
$\cos x$	$\sin x$
$\sin x$	$-\cos x$
$\sec^2 x$	$\tan x$
$\operatorname{cosec}^2 x$	$\cot x$

The following methods are general guide lines; alternative approaches can be better in individual cases.

Integrating products can be done by
a) recognition: in particular

$$\int f'(x)e^{f(x)}\,dx=e^{f(x)}+K$$

$$\int \sin^p x\cos x\,dx=\frac{1}{p+1}\sin^{p+1}x+K\ (p\neq -1)$$

$$\int \cos^p x\sin x\,dx=-\frac{1}{p+1}\cos^{p+1}x+K\ (p\neq -1)$$

$$\int \tan^p x\sec^2 x\,dx=\frac{1}{p+1}\tan^{p+1}x+K\ (p\neq -1)$$

b) change of variable: suitable for $f'(x)g\{f(x)\}$

c) by parts: $\int v\frac{du}{dx}dx = uv - \int u\frac{dv}{dx}dx$

Integration by parts can be used also to integrate $\ln x$.

Integrating fractions can be done by

a) recognition: in particular

$$\int\frac{f'(x)}{f(x)}dx = \ln f(x) + K$$

b) change of variable: suitable for $\dfrac{f'(x)}{g\{f(x)\}}$

c) using partial fractions.

DIFFERENTIAL EQUATIONS

A first order linear differential equation is a relationship between x, y and dy/dx. It can be solved by collecting all the x terms, along with dx, on one side, with all y terms and dy on the other side. Then each side is integrated with respect to its own variable. A constant of integration called an arbitrary constant is introduced on one side only to give a general solution which is a family of lines or curves.

If extra information provides the value of this constant we have a particular solution, i.e. one member of the family.

Volume by Integration

When the area bounded by the x-axis, the ordinates at a and b, and part of the curve $y = f(x)$ rotates completely about the x-axis, the volume generated is

given by $\lim_{\delta x\to 0}\sum_{x=a}^{x=b}\pi y^2\delta x = \int_a^b \pi y^2 dx$

and about the y-axis, by $\lim_{\delta y\to 0}\sum_{y=a}^{y=b}\pi x^2\delta y = \int_a^b \pi x^2 dy$

MULTIPLE CHOICE EXERCISE F

In Questions 1 to 15 write down the letter or letters corresponding to a correct answer.

1 A point P moves so that it is equidistant from A and B. The locus of the set of points P is

A a circle on AB as diameter

B a line parallel to AB

C the perpendicular bisector of AB

D a parabola

2 The point $P(x,y)$ lies on the curve whose equation is $x^2+y^2-2x = 0$.

A P is on a circle, centre the origin.

B The curve is not a circle.

C P is always 1 unit from the point $(1,0)$.

D The origin is a possible position of P.

3

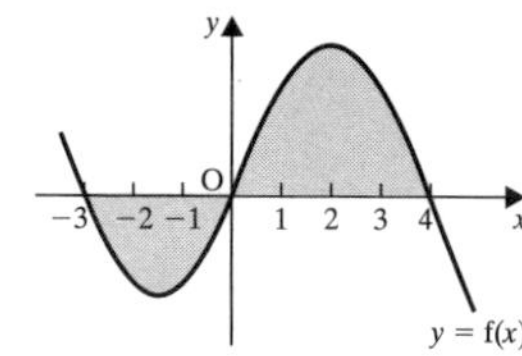

The volume formed when the shaded area is rotated about Ox can be found from

A $\pi\int_{-3}^{4}[f(x)]^2dx$

B $\pi\int_{-3}^{0}[f(x)]^2dx + \pi\int_{0}^{4}[f(x)]^2dx$

C $\pi\int_{-3}^{1}[f(x)]^2dx + \pi\int_{1}^{4}[f(x)]^2dx$

D $\pi\int_{-3}^{4}x^2dx$

4 e^{x^2} could be the integral w.r.t. x of

A e^{2x} **D** $x^2e^{x^2} - 1$

B $2xe^{x^2}$ **E** none of these

C $\dfrac{e^{x^2}}{2x}$

5 If $\displaystyle\int_1^5 \frac{dx}{2x-1} = \ln K$, the value of K is

A 9 **D** 81

B 3 **E** 8

C undefined

6 $I = \displaystyle\int_1^2 x\sqrt{(x^2-1)}\,dx$ is found as follows. Where does an error first occur?

A Let $u \equiv x^2 - 1$

B $\ldots du \equiv \ldots 2x\,dx$

C $I = \frac{1}{2}\displaystyle\int_1^2 u^{1/2}du$

D $I = \frac{3}{4}\left[u^{3/2}\right]_1^2$

7 The value of $\displaystyle\int_0^2 2e^{2x}\,dx$ is

A e^4 **C** ∞ **E** $\frac{1}{2}e^4$

B $e^4 - 1$ **D** $4e^4$

8 $x^3 - 3x^2 + 2x - 6$ has a factor

A $x-3$ **C** $x-4$ **E** $x+2$

B $x-2$ **D** $x+3$

9 $x^3 - 3x^2 + 6x - 2$ has remainder 2 when divided by

A $x-1$ **C** x **E** $2x-1$

B $x+1$ **D** $x+2$

10 If the equation $ax^2 + by^2 + 2gx + 2fy + c = 0$ represents a circle through the origin,

A $g = 0$ and $f = 0$

B $c = 0$

C $a = b$

D $a = -b$

11 $f(x) = (3-5x)^4$

A $f(x)$ has a remainder 16 when divided by $x-1$

B $f(x) = 3^4 - 5x^4$

C the equation $f(x) = 0$ is satisfied by only one value of x

12 $f(x) \equiv 2x^2 + 3x - 2$

A $f(x)$ can be expressed as the sum of two partial fractions

B the equation $f(x) = 0$ has two real distinct roots

C $x+2$ is a factor of $f(x)$

13 Using $x = \sin\theta$ transforms

$$\int \frac{x^2}{\sqrt{(1-x^2)}}\,dx \text{ into}$$

A $\displaystyle\int \frac{\sin^2\theta}{\cos\theta}\,d\theta$

B $\frac{1}{2}\displaystyle\int (1-\cos 2\theta)\,d\theta$

C $\displaystyle\int \sin^2\theta\,d\theta$

D $\frac{1}{2}\displaystyle\int (1+\cos 2\theta)\,d\theta$

14 Which of the following differential equations can be solved by separating the variables?

A $x\dfrac{dy}{dx} = y + x$

B $xy\dfrac{dy}{dx} = x + 1$

C $e^{x+y} = y\dfrac{dy}{dx}$

D $x + \dfrac{dy}{dx} = \ln y$

15 $\displaystyle\int_1^2 xe^x\,dx$

A is a definite integral

B is equal to $xe^x - e^x$

C is equal to $\left[\frac{1}{2}e^{x^2}\right]_1^2$

D can be integrated by parts.

In Questions 16 to 21 a single statement is made. Write T if it is true, F if it is false.

16 $\displaystyle\sum_{x=a}^{x=b} y\,\delta x = \int_a^b y\,dx$

17 $x^2 + 2y^2 = 1$ is the equation of a circle.

18 $\displaystyle\int \tan x\,dx = \sec^2 x + K$

19 $\displaystyle\int_0^a f(y)\,dy = \lim_{\delta y \to 0} \sum_{y=0}^{y=a} f(y)\,\delta y$

20 $\left[f(x)\right]_0^a = f(a) - 0$

21 $x^2 + y^2 - 2x - 4y + 6 = 0$ is the equation of a circle.

Examination Questions F

1 The polynomial $x^3 + 2x^2 + ax + b$

i is divisible by $(x - 1)$ **ii** leaves a remainder 12 when divided by $(x - 2)$.

Find the values of a and b. (AQA)

2 Find $\int (2 + e^{-x})\,dx$. (OCR)

3 Sketch the curve defined by the equations

$$x = 3a\cos t,\ y = 2a\sin t,$$

where a is a positive constant, for $0 \leqslant t \leqslant 2\pi$. (AQA)

4 Use integration by parts to show that $\int_2^4 x\ln x\,dx = 7\ln 4 - 3$ (Edexcel)

5
$$\frac{5x-1}{(1+x)(1+x+3x^2)} = \frac{Ax+B}{1+x+3x^2} + \frac{C}{1+x}$$

Find the values of the constants A, B and C. (Edexcel)

6
$$f(x) = \frac{x-8}{(x+2)(2x-1)}$$

a Express $f(x)$ in partial fractions. **b** Find $f'(0)$. (Edexcel)

7 Using the substitution $2x = \sin\theta$, or otherwise, find the exact value of $\int_0^{\frac{1}{4}} \frac{1}{\sqrt{1-4x^2}}\,dx$. (AQA)

8

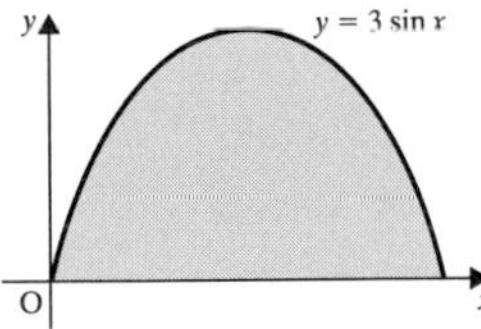

The graph of $y = 3\sin x$ is sketched above.
Write down an expression for the shaded area. Hence find the value of the shaded area. (AQA)

9 Find **i** $\int \frac{1}{2x+1}\,dx$ **ii** $\int (2x+1)^7\,dx$ (OCR)

10 The points $(5, 5)$ and $(-3, -1)$ are the ends of a diameter of the circle C with centre A. Write down the coordinates of A and show that the equation of C is

$$x^2 + y^2 - 2x - 4y - 20 = 0.$$

The line L with equation $y = 3x - 16$ meets C at the points P and Q.
Show that the x-coordinates of P and Q satisfy the equation

$$x^2 - 11x + 30 = 0$$

Hence find the coordinates of P and Q. (WJEC)

11 The polynomial

$$p(x) = x^3 + cx^2 + 7x + d$$

has a factor of $x + 2$, and leaves a remainder of 3 when divided by $x - 1$.

a Determine the value of each of the constants c and d.

b Find the exact values of the three roots of the equation $p(x) = 0$. (AQA)

12 Express $\dfrac{x}{(x+1)(x+2)^2}$ in partial fractions. (AQA)

13 a Show that $2x - 1$ is a factor of $8x^4 - 4x^3 + 14x^2 - 3x - 2$.

b Find the remainder when $8x^4 - 4x^3 + 14x^2 - 3x - 2$ is divided by $x + 1$. (OCR)

14 a Find the radius and the coordinates of the centre of the circle with equation

$$x^2 + y^2 - 2x - 8y = 0.$$

b Determine, by calculation, whether the point $(2.9, 1.7)$ lies inside or outside the circle. (Edexcel)

15 Evaluate, in terms of e, $\displaystyle\int_0^1 (e^{2x} - 1)^2\,dx$ (Edexcel)

16 $f(n) \equiv n^2 + n + 1$, where n is a positive integer.
Classify the following statements about $f(n)$ as true or false. If a statement is true, prove it; if it is false, provide a counter-example.

a $f(n)$ is always a prime number.

b $f(n)$ is always an odd number. (Edexcel)

17 The curve shown has parametric equations

$$x = t^2,\ y = t^3,$$

where $t \geqslant 0$ is a parameter.
Also shown is part of the normal to the curve at the point where $t = 1$.

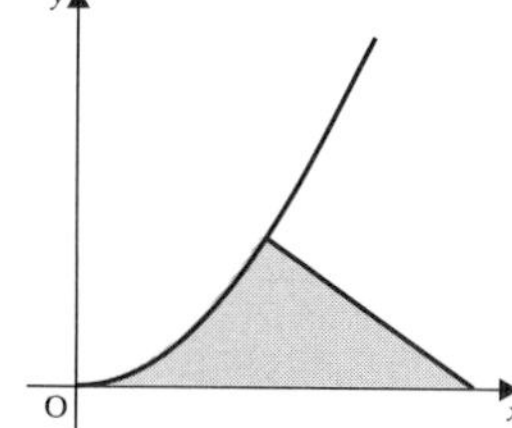

a Find an equation of this normal.

b Find the area of the finite region bounded by the curve, the x-axis and this normal. (Edexcel)

18 Use the identity $\cos(A + B) \equiv \cos A \cos B - \sin A \sin B$ to prove that $\sin^2\theta \equiv \frac{1}{2}(1 - \cos 2\theta)$.

The finite region R is bounded by the curve with equation $y = \sin^2 2x$, the lines $x = \dfrac{\pi}{8}$, $x = \dfrac{\pi}{4}$ and $y = 0$.

Find the area of R giving your answer in the form $p + q\pi$, where p and q are numbers to be found. (Edexcel)

19 Find $\int x \cos 2x \, dx$. (OCR)

20 Use the substitution $u = 3x - 1$ to express $\int x(3x-1)^4 \, dx$ as an integral in terms of u.

Hence, or otherwise, find $\int x(3x-1)^4 \, dx$, giving your answer in terms of u. (OCR)

21 A circle C has equation

$$x^2 + y^2 - 2x - 8y - 8 = 0.$$

a Find the coordinates of the centre of C.

b Find the radius of C.

c Find the equation of the tangent to C at the point (5, 1). (WJEC)

22

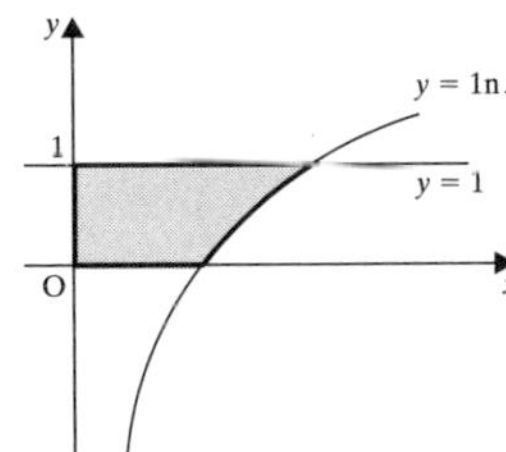

The region R in the first quadrant is bounded by the curve $y = \ln x$, the x-axis, the y-axis and the line $y = 1$, as shown in the diagram. Show that the volume of the solid formed when R is rotated completely about the y-axis is $\frac{1}{2}\pi(e^2 - 1)$. (OCR)

23 A class is asked to prove the result that for all positive integers n, if

$$S_n = 1 + 3 + 5 + \ldots + (2n-1), \text{ then } S_n = n^2.$$

a The following is a student's attempt at a proof.

$$S_1 = 1 \qquad\qquad = 1^2$$
$$S_2 = 1 + 3 = 4 \qquad = 2^2$$
$$S_3 = 1 + 3 + 5 = 9 = 3^2$$
etc.

so the formula is true for all positive integers n.

Why is this student's argument incorrect?

b Another student begins to prove the result in the following way.

$$S_n = 1 \quad + 3 \quad + 5 \quad + \ldots + (2n-1).$$
Also, $S_n = (2n-1) + (2n-3) + (2n-5) + \ldots + 1.$

Adding each term, we have ...

Complete this student's proof correctly to show that $S_n = n^2$ for all positive integers n. (AQA)

24 Determine the coordinates of the centre C and the radius of the circle with equation

$$x^2 + y^2 + 4x - 10y + 13 = 0.$$

Find the distance from the point P(2, 3) to the centre of the circle.
Hence find the length of the tangents from P to the circle. (AQA)

25

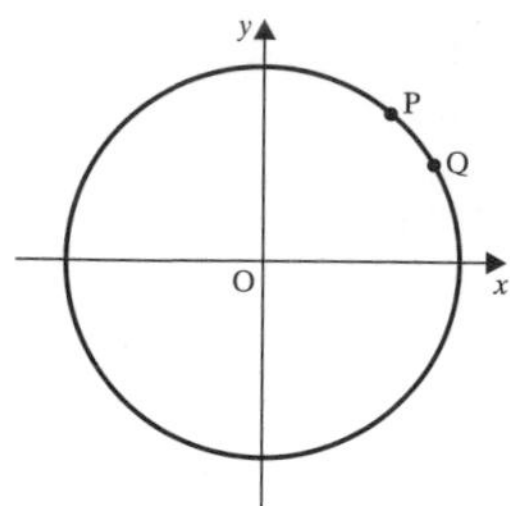

The diagram shows the circle with equation $x^2 + y^2 = 25$. Find the length of the minor arc between the points P(3, 4) and Q(4, 3), giving your answer correct to 3 significant figures. (OCR)

26 Express $\dfrac{3}{(2x+1)(x-1)}$ in partial fractions.

Hence find the exact value of $\displaystyle\int_2^3 \frac{3}{(2x+1)(x-1)}\,dx$ giving your answer as a single logarithm. (OCR)

27 $f(x) \equiv \dfrac{5x^2 - 8x + 1}{2x(x-1)^2} \equiv \dfrac{A}{x} + \dfrac{B}{x-1} + \dfrac{C}{(x-1)^2}$.

a Find the values of the constants A, B and C.

b Hence find $\displaystyle\int f(x)\,dx$.

c Hence show that

$$\int_4^9 f(x)\,dx = \ln\left(\tfrac{32}{3}\right) - \tfrac{5}{24}.$$

(Edexcel)

28

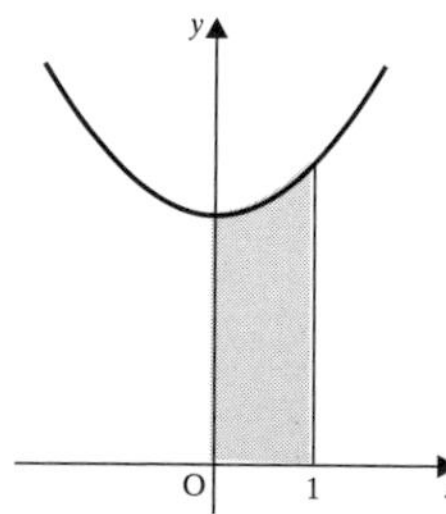

The figure shows the finite shaded region bounded by the curve with equation $y = x^2 + 3$, the lines $x = 1$, $x = 0$ and the x-axis. This region is rotated through $360°$ about the y-axis. Find the volume generated. (Edexcel)

29 a i Find $\displaystyle\int \frac{1}{x(x+1)}\,dx,\ x > 0$.

Using the substitution $u = e^x$ and the answer to **i**, or otherwise,

ii find $\displaystyle\int \frac{1}{1+e^x}\,dx$.

b Use integration by parts to find $\int x^2 \sin x\,dx$. (Edexcel)

30 a

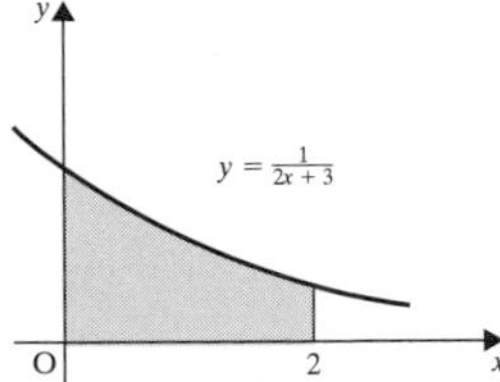

Calculate the volume of the solid formed when the area between $y = \dfrac{1}{2x+3}$, the axes and the line $x = 2$ is rotated through an angle of 2π radians about the x-axis.

b Use the substitution $u = 2x + 3$ to find $\displaystyle\int \frac{x}{(2x+3)^2}\,dx$. (AQA)

31 Find

a $\displaystyle\int \sin 2y\,dy$,

b $\displaystyle\int x e^x\,dx$.

c Hence find the general solution of the differential equation

$$\frac{dy}{dx} = \frac{x e^x}{\sin y \cos y}, \quad 0 < y < \frac{\pi}{2}.$$ (Edexcel)

32 a Obtain the general solution of the differential equation

$$\frac{dy}{dx} = xy^2, \quad y > 0$$

b Given also that $y = 1$ at $x = 1$ show that

$$y = \frac{2}{3 - x^2}, \quad -\sqrt{3} < x < \sqrt{3},$$

is a particular solution of the differential equation.

The curve C has equation $y = \dfrac{2}{3 - x^2}$, $x \neq -\sqrt{3}$, $x \neq \sqrt{3}$.

c Write down the gradient of C at the point $(1, 1)$.

d Deduce that the line which is a tangent to C at the point $(1, 1)$ has equation $y = x$.

e Find the coordinates of the point where the line $y = x$ again meets the curve C. (Edexcel)

33 Sketch on the same axes the curves with equations $y = 1 + x^2$ and $y = 9 - x^2$ and determine their points of intersection.
The finite region bounded by the two curves is rotated through π radians about the y-axis.
Calculate, in terms of π, the volume of the solid of revolution. (AQA)

34 At time t hours the rate of decay of the mass of a radioactive substance is proportional to the mass x kg of the substance at that time. At time $t = 0$ the mass of the substance is A kg.

a By forming and integrating a differential equation, show that $x = Ae^{-kt}$, where k is a constant.

It is observed that $x = \frac{1}{3}A$ at time $t = 10$.

b Find the value of t when $x = \frac{1}{2}A$, giving your answer to 2 decimal places. (Edexcel)

35 The curve C is given by the equations

$$x = 2t, y = t^2,$$

where t is a parameter.

a Find an equation of the normal to C at the point P on C where $t = 3$.

The normal meets the y-axis at the point B. The finite region R is bounded by the part of the curve C between the origin O and P, and the lines OB and BP.

b Show the region R, together with its boundaries, in a sketch.

The region R is rotated through 2π about the y-axis to form a solid S.

c Using integration, and explaining each step in your method, find the volume of S, giving your answer in terms of π. (Edexcel)

36 Showing your method clearly in each case, find

a $\int \sin^2 x \cos x \, dx$

b $\int x \ln x \, dx$

c Use the substitution $t^2 = x + 1$, where $x > -1$, $t > 0$, to find $\int \frac{x}{\sqrt{x+1}} dx$.

d Hence evaluate $\int_0^3 \frac{x}{\sqrt{x+1}} dx$. (Edexcel)

37 The rate, in $cm^3 s^{-1}$, at which oil is leaking from an engine sump at any time t seconds is proportional to the volume of oil, V cm^3, in the sump at that instant. At time $t = 0$, $V = A$.

a By forming and integrating a differential equation, show that

$$V = Ae^{-kt},$$

where k is a positive constant.

b Sketch a graph to show the relation between V and t.

Given further that $V = \frac{1}{2}A$ at $t = T$,

c show that $kT = \ln 2$. (Edexcel)

38 During a spell of freezing weather, the ice on a pond has thickness x mm at time t hours after the start of freezing. At 3.00 pm, after one hour of freezing weather, the ice is 2 mm thick and it is desired to predict when it will be 4 mm thick.

i In a simple model, the rate of increase of x is assumed to be constant. For this model, express x in terms of t and hence determine when the ice will be 4 mm thick.

ii In a more refined model, the rate of increase of x is taken to be proportional to $\frac{1}{x}$. Set up a differential equation for x, involving a constant of proportionality k.
Solve the differential equation and hence show that the thickness of ice is proportional to the square root of the time elapsed from the start of freezing.
Determine the time at which the second model predicts that the ice will be 4 mm thick.

iii What assumption about the weather underlies both models? (OCR)

39 i Express $\sin^2 x$ in terms of $\cos 2x$.

ii The region R is bounded by the part of the curve $y = \sin x$ between $x = 0$ and $x = \pi$ and the x-axis. Show that the volume of the solid formed when R is rotated completely about the x-axis is $\frac{1}{2}\pi^2$. (OCR)

40 i Find $\int x e^{-x} \, dx$.

ii A curve is such that at every point (x, y) on it the gradient, $\frac{dy}{dx}$, satisfies the differential equation

$$\frac{dy}{dx} = x e^{-x-y}.$$

Show that

$$e^y = \int x e^{-x} \, dx.$$

Given that the curve is defined for all values of x and that it passes through the origin, show that it also passes through the point $(-1, \ln 2)$. (AQA)

41 The graph of $y = 2\sqrt{x-1}$ is shown in the sketch, together with the line $y = 4$.

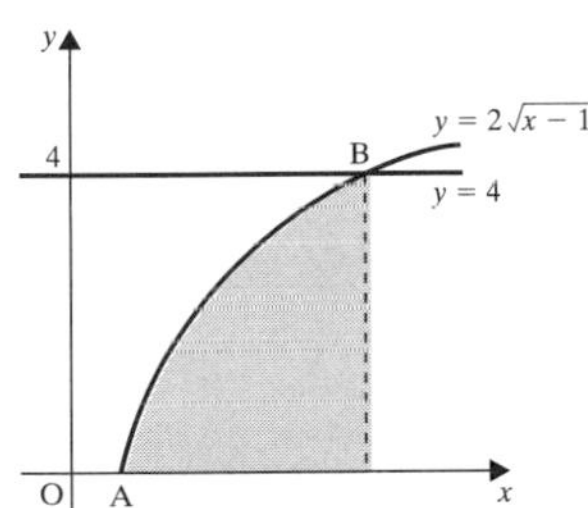

a Find the coordinates of the points A and B.

b Find the **exact** value of the shaded area shown.

c Calculate the **exact** volume of the solid formed by rotating the shaded area through 360° about the x-axis, leaving your answer in terms of π. (AQA)

42 a Use integration by parts to find $\int x \cos 2x \, dx$.

b By means of the substitution $u = 3x^2 + 1$, or otherwise, evaluate

$$\int_0^1 x\sqrt{3x^2 + 1} \, dx.$$

No credit will be given for a numerical approximation or for a numerical answer without supporting working. (AQA)

43 A curve has equation $y = \dfrac{3x+4}{(x-2)(2x+1)}$

a Express $\dfrac{3x+4}{(x-2)(2x+1)}$ in partial fractions.

b Show that $\dfrac{dy}{dx} = \dfrac{2}{(2x+1)^2} - \dfrac{2}{(x-2)^2}$ and hence, or otherwise, show that the curve has a turning point when $x = -3$.

Determine the value of x at the other stationary point of the curve.

c Find $\dfrac{d^2y}{dx^2}$ and hence determine the nature of the turning point when $x = -3$.

d Find $\displaystyle\int \frac{3x+4}{(x-2)(2x+1)}\,dx$.

Hence show that the area of the region bounded by the curve, the x-axis and the lines $x = 4$ and $x = 12$ is equal to ln 15. (AQA)

44 The circle C has equation $x^2 + y^2 - 2x - 4y - 20 = 0$.

a Find the radius and the coordinates of the centre of C.

b Find the equation of the tangent to C at the point (4, 6), giving your answer in the form $ax + by + c = 0$. (AQA)

45 a Use the substitution $v = \sqrt{x^2+5}$ to evaluate: $\displaystyle\int_0^2 \frac{x\,dx}{\sqrt{x^2+5}}$

b Use integration by parts to find $\displaystyle\int_1^3 x e^{-2x}\,dx$. (OCR)

46 At time t hours, the rate of decay of the mass of a radioactive substance is proportional to the mass x kg at that time.

a Write down a differential equation satisfied by x.

b Given that $x = C$ when $t = 0$, show that $x = Ce^{-at}$ where a is a positive constant.

c Find the value of a if the mass halves every 2.5 hrs. (AQA)

47 The rate of destruction of a drug by the kidneys is proportional to the amount of the drug present in the body. The constant of proportionality is denoted by k. At time t the quantity of drug in the body is x. Write down a differential equation relating x and t, and show that the general solution is $x = Ae^{-kt}$, where A is an arbitrary constant.

Before $t = 0$ there is no drug in the body, but at $t = 0$ a quantity Q of the drug is administered. When $t = 1$ the amount of drug in the body is $Q\alpha$, where α is a constant such that $0 < \alpha < 1$. Show that $x = Q\alpha^t$.

Sketch the graph of x against t for $0 < t < 1$.

When $t = 1$ and again when $t = 2$ another dose Q is administered. Show that the amount of drug in the body immediately after $t = 2$ is $Q(1 + \alpha + \alpha^2)$. (OCR)

48 The parametric equations of a curve are

$$x = 3\cos\theta,\ y = 2\sin\theta \quad \text{for} \quad 0 \leqslant \theta < 2\pi.$$

i By eliminating θ between these two equations, find the Cartesian equation of the curve.

ii Draw a sketch of the curve, giving the coordinates of the points where it cuts the axes. On your sketch show the pair of tangents which pass through the point (6, 2).

iii Use the parametric equations to calculate $\frac{dy}{dx}$ in terms of θ.

You are given that the equation of the tangent to the curve at $(3\cos\theta, 2\sin\theta)$ is

$$2x\cos\theta + 3y\sin\theta = 6.$$

iv Show that, for tangents to the curve which pass through the point (6, 2),

$$2\cos\theta + \sin\theta = 1.$$

v Solve the equation in **iv** to find the two values of θ (in radians correct to 2 decimal places) corresponding to the two tangents. (OCR)

49 A patch of oil pollution in the sea is approximately circular in shape. When first seen its radius was 100 m and its radius was increasing at a rate of 0.5 m per minute. At a time t minutes later, its radius is r metres. An expert believes that, if the patch is untreated, its radius will increase at a rate which is proportional to $1/r^2$.

i Write down a differential equation for this situation, using a constant of proportionality, k.

ii Using the initial conditions, find the value of k. Hence calculate the expert's prediction of the radius of the oil patch after 2 hours.

The expert thinks that if the oil patch is treated with chemicals then its radius will increase at a rate which is proportional to $\frac{1}{r^2(2+t)}$.

iii Write down a differential equation for this new situation and, using the same initial conditions as before, find the value of the new constant of proportionality.

iv Calculate the expert's prediction of the radius of the treated oil patch after 2 hours. (OCR)

50 i Use integration by parts to evaluate

$$\int 4x\cos 2x\,dx.$$

ii Use part **i**, together with a suitable expression for $\cos^2 x$, to show that

$$\int 8x\cos^2 x\,dx = 2x^2 + 2x\sin 2x + \cos 2x + c.$$

iii Find the solution of the differential equation

$$\frac{dy}{dx} = \frac{8x\cos^2 x}{y}$$

which satisfies $y = \sqrt{3}$ when $x = 0$.

iv Show that any point (x, y) on the graph of this solution which satisfies $\sin 2x = 1$ also lies on one of the lines $y = 2x + 1$ or $y = -2x - 1$. (OCR)

CHAPTER 34 Series

Power Series

A series such as $x + x^2 + x^3 + \dots$ is called a power series because the terms involve powers of a variable quantity. Series, such as those considered in Chapter 19, each of whose terms has a fixed numerical value, are called number series.

The Binomial Theorem

We saw in Chapter 1 that when an expression such as $(1+x)^4$ is expanded, the coefficients of the terms in the expansion can be obtained from Pascal's Triangle. Now $(1+x)^{20}$ could be expanded in the same way but, as the construction of the triangular array would be tedious, we need a more general method to expand powers of $(1+x)$.

This general method uses the binomial theorem which states that, if n is a positive integer,

$$(1+x)^n \equiv 1 + nx + \frac{n(n-1)}{(2)(1)}x^2 + \frac{n(n-1)(n-2)}{(3)(2)(1)}x^3$$

$$+ \frac{n(n-1)(n-2)(n-3)}{(4)(3)(2)(1)}x^4 + \dots + nx^{n-1} + x^n.$$

The right-hand side of this identity is called the *series expansion* of $(1+x)^n$.
The coefficients of the powers of x are called *binomial coefficients*.
The denominators of these coefficients involve the products of all the positive integers from the power of x in that term down to 1 and we can write these more concisely using *factorial notation*.

The product $(4)(3)(2)(1)$ is called '4 factorial' and is written 4!
Similarly, 8! means the product of all the positive integers from 8 down to 1.

In general, when r is a positive integer,

***r*! means the product of all the positive integers from *r* down to 1.**

Consider a general term somewhere in the middle of the series, involving x^r where r is an integer between 1 and n.

Looking at the pattern of the binomial coefficients, we can see that the coefficient of x^r is $\dfrac{n(n-1)(n-2)\dots(n-r+1)}{r!}$ which is denoted by $\dbinom{n}{r}$,

i.e. $$\binom{n}{r} \equiv \frac{n(n-1)(n-2)\dots(n-r+1)}{r!}$$

Hence the binomial theorem states that, if n is a positive integer,

$$(1+x)^n \equiv 1 + nx + \frac{n(n-1)}{2!}x^2 + \frac{n(n-1)(n-2)}{3!}x^3 + \frac{n(n-1)(n-2)(n-3)}{4!}x^4 + \ldots + nx^{n-1} + x^n.$$

$$\equiv 1 + \binom{n}{1}x + \binom{n}{2}x^2 + \binom{n}{3}x^3 + \ldots + \binom{n}{r}x^r + \ldots + x^n.$$

Notice that

1) the expansion of $(1+x)^n$ is a finite series with $n+1$ terms,

2) the coefficient of x^r, i.e. $\frac{n(n-1)(n-2)\ldots(n-r+1)}{r!}$, has r factors in the numerator,

3) the term containing x^2 is the third term, the term in x^3 is the fourth term, and so on.

Now consider $(a+bx)^n$, where n is a positive integer.

$$(a+bx)^n \equiv a^n\left(1+\frac{b}{a}x\right)^n$$

Replacing x by $\frac{b}{a}x$ in the binomial series gives

$$\begin{aligned}(a+bx)^n &= a^n\left[1+\binom{n}{1}\left(\frac{b}{a}x\right)+\binom{n}{2}\left(\frac{b}{a}x\right)^2+\ldots+\binom{n}{r}\left(\frac{b}{a}x\right)^r+\ldots+\binom{n}{n}\left(\frac{b}{a}x\right)^n\right]\\ &= a^n+\binom{n}{1}a^{n-1}bx+\binom{n}{2}a^{n-2}b^2x^2+\ldots+\binom{n}{r}a^{n-r}b^rx^r+\ldots+\binom{n}{n}b^nx^n\\ &= a^n+na^{n-1}bx+\frac{n(n-1)}{2!}a^{n-2}b^2x^2+\ldots+b^nx^n\end{aligned}$$

EXAMPLES 34A

1 Write down the first three terms in the expansion in ascending powers of x of

a $\left(1-\frac{x}{2}\right)^{10}$ **b** $(3-2x)^8$

a

> Use $(1+x)^n = 1+nx+\frac{n(n-1)}{2!}x^2+\ldots$ and replace x by $-\frac{x}{2}$ and n by 10.

$$\left(1-\frac{x}{2}\right)^{10} = 1+(10)\left(-\frac{x}{2}\right)+\frac{10\times 9}{2!}\left(-\frac{x}{2}\right)^2+\ldots = 1-5x+\tfrac{45}{4}x^2+\ldots$$

b $(3-2x)^8 = 3^8(1-\frac{2}{3}x)^8$.

> Replace x by $-\frac{2}{3}x$ and n by 8 in the expansion of $(1+x)^n$.

$$(3-2x)^8 = 3^8\left(1+8(-\tfrac{2}{3}x)+\left(\frac{8\times 7}{2}\right)(-\tfrac{2}{3}x)^2+\ldots\right)$$

Therefore the first three terms of this series are $3^8-(16)(3^7)x+(112)(3^6)x^2$

2 Find the fourth term in the expansion of $(a-2b)^{20}$ as a series in ascending powers of b.

$$(a-2b)^{20} = a^{20}\left(1-\frac{2b}{a}\right)^{20}$$

The fourth term in the expansion of $(1+x)^n$ is $\binom{n}{3}x^3$. Replace x by $-\frac{2b}{a}$ and n by 20.

$$\text{The fourth term is } a^{20}\binom{20}{3}\left(-\frac{2b}{a}\right)^3 = -\frac{(20)(19)(18)}{3!}(a)^{17}(8)(b)^3$$
$$= -9120a^{17}b^3$$

3 Write down the first three terms in the binomial expansion of

$$(1-2x)(1+\tfrac{1}{2}x)^{10}$$

The third term in the binomial expansion is the term containing x^2, so start by expanding $(1+\frac{1}{2}x)^{10}$ as far as the term in x^2.

$$(1+\tfrac{1}{2}x)^{10} = 1+(10)(\tfrac{1}{2}x)+\frac{(10)(9)}{2!}(\tfrac{1}{2}x)^2+\dots$$
$$= 1+5x+\tfrac{45}{4}x^2+\dots$$
$$\therefore \quad (1-2x)(1+\tfrac{1}{2}x)^{10} = (1-2x)(1+5x+\tfrac{45}{4}x^2+\dots)$$
$$= 1+5x+\tfrac{45}{4}x^2+\dots-2x-10x^2+\dots$$
$$= 1+3x+\tfrac{5}{4}x^2+\dots$$

Notice that we do not write down the product of $-2x$ and $\frac{45}{4}x^2$, as terms in x^3 are not required.

EXERCISE 34A

1 Write down the first four terms in the binomial expansion of

a $(1+3x)^{12}$ **b** $(1-2x)^9$ **c** $(2+x)^{10}$

d $\left(1-\frac{x}{3}\right)^{20}$ **e** $\left(2-\frac{3}{2}x\right)^7$ **f** $\left(\frac{3}{2}+2x\right)^9$

2 Write down the term indicated in the binomial expansion of each of the following functions.

a $(1-4x)^7$, 3rd term **b** $\left(1-\frac{x}{2}\right)^{20}$, 2nd term

c $(2-x)^{15}$, 12th term **d** $(p-2q)^{10}$, 5th term

e $(3a+2b)^8$, 2nd term **f** $(1-2x)^{12}$, the term in x^4

g $\left(2+\frac{x}{2}\right)^9$, the term in x^5 **h** $(a+b)^8$, the term in a^3

3 Write down the binomial expansion of each function as a series of ascending powers of x as far as, and including, the term in x^2.

a $(1+x)(1-x)^9$ **b** $(1-x)(1+2x)^{10}$

c $(2+x)\left(1-\frac{x}{2}\right)^{20}$ **d** $(1+x)^2(1-5x)^{14}$

Using Series to Find Approximations

Consider $(1+x)^{20}$ and its binomial expansion,

$$(1+x)^{20} = 1+20x+\frac{(20)(19)}{2!}x^2+\frac{(20)(19)(18)}{3!}x^3+\ldots+x^{20}$$

This is valid for all values of x so if, for example, $x=0.01$ we have

$$(1.01)^{20} = 1+20(0.01)+\frac{(20)(19)}{2!}(0.01)^2+\frac{(20)(19)(18)}{3!}(0.01)^3+\ldots+(0.01)^{20}$$

i.e. $(1.01)^{20} = 1+0.2+0.019+0.001\,14+0.000\,048\,45+\ldots+10^{-40}$

Because the value of x (i.e. 0.01) is small, we see that adding successive terms of the series makes progressively smaller contributions to the accuracy of $(1.01)^{20}$.
In fact, taking only the first four terms gives $(1.01)^{20} \approx 1.220\,14$

This approximation is correct to three decimal places as the fifth and succeeding terms do not add anything to the first four decimal places.

In general, if x is small so that successive powers of x quickly become negligible in value, then the sum of the first few terms in the expansion of $(1+x)^n$ gives an approximate value for $(1+x)^n$.

The number of terms required to obtain a good approximation depends on two considerations:

the value of x (the smaller x is, the fewer are the terms needed to obtain a good approximation),

the accuracy required (an answer correct to 3 s.f. needs fewer terms than an answer correct to 6 s.f.).

EXAMPLES 34B

1 By substituting 0.001 for x in the expansion of $(1-x)^7$ find the value of $(1.998)^7$ correct to five significant figures.

Now $(1.998)^7 = (2-0.002)^7 = 2^7(1-0.001)^7$

$= 2^7(1-x)^7$ with $x = 0.001$

So $(1.998)^7 = 2^7\left[1-7(0.001)+\frac{(7)(6)}{2!}(0.001)^2-\frac{(7)(6)(5)}{3!}(0.001)^3+\ldots\right]$

> To give an answer correct to 5 s.f. we will work to 7 s.f. so only need the first three terms.

$\therefore \quad (1.998)^7 = 128(1-0.007+0.000\,021\,0)$ to 7 s.f.

$= 127.11$ correct to 5 s.f.

In the last example, a calculator will give the value of $(1.998)^7$ to about 8 s.f. (depending on the particular calculator). If, however, the value is required to, say, 15 s.f., the method used in the worked example will give the extra accuracy.

Even greater accuracy can be achieved by using a spreadsheet. This is an extract from a spreadsheet that allows numbers to be shown to a maximum of 15 decimal places. The second column gives the values of the individual terms of the series. The last column gives the sum of the terms from the first to the $(r+1)$th term and it shows how the approximation improves as more terms are added in.

r	$128\binom{7}{r}(0.001)^r$	sum of first $(r+1)$ terms
0	128	128.000000000000000
1	-0.896	127.104000000000000
2	0.002688	127.106688000000000
3	-4.48E-06	127.106683520000000
4	4.48E-09	127.106683524480000
5	-2.688E-12	127.106683524477000
6	8.96E-16	127.106683524477000
7	-1.28E-19	127.106683524477000

The next worked example illustrates how a series expansion enables us to find a simple function which can be used as an approximation to a given function when x has values that are close to zero.

2 If x is so small that x^2 and higher powers can be neglected show that

$$(1-x)^5\left(2+\frac{x}{2}\right)^{10} \approx 2^9(2-5x)$$

> Use the binomial expression of $(1-x)^5$ and neglect terms containing x^2 and higher powers of x.

$$(1-x)^5 \approx 1-5x$$

Similarly

$$\left(2+\frac{x}{2}\right)^{10} \equiv 2^{10}\left(1+\frac{x}{4}\right)^{10}$$

$$\approx 2^{10}\left[1+10\left(\frac{x}{4}\right)\right]$$

Therefore

$$(1-x)^5\left(2+\frac{x}{2}\right)^{10} \approx 2^{10}(1-5x)\left(1+\frac{5x}{2}\right)$$

$$= 2^9(1-5x)(2+5x)$$

$$\approx 2^9(2-5x) \quad \text{again neglecting the term in } x^2.$$

The graphical significance of the approximation in the last example is interesting.

If $\qquad y = (1-x)^5\left(2+\frac{x}{2}\right)^{10}$

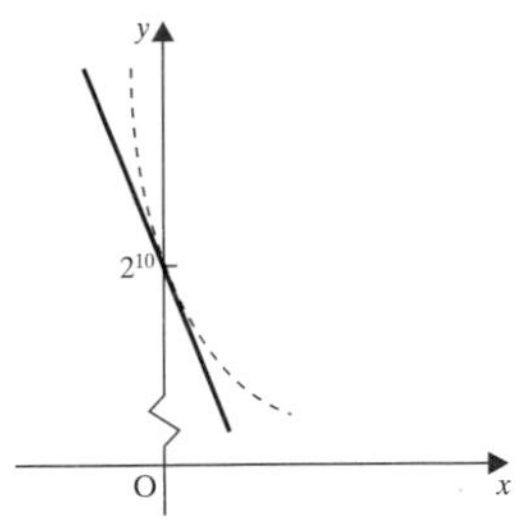

then, for values of x close to zero, $y \approx 2^9(2-5x)$ which is the equation of a straight line, i.e.

$y = 2^9(2-5x)$ is the tangent to $y = (1-x)^5\left(2+\frac{x}{2}\right)^{10}$ at the point where $x = 0$.

Note that the function $2^9(2-5x)$ is called a *linear approximation* for the function $(1-x)^5\left(2+\frac{x}{2}\right)^{10}$ in the region where $x \approx 0$.

EXERCISE 34B

1 By substituting 0.01 for x in the binomial expansion of $(1-2x)^{10}$, find the value of $(0.98)^{10}$ correct to four decimal places.

2 By substituting 0.05 for x in the binomial expansion of $\left(1+\frac{x}{5}\right)^6$, find the value of $(1.01)^6$ correct to four significant figures.

3 Use the binomial expansion of $(2+x)^7$ to show that, correct to 3 d.p., $(2.08)^7 = 168.439$.

4 Show that, if x is small enough for x^2 and higher powers of x to be neglected, the function $(x-2)(1+3x)^8$ has a linear approximation of $-2-47x$.

5 If x is so small that x^3 and higher powers of x are negligible, show that $(2x+3)(1-2x)^{10} \approx 3-58x+500x^2$.

6 By neglecting x^2 and higher powers of x find linear approximations for the following functions in the immediate neighbourhood of $x = 0$.

a $(1-5x)^{10}$ **b** $(2-x)^8$ **c** $(1+x)(1-x)^{20}$

The Binomial Theorem for any Value of n

The expansion of $(1+x)^n$ as a series when n is not a positive integer is very similar to the series when n is a positive integer with two important differences:

- the series does not terminate but carries on to infinity,
- the series is valid only for values of x in the range $-1 < x < 1$,

i.e. $\quad (1+x)^n \equiv 1 + \binom{n}{1}x + \binom{n}{2}x^2 + \binom{n}{3}x^3 + \dots$

for *any* value of *n* provided that $|x| < 1$.

Similarly, $(a+bx)^n = a^n\left(1+\frac{bx}{a}\right)^n$

$= a^n\left[1+\binom{n}{1}\left(\frac{bx}{a}\right)+\binom{n}{2}\left(\frac{bx}{a}\right)^2+\binom{n}{3}\left(\frac{bx}{a}\right)^3+\dots\right]$ provided that $\left|\frac{bx}{a}\right| < 1$.

Notice that this is true only when x has a value within the stated range, therefore this *range must be stated* for every expansion.

The proof of this expansion is in the next section of this chapter, but there are two particular expansions that can be verified.
Using the expansion of $(1+x)^n$ with $n = -1$ gives

$$(1+x)^{-1} = 1+(-1)x+\frac{(-1)(-2)}{2!}x^2+\frac{(-1)(-2)(-3)}{3!}x^3+\frac{(-1)(-2)(-3)(-4)}{4!}x^4+\dots$$
$$= 1-x+x^2-x^3+x^4-\dots$$

The RHS is a GP with common ratio $-x$, and so has a sum to infinity of

$$\frac{1}{1-(-x)} = \frac{1}{1+x} = (1+x)^{-1} \text{ provided that } |x|<1, \text{ i.e.}$$

$$\mathbf{(1+x)^{-1} = 1-x+x^2-x^3+x^4-\dots} \quad \textbf{provided that } |x|<1.$$

Replacing x with $-x$ gives

$$\mathbf{(1-x)^{-1} = 1+x+x^2+x^3+x^4-\dots} \quad \textbf{provided that } |x|<1.$$

These two expansions are well worth memorising.

EXAMPLES 34C

1 Expand each of the following functions as a series of ascending powers of x up to and including the term in x^3 stating the set of values of x for which each expansion is valid.

a $(1+x)^{1/2}$ **b** $(1-2x)^{-3}$ **c** $(2-x)^{-2}$

For $|x|<1$

$$(1+x)^n = 1+nx+\frac{n(n-1)}{2!}x^2+\frac{n(n-1)(n-2)}{3!}x^3+\dots \qquad [1]$$

a Replacing n by $\frac{1}{2}$ in [1] gives

$$(1+x)^{1/2} = 1+\tfrac{1}{2}x+\frac{\frac{1}{2}(\frac{1}{2}-1)}{2!}x^2+\frac{\frac{1}{2}(\frac{1}{2}-1)(\frac{1}{2}-2)}{3!}x^3+\dots$$
$$= 1+\tfrac{1}{2}x+\frac{\frac{1}{2}(-\frac{1}{2})}{2!}x^2+\frac{\frac{1}{2}(-\frac{1}{2})(-\frac{3}{2})}{3!}x^3+\dots$$
$$= 1+\frac{x}{2}-\frac{x^2}{8}+\frac{x^3}{16}-\dots \text{ for } |x|<1$$

b Replacing n by -3 and x by $-2x$ in [1] gives

$$(1-2x)^{-3} = 1+(-3)(-2x)+\frac{(-3)(-4)}{2!}(-2x)^2+\frac{(-3)(-4)(-5)}{3!}(-2x)^3+\dots$$
$$= 1+6x+24x^2+80x^3+\dots$$

provided that $|2x|<1$, i.e. $-\frac{1}{2}<x<\frac{1}{2}$.

c $(2-x)^{-2} = 2^{-2}(1-\frac{1}{2}x)^{-2}$

Replacing n by -2 and x by $-\frac{1}{2}x$ in [1] gives

$$(2-x)^{-2} = \frac{1}{4}\left[1+(-2)(-\tfrac{1}{2}x)+\frac{(-2)(-3)}{2!}(-\tfrac{1}{2}x)^2 + \frac{(-2)(-3)(-4)}{3!}(-\tfrac{1}{2}x)^3+\ldots\right]$$

$$= \tfrac{1}{4}(1+x+\tfrac{3}{4}x^2+\tfrac{1}{2}x^3+\ldots)$$

$$= \tfrac{1}{4}+\tfrac{1}{4}x+\tfrac{3}{16}x^2+\tfrac{1}{8}x^3+\ldots$$

The expansion of $(1-\frac{1}{2}x)^{-2}$ is valid for $|\frac{1}{2}x|<1$, i.e. for $-2<x<2$.

Therefore the expansion $(2-x)^{-\frac{1}{2}}$ also is valid for $-2<x<2$.

2 Express $\dfrac{5}{(1+3x)(1-2x)}$ in partial fractions.

Hence expand $\dfrac{5}{(1+3x)(1-2x)}$ as a series of ascending powers of x giving the first four terms and the range of values of x for which the expansion is valid.

Expressing $\dfrac{5}{(1+3x)(1-2x)}$ in partial fractions gives

$$\frac{5}{(1+3x)(1-2x)} = \frac{3}{(1+3x)}+\frac{2}{(1-2x)} = 3(1+3x)^{-1}+2(1-2x)^{-1}$$

Now $(1+x)^{-1} = 1-x+x^2-x^3+\ldots$ for $-1<x<1$

Replacing x by $3x$ gives

$$(1+3x)^{-1} = 1-3x+(3x)^2-(3x)^3+\ldots$$

$$= 1-3x+9x^2-27x^3+\ldots \quad \text{for} \quad -1<3x<1$$

Also $(1-x)^{-1} = 1+x+x^2+\ldots$ and replacing x by $2x$ gives

$$(1-2x)^{-1} = 1+(2x)+(2x)^2+(2x)^3+\ldots$$

$$= 1+2x+4x^2+8x^3+\ldots \quad \text{for} \quad -1<-2x<1$$

Hence $\dfrac{5}{(1+3x)(1-2x)} = 3(1+3x)^{-1}+2(1-2x)^{-1}$

$$= (3+2)+(-9+4)x+(27+8)x^2+(-81+16)x^3+\ldots$$

provided that $-\frac{1}{3}<x<\frac{1}{3}$ *and* $-\frac{1}{2}<x<\frac{1}{2}$.

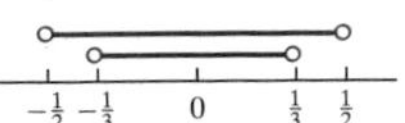

Therefore the first four terms of the series are $5-5x+35x^2-65x^3$. The expansion is valid for the range of values of x satisfying both $-\frac{1}{3}<x<\frac{1}{3}$ *and* $-\frac{1}{2}<x<\frac{1}{2}$,

i.e. for $-\frac{1}{3}<x<\frac{1}{3}$.

3 Expand $\sqrt{\dfrac{1+x}{1-2x}}$ as a series of ascending powers of x up to and including the term containing x^2.

$$\sqrt{\frac{1+x}{1-2x}} \equiv (1+x)^{1/2}(1-2x)^{-\frac{1}{2}}$$

Now $(1+x)^{1/2} = \left[1+\tfrac{1}{2}x+\dfrac{(\frac{1}{2})(-\frac{1}{2})}{2!}x^2+\dots\right]$ for $-1<x<1$

and $(1-2x)^{-1/2} = \left[1+(-\tfrac{1}{2})(-2x)+\dfrac{(-\frac{1}{2})(-\frac{3}{2})}{2!}(-2x)^2+\dots\right]$

for $-1 < 2x < 1$

Hence
$$\begin{aligned}\sqrt{\frac{1+x}{1-2x}} &\equiv (1+x)^{1/2}(1-2x)^{-1/2}\\ &= (1+\tfrac{1}{2}x-\tfrac{1}{8}x^2+\dots)(1+x+\tfrac{3}{2}x^2+\dots)\\ &= 1+(\tfrac{1}{2}x+x)+(\tfrac{1}{2}x^2-\tfrac{1}{8}x^2+\tfrac{3}{2}x^2)+\dots\\ &= 1+\tfrac{3}{2}x+\tfrac{15}{8}x^2+\dots\end{aligned}$$

provided that $-1<x<1$ *and* $-\frac{1}{2}<x<\frac{1}{2}$, i.e. $-\frac{1}{2}<x<\frac{1}{2}$.

$-1 \quad -\frac{1}{2} \quad 0 \quad \frac{1}{2} \quad 1$

It is interesting to compare the methods used in the last two examples.

In Example 2, the function is expressed as the sum of two binomials and the series is obtained by adding two binomial expansions.

In Example 3 the function is expressed as a product of two binomials and the series is obtained by multiplying two binomial expansions.

The first method has the advantage that it is very much easier to add the terms of two series than it is to multiply them.

Therefore, *whenever possible, a compound function should be expressed as a sum of simpler functions before it is expanded as a series* and, when this is not possible, a compound function should be expressed as a product of simpler functions.

Further Approximations

We have seen how a series can be used to find an approximate value of a rational number without having to calculate its exact value.

The next example illustrates how a series can be used to find the decimal value, to any required degree of accuracy, of an irrational quantity.

EXAMPLES 34C (continued)

4 Use the expansion of $(1-x)^{1/2}$ with $x = 0.02$ to find the decimal value of $\sqrt{2}$ correct to nine decimal places.

$$(1-x)^{1/2} = 1 - \tfrac{1}{2}x + \frac{(\frac{1}{2})(-\frac{1}{2})}{2!}(-x)^2 + \frac{(\frac{1}{2})(-\frac{1}{2})(-\frac{3}{2})}{3!}(-x)^3$$
$$+ \frac{(\frac{1}{2})(-\frac{1}{2})(-\frac{3}{2})(-\frac{5}{2})}{4!}(-x)^4 + \frac{(\frac{1}{2})(-\frac{1}{2})(-\frac{3}{2})(-\frac{5}{2})(-\frac{7}{2})}{5!}(-x)^5 \ldots$$
$$= 1 - \tfrac{1}{2}x - \tfrac{1}{8}x^2 - \tfrac{1}{16}x^3 - \tfrac{5}{128}x^4 - \tfrac{7}{256}x^5 - \ldots$$

This is valid for $-1 < x < 1$ and so is valid when $x = 0.02$.

Replacing x by 0.02 gives
$(0.98)^{1/2} = 1 - 0.01 - 0.000\,05 - 0.000\,000\,5 - 0.000\,000\,006\,25$
$- 0.000\,000\,000\,087\,5 - \ldots$

> The next term in the series is 1.3125×10^{-12} and as this does not contribute to the first ten decimal places we do not need it, or any further terms.

i.e. $\sqrt{\frac{98}{100}} = 0.989\,949\,493\,7$ to 10 d.p.

$\Rightarrow$ $\frac{7}{10}\sqrt{2} = 0.989\,949\,493\,7$ to 10 d.p.

$\therefore$ $\sqrt{2} = 1.414\,213\,562$ correct to 9 d.p.

EXERCISE 34C

Expand the following functions as series of ascending powers of x up to and including the term in x^3. In each case give the range of values of x for which the expansion is valid.

1 $(1-2x)^{1/2}$

2 $(1+5x)^{-2}$

3 $(1-\frac{1}{2}x)^{-3}$

4 $(1+x)^{3/2}$

5 $(3+x)^{-1}$

6 $\left(1+\dfrac{x}{2}\right)^{-1/2}$

7 $\dfrac{1}{(1-x)^2}$

8 $\sqrt{\dfrac{1}{1+x}}$

9 $(1+x)\sqrt{1-x}$

10 $\dfrac{x+2}{x-1}$

11 $\dfrac{2-x}{\sqrt{1-3x}}$

12 $\dfrac{1}{(2-x)(1+2x)}$

13 $\sqrt{\dfrac{1+x}{1-x}}$

14 $\left(1+\dfrac{x^2}{9}\right)^{-1}$

15 $\dfrac{x}{(1+x)(1-2x)}$

16 $\left(1+\dfrac{1}{x}\right)^{-1}$ $\left[\textit{Hint}\ \left(1+\dfrac{1}{x}\right)^{-1} \equiv \left(\dfrac{x+1}{x}\right)^{-1} \equiv \dfrac{x}{1+x}\right]$

17 Expand $\left(1+\dfrac{1}{p}\right)^{-3}$ as a series of descending powers of p, as far as and including the term containing p^{-4}. State the range of values of p for which the expansion is valid.

$\left(\textit{Hint}\ \text{Replace } x \text{ by } \dfrac{1}{p} \text{ in } (1+x)^{-3}\right)$

18 By substituting 0.08 for x in $(1+x)^{1/2}$ and its expansion find $\sqrt{3}$ correct to four significant figures.

19 By substituting $\frac{1}{10}$ for x in $(1-x)^{-1/2}$ and its expansion find $\sqrt{10}$ correct to six significant figures.

20 Expand $\sqrt{\dfrac{1+2x}{1-2x}}$ as a series of ascending powers of x up to and including the term in x^2.

21 If x is so small that x^2 and higher powers of x may be neglected show that
$\dfrac{1}{(x-1)(x+2)} \approx -\frac{1}{2} - \frac{1}{4}x.$

22 By neglecting x^3 and higher powers of x, find a quadratic function that approximates to the function $\dfrac{1-2x}{\sqrt{1+2x}}$ in the region close to $x = 0$.

23 Find a quadratic function that approximates to

$$f(x) = \frac{1}{\sqrt[3]{(1-3x)^2}}$$

for values of x close to zero.

24 Use partial fractions and the binomial series to find a linear approximation for

$$\frac{3}{(1-2x)(2-x)}$$

25 If terms containing x^4 and higher powers of x can be neglected, show that

$$\frac{2}{(x+1)(x^2+1)} \approx 2(1-x)$$

26 Show that

$$\frac{12}{(3+x)(1-x)^2} \approx 4 + \frac{20}{3}x + \frac{88}{9}x^2$$

provided that x is small enough to neglect powers higher than 2.

27 If x is very small, find a cubic approximation for

$$\frac{1}{(3-x)^3}$$

Maclaurin's Theorem

We have seen that the function $(1+x)^n$ can be expressed as a series of ascending powers of x. It is possible to express many functions of x as power series.
Before we look at any individual expansions, however, we will examine the series for $f(x)$, where $f(x)$ is a general function of x.

Suppose that f(x) can be expanded as a series of ascending powers of x, so that

$$f(x) = a_0 + a_1x + a_2x^2 + a_3x^3 + a_4x^4 + \dots \qquad [1]$$

where $a_0, a_1, a_2 \dots$ are unknown constants.

Now if we substitute 0 for x in [1] we get $f(0) = a_0$

i.e. $a_0 = f(0)$

Using $f'(x)$ to denote the first derivative of $f(x)$ with respect to x, differentiating $f(x)$ with respect to x gives

$$f'(x) = a_1 + 2a_2x + 3a_3x^2 + 4a_4x^3 + 5a_5x^4 + \dots \qquad [2]$$

i.e. $a_1 = f'(0)$

Differentiating [2] gives the second derivative of $f(x)$ w.r.t. x, i.e.

$$f''(x) = 2a_2 + (3)(2)a_3x + (4)(3)a_4x^2 + (5)(4)a_5x^3 + \dots \qquad [3]$$

$\Rightarrow$ $f''(0) = 2a_2$

i.e $a_2 = \dfrac{f''(0)}{2}$

Similarly

$$f'''(x) = (3)(2)a_3 + (4)(3)(2)a_4x + (5)(4)(3)a_5x^2 + \dots$$

$\Rightarrow$ $f'''(0) = 3!a_3$

i.e $a_3 = \dfrac{f'''(0)}{3!}$

and $f''''(x) = (4)(3)(2)a_4 + (5)(4)(3)(2)a_5x + \dots$

$\Rightarrow$ $f''''(0) = 4!a_4$

i.e. $a_4 = \dfrac{f''''(0)}{4!}$

and so on.

Now using these values in [1] we have

$$\mathbf{f(x) = f(0) + f'(0)x + \frac{f''(0)}{2!}x^2 + \frac{f'''(0)}{3!}x^3 + \frac{f''''(0)}{4!}x^4} + \dots$$

This is known as *Maclaurin's Theorem.*

The right-hand side is the Maclaurin's Series for $f(x)$ and it can be used to find an expansion for a function of x, in ascending powers of x, provided that:

1) it is possible to find *all* the derivatives of $f(x)$,

2) all the derivatives are defined when $x = 0$,

3) the series obtained is convergent (this condition is stated, and will be used here, but without proof).

We will now use Maclaurin's Theorem to expand some familiar functions.

EXAMPLES 34D

1 Express e^x as a series of ascending powers of x.

When $f(x) = e^x$, $f'(x) = e^x$, $f''(x) = e^x$, $f'''(x) = e^x$, ...

So condition (1) is satisfied.

Then $f(0) = e^0 = 1$, $f'(0) = e^0 = 1$, $f''(0) = 1$, $f'''(0) = 1$, ...

So condition (2) is satisfied.

Now using

$$f(x) = f(0) + f'(0)x + \frac{f''(0)}{2!}x^2 + \frac{f'''(0)}{3!}x^3 + \frac{f''''(0)}{4!}x^4 + \dots$$

gives $$e^x = 1 + x + \frac{x^2}{2} + \frac{x^3}{3!} + \frac{x^4}{4!} + \dots$$

This series is always convergent so it is a valid expansion of e^x for all values of x.

The power series for e^x can be used to derive similar series for other powers of e. For example, we can express e^{3x} as a power series by replacing x by $3x$ in the series found above for e^x.

From $$e^x = 1 + x + \frac{x^2}{2!} + \frac{x^3}{3!} + \frac{x^4}{4!} + \dots$$

we get $$e^{3x} = 1 + (3x) + \frac{(3x)^2}{2!} + \frac{(3x)^3}{3!} + \frac{(3x)^4}{4!} + \dots$$

2 Show that $\sin x$ can be expressed as a series of ascending powers of x.

If $f(x) = \sin x$, $f'(x) = \cos x$, $f''(x) = -\sin x$, $f''''(x) = -\cos x$, ...

So condition (1) is satisfied.

Then $f(0) = \sin 0 = 0$, $f'(0) = \cos 0 = 1$, $f''(0) = -\sin 0 = 0$,

$f'''(0) = -\cos 0 = -1$, ...

So condition (2) is satisfied.

From $$f(x) = f(0) + f'(0)x + \frac{f''(0)}{2!}x^2 + \frac{f'''(0)}{3!}x^3 + \frac{f''''(0)}{4!}x^4 + \dots$$

we get $$f(x) = 0 + x + 0x^2 - \frac{x^3}{3!} - 0\frac{x^4}{4!} + \frac{x^5}{5!} + \dots$$

This series is always convergent so, for all values of x,

$$\sin x = x - \frac{x^3}{3!} + \frac{x^5}{5!} - \frac{x^7}{7!} + \dots$$

3 Use Maclaurin's Theorem to prove that

$$(1+x)^n = 1 + \binom{n}{1}x + \binom{n}{2}x^2 + \binom{n}{3}x^3 + \dots + x^n$$ when n is a positive integer

and $$(1+x)^n = 1 + \binom{n}{1}x + \binom{n}{2}x^2 + \binom{n}{3}x^3 + \dots$$ for *any* real value of n.

When $f(x) = (1+x)^n$,

$f'(x) = n(1+x)^{n-1}$, $f''(x) = n(n-1)(1+x)^{n-2}$,

$f'''(x) = n(n-1)(n-2)(1+x)^{n-3}$, ...

so condition (1) is satisfied.

Also $f(0) = 1$, $f'(0) = n$, $f''(0) = n(n-1)$, $f'''(0) = n(n-1)(n-2)$, ...

so condition (2) is satisfied,

$\therefore$ using $f(x) = f(0) + f'(0)x + \frac{f''(0)}{2!}x^2 + \frac{f'''(0)}{3!}x^3 + \ldots$

gives $(1+x)^n \equiv 1 + nx + \frac{n(n-1)}{2!}x^2 + \frac{n(n-1)(n-2)}{3!}x^3 + \ldots$

$$= 1 + \binom{n}{1}x + \binom{n}{2}x^2 + \binom{n}{3}x^3 + \ldots$$

Now if n is a positive integer, the $(n+1)$th derivative of $f(x)$ is zero for all values of x as the last factor in the numerator is $(n-n)$, and all subsequent derivatives are also zero. So the series terminates with x^n and is valid for all values of x.

When n is not a positive integer, there is no value of n for which any of the factors in any derivative is zero, so the series is infinite.

> At this stage, it is not possible to prove that the infinite series converges only if $|x| < 1$, but it is intuitively obvious that the individual terms must approach zero for their sum to approach a finite value. For this to happen, the numerical value of x must be less than 1.

4 Show that $\ln x$ can *not* be expanded as a power series in x.

If $f(x) = \ln x$, $f'(x) - \frac{1}{x} \Rightarrow f'(0) = \frac{1}{0}$ which is undefined.

Condition (2) is not satisfied so $\ln x$ cannot be expressed as a power series in x.

> Note, however, that $\ln(1+x)$ *can* be expanded as a power series as will be found in the next Exercise, Question 3.

EXERCISE 34D

Use Maclaurin's Theorem to express each function as a series of ascending powers of x, up to the term containing the given power of x. Some of these series are convergent only for limited ranges of values of x; in each of these cases the range is given.

1 e^{-x}; x^4

2 $\cos x$; x^6

3 $\ln(1+x)$; x^4 for $-1 < x \leqslant 1$

4 $(1+x)^{12}$; x^3

5 e^{2x}; x^3

6 $\sin 3x$; x^5

7 $(1-x)^{-1}$; x^4 for $|x|<1$

8 $(1+x)^{1/2}$; x^4 for $|x|<1$

9 $\dfrac{e^{2x}-1}{e^x}$; x^4

$\left(\text{Hint: express } \dfrac{e^{2x}-1}{e^x} \text{ as } \dfrac{e^{2x}}{e^x} - \dfrac{1}{e^x}\right)$

10 $\ln(1-x)$; x^4 for $-1 \leqslant x < 1$

Further Approximations

We have seen how a binomial series can be used to find an approximate value of a number. The following example illustrates how a power series can be used to find the decimal value, to any degree of accuracy required, of an irrational number.

EXAMPLES 34E

1 Use the Maclaurin expansion of $\sin x$ to show that

a for small values of x, $\sin x \approx x$.

b $\sin 0.1 \text{ rad} = 0.10$ correct to 2 decimal places.

Using Example 34d, number 2

$$\sin x = x - \frac{x^3}{3!} + \frac{x^5}{5!} - \frac{x^7}{7!} + \dots \quad \text{for all values of } x.$$

a If x is small enough for x^3 and higher powers of x to be negligible in value then

$$\sin x \approx x$$

b $\sin(0.1) = (0.1) - \dfrac{(0.1)^3}{3!} + \dots$

> The next term, $\dfrac{(0.1)^5}{5!}$, is so small that it will not affect the first 3 decimal places.

$$\therefore \quad \sin(0.1) \approx (0.1) - 0.000\,16 + \dots$$
$$= 0.0998$$
$$= 0.10 \quad \text{correct to 2 d.p.}$$

2 Show that, if x is small enough for x^2 and higher powers of x to be ignored, $e^x + e^{2x} \approx 2 + 3x$.

$$e^x + e^{2x} = \left[1 + x + \frac{x^2}{2!} + \dots\right] + \left[1 + 2x + \frac{(2x)^2}{2!} + \dots\right]$$
$$= 2 + 3x + \dots$$
$$\approx 2 + 3x \quad \text{when } x \text{ is small.}$$

> Note that $2+3x$ is a linear approximation to $e^x + e^{2x}$ and further, the line $y = 2+3x$ is the tangent to the curve $y = e^x + e^{2x}$ at the point where $x = 0$.

EXERCISE 34E

Find a linear approximation for each function when the value of x is small.

1 $\sin 3x$ **2** $\ln(1+3x)$ **3** $(1+x)e^x$

Find a quadratic approximation for each function when the value of x is small.

4 $\cos x$ **5** e^{-2x} **6** $x\ln(1+x)$

7 Use the expansion of e^x to find the value of $e^{0.1}$ correct to 2 significant figures.

8 Write down u_n and u_{n+1} in terms of n where u_n and u_{n+1} are the nth and $(n+1)$th terms in the Maclaurin expansion of e^x. Hence find a recurrence relation between u_n and u_{n+1}.

9 This question demonstrates how a spreadsheet can be used to find the sum of the first n terms of a series accurate to however many significant figures the software allows. In particular it shows how to find successfully better approximations to the value of $e^{0.2}$. *You will need to use a computer with spreadsheet software.*

a Enter labels n, u_n, Σu_n in cells A1, A2, A3.

Then enter 1 in cell B1, the formula B1 + 1 in B2 and replicate this across the sheet. (This generates the sequence 1, 2, 3, ...)

b Enter 1 in C1, 0.2 in C2, the formula $\dfrac{\text{C2} \times 0.2}{\text{B2}}$ in C3 and replicate across

(this generates the sequence $1 + 0.2 + \dfrac{0.2^2}{2!} + \ldots$).

c Finally enter the formula for the sum of cells C1 to C1 in D1, C1 to C2 in D2 and again replicate this across the sheet. This generates the sum of the series to n terms for $n = 1, 2, 3, \ldots$

d How many terms of the series are needed to give an answer correct to 4 decimal places?

e How many terms are needed to give $e^{0.2}$ correct to 15 decimal places?

f In the C line of cells replace 0.2 by other values and hence find e^x for other values of x, giving these values correct to 3 significant figures. Compare the rates at which the sequence of values for the sums to n terms of the series converge.

10 a Write down the Maclaurin expansion of $(1-x)^{1/2}$.

b Deduce a recurrence relation between successive terms of this series.

c Repeat Question 9 for the expansion of $(1-x)^{1/2}$ with $x = 0.5$ and the appropriate formula in the second row of cells (derive this from the result of part **b**) to find $\sqrt{0.5}$ to as many significant figures as possible.

d Find the square roots of some other numbers by using other values for x.

e What happens if 1 is used as a value for x?

f What happens if 2 is taken as the value for x?

MIXED EXERCISE 34

1 Find the first three terms and the last term in the expansion of $(1+2x)^9$ as a series of ascending powers of x. For what values of x is this expansion valid?

2 Expand $(1+3x)^5$ as a series of ascending powers of x.
Hence write down the expansion of $(1-3x)^5$.
Hence show that $(1+3\sqrt{3})^5+(1-3\sqrt{3})^5 = 7832$.

3 When a sum of money is invested at 2% pa compound interest, it grows by a factor of 1.02^{25}. By using a suitable series expansion together with a suitable value of x, show that $1.02^{25} = 1.64$ correct to 3 significant figures.

4 Expand $\dfrac{1}{1+2x}$ as a series of ascending powers of x, giving the first three terms.

5 Express $\dfrac{1}{(x-1)(2x-1)}$ in partial fractions.

Hence expand $\dfrac{1}{(x-1)(2x-1)}$ as a series of ascending powers of x up to and including the term in x^3 and state the range of values of x for which it is valid.

6 Find the first three terms in the expansion of $\dfrac{1}{(1+x)(1+x^2)}$ as a series of ascending powers of x and state the range of values of x for which it is valid.

7 Expand $\dfrac{2-x-3x^2}{(1-2x)(1+x)^2}$ as a series of ascending powers of x up to and including the term in x^2 and state the range of values of x for which it is valid.

8 Find the first four terms in the expansion of $\sqrt{1-2x}$ as a series of ascending powers of x.

9 The coefficient of x^2 in the expansion of $(4-3x)^n$ as a series of ascending powers of x is $-\frac{9}{64}$. Show that n satisfies the equation $4^{n+1} = \dfrac{2}{n(1-n)}$ and hence verify that $n = \frac{1}{2}$.

10 Find a linear approximation to the curve $y = \dfrac{1}{(1-2x)^2}$ near the origin.

11 Express $\cos 2x$ as a series of ascending powers of x, as far as the term in x^2.

12 Show that $\dfrac{e^{2x}-1}{e^x} \approx 2x$ if x is small enough for x^2 and higher powers of x to be ignored.

13 Find the first two terms in the expansion of $e^x + \ln(1-x)$ as a series of ascending powers of x.

14 Show that $2^x = e^{x\ln 2}$. Hence find the first three terms in the expansion of 2^x as a series of ascending powers of x. Use your series with $x = \frac{1}{3}$ to find an approximate value for $\sqrt[3]{2}$.

CHAPTER 35 Equations 3

This chapter brings together some categories of equation that have exact solutions and, by looking at the form of the equation, suggests possible ways of solving them. Remember though, that it is not always possible to find exact solutions, even when they exist, so the chapter ends with some methods for finding approximate solutions.

Trigonometric Equations

Successful solutions of trig equations depend on three factors; recognising and knowing the trig identities, correctly classifying the equation so that the first attempt at solution is likely to be successful and, finally, experience.

There are two general approaches to trig equations. First see whether the equation can be factorised, either in its given form or by applying an appropriate identity. If this is not possible (and often it is not) try to reduce the equation to a form which involves only one variable, i.e. one trig ratio of one angle.

In this section we give most of the common categories of trig equation followed by an appropriate method of solution. This list is neither exhaustive nor infallible, but it covers most forms of equation met at this level.

A. Equations Containing One Angle Only

Form of Equation	Method
1. $a\cos\theta + b\sin\theta = 0$	Divide by $\cos\theta$, provided that $\cos\theta \neq 0$.
2. $a\cos\theta + b\sin\theta = c$	Write LHS as $R\cos(\theta+\alpha)$.
3. $\left.\begin{aligned} a\cos^2\theta + b\sin\theta &= c \\ a\sin^2\theta + b\cos\theta &= c \\ a\tan^2\theta + b\sec\theta &= c \end{aligned}\right\}$	Use the Pythagorean identities to express in terms of one ratio only.
4. $\left.\begin{aligned} a\cos\theta + b\tan\theta &= 0 \\ a\sin\theta + b\tan\theta &= 0 \end{aligned}\right\}$	Multiply by $\cos\theta$ and exclude $\cos\theta = 0$ from solution.

B. Equations Containing Multiples of One Angle

Form of Equation	Method
1. $\left.\begin{aligned} a\cos\theta + b\cos 2\theta &= c \\ a\sin\theta + b\cos 2\theta &= c \end{aligned}\right\}$	Use the double angle formulae to reduce to Section A type.
2. $\left.\begin{aligned} \cos a\theta &= c \\ \sin a\theta &= c \end{aligned}\right\}$	Solve for $a\theta$ and divide by a.

It is important to realise that a method given in this list does not represent the only way of solving a particular equation, nor does it always lead to the quickest solution. Sometimes an equation can be simplified quickly when part of it is recognised as part of a trig identity. Sometimes it may be necessary to classify each side of an equation independently.

EXAMPLE 35A Find the solution of the equation in the interval $0 \leqslant \theta \leqslant \frac{1}{2}\pi$

$$\cos 2\theta \sin\theta + \sin 2\theta \cos\theta = \cos 3\theta$$

$\cos 2\theta \sin\theta + \sin 2\theta \cos\theta = \cos 3\theta$

> The LHS is recognised as the expansion of $\sin(2\theta + \theta)$, i.e. $\sin 3\theta$.

$\sin 3\theta = \cos 3\theta$

> Divide both sides by $\cos 3\theta$ ($\cos 3\theta \neq 0$).

$\tan 3\theta = 1$

In the interval $0 \leqslant 3\theta \leqslant \frac{3}{2}\pi$, $3\theta = \dfrac{\pi}{4}, \dfrac{5\pi}{4}$

In the interval $0 \leqslant \theta \leqslant \frac{1}{2}\pi$, $\theta = \dfrac{\pi}{12}, \dfrac{5\pi}{12}$

> Check: $\cos\frac{\pi}{4} \neq 0$, $\cos\frac{3\pi}{4} \neq 0$

EXERCISE 35A Solve each equation for values of x in the range $0 \leqslant x \leqslant \pi$.

1 $\sin 2x \cos x + \cos 2x \sin x = 1$

2 $2\sin^2 x + \cos x = 1$

3 $5\cos x + 12\sin x = 13$

4 $2\sin x + \sin 2x = 0$

5 $\cos^2 x + 2\sin^2 x = 2$

6 $4\sin x - 5 = 3\cos x$

7 $\sin^2 x = 2\cos x + 1$

8 $\cos x = 2\tan x$

Solve each equation for values of x in the range $0 \leqslant x < 360°$ giving answers correct to 1 decimal place where necessary.

9 $\cos 2x + \sin x = 1$

10 $2\sin x - \cos x = 1$

11 $\sin x + \tan x = 0$

12 $\cos 2x + 2\sin x = 0$

13 $\tan x + 2\sec^2 x = 3$

14 $2\cos x + \cos 2x = 2$

15 $\cos(x + 30°) + \sin x = 0$

16 $\cos x \cos 2x - \sin x \sin 2x = \frac{1}{2}$

Polynomial Equations

If a quadratic equation has real roots then these roots can always be found either by factorisation or by using the formula.

If the equation cannot be reduced to a quadratic, real roots can *sometimes* be found by using the factor theorem but only if these roots are integers or simple rational numbers.

When the factor theorem fails to find *exact* solutions, they can be obtained by other methods in some special cases, one of which we look at now.

Equations with a Repeated Root

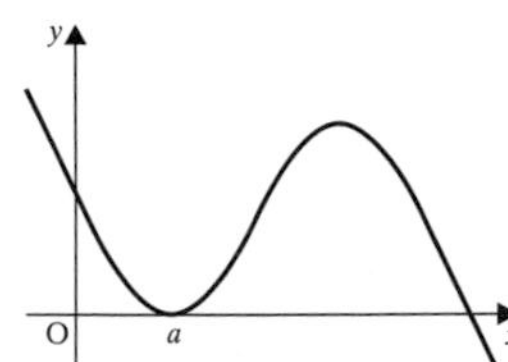

If the equation $f(x) = 0$ has a repeated root, a, then $(x-a)^2$ is a factor of $f(x) = 0$. This means that the curve $y = f(x)$ *touches* the x-axis at $x = a$.

As $y = f(x)$ has a turning point at $x = a$, $f'(x) = 0$ when $x = a$, i.e. $f'(a) = 0$, therefore

$f(a) = 0$ *and* $f'(a) = 0$ $\Leftrightarrow$ $f(x)$ has a repeated factor.

EXAMPLES 35B

1 Solve the equation $18x^3 - 111x^2 + 224x - 147 = 0$ given that it has two equal roots.

The repeated root of the given equation

$$f(x) = 18x^3 - 111x^2 + 224x - 147 = 0 \quad [1]$$

also satisfies the equation

$$f'(x) = 54x^2 - 222x + 224 = 0$$

$$\Rightarrow \quad 27x^2 - 111x + 112 = 0$$

$$\Rightarrow \quad (9x-16)(3x-7) = 0$$

So *either* $x = \frac{16}{9}$ or $x = \frac{7}{3}$ is a solution of the given equation.

> To check which of these values is the repeated root of the given equation, each is substituted in turn into equation [1].

We find that $f(\frac{16}{9}) \neq 0$ and that $f(\frac{7}{3}) = 0$.

$f(x)$ and $f'(x)$ are both zero when $x = \frac{7}{3}$ so $x = \frac{7}{3}$ is the repeated root of the given equation.

Hence $\quad 18x^3 - 111x^2 + 224x - 147 \equiv (3x-7)^2(ax+b)$

Comparing coefficients of x^3 gives $18 = 9a \quad \Rightarrow \quad a = 2$

Comparing constants gives $-147 = 49b \quad \Rightarrow \quad b = -3$

Therefore $18x^3 - 111x^2 + 224x - 147 \equiv (3x-7)^2(2x-3)$

Then $(3x-7)^2(2x-3) = 0 \quad \Rightarrow \quad x = \frac{7}{3}$ or $\frac{3}{2}$

Equations Involving Square Roots

When an equation contains square roots involving the unknown, we eliminate those square roots by squaring. However, we saw in Chapter 5 that, when both sides of an equation are squared, an extra equation, and hence extra solutions, is introduced.

For example if $x = 2$ then squaring gives $x^2 = 4$.
But $x^2 = 4$ includes both $x = 2$ and $x = -2$, therefore remember that whenever a solution involves squaring, all solutions must be checked in the original equation.

EXAMPLES 35B (continued)

2 Solve the equation $\sqrt{x+8} - \sqrt{x+3} = \sqrt{2x-1}$

Squaring both sides of the equation gives

$x+8-2\sqrt{x+8}\sqrt{x+3}+x+3 = 2x-1 \quad \Rightarrow \quad 6 = \sqrt{x+8}\sqrt{x+3}$

Squaring again gives $36 = (x+8)(x+3)$

$\Rightarrow \quad x^2+11x-12 = 0 \quad \Rightarrow \quad x = 1$ or -12

Checking these values of x in the original equation shows that when $x = 1$, LHS $= \sqrt{9} - \sqrt{4} = 1$ and RHS $= \sqrt{1} = 1$,

so $x = 1$ is a solution of the given equation,

when $x = -12$, RHS $= \sqrt{-25}$ which is not real,

so $x = -12$ is not a solution of the given equation.

Therefore, $x = 1$ is the only solution.

Note that we have looked at a very small sample of the possible types of equation that have exact solutions. Although there are rules for solving a few particular forms of equation, in most cases such rules do not exist. The solution of equations is an art; experience will suggest a likely form of attack but there is never any guarantee that the method will produce a solution and other methods have to be tried. In the following exercise, all the equations have exact solutions and most of them are of a type that has been discussed either in this chapter or earlier in the book. Just a few may be unfamiliar.

EXERCISE 35B

1 Solve the equation $80x^3+88x^2-3x-18 = 0$ given that it has a repeated root.

2 Show that $x = 0$ is not a root of the equation

$$5x^4-16x^3-42x^2-16x+5 = 0$$

Divide the equation by x^2, then use $y = x+1/x$ to show that

$$5(y^2-2)-16y-42 = 0$$

Hence solve the given equation.

3 The equation $20x^3-52x^2+21x+18 = 0$ has a repeated root. Solve it.

Solve the given equations using any suitable methods.

4 $x^3-6x^2+11x-6 = 0$

5 $1+\sqrt{x} = \sqrt{3x-3}$

6 $\sqrt{2x-5} - \sqrt{x-2} = 1$

7 $x^2-3x = 8$

8 $1-\sqrt{x}+\sqrt{x-3} = 0$

9 $\sqrt{3x+1}+\sqrt{x-1} = \sqrt{7x+1}$

10 $|x| = 3-|1-x|$

11 $2x^3-x^2+20 = 0$

12 $\dfrac{x^2}{4}+y^2 = 1$ and $xy = 1$

13 $x^2+y^2+4x-6y = 3$ and $y = x+1$

14 $x^2+y^2+8x-4y+15 = 0$ and $x^2+y^2+6x+2y-15 = 0$

15 $x^4-x^3-12x^2-4x+16 = 0$ (*Hint* Use $y = x+4/x$)

Approximate Solutions

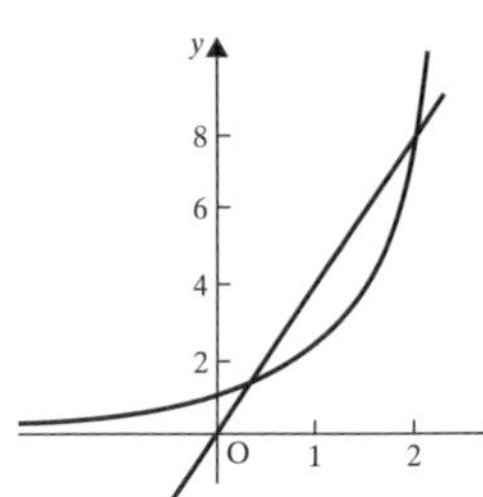

When the roots of an equation cannot be found exactly, we can look for approximate solutions. The first step is to locate the roots roughly and this can be done graphically.

Consider, for example, the equation $e^x = 4x$.

The roots of this equation are the values of x where the curve $y = e^x$ and the line $y = 4x$ intersect.

From the sketch we can see that there is one root between 0 and 0.5 and another root somewhere near 2.

We now need a way to locate the roots more accurately.

Locating the Roots of an Equation

Suppose that we have very roughly located the roots of an equation $f(x) = 0$.

Now consider the curve $y = f(x)$.

The roots of the equation $f(x) = 0$ are the values of x where this curve crosses the x-axis, e.g.

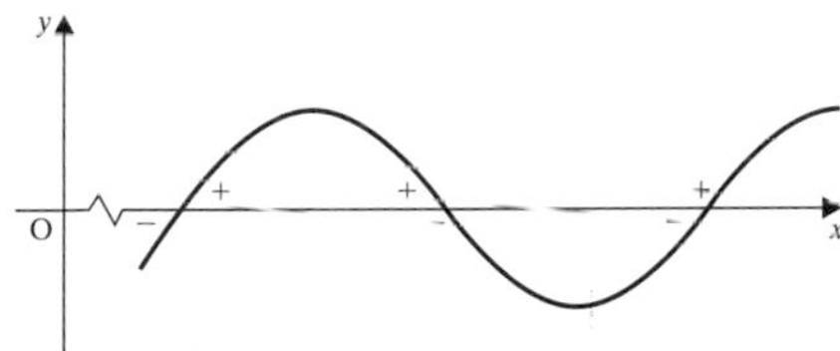

Each time that the curve crosses the x-axis, the sign of y changes. So

if one root only of the equation $f(x) = 0$ lies between x_1 and x_2, and if the curve $y = f(x)$ is unbroken between the points where $x = x_1$ and $x = x_2$, then $f(x_1)$ and $f(x_2)$ are opposite in sign.

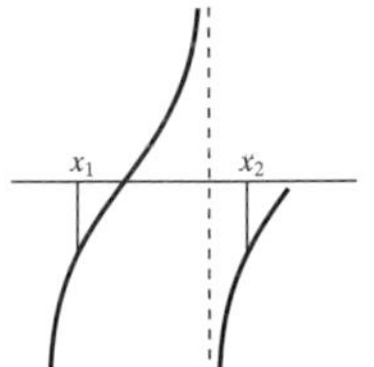

The condition that the curve $y = f(x)$ must be unbroken between x_1 and x_2 is essential as we can see from the curve on the left.

This curve crosses the x-axis between x_1 and x_2 but $f(x_1)$ and $f(x_2)$ have the same sign because the curve is broken between these values.

Returning to the equation $e^x = 4x$, we will now locate the larger root a little more precisely.

First we write the equation in the form $f(x) = 0$, i.e. $e^x - 4x = 0$, then we find where there is a change in the sign of $f(x)$.

We know that there is a root in region of $x = 2$, so we will see if it lies between 1.8 and 2.2.

Using $f(x) = e^x - 4x$, gives $f(1.8) = e^{1.8} - 4(1.8) = -1.1\ldots$

and $f(2.2) = e^{2.2} - 4(2.2) = 0.2\ldots$

Therefore the larger root of the equation lies between 1.8 and 2.2 (and is likely to be nearer to 2.2 as $f(2.2)$ is nearer to zero then $f(1.8)$ is).

EXAMPLE 35C

Find the turning points on the curve $y = x^3 - 2x^2 + x + 1$. Hence sketch the curve and use the sketch to show that the equation $x^3 - 2x^2 + x + 1 = 0$ has only one real root. Find two consecutive integers between which this root lies.

At turning points $\dfrac{dy}{dx} = 0$, i.e. $3x^2 - 4x + 1 = 0$

$\Rightarrow \qquad (3x - 1)(x - 1) = 0$ so $x = \frac{1}{3}$ and 1

When $x = \frac{1}{3}$, $y = \frac{31}{27}$ and when $x = 1$, $y = 1$.

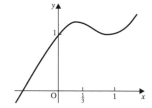

As the curve is a cubic, and as $y \to \infty$ as $x \to \infty$, we deduce that the curve has a maximum point at $(\frac{1}{3}, \frac{31}{27})$ and a minimum point at $(1, 1)$.

From the sketch, we see that the curve $y = x^3 - 2x^2 + x + 1$ crosses the x-axis at one point only, therefore the equation $x^3 - 2x^2 + x + 1 = 0$ has only one real root.

Also from the sketch it appears that this root lies between $x = -1$ and $x = 0$.

Using $f(x) = x^3 - 2x^2 + x + 1$ gives $f(-1) = -3$ and $f(0) = 1$.

As $f(-1)$ and $f(0)$ are opposite in sign, the one real root of $x^3 - 2x^2 + x + 1 = 0$ lies between $x = -1$ and $x = 0$.

EXERCISE 35C

1 Use sketch graphs to determine the number of real roots of each equation. (Some may have an infinite set of roots.)

a $\sin x = \dfrac{1}{x}$ **b** $\cos x = x^2 - 1$ **c** $2^x = \tan x$

d $2^x \sin x = 1$ **e** $(x^2 - 4) = \dfrac{1}{x}$ **f** $x2^x = 1$

g $x \ln x = 1$ **h** $\sin x = x^2$ **i** $\ln x + 2^x = 0$

2 For each equation in Question 1 with a finite number of roots, locate the root, or the larger root where there is more than one, within an interval of half a unit.

3 Find the turning points on the curve whose equation is $y = x^3 - 3x^2 + 1$. Hence sketch the curve and use your sketch to find the number of real roots of the equation $x^3 - 3x^2 + 1 = 0$.

4 Using a method similar to that given in Question 3, or otherwise, determine the number of real roots of each equation.

a $x^4 - 3x^3 + 1 = 0$ **b** $x^3 - 24x + 1 = 0$ **c** $x^5 - 5x^2 + 4 = 0$

5 Show that the equation $3^{-x} = x^2 + 2$ has just one root and find this root to the nearest integer.

6 Find the successive integers between which the smallest root of the equation $2^x = \frac{1}{2}(x + 3)$ lies

Approximations

There are many ways in which successive numerical approximations can be used to find a root of an equation to any degree of accuracy required. Those given below are common methods.

Whatever procedure is used we must first find an interval in which the required root lies. This can be done using the methods described in the last section.

Method 1: The Interval Bisection Method

Suppose that we want to find the larger root of the equation

$$e^x - 3x = 0.$$

The first step is to find two consecutive integer values of x between which the required root lies.

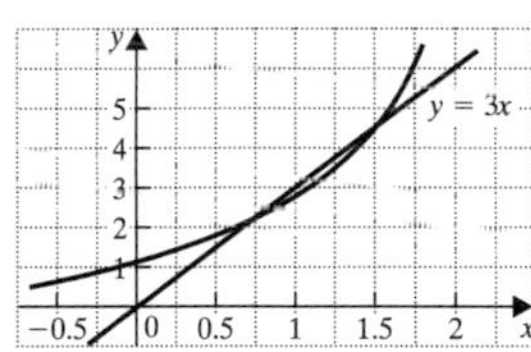

To do this, the equation is written in the form $e^x = 3x$. Simple sketches of the curve $y = e^x$ and the line $y = 3x$ can then be drawn and their points of intersection give the approximate location of the roots of the given equation. A graphics calculator can be used to draw the curves easily and to check results.

We see that the given equation has two roots and that the larger is between 1 and 2. So if the required root is α, we now know that $1 < \alpha < 2$.

Now consider $f(x) = e^x - 3x$

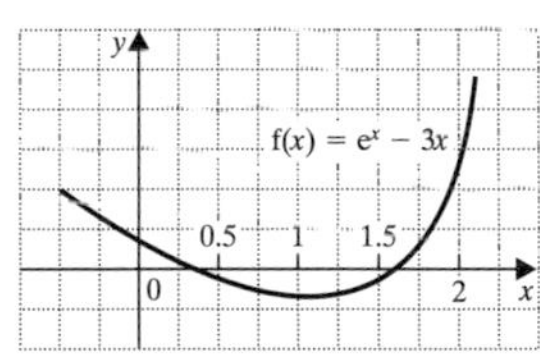

A sketch of $f(x)$ shows that the graph of $y = e^x - 3x$ crosses the x-axis somewhere in the interval from 1 to 2 and at this point $x = \alpha$. We also see that $f(1) < 0$ and $f(2) > 0$, confirming that $1 < \alpha < 2$.

At this stage we cannot estimate the value of α even to the nearest integer, but such an estimate can be made if the interval in which α lies is reduced. So we now consider $f(1.5)$.

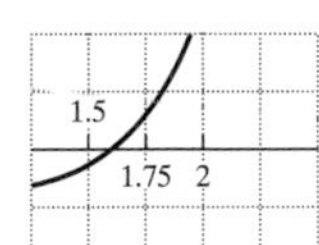

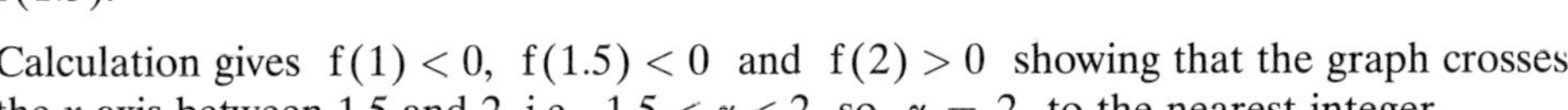

Calculation gives $f(1) < 0$, $f(1.5) < 0$ and $f(2) > 0$ showing that the graph crosses the x-axis between 1.5 and 2, i.e. $1.5 < \alpha < 2$ so $\alpha = 2$ to the nearest integer.

Now halving this interval, we investigate $f(1.75)$ and find that

$$f(1.5) < 0, \quad f(1.75) > 0, \quad f(2) > 0 \quad \Rightarrow \quad 1.5 < \alpha < 1.75$$

This is not accurate enough to give α correct to one decimal place but succesive bisections of the interval can give the value of α to any specified degree of accuracy, e.g.

$$f(1.5) < 0, \quad f(1.625) > 0 \quad \Rightarrow \quad 1.5 < \alpha < 1.625$$

$$f(1.5) < 0, \quad f(1.5625) > 0 \quad \Rightarrow \quad 1.5 < \alpha < 1.5625$$

(We still do not know whether x is 1.5 or 1.6 to 1 dp, so we continue.)

$$f(1.5) < 0, \quad f(1.531\,25) > 0 \quad \Rightarrow \quad 1.5 < \alpha < 1.531\,25$$

Now we *do* know that $\alpha = 1.5$ correct to 1 dp.
At this stage the interval is less than 0.05. If the value of α were needed correct to 2 dp we would continue halving the interval until it becomes less than 0.005. (In this case two further bisections are needed.)

EXERCISE 35D

If you have a graphics calculator you can use it to sketch the graphs and to check the accuracy of the answers, but for exam purposes the *method* of interval bisection *must* be used for the intermediate working.

1 On the same axes, draw the graphs of $y = x^3$ and $y = 5 - x^2$ for values of x from -2 to 2. From your graphs find two consecutive integer values of x between which there is a root of the equation $x^3 + x^2 - 5 = 0$.

Use the interval bisection method to find this root correct to 1 dp.

2 Use sketch graphs to show that there is only one positive root of the equation $3\sin x - x = 0$ where the angle is measured in radians. Use a calculator to show that, if this root is α, then $2.2 < \alpha < 2.3$. By applying the interval bisection method, find this root correct to 2 dp.

3 On the same axes, draw the graphs of $y = 4x$ and $y = 2^x$ for values of x from -1 to 4. Hence show that one root of the equation $2^x - 4x = 0$ is 4, and that there is another root between 0 and 1. Find the value of this root correct to 1 dp.

4 By drawing suitable sketch graphs for the range $0 < x < 4$, find two consecutive integers between which there lies a value of x that satisfies the equation $\ln(x - 2) = 1/x$. Find this value of x correct to 2 sf.

Method 2: x = g(x)

This method can often be used to find a root of an equation $f(x) = 0$ which can be written in the form $x = g(x)$. The roots of the equation $x = g(x)$ are the values of x at the points of intersection of the line $y = x$ and the curve $y = g(x)$.

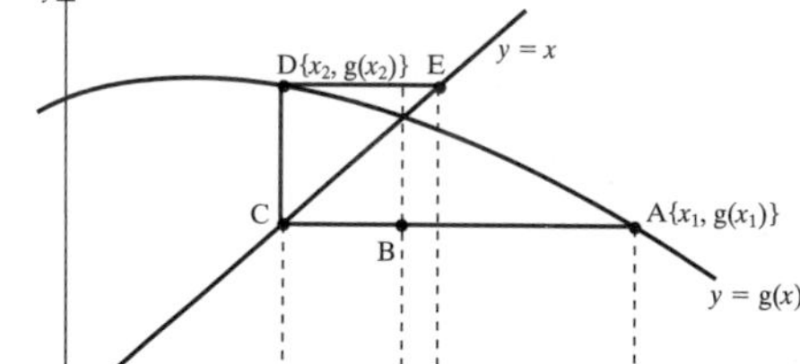

Taking x_1 as a first approximation to a root α then in the diagram,

A is the point on the *curve* where $x = x_1,\ y = g(x_1)$

B is the point where $x = \alpha,\ y = g(x_1)$

C is the point on the *line* where $x = x_2,\ y = g(x_1)$

If, in the region of α, the slope of $y = g(x)$ is less steep than that of the line $y = x$, i.e. provided that $|g'(x)| < 1$,

then $\quad CB < BA$

so x_2 is closer to α than is x_1 and x_2 is a better approximation to α

But C is on the line $y = x$,

therefore $\quad x_2 = g(x_1)$

Now taking the point D on the curve where $x = x_2,\ y = g(x_2)$ and repeating the argument above we find that x_3 is a better approximation to α than is x_2

where $\quad x_3 = g(x_2)$

This process can be repeated as often as necessary to achieve the required degree of accuracy and is called iteration.

The rate at which these approximations converge to α depends on the value of $|g'(x)|$ near α. The smaller $|g'(x)|$ is, the more rapid is the convergence.

It should be noted that this method fails if $|g'(x)| > 1$ near α.
The following diagrams illustrate some of the factors which determine the success, or otherwise, of this method.

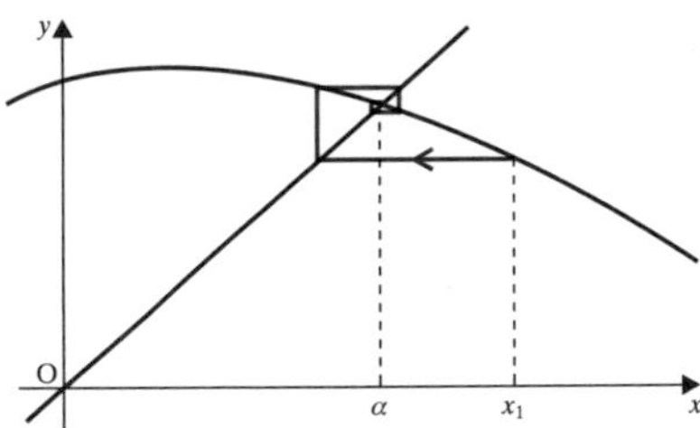

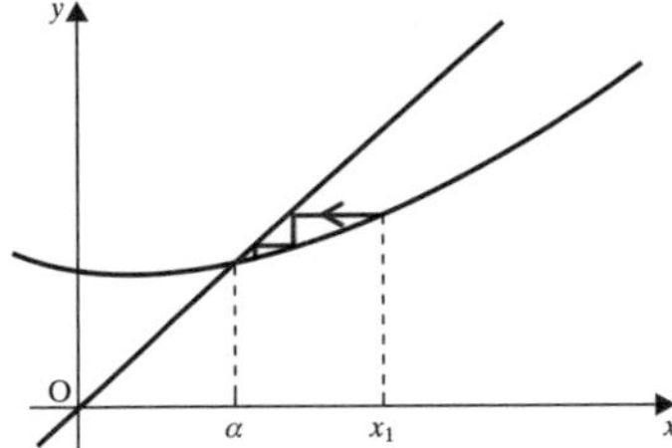

Rapid rate of convergence ($|g'(x)|$ small).

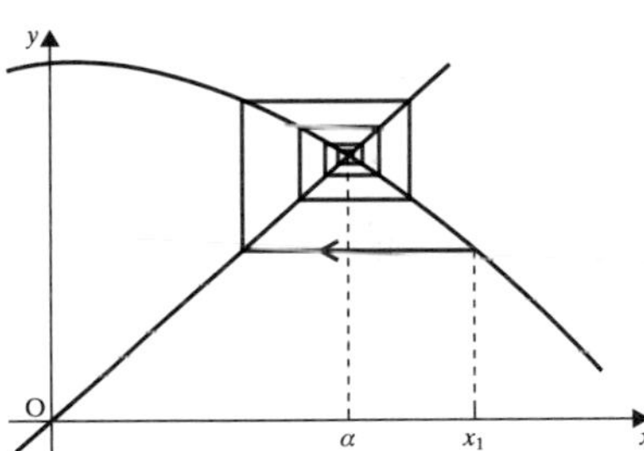

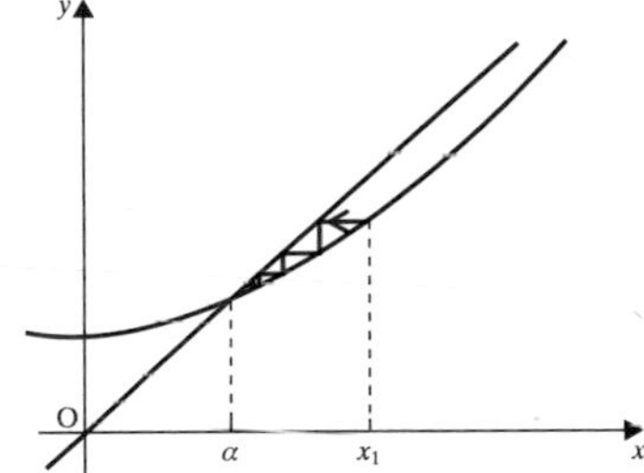

Slow rate of convergence ($|g'(x)| < 1$ but close to 1).

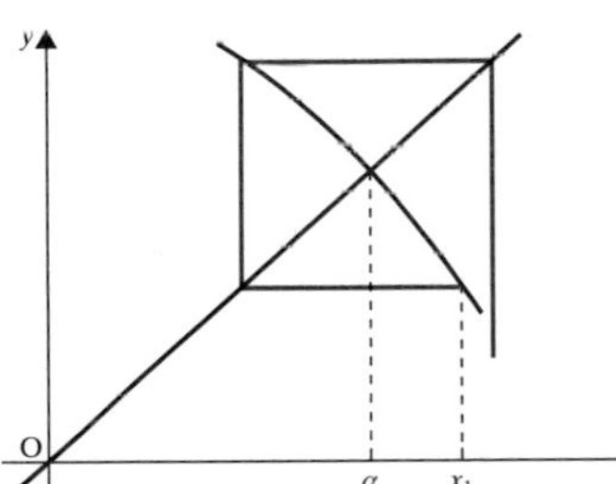

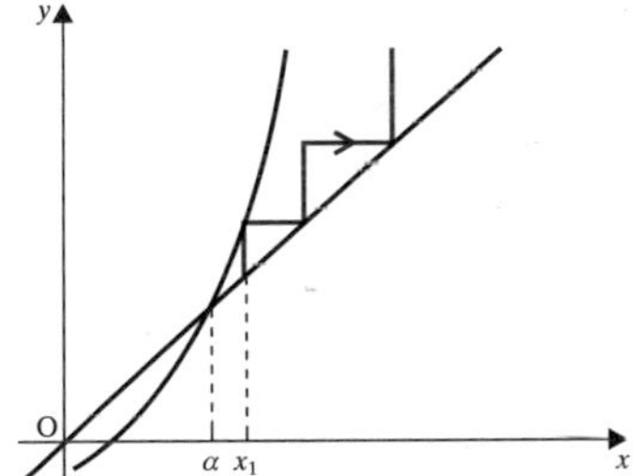

Divergence, i.e. failure, ($|g'(x)| > 1$).

As an example consider the equation $x^3 + 2x^2 + 5x - 1 = 0$

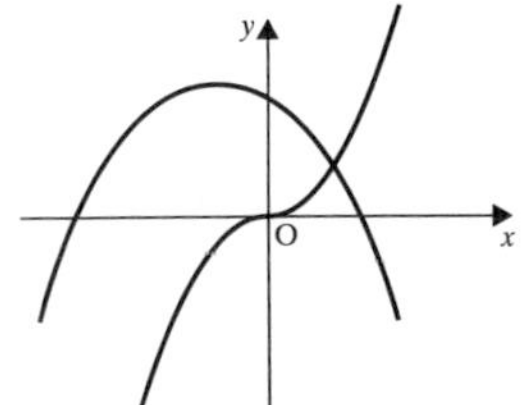

The equation can be written $x^3 = 1 - 5x - 2x^2$ so that a sketch of $y = x^3$ and $y = 1 - 5x - 2x^2$ shows the number of roots.

From the sketch we see that there is only one root and it is near the origin.

The given equation can be written in the form $x = g(x)$

where $\quad g(x) = -\frac{1}{5}(x^3 + 2x^2 - 1)$

We will take $x_1 = 0$ as our first approximation.

A better approximation, x_2, is found from

$$x_2 = g(x_1) = -\tfrac{1}{5}[0^3 + 2(0)^2 - 1] = 0.2$$

Further improvements are obtained by repeating this step,

i.e. $x_3 = g(x_2) = -\frac{1}{5}[(0.2)^3 + 2(0.2)^2 - 1] = 0.1824$

$x_4 = g(x_3) = -\frac{1}{5}[0.1824)^3 + 2(0.1824)^2 - 1] = 0.1854\ldots$

$x_5 = g(x_4) = -\frac{1}{5}[(0.1854\ldots)^3 + 2(0.1854\ldots)^2 - 1] = 0.1849\ldots$ and so on.

The degree of accuracy at any stage can be checked by determining the sign of $f(x)$ on either side of the value so far obtained for the root, e.g., taking $x \approx 0.1850$ we find that $f(0.1846)$ is negative and $f(0.1854)$ is positive, so $x = 0.185$ correct to 3 d.p.

Note This does *not* show that $x = 0.1850$ to 4 d.p.

Method 3: The Newton–Raphson or Newton's Method

This method is based on determining a linear approximation for a function. Suppose that the equation $f(x) = 0$ has a root α and that a is an approximation for α.

The curve $y = f(x)$ cuts the x-axis where $x = \alpha$. If we consider the tangent to $y = f(x)$ at the point where $x = a$ then the point B where this tangent cuts the x-axis will, in most circumstances, be nearer to the point $x = \alpha$.

i.e.

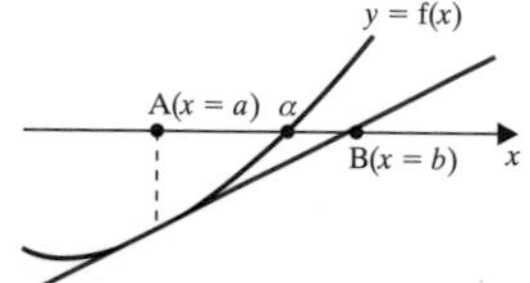

So if this tangent cuts the x-axis at B where $x = b$, then b is a better approximation to α than a is.

The gradient of $y = f(x)$ at the point A is $f'(a)$.
The coordinates of A are $(a, f(a))$.

So the equation of the tangent at A is $y - f(a) = f'(a)(x - a)$

This line cuts the x-axis where $y = 0$, i.e. where $x = a - \dfrac{f(a)}{f'(a)}$

So if a is an approximation for a root of the equation $f(x) = 0$, then $b = a - \dfrac{f(a)}{f'(a)}$ is a better approximation.

As an example we will use Newton's Method to find the root of $xe^x = 3$ correct to three decimal places.

A first approximation to the root is found by drawing the graphs of $y = \dfrac{3}{x}$ and $y = e^x$.

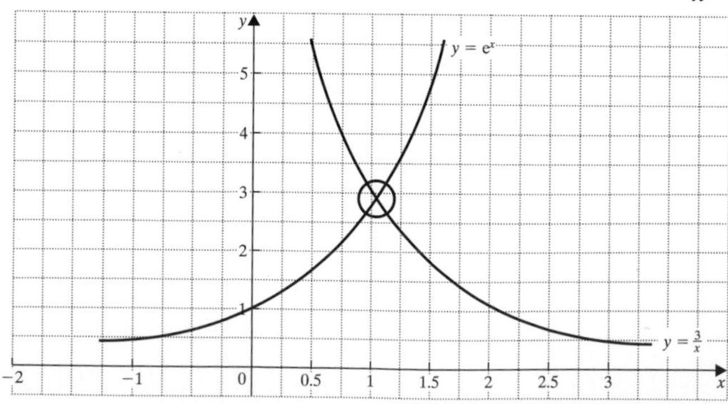

From these graphs we see that $xe^x = 3$ has a root α which is approximately 1.

Now $\qquad f(x) = xe^x - 3$

$\Rightarrow \qquad f'(x) = (x+1)e^x$

Using $b = a - \dfrac{f(a)}{f'(a)}$, and taking $a = 1$ as our first approximation to α, the second approximation is

$$1 - \frac{e-3}{2e} = 1.0518 \text{ (to 4 d.p.)}$$

Taking $a = 1.0518$ and repeating the procedure, the third approximation is

$$1.0518 - \frac{(1.0518)e^{1.0518} - 3}{(2.0518)e^{1.0518}} = 1.0499 \text{ (to 4 d.p.)}$$

So, to three decimal places, the root is likely to be 1.050 and we can check this by calculating f(1.0495) and f(1.0504).
Now f(1.0495) is negative and f(1.0504) is positive. Thus the root lies between these values and, correct to 3 d.p., is 1.050.

The rate of convergence using Newton's Method depends on the crudeness, or otherwise, of the first approximation and on the shape of the curve in the neighbourhood of the root. In extreme cases these factors may lead to failure. Some of these cases are illustrated by the following graphs.

In the following diagrams, α is a root of $f(x) = 0$; a is the first approximation for α and b is the second approximation for α given by Newton's Method. In each case we see that b is *not* a better approximation to the root than a is.

a $f'(a)$ is too small.

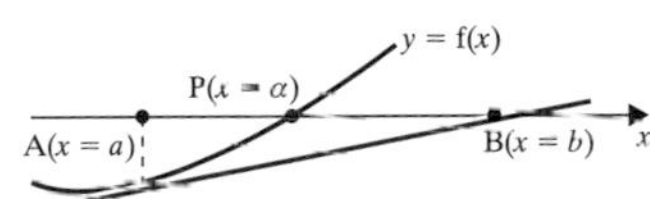

b $f'(x)$ increases too rapidly.

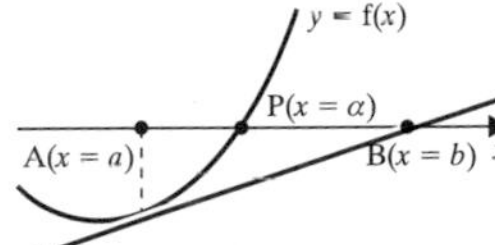

c A is too far from P.

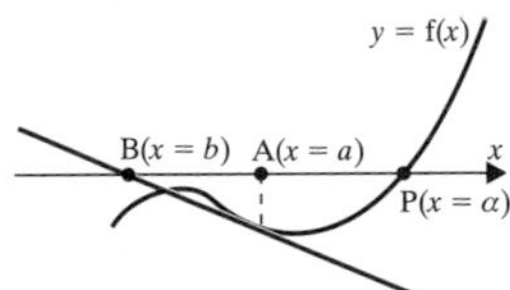

EXERCISE 35E

Show that each of the following equations has a root between $x = 0$ and $x = 1$. Using $x = g(x)$ find this root correct to 2 decimal places.

1 $x^3 - x^2 + 10x - 2 = 0$

2 $3x^3 - 2x^2 - 9x + 2 = 0$

3 $2x^3 + x^2 + 6x - 1 = 0$

4 $x^2 + 8x - 8 = 0$

Use the change of sign of $f(x)$ to find, correct to 2 significant figures, the smallest root of each of the following equations.

5 $4+5x^2-x^3 = 0$

6 $x^4-4x^3-x^2+4x-10 = 0$

Use a graphical method to find a first approximation to the root(s) of each equation. Then apply two stages of Newton's Method to give a better approximation. State the accuracy of each of your results.

7 $\tan x = 2x$
(the smallest positive root)

8 $x^3-6x+3 = 0$
(the negative root)

9 $e^x = 2x+1$

10 $\sin x = 1-x$

11 $x^2 = \ln(x+1)$

12 $e^x(1+x) = 2$

13 $e^x = 3x+1$

14 $x = 1+\ln x$

15 $3+x-2x^2 = e^x$

16 $x^3-3x^2-1 = 0$

17 $e^x = 2\cos x$
(the roots between $-\frac{1}{2}\pi$ and $\frac{1}{2}\pi$)

MIXED EXERCISE 35

Solve the following equations giving exact roots where possible.

1 $3^{2x} = 10$

2 $3\sin x-4\cos x = 2$
for $0 \leqslant x \leqslant 360°$.

3 $x^2+y^2 = 4$ and $x+y = 1$

4 $x^3-3x^2+3x-1 = 0$

5 $|x| = 1-2|x|$

6 $\cos 2x+\cos x = 2$
for all values of x.

7 $2x^4-15x^3+38x^2-39x+18 = 0$ given that it has a repeated root.

8 Show that the equation $x^3-2x^2-1 = 0$ has a root between 2 and 3. Taking 2 as the first approximation to this root, use the Newton–Raphson method twice to obtain a better approximation.

9 Show that the equation $e^x = x^2+2$ has only one root and use the interval bisection method to find this root correct to two significant figures.

10 Show that the equation $x^4+x^2-x = 0$ has two roots. By writing the equation in the form $x = \dfrac{x}{x^3+x}$ find the larger root correct to one significant figure.

CHAPTER 36 Vectors

Vectors

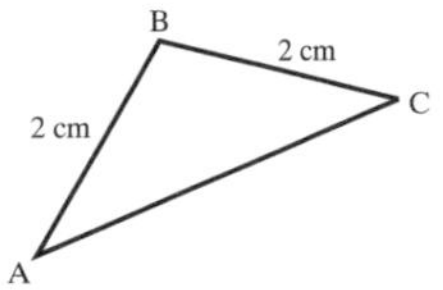

Although we usually assume that two and two make four, this is not always the case.

If, for example a point B is 2 cm from a point A and C is 2 cm from B then, in general, C is *not* 4 cm from A.

AB, BC and AC are displacements. Each of them has a magnitude and is related to a definite direction in space and so is called a *vector*.

A vector is a quantity which has both magnitude and a specific direction in space.

A scalar quantity is one that is fully defined by magnitude alone. Length, for example, is a scalar quantity, as the length of a piece of string does not depend on its direction when it is measured.

Vector Representation

A vector can be represented by a section of a straight line, whose length represents the magnitude of the vector and whose direction, indicated by an arrow, represents the direction of the vector. Such vectors can be denoted by a letter in bold type, e.g. **a** or, when hand-written, by $\underline{a}$.

Alternatively we can represent a vector by the magnitude and direction of a line joining A to B. When we denote the vector by $\overrightarrow{AB}$ or **AB**, the vector in the opposite direction, i.e from B to A, is written $\overrightarrow{BA}$ or **BA**.

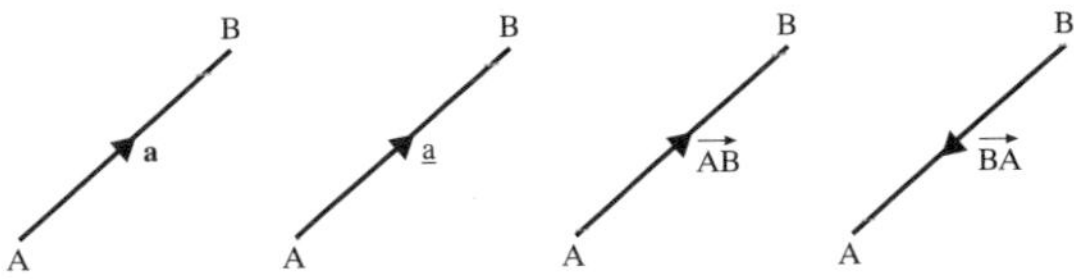

Equivalent Displacements

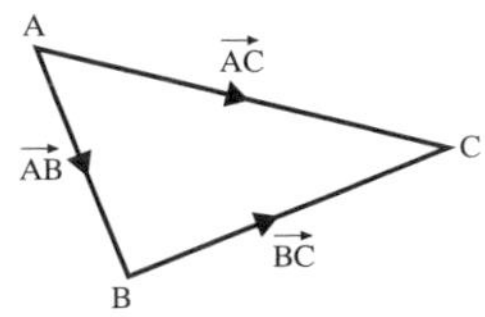

The displacement from A to B, followed by the displacement from B to C, is equivalent to the displacement from A to C.

This is written as the vector equation

$$\overrightarrow{AB} + \overrightarrow{BC} = \overrightarrow{AC}$$

Note that, in vector equations like the one above,
$+$ means 'together with' and $=$ means 'is equivalent to'

Many quantities other than displacements behave in the same way and all of them are vectors.

Properties of Vectors

The Modulus of a Vector

The *modulus* of a vector **a** is its magnitude and is written $|\mathbf{a}|$ or a i.e. $|\mathbf{a}|$ is the length of the line representing **a**.

Equal Vectors

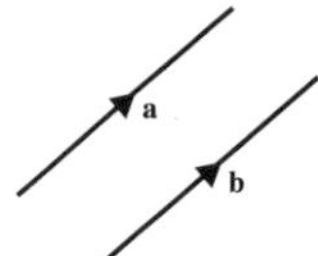

Two vectors with the same magnitude and the same direction are equal.

i.e. $\mathbf{a} = \mathbf{b} \iff \begin{cases} |\mathbf{a}| = |\mathbf{b}| \quad \text{and} \\ \text{the directions of } \mathbf{a} \text{ and } \mathbf{b} \text{ are the same.} \end{cases}$

It follows that a vector can be represented by *any* line of the right length and direction, regardless of position, i.e. each of the lines in the diagram below represents the vector **a**.

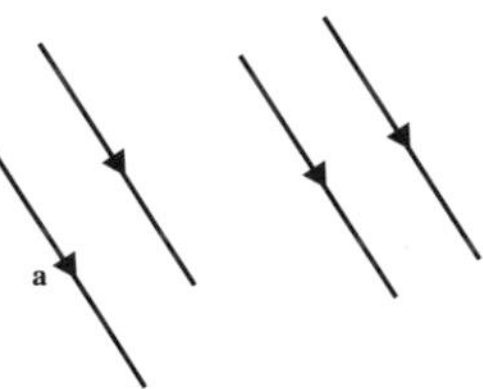

Negative Vectors

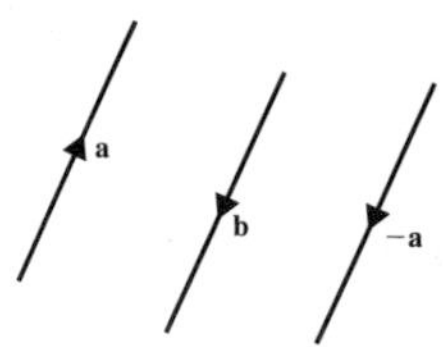

If two vectors, **a** and **b**, have the same magnitude but opposite directions we say

$$\mathbf{b} = -\mathbf{a}$$

i.e. $-\mathbf{a}$ is a vector of magnitude $|\mathbf{a}|$ and in the direction opposite to that of **a**.

We also say that **a** and **b** are *equal and opposite* vectors.

Multiplication of a Vector by a Scalar

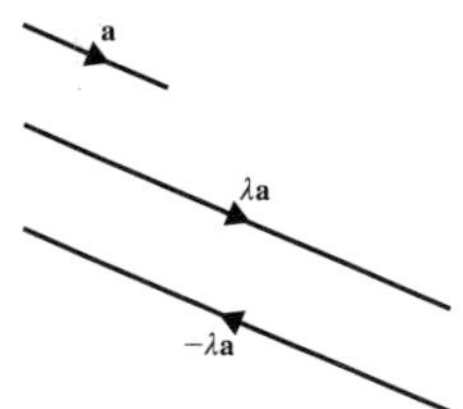

If λ is a positive real number, then $\lambda\mathbf{a}$ is a vector in the same direction as **a** and of magnitude $\lambda|\mathbf{a}|$.

It follows that $-\lambda\mathbf{a}$ is a vector in the opposite direction, with magnitude $\lambda|\mathbf{a}|$.

Addition of Vectors

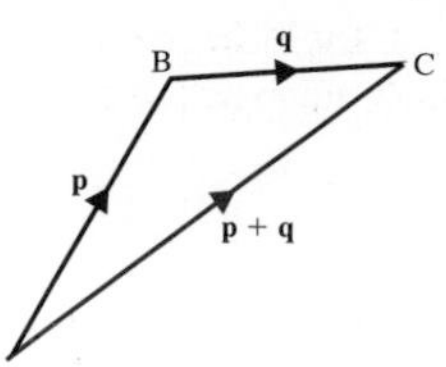

If the sides AB and BC of a triangle ABC represent the vectors **p** and **q** then the third side AC represents the vector sum, or resultant, of **p** and **q**, which is denoted by $\mathbf{p} + \mathbf{q}$.

(This property was demonstrated for displacement vectors at the beginning of the chapter.)

Note that **p** and **q** follow each other round the triangle (in this case in the clockwise sense), whereas the resultant, $\mathbf{p} + \mathbf{q}$, goes the opposite way round (anticlockwise in the diagram).

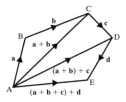

This is known as *the triangle law* for addition of vectors. It can be extended to cover the addition of more than two vectors.

Let $\overrightarrow{AB}$, $\overrightarrow{BC}$, $\overrightarrow{CD}$ and $\overrightarrow{DE}$ represent the vectors **a**, **b**, **c** and **d** respectively.

The triangle law gives $\overrightarrow{AB} + \overrightarrow{BC} = \mathbf{a} + \mathbf{b} = \overrightarrow{AC}$

then $\overrightarrow{AC} + \overrightarrow{CD} = (\mathbf{a} + \mathbf{b}) + \mathbf{c} = \overrightarrow{AD}$

and $\overrightarrow{AD} + \overrightarrow{DE} = (\mathbf{a} + \mathbf{b} + \mathbf{c}) + \mathbf{d} = \overrightarrow{AE}$

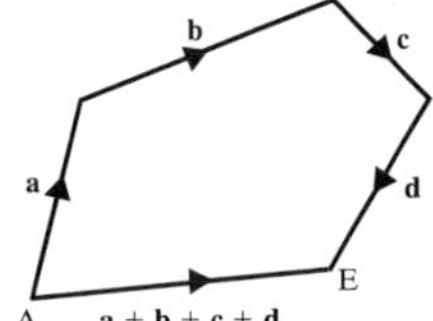

Now AE completes the polygon of which AB, BC, CD and DE are four sides taken in order, (i.e. they follow each other round the polygon in the *same sense*).

Note that, again, the side representing the resultant closes the polygon in the *opposite* sense.

Note that the vectors **a**, **b**, **c** and **d** are not necessarily coplanar so the polygon may not be a plane figure.

The order in which the addition is performed does not matter as we can see by considering a parallelogram ABCD.

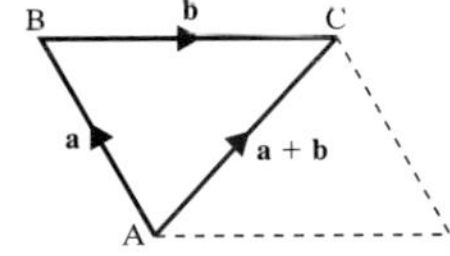

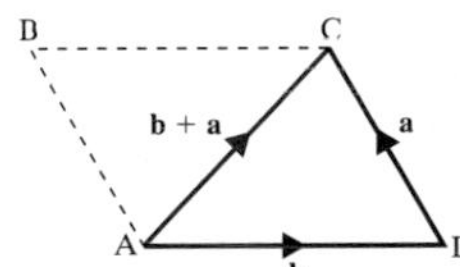

Because the opposite sides of a parallelogram are equal and parallel, $\overrightarrow{AB}$ and $\overrightarrow{DC}$ both represent **a** and $\overrightarrow{BC}$ and $\overrightarrow{AD}$ both represent **b**.

In $\triangle ABC$ $\overrightarrow{AC} = \mathbf{a} + \mathbf{b}$ and in $\triangle ADC$ $\overrightarrow{AC} = \mathbf{b} + \mathbf{a}$

Therefore $\mathbf{a} + \mathbf{b} = \mathbf{b} + \mathbf{a}$

The Angle between Two Vectors

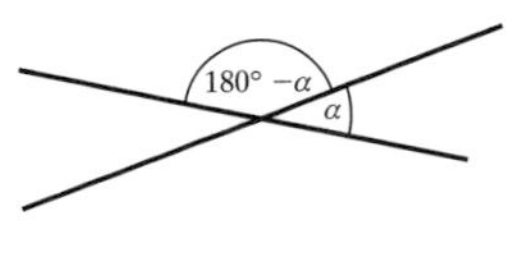

There are two angles between two lines i.e. α and $180° - \alpha$.

The angle between two vectors, however, is defined uniquely. It is the angle between their directions when the lines representing them *both converge* or *both diverge* (see diagrams i and ii).

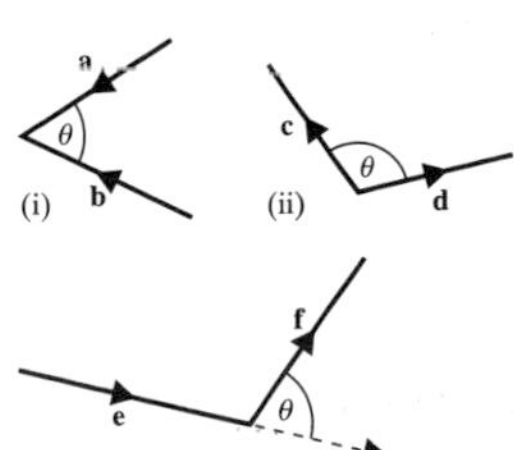

In some cases one of the lines may have to be produced in order to mark the correct angle (see diagram iii).

Position Vectors

In general a vector has no specific location in space and is called a *free vector*. Some vectors, however, are constrained to a specific position, e.g. the vector $\overrightarrow{OA}$ where O is a fixed origin.

$\overrightarrow{OA}$ is called the position of A relative to O.

This displacement is unique and *cannot* be represented by any other line of equal length and direction.

Vectors such as $\overrightarrow{OA}$, representing quantities that have a specific location, are called *position* vectors or *tied* vectors.

The Position Vector of the Midpoint of a Line

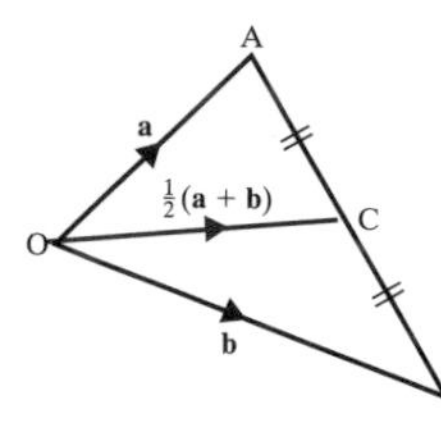

Consider a line AB where the position vectors relative to O of A and B are **a** and **b** respectively and C is the midpoint of AB.

In the diagram, $\overrightarrow{AB} = \mathbf{b} - \mathbf{a}$

$\therefore \quad \overrightarrow{AC} = \frac{1}{2}(\mathbf{b} - \mathbf{a})$

$\Rightarrow \quad \overrightarrow{OC} = \overrightarrow{OA} + \overrightarrow{AC} = \mathbf{a} + \frac{1}{2}(\mathbf{b} - \mathbf{a}) = \frac{1}{2}\mathbf{a} + \frac{1}{2}\mathbf{b}$

i.e. the position vector of C is $\frac{1}{2}(\mathbf{a} + \mathbf{b})$.

EXAMPLES 36A

1 Two vectors, **a** and **b**, are such that $|\mathbf{a}| = 3$, $|\mathbf{b}| = 5$ and the angle between **a** and **b** is $\frac{1}{3}\pi$. If the line OP represents the vector $3\mathbf{a} - 2\mathbf{b}$, find, correct to 1 d.p., the angle between $\overrightarrow{OP}$ and **a**.

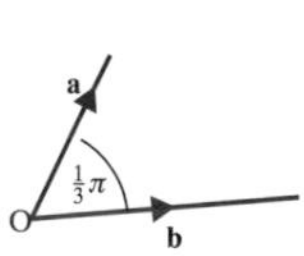

The line OP is found by drawing OQ parallel to **a** such that $\overrightarrow{OQ} = 3\mathbf{a}$, followed by QP parallel to **b** such that $\overrightarrow{QP} = -2\mathbf{b}$

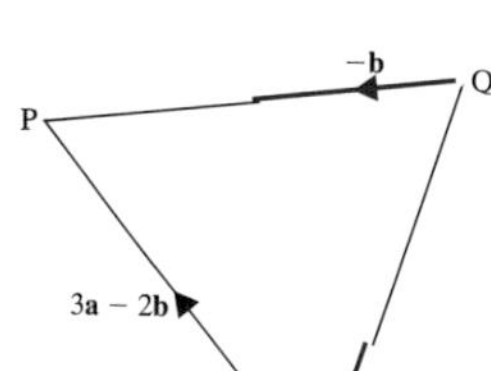

Thus $\overrightarrow{OP} = 3\mathbf{a} - 2\mathbf{b}$

Now $OQ = 3|\mathbf{a}| = 9$ and $QP = 2|\mathbf{b}| = 10$

The angle between $\overrightarrow{OP}$ and **a** is α where

$$\frac{\sin\alpha}{10} = \frac{\sin\frac{1}{3}\pi}{OP} \quad \text{(using the sine formula)}$$

So first we must find OP.

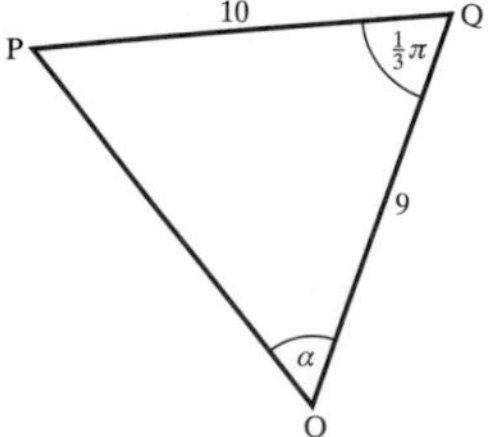

Using the cosine formula in OPQ gives

$$OP^2 = 81 + 100 - (2)(9)(10)\cos\tfrac{1}{3}\pi = 91 \quad \Rightarrow \quad OP = \sqrt{91}$$

Then $\dfrac{\sin\alpha}{10} = \dfrac{\sin\frac{1}{3}\pi}{\sqrt{91}} \quad \Rightarrow \quad \sin\alpha = 0.9078$

$\therefore$ OP is inclined at 65.2° to **a**.

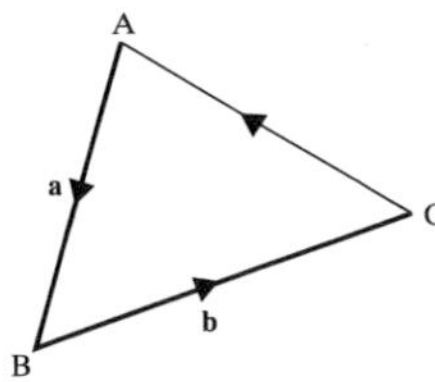

2 In a triangle ABC, $\overrightarrow{AB}$ represents $\mathbf{a}$ and $\overrightarrow{BC}$ represents $\mathbf{b}$. If D is the midpoint of AB express in terms of $\mathbf{a}$ and $\mathbf{b}$ the vectors $\overrightarrow{CA}$ and $\overrightarrow{DC}$.

$\overrightarrow{CA} = \overrightarrow{CB} + \overrightarrow{BA}$

Now $\quad \overrightarrow{CB} = -\overrightarrow{BC} = -\mathbf{b}$

and $\quad \overrightarrow{BA} = -\overrightarrow{AB} = -\mathbf{a}$

$\therefore \quad \overrightarrow{CA} = -\mathbf{b} - \mathbf{a} = -(\mathbf{a} + \mathbf{b})$

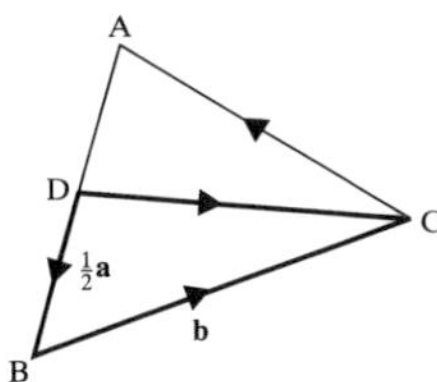

$\overrightarrow{DC} = \overrightarrow{DB} + \overrightarrow{BC}$

Now $\quad DB = \frac{1}{2}AB$

$\Rightarrow \quad \overrightarrow{DB} = \frac{1}{2}\overrightarrow{AB} = \frac{1}{2}\mathbf{a}$

$\therefore \quad \overrightarrow{DC} = \frac{1}{2}\mathbf{a} + \mathbf{b}$

3 If D is the midpoint of the side BC of a triangle ABC, show that $\overrightarrow{AB} + \overrightarrow{AC} = 2\overrightarrow{AD}$.

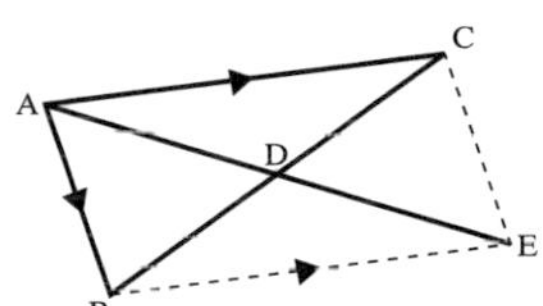

Completing the parallelogram ABEC we see that $\overrightarrow{BE} = \overrightarrow{AC}$

Therefore $\overrightarrow{AB} + \overrightarrow{AC} = \overrightarrow{AB} + \overrightarrow{BE} = \overrightarrow{AE}$

The diagonals of a parallelogram bisect each other.

Therefore $\quad \overrightarrow{AE} = 2\overrightarrow{AD}$

$\Rightarrow \quad \overrightarrow{AB} + \overrightarrow{AC} = 2\overrightarrow{AD}$

4 Four points O, A, B and C are such that $\overrightarrow{OA} = 10\mathbf{a}$, $\overrightarrow{OB} = 5\mathbf{b}$ and $\overrightarrow{OC} = 4\mathbf{a} + 3\mathbf{b}$. Show that A, B and C are collinear.

> If A, B and C are collinear then AB and BC have the same direction so this is what we must show.

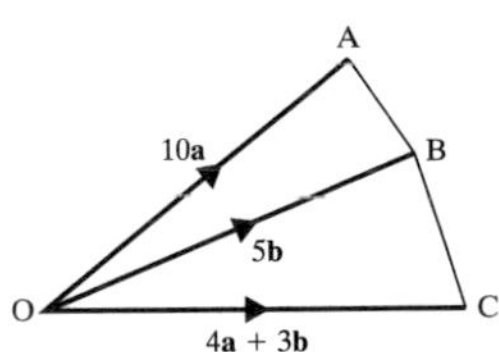

$\overrightarrow{AB} = \overrightarrow{AO} + \overrightarrow{OB} = -10\mathbf{a} + 5\mathbf{b} = 5(\mathbf{b} - 2\mathbf{a})$

$\overrightarrow{BC} = \overrightarrow{BO} + \overrightarrow{OC} = -5\mathbf{b} + 4\mathbf{a} + 3\mathbf{b}$

$= 4\mathbf{a} - 2\mathbf{b} = -2(\mathbf{b} - 2\mathbf{a})$

AB and BC both have a direction given by $\lambda(\mathbf{b} - 2\mathbf{a})$ so they are parallel. Hence, since C is a common point, A, B and C are collinear.

Note that $\overrightarrow{BC} = -\frac{2}{5}\overrightarrow{AB}$ so, although $\overrightarrow{AB}$ and $\overrightarrow{BC}$ are parallel, they are in opposite directions, showing that the diagram really looks like this.

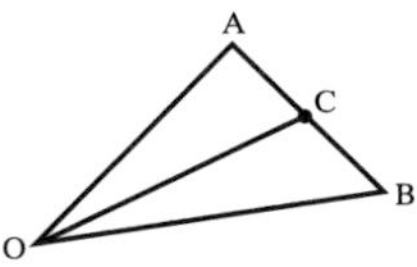

EXERCISE 36A

1 ABCD is a quadrilateral. Find the single vector which is equivalent to

a $\overrightarrow{AB} + \overrightarrow{BC}$ **b** $\overrightarrow{BC} + \overrightarrow{CD}$

c $\overrightarrow{AB} + \overrightarrow{BC} + \overrightarrow{CD}$ **d** $\overrightarrow{AB} + \overrightarrow{DA}$

2 ABCDEF is a regular hexagon in which $\overrightarrow{BC}$ represents $\mathbf{b}$ and $\overrightarrow{FC}$ represents $2\mathbf{a}$. Express the vectors $\overrightarrow{AB}$, $\overrightarrow{CD}$ and $\overrightarrow{BE}$ in terms of $\mathbf{a}$ and $\mathbf{b}$.

3 Draw diagrams representing the following vector equations.

a $\overrightarrow{AB} - \overrightarrow{CB} = \overrightarrow{AC}$ **b** $\overrightarrow{AB} = 2\overrightarrow{PQ}$

c $\overrightarrow{AB} + \overrightarrow{BC} = 3\overrightarrow{AD}$ **d** $2\overrightarrow{AB} + \overrightarrow{PQ} = 0$

4 If A, B, C, D are four points such that $\overrightarrow{AB} = \overrightarrow{DC}$ and $\overrightarrow{BC} + \overrightarrow{DA} = \mathbf{0}$ prove that ABCD is a parallelogram.

5 O, A, B, C, D are five points such that
$\overrightarrow{OA} = \mathbf{a}$, $\overrightarrow{OB} = \mathbf{b}$, $\overrightarrow{OC} = \mathbf{a} + 2\mathbf{b}$, $\overrightarrow{OD} = 2\mathbf{a} - \mathbf{b}$.
Express $\overrightarrow{AB}$, $\overrightarrow{BC}$, $\overrightarrow{CD}$, $\overrightarrow{AC}$, $\overrightarrow{BD}$ in terms of $\mathbf{a}$ and $\mathbf{b}$.

6

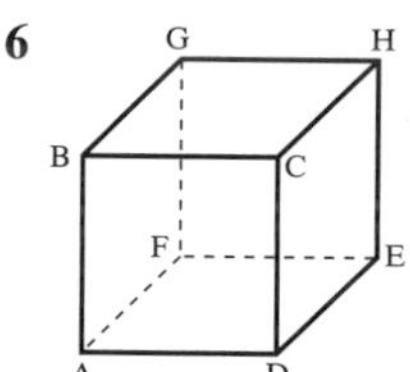

If $\mathbf{a}$, $\mathbf{b}$, $\mathbf{c}$ are represented by the edges $\overrightarrow{AB}$, $\overrightarrow{AD}$, $\overrightarrow{AF}$ of the cube in the diagram, find, in terms of $\mathbf{a}$, $\mathbf{b}$ and $\mathbf{c}$, the vectors represented by the remaining edges.

7 If O, A, B, C are four points such that $\overrightarrow{OA} = \mathbf{a}$, $\overrightarrow{OB} = 2\mathbf{a} - \mathbf{b}$, $\overrightarrow{OC} = \mathbf{b}$ show that A, B and C are collinear.

8

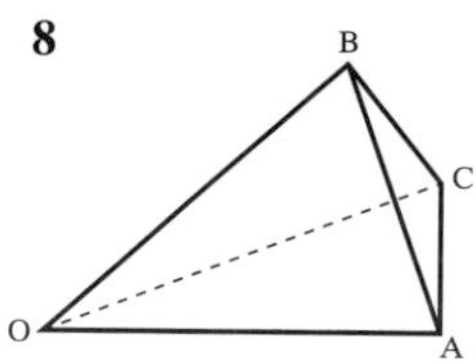

If OABC is a tetrahedron and $\overrightarrow{OA} = \mathbf{a}$, $\overrightarrow{OB} = \mathbf{b}$, $\overrightarrow{OC} = \mathbf{c}$, find $\overrightarrow{AC}$, $\overrightarrow{AB}$, $\overrightarrow{CB}$ in terms of $\mathbf{a}$, $\mathbf{b}$, $\mathbf{c}$.

9 For the cube defined in Question 6, find, in terms of $\mathbf{a}$, $\mathbf{b}$ and $\mathbf{c}$, the vectors $\overrightarrow{BE}$, $\overrightarrow{GD}$, $\overrightarrow{AH}$, $\overrightarrow{FC}$.

10 If $\mathbf{a}$ and $\mathbf{b}$ are vectors such that $|\mathbf{a}| = 2$, $|\mathbf{b}| = 4$ and the angle between $\mathbf{a}$ and $\mathbf{b}$ is $\frac{1}{3}\pi$, find the angle between

a $\mathbf{a}$ and $\mathbf{a} - \mathbf{b}$ **b** $\mathbf{b}$ and $\mathbf{a} + \mathbf{b}$ **c** $3\mathbf{a} - \mathbf{b}$ and $\mathbf{b}$

In Questions 11 to 13 the position vectors, relative to O, of A, B, C and D are $\mathbf{a}$, $\mathbf{b}$, $\mathbf{c}$ and $\mathbf{d}$ respectively. P, Q and R are the midpoints of AB, BC and CD respectively.

In Questions 11 and 12 find the position vector of each given point.

11 a The midpoint of AC **b** The midpoint of BD

12 a The midpoint of PQ **b** The midpoint of QR

13 Show that PQ is parallel to AC.

The Location of a Point in Space

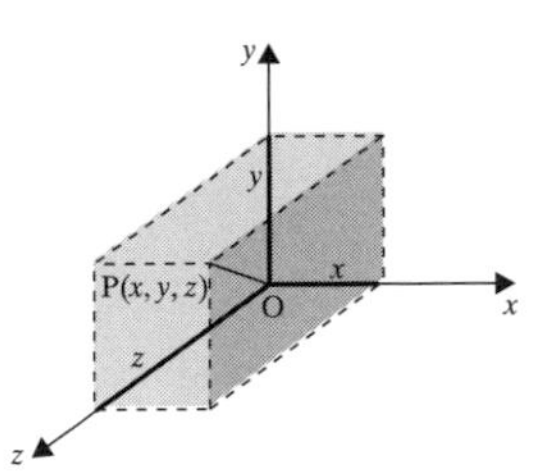

We saw in Chapter 5 that any point P in a plane can be located by giving its distances from a fixed point O, in each of two perpendicular directions. These distances are the Cartesian coordinates of the point.

Now we consider locating a point in three-dimensional space.

If we have a fixed point, O, then any other point can be located by giving its distances from O in each of *three* mutually perpendicular directions, i.e. we need *three* coordinates to locate a point in space. So we use the familiar x and y axes, together with a third axis Oz. Then any point has coordinates (x, y, z) relative to the origin O.

Cartesian Unit Vectors

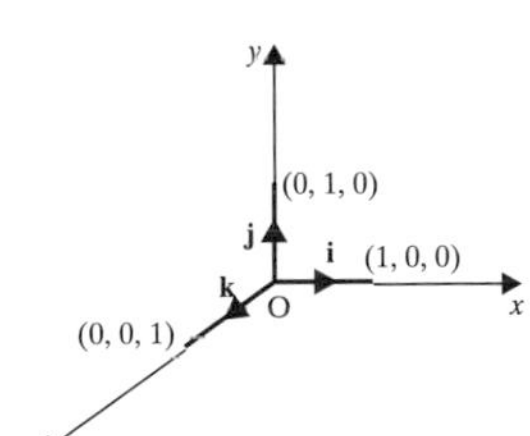

A unit vector is a vector whose magnitude is one unit.

Now if
i is a unit vector in the direction of Ox
j is a unit vector in the direction of Oy
k is a unit vector in the direction of Oz

then the position vector, relative to O, of any point P can be given in terms of **i**, **j** and **k**,

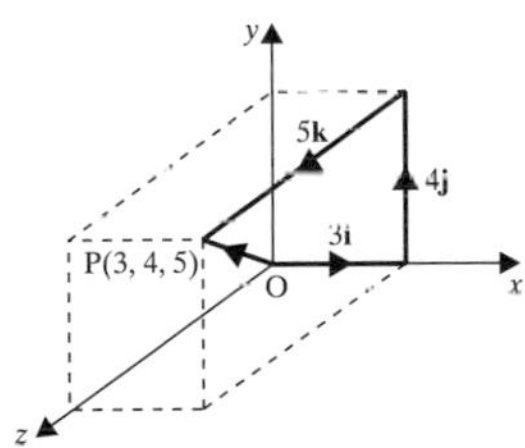

e.g. the point distant

3 units from O in the direction Ox
4 units from O in the direction Oy
5 units from O in the direction Oz

has coordinates (3, 4, 5) and $\overrightarrow{OP} = 3\mathbf{i} + 4\mathbf{j} + 5\mathbf{k}$.

This can also be written as $\overrightarrow{OP} = \begin{pmatrix} 3 \\ 4 \\ 5 \end{pmatrix}$

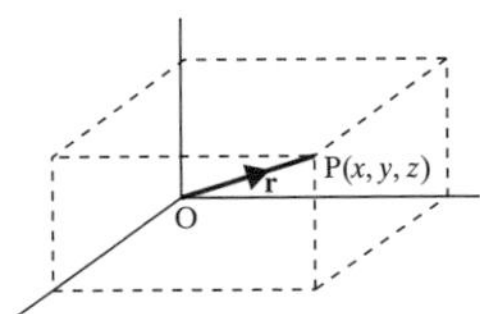

In general, if P is a point, (x, y, z) and $\overrightarrow{OP} = \mathbf{r}$,

then $\quad \mathbf{r} = x\mathbf{i} + y\mathbf{j} + z\mathbf{k}$

or $\quad \mathbf{r} = \begin{pmatrix} x \\ y \\ z \end{pmatrix}$

and **r** is the position vector of P.

Free vectors can be given in the same form. For example, the vector $3\mathbf{i} + 4\mathbf{j} + 5\mathbf{k}$ *can* represent the position vector of the point P(3, 4, 5) but it can equally well represent *any* vector of length and direction equal to those of OP.

Note that, unless a vector is *specified* as a position vector it is taken to be free.

Operations on Cartesian Vectors

Addition and Subtraction

To add or subtract vectors given in **i j k** form, the coefficients of **i**, **j** and **k** are collected separately,

e.g. if $\mathbf{v}_1 = 3\mathbf{i} + 2\mathbf{j} + 2\mathbf{k}$ and $\mathbf{v}_2 = \mathbf{i} + 2\mathbf{j} - 3\mathbf{k}$

then
$$\begin{aligned}\mathbf{v}_1 + \mathbf{v}_2 &= (3\mathbf{i} + 2\mathbf{j} + 2\mathbf{k}) + (\mathbf{i} + 2\mathbf{j} - 3\mathbf{k})\\ &= (3+1)\mathbf{i} + (2+2)\mathbf{j} + (2-3)\mathbf{k}\\ &= 4\mathbf{i} + 4\mathbf{j} - \mathbf{k}\end{aligned}$$

and
$$\begin{aligned}\mathbf{v}_1 - \mathbf{v}_2 &= (3-1)\mathbf{i} + (2-2)\mathbf{j} + (2-\{-3\})\mathbf{k}\\ &= 2\mathbf{i} + 5\mathbf{k}\end{aligned}$$

Modulus

The modulus of **v**, where $\mathbf{v} = 12\mathbf{i} - 3\mathbf{j} + 4\mathbf{k}$, is the length of OP where P is the point (12, −3, 4).

Using Pythagoras twice we have

$OB^2 = OA^2 + AB^2 = 12^2 + 4^2$

$OP^2 = OB^2 + BP^2 = (12^2 + 4^2) + (-3)^2$

$\therefore \quad OP = \sqrt{12^2 + 4^2 + 3^2} = 13$

In general, if $\mathbf{v} = a\mathbf{i} + b\mathbf{j} + c\mathbf{k}$,

$$|\mathbf{v}| = \sqrt{a^2 + b^2 + c^2}$$

Parallel Vectors

Two vectors $\mathbf{v}_1$ and $\mathbf{v}_2$ are parallel if $\mathbf{v}_1 = \lambda\mathbf{v}_2$

e.g. $2\mathbf{i} - 3\mathbf{j} - \mathbf{k}$ is parallel to $4\mathbf{i} - 6\mathbf{j} - 2\mathbf{k}$ $\quad (\lambda = 2)$

and $\mathbf{i} + \mathbf{j} + \mathbf{k}$ is parallel to $-3\mathbf{i} - 3\mathbf{j} - 3\mathbf{k}$ $\quad (\lambda = -3)$

Equal Vectors

If two vectors $\mathbf{v}_1 = a_1\mathbf{i} + b_1\mathbf{j} + c_1\mathbf{k}$ and $\mathbf{v}_2 = a_2\mathbf{i} + b_2\mathbf{j} + c_2\mathbf{k}$ are equal then

$$a_1 = a_2 \text{ and } b_1 = b_2 \text{ and } c_1 = c_2$$

EXAMPLES 36B

1 Given the vector **v** where $\mathbf{v} = 5\mathbf{i} - 2\mathbf{j} + 4\mathbf{k}$, state whether each of the following vectors is parallel to **v**, equal to **v** or neither.

a $10\mathbf{i} - 4\mathbf{j} + 8\mathbf{k}$ **b** $-\frac{1}{2}(-10\mathbf{i} + 4\mathbf{j} - 8\mathbf{k})$

c $-5\mathbf{i} + 2\mathbf{j} - 4\mathbf{k}$ **d** $4\mathbf{i} - 2\mathbf{j} + 5\mathbf{k}$

a $10\mathbf{i} - 4\mathbf{j} + 8\mathbf{k} = 2(5\mathbf{i} - 2\mathbf{j} + 4\mathbf{k}) \quad (\lambda = 2)$

$\therefore$ $10\mathbf{i} - 4\mathbf{j} + 8\mathbf{k}$ is parallel to **v**.

b $-\frac{1}{2}(-10\mathbf{i}+4\mathbf{j}-8\mathbf{k}) = 5\mathbf{i}-2\mathbf{j}+4\mathbf{k}$

$\therefore\ -\frac{1}{2}(-10\mathbf{i}+4\mathbf{j}-8\mathbf{k})$ is equal to $\mathbf{v}$.

c $-5\mathbf{i}+2\mathbf{j}-4\mathbf{k} = -(5\mathbf{i}-2\mathbf{j}+4\mathbf{k}) \qquad (\lambda = -1)$

$\therefore\ -5\mathbf{i}+2\mathbf{j}-4\mathbf{k}$ is parallel to $\mathbf{v}$.

d $4\mathbf{i}-2\mathbf{j}+5\mathbf{k}$ is not a multiple of $5\mathbf{i}-2\mathbf{j}+4\mathbf{k}$

$\therefore\ 4\mathbf{i}-2\mathbf{j}+5\mathbf{k}$ is not equal or parallel to $\mathbf{v}$.

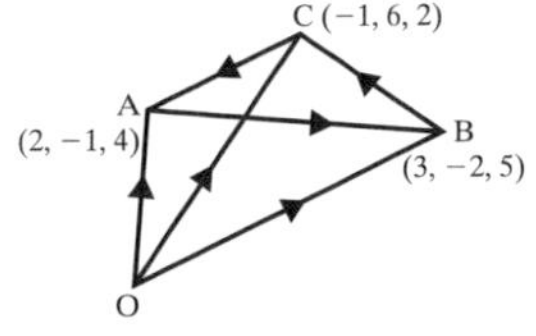

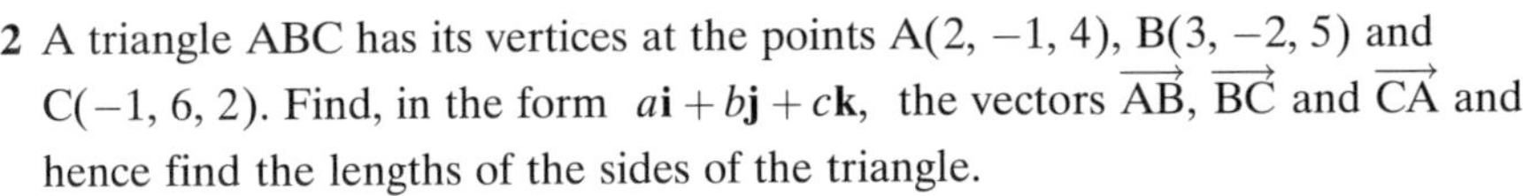

2 A triangle ABC has its vertices at the points A(2, −1, 4), B(3, −2, 5) and C(−1, 6, 2). Find, in the form $a\mathbf{i}+b\mathbf{j}+c\mathbf{k}$, the vectors $\overrightarrow{AB}$, $\overrightarrow{BC}$ and $\overrightarrow{CA}$ and hence find the lengths of the sides of the triangle.

> The coordinate axes are not drawn in this diagram as they tend to cause confusion when two or more points are illustrated. The origin should always be included however as it provides a reference point.

$$\begin{aligned}\overrightarrow{AB} &= \overrightarrow{OB}-\overrightarrow{OA}\\ &= (3\mathbf{i}-2\mathbf{j}+5\mathbf{k})-(2\mathbf{i}-\mathbf{j}+4\mathbf{k})\\ &= \mathbf{i}-\mathbf{j}+\mathbf{k}\end{aligned}$$

$$\begin{aligned}\overrightarrow{BC} &= \overrightarrow{OC}-\overrightarrow{OB}\\ &= (-\mathbf{i}+6\mathbf{j}+2\mathbf{k})-(3\mathbf{i}-2\mathbf{j}+5\mathbf{k})\\ &= -4\mathbf{i}+8\mathbf{j}-3\mathbf{k}\end{aligned}$$

$$\begin{aligned}\overrightarrow{CA} &= \overrightarrow{OA}-\overrightarrow{OC}\\ &= (2\mathbf{i}-\mathbf{j}+4\mathbf{k})-(-\mathbf{i}+6\mathbf{j}+2\mathbf{k})\\ &= 3\mathbf{i}-7\mathbf{j}+2\mathbf{k}\end{aligned}$$

Hence $\quad AB = |\overrightarrow{AB}| = \sqrt{(1)^2+(-1)^2+(1)^2} = \sqrt{3}$

$BC = |\overrightarrow{BC}| = \sqrt{(-4)^2+(8)^2+(-3)^2} = \sqrt{89}$

$CA = |\overrightarrow{CA}| = \sqrt{(3)^2+(-7)^2+(2)^2} = \sqrt{62}$

Two-dimensional problems can be solved by using the same principles as for three-dimensional cases but the working tends to be easier because it involves fewer terms.

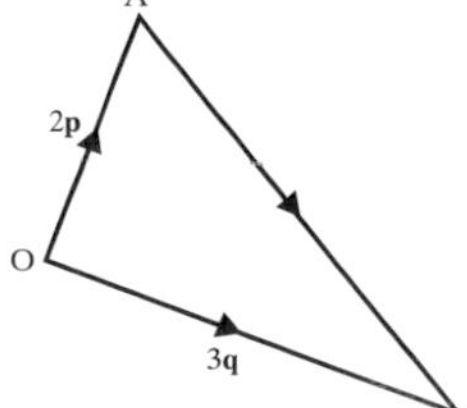

3 Given that $\mathbf{p} = \mathbf{i}+3\mathbf{j}$, $\mathbf{q} = 4\mathbf{i}-2\mathbf{j}$, $\mathbf{OA} = 2\mathbf{p}$ and $\mathbf{OB} = 3\mathbf{q}$, find

a $|\mathbf{OA}|$ **b** $|\mathbf{OB}|$ **c** $|\mathbf{AB}|$

a $|\mathbf{OA}| = 2|\mathbf{i}+3\mathbf{j}| = 2\sqrt{1^2+3^2} = 2\sqrt{10}$

b $|\mathbf{OB}| = 3|4\mathbf{i}-2\mathbf{j}| = 3\sqrt{4^2+(-2)^2} = 6\sqrt{5}$

c $\mathbf{AB} = \mathbf{OB}-\mathbf{OA} = 3(4\mathbf{i}-2\mathbf{j})-2(\mathbf{i}+3\mathbf{j}) = 10\mathbf{i}-12\mathbf{k}$

$|\mathbf{AB}| = \sqrt{10^2+12^2} = 2\sqrt{61}$

EXERCISE 36B

1 Write down, in the form $a\mathbf{i}+b\mathbf{j}+c\mathbf{k}$, the vector represented by $\overrightarrow{OP}$ if P is a point with coordinates

a (3, 6, 4) b (1, –2, –7) c (1, 0, –3)

2 $\overrightarrow{OP}$ represents a vector $\mathbf{r}$. Write down the coordinates of P if

a $\mathbf{r} = 5\mathbf{i} - 7\mathbf{j} + 2\mathbf{k}$ b $\mathbf{r} = \mathbf{i} + 4\mathbf{j}$ c $\mathbf{r} = \mathbf{j} - \mathbf{k}$

3 Find the length of the line OP if P is the point

a (2, –1, 4) b (3, 0, 4) c (–2, –2, 1)

4 Find the modulus of the vector $\mathbf{V}$ if

a $\mathbf{V} = 2\mathbf{i} - 4\mathbf{j} + 4\mathbf{k}$ b $\mathbf{V} = 6\mathbf{i} + 2\mathbf{j} - 3\mathbf{k}$ c $\mathbf{V} = 11\mathbf{i} - 7\mathbf{j} - 6\mathbf{k}$

5 If $\mathbf{a} = \mathbf{i}+\mathbf{j}+\mathbf{k}$, $\mathbf{b} = 2\mathbf{i}-\mathbf{j}+3\mathbf{k}$, $\mathbf{c} = -\mathbf{i}+3\mathbf{j}-\mathbf{k}$ find

a $\mathbf{a}+\mathbf{b}$ b $\mathbf{a}-\mathbf{c}$ c $\mathbf{a}+\mathbf{b}+\mathbf{c}$ d $\mathbf{a}-2\mathbf{b}+3\mathbf{c}$

In Questions 6 to 8, $\overrightarrow{OA} = \mathbf{a} = 4\mathbf{i} - 12\mathbf{j}$ and $\overrightarrow{OB} = \mathbf{b} = \mathbf{i} + 6\mathbf{j}$.

6 Which of the following vectors are parallel to $\mathbf{a}$?

a $\mathbf{i}+3\mathbf{j}$ b $-\mathbf{i}+3\mathbf{j}$ c $12\mathbf{i}-4\mathbf{j}$ d $-4\mathbf{i}+12\mathbf{j}$ e $\mathbf{i}-3\mathbf{j}$

7 Which of the following vectors are equal to $\mathbf{b}$?

a $2\mathbf{i}+12\mathbf{j}$ b $-\mathbf{i}-6\mathbf{j}$

c $\overrightarrow{AE}$ if E is (5, –6) d $\overrightarrow{AF}$ if F is (6, 0)

8 If $\overrightarrow{OD} = \lambda\overrightarrow{OA}$, find the value of λ for which $\overrightarrow{OD} + \overrightarrow{OB}$ is parallel to the x-axis.

9 Which of the following vectors are parallel to $3\mathbf{i}-\mathbf{j}-2\mathbf{k}$?

a $6\mathbf{i}-3\mathbf{j}-4\mathbf{k}$ b $-9\mathbf{i}+3\mathbf{j}+6\mathbf{k}$ c $-3\mathbf{i}-\mathbf{j}-2\mathbf{k}$

d $-2(3\mathbf{i}+\mathbf{j}+2\mathbf{k})$ e $\frac{3}{2}\mathbf{i}-\frac{1}{2}\mathbf{j}-\mathbf{k}$ f $-\mathbf{i}+\frac{1}{3}\mathbf{j}+\frac{2}{3}\mathbf{k}$

10 Given that $\mathbf{a} = 4\mathbf{i}+\mathbf{j}-6\mathbf{k}$, state whether each of the following vectors is parallel or equal to $\mathbf{a}$ or neither.

a $8\mathbf{i}+2\mathbf{j}-10\mathbf{k}$ b $-4\mathbf{i}-\mathbf{j}+6\mathbf{k}$ c $2(2\mathbf{i}+\frac{1}{2}\mathbf{j}-3\mathbf{k})$

11 The triangle ABC has its vertices at the points A(–1, 3, 0), B(–3, 0, 7), C(–1, 2, 3). Find in the form $a\mathbf{i}+b\mathbf{j}+c\mathbf{k}$ the vectors representing

a $\overrightarrow{AB}$ b $\overrightarrow{AC}$ c $\overrightarrow{CB}$

12 Find the lengths of the sides of the triangle described in Question 11.

13 Find $|\mathbf{a}-\mathbf{b}|$ where $\mathbf{a} = \mathbf{i}-\mathbf{j}+2\mathbf{k}$, $\mathbf{b} = 2\mathbf{i}-\mathbf{j}$.

14 A, B, C and D are the points (0, 0, 2), (–1, 3, 2), (1, 0, 4) and (–1, 2, –2) respectively. Find the vectors representing $\overrightarrow{AB}$, $\overrightarrow{BD}$, $\overrightarrow{CD}$, $\overrightarrow{AD}$.

Finding a Unit Vector

Consider the vector $\mathbf{v} = 6\mathbf{i} + 2\mathbf{j} + 3\mathbf{k}$, represented by $\overrightarrow{OP}$ where P is the point (6, 2, 3).

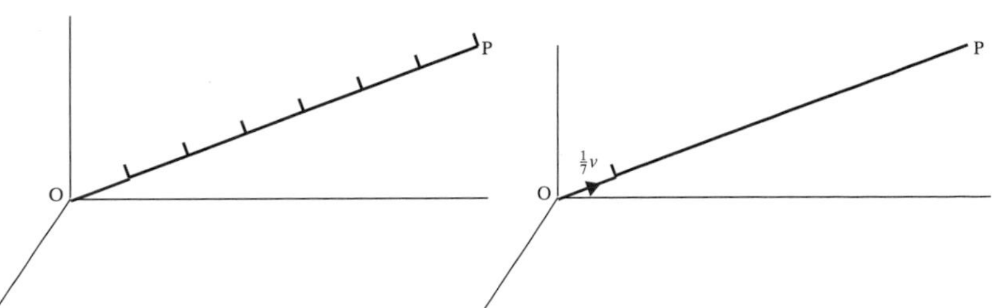

Now $|\mathbf{v}|$ is $\sqrt{6^2 + 2^2 + 3^2}$, i.e. OP is 7 units long.

Therefore $\frac{1}{7}\mathbf{v}$ is a vector of unit magnitude and is denoted by $\hat{\mathbf{v}}$

i.e. $$\hat{\mathbf{v}} = \frac{\mathbf{v}}{|\mathbf{v}|}$$

In general **a unit vector in the direction of v is given by** $\dfrac{\mathbf{v}}{|\mathbf{v}|}$

EXAMPLE 36C

Find the coordinates of P if OP is of length 5 units and is parallel to the vector $2\mathbf{i} - \mathbf{j} + 4\mathbf{k}$.

A vector parallel to $2\mathbf{i} - \mathbf{j} + 4\mathbf{k}$ can be in either of the directions given by $\pm(2\mathbf{i} - \mathbf{j} + 4\mathbf{k})$

$\therefore \qquad \mathbf{OP} = \pm k(2\mathbf{i} - \mathbf{j} + 4\mathbf{k})$

Now $\qquad |\mathbf{OP}| = 5, \quad \Rightarrow \quad 5^2 = k^2(4 + 1 + 16)$

i.e. $\qquad k^2 = \dfrac{25}{21} \quad \Rightarrow \quad k = \pm\dfrac{5}{\sqrt{21}}$

$$\mathbf{OP} = (5)\left\{\frac{\pm(2\mathbf{i} - \mathbf{j} + 4\mathbf{k})}{\sqrt{21}}\right\} = \pm\left(\frac{10}{\sqrt{21}}\mathbf{i} - \frac{5}{\sqrt{21}}\mathbf{j} + \frac{20}{\sqrt{21}}\mathbf{k}\right)$$

The coordinates of P are

either $\left(\dfrac{10}{\sqrt{21}}, \dfrac{-5}{\sqrt{21}}, \dfrac{20}{\sqrt{21}}\right)$ or $\left(\dfrac{-10}{\sqrt{21}}, \dfrac{5}{\sqrt{21}}, \dfrac{-20}{\sqrt{21}}\right)$

EXERCISE 36C

1 Find a unit vector in the direction of each of the vectors.

a $2\mathbf{i} + 2\mathbf{j} - \mathbf{k}$ **b** $6\mathbf{i} - 2\mathbf{j} - 3\mathbf{k}$

c $3\mathbf{i} + 4\mathbf{k}$ **d** $\mathbf{i} + 8\mathbf{j} + 4\mathbf{k}$

2 Find the coordinates of Q if $|\mathbf{OQ}| = 1$ and $\mathbf{OQ}$ is in the direction of

a $\mathbf{i} + 2\mathbf{j} - 2\mathbf{k}$ **b** $3\mathbf{i} + 2\mathbf{j} + 6\mathbf{k}$

c $8\mathbf{i} - \mathbf{j} - 4\mathbf{k}$ **d** $\mathbf{i} - \mathbf{j} - \mathbf{k}$

3 Find the vector **V** if

a $\mathbf{V} = \mathbf{OP}$ where P is the point (0, 4, 5)

b $|\mathbf{V}| = 24$ units and $\widehat{\mathbf{V}} = \frac{2}{3}\mathbf{i} - \frac{2}{3}\mathbf{j} - \frac{1}{3}\mathbf{k}$

c **V** is parallel to the vector $8\mathbf{i} + \mathbf{j} + 4\mathbf{k}$ and equal in magnitude to the vector $\mathbf{i} - 2\mathbf{j} + 2\mathbf{k}$

4 Find $\widehat{\mathbf{r}}$ in the form $a\mathbf{i} + b\mathbf{j} + c\mathbf{k}$ if

a $\mathbf{r} = \mathbf{i} - \mathbf{j} + \mathbf{k}$ **b** $\mathbf{r} = 5\mathbf{j} - 12\mathbf{k}$ **c** $\mathbf{r} = \mathbf{i}$

Properties of a Line Joining Two Points

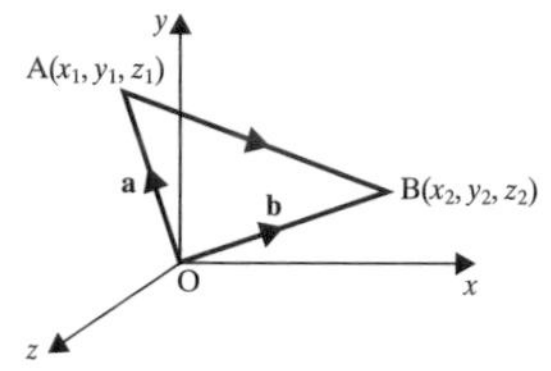

Consider the line joining the points $A(x_1, y_1, z_1)$ and $B(x_2, y_2, z_2)$.

$$\overrightarrow{OA} = x_1\mathbf{i} + y_1\mathbf{j} + z_1\mathbf{k} \quad \text{and} \quad \overrightarrow{OB} = x_2\mathbf{i} + y_2\mathbf{j} + z_2\mathbf{k}$$

and $\quad \overrightarrow{AB} = \overrightarrow{AO} + \overrightarrow{OB} = \overrightarrow{OB} - \overrightarrow{OA}$

hence $\quad \overrightarrow{AB} = (x_2 - x_1)\mathbf{i} + (y_2 - y_1)\mathbf{j} + (z_2 - z_1)\mathbf{k}$

The Length of AB

$$AB = |(x_2 - x_1)\mathbf{i} + (y_2 - y_1)\mathbf{j} + (z_2 - z_1)\mathbf{k}|$$

so **the length of the line joining (x_1, y_1, z_1) and (x_2, y_2, z_2) is $\sqrt{(x_2 - x_1)^2 + (y_2 - y_1)^2 + (z_2 - z_1)^2}$**

The Position Vector of the Midpoint of AB

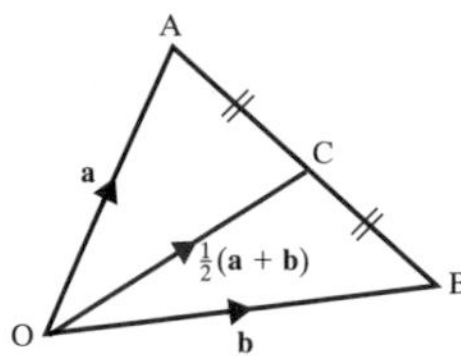

We saw on p. 452 that if C is the midpoint of AB then the position vector of C is $\frac{1}{2}(\overrightarrow{OA} + \overrightarrow{OB})$.

If A is the point (x_1, y_1, z_1) and B is the point (x_2, y_2, z_2) the coordinates of C are

$$\left(\tfrac{1}{2}(x_1 + x_2),\ \tfrac{1}{2}(y_1 + y_2),\ \tfrac{1}{2}(z_1 + z_2)\right)$$

i.e. **the coordinates of the midpoint are the averages of the respective coordinates of the end points.**

EXAMPLES 36D

1 The coordinates of the midpoint R of a line PQ are (3, − 2, 6). If P is the point (4, 1, − 3) and Q is the point (a, b, c), find the values of a, b and c.

R is the point (3, −2, 6)

But R is the midpoint of PQ, so R is the point $(\frac{1}{2}\{4 + a\}, \frac{1}{2}\{1 + b\}, \frac{1}{2}\{-3 + c\})$

$\therefore \quad \frac{1}{2}(4 + a) = 3, \ \frac{1}{2}(1 + b) = -2, \ \frac{1}{2}(-3 + c) = 6$

$\therefore \quad a = 2, \ b = -5, \ c = 15.$

2 Find the length of the median through O of the triangle OAB, where A is the point $(2, 7, -1)$ and B is the point $(4, 1, 2)$.

The coordinates of M, the midpoint of AB, are

$\left(\frac{1}{2}\{2+4\}, \frac{1}{2}\{7+1\}, \frac{1}{2}\{-1+2\}\right)$

i.e. $\left(3, 4, \frac{1}{2}\right)$

So the length of OM is $\sqrt{3^2+4^2+\frac{1}{2}^2} = \frac{1}{2}\sqrt{101}$

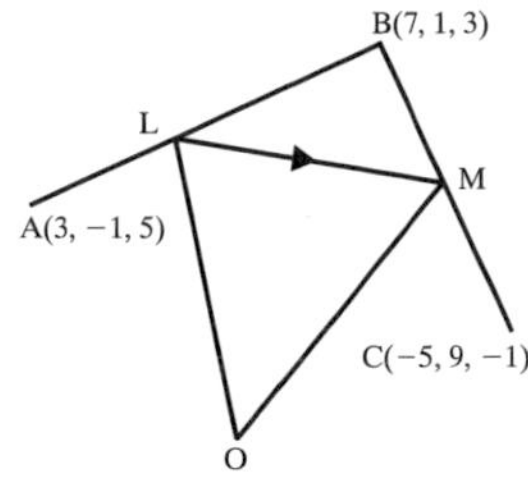

3 The points A, B and C have coordinates $(3, -1, 5)$, $(7, 1, 3)$ and $(-5, 9, -1)$ respectively. If L is the midpoint of AB and M is the midpoint of BC, find the length of LM.

L is the point $\left(\frac{1}{2}\{3+7\}, \frac{1}{2}\{-1+1\}, \frac{1}{2}\{5+3\}\right)$, i.e. $(5, 0, 4)$

M is the point $\left(\frac{1}{2}\{7-5\}, \frac{1}{2}\{1+9\}, \frac{1}{2}\{3-1\}\right)$, i.e. $(1, 5, 1)$

$\therefore \quad LM = \sqrt{(1-5)^2+(5-0)^2+(1-4)^2} = 5\sqrt{2}$

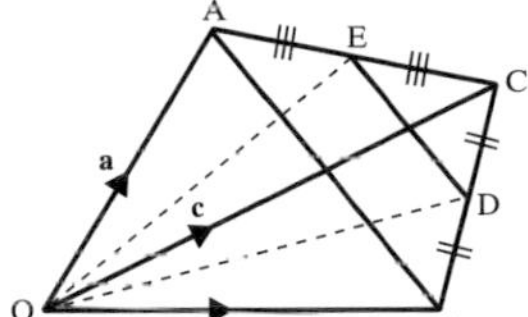

4 A, B and C are the points with position vectors $2\mathbf{i}-\mathbf{j}+5\mathbf{k}$, $\mathbf{i}-2\mathbf{j}+\mathbf{k}$ and $3\mathbf{i}+\mathbf{j}-2\mathbf{k}$ respectively. If D and E are the respective midpoints of BC and AC, show that DE is parallel to AB.

Using $\mathbf{a} = 2\mathbf{i}-\mathbf{j}+5\mathbf{k}$, $\mathbf{b} = \mathbf{i}-2\mathbf{j}+\mathbf{k}$ and $\mathbf{c} = 3\mathbf{i}+\mathbf{j}-2\mathbf{k}$ we have

In $\triangle$OBC, $\quad \overrightarrow{OD} = \frac{1}{2}(\mathbf{b}+\mathbf{c}) = \frac{1}{2}(4\mathbf{i}-\mathbf{j}-\mathbf{k})$

and in $\triangle$OAC $\quad \overrightarrow{OE} = \frac{1}{2}(\mathbf{a}+\mathbf{c}) = \frac{1}{2}(5\mathbf{i}+3\mathbf{k})$

$\therefore \quad \overrightarrow{DE} = \overrightarrow{OE} - \overrightarrow{OD} = \frac{1}{2}(\mathbf{i}+\mathbf{j}+4\mathbf{k})$

Also $\quad \overrightarrow{AB} = \mathbf{b}-\mathbf{a} = -\mathbf{i}-\mathbf{j}-4\mathbf{k}$

So $\quad \overrightarrow{AB} = -2\overrightarrow{DE}$

$\therefore \quad$ AB and DE are parallel.

EXERCISE 36D

In this exercise A, B, C and D are the points with position vectors $\mathbf{i}+\mathbf{j}-\mathbf{k}$, $\mathbf{i}-\mathbf{j}+2\mathbf{k}$, $\mathbf{j}+\mathbf{k}$ and $2\mathbf{i}+\mathbf{j}$ respectively.

1 Find $|\mathbf{AB}|$ and $|\mathbf{BD}|$.

2 Determine whether any of the following pairs of lines are parallel.

a AB and CD

b AC and BD

c AD and BC

3 If L and M are the position vectors of the midpoints of AD and BD respectively, show that $\overrightarrow{LM}$ is parallel to $\overrightarrow{AB}$.

4 If H and K are the midpoints of AC and CD respectively show that $\overrightarrow{HK} = \frac{1}{2}\overrightarrow{AD}$.

5 If L, M, N and P are the midpoints of AD, BD, BC and AC respectively, show that $\overrightarrow{LM}$ is parallel to $\overrightarrow{NP}$.

The Scalar Product

We are now going to look at an operation involving two vectors and the angle between them. This operation is called a product but, because it involves vectors, it is in no way related to the product of real numbers.

The Definition of the Scalar Product

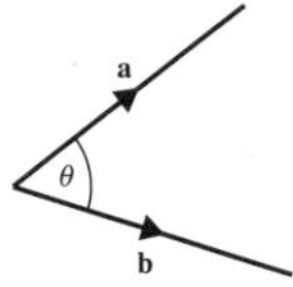

The scalar product of two vectors **a** and **b** is denoted by **a . b** and defined as $ab\cos\theta$ where θ is the angle between **a** and **b**.

i.e. $\quad \mathbf{a}\,.\,\mathbf{b} = ab\cos\theta$

Note that as $ab\cos\theta = ba\cos\theta$, $\mathbf{a}\,.\,\mathbf{b} = \mathbf{b}\,.\,\mathbf{a}$

Properties of the Scalar Product

Parallel Vectors

If **a** and **b** are parallel then

either $\quad \mathbf{a}\,.\,\mathbf{b} = ab\cos 0 \qquad$ or $\qquad \mathbf{a}\,.\,\mathbf{b} = ab\cos\pi$

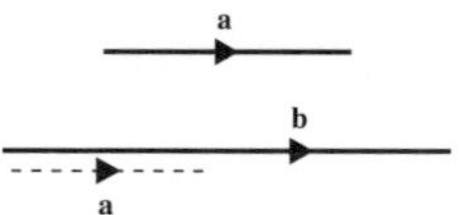

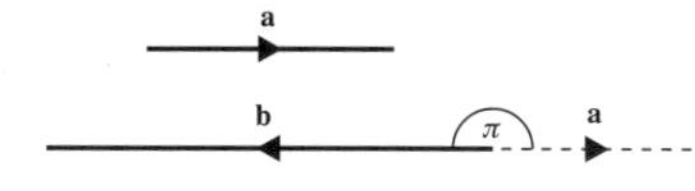

i.e. **for like parallel vectors** $\quad \mathbf{a}\,.\,\mathbf{b} = ab$

and **for unlike parallel vectors** $\quad \mathbf{a}\,.\,\mathbf{b} = -ab$

In the special case when $\mathbf{a} = \mathbf{b}$

$$\mathbf{a}\,.\,\mathbf{b} = \mathbf{a}\,.\,\mathbf{a} = a^2 \quad \text{(sometimes } \mathbf{a}\,.\,\mathbf{a} \text{ is written } \mathbf{a}^2\text{)}$$

In particular, for the Cartesian unit vectors **i**, **j** and **k**

$$\mathbf{i}\,.\,\mathbf{i} = \mathbf{j}\,.\,\mathbf{j} = \mathbf{k}\,.\,\mathbf{k} = 1$$

Perpendicular Vectors

If **a** and **b** are perpendicular then $\theta = \frac{1}{2}\pi, \quad \Rightarrow \quad \mathbf{a}\,.\,\mathbf{b} = ab\cos\frac{1}{2}\pi = 0$

i.e. **for perpendicular vectors** $\quad \mathbf{a}\,.\,\mathbf{b} = 0$

For the unit vectors **i**, **j** and **k** we have

$$\mathbf{i}\,.\,\mathbf{j} = \mathbf{j}\,.\,\mathbf{k} = \mathbf{k}\,.\,\mathbf{i} = 0$$

Calculating a . b in Cartesian Form

If $\mathbf{a} = x_i\mathbf{i} + y_1\mathbf{j} + z_1\mathbf{k}$ and $\mathbf{b} = x_2\mathbf{i} + y_2\mathbf{j} + z_2\mathbf{k}$ then we can find **a . b** from

$$\mathbf{a}\cdot\mathbf{b} = (x_1x_2 + y_1y_2 + z_1z_2)$$

i.e. $\quad \mathbf{(x_1 i + y_1 j + z_1 k) . (x_2 i + y_2 j + z_2 k) = x_1x_2 + y_1y_2 + z_1z_2}$

So $(2\mathbf{i} - 3\mathbf{j} + 4\mathbf{k}).(\mathbf{i} + 3\mathbf{j} - 2\mathbf{k}) = (2)(1) + (-3)(3) + (4)(-2) = -15$

EXAMPLES 36E

1 Find the scalar product of $\mathbf{a} = 2\mathbf{i} - 3\mathbf{j} + 5\mathbf{k}$ and $\mathbf{b} = \mathbf{i} - 3\mathbf{j} + \mathbf{k}$ and hence find the cosine of the angle between **a** and **b**.

$$\mathbf{a}\cdot\mathbf{b} = (2)(1) + (-3)(-3) + (5)(1) = 16$$

But $\quad \mathbf{a}\cdot\mathbf{b} = |\mathbf{a}||\mathbf{b}|\cos\theta$

$|\mathbf{a}| = \sqrt{4+9+25} = \sqrt{38}$ and $|\mathbf{b}| = \sqrt{1+9+1} = \sqrt{11}$

Hence $\quad \cos\theta = \dfrac{\mathbf{a}\cdot\mathbf{b}}{|\mathbf{a}||\mathbf{b}|} = \dfrac{16}{\sqrt{11}\sqrt{38}} = \dfrac{16}{\sqrt{418}}$

2 If $\mathbf{a} = 10\mathbf{i} - 3\mathbf{j} + 5\mathbf{k}$, $\mathbf{b} = 2\mathbf{i} + 6\mathbf{j} - 3\mathbf{k}$ and $\mathbf{c} = \mathbf{i} + 10\mathbf{j} - 2\mathbf{k}$, verify that $\mathbf{a}\cdot\mathbf{b} + \mathbf{a}\cdot\mathbf{c} = \mathbf{a}\cdot(\mathbf{b} + \mathbf{c})$.

$$\mathbf{a}\cdot\mathbf{b} = (10)(2) + (-3)(6) + (5)(-3) = -13$$
$$\mathbf{a}\cdot\mathbf{c} = (10)(1) + (-3)(10) + (5)(-2) = -30$$
$$\mathbf{b} + \mathbf{c} = 3\mathbf{i} + 16\mathbf{j} - 5\mathbf{k}$$

Hence $\quad \mathbf{a}\cdot(\mathbf{b} + \mathbf{c}) = (10)(3) + (-3)(16) + (5)(-5) = -43$

But $\quad \mathbf{a}\cdot\mathbf{b} + \mathbf{a}\cdot\mathbf{c} = -13 - 30 = -43$

Therefore $\mathbf{a}\cdot\mathbf{b} + \mathbf{a}\cdot\mathbf{c} = \mathbf{a}\cdot(\mathbf{b} + \mathbf{c})$

3 Find a unit vector which is perpendicular to AB and AC, if $\overrightarrow{AB} = \mathbf{i} + 2\mathbf{j} + 3\mathbf{k}$ and $\overrightarrow{AC} = 4\mathbf{i} - \mathbf{j} + 2\mathbf{k}$.

Let $a\mathbf{i} + b\mathbf{j} + c\mathbf{k}$ be a vector perpendicular to both AB and AC.

It is perpendicular to AB so $(a\mathbf{i} + b\mathbf{j} + c\mathbf{k}).(\mathbf{i} + 2\mathbf{j} + 3\mathbf{k}) = 0$

It is perpendicular to AC so $(a\mathbf{i} + b\mathbf{j} + c\mathbf{k}).(4\mathbf{i} - \mathbf{j} + 2\mathbf{k}) = 0$

Therefore $\begin{cases} a + 2b + 3c = 0 \\ 4a - b + 2c = 0 \end{cases}$

Eliminating b gives $a = -\frac{7}{9}c$ and eliminating a gives $b = -\frac{10}{9}c$

Hence $\quad a\mathbf{i} + b\mathbf{j} + c\mathbf{k} = -\frac{7}{9}c\mathbf{i} - \frac{10}{9}c\mathbf{j} + c\mathbf{k} = \frac{1}{9}c(-7\mathbf{i} - 10\mathbf{j} + 9\mathbf{k})$

Thus $-7\mathbf{i} - 10\mathbf{j} + 9\mathbf{k}$ is perpendicular to both AB and AC.

A unit vector perpendicular to AB and AC is therefore

$$(-7\mathbf{i} - 10\mathbf{j} + 9\mathbf{k})/\sqrt{230}$$

EXERCISE 36E

1 Calculate $\mathbf{a}.\mathbf{b}$ if

a $\mathbf{a} = 2\mathbf{i} - 4\mathbf{j} + 5\mathbf{k}$, $\mathbf{b} = \mathbf{i} + 3\mathbf{j} + 8\mathbf{k}$

b $\mathbf{a} = 3\mathbf{i} - 7\mathbf{j} + 2\mathbf{k}$, $\mathbf{b} = 5\mathbf{i} + \mathbf{j} - 4\mathbf{k}$

c $\mathbf{a} = 2\mathbf{i} - 3\mathbf{j} + 6\mathbf{k}$, $\mathbf{b} = \mathbf{i} + \mathbf{j}$

What conclusion can you draw in part **b**?

2 Find $\mathbf{p}.\mathbf{q}$ and the cosine of the angle between $\mathbf{p}$ and $\mathbf{q}$ if

a $\mathbf{p} = 2\mathbf{i} + 4\mathbf{j} + \mathbf{k}$, $\mathbf{q} = \mathbf{i} + \mathbf{j} + \mathbf{k}$

b $\mathbf{p} = -\mathbf{i} + 3\mathbf{j} - 2\mathbf{k}$, $\mathbf{q} = \mathbf{i} + \mathbf{j} - 6\mathbf{k}$

c $\mathbf{p} = -2\mathbf{i} + 5\mathbf{j}$, $\mathbf{q} = \mathbf{i} + \mathbf{j}$

d $\mathbf{p} = 2\mathbf{i} + \mathbf{j}$, $\mathbf{q} = \mathbf{j} - 2\mathbf{k}$

3 The cosine of the angle between two vectors $\mathbf{v}_1$ and $\mathbf{v}_2$ is $\frac{4}{21}$.
If $\mathbf{v}_1 = 6\mathbf{i} + 3\mathbf{j} - 2\mathbf{k}$ and $\mathbf{v}_2 = -2\mathbf{i} + \lambda\mathbf{j} - 4\mathbf{k}$, find the positive value of λ.

4 In a triangle ABC, $\overrightarrow{AB} = \mathbf{i} + 2\mathbf{j} + 3\mathbf{k}$ and $\overrightarrow{BC} = -\mathbf{i} + 4\mathbf{j}$.
Find the cosine of angle ABC.
Find the vector $\overrightarrow{AC}$ and use it to calculate the angle BAC.

5 Show that $\mathbf{i} + 7\mathbf{j} + 3\mathbf{k}$ is perpendicular to both $\mathbf{i} - \mathbf{j} + 2\mathbf{k}$ and $2\mathbf{i} + \mathbf{j} - 3\mathbf{k}$.

6 Show that $13\mathbf{i} + 23\mathbf{j} + 7\mathbf{k}$ is perpendicular to both $2\mathbf{i} + \mathbf{j} - 7\mathbf{k}$ and $3\mathbf{i} - 2\mathbf{j} + \mathbf{k}$.

7 The magnitudes of two vectors $\mathbf{p}$ and $\mathbf{q}$ are 5 and 4 units respectively.
The angle between $\mathbf{p}$ and $\mathbf{q}$ is 30°.
Find **a** $\mathbf{p}.\mathbf{q}$ **b** the magnitude of the vector $\mathbf{p} - \mathbf{q}$

8 Calculate the acute angle between the vectors

$$\begin{pmatrix} 2 \\ -1 \\ 3 \end{pmatrix} \text{ and } \begin{pmatrix} 0 \\ -1 \\ -1 \end{pmatrix}.$$

9

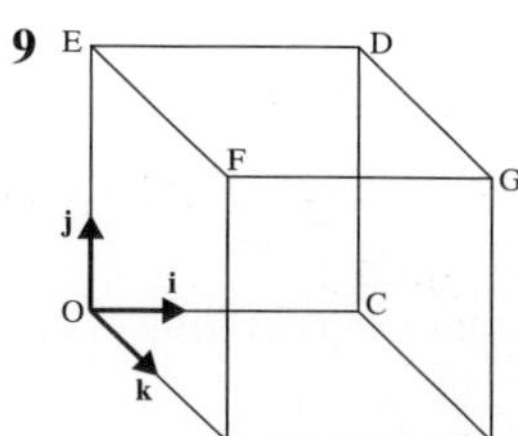

The diagram shows a cube where the length of each edge is 4 cm.

a Express in terms of **i**, **j**, and **k**, the vectors
i $\overrightarrow{AE}$ **ii** $\overrightarrow{AG}$

b Given that H is the midpoint of AB, find the angles of the triangle AEH, giving your answers to the nearest degree.

10 Is it true to say that if $\mathbf{a}.\mathbf{b} = 0$ then either $\mathbf{a} = \mathbf{0}$ or $\mathbf{b} = \mathbf{0}$?
Justify your answer.

11 In triangle OAB, O is the origin, $\overrightarrow{OA} = 4\mathbf{i} - 3\mathbf{j} + 4\mathbf{k}$ and $\overrightarrow{OB} = \mathbf{i} + 6\mathbf{j} - 2\mathbf{k}$.

a Show that triangle OAB is isosceles.

b Find angle AOB correct to the nearest degree.

c Hence or otherwise find the area of triangle OAB.

12 The points A, B, C have position vectors

$$\mathbf{a} = 2\mathbf{i} + \mathbf{j} - \mathbf{k}, \quad \mathbf{b} = 3\mathbf{i} + 4\mathbf{j} - 2\mathbf{k}, \quad \mathbf{c} = 5\mathbf{i} - \mathbf{j} + 2\mathbf{k},$$

respectively, relative to a fixed origin O.

a Evaluate the scalar product $(\mathbf{a} - \mathbf{b}).(\mathbf{c} - \mathbf{b})$.
Hence calculate the size of angle ABC, giving your answer to the nearest 0.1°.

b Given that ABCD is a parallelogram:
i determine the position vector of D;
ii calculate the area of ABCD.

c The point E lies on BA produced so that $\overrightarrow{BE} = 3\overrightarrow{BA}$.
Write down the position vector of E.
The line CE cuts the line AD at X. Find the position vector of X.

The Equation of a Straight Line

A particular line is uniquely located in space if

a it has a known direction and passes through a known fixed point, or

b it passes through two known fixed points.

A Line with Known Direction Passing through a Fixed Point

Consider a line that is parallel to a vector **m** and which passes through a fixed point A with position vector **a**.

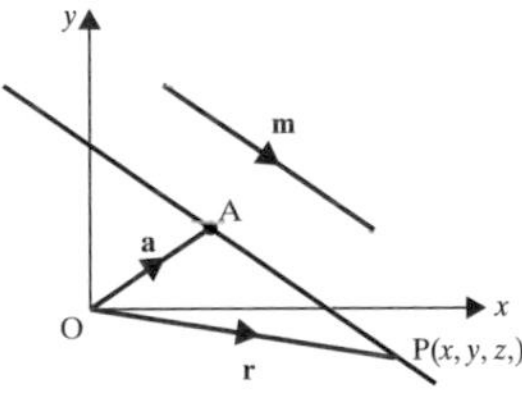

If **r** is the position vector, $\overrightarrow{OP}$, of a point P then

$$\text{P is a point on this line} \iff \overrightarrow{AP} = \lambda\mathbf{m}$$

where λ is a variable scalar, i.e. a parameter.

Now $\quad \overrightarrow{OP} = \overrightarrow{OA} + \overrightarrow{AP}$

i.e. $\quad \mathbf{r} = \mathbf{a} + \lambda\mathbf{m}$

i.e. $\quad$ P is on the line $\iff \mathbf{r} = \mathbf{a} + \lambda\mathbf{m}$.

For each value of the parameter λ this equation gives the position vector of one point on the line and it is called the vector equation of the line.

For example, the line whose vector equation is

$$\mathbf{r} = (5\mathbf{i} - 2\mathbf{j} + 4\mathbf{k}) + \lambda(2\mathbf{i} - \mathbf{j} + 3\mathbf{k}) \qquad [1]$$

is *parallel* to the vector $2\mathbf{i} - \mathbf{j} + 3\mathbf{k}$ and passes through the point whose position vector is $5\mathbf{i} - 2\mathbf{j} + 4\mathbf{k}$.
Now $\mathbf{r}$ is the position vector of any point $P(x, y, z)$ on the line,

i.e. $$x\mathbf{i} + y\mathbf{j} + z\mathbf{k} = 5\mathbf{i} - 2\mathbf{j} + 4\mathbf{k} + \lambda(2\mathbf{i} - \mathbf{j} + 3\mathbf{k})$$
$$= (5 + 2\lambda)\mathbf{i} + (-2 - \lambda)\mathbf{j} + (4 + 3\lambda)\mathbf{k}$$

$$\Rightarrow \quad \left.\begin{aligned} x &= 5 + 2\lambda \\ y &= -2 - \lambda \\ z &= 4 + 3\lambda \end{aligned}\right\} \qquad [2]$$

Isolating λ in each of these equations gives

$$\frac{x-5}{2} = \frac{y+2}{-1} = \frac{z-4}{3} \quad (=\lambda) \qquad [3]$$

So there are three ways of expressing the relationships between the coordinates of any point P on this line:

[1] is the vector equation of the line,

[2] are the parametric equations of the line.

[3] are the Cartesian equations of the line.

Note that the line is parallel to $2\mathbf{i} - \mathbf{j} + 3\mathbf{k}$; so the vector $2\mathbf{i} - \mathbf{j} + 3\mathbf{k}$ defines the direction of the line, and the coefficients of **i**, **j** and **k**, i.e. 2, −1, and 3, appear in all three forms of its equation.

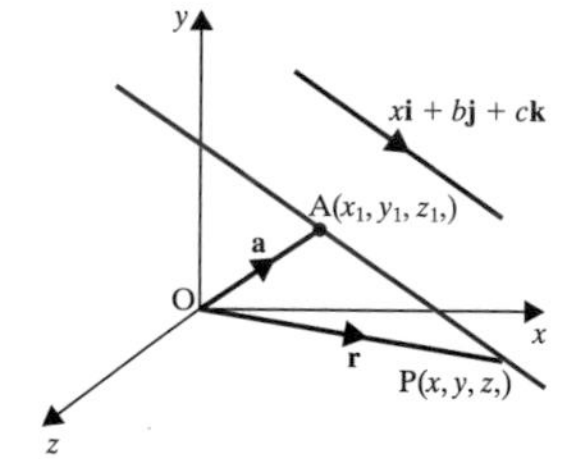

In general, if a line passes through $A(x_1, y_1, z_1)$ and is parallel to $a\mathbf{i} + b\mathbf{j} + c\mathbf{k}$ its equations may be written

$$\mathbf{r} = (x_1\mathbf{i} + y_1\mathbf{j} + z_1\mathbf{k}) + \lambda(a\mathbf{i} + b\mathbf{j} + c\mathbf{k})$$

or $$\left\{\begin{aligned} x &= x_1 + \lambda a \\ y &= y_1 + \lambda b \\ z &= z_1 + \lambda c \end{aligned}\right.$$

or $$\frac{x - x_1}{a} = \frac{y - y_1}{b} = \frac{z - z_1}{c} \quad (=\lambda).$$

Note that the point (x_1, y_1, z_1) is only one of an infinite set of particular points on the line. Hence the equations representing a given line are not unique.

EXAMPLES 36F

1 Find a vector equation of the line that passes through the point with position vector $2\mathbf{i} - \mathbf{j} + 4\mathbf{k}$ and is parallel to the vector $\mathbf{i} + \mathbf{j} - 2\mathbf{k}$.

> The vector equation of a line is $\mathbf{r} = \mathbf{a} + \lambda\mathbf{m}$ where **a** is the position vector of a point on the line and **m** is parallel to the line. For this line, $\mathbf{a} = 2\mathbf{i} - \mathbf{j} + 4\mathbf{k}$ and $\mathbf{m} = \mathbf{i} + \mathbf{j} - 2\mathbf{k}$.

A vector equation of the line is $\mathbf{r} = 2\mathbf{i} - \mathbf{j} + 4\mathbf{k} + \lambda(\mathbf{i} + \mathbf{j} - 2\mathbf{k})$.

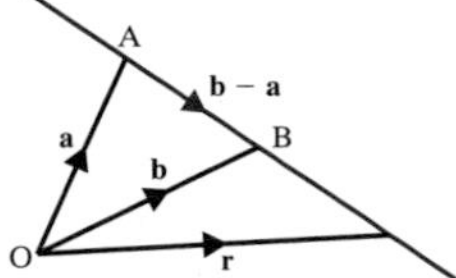

2 Find a vector equation for the line through the points A(3, 4, −7) and B(1, −1, 6).

> To find a vector equation of a line we need a point on the line (we can use either A or B) and a vector parallel to the line.

As A and B are on the line, $\overrightarrow{AB}$ is parallel to the line and $\overrightarrow{AB} = \overrightarrow{OB} - \overrightarrow{OA}$.

$\overrightarrow{OA} = \mathbf{a} = 3\mathbf{i} + 4\mathbf{j} - 7\mathbf{k}$ and $\overrightarrow{OB} = \mathbf{b} = \mathbf{i} - \mathbf{j} + 6\mathbf{k}$

$\therefore \quad \overrightarrow{AB} = \mathbf{b} - \mathbf{a} = (\mathbf{i} - \mathbf{j} + 6\mathbf{k}) - (3\mathbf{i} + 4\mathbf{j} - 7\mathbf{k})$

$= -2\mathbf{i} - 5\mathbf{j} + 13\mathbf{k}$

A vector equation of the line is $\mathbf{r} = 3\mathbf{i} + 4\mathbf{j} - 7\mathbf{k} + \lambda(-2\mathbf{i} - 5\mathbf{j} + 13\mathbf{k})$

> Remember that this equation is not unique; we could have used **b** instead of **a** and also any multiple of $\mathbf{b} - \mathbf{a}$, all of which are parallel to the line, e.g. $\mathbf{r} = \mathbf{i} - \mathbf{j} + \mathbf{k} + \lambda(2\mathbf{i} + 5\mathbf{j} - 13\mathbf{k})$ is an equally valid vector equation for this line.

3 State whether or not the lines with equations

$\mathbf{r} = 2\mathbf{i} - 3\mathbf{j} + 2\mathbf{k} + \lambda(\mathbf{i} - \mathbf{j} + 4\mathbf{k})$ and $\mathbf{r} = (3 - \mu)\mathbf{i} - (3 - \mu)\mathbf{j} + (2 - 4\mu)\mathbf{k}$

are parallel.

> To determine whether the lines are parallel we need to find a vector in the direction of each line. This can be 'read' from the equation of the first line but the equation of the second line needs rearranging first.

$\mathbf{r} = 2\mathbf{i} - 3\mathbf{j} + 2\mathbf{k} + \lambda(\mathbf{i} - \mathbf{j} + 4\mathbf{k})$ is in the direction of the vector $(\mathbf{i} - \mathbf{j} + 4\mathbf{k})$.

$\mathbf{r} = (3 - \mu)\mathbf{i} - (3 - \mu)\mathbf{j} + (2 - 4\mu)\mathbf{k}$

$= 3\mathbf{i} - 3\mathbf{j} + 2\mathbf{k} + \mu(-\mathbf{i} + \mathbf{j} - 4\mathbf{k})$ so is in the direction of the vector $(-\mathbf{i} + \mathbf{j} - 4\mathbf{k})$.

But $-\mathbf{i} + \mathbf{j} - 4\mathbf{k} = -(\mathbf{i} - \mathbf{j} + 4\mathbf{k})$, $\therefore$ the lines are parallel.

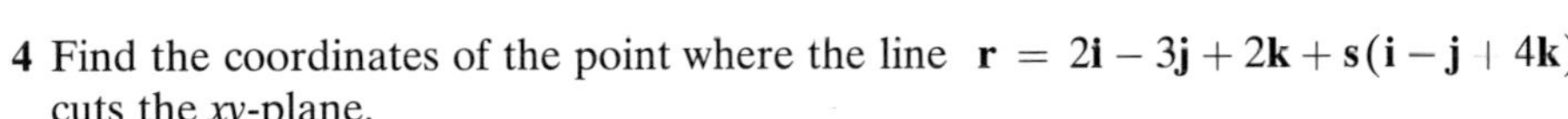

4 Find the coordinates of the point where the line $\mathbf{r} = 2\mathbf{i} - 3\mathbf{j} + 2\mathbf{k} + s(\mathbf{i} - \mathbf{j} + 4\mathbf{k})$ cuts the *xy*-plane.

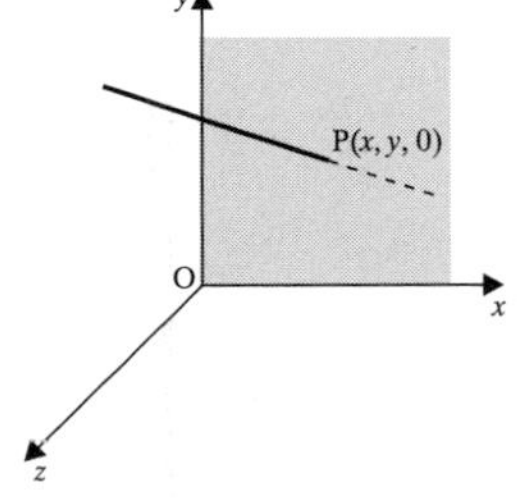

> The *z*-coordinate of any point P on the *xy*-plane is zero,
> i.e. **OP** has zero **k** component.
> We can identify the **k** component if we collect the components of **i**, **j** and **k**.

Rearranging the equation of the line as $\mathbf{r} = (2 + s)\mathbf{i} + (-3 - s)\mathbf{j} + (2 + 4s)\mathbf{k}$

shows that it cuts the *xy*-plane where $2 + 4s = 0$, i.e. where $s = -\frac{1}{2}$.

When $s = -\frac{1}{2}$, $\mathbf{r} = \frac{3}{2}\mathbf{i} - \frac{7}{2}\mathbf{j}$,

$\therefore$ the line cuts the *xy*-plane at the point $(\frac{3}{2}, -\frac{7}{2}, 0)$.

EXERCISE 36F

1 Write down a vector which is parallel to each of these lines.

a $\mathbf{r} = \mathbf{i} - 2\mathbf{j} + 4\mathbf{k} + t(2\mathbf{i} - \mathbf{j} - 5\mathbf{k})$ b $\mathbf{r} = 2\mathbf{i} - \mathbf{k} + s(3\mathbf{j} - 5\mathbf{k})$

c $\mathbf{r} = (1 - 2s)\mathbf{i} + (4s - 3)\mathbf{j} + (1 + s)\mathbf{k}$

d $\mathbf{r} = t\mathbf{i} + 3\mathbf{j} - (1 - t)\mathbf{k}$ e $\frac{x-1}{2} = \frac{y+1}{3} = \frac{z}{2}$

f $\frac{x+3}{4} = \frac{y}{3} = 2z$

2 Write down a Cartesian equation of the line whose vector equation is

a $\mathbf{r} = \mathbf{i} - 5\mathbf{j} + 2\mathbf{k} + t(\mathbf{i} + 3\mathbf{j} + \mathbf{k})$ b $\mathbf{r} = \mathbf{i} + (3 - 2\lambda)\mathbf{j} + (3 + \lambda)\mathbf{k}$

3 Write down a vector equation of the line whose Cartesian equation is

a $\frac{x-3}{2} = \frac{y+1}{3} = \frac{z-4}{2}$ b $\frac{x}{2} = y = \frac{z+1}{3}$

4 Write down equations in vector form for the line through a point A with position vector **a** and in the direction of vector **b** where

a $\mathbf{a} = \mathbf{i} - 3\mathbf{j} + 2\mathbf{k}$ $\mathbf{b} = 5\mathbf{i} + 4\mathbf{j} - \mathbf{k}$ b $\mathbf{a} = 2\mathbf{i} + \mathbf{j}$ $\mathbf{b} = 3\mathbf{j} - \mathbf{k}$

c A is the origin $\mathbf{b} = \mathbf{i} - \mathbf{j} - \mathbf{k}$

5 State whether or not the following pairs of lines are parallel:

a $\mathbf{r} = \mathbf{i} + \mathbf{j} - \mathbf{k} + \lambda(2\mathbf{i} - 3\mathbf{j} + \mathbf{k})$
$\mathbf{r} = 2\mathbf{i} - 4\mathbf{j} + 5\mathbf{k} + \lambda(\mathbf{i} + \mathbf{j} - \mathbf{k})$

b $\mathbf{r} = 2\mathbf{i} - \mathbf{j} + 5\mathbf{k} + s(\mathbf{i} + \mathbf{j} - \mathbf{k})$
$\mathbf{r} = (3 + t)\mathbf{i} + (t - 1)\mathbf{j} + (5 - t)\mathbf{k}$

c $\mathbf{r} = 2\mathbf{i} - \mathbf{j} + 4\mathbf{k} + \lambda(\mathbf{i} + \mathbf{j} + 3\mathbf{k})$
$\mathbf{r} = \mu(2\mathbf{i} + 2\mathbf{j} + 6\mathbf{k})$

d $\mathbf{r} = \lambda(3\mathbf{i} - 3\mathbf{j} + 6\mathbf{k})$
$\mathbf{r} = 4\mathbf{j} + \lambda(-\mathbf{i} + \mathbf{j} - 2\mathbf{k})$

6 The points A(4, 5, 10), B(2, 3, 4) and C(1, 2, −1) are three vertices of a parallelogram ABCD. Find vector equations for the sides AB, BC and AD.

7 Write down a vector equation for the line through A and B if

a $\overrightarrow{AB}$ is $3\mathbf{i} + \mathbf{j} - 4\mathbf{k}$ and $\overrightarrow{OB}$ is $\mathbf{i} + 7\mathbf{j} + 8\mathbf{k}$.

b A and B have coordinates (1, 1, 7) and (3, 4, 1).

Find, in each case, the coordinates of the points where the line crosses the *xy*-plane, the *yz*-plane and the *zx*-plane.

8 A line has Cartesian equations $\frac{x-1}{3} = \frac{y+2}{4} = \frac{z-4}{5}$.

a Write down a vector parallel to this line.

b Find a vector equation for a parallel line passing through the point with position vector $5\mathbf{i} - 2\mathbf{j} - 4\mathbf{k}$ and find the coordinates of the point on this line where $y = 0$.

Pairs of Lines

The location of two lines in space may be such that:

a the lines are parallel

b the lines are not parallel and intersect

c the lines are not parallel and do not intersect. Such lines are called *skew*.

Parallel Lines

If two lines are parallel, this property can be observed from their equations by comparing the vectors parallel to the lines.

Non-parallel Lines

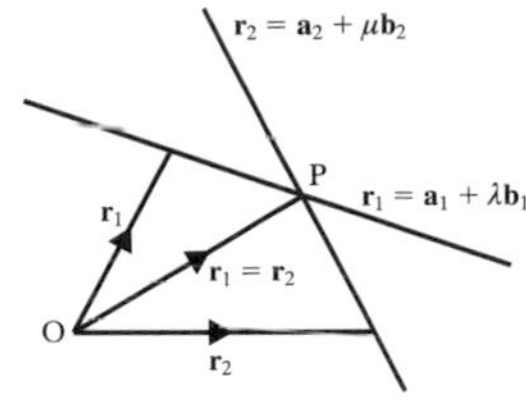

Consider two lines whose vector equations are $\mathbf{r}_1 = \mathbf{a}_1 + \lambda\mathbf{b}_1$ and $\mathbf{r}_2 = \mathbf{a}_2 + \mu\mathbf{b}_2$.

If these lines intersect, there must be unique values of λ and μ such that

$$\mathbf{a}_1 + \lambda\mathbf{b}_1 = \mathbf{a}_2 + \mu\mathbf{b}_2.$$

If no such values can be found, the lines do not intersect.

EXAMPLES 36G

1 Find out whether the following pairs of lines are parallel, non-parallel and intersecting, or non-parallel and non-intersecting.

a $\mathbf{r}_1 = \mathbf{i} + \mathbf{j} + 2\mathbf{k} + \lambda(3\mathbf{i} - 2\mathbf{j} + 4\mathbf{k})$ and $\mathbf{r}_2 = 2\mathbf{i} - \mathbf{j} + 3\mathbf{k} + \mu(-6\mathbf{i} + 4\mathbf{j} - 8\mathbf{k})$

b $\mathbf{r}_1 = \mathbf{i} - \mathbf{j} + 3\mathbf{k} + \lambda(\mathbf{i} - \mathbf{j} + \mathbf{k})$ and $\mathbf{r}_2 = 2\mathbf{i} + 4\mathbf{j} + 6\mathbf{k} + \mu(2\mathbf{i} + \mathbf{j} + 3\mathbf{k})$

c $\mathbf{r}_1 = \mathbf{i} + \mathbf{k} + \lambda(\mathbf{i} + 3\mathbf{j} + 4\mathbf{k})$ and $\mathbf{r}_2 = 2\mathbf{i} + 3\mathbf{j} + \mu(4\mathbf{i} - \mathbf{j} + \mathbf{k})$

a Checking first whether the lines are parallel we compare their directions.

The first line is parallel to $3\mathbf{i} - 2\mathbf{j} + 4\mathbf{k}$
The second line is parallel to $-6\mathbf{i} + 4\mathbf{j} - 8\mathbf{k} = -2(3\mathbf{i} - 2\mathbf{j} + 4\mathbf{k})$
Therefore these two lines are parallel.

b In this case the directions of the lines are $\mathbf{i} - \mathbf{j} + \mathbf{k}$ and $2\mathbf{i} + \mathbf{j} + 3\mathbf{k}$
These are not equal, so these two lines are not parallel.
Now if the lines intersect it will be at a point where $\mathbf{r}_1 = \mathbf{r}_2$, i.e. where

$$(1+\lambda)\mathbf{i} - (1+\lambda)\mathbf{j} + (3+\lambda)\mathbf{k} = 2(1+\mu)\mathbf{i} + (4+\mu)\mathbf{j} + (6+3\mu)\mathbf{k}.$$

Equating the coefficients of **i** and **j**, we have

$$1 + \lambda = 2(1+\mu) \quad \text{and} \quad -(1+\lambda) = 4 + \mu$$

Hence $\mu = -2, \lambda = -3$

With these values for λ and μ, the coefficients of **k** become

$$\left.\begin{array}{ll}\text{first line} & 3 + \lambda = 0 \\ \text{second line} & 6 + 3\mu = 0\end{array}\right\} \text{ equal values.}$$

So $\mathbf{r}_1 = \mathbf{r}_2$ when $\lambda = -3$ and $\mu = -2$.
Therefore the lines *do* intersect at the point with position vector
$(1-3)\mathbf{i} - (1-3)\mathbf{j} + (3-3)\mathbf{k}$, ($\lambda = -3$ in $\mathbf{r}_1$) i.e. $-2\mathbf{i} + 2\mathbf{j}$.

c The directions of these two lines are not equal so the lines are not parallel. If the lines intersect it will be where $\mathbf{r}_1 = \mathbf{r}_2$, i.e. where $(1+\lambda)\mathbf{i} + 3\lambda\mathbf{j} + (1+4\lambda)\mathbf{k} = (2+4\mu)\mathbf{i} + (3-\mu)\mathbf{j} + \mu\mathbf{k}$.

Equating the coefficients of **i** and **j** we have

$$\left.\begin{aligned} 1+\lambda &= 2+4\mu \\ 3\lambda &= 3-\mu \end{aligned}\right\} \Rightarrow \quad \mu = 0, \quad \lambda = 1.$$

With these values of λ and μ, the coefficients of **k** become

$$\left.\begin{aligned} &\text{first line} \quad & 1+4\lambda &= 5 \\ &\text{second line} \quad & \mu &= 0 \end{aligned}\right\} \quad \text{unequal values.}$$

So there are no values of λ and μ for which $\mathbf{r}_1 = \mathbf{r}_2$ and these lines do not intersect and are therefore skew.

2 Find the angle between the lines

$\mathbf{r} = 2\mathbf{i} - \mathbf{j} + 4\mathbf{k} + \lambda(\mathbf{i} + \mathbf{j} - 2\mathbf{k})$ and $\mathbf{r} = (3-2\mu)\mathbf{i} + (4-5\mu)\mathbf{j} + (-7+13\mu)\mathbf{k}$.

> The angle between the lines is the angle between their directions, i.e. between the vectors $\mathbf{i} + \mathbf{j} - 2\mathbf{k}$ and $-2\mathbf{i} - 5\mathbf{j} + 13\mathbf{k}$, which we can find by using the scalar product.

If θ is the angle between the lines, then

$$(\mathbf{i} + \mathbf{j} - 2\mathbf{k}).(-2\mathbf{i} - 5\mathbf{j} + 13\mathbf{k}) = |\mathbf{i} + \mathbf{j} - 2\mathbf{k}||-2\mathbf{i} - 5\mathbf{j} + 13\mathbf{k}|\cos\theta$$

$$\Rightarrow \quad -33 = (\sqrt{6})(\sqrt{198})\cos\theta$$

$$\Rightarrow \quad \cos\theta = -\frac{33}{6\sqrt{33}}, \quad \text{i.e.} \quad \theta = \cos^{-1}\frac{\sqrt{33}}{6}$$

EXERCISE 36G

1 Find whether the following pairs of lines are parallel, intersecting or skew, in the case of intersection state the position vector of the common point, and find the angle between the lines.

a $\mathbf{r} = \mathbf{i} - \mathbf{j} + \mathbf{k} + \lambda(3\mathbf{i} - 4\mathbf{j} + \mathbf{k})$, $\mathbf{r} = \mu(-9\mathbf{i} + 12\mathbf{j} - 3\mathbf{k})$

b $\mathbf{r} = (4-t)\mathbf{i} + (8-2t)\mathbf{j} + (3-t)\mathbf{k}$, $\mathbf{r} = (7+6s)\mathbf{i} + (6+4s)\mathbf{j} + (5+5s)\mathbf{k}$

c $\mathbf{r} = \mathbf{i} + 3\mathbf{k} + \lambda(2\mathbf{i} + \mathbf{j} + \mathbf{k})$, $\mathbf{r} = 2\mathbf{i} - \mathbf{j} + \mathbf{k} + \mu(\mathbf{i} - 2\mathbf{j})$

2 Two lines which intersect have equations

$$\mathbf{r} = 2\mathbf{i} + 9\mathbf{j} + 13\mathbf{k} + \lambda(\mathbf{i} + 2\mathbf{j} + 3\mathbf{k})$$

and $$\mathbf{r} = a\mathbf{i} + 7\mathbf{j} - 2\mathbf{k} + \mu(-\mathbf{i} + 2\mathbf{j} - 3\mathbf{k}).$$

Find the value of a, the position vector of the point of intersection, and the angle between the lines.

3 Show that the lines

$$\mathbf{r} = 2\mathbf{i} - \mathbf{j} + \mathbf{k} + \lambda(\mathbf{i} - 2\mathbf{j} + 2\mathbf{k})$$

and $$\mathbf{r} = \mathbf{i} - 3\mathbf{j} + 4\mathbf{k} + \mu(2\mathbf{i} + 3\mathbf{j} - 6\mathbf{k})$$

are skew.

Equation of a Plane

A particular plane can be specified in several ways, for example:

a one and only one plane can be drawn through three non-collinear points, therefore three given points specify a particular plane,

b one and only one plane can be drawn to contain two concurrent lines, therefore two given concurrent lines specify a particular plane,

c one and only one plane can be drawn perpendicular to a given direction at a given distance from the origin, therefore the normal to a plane and the distance of the plane from the origin specify a particular plane,

d one and only one plane can be drawn through a given point and perpendicular to a given direction, therefore a point on the plane and a normal to the plane specify a particular plane.

There are many other ways of specifying a particular plane but those described in **c** and **d** are particularly suitable for deriving the vector equation of a plane.

The Vector Equation of a Plane

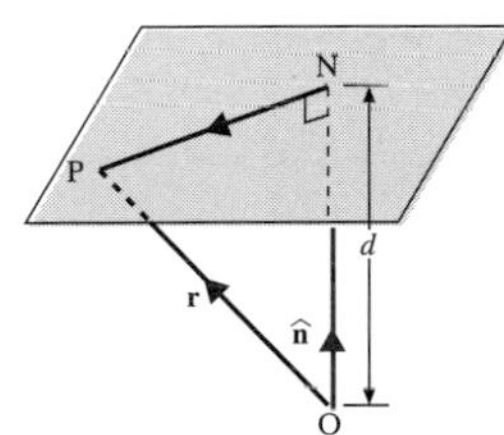

Consider the plane which is at a distance d from the origin and which is perpendicular to the unit vector $\hat{\mathbf{n}}$ ($\hat{\mathbf{n}}$ being directed *away* from O).

If ON is the perpendicular from the origin to the plane then $\overrightarrow{ON} = d\hat{\mathbf{n}}$. For any point, P, on the plane, NP is perpendicular to ON. Conversely if P is not on the plane, NP is not perpendicular to ON. So

$$\text{P is on the plane} \iff \overrightarrow{NP}\cdot\overrightarrow{ON} = 0. \qquad [1]$$

This equation is called the scalar product form of the vector equation of the plane.

If $\mathbf{r}$ is the position vector of P, $\overrightarrow{NP} = \mathbf{r} - d\hat{\mathbf{n}}$.

Therefore [1] becomes $(\mathbf{r} - d\hat{\mathbf{n}})\cdot d\hat{\mathbf{n}} = 0 \quad \Rightarrow \quad \mathbf{r}\cdot\hat{\mathbf{n}} - d\hat{\mathbf{n}}\cdot\hat{\mathbf{n}} = 0$

But $\hat{\mathbf{n}}\cdot\hat{\mathbf{n}} = 1$ so $\mathbf{r}\cdot\hat{\mathbf{n}} = d$ [2]

The equation $\mathbf{r}\cdot\hat{\mathbf{n}} = d$ is the standard form of the vector equation of the plane, where
$\mathbf{r}$ is the position vector of any point on the plane,
$\hat{\mathbf{n}}$ is the unit vector perpendicular to the plane,
d is the distance of the plane from the origin.

This standard form of the vector equation of a plane can be multiplied by any scalar quantity, so

any equation of the form $\mathbf{r}\cdot\mathbf{n} = D$ represents a plane that is perpendicular to $\mathbf{n}$.

Converting back to standard form we have $\mathbf{r}\cdot\hat{\mathbf{n}} = \dfrac{D}{|\mathbf{n}|}$ **where** $\dfrac{D}{|\mathbf{n}|}$ **is the distance of** the plane from the origin.

Now consider the plane which contains the point A whose position vector is $\mathbf{a}$ and which is perpendicular to the unit vector $\hat{\mathbf{n}}$.

The distance of the plane from the origin is ON and $\text{ON} = |\mathbf{a}|\cos\theta = \mathbf{a}.\hat{\mathbf{n}}$.
So equation [2] becomes $\mathbf{r}.\hat{\mathbf{n}} = \mathbf{a}.\hat{\mathbf{n}}$.

Thus $\mathbf{r}.\hat{\mathbf{n}} = \mathbf{a}.\hat{\mathbf{n}}$ is the vector equation of a plane that is perpendicular to $\hat{\mathbf{n}}$ and that contains the point with position vector $\mathbf{a}$.

The Cartesian Equation of a Plane

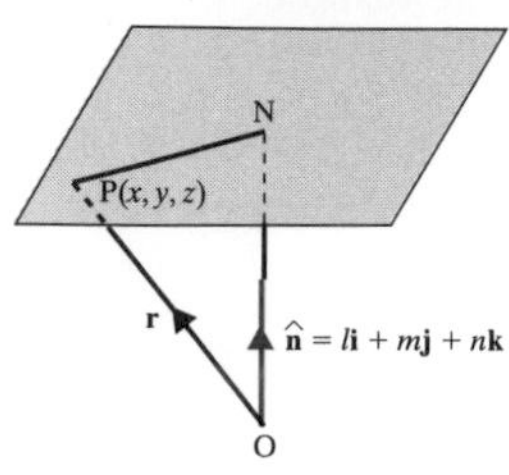

Consider a plane whose vector equation in standard form is

$$\mathbf{r}.\hat{\mathbf{n}} = d \qquad [1]$$

where $\hat{\mathbf{n}} = l\mathbf{i} + m\mathbf{j} + n\mathbf{k}$.
Now if a point $P(x, y, z)$ is on this plane its position vector, $\mathbf{r} = x\mathbf{i} + y\mathbf{j} + z\mathbf{k}$, satisfies equation [1], so

$$(x\mathbf{i} + y\mathbf{j} + z\mathbf{k}).(l\mathbf{i} + m\mathbf{j} + n\mathbf{k}) = d \quad \Rightarrow \quad lx + my + nz = d$$

i.e. **$lx + my + nz = d$ is the Cartesian equation of the plane.**

Now if each term in this equation is multiplied by a constant we have

$$Ax + By + Cz = D$$

which is equivalent to the vector equation $\mathbf{r}.(A\mathbf{i} + B\mathbf{j} + C\mathbf{k}) = D$

It is easy to convert back to standard form however, by dividing by $\sqrt{A^2 + B^2 + C^2}$, e.g. a plane whose equation is

$$2x + 3y + 6z = 21 \quad \text{or} \quad \mathbf{r}.(2\mathbf{i} + 3\mathbf{j} + 6\mathbf{k}) = 21$$

can be converted to the standard form

$$\tfrac{2}{7}x + \tfrac{3}{7}y + \tfrac{6}{7}z = 3 \quad \text{or} \quad \mathbf{r}.(\tfrac{2}{7}\mathbf{i} + \tfrac{3}{7}\mathbf{j} + \tfrac{6}{7}\mathbf{k}) = 3$$

So this plane is 3 units from the origin and it is perpendicular to the unit vector $\tfrac{2}{7}\mathbf{i} + \tfrac{3}{7}\mathbf{j} + \tfrac{6}{7}\mathbf{k}$.

Notice that the coefficients of $\mathbf{i}$, $\mathbf{j}$ and $\mathbf{k}$ in the unit vector $\hat{\mathbf{n}}$ appear in both forms of the equation; this means that a vector perpendicular to a plane can be read from its equation.

EXAMPLES 36H

1 Show that the line L whose vector equation is $\mathbf{r} = 2\mathbf{i} - 2\mathbf{j} + 3\mathbf{k} + t(\mathbf{i} - \mathbf{j} + 4\mathbf{k})$ is parallel to the plane Π whose vector equation is $\mathbf{r}.(\mathbf{i} + 5\mathbf{j} + \mathbf{k}) = 5$.

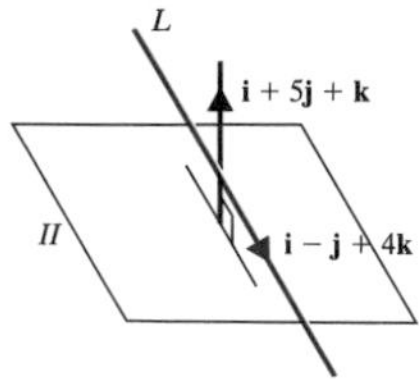

> The line L is parallel to the vector $\mathbf{i} - \mathbf{j} + 4\mathbf{k}$ and the plane Π is perpendicular to the vector $\mathbf{i} + 5\mathbf{j} + \mathbf{k}$ so L is parallel to Π if and only if $\mathbf{i} - \mathbf{j} + 4\mathbf{k}$ and $\mathbf{i} + 5\mathbf{j} + \mathbf{k}$ are perpendicular.

L is parallel to Π if and only if $(\mathbf{i} - \mathbf{j} + 4\mathbf{k}).(\mathbf{i} + 5\mathbf{j} + \mathbf{k}) = 0$

$(\mathbf{i} - \mathbf{j} + 4\mathbf{k}).(\mathbf{i} + 5\mathbf{j} + \mathbf{k}) = 1 - 5 + 4 = 0$,

$\therefore \quad L$ is parallel to Π.

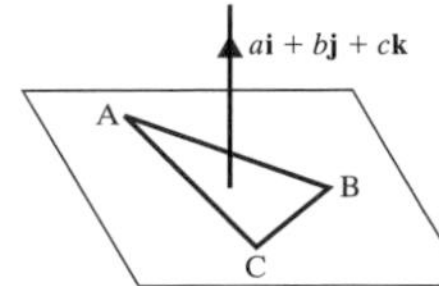

2 Find the Cartesian equation of the plane containing the points A(0, 1, 1), B(2, 1, 0) and C(−2, 0, 3).

If the equation of the plane is $ax + by + cz = d$ then the coordinates of A, B and C satisfy this equation

$\Rightarrow \quad b + c = d \quad [1]$ — using A

$\Rightarrow \quad 2a + b = d \quad [2]$ — using B

$\Rightarrow \quad -2a + 3c = d \quad [3]$ — using C

> These three equations are enough to find *a*, *b* and *c* in terms of *d*. This is all we need because any multiple of $ax + by + cz = d$ is also the equation of the plane.

$[2] - [1] \quad \Rightarrow \quad 2a - c = 0 \quad [4]$

$[3] + [4] \quad \Rightarrow \quad 2c = d$

$\therefore c = \frac{1}{2}d, \; a = \frac{1}{4}d$ and $b = \frac{1}{2}d$

$\therefore$ the equation of the plane is $\frac{1}{4}dx + \frac{1}{2}dy + \frac{1}{2}dz = d \quad \Rightarrow \quad x + 2y + 2z = 4$

3 The equation of a line is $\mathbf{r} = (1 - s)\mathbf{i} + (2 + 3s)\mathbf{j} + 6s\mathbf{k}$ and the equation of a plane is $\mathbf{r}.(2\mathbf{i} - \mathbf{j} + \mathbf{k}) = 3$. Find

a the coordinates of the point where the line cuts the plane,

b the angle between the line and the plane in degrees correct to 1 decimal place.

a

> The line cuts the plane at the point whose position vector satisfies both the equation of the line and the equation of the plane.

The line cuts the plane where $[(1 - s)\mathbf{i} + (2 + 3s)\mathbf{j} + 6s\mathbf{k}].(2\mathbf{i} - \mathbf{j} + \mathbf{k}) = 3$

$\Rightarrow \quad 2(1 - s) - (2 + 3s) + 6s = 3 \quad \Rightarrow \quad s = 3$

When $s = 3$, $\mathbf{r} = -2\mathbf{i} + 11\mathbf{j} + 18\mathbf{k}$,

$\therefore$ the line and the plane meet at the point (−2, 11, 18).

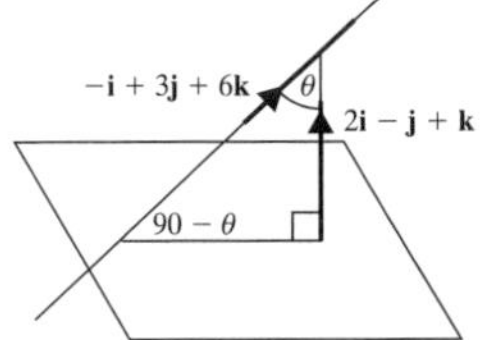

b Rearranging the equation of the line as $\mathbf{r} = \mathbf{i} + 2\mathbf{j} + s(-\mathbf{i} + 3\mathbf{j} + 6\mathbf{k})$ shows that the line is parallel to the vector $-\mathbf{i} + 3\mathbf{j} + 6\mathbf{k}$.
The plane is perpendicular to the vector $2\mathbf{i} - \mathbf{j} + \mathbf{k}$.

> The diagram shows that we can get the angle between the line and the plane by using the scalar product to find the angle between the line and the normal to the plane.

$(-\mathbf{i} + 3\mathbf{j} + 6\mathbf{k}).(2\mathbf{i} - \mathbf{j} + \mathbf{k}) = |-\mathbf{i} + 3\mathbf{j} + 6\mathbf{k}||2\mathbf{i} - \mathbf{j} + \mathbf{k}|\cos\theta$

$\Rightarrow \quad 1 = \sqrt{46}\sqrt{6}\cos\theta \quad \Rightarrow \quad \theta = 86.54\ldots°$

$\therefore$ the angle between the line and the plane is $90° - 86.54\ldots°$
i.e. 3.5° correct to 1 decimal place.

EXERCISE 36H

1 Write down the Cartesian equations of these planes.

a $\mathbf{r}.(\mathbf{i}+\mathbf{j}-\mathbf{k}) = 2$

b $\mathbf{r}.(2\mathbf{i}+3\mathbf{j}-4\mathbf{k}) = 1$

2 Write down the vector equations of these planes.

a $3x - 2y + z = 5$

b $5x - 3y - 4z = 7$

3 Write down a vector which is perpendicular to each plane given in Questions 1 and 2.

4 Find the distance from the origin of each plane given in Questions 1 and 2.

5 Two planes Π_1 and Π_2 have vector equations $\mathbf{r}.(2\mathbf{i}+\mathbf{j}-2\mathbf{k}) = 3$ and $\mathbf{r}.(2\mathbf{i}+\mathbf{j}-2\mathbf{k}) = 9$. Explain why Π_1 and Π_2 are parallel and hence find the distance between them.

6 Find the vector equation of the line through the origin which is perpendicular to the plane $\mathbf{r}.(\mathbf{i}-2\mathbf{j}+\mathbf{k}) = 3$.

7 Find the vector equation of the line through the point (2, 1, 1) which is perpendicular to the plane $\mathbf{r}.(\mathbf{i}+2\mathbf{j}-3\mathbf{k}) = 6$.

8 A plane goes through the three points whose position vectors are **a**, **b** and **c** where

$\mathbf{a} = \mathbf{i}+\mathbf{j}+2\mathbf{k}$
$\mathbf{b} = 2\mathbf{i}-\mathbf{j}+3\mathbf{k}$
$\mathbf{c} = -\mathbf{i}+2\mathbf{j}-2\mathbf{k}$

Find the vector of this plane in scalar product form and hence find the distance of the plane from the origin.

9 Find the vector equation of the plane which contains the points A(0, 1, 1), B(−1, 2, 1) and C(2, 0, 2).

10 Find the vector equation of the plane that contains the lines
$\mathbf{r} = -3\mathbf{i}-2\mathbf{j}+t(\mathbf{i}-2\mathbf{j}+\mathbf{k})$ and
$\mathbf{r} = \mathbf{i}-11\mathbf{j}+4\mathbf{k}+s(2\mathbf{i}-\mathbf{j}+2\mathbf{k})$.

11 Find the point of intersection of the line $\mathbf{r} = (\mathbf{i}+\mathbf{j}-2\mathbf{k})+\lambda(\mathbf{i}-\mathbf{j}+\mathbf{k})$ and the plane $\mathbf{r}.(\mathbf{i}+2\mathbf{j}-\mathbf{k}) = 2$.

12 Find the point of intersection of the line $x-2 = 2y+1 = 3-z$ and the plane $x+2y+z = 3$.

13 One flat face of a metal disc has equation $3x - y + 2z = 5$. A laser beam is used to burn a hole in the disc. The beam passes along the line whose equation is

$$\frac{x-2}{1} = \frac{y-3}{4} = \frac{z}{2}.$$

Find

a the coordinates of the point where the beam meets the face of the disc

b the angle between the beam and the face of the disc.

14

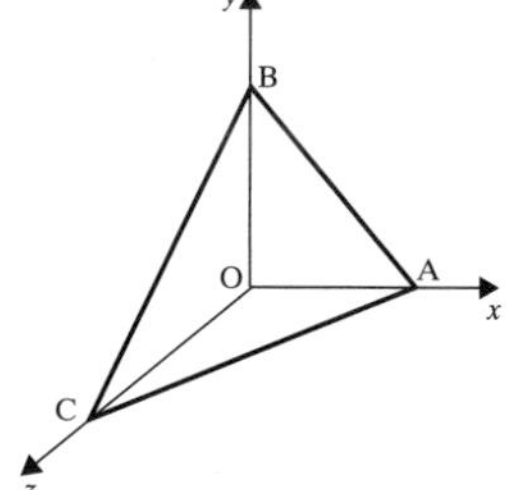

OABC is a tetrahedron. The coordinates of A, B and C are (2, 0, 0), (0, 3, 0) and (0, 0, 4) respectively. Find

a a vector equation for the plane ABC

b the distance of the plane ABC from O

c the angle between the edge OB and the plane ABC.

SUMMARY G

POWER SERIES

The terms of a power series involve powers of a variable, e.g.

$$1+x+x^2+x^3+\dots$$

The Binomial Theorem

If n is a positive integer then $(1+x)^n$ can be expanded as a finite series, where

$$(1+x)^n = 1+nx+\binom{n}{2}x^2+\binom{n}{3}x^3+\dots+x^n$$

and where

$$\binom{n}{r} = \frac{n(n-1)(n-2)\dots(n-r+1)}{r!}$$

Further, $(a+x)^n = a^n+na^{n-1}x+\binom{n}{2}a^{n-2}x^2 + \binom{n}{3}a^{n-3}x^3+\dots+x^n$

If n is *any* number then $(1+x)^n$ can be expanded as the *infinite* series given by

$$(1+x)^n = 1+nx+\binom{n}{2}x^2+\binom{n}{3}x^3+\dots$$

provided that $|x|<1$.

To expand $(a+bx)^n$, express $(a+bx)^n$ as $a^n\left(1+\frac{b}{a}x\right)^n$, then use the expansions above with $\frac{b}{a}x$ replacing x.

MACLAURIN'S THEOREM

A function of x, $f(x)$, can be expanded as a power series in x using

$$f(x) = f(0)+f'(0)x+\frac{f''(0)}{2!}x^2+\frac{f'''(0)}{3!}x^3+\dots$$

provided that it is possible
to find all the derivatives of $f(x)$
all the derivatives are defined when $x=0$
the series obtained is convergent.

SOLUTION OF EQUATIONS

Trigonometric Equations

Listed below are some of the trig identities useful in solving equations.

$$\left.\begin{aligned} \cos^2\theta+\sin^2\theta &\equiv 1 \\ \tan^2\theta+1 &\equiv \sec^2\theta \\ \cot^2\theta+1 &\equiv \csc^2\theta \end{aligned}\right\}$$

Use in an equation containing two ratios of one angle, at least one ratio being squared.

$$\left.\begin{aligned} \cos 2\theta &\equiv 2\cos^2\theta-1 \\ &\equiv 1-2\sin^2\theta \end{aligned}\right\}$$

Use to express an equation in terms of trig ratios of θ only.

$a\sin\theta+b\cos\theta \equiv r\sin(\theta+\alpha)$
and variations of this form

Use to reduce $a\sin\theta+b\cos\theta = c$
to $\sin(\theta+\alpha) = k$.

Exponential Equations

If there is only one term on each side, take logs.

If a sum or difference of terms is involved, suspect a disguised quadratic equation, e.g.

$$e^{2x}-2e^x-3=0 \quad \Rightarrow \quad y^2-2y-3=0$$

where $y = e^x$

Polynomial Equations

If $f(a)=0$ and $f'(a)=0$ then $y=f(x)$ has a repeated root $x=a$.

Approximate Solutions

Read solutions from a graph.

Use $f(x)=g(x)$ where the curves $y=f(x)$ and $y=g(x)$ cut.

Use $f(x)=0$ between two points on the x-axis where $f(x)<0$ and $f(x)>0$.

Iteration

Rearrange $f(x) = 0$ as $x = g(x)$ then use the iteration $x_{n+1} = g(x_n)$

Newton–Raphson Method

If $x = a$ is an approximate solution of the equation $f(x) = 0$ then a better approximation is $x = b$ where $b = a - \dfrac{f(a)}{f'(a)}$.

VECTORS

A vector is a quantity with both magnitude and direction and can be represented by a line segment.

If lines representing several vectors are drawn 'head to tail' in order, then the line (in the opposite sense) which completes a closed polygon represents the sum of the vectors (or the resultant vector).

A position vector has a fixed location in space.

Cartesian Unit Vectors

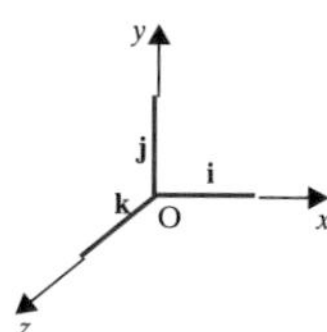

i, **j** and **k** are unit vectors in the directions of O*x*, O*y* and O*z* respectively.

Any vector can be given in the form $a\mathbf{i} + b\mathbf{j} + c\mathbf{k}$.

$$(a_1\mathbf{i} + b_1\mathbf{j} + c_1\mathbf{k}) \pm (a_2\mathbf{i} + b_2\mathbf{j} + c_2\mathbf{k}) = (a_1 \pm a_2)\mathbf{i} + (b_1 \pm b_2)\mathbf{j} + (c_1 \pm c_2)\mathbf{k}$$

$$|a\mathbf{i} + b\mathbf{j} + c\mathbf{k}| = \sqrt{a^2 + b^2 + c^2}$$

For two vectors $\mathbf{v}_1 = a_1\mathbf{i} + b_1\mathbf{j} + c_1\mathbf{k}$ and $\mathbf{v}_2 = a_2\mathbf{i} + b_2\mathbf{j} + c_2\mathbf{k}$

$\mathbf{v}_1$ and $\mathbf{v}_2$ are parallel if $\mathbf{v}_1 = \lambda\mathbf{v}_2$,

i.e. $a_1 = \lambda a_2,\ b_1 = \lambda b_2,\ c_1 = \lambda c_2$

$\mathbf{v}_1$ and $\mathbf{v}_2$ are equal if $a_1 = a_2,\ b_1 = b_2,\ c_1 = c_2$.

If $\mathbf{v} = a\mathbf{i} + b\mathbf{j} + c\mathbf{k}$ then the unit vector in the direction of **v** is $\hat{\mathbf{v}}$ where $\hat{\mathbf{v}} = \dfrac{\mathbf{v}}{|\mathbf{v}|}$.

Equations for a Line

For a line in the direction of the vector $\mathbf{d} = a\mathbf{i} + b\mathbf{j} + c\mathbf{k}$ and passing through a point with position vector $\mathbf{a} = x_1\mathbf{i} + y_1\mathbf{j} + z_1\mathbf{k}$, a vector equation in standard form is $\mathbf{r} = \mathbf{a} + \lambda\mathbf{d}$

Cartesian equations are $\dfrac{x - x_1}{a} = \dfrac{y - y_1}{b} = \dfrac{z - z_1}{c}$.

Two lines with equations $\mathbf{r}_1 = \mathbf{a}_1 + \lambda\mathbf{d}_1$ and $\mathbf{r}_2 = \mathbf{a}_2 + \mu\mathbf{d}_2$

are parallel if $\mathbf{d}_1$ is a multiple of $\mathbf{d}_2$

intersect if there are values of λ and μ for which $\mathbf{r}_1 = \mathbf{r}_2$

are skew in all other cases.

The Scalar Product of Two Vectors

If θ is the angle between two vectors **a** and **b** then

$$\mathbf{a}.\mathbf{b} = |\mathbf{a}||\mathbf{b}|\cos\theta$$

a and **b** are perpendicular $\Rightarrow$ $\mathbf{a}.\mathbf{b} = 0$.

If $\mathbf{a} = x_1\mathbf{i} + y_1\mathbf{j} + z_1\mathbf{k}$ and $\mathbf{b} = x_2\mathbf{i} + y_2\mathbf{j} + z_2\mathbf{k}$ then

$$\mathbf{a}.\mathbf{b} = x_1x_2 + y_1y_2 + z_1z_2$$

Equations of a Plane

A plane perpendicular to a vector **n** has a vector equation in standard form of $\mathbf{r}.\mathbf{n} = D$ where $\dfrac{D}{\hat{\mathbf{n}}}$ is the distance of the plane from O.

Taking $\mathbf{n} = A\mathbf{i} + B\mathbf{j} + C\mathbf{k}$ gives the Cartesian equation for the plane

i.e. $Ax + By + Cz = D$

MULTIPLE CHOICE EXERCISE G

In Questions 1 to 11 write down the letter or letters corresponding to a correct answer.

1 $(\mathbf{i} - \mathbf{j} + \mathbf{k}).(\mathbf{i} + \mathbf{j} - \mathbf{k}) =$

A $2\mathbf{i}$ **B** 1 **C** -1 **D** 2

2 The coefficient of x^2 in the expansion of $(1 - 2x)^{-1}$ is

A 0 **B** 4 **C** 2 **D** -4

3 $f(x) = x - \cos x$

A $f(x) = 0$ has no roots.

B $f(x) = 0$ has a root between $x = 0$ and $x = \pi/2$.

C If a is an approximate root of the equation $f(x) = 0$, then

$a - \dfrac{a - \cos a}{\sin a}$ is a better one.

4 The first two terms in the expansion of $(1 - 2x)^{-3}$ are

A $1 - 6x$ **D** $3 + 6x$

B $1 + 6x$ **E** $1 + x$

C $6x + 24x^2$

5 The sum to infinity of the series $1 + 2x + 4x^2 + 8x^3 + \ldots$ for $|x| < \frac{1}{2}$ is

A $\dfrac{2x}{1 - 2x}$ **D** $\dfrac{2}{1 - x}$

B $\dfrac{1}{1 + 2x}$ **E** $1 - 2x$

C $\dfrac{1}{1 - 2x}$

6 The third term of the series $\sum_{r=0}^{\infty} (-1)^{r+1} 2^r x^{-r}$ is

A $\dfrac{4}{x^2}$ **C** $-\dfrac{4}{x^2}$

B $\dfrac{8}{x^3}$ **D** $-4x^2$

7 The line
$\mathbf{r} = \mathbf{i} + 2\mathbf{j} + \mathbf{k} + \lambda(4\mathbf{i} - \mathbf{j} + 7\mathbf{k})$

A is in the direction $4\mathbf{i} - \mathbf{j} + 7\mathbf{k}$

B has length $\sqrt{66}$

C is parallel to the line
$\mathbf{r} = (2 + 4t)\mathbf{i} + (2 - t)\mathbf{j} + (1 + 7t)\mathbf{k}$

D does not intersect the line
$\mathbf{r} = \mathbf{k} + \lambda(4\mathbf{i} - \mathbf{j} + 7\mathbf{k})$

8 The coefficient of x^3 in the expansion of $(2 - x)^8$ is

A 1792 **C** −448 **E** −2000

B −1792 **D** 56

9 A linear approximation for $(1 + 2x)^{-2}$ where $x \approx 0$ is

A $1 + 2x$ **C** $1 + 4x$

B $1 - 4x$ **D** $1 - 2x$

10 The equation of the line through the points (1, 0, 1) and (1, 3, 2) is

A $\mathbf{r} = \mathbf{i} + \mathbf{k} + \lambda(3\mathbf{i} + \mathbf{k})$

B $\mathbf{r} = \mathbf{i} + (3 - 3t)\mathbf{j} + (2 - t)\mathbf{k}$

C $x = 1,\ \dfrac{y}{3} = \dfrac{z - 1}{1}$

11 The cosine of the angle between the lines $\mathbf{r} = \mathbf{a}_1 + \lambda\mathbf{b}_1$ and $\mathbf{r} = \mathbf{a}_2 + \lambda\mathbf{b}_2$ is

A $\mathbf{b}_1 . \mathbf{b}_2$ **C** $\dfrac{\mathbf{b}_1 . \mathbf{b}_2}{b_1 b_2}$

B $\mathbf{a}_1 . \mathbf{a}_2$ D $\dfrac{\mathbf{a}_1 . \mathbf{a}_2}{a_1 a_2}$

In Questions 12 to 16 a single statement is made. Write T if it is true, F if it is false.

12 $\mathbf{a} . \mathbf{b} = 0 \quad \Rightarrow \quad \mathbf{a} = 0$ or $\mathbf{b} = 0$.

13 A root of the equation $x^2 + 2x - 6 = 0$ can be found from $x_{n+1} = \frac{1}{2}(1 - x_n{}^2)$ with $x_1 = 1$.

14 A linear approximation to $(1 + x)^{-1}$ near the origin is $1 - x$.

15 The line $\dfrac{x - 1}{2} = \dfrac{y + 1}{3}$ lies in the xy-plane.

16 The magnitude of the vector $6\mathbf{i} - 2\mathbf{j} - 3\mathbf{k}$ is 49.

EXAMINATION QUESTIONS G

1 Write down the expansions of the following expressions in ascending powers of x, as far as the term containing x^3. In each case state the values of x for which the expansion is valid.

i $(1-x)^{-1}$ **ii** $(1+2x)^{-2}$ **iii** $\dfrac{1}{(1-x)(1+2x)^2}$ (OCR)

2 Referred to an origin O, the position vectors of the points A and B are $-3\mathbf{i}+\mathbf{j}-7\mathbf{k}$ and $5\mathbf{i}+3\mathbf{j}+5\mathbf{k}$ respectively. By using a scalar product, calculate, in degrees to one decimal place, the size of $\angle$AOB. (Edexcel)

3 The sequence given by the iteration formula

$$x_{n+1} = 100 + \ln x_n,$$

with $x_1 = 100$, converges to α. Find α correct to 2 decimal places, and write down an equation of which α is a root. (OCR)

4 Determine the coefficient of x^6 in the binomial expansion of

i $(1-x)^{15}$ **ii** $(1-x^2)^{15}$. (OCR)

5 The coefficient of x^3 in the expansion of $(1+2x)^n$ is $8n$, where n is a positive integer. Find the value of n. (OCR)

6 Expand $\left(x-\dfrac{1}{x}\right)^5$, simplifying the coefficients. (Edexcel)

7 Three points P, Q and R have position vectors, $\mathbf{p}$, $\mathbf{q}$ and $\mathbf{r}$ respectively, where

$$\mathbf{p} = 7\mathbf{i}+10\mathbf{j},\ \ \mathbf{q} = 3\mathbf{i}+12\mathbf{j},\ \ \mathbf{r} = -\mathbf{i}+4\mathbf{j}.$$

i Write down the vectors $\overrightarrow{PQ}$ and $\overrightarrow{RQ}$, and show that they are perpendicular.

ii Using a scalar product, or otherwise, find the angle PRQ.

iii Find the position vector of S, the midpoint of PR.

iv Show that $|\overrightarrow{QS}| = |\overrightarrow{RS}|$. Using your previous results, or otherwise, find the angle PSQ. (OCR)

8 a Obtain the first 4 non-zero terms of the binomial expansion in ascending powers of x of

$$(1-x^2)^{-\frac{1}{2}},\text{ given that } |x| < 1,$$

b Show that, when $x = \frac{1}{3}$, $(1-x^2)^{-\frac{1}{2}} = \frac{3}{4}\sqrt{2}$.

c Substitute $x = \frac{1}{3}$ into your expansion and hence obtain an approximation to $\sqrt{2}$, giving your answer to 5 decimal places. (Edexcel)

9 Given that

$$(2-x)^{13} \equiv A + Bx + Cx^2 + \ldots,$$

find the values of the integers A, B and C. (Edexcel)

10 a By sketching the curves with equations $y = 4 - x^2$ and $y = e^x$, show that the equation $x^2 + e^x - 4 = 0$ has one negative root and one positive root.

b Use the iteration formula $x_{n+1} = -(4 - e^{x_n})^{\frac{1}{2}}$ with $x_0 = -2$ to find in turn x_1, x_2, x_3 and x_4 and hence write down an approximation to the negative root of the equation, giving your answer to 4 decimal places.

An attempt to evaluate the positive root of the equation is made using the iteration formula $x_{n+1} = (4 - e^{x_n})^{\frac{1}{2}}$ with $x_0 = 1.3$.

c Describe the result of such an attempt. (Edexcel)

11 a Rearrange the cubic equation $x^3 - 6x - 2 = 0$ into the form

$$x = \pm\sqrt{a + \frac{b}{x}}.$$

State the values of the constants a and b.

b Use the iterative formula $x_{n+1} = \sqrt{a + \frac{b}{x_n}}$ with $x_0 = 2$ and your values of a and b to find the approximate positive solution x_4 of the equation to an appropriate degree of accuracy. Show all your intermediate answers. (Edexcel)

12 a Expand $(1 - 2x)^{10}$ in ascending powers of x up to and including the term in x^3, simplifying each coefficient in the expansion.

b Use your expansion to find an approximation to $(0.98)^{10}$, stating clearly the substitution which you have used for x. (Edexcel)

13 a Expand $(3 + 2x)^4$ in ascending powers of x, giving each coefficient as an integer.

b Hence, or otherwise, write down the expansion of $(3 - 2x)^4$ in ascending powers of x.

c Hence by choosing a suitable value for x show that $(3 + 2\sqrt{2})^4 + (3 - 2\sqrt{2})^4$ is an integer and state its value. (Edexcel)

14 a Use the iteration $x_{n+1} = (3x_n + 3)^{\frac{1}{3}}$, with $x_0 = 2$, to find, to 3 significant figures, x_4.

The only real root of the equation $x^3 - 3x - 3 = 0$ is α. It is given that, to 3 significant figures, $\alpha = x_4$.

b Use the substitution $y = 3^x$ to express

$$27^x - 3^{x+1} - 3 = 0$$

as a cubic equation.

c Hence, or otherwise, find an approximate solution to the equation

$$27^x - 3^{x+1} - 3 = 0,$$

giving your answer to 2 significant figures. (Edexcel)

15

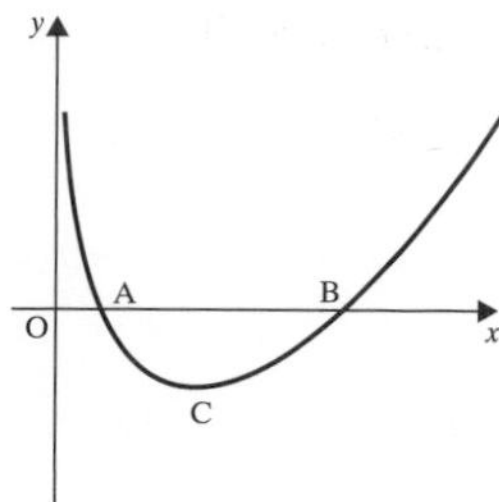

The figure shows the curve with equation $y = x - \ln x - 2$. The curve crosses the x-axis at A and B, and C is the minimum point on the curve.

a Find the coordinates of point C.

b Given the iteration $u_{n+1} = \ln u_n + 2$, and $u_0 = 3$, find, to three significant figures, u_4. (You are advised to show all stages of your working.)

c Given the iteration $v_{n+1} = e^{v_n - 2}$, and $v_0 = 0.5$, find, to three significant figures, v_4.

d Given that $u_n \to u$ as $n \to \infty$, show that u satisfies the equation $x - \ln x - 2 = 0$.

e Given that $v_n \to v$ as $n \to \infty$, show that v satisfies the equation $x - \ln x - 2 = 0$.

f Hence estimate the coordinates of points A and B, giving your answers to 3 significant figures.

(Edexcel)

16 i Write down the expansion of $(2 - x)^4$.

ii Find the first four terms in the expansion of $(1 + 2x)^{-3}$ in ascending powers of x. For what range of values of x is this expansion valid?

iii When the expansion is valid,

$$\frac{(2-x)^4}{(1+2x)^3} = 16 + ax + bx^2 + \ldots$$

Find the values of a and b.

(OCR)

17 A golden rectangle has one side of length 1 unit and a shorter side of length ψ units, where ψ is called the golden section.

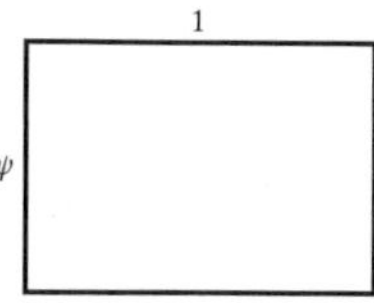

ψ can be found using the iterative formula

$$x_{n+1} = \sqrt[3]{x_n(1 - x_n)}.$$

Choosing a suitable value for x_1 and showing intermediate values, use this iterative formula to obtain the value of ψ to 2 decimal places.

(AQA)

18 The figure shows a circle of radius 1 and centre O.

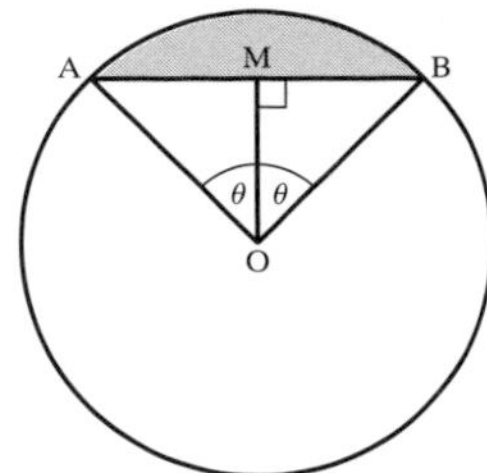

Given that the shaded area is $\frac{1}{6}$ of the area of the complete circle, show that

$$\theta - \frac{1}{2}\sin 2\theta = \frac{\pi}{6}.$$

Show that θ lies between 0.9 and 1.1 radians.

An iterative procedure that can be used to find θ is based on the sequence

$$\theta_{n+1} = \frac{1}{2}\sin 2\theta_n + \frac{\pi}{6}.$$

Taking $\theta_0 = 1$, find θ correct to 2 decimal places. (WJEC)

19 Three points have coordinates A(9, 2, −4), B(3, 1, −4) and C(2, 7, 6).

a **i** Write down the vector $\overrightarrow{AB}$.

ii Write down the vector $\overrightarrow{BC}$.

iii Show that AB is perpendicular to BC.

b **i** Write down a vector equation for the line L which is parallel to BC and which passes through the point (−4, 6, 6).

ii Show that L intersects the line AB. (AQA)

20 **a** Illustrate, by sketching graphs of

$$y = \ln(5x) \text{ and } y = \frac{10}{x}, \quad x > 0$$

on the same diagram, that the equation

$$x\ln(5x) - 10 = 0$$

has just one real root.

b Show, by using Newton's method, that when a is an approximation to this root then a better approximation is usually given by

$$\frac{a+10}{1+\ln(5a)}.$$

c Use this result twice, starting with $a = 3$, to find a further approximation giving three decimal places in your answer. (AQA)

21 **a** Express $\dfrac{1-x-x^2}{(1-2x)(1-x)^2}$ as the sum of three partial fractions.

b Hence, or otherwise, expand this expression in ascending powers of x up to and including the term in x^3.

c State the range of values of x for which the full expansion is valid. (AQA)

22 A pipe runs from the point A(2, 1, 5) to the point B(6, 0, 6). At B the pipe bends and then runs to the point C(7, 0, 3).

a Find a vector equation of the line AB.

b Find the angle between the vectors

$$\begin{pmatrix} 4 \\ -1 \\ 1 \end{pmatrix} \text{ and } \begin{pmatrix} 1 \\ 0 \\ -3 \end{pmatrix}.$$

c Find the angle ABC formed by the pipe. (AQA)

23 Given that $|x| < 1$, expand $\sqrt{1+x}$ as a series of ascending powers of x, up to and including the term in x^2.

Show that, if x is small, then

$$(2-x)\sqrt{1+x} \approx a + bx^2,$$

where the values of a and b are to be stated. (OCR)

24 A curve C has equation $y = \dfrac{\sin x}{x}$, where $x > 0$.

i Find $\dfrac{dy}{dx}$, and hence show that the x-coordinate of any stationary point of C satisfies the equation $x = \tan x$.

ii Use the iteration $x_{n+1} = \pi + \tan^{-1} x_n$ to find, correct to 2 decimal places, the root of the equation $x = \pi + \tan^{-1} x$ which lies between 4 and 5.

iii Show that every root of $x = \pi + \tan^{-1} x$ is also a root of $x = \tan x$. (OCR)

25 Find, in their simplest form, the first three terms in the expansion of

$$(1+3t)^{\frac{2}{3}}$$

in ascending powers of t, where $|t| < \frac{1}{3}$. (AQA)

26 Referred to a fixed origin O, the points A and B have position vectors $3\mathbf{i} - \mathbf{j} + 2\mathbf{k}$ and $-\mathbf{i} + \mathbf{j} + 9\mathbf{k}$ respectively.

a Show that OA is perpendicular to AB.

b Find, in vector form, an equation of the line L_1 through A and B.

The line L_2 has equation $\mathbf{r} = (8\mathbf{i} + \mathbf{j} - 6\mathbf{k}) + \lambda(\mathbf{i} - 2\mathbf{j} - 2\mathbf{k})$, where λ is a scalar parameter.

c Show that the lines L_1 and L_2 intersect and find the position vector of their point of intersection. (Edexcel)

27 Write down and simplify the binomial expansion of $(1-2x)^{\frac{1}{3}}$ up to and including the x^3 term. By putting $x = \frac{1}{10}$, use your expansion to find an approximation to the cube root of 10, giving your answer as the ratio of two integers. (AQA)

28

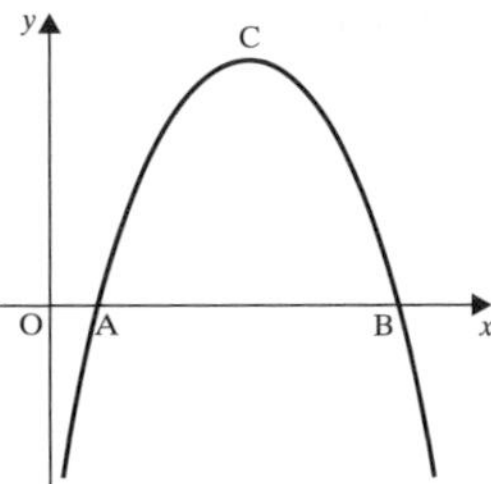

The function f is defined by

$$f(x) = 18\ln x - x^2, \ x > 0$$

The diagram shows a sketch of the curve with equation $y = f(x)$.

i The point C is the maximum point of the curve. Find the x-coordinate of C.

ii The curve crosses the x-axis at the points A and B. Show by calculation that the x-coordinate of A lies between 1.0 and 1.1.

iii Using $x = 1.0$ as a first approximation to the x-coordinate of A, apply the Newton–Raphson method twice to $f(x) = 0$ to obtain a third approximation, giving your answer correct to three decimal places.

iv The normal at the point $P(2, 18\ln 2 - 4)$ meets the x-axis at the point N. Find the area of the triangle OPN, where O is the origin, giving your answer correct to three significant figures.

(OCR)

29 On a single diagram, sketch the graphs of $y = \ln(10x)$ and $y = \frac{6}{x}$, and explain how you can deduce that the equation $\ln(10x) = \frac{6}{x}$ has exactly one real root.

Given that the root is close to 2, use the iteration

$$x_{n+1} = \frac{6}{\ln(10x_n)}$$

to evaluate the root correct to three decimal places.

The same equation may be written in the form $x\ln(10x) - 6 = 0$.
Taking $f(x)$ to be $x\ln(10x) - 6$, find $f'(x)$, and show that the Newton–Raphson iteration for the root of $f(x) = 0$ may be simplified to the form

$$x_{n+1} = \frac{x_n + 6}{1 + \ln(10x_n)}$$

(OCR)

30 Show that the equation $x^3 - x^2 - 2 = 0$ has a root a which lies between 1 and 2.

a Using 1.5 as a first approximation for a, use the Newton–Raphson method once to obtain a second approximation for a, giving your answer to 3 decimal places.

b Show that the equation $x^3 - x^2 - 2 = 0$ can be arranged in the form $x = \sqrt[3]{f(x)}$ where $f(x)$ is a quadratic function.

Use an iteration of the form $x_{n+1} = g(x_n)$ based on this rearrangement and, with $x_1 = 1.5$, to find x_2 and x_3, giving your answers to 3 decimal places. (AQA)

31 Find the term independent of x in the expansion of $\left(x^2 - \frac{2}{x}\right)^6$. (OCR)

32

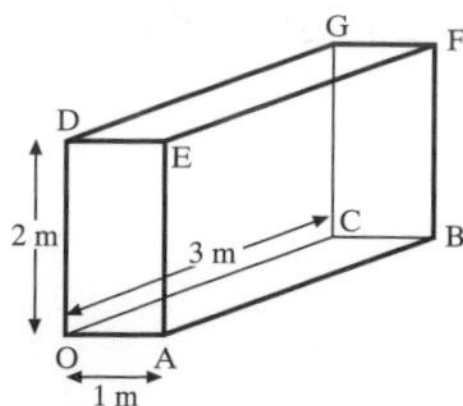

The figure shows a cuboid in which OA = 1 m, OC = 3 m and OD = 2 m. Taking O as origin and unit vectors **i**, **j**, **k** in the directions OA, OC, OD respectively, express in terms of **i**, **j**, **k** the vectors

i **OF** and **ii** **AG**

By considering an appropriate scalar product, find the acute angle between the diagonals OF and AG.

(WJEC)

33

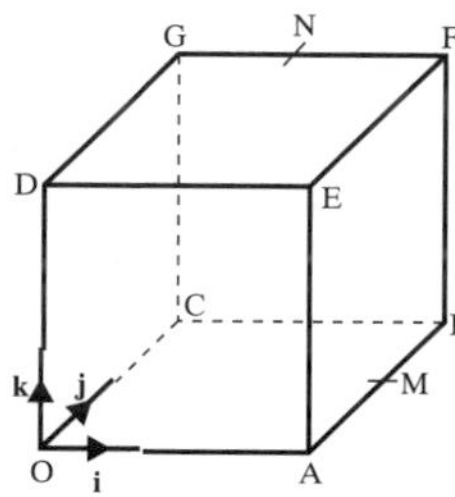

In the diagram OABCDEFG is a cube in which the length of each edge is 2 units. Unit vectors **i**, **j**, **k** are parallel to $\overrightarrow{OA}$, $\overrightarrow{OC}$, $\overrightarrow{OD}$ respectively. The midpoints of AB and FG are M and N respectively.

i Express each of the vectors $\overrightarrow{ON}$ and $\overrightarrow{MG}$ in terms of **i**, **j** and **k**.

ii Show that the acute angle between the direction of $\overrightarrow{ON}$ and $\overrightarrow{MG}$ is 63.6°, correct to the nearest 0.1°.

(OCR)

34 Given that $y = \ln(4 + 3x)$, find $\dfrac{dy}{dx}$ and show that $\dfrac{d^2y}{dx} = -\frac{9}{16}$ when $x = 0$.

Hence, or otherwise, obtain the Maclaurin series for $\ln(4 + 3x)$, up to and including the term in x^2.

(OCR)

35 Given that $y = \cos(\frac{1}{3}\pi + 2x)$, find $\dfrac{d^2y}{dx^2}$.

Hence obtain the Maclaurin series for $\cos(\frac{1}{3}\pi + 2x)$, up to and including the term in x^2. (OCR)

36 Obtain the first three terms in the Maclaurin series for $\ln(3 + x)$. (OCR)

37 A curve is defined by the parametric equations $x = \cos^{-1}(t)$, $y = -\ln t$.

a Find $\dfrac{dt}{dx}$. Hence show that $\dfrac{dy}{dx} = \tan x$.

b Use the Maclaurin Series to find an approximation to the curve near the origin, of the form

$$y = a + bx + cx^2.$$

(AQA)

38 a Given that $|x| < \frac{1}{3}$, write down the expansion of $(1 - 3x)^{-2}$ in ascending powers of x up to and including the term x^3.

b Write down the series expansions of

i e^{4x} **ii** $\sin 2x$

in ascending powers of x up to and including the term in x^3 simplifying the coefficients as much as possible.

c Given that x is small enough for x^4 and higher powers to be ignored, show that $(1 - 3x)^{-2} - e^{4x} - \sin 2x = ax^2 + bx^3$ and state the values of the constants a and b. (AQA)

39 i Show that $\frac{d}{dx}(e^{-x^2}) = -2xe^{-x^2}$.

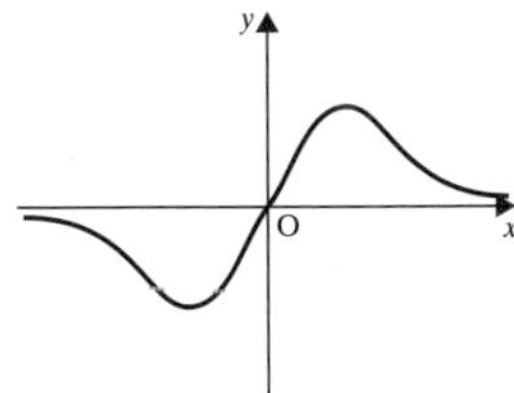

The sketch above shows the curve with equation $y = xe^{-x^2}$.

ii Differentiate xe^{-x^2} and find the coordinates of the two stationary points on the curve.

iii Find the area of the region between the curve and the x-axis for $0 \leqslant x \leqslant 0.4$.

iv Give the first two terms of the series expansion for e^{-x^2} and hence find an approximation for $y = xe^{-x^2}$ when x is small.

v Use your answer to part **iv** to find an approximation for the area which you calculated in part **iii**. (OCR)

40 Referred to a three-dimensional Cartesian coordinate system with origin O, the points A, B and C have coordinates (4, 0, 0), (0, 4, 0) and (0, 0, 5) respectively.
Calculate the acute angle between the planes OAB and CAB, giving your answer in degrees to 1 decimal place. (Edexcel)

41 A plane Π has equation $ax + by + z = d$.

i Write down, in terms of a and b, a vector which is perpendicular to Π.

Points A(2, −1, 2), B(4, −4, 2), C(5, −6, 3) lie on Π.

ii Write down the vectors $\overrightarrow{AB}$ and $\overrightarrow{AC}$.

iii Use scalar products to obtain two equations for a and b.

iv Find the equation of the plane Π.

v Find the angle which the plane Π makes with the plane $x = 0$.

vi Point D is the midpoint of AC. Point E is on the line between D and B such that DE : EB = 1 : 2. Find the coordinates of E. (OCR)

42 The figure shows an arrow embedded in a target. The line of the arrow passes through the point A(2, 3, 5) and has direction vector $3\mathbf{i} + \mathbf{j} - 2\mathbf{k}$. The arrow intersects the target at the point B. The plane of the target has equation $x + 2y - 3z = 4$. The units are metres.

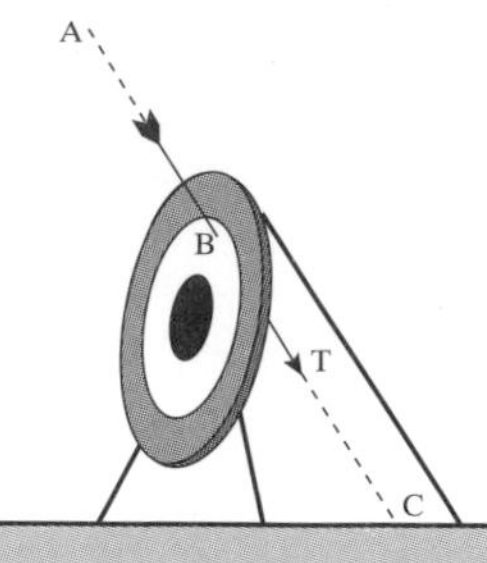

i Write down the vector equation of the line of the arrow in the form

$$\mathbf{r} = \mathbf{p} + \lambda\mathbf{q}.$$

ii Find the value of λ which corresponds to B. Hence write down the coordinates of B.

iii The point C is where the line of the arrow meets the ground, which is the plane $z = 0$. Find the coordinates of C.

iv The tip, T, of the arrow is one-third of the way from B to C. Find the coordinates of T and the length of BT.

v Write down a normal vector to the plane of the target. Find the acute angle between the arrow and this normal. (OCR)

43 The position vectors of three points A, B C on a plane ski-slope are

$$\mathbf{a} = 4\mathbf{i} + 2\mathbf{j} - \mathbf{k}, \quad \mathbf{b} = -2\mathbf{i} + 26\mathbf{j} + 11\mathbf{k}, \quad \mathbf{c} = 16\mathbf{i} + 17\mathbf{j} + 2\mathbf{k}$$

where the units are metres.

i Show that the vector $2\mathbf{i} - 3\mathbf{j} + 7\mathbf{k}$ is perpendicular to $\overrightarrow{AB}$ and also perpendicular to $\overrightarrow{AC}$. Hence find the equation of the plane of the ski-slope.

The track for an overhead railway lies along the straight line DEF, where D and E have position vectors $\mathbf{d} = 130\mathbf{i} - 40\mathbf{j} + 20\mathbf{k}$ and $\mathbf{e} = 90\mathbf{i} - 20\mathbf{j} + 15\mathbf{k}$, and F is a point on the ski-slope.

ii Find the equation of the straight line DE.

iii Find the position vector of the point F.

iv Show that $\overrightarrow{FD} = 15(8\mathbf{i} - 4\mathbf{j} + \mathbf{k})$ and hence find the length of the track. (OCR)

44 In a crystal structure one of the planes of the crystal has equation

$$2x - 3y + z = 5.$$

An X-ray beam passes through the crystal along the line having equation

$$\begin{bmatrix} x \\ y \\ z \end{bmatrix} = \begin{bmatrix} 1 \\ 1 \\ 2 \end{bmatrix} + \lambda \begin{bmatrix} 3 \\ 1 \\ 1 \end{bmatrix}.$$

Find the angle between the X-ray beam and the given plane. (AQA)

45

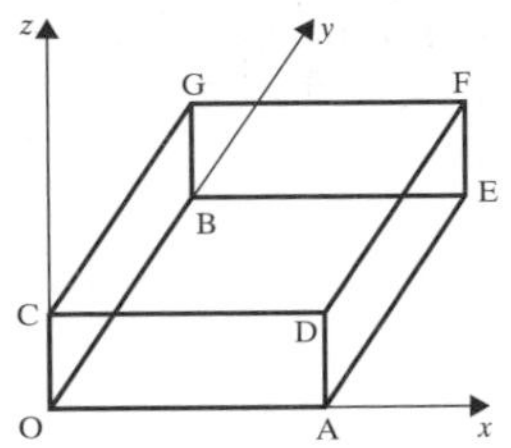

A rectangular-faced box OADCBEFG, has sides of length 5 cm, 4 cm, 3 cm. A rectangular Cartesian coordinate system has its origin O, as shown above. The following are the coordinates of A, B and C:

$$A : (5, 0, 0); \quad B : (0, 4, 0); \quad C : (0, 0, 3).$$

a Write down the coordinates of the point G.

b Calculate the length of the line AG.

c Find, in an appropriate form, the equations of the line AG.

d The plane with equation $24x - 16y + 20z = 244$ meets the line AG in a point P. Prove that the coordinates of P are $(10, -4, -3)$. (OCR)

46 The Cartesian coordinates of three points in three-dimensional space are

$$A = (1, 0, 2), \quad B = (3, 1, 0) \quad \text{and} \quad C = (2, -1, 1).$$

a Find, in the form $ax + by + cz = d$, the equation of the plane containing the points A, B and C.

b Find, in the form $\frac{x - \alpha}{l} = \frac{y - \beta}{m} = \frac{z - \gamma}{n}$, the equations of the line that passes through A and B. (OCR)

ANSWERS

Answers to questions taken from past examination papers are the sole responsibility of the authors and have not been approved by the Examining Boards.

Chapter 1

Exercise 1A – p.1

1. $15x$
2. $2x^2$
3. $4x^2$
4. $10pq$
5. $8x^2$
6. $10p^2qr$
7. $9a^2$
8. $63ab$
9. $24st^2$
10. $8a^3$
11. $\frac{5}{3}x$
12. $2m$
13. $4ab^3$
14. $5xy$
15. $196p^4q^2$
16. $2a$
17. $6ax$
18. $2x$
19. $9b/5a$
20. $\frac{2}{5}x$
21. x^3/y^2

Exercise 1B – p.2

1. $3x^2 - 4x$
2. $a - 12$
3. $2y - xy + y^2$
4. $5pq - 9p^2$
5. $3xy + y^2$
6. $x^3 - x^2 + x + 7$
7. $5 + t - t^2$
8. $a^2 - ab - 2b$
9. $7 - x$
10. $4x - 9$
11. $3x^2 + 18x - 20$
12. $ab - 2ac + cb$
13. $11cT - 2cT^2 - 55T^2$
14. $-x^3 + 7x^2 - 7x$
15. $-4y^2 + 24y - 10$
16. $5RS + 5RF - R^2$

Exercise 1C – p.3

1. -7
2. 2
3. **a** 1 **b** -5 **c** -1
4. **a** 1 **b** 0 **c** -3

Exercise 1D – p.3

1. $x^2 + 6x + 8$
2. $x^2 + 8x + 15$
3. $a^2 + 13a + 42$
4. $t^2 + 15t + 56$
5. $s^2 + 17s + 66$
6. $2x^2 + 11x + 5$
7. $5y^2 + 28y + 15$
8. $6a^2 + 17a + 12$
9. $35t^2 + 86t + 48$
10. $99s^2 + 49s + 6$
11. $x^2 - 5x + 6$
12. $y^2 - 5y + 4$
13. $a^2 - 11a + 24$
14. $b^2 - 17b + 72$
15. $p^2 - 15p + 36$
16. $2y^2 - 13y + 15$
17. $3x^2 - 13x + 4$
18. $6r^2 - 25r + 14$
19. $20x^2 - 19x + 3$
20. $6a^2 - 7ab + 2b^2$
21. $x^2 - x - 6$
22. $a^2 + a - 56$
23. $y^2 + 2y - 63$
24. $s^2 + s - 30$
25. $q^2 + 8q - 65$
26. $2t^2 + 3t - 20$
27. $4x^2 + 11x - 3$
28. $6q^2 - q - 15$
29. $x^2 - xy - 2y^2$
30. $2s^2 + st - 6t^2$

Exercise 1E – p.4

1. $x^2 - 4$
2. $25 - x^2$
3. $x^2 - 9$
4. $4x^2 - 1$
5. $x^2 - 64$
6. $x^2 - a^2$
7. $x^2 - 1$
8. $9b^2 - 16$
9. $4y^2 - 9$
10. $a^2b^2 - 36$
11. $25x^2 - 1$
12. $x^2y^2 - 16$

Exercise 1F – p.4

1. $x^2 + 8x + 16$
2. $x^2 + 4x + 4$
3. $4x^2 + 4x + 1$
4. $9x^2 + 30x + 25$
5. $4x^2 + 28x + 49$
6. $x^2 - 2x + 1$
7. $x^2 - 6x + 9$
8. $4x^2 - 4x + 1$
9. $16x^2 - 24x + 9$
10. $25x^2 - 20x + 4$
11. $9t^2 - 42t + 49$
12. $x^2 + 2xy + y^2$
13. $4p^2 + 36p + 81$
14. $9q^2 - 66q + 121$
15. $4x^2 - 20xy + 25y^2$

Exercise 1G – p.5

1. $11x - 2x^2 - 12$
2. $x^2 - 49$
3. $6 - 25x + 4x^2$
4. $14p^2 - 3p - 2$
5. $9p^2 - 6p + 1$
6. $15t^2 + t - 2$
7. $16 - 8p + p^2$
8. $14t - 3 - 8t^2$
9. $x^2 + 4xy + 4y^2$
10. $16x^2 - 9$
11. $9x^2 + 42x + 49$
12. $15 - R - 2R^2$
13. $a^2 - 6ab + 9b^2$
14. $4x^2 - 20x + 25$
15. $49a^2 - 4b^2$
16. $9a^2 + 30ab + 25b^2$
17. **a** $6, -22$ **b** $15, 31$ **c** $14, -31$ **d** $81, 18$

Exercise 1H – p.6

1. $x^3 - x^2 - x - 2$
2. $3x^3 - 5x^2 - x + 2$
3. $4x^3 - 8x^2 + 13x - 5$
4. $x^3 - 2x^2 + 1$
5. $2x^3 - 9x^2 - 24x - 9$
6. $x^3 + 6x^2 + 11x + 6$
7. $x^3 + 4x^2 - x - 4$

8 $x^3 - 4x^2 + x + 6$
9 $2x^3 + 7x^2 + 7x + 2$
10 $x^3 + 4x^2 + 5x + 2$
11 $4x^3 + 4x^2 - 7x + 2$
12 $27x^3 - 27x^2 + 9x - 1$
13 $4x^3 - 9x^2 - 25x - 12$
14 $4x^3 - 4x^2 - x + 1$
15 $6x^3 + 13x^2 + x - 2$
16 $x^3 + 3x^2 + 3x + 1$
17 $x^3 + x^2 - 4x - 4$
18 $2x^3 + 7x^2 - 9$
19 $24x^3 + 38x^2 - 51x + 10$
20 $4x^3 - 42x^2 + 68x + 210$
21 $3x^3 - 16x^2 + 28x - 16$; -16, 28
22 6, -17
23 $x^3 + 3x^2y + 3xy^2 + y^3$
24 $x^4 + 4x^3y + 6x^2y^2 + 4xy^3 + y^4$

Exercise 1I – p.7

1 $x^3 + 9x^2 + 27x + 27$
2 $x^4 - 8x^3 + 24x^2 - 32x + 16$
3 $x^4 + 4x^3 + 6x^2 + 4x + 1$
4 $8x^3 + 12x^2 + 6x + 1$
5 $x^5 - 15x^4 + 90x^3 - 270x^2 + 405x - 243$
6 $p^4 - 4p^3q + 6p^2q^2 - 4pq^3 + q^4$
7 $8x^3 + 36x^2 + 54x + 27$
8 $x^5 - 20x^4 + 160x^3 - 640x^2 + 1280x - 1024$
9 $81x^4 - 108x^3 + 54x^2 - 12x + 1$
10 $1 + 20a + 150a^2 + 500a^3 + 625a^4$
11 $64a^6 - 192a^5b + 240a^4b^2 - 160a^3b^3 + 60a^2b^4 - 12ab^5 + b^6$
12 $8x^3 - 60x^2 + 150x - 125$

Exercise 1J – p.9

1 $(x + 5)(x + 3)$
2 $(x + 7)(x + 4)$
3 $(x + 6)(x + 1)$
4 $(x + 4)(x + 3)$
5 $(x - 1)(x - 9)$
6 $(x - 3)^2$
7 $(x + 6)(x + 2)$
8 $(x - 8)(x - 1)$
9 $(x + 7)(x - 2)$
10 $(x + 4)(x - 3)$
11 $(x - 5)(x + 1)$
12 $(x - 12)(x + 2)$
13 $(x + 7)(x + 2)$
14 $(x - 1)^2$
15 $(x - 3)(x + 3)$
16 $(x + 8(x - 3)$
17 $(x + 2)^2$
18 $(x - 1)(x + 1)$
19 $(x - 6)(x + 3)$
20 $(x + 5)^2$
21 $(x - 4)(x + 4)$
22 $(4 + x)(1 + x)$
23 $(2x - 1)(x - 1)$
24 $(3x + 1)(x + 1)$
25 $(3x - 1)^2$
26 $(3x + 1)(2x - 1)$
27 $(3 + x)^2$
28 $(2x - 3)(2x + 3)$
29 $(x + a)^2$
30 $(xy - 1)^2$

Exercise 1K – p.9

1 $(3x - 4)(2x + 3)$
2 $(4x - 3)(x - 2)$
3 $(4x - 1)(x + 1)$
4 $(3x - 2)(x - 5)$
5 $(2x - 3)^2$
6 $(1 - 2x)(3 + x)$
7 $(5x - 4)(5x + 4)$
8 $(3 + x)(1 - x)$
9 $(5x - 1)(x - 12)$
10 $(3x + 5)^2$
11 $(3 - x)(1 + x)$
12 $(3 + 4x)(4 - 3x)$
13 $(1 + x)(1 - x)$
14 $(3x + 2)^2$
15 $(x + y)^2$
16 $(1 - 2x)(1 + 2x)$
17 $(2x - y)^2$
18 $(3 - 2x)(3 + 2x)$
19 $(6 + x)^2$
20 $(5x - 4)(8x + 3)$
21 $(7x + 30)(x - 5)$
22 $(6 - 5x)(6 + 5x)$
23 $(x - y)(x + y)$
24 $(9x - 2y)^2$
25 $(7 - 6x)^2$
26 $(5x - 2y)(5x + 2y)$
27 $(6x + 5y)^2$
28 $(2x - 3y)(2x + y)$
29 $(3x + 4y)(2x + y)$
30 $(7pq - 2)^2$

Exercise 1L – p.10

1 not possible
2 $2(x + 1)^2$
3 $(x + 2)(x + 1)$
4 $3(x + 5)(x - 1)$
5 not possible
6 not possible
7 not possible
8 $2(x - 2)^2$
9 $3(x - 2)(x + 1)$
10 $2(x^2 - 3x + 4)$
11 $3(x - 4)(x + 2)$
12 $(x - 6)(x + 2)$
13 not possible
14 $4(x - 5)(x + 5)$
15 $5(x^2 - 5)$
16 not possible
17 not possible
18 not possible

Exercise 1M – p.12

1 yes
2 no
4 neither are
6 **a** $(x - 1)(x + 1)(x + 2)$
b $(x - 2)(x^2 + x + 1)$
c $(2x - 1)(x^2 + 1)$
d $(x + 3)(x - 3)(x^2 + 9)$
e $(x + 3)(x^2 - 3x + 9)$
f $(x - 2)(x + 2)(x^2 + x + 1)$
7 20
8 $(2x - 1)$; 2, -1, 1, 1, 5
9 3, -6
10 5

Exercise 1N – p.14

1 $\frac{1}{4}$
2 $\dfrac{2(x + 2)}{3(x - 2)}$
3 $\frac{2}{3}$
4 $\frac{3}{5}$
5 $\dfrac{x}{y}$
6 not possible
7 not possible
8 $\dfrac{5x(x + y)}{5y + 2x}$
9 $\dfrac{2a - 6b}{6a + b}$
10 $\dfrac{b - 4}{3x(b + 4)}$
11 $\dfrac{x - 3}{x + 4}$
12 $\dfrac{4y^2 + 3}{(y + 3)(y - 3)}$
13 $\dfrac{1}{3(x + 3)}$
14 $\dfrac{x + 2}{2x + 1}$
15 $\dfrac{x - 2}{x - 1}$
16 $\dfrac{1}{2(a - 5)}$

17 $\dfrac{3}{p+3q}$

18 $\dfrac{a^2+2a+4}{(a+5)(a+2)}$

19 $\dfrac{x+1}{3(x+3)}$

20 $\dfrac{4(x-3)}{(x+1)^2}$

Exercise 1P – p.15

1 $\dfrac{2x^2}{3y^2}$

2 $\dfrac{6t^2}{s}$

3 $\dfrac{8v^2}{3}$

4 $\dfrac{2r}{3}$

5 $3x$

6 $\dfrac{9x}{4y^2}$

7 $\dfrac{x^2}{24}$

8 $\dfrac{2}{a(a+b)}$

9 $\dfrac{1}{x+1}$

10 $\dfrac{1}{2a}$

11 $\dfrac{1}{x-1}$

12 $\dfrac{3}{2(x+3)}$

13 $\dfrac{a^4}{27}$

14 $\frac{2}{3}$

15 $\dfrac{2r}{3s^2}$

16 $\dfrac{3x^2}{2(y-2)}$

17 $\dfrac{b^2}{c^2}$

18 $2(x+3)$

19 $\dfrac{x-3}{3}$

20 $\dfrac{x(2x-3)}{x-1}$

Exercise 1Q – p.15

1 $\dfrac{b-a}{ab}$

2 $\dfrac{8}{15x}$

3 $\dfrac{q-p}{pq}$

4 $\dfrac{11}{10x}$

5 $\dfrac{x^2+1}{x}$

6 $\dfrac{x^2-y^2}{xy}$

7 $\dfrac{2p^2-1}{p}$

8 $\dfrac{7x+3}{12}$

9 $\dfrac{5x-1}{6}$

10 $\dfrac{11-7x}{15}$

11 $\dfrac{\sin B+\sin A}{\sin A\sin B}$

12 $\dfrac{\sin A+\cos A}{\cos A\sin A}$

13 $\dfrac{12x^2+1}{4x}$

14 $\dfrac{2x^2+x-2}{2x+1}$

15 $\dfrac{x^2+2x+2}{x+1}$

16 $\dfrac{2x+3}{2x}$

17 $\dfrac{1+x-x^2}{x}$

18 $\dfrac{n+1}{n^2}$

19 $\dfrac{x(b^2+a^2)}{a^2b^2}$

20 $\dfrac{a^2+3a+1}{a(a+1)}$

Exercise 1R – p.16

1 $\dfrac{2x}{(x+1)(x-1)}$

2 $\dfrac{2x-1}{(x+1)(x-2)}$

3 $\dfrac{7x+18}{(x+2)(x+3)}$

4 $\dfrac{x}{(x-1)(x+1)}$

5 $\dfrac{-1-3a}{(a-1)(a+1)}=\dfrac{1+3a}{(1-a)(1+a)}$

6 $\dfrac{x+2}{(x+1)^2}$

7 $\dfrac{1-4x}{(2x+1)^2}$

8 $\dfrac{-3x-10}{(x+1)(x+4)}=-\dfrac{3x+10}{(x+1)(x+4)}$

9 $\dfrac{2x+6}{(x+1)^2}$

10 $\dfrac{8-x-x^2}{(x+2)^2(x+4)}$

11 $\dfrac{7x+8}{6(x-1)(x+4)}$

12 $\dfrac{8-3x}{5(x+2)(x+4)}$

13 $\dfrac{15x-58}{6(x+2)(3x-5)}$

14 $\dfrac{5x^2-9x-32}{(x+1)(x-2)(x+3)}$

15 $\dfrac{2x^2+6x+6}{(x+1)(x+2)(x+3)}$

$=\dfrac{2(x^2+3x+3)}{(x+1)(x+2)(x+3)}$

16 $-\dfrac{1}{x(x+1)^2}$

17 $\dfrac{7t+3}{(t+1)^2}$

18 $\dfrac{-t^4+2t^3-2t^2-2t-1}{(t^2+1)(t^2-1)}$

19 $\dfrac{1+3y-3x}{(y-x)(y+x)}$

20 $\dfrac{n^3+6n^2+8n+2}{n(n+1)(n+2)}$

Exercise 1S – p.18

1 quotient $2x+1$, remainder -5
2 quotient $x-2$, remainder 6
3 quotient $2x^2+x+1$, remainder 0
4 quotient $2x^2+3x+6$, remainder 14

Mixed Exercise 1 – p.18

1 -23
2 $115x - 105x^2 - 30$
3 108
4 $3(x-2)(x-1)$
5 250
6 $4(x-3)(x+3)$
7 $4x^3 - 4x^2 + x$
8 $(x-5)^2$
9 7
10 $(15a^2 - ab - 6b^2)$
11 $x^3 - 7x^2 + 13x - 6$
12 $6x^3 + 17x^2 - 5x - 6$
13 -5
15 $(2x-1)(x+1)(x-1)$
16 1, 2, -1
17 -6
18 $(x-1)(x+1)(x-2)(x-4)$
19 **a** $(x-1)(x^2+x+1)$
b $(x+y)(x^2-xy+y^2)$
c $(2x+3)(4x^2-6x+9)$
20 4, 1
21 **a** $\frac{x^2-9}{2(x-3)} = \frac{x+3}{2}$ **b** $\frac{1}{x-3}$
22 **a** $\frac{4a}{rp}$ **b** $\frac{2p^2-3r}{pr}$
23 **a** $\frac{2}{3(n+2)}$ **b** $\frac{5x^2+x-1}{x(x+1)(2x-1)}$
24 **a** $\frac{2x-5}{2x+5}$ **b** $\frac{2t}{t^2-1}$
25 **a** $\frac{(x-1)^3}{x+1}$ **b** $\frac{ab+bc+ac}{abc}$
26 quotient $2x+7$, remainder 18
27 $3x - 14 + \frac{43}{x+3}$
28 quotient $x^2 - 3x - 3$, remainder 2; $x^2 - 3x - 3 + \frac{2}{x-1}$

Chapter 2

Exercise 2A – p.21

1 $2\sqrt{3}$
2 $4\sqrt{2}$
3 $3\sqrt{3}$
4 $5\sqrt{2}$
5 $10\sqrt{2}$
6 $6\sqrt{2}$
7 $9\sqrt{2}$
8 $12\sqrt{2}$
9 $5\sqrt{3}$
10 $4\sqrt{3}$
11 $10\sqrt{5}$
12 $2\sqrt{5}$

Exercise 2B – p.22

1 $2\sqrt{3} - 3$
2 $5\sqrt{2} + 8$
3 $2\sqrt{5} + 5\sqrt{15}$
4 4
5 $\sqrt{6} + \sqrt{2} - \sqrt{3} - 1$
6 $13 + 7\sqrt{3}$
7 4
8 $5 - 3\sqrt{2}$
9 $22 - 10\sqrt{5}$
10 9
11 $10 - 4\sqrt{6}$
12 $31 + 12\sqrt{3}$
13 $(4+\sqrt{5})(4-\sqrt{5}) = 11$
14 $(\sqrt{11}-3)(\sqrt{11}+3) = 2$
15 $(2\sqrt{3}+4)(2\sqrt{3}-4) = -4$
16 $(\sqrt{6}+\sqrt{5})(\sqrt{6}-\sqrt{5}) = 1$
17 $(3+2\sqrt{3})(3-2\sqrt{3}) = -3$
18 $(2\sqrt{5}+\sqrt{2})(2\sqrt{5}-\sqrt{2}) = 18$

Exercise 2C – p.23

1 $\frac{3}{7}\sqrt{2}$
2 $\frac{1}{7}\sqrt{7}$
3 $\frac{2}{11}\sqrt{11}$
4 $\frac{3}{5}\sqrt{10}$
5 $\frac{1}{9}\sqrt{3}$
6 $\frac{1}{2}\sqrt{2}$
7 $\sqrt{2}+1$
8 $\frac{1}{23}(15\sqrt{2}-6)$
9 $\frac{1}{3}(4\sqrt{3}+6)$
10 $-5(2+\sqrt{5})$
11 $\frac{1}{4}(\sqrt{7}+\sqrt{3})$
12 $4(2+\sqrt{3})$
13 $\sqrt{5}-2$
14 $\frac{1}{13}(7\sqrt{3}+2)$
15 $3+\sqrt{5}$
16 $3(\sqrt{3}+\sqrt{2})$
17 $\frac{3}{19}(10-\sqrt{5})$
18 $3+2\sqrt{2}$
19 $\frac{2}{3}(7-2\sqrt{7})$
20 $\frac{1}{2}(1+\sqrt{5})$
21 $\frac{1}{4}(\sqrt{11}+\sqrt{7})$
22 $\frac{1}{6}(9+\sqrt{3})$
23 $\frac{1}{14}(9\sqrt{2}-20)$
24 $\frac{1}{6}(3\sqrt{2}+2\sqrt{3})$
25 $\frac{1}{2}(2+\sqrt{2})$
26 $\frac{1}{42}(3\sqrt{7}-\sqrt{21})$
27 $\frac{1}{9}(\sqrt{30}+2\sqrt{3})$

Exercise 2D – p.26

1 $\frac{1}{2^4} = \frac{1}{16}$
2 $\frac{1}{2^2} = \frac{1}{4}$
3 $3^2 = 9$
4 x^2
5 1
6 t^4
7 1
8 2
9 $y^{3/2}$
10 x^5
11 $\frac{1}{y^{3/4}}$
12 p
13 3
14 $\frac{1}{32}$
15 $\frac{1}{2}$
16 2
17 27
18 $\frac{9}{4}$
19 1
20 16
21 $\frac{5}{4}$
22 -5
23 1331
24 $\frac{3}{5}$
25 6
26 $\frac{16}{27}$
27 8
28 5
29 1
30 1

Exercise 2E – p.28

1 $\log_{10} 1000 = 3$
2 $\log_2 16 = 4$
3 $\log_{10} 10\,000 = 4$
4 $\log_3 9 = 2$
5 $\log_4 16 = 2$
6 $\log_5 25 = 2$
7 $\log_{10} 0.01 = -2$

8 $\log_9 3 = \frac{1}{2}$
9 $\log_5 1 = 0$
10 $\log_4 2 = \frac{1}{2}$
11 $\log_{12} 1 = 0$
12 $\log_8 2 = \frac{1}{3}$
13 $\log_q p = 2$
14 $\log_x 2 = y$
15 $\log_p r = q$
16 $\ln 4 = x$
17 $\ln y = 2$
18 $\ln b = a$
19 $10^5 = 100\,000$
20 $4^3 = 64$
21 $10^1 = 10$
22 $2^2 = 4$
23 $2^5 = 32$
24 $10^3 = 1000$
25 $5^0 = 1$
26 $3^2 = 9$
27 $4^2 = 16$
28 $3^3 = 27$
29 $36^{\frac{1}{2}} = 6$
30 $a^0 = 1$
31 $x^z = y$
32 $a^b = 5$
33 $p^r = q$
34 $e^4 = x$
35 $e^a = 0.5$
36 $e^b = a$

Exercise 2F – p.29

1 2
2 6
3 6
4 4
5 2
6 3
7 $\frac{1}{2}$
8 −2
9 −1
10 $\frac{1}{2}$
11 0
12 1
13 $\frac{1}{3}$
14 0
15 $\frac{1}{3}$
16 3
17 1
18 2
19 3
20 1.5
21 **a** 7.39 **b** 4.48 **c** 0.135 **d** 1.05
22 **a** 1.10 **b** 0.875 **c** −1.60 **d** 2.85 **e** 0.748 **f** 2.40

Exercise 2G – p.31

1 $\log p + \log q$
2 $\log p + \log q + \log r$
3 $\log p - \log q$
4 $\log p + \log q - \log r$
5 $\log p - \log q - \log r$
6 $2\log p + \log q$
7 $\log q - 2\log r$
8 $\log p + \frac{1}{2}\log q$
9 $2\log p + 3\log q - \log r$
10 $\frac{1}{2}\log q - \frac{1}{2}\log r$
11 $n \log q$
12 $n \log p + m \log q$
13 $\log pq$
14 $\log p^2 q$
15 $\log q/r$
16 $\log q^3 p^4$
17 $\log p^n/q$
18 $\log pq^2/r^3$
19 $\ln 5 + \ln x$
20 $\ln 5 + 2\ln x$
21 $\ln 3 + \ln(x+1)$
22 $\ln x - \ln(x+1)$
23 $\ln 2 + \ln x - \ln(x-1)$
24 $\ln x + 2\ln y$
25 $\frac{1}{2}\ln(x+1)$
26 $\ln x + \ln(x+4)$
27 $\ln(x+1) + \ln(x-1)$
28 $2\ln x + \ln(x+y)$
29 $1 + \ln x$
30 $2 + \ln x + \ln(x-e)$
31 $\ln 2x$
32 $\ln(3/x)$
33 $\ln(x^2/4)$
34 $\ln(x/(1-x)^2)$
35 $\ln(e/x)$
36 $\ln(e^2 x)$
37 $\ln(x^2/\sqrt{x-1})$
38 $\ln(x^2/y^3)$

Mixed Exercise 2 – p.32

1 **a** $2\sqrt{21}$ **b** $10\sqrt{3}$ **c** $3\sqrt{5}$
2 **a** $8-2\sqrt{2}$ **b** $7-2\sqrt{10}$
3 **a** $(7+\sqrt{3})(7-\sqrt{3}) = 46$
b $(2\sqrt{2}-1)(2\sqrt{2}+1) = 7$
c $(\sqrt{7}+\sqrt{5})(\sqrt{7}-\sqrt{5}) = 2$
4 **a** $\frac{5}{7}\sqrt{7}$ **b** $\frac{1}{3}(\sqrt{13}+2)$
c $4(\sqrt{3}+\sqrt{2})$ **d** $2-\sqrt{3}$
5 **a** 1 **b** 1
6 **a** $\frac{1}{4}$ **b** $\frac{4}{7}$ **c** $\frac{16}{243}\sqrt{6}$
7 **a** 1 **b** $\frac{1}{5}\sqrt{15}$
8 **a** 7 **b** $\frac{1}{2}$ **c** 0 **d** 5
9 **a** $3\log a - \log b - 2\log c$
b $n\log a - \log b$
c $\log a + \log b - \log c$
d $\log a + \frac{1}{2}\log(1+b)$
10 **a** $\log\dfrac{a^3}{b}$ **b** $-\log a$
11 **a** $\ln x/y$ **b** $\ln e^2(x+1)$
c $\ln Ax$ **d** $\ln y$
12 **b** **i** £2.7145...
ii £2.7181...
iii £2.7182... **c** £e

Chapter 3

Exercise 3A – p.33

1 =
2 ≡
3 ≡
4 =
5 ≡
6 =
7 ≡
8 =
9 =

Exercise 3B – p.34

1 $x = -2$ or $x = -3$
2 $x = 2$ or $x = -3$
3 $x = 3$ or $x = -2$
4 $x = -2$ or $x = -4$
5 $x = 1$ or $x = 3$
6 $x = 1$ or $x = -3$
7 $x = -1$ or $x = -\frac{1}{2}$
8 $x = 2$ or $x = \frac{1}{4}$
9 $x = 1$ or $x = -5$
10 $x = 8$ or $x = -9$
11 −1, 3
12 −1, −4
13 1, 5
14 2, −5
15 −2, 7
16 2, 7

Exercise 3C – p.35

1 $x = 2$ or $x = 5$
2 $x = 3$ or $x = -5$
3 $x = 4$ or $x = -1$
4 $x = 3$ or $x = 4$
5 $x = \frac{1}{3}$ or $x = -1$
6 $x = -1$ or $x = -6$
7 $x = 0$ or $x = 2$
8 $x = -1$ or $x = -\frac{1}{4}$
9 $x = \frac{2}{3}$ or $x = -1$
10 $x = 0$ or $x = -\frac{1}{2}$
11 $x = 0$ or $x = -6$
12 $x = 0$ or $x = 10$
13 $x = 0$ or $x = \frac{1}{2}$
14 $x = 5$ or $x = -4$
15 $x = 2$ or $x = -\frac{4}{3}$
16 $x = 2$ or $x = -1$
17 $x = 0$ or $x = 1$
18 $x = 0$ or $x = 2$
19 $x = 3$ or $x = -1$
20 $x = -1$ or $x = \frac{1}{2}$

Exercise 3D – p.37

1 4
2 1
3 9
4 25
5 2
6 $\frac{25}{4}$
7 192
8 81
9 200
10 $\frac{1}{4}$
11 $\frac{1}{3}$
12 $\frac{9}{8}$
13 $x = -4 \pm \sqrt{17}$
14 $x = 1 \pm \sqrt{3}$
15 $x = -\frac{1}{2}(1 \pm \sqrt{5})$
16 $x = -\frac{1}{2}(1 \pm \sqrt{3})$
17 $x = -\frac{1}{2}(3 \pm \sqrt{5})$
18 $x = \frac{1}{4}(1 \pm \sqrt{17})$
19 $x = -2 \pm \sqrt{6}$
20 $x = -\frac{1}{6}(1 \pm \sqrt{13})$
21 $x = \frac{1}{2}(-2 \pm 3\sqrt{2})$
22 $x = \frac{1}{2}(1 \pm \sqrt{13})$
23 $x = -\frac{1}{8}(1 \pm \sqrt{17})$
24 $x = \frac{1}{4}(3 \pm \sqrt{41})$

Exercise 3E – p.38

1 $x = -2 \pm \sqrt{2}$
2 $x = \frac{1}{4}(1 \pm \sqrt{17})$
3 $x = \frac{1}{2}(-5 \pm \sqrt{21})$
4 $x = \frac{1}{4}(1 \pm \sqrt{33})$
5 $x = 2 \pm \sqrt{3}$
6 $x = \frac{1}{4}(1 \pm \sqrt{41})$
7 $x = \frac{1}{6}(1 \pm \sqrt{13})$
8 $x = -\frac{1}{6}(1 \pm \sqrt{13})$
9 $x = -0.260$ or -1.540
10 $x = 2.781$ or 0.719
11 $x = 1.883$ or -0.133
12 $x = 0.804$ or -1.554
13 $x = 0.804$ or -1.554
14 $x = 0.724$ or 0.276
15 $x = 7.873$ or 0.127
16 $x = 3.303$ or -0.303

Exercise 3F – p.39

1 $3, 2\frac{1}{2}, -\frac{1}{3}$
2 $-2\frac{1}{2}, -4, 1$
3 $1, 2, 3$
4 $2, 5, \frac{1}{3}$
5 $\frac{1}{2}, 4, -2$
6 $2, 1 + \sqrt{5}, 1 - \sqrt{5}$
7 $1, 1, -2$
8 $1, \frac{1}{2}(\sqrt{5} - 5), -\frac{1}{2}(\sqrt{5} + 5)$
9 $a = 3$; $1, -2$

Exercise 3G – p.41

1 $x = 2, y = 1, z = 1$
2 $x = 3, y = 4, z = 1$
3 $x = 1, y = -1, z = 2$
4 $x = 3, y = 2, z = -1$
5 $x = 1, y = 2, z = 3$
6 $x = 4, y = -1, z = 2$

Exercise 3H – p.42

1

x	-2	1
y	-1	2

2 $x = -1, y = 3$
3 $x = 2, y = 3$

4

x	-1	$\frac{1}{2}$
y	4	1

5

x	2	$-\frac{1}{2}$
y	-3	2

6

x	$\frac{7}{2}$	-2
y	$-\frac{1}{2}$	5

7

x	1	2
y	2	1

8

x	-1	3
y	-4	4

9 $x = 1, y = 5$

10

x	6	-6
y	2	-4

11 $x = \frac{1}{2}, y = -1$

12

x	1	0
y	$\frac{1}{3}$	$\frac{2}{3}$

13 $x = -1, y = -\frac{1}{2}$
14 $x = 1, y = -\frac{1}{3}$

15

x	$-\frac{1}{3}$	$\frac{2}{3}$
y	$-\frac{1}{2}$	$\frac{1}{4}$

16 $x = 1, y = \frac{1}{2}$

17

x	-3	6
y	-3	$\frac{3}{2}$

18

x	1	$-\frac{1}{4}$
y	-2	3

19

x	-1	2
y	$\frac{1}{3}$	$-\frac{1}{6}$

20

x	$\frac{1}{2}$	0
y	$\frac{1}{2}$	1

21

x	1	$3\frac{1}{2}$
y	1	-4

22

x	-1	$7\frac{1}{2}$
y	2	$-\frac{7}{5}$

Mixed Exercise 3 – p.43

1 $-1, 6$
2 $3 \pm \sqrt{14}$
3 $\frac{1}{4}(-3 \pm \sqrt{17})$
4 $\frac{1}{3}(-2 \pm \sqrt{19})$
5 $1, 1$
6 $-\frac{1}{4}, 3$
7 $\frac{1}{2}(-1 \pm \sqrt{13})$
8 $-6, 2$
9 $-1 \pm \sqrt{3}$
10 $-2, -2$
11 $0, 2$
12 $-4, 1$
13 $4, -3, -2$
14 $0, -1$
15 $\frac{1}{2}(7 \pm \sqrt{89})$
16 1
17 $-3, 5, -\frac{1}{3}$
18 $2, 1.19, -4.19$
19 $x = 4, y = 5$ or $x = -22, y = 31$
20 $x = 2, y = 5$ or $x = 4, y = 9$
21 $5, \frac{2}{3}$
a rational **b** it factorises

Chapter 4

Exercise 4A – p.46

1 4
2 $-\frac{5}{3}$
3 1
4 $\frac{4}{3}$
5 -3
6 $\frac{2}{5}$
7 real and different
8 not real
9 real and different
10 real and equal
11 real and different
12 real and equal
13 real and different
14 not real
15 real and different
16 real and equal
17 $k = \pm 12$
18 $a = 2\frac{1}{4}$
19 $p = 2$
22 $q^2 = 4p$

Exercise 4B – p.47

1 2
2 -2
3 1.5
4 1.63
5 1.16
6 0.861
7 2.77
8 $\frac{1}{4}$
9 1
10 16
11 $1, 4$
12 $\frac{3}{2}$
13 $\log_x(5/9), \frac{1}{3}\sqrt{5}$
14 1
15 $\log_3(y/x^2), y = 3x^2$
16 $y^2, 1$

Exercise 4C – p.49

1 $2, -1$
2 0
3 3
4 $\pm 3, \pm\sqrt{3}$
5 $-\frac{1}{2}, 1$
6 1
7 $\pm 8, \pm 1$
8 $\frac{1}{2}(1 \pm \sqrt{3})$
9 1
10 $-1, 32$
11 4
12 no solutions

Mixed Exercise 4 – p.49

1 **a** $6; 4, 2$
b $-\frac{5}{4}; -\frac{5}{8} \pm \frac{\sqrt{73}}{8}$
2 **a** not real
b real and different
c real and equal
d real and different
3 $4, -1$
5 2
7 1
8 3
9 2
10 2.10
11 2
12 $\frac{1}{2}\ln 2$
13 $8, 1$
14 $0, 9$
15 $1 \pm \sqrt{2}$
16 $\frac{1}{2}(\sqrt{5} - 1)$
17 2
18 $1, -1$
19 2
20 $\log_4 \frac{x-1}{y^{1/2}}, y = 4(x-1)^2,$
$x = 2, y = 4$
21 $x = 2, y = 4$
22 $x = 0, y = 0$ or $x = \frac{1}{2}, y = 1$
23 $x = 8, y = 2$

Summary A

Multiple Choice Exercise A – p.51

1 A
2 E
3 D
4 D
5 B
6 A, D
7 B
8 E
9 A
10 E
11 C
12 C
13 B, C
14 A, E
15 B
16 B, C
17 A
18 A
19 A, B
20 T
21 F
22 T
23 F

Examination Questions A – p.53

1 11/12
2 (3, 1), (−11/5, −8/5)
3 $1/a^2$
4 (2, −1), (−1, 2)
5 $a^{-2/3}$
6 (2, −1), (−5/4, 11/2)
7 a $(x+2)$
b $(x+2)(5x^2+14x+1)$
c $-2, \frac{1}{5}(-7\pm 2\sqrt{11})$
8 4
9 a $11\sqrt{7}-32$ b $11+4\sqrt{7}$
10 a $x = 3, y = 2$
b line touches curve at (3, 2)
11 $\dfrac{2(x+1)}{(x+3)}$
12 ln 4
13 a 3/2 b 1.431
14 a $1+x^2+x^3+x^5$
b 1.000 001 001 000 001
15 $y = \pm x^2$
16 $(x-4)^2 - 19, x = 4\pm\sqrt{19}$
17 −2, −1.5, 0.5
18 $-12, \frac{1}{2}(-1\pm 3\sqrt{5})$
19 $3\pm 2\sqrt{2}$
20 a pq b p^2/q
21 $y = \log_b \frac{1}{5}, -1$
22 $\log_2 \dfrac{x+2}{x}$, 2/7
23 c $y^2 - 10\,001\left(\dfrac{y}{10}\right) + 100 = 0$
d −1, 3
24 $a = -3, b = 1$
25 i 2 ii $2, \pm\sqrt{2}$
26 $\frac{13}{3} - \frac{1}{6}\log_2 3$
27 b $(x+3)(2x^2-x-5)$
c 1.85, −1.35
29 $(x-2)(2x+1)(x-3)$
30 a $p = 3, q = 2, r = -7$
b −7 c −0.5, −3.5
31 b ln 4
32 $(2x-5)(x+1)$
33 (3, −1)(−1/3, 7/3)
35 $a = -13, b = 12, (x+4)$
36 $a = 1, b = -4, (x+1)$
37 $(x^2-3)(x^2+x+2)$
38 b −0.443

Chapter 5

Exercise 5A – p.57

1 T
2 T
3 F: $\ln x + \ln y = \ln xy$ not $\ln(x+y)$
4 F: x can be −4
5 T
6 T
7 F: $x^2 = 1 \Rightarrow x = \pm 1$
8 T
9 F: $(-a)^3 + 1 = 0$
10 T
11 $\Rightarrow$
12 $\Leftrightarrow$
13 $\Leftrightarrow$
14 $\Leftarrow$
15 $\Leftarrow$
16 $\Rightarrow$
17 $\Rightarrow$
18 $\Rightarrow$
19 $\Leftarrow$
20 $\Leftrightarrow$

Chapter 6

Exercise 6A – p.63

1 a 5 b $\sqrt{2}$ c $\sqrt{13}$
2 a $(\frac{5}{2}, 4)$ b $(\frac{5}{2}, \frac{1}{2})$ c $(3, \frac{7}{2})$
3 a $\sqrt{109}, (\frac{1}{2}, 1)$
b $\sqrt{5}, (-\frac{1}{2}, -1)$
c $2\sqrt{2}, (-2, -3)$
4 $\sqrt{65}$
5 $\sqrt{13}$
6 (2, −4)
8 b $(-3\frac{1}{2}, -\frac{1}{2})$ c $17\frac{1}{2}$ sq units
9 a $\sqrt{5}(2+\sqrt{2})$ b $(0, 4\frac{1}{2})$
c $2\frac{1}{2}$
11 (−5, −3)
12 (1, 5)
13 (6, 9.5)

Exercise 6B – p.66

1 a 3 b $\frac{3}{2}$ c $\frac{1}{3}$ d $\frac{3}{4}$
e −4 f 6 g $-\frac{7}{3}$ h $-\frac{3}{2}$
i $\dfrac{k}{h}$
2 a yes b no c yes d yes
3 a parallel b perpendicular
c perpendicular
d neither e parallel

Exercise 6C – p.67

1 $a = 0, b = 4$
2 b $22\frac{1}{2}$ square units
5 $\sqrt{a^2+4b^2}$
8 $\left(\dfrac{p+q}{2}, \dfrac{p+q}{2}\right)$
9 $(a-2)^2 + (b-1)^2 = 9$
10 8
11 $b(d-b) = ac$
12 $b^2 = 8a - 16$

Chapter 7

Exercise 7A – p.68

1 $\sin A = \frac{12}{13}, \cos A = \frac{5}{13}$
2 $\tan X = \frac{3}{4}, \sin X = \frac{3}{5}$
3 $\cos P = \frac{9}{41}, \tan P = \frac{40}{9}$
4 $\sin A = \dfrac{1}{\sqrt{2}} = \cos A$
5 $\sin Y = \frac{1}{3}\sqrt{5}, \tan Y = \frac{1}{2}\sqrt{5}$
6 $\cos A = \frac{1}{2}\sqrt{3}$; 30°
7 $\cos X = \frac{24}{25}$
8 $\cos X = \frac{4}{5}$;
$\cos^2 X - \sin^2 X = 0.28 = \cos 2X$
9 1
10 $2\sin X\cos X = 0.96, \sin 2X = 0.96,$
$\sin 2X = 2\sin X\cos X$

Exercise 7B – p.70

1 $\frac{3}{5}, -\frac{4}{5}, -\frac{3}{4}$
2 $\frac{5}{13}, -\frac{12}{13}, -\frac{5}{12}$
3 $\frac{3}{5}, -\frac{4}{5}, -\frac{3}{4}$
4 $\frac{5}{\sqrt{34}}, \frac{-3}{\sqrt{34}}, \frac{-5}{3}$
5 80° or 100°
6 105°
7 52° or 128°
8 150°
9 81° or 99°
10 57°
11 90°
12 89°
13 $\frac{5}{13}$
14 53° or 127°
15 150°
16 **a** yes, 90° **b** yes, 0 **c** no
17 $A + B = 180°$

Exercise 7C – p.73

Answers are correct to 3 s.f.

1 11.1 cm
2 10.2 cm
3 156 cm
4 113 cm
5 7.01 cm
6 141 cm
7 16.3 cm
8 no; an angle and the side opposite to it are not known

Exercise 7D – p.75

Answers are correct to the nearest degree.

1 18°

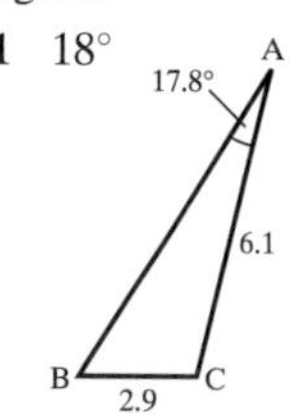

2 58° or 122°

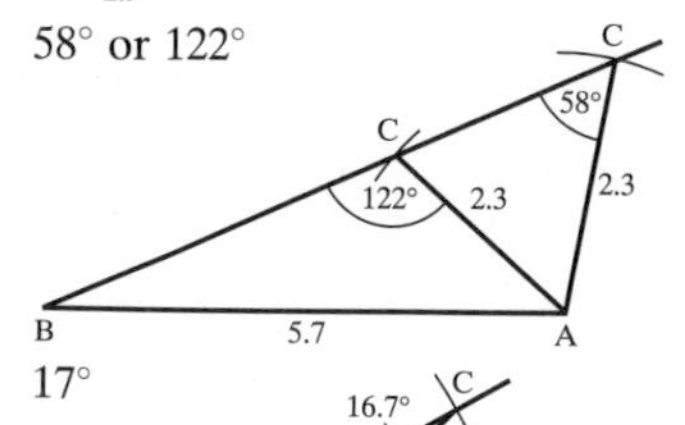

3 17°

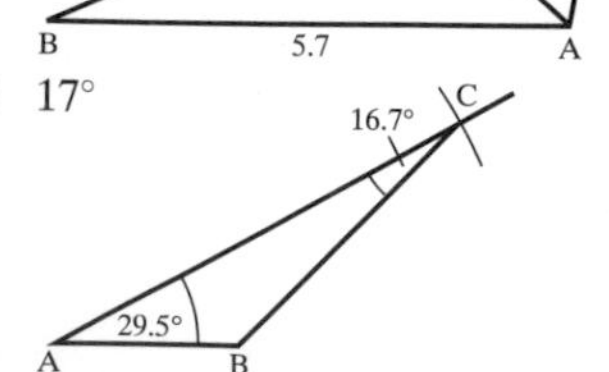

4 35°

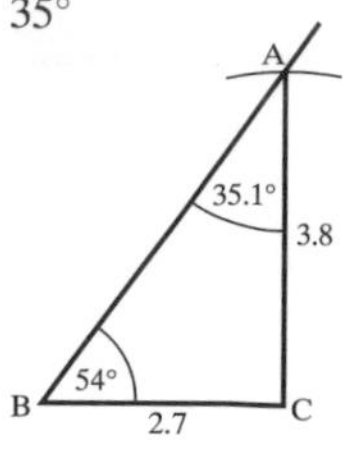

5 57° or 123°

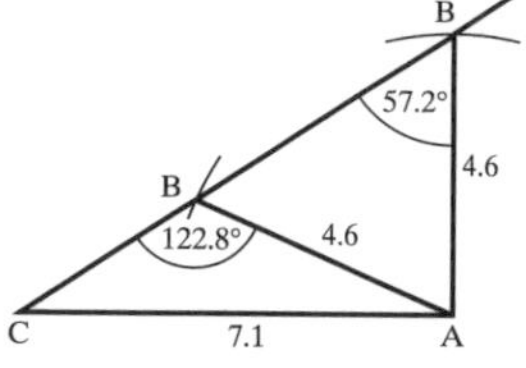

6 30°

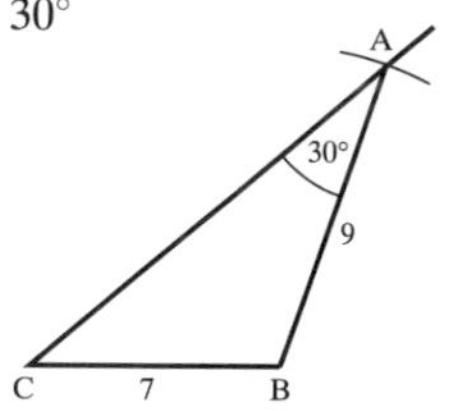

Exercise 7E – p.77

1 5.29 cm
2 12.9 cm
3 53.9 cm
4 4.04 cm
5 101 cm
6 12.0 cm
7 64.0 cm
8 31.8 cm

Exercise 7F – p.78

1 38°
2 55°
3 45°
4 94°
5 **a** 18°
b 126°
6 29°
7 11.4 cm, 68°

Exercise 7G – p.79

1 87.4 cm
2 23.8 cm
3 17.5 cm
4 ∠B = 81°; a = 112 cm
5 a = 164 cm; c = 272 cm
6 ∠B = 34°; a = 37.0 cm
7 ∠C = 43°; b = 19.4 cm; c = 13.5 cm
8 ∠A = 52°; a = 33.2 cm; c = 41.5 cm
9 ∠B = 43°; ∠C = 60°; a = 27.1 cm
10 ∠A = 22°; ∠C = 33°; b = 30.3 cm
11 14.1 m
12 40°, 53°, 87°

Exercise 7H – p.80

1 12 300 cm^2
2 2190 cm^2
3 1680 cm^2
4 453 square units
5 42.9 square units
6 51°, 21.0 cm^2
7 10.6 cm, 59.8 cm^2
8 52°, 151 cm (or 150 cm)
9 5.25 cm
10 $h = c \sin A$

Exercise 7I – p.82

1 11.0 cm, 67.7 cm^2 (or 67.8 cm^2)
2 58.5 km
3 477 m
4 ∠BAO = 74°, ∠CAO = 52°; 22 cm^2
5 **a** 60.8 cm **b** 35° **c** 42.8 cm
d 2140 cm^2 **e** 427 000 cm^3

Mixed Exercise 7 – p.83

1 **a** 116° **b** 86°
2 $-\frac{24}{25}$
3 **a** $\frac{5}{\sqrt{39}}$ **b** $\frac{-5}{\sqrt{39}}$
4 **a** $\frac{1}{\sqrt{2}}, \frac{-1}{\sqrt{2}}$ **b** $\frac{3}{\sqrt{13}}, \frac{-2}{\sqrt{13}}$
5 $-\frac{5}{13}$
6 9.05 cm
7 4.82
8 83°
9 54° or 126°
10 108°, 50°, 22°
12 75.8 cm^2
13 $\frac{1}{2}$; yes, ∠A can be 30° or 150°
14 **a** 98°
b 19.8 cm^2
c 3.96 cm

Chapter 8

Exercise 8A – p.86

1 **a** $y = 2x$ **b** $x + y = 0$
c $3y = x$ **d** $4y + x = 0$
e $y = 0$ **f** $x = 0$
2 **a** $2y = x + 2$ **b** $3y + 2x = 0$
c $y = 4x$
3 **a** $2y = x$ **b** $y + 2x - 6 = 0$
c $y = 2x + 5$
4 **a** $2y + x = 0$ **b** $2x - 3y = 0$
c $2x + y = 0$
5 **a** $x - 3y + 1 = 0$
b $2x + y - 5 = 0$
6 **a** $5x - y - 17 = 0$
b $x + 7y + 11 = 0$
7 $3x + 4y - 48 = 0$, 5

Exercise 8B – p.89

1 **a** $y = 3x - 3$
b $5x + y - 6 = 0$
c $x - 4y - 4 = 0$
d $y = 5$
e $2x + 5y - 21 = 0$
f $15x + 40y + 34 = 0$
2 **a** $3x - 2y + 2 = 0$
b $3x - 2y + 7 = 0$
c $x = 3$
3 **a**, **c** and **d**
4 $x + y - 7 = 0$
5 **a** $x + 2y - 5 = 0$
b $16x - 6y + 19 = 0$
c $10x - 16y + 23 = 0$
6 $2x + y = 0$
7 $4x + 5y = 0$
8 $5x + 4y = 0$
9 $x + 2y - 11 = 0$
10 $3x - 4y + 19 = 0$
11 $y = x + 2$
12 $x + y - 1 = 0$
13 50°
14 $y + 3x = 0$, $3y - x = 0$

Exercise 8C – p.90

2 $\frac{1}{4}$ sq. units
3 $(-\frac{4}{3}, \frac{11}{3})$, $(-5, 0)$, $(6, 0)$
4 $10x - 26y - 1 = 0$
5 $x + 3y - 11 = 0$, $(\frac{13}{5}, \frac{14}{5})$
6 $[\frac{2}{5}(2 + 2a - b), \frac{1}{5}(8 - 2a + b)]$
7 $y = 2x - 3$
8 $y = x - 4$ (or $y + x = 10$ if line is inclined at −45° to Ox)
9 $8by - 2ax + 8b^2 + 3a^2 = 0$
10 **a** $\sqrt{20}$ **b** $x - 2y + 1 = 0$
11 $(\frac{9}{10}, \frac{17}{10})$ and $(-\frac{18}{10}, \frac{26}{10})$, $(-\frac{27}{10}, -\frac{1}{10})$ or $(\frac{36}{10}, \frac{8}{10})$, $(\frac{27}{10}, -\frac{19}{10})$
12 **a** (1, 2), (5, 2), (3, 6) **b** 8

Exercise 8D – p.95

1 $a = 2, b = -4$ (exactly)
2 $a = 6, b = 4$
3 $a = 30, b = 2$
4 $k = 4, n = 0.07$
5 $k = 5500, n = 1.5$
6 **a** 10 000, 1.2
c gradient and intercept different so values of constants different.
7 **a** $\frac{1}{30}$
d $s = 780$

Chapter 9

Exercise 9A – p.98

1 **a** ∠ADE, ∠ACE **b** ∠AFE
c ∠AOC **d** ∠ABC
e ∠COE **f** ∠CAD, ∠CED
g ∠BAC
3 **c** 106.3° **d** (5, 8), 53.1°

Exercise 9B – p.100

1 **a** 50° **b** 40°, 40°
c 60° **d** 54°, 108°
2 $2x$, $2y$
4 (4, 4)
7 $a^2 + b^2 - 9a - b + 14 = 0$
9 $(-\frac{115}{18}, -\frac{29}{6})$, 10.8 units to 3 s.f.
10 60°, 60°

Exercise 9C – p.102

2 61.9°
3 5 units
4 67.4°
5 7.22 cm
6 60°, 30°, 90°
7 8.43
8 16 sq. units
9 14.9 cm
10 2.85
11 $\frac{2\sqrt{5}}{5}, -\frac{2\sqrt{5}}{5}$ or $\frac{2}{\sqrt{5}}, -\frac{2}{\sqrt{5}}$
13 $(-\frac{3}{4}, -\frac{9}{4})$
14 **a** $2x + y = 0$ **b** $(-\frac{4}{5}, \frac{8}{5})$
15 $x + y - 14 = 0$

Chapter 10

Exercise 10A – p.105

1 **a** $\frac{\pi}{4}$ **b** $\frac{5\pi}{6}$ **c** $\frac{\pi}{6}$
d $\frac{\pi}{2}$ **e** $\frac{3\pi}{2}$ **f** $\frac{2\pi}{3}$
g $\frac{\pi}{3}$ **h** $\frac{\pi}{8}$ **i** $\frac{4\pi}{3}$
j $\frac{5\pi}{3}$ **k** $\frac{7\pi}{4}$ **l** $\frac{3\pi}{4}$
m $\frac{7\pi}{6}$ **n** $\frac{5\pi}{4}$
2 **a** 30° **b** 180° **c** 18°
d 60° **e** 150° **f** 15°
g 210° **h** 135° **i** 20°
j 270° **k** 80° **l** 45°
m 108° **n** 22.5°
3 **a** 0.61 **b** 0.82 **c** 1.62
d 4.07 **e** 0.25 **f** 2.04
g 6.46
4 **a** 97.4° **b** 190.2° **c** 57.3°
d 119.7° **e** 286.5° **f** 360.0°
5 **a** 0.8660 **b** 0.5 **c** 0
d 0.5 **e** 1 **f** 0
g 1 **h** −1 **i** −1
j −0.5

6 a $0, 2\pi$ b $\frac{\pi}{4}, \frac{5\pi}{4}$ c $\frac{\pi}{6}, \frac{5\pi}{6}$
d $\frac{\pi}{3}, \frac{5\pi}{3}$ e $\frac{3\pi}{2}$ f π
g $\frac{\pi}{2}$ h $\frac{3\pi}{4}, \frac{7\pi}{4}$
i $\frac{7\pi}{6}, \frac{11\pi}{6}$ j $0, \pi$
k $\frac{\pi}{2}, \frac{3\pi}{2}$ l $\frac{5\pi}{4}, \frac{7\pi}{4}$
7 a 0.932 b 0.939
c 9.89 d −0.801
8 a 0.284 b 0.929
c 0.644 d 0.0226

Exercise 10B – p.108

1 $\frac{2\pi}{3}$ cm
2 $\frac{25\pi}{2}$ cm
3 2.4 rad
4 0.692 rad
5 $\frac{15}{\pi}$ cm
6 $\frac{25}{\pi}$ cm
7 4π cm
8 $\frac{60}{\pi}$ cm
9 $\frac{360°}{\pi}$
10 146.4°
11 182.1 mm
12 0.333 rad
13 85.6 cm

Exercise 10C – p.109

1 4.19 cm^2
2 75.4 cm^2
3 $\frac{\pi}{2}$
4 0.96 rad
5 $\frac{125\pi}{3}$ cm^2
6 $\frac{15}{\pi}$ cm, $\frac{225}{2\pi}$ cm^2
7 6 cm
8 $4\sqrt{3}$ cm
9 8 cm
10 0.283 rad
11 a 12 cm^2 b 23.2 cm^2
12 14.5 mm^2, 139 mm^2
13 a 15.2 cm b 32.5 cm^2
14 19.6 cm, 108 cm^2

Exercise 10D – p.112

3 $32(\tan\theta - \theta)$ cm^2
6 a $\frac{\pi}{3}$
7 a $\frac{(50-2r)}{r}$
8 0.979 cm^2
9 a 2.35 rad b 9.41 cm
c 35.4 cm d 68.8 cm
10 10.2 cm
11 182 cm

Summary B

Multiple Choice Exercise B – p.115

1 C
2 E
3 C
4 E
5 A, B, C
6 C
7 D
8 B
9 A
10 A
11 A, B
12 A
13 D
14 D
15 D
16 C, D
17 A, B
18 B
19 A, C
20 B
21 A, B
22 C, D
23 A, C
24 F
25 T
26 F
27 T
28 T
29 T
30 F

Examination Questions B – p.118

1 a i $-\frac{2}{3}$ ii $3x - 2y - 29 = 0$
b $(5, -7)$
2 4.85 cm^2
3 a $(2, 6)$ b $3x + y - 12 = 0$
5 $9x + 13y + 14 = 0$
6 63.1°, 116.9°
7 a $y = -\frac{3}{4}x + \frac{5}{2}$
8 a 10 cm b 0.7896 c 13.4 cm^2
9 b $2x + y - 11 = 0$,
$x - 2y - 13 = 0$, $(7, -3)$
c 15 sq units
10 a $y = -\frac{3}{4}x + \frac{5}{2}$
11 a $5x + 12y - 22 = 0$
b −22 c 26
12 a 120 cm^2 b 2.16
c 161.07 cm^2
13 a $3x - 2y - 3 = 0$
b $(3, 3)$
14 c $\frac{1}{2}r^2\sqrt{3} - \frac{1}{6}\pi r^2$
15 $x - 2y = 8$, $(2, -3)$, $4\sqrt{5}$
16 a $\pi - 2\theta$ b $2r^2 + 2r^2\cos 2\theta$
c $2r\cos\theta$
17 a 5 b $\frac{1}{8}$
18 b 80.9 m c 26.7 m d 847 m^2
19 $(8, 2)$
20 17 cm^2
21 a 29.7 b $11x - 10y + 19 = 0$
22 a $7x + 5y - 18 = 0$
b 162/35 sq units
23 $\theta = 41.4°$, $\angle ACB > \theta$
24 $3x + 5y - 4 = 0$, $(\frac{4}{3}, 0)$
25 24.6 m
26 $6y - 5x + 2 = 0$
27 a $y = \frac{1}{2}x + 3$ b $3\sqrt{5}$
28 a $y = -\frac{1}{2}x + 4$
b $y = -\frac{1}{2}x + \frac{3}{2}$; $(1, 1)$
29 a $\ln y + \frac{6}{7}\ln x = 3$
b $y = 20.1x^{-0.857}$
30 a $\ln y = 3 - 9x/13$
b $a = 20, b = 0.5$
31 $m = 0.8$, 12 kg–13 kg
32 b i 3.3
ii $a = 5, b = 0.3$

Chapter 11

Exercise 11A – p.125

1 **a** yes **b** yes **c** yes, $x \neq 1$
d no **e** yes, $x \geqslant 0$
f yes **g** yes **h** no

2 $-4, -24$

3 25, 217

4 1, not defined, 12

5 $1, \frac{\sqrt{3}}{2}$

Exercise 11B – p.127

1 **a** $f(x) \geqslant -3$ **b** $f(x) \geqslant -5$
c $f(x) \geqslant 0$ **d** $0 < f(x) \leqslant \frac{1}{2}$

2 **a**

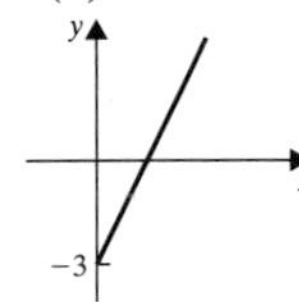

b

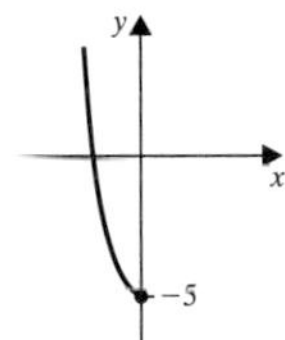

c

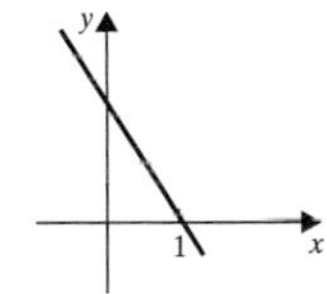

d

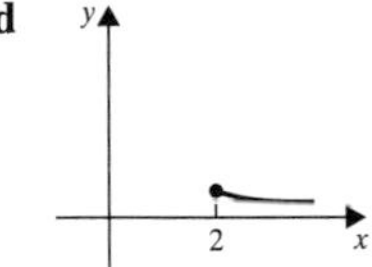

3 **a** 5, 4, 2, 0

b

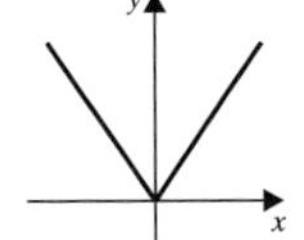

4 **a** 0, 2, 4, 5, 5

b

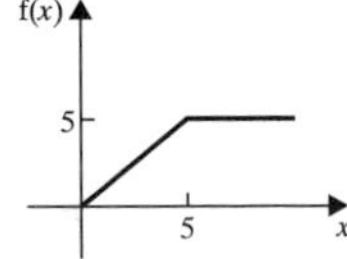

c $0 \leqslant f(x) \leqslant 5$

5 **a** 0, £5000

b $f(x) = 0$ for $0 \leqslant x \leqslant 20\,000$
$f(x) = \frac{1}{5}(x - 20\,000)$ for $x > 20\,000$

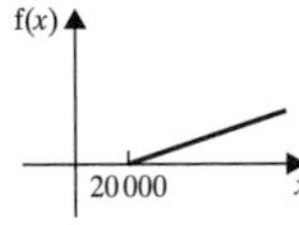

domain $x \geqslant 0$ (but $x <$ GNP!)
range $f(x) \geqslant 0$

Exercise 11C – p.130

1 **a** $\frac{11}{4}$ **b** 3 **c** 4

2 **a** $f(x) \leqslant \frac{29}{4}$ **b** $f(x) \geqslant -2$
c $f(x) \leqslant 1$

3 **a**

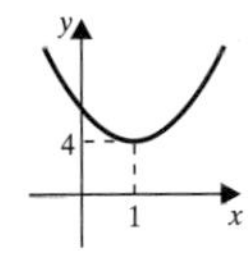

b

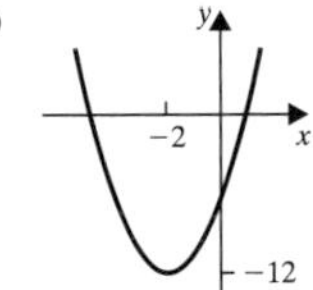

c

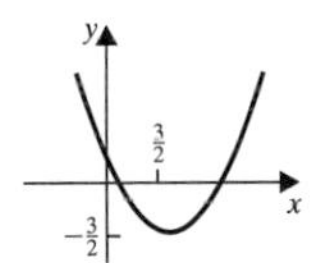

d

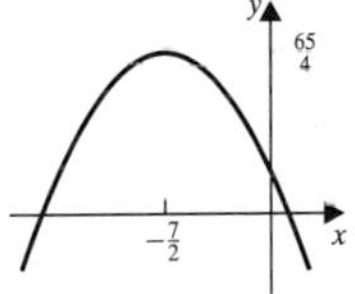

e

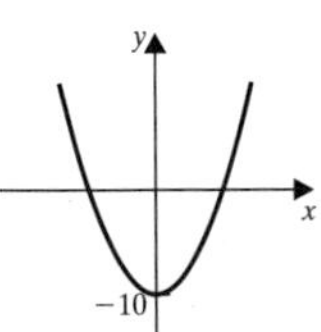

f

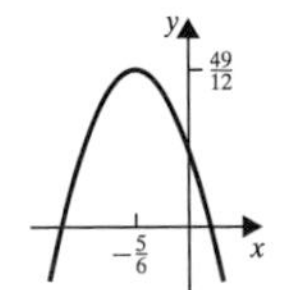

4 **a**

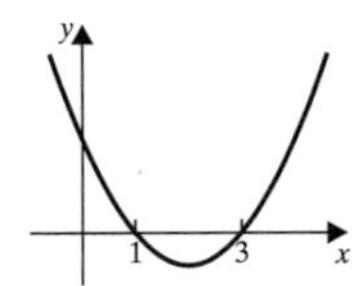

b

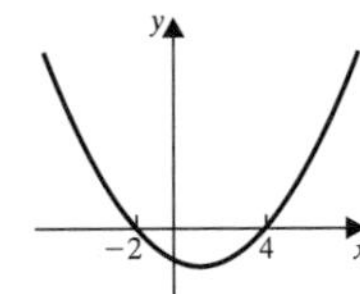

c

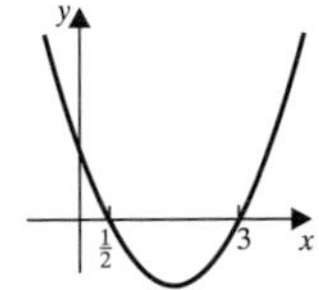

d

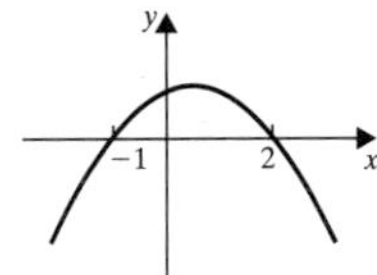

e

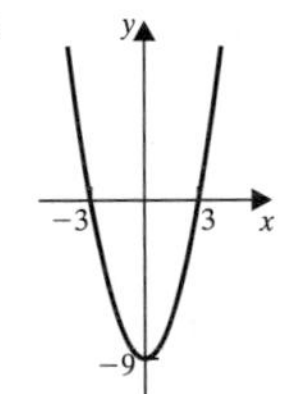

f

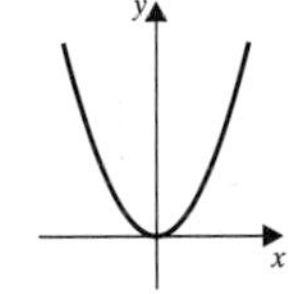

Exercise 11D – p.133

1 **a**

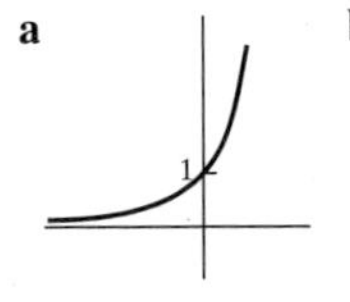

b

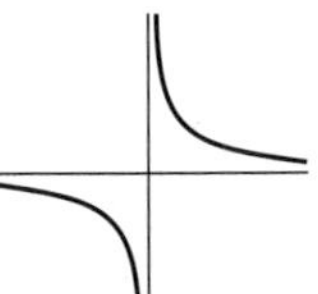

c

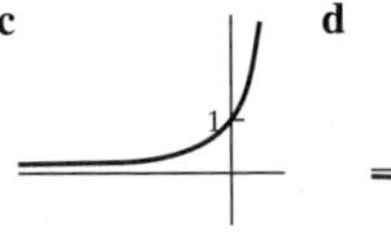

d

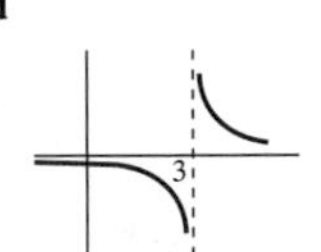

2 16, 8, 4, 2, 1, $\frac{1}{2}$, $\frac{1}{4}$, $\frac{1}{8}$, $\frac{1}{16}$
as $x \to \infty$, $f(x) \to 0$ and
as $x \to -\infty$, $f(x) \to \infty$

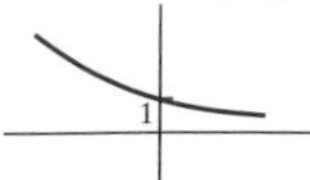

3 -2, as $x \to -2$ from below,
$f(x) \to -\infty$
as $x \to -2$ from above, $f(x) \to \infty$
as $x \to \infty$, $f(x) \to 0$ from above
as $x \to -\infty$, $f(x) \to 0$ from below

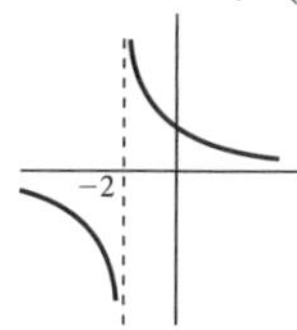

4 a

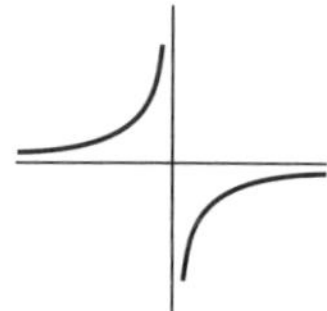

b

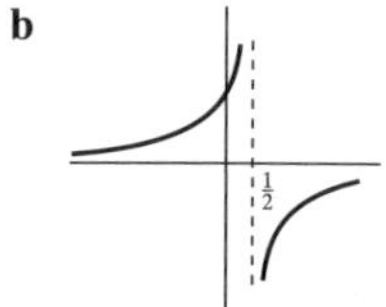

c

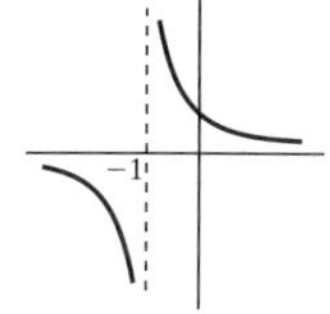

d

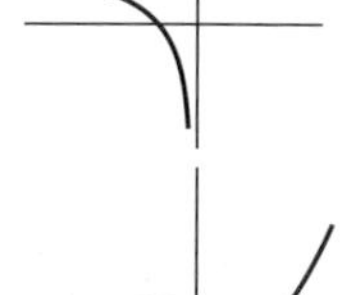

5 a

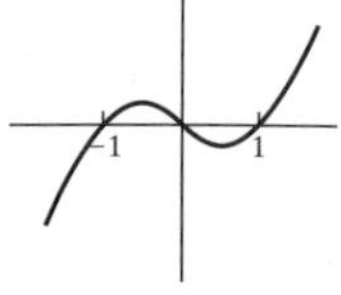

b

c

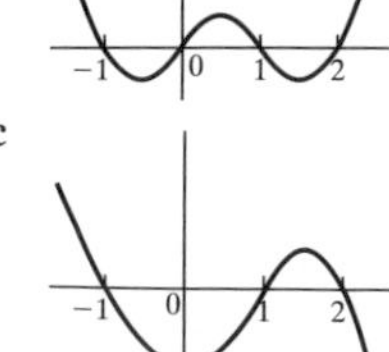

d

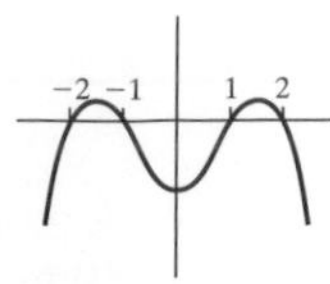

Exercise 11E – p.134

1 b translation $\begin{pmatrix} 0 \\ c \end{pmatrix}$

2 translation $\begin{pmatrix} -c \\ 0 \end{pmatrix}$

3 a reflection Ox
b reflection Oy

Exercise 11F – p.136

11 Reflection in line $y = x$
(there are several alternatives).
12 Reflection in line $y = x$
(there are several alternatives).
13 (5, 2)
14 (b, a)

Exercise 11G – p.139

1 a

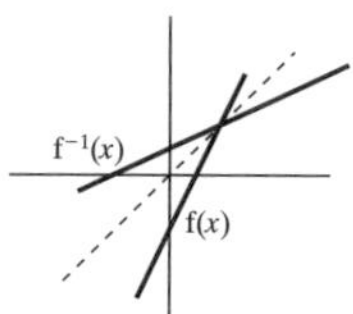

b

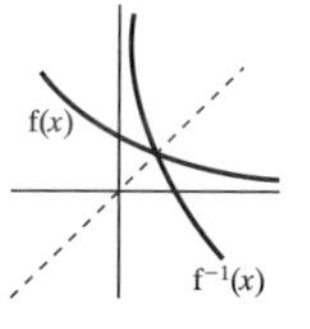

c

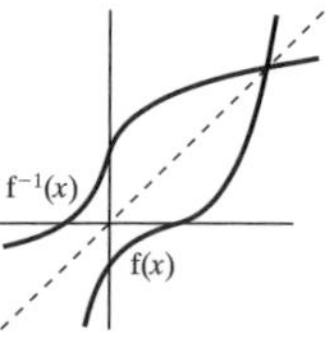

d

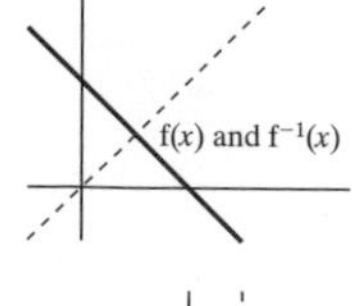

e

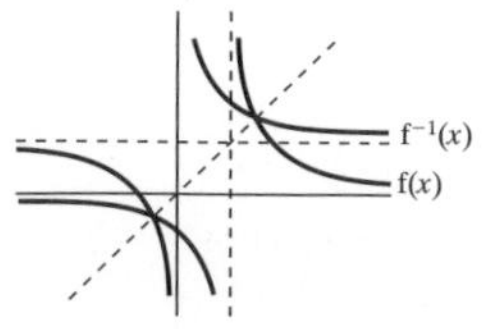

f

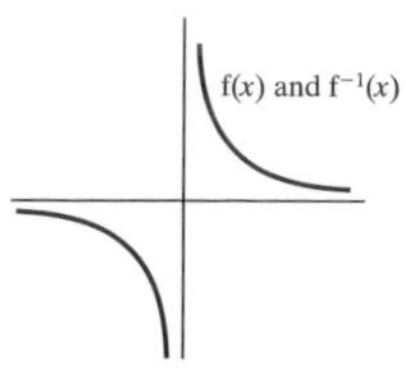

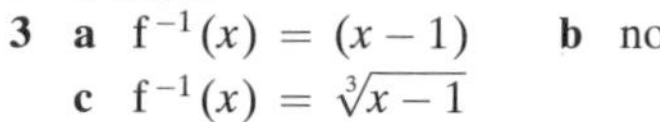

2 **d** and **f**
3 a $f^{-1}(x) = (x - 1)$ **b** no
c $f^{-1}(x) = \sqrt[3]{x - 1}$
d $f^{-1}(x) = \sqrt{x + 4}, x \geqslant -4$
e $f^{-1}(x) = \sqrt[4]{x} - 1, x \geqslant 0$
4 a $-\frac{1}{3}$ **b** $\frac{1}{2}$
c there isn't one
5 a 9 **b** 2 **c** -1

Exercise 11H – p.140

1 a $\dfrac{1}{x^2}$ **b** $(1 - x)^2$ **c** $1 - \dfrac{1}{x}$
d $1 - x^2$ **e** $\dfrac{1}{x^2}$

2 a 125 **b** 15 **c** -1
d -1
3 a $(1 + x)^2$ **b** $2(1 + x)^2$
c $1 + 4x^2$
4 $g(x) = x^2$, $h(x) = 2 - x$
5 $g(x) = x^4$, $h(x) = (x + 1)$
6 a $f(x) = gh(x)$, $g(x) = 10^x$,
$h(x) = x + 1$
b $f(x) = gh(x)$, $g(x) = \dfrac{1}{x^2}$
$h(x) = 3x - 2$
c $f(x) = g(x) + h(x)$,
$g(x) = 2^x$, $h(x) = x^2$
d $f(x) = \dfrac{g(x)}{h(x)}$, $g(x) = 2x + 1$,
$h(x) = x$
e $f(x) = gh(x)$, $g(x) = x^4$,
$h(x) = 5x - 6$
f $f(x) = g(x)\,h(x)$,
$g(x) = x - 1$, $h(x) = x^2 - 2$

Mixed Exercise 11 – p.140

1 a $\frac{1}{0}$ is meaningless **b** $\frac{1}{4}$
c

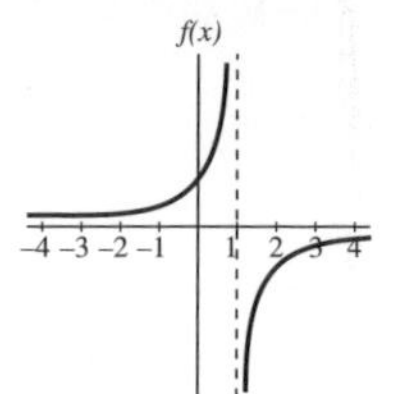

d $f^{-1}(x) = 1 - \dfrac{1}{x}, x \neq 0$

2 **a** $\frac{11}{4}$ when $x = \frac{3}{2}$

b $-\frac{41}{8}$ when $x = \frac{7}{4}$

c -9 when $x = -2$

3

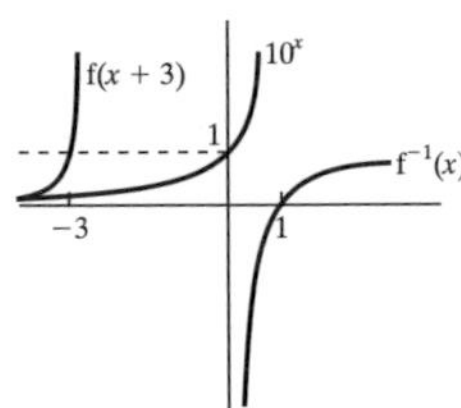

4 **a** 10 000, $\frac{1}{9}$, $\frac{1}{100}$

b 10^{-x^2}, $10^{2/x}$

c $\log x$, g^{-1} does not exist, $\dfrac{1}{x}$

d $\pm\frac{1}{3}$

e no for all x, yes for $x > 0$

5 **a**

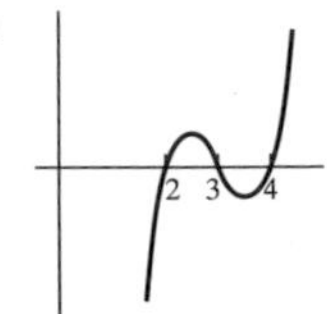

b

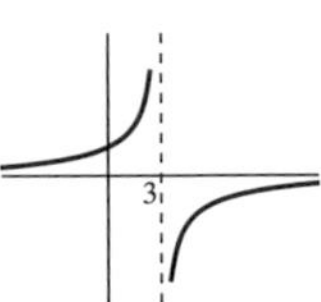

c

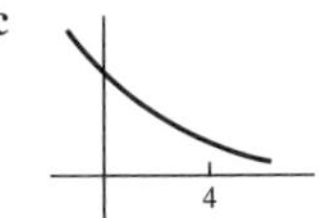

6 **a** $g(x) = 2^x, h(x) = 3x - 2$

b 2^{46}, 1

7 **a**

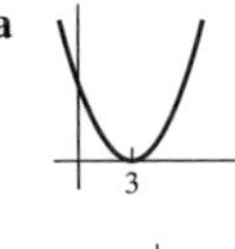

b 0, $\frac{1}{6}$

8 **a** $\dfrac{3}{x}$ **b** $\dfrac{1}{3(x^2-1)}$

c $\dfrac{3}{x}$ **d** $\dfrac{1}{3x}$

Chapter 12

Exercise 12A – p.143

1. $x < \frac{7}{2}$
2. $x > 4$
3. $x > -3$
4. $x > -2$
5. $x < \frac{1}{2}$
6. $x < -3$
7. $x < -\frac{1}{4}$
8. $x > \frac{8}{3}$
9. $x > \frac{3}{8}$

Exercise 12B – p.144

1. $x > 2$ and $x < 1$
2. $x \geqslant 5$ and $x \leqslant -3$
3. $-4 < x < 2$
4. $x \geqslant \frac{1}{2}$ and $x \leqslant -1$
5. $x > 2 + \sqrt{7}$ and $x < 2 - \sqrt{7}$
6. $-\frac{1}{2} < x < \frac{1}{2}$
7. $-4 \leqslant x \leqslant 2$
8. $x > 1$ and $x < -\frac{2}{5}$
9. $x \geqslant \frac{3}{2}$ and $x \leqslant -5$
10. $x > 4$ and $x < -2$
11. $\frac{1}{2}(-3 - \sqrt{17}) \leqslant x \leqslant \frac{1}{2}(-3 + \sqrt{17})$
12. $x > 7$ and $x < -1$

Exercise 12C – p.146

1. $x > -2$
2. $x < -2$
3. $1 < x < \frac{5}{2}$
4. $4 < x < \frac{24}{5}$
5. $-3 < x < 3$ and $x > 5$
6. $1 < x < 2$ and $x < 0$
7. $x > 3$ and $1 < x < 2$
8. $2 < x < 8$
9. $-1 < x < 1$ and $2 < x < 3$
10. $\frac{1}{6} < x < 1$
11. $x > -1$ and $-5 < x < -3$
12. $-4 < x < -\frac{2}{3}$
13. $-1 < x < 2$
14. $3 < x < 4$ and $-4 < x < -1$
15. $4 < x < 5$ and $-1 < x < 1$
16. $2 < x < 3$

Exercise 12D – p.148

1 **a** $p \geqslant 9$ and $p \leqslant 1$

b $p \geqslant 5$ and $p \leqslant 1$

2 $-2 < a < 6$

3 $-1 < p < \frac{7}{2}$

4 $-\frac{1}{7} \leqslant f(x) \leqslant 1$

5 $f(x) \geqslant 2$ and $f(x) \leqslant -2$

6 all values

7 $-\frac{1}{2} \leqslant f(x) \leqslant \frac{1}{2}$

8 $f(x) \geqslant 3 + \sqrt{8}$ and $f(x) \leqslant 3 - \sqrt{8}$

9 $f(x) \geqslant 2$ and $f(x) \leqslant -2$

10 $0 < k < \frac{4}{9}$

12 $-\frac{1}{2} \leqslant f(x) \leqslant 1$

Mixed Exercise 12 – p.149

1 $x < 1$

2 $x > \frac{3}{2}$

3 $x > \frac{3}{2}$

4 $x > 3$ and $x < -2$

5 $-\frac{2}{3} < x < \frac{3}{2}$

6 $-\sqrt{13} < x < \sqrt{13}$

7 $x > 3 + \sqrt{2}$ and $x < 3 - \sqrt{2}$

8 $-2 < x < 7$

9 all values of x

10 $x > 6$ and $x < 1$

11 $1 < x < 4$

12 $-1 < x < 1$

13 $1 < x < 3$

14 $x > -1$

15 $x \geqslant 1$ and $x \leqslant -3$

16 $-2 < x \leqslant -1$ and $1 \leqslant x < 2$
(Note: $x \neq 2$ or -2)

18 $k \leqslant 3$ and $k \geqslant 4$

19 $\frac{2}{3} \leqslant f(x) \leqslant 2$

Chapter 13

Exercise 13A – p.153

1 4; 4
2 1; 1
3 $3x^2$; 3
4 $2x$; 4
5 $2x - 1$; 1

Exercise 13B – p.154

1 $-\frac{2}{x^3}$
2 $-\frac{2}{x^2}$

Exercise 13C – p.155

1 $5x^4$
2 $-3x^{-4}$
3 $\frac{4}{3}x^{1/3}$
4 $-\frac{1}{x^2}$
5 $10x^9$
6 $-\frac{2}{x^3}$
7 $\frac{3}{2}\sqrt{x}$
8 $-\frac{1}{2}x^{-3/2}$
9 $-\frac{4}{x^5}$
10 $\frac{1}{3}x^{-2/3}$
11 $-\frac{1}{4}x^{-5/4}$
12 1
13 $\frac{7}{2}\sqrt{x^5}$
14 $-\frac{7}{x^8}$
15 $\frac{1}{7}x^{-6/7}$
16 $3x^2$

Exercise 13D – p.156

1 $3x^2 - 2x + 5$
2 $6x + \frac{4}{x^2}$
3 $\frac{1}{2\sqrt{x}} - \frac{1}{2x\sqrt{x}}$
4 $8x^3 - 8x$
5 $3x^2 - 4x - 8$
6 $2x + \frac{5}{2\sqrt{x}}$
7 $-\frac{3}{4}x^{-7/4} - \frac{3}{4}x^{-1/4} + 1$
8 $9x^2 - 8x + 9$
9 $\frac{3}{2}x^{1/2} - \frac{1}{2}x^{-1/2} - \frac{1}{2}x^{-3/2}$
10 $\frac{1}{2\sqrt{x}} + \frac{3\sqrt{x}}{2}$
11 $-\frac{2}{x^3} + \frac{3}{x^4}$
12 $\frac{-1}{2\sqrt{x^3}} + \frac{2}{x^2}$
13 $-\frac{1}{2}x^{-3/2} + \frac{9}{2}x^{1/2}$
14 $\frac{1}{4}x^{-3/4} - \frac{1}{5}x^{-4/5}$
15 $-\frac{12}{x^4} + \frac{3x^2}{4}$
16 $-\frac{4}{x^2} - \frac{10}{x^3} + \frac{18}{x^4}$
17 $\frac{3}{2\sqrt{x}} - 3$
18 $1 + 2x^{-2} + 9x^{-4}$
19 $\frac{3}{2}\sqrt{x} - \frac{5}{2}x\sqrt{x}$
20 $\frac{-3\sqrt{x}}{2x^3} + \frac{3\sqrt{x}}{2}$

Exercise 13E – p.157

1 $\frac{dy}{dx} = 2x + 2$
2 $\frac{dz}{dx} = -4x^{-3} + x^{-2}$
3 $\frac{dy}{dx} = 6x + 11$
4 $\frac{dy}{dz} = 2z - 8$
5 $\frac{ds}{dt} = -\frac{3}{2t^4}$
6 $\frac{ds}{dt} = \frac{1}{2}$
7 $\frac{dy}{dx} = 1 - \frac{1}{x^2}$
8 $\frac{dy}{dz} = \frac{5z^2 - 1}{2\sqrt{z}}$
9 $\frac{dy}{dx} = 18x^2 - 8$
10 $\frac{ds}{dt} = 2t$
11 $\frac{ds}{dt} = 1 - \frac{7}{t^2}$
12 $\frac{dy}{dx} = -\frac{3\sqrt{x} + 28}{2x^3}$

Exercise 13F – p.158

1 2; $-\frac{1}{2}$
2 $-\frac{1}{3}$; 3
3 $\frac{1}{4}$; -4
4 6; $-\frac{1}{6}$
5 1; -1
6 5; $-\frac{1}{5}$
7 11; $-\frac{1}{11}$
8 -11; $\frac{1}{11}$
9 4; $-\frac{1}{4}$
10 $\frac{4}{27}$; $-\frac{27}{4}$
11 $\frac{5}{4}$; $-\frac{4}{5}$
12 2; $-\frac{1}{2}$
13 (2, 2) and (−2, 4)
14 (1, 0) and $(-\frac{1}{3}, \frac{4}{27})$
15 (3, 0) and (−3, 18)
16 (−1, −2) and (1, 2)
17 (1, −16)
18 $(-2, \frac{1}{4})$
19 (0, −5)
20 (1, −2) and (−1, 2)

Mixed Exercise 13 – p.159

1 $6x + 1$
2 a $-3x^{-4} - 3x^2$
b $\frac{1}{2}x^{-1/2} + \frac{1}{2}x^{-3/2}$
c $-\frac{2}{x^3} - \frac{6}{x^4}$
3 a $\frac{dy}{dx} = \frac{3}{2}x^{1/2} - \frac{2}{3}x^{-1/3} - \frac{1}{3}x^{-4/3}$
b $\frac{dy}{dx} = \frac{1}{2\sqrt{x}} + \frac{1}{x^2} - \frac{3}{x^4}$
c $-\frac{3}{4x^{7/4}} + \frac{1}{4x^{5/4}}$
4 a 5 b 5 c 17
5 a 5 b 6
6 a 1 b 7 and −7
7 a −3 b $(-\frac{1}{2}, 0)$ and (2, 0)
c −5 and 5
8 a (1, 10) and (−1, 6)
b $(\frac{1}{3}, 7\frac{7}{9})$ and $(-\frac{1}{3}, 8\frac{2}{9})$
9 a $4x^3 - 2x$ b $6(3x + 4)$
c $\frac{x + 3}{2x\sqrt{x}}$
10 $1 - \sqrt{2}$
11 (−2, 4)
12 $\frac{1}{5}$ and $-\frac{1}{5}$
13 (1, 9) and (3, 11)
14 1
15 b, d

Chapter 14

Exercise 14A – p.161

1 **a** $y = 2x - 5$
b $2y + x + 5 = 0$
2 **a** $y = 4x - 2$
b $4y + x + 8 = 0$
3 **a** $y + x + 2 = 0$ **b** $y = x$
4 **a** $y = 5$ **b** $x = 0$
5 **a** $y + x = 3$ **b** $y = x - 1$
6 **a** $y = 19x + 26$
b $19y + x + 230 = 0$
7 $4y + x + 12 = 0$
8 $y = 7x - 29$
9 $y + x = 1, 2y = 2x - 3; (\frac{5}{4}, -\frac{1}{4})$
10 $4y - x + 1 = 0, 4y + x - 5 = 0$
11 $y = 5x - 1, 3y + 9x + 19 = 0$
12 $y = 7x - 4, y + 5x + 28 = 0$
13 $(2, 8), y = 8x - 8$
14 $(\frac{1}{2}, -\frac{1}{4})$
15 $y + x + 1 = 0$
16 $2y = x + 2$
17 $k = -\frac{7}{2}$
18 $8y + 121 = 0$
19 $(1, -1)$
20 $p = 12, q = 8; (-2, 24)$

Exercise 14B – p.163

1 $x = 0$
2 $x = \frac{3}{4}$
3 $x = 0, x = \frac{8}{3}$
4 $x = \pm\frac{1}{2}$
5 $x = 0, x = \frac{4}{3}$
6 $x = \pm 1$
7 $x = 4$
8 $x = \pm 3$
9 $x = 1, x = -\frac{4}{3}$
10 $x = \pm\frac{5}{9}\sqrt{3}$
11 $x = 1, x = -4$
12 $x = \pm\frac{2}{3}\sqrt{3}$
13 $(3, 3), (-3, -3)$
14 $(1, -7), (\frac{1}{3}, -\frac{185}{27})$
15 $(\frac{1}{2}, -\frac{25}{4})$
16 $(\frac{1}{3}, \frac{-2}{9}\sqrt{3})$
17 $(1, 2)$
18 $(4, 10), (-4, 6)$

Exercise 14C – p.168

1 $(1, 1)$ max
2 $(-1, -2)$ min; $(1, 2)$ max
3 $(3, 6)$ min; $(-3, -6)$ max
4 $(0, 0)$ max; $(\frac{10}{3}, -\frac{500}{27})$ min
5 $(0, 0)$ min
6 $(1, \frac{3}{2})$ min
7 $(-1, 1)$ max; $(0, 0)$ min; $(1, 1)$ max
8 $(0, 0)$ min
9 $(\frac{5}{4}, -\frac{49}{8})$ min
10 $(-1, 4)$ max; $(1, -4)$ min
11 $(-2, -16)$ min; $(0, 0)$ max; $(2, -16)$ min
12 $(-2, 8)$ min; $(2, 8)$ min
13 -2 max; 2 min
14 $2\frac{3}{4}$ min
15 0 inflex; 27 max
16 8 inflex
17 7 inflex
18 $-\frac{5}{16}$ min, 0 max, -2 min
19 $5x^4 + 3x^2 + 40 = 0$ has no real roots so there is no stationary point; $20x^3 + 6x = 0$ when $x = 0$ so there is a point of inflexion at $(0, -3)$; $y = 4x - 3$.

Exercise 14D – p.170

1 $800\,\text{m}^2$; $20\,\text{m} \times 40\,\text{m}$
2 $20\,\text{cm} \times 20\,\text{cm} \times 10\,\text{cm}$
3 $r = \sqrt{9 - h^2}$; $12\pi\sqrt{3}\,\text{cm}^3$
4 5 cm square
5 $\sqrt{35}$ cm square
6 $a = 1, b = -2, c = 3$
7 $p = q = 1, r = 2; (-1, 0)$
8 $5y = x^2 + 4x + 9$

Mixed Exercise 14 – p.171

1 $11; y = 11x - 6; (\frac{2}{3}, \frac{4}{3})$
2 $2y = x - 1; (-\frac{3}{2}, -\frac{5}{4})$
3 $(2, 14), (-2, -14)$
4 $y = 3x + 6, y = 3x + 2$
5 $x + 2y + 9 = 0$
6 $(-2, 16)$ max, $(2, -16)$ min

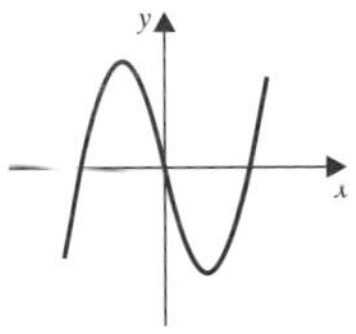

7 min 2, max -2;

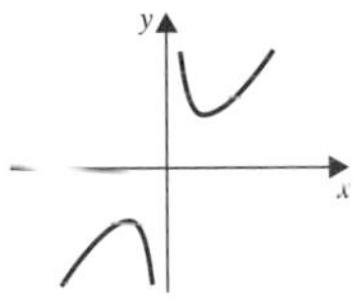

9 $h = \frac{1}{2}(7 - 2r - \pi r)$
10 $4\,\text{m} \times 4\,\text{m} \times 2\,\text{m}$
11 $12.5\,\text{cm}^2$

Chapter 15

Exercise 15A – p.175

1 $\frac{1}{2}\sqrt{3}$
2 0
3 $-\frac{1}{2}\sqrt{3}$
4 $\frac{1}{2}$
5 $\frac{1}{2}\pi, \frac{5}{2}\pi, \frac{9}{2}\pi$
6 $-\frac{1}{2}\pi, -\frac{5}{2}\pi$
7 $\sin 55°$
8 $-\sin 70°$
9 $-\sin 60°$
10 $-\sin\frac{1}{6}\pi$
11

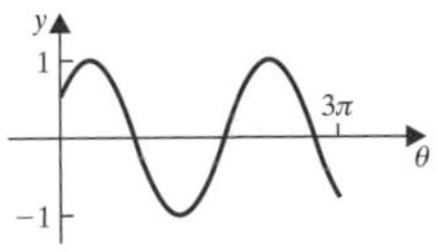

12

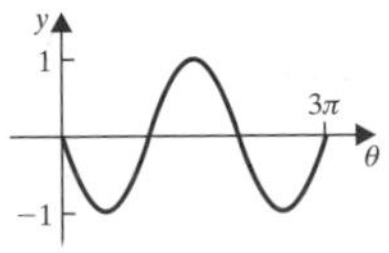

13

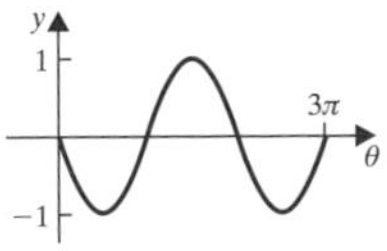

14

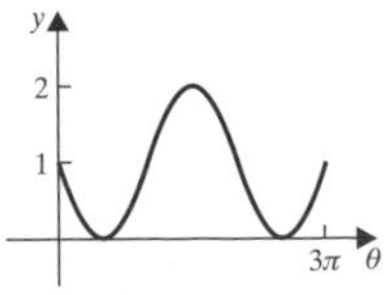

15

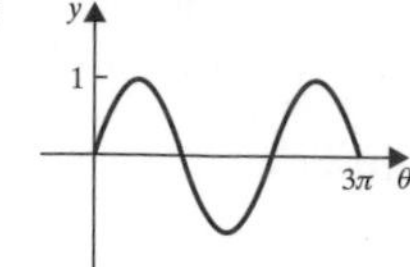

16

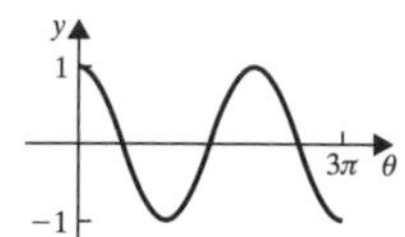

17 The curve $y = a\sin\theta$ is a one-way stretch of the curve $y = \sin\theta$ by a factor a parallel to the y-axis.

19 The curve $y = \sin 3\theta$ is a one-way shrinkage of the curve $y = \sin\theta$ by a factor $\frac{1}{3}$ parallel to the x-axis.

20 a

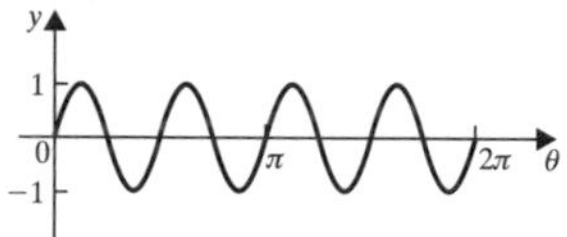

b

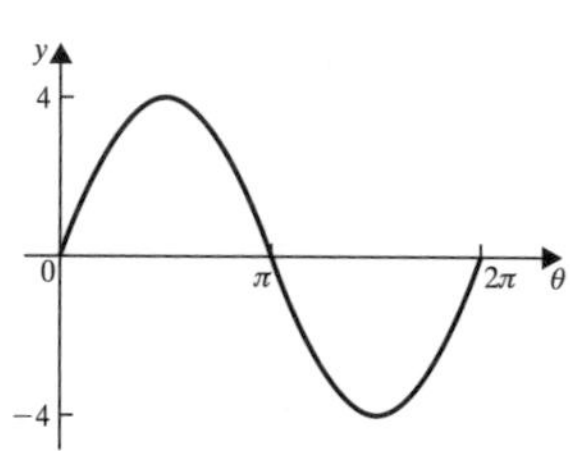

Exercise 15B – p.177

1 a $-\cos 57°$ **b** $-\cos 70°$
c $\cos 20°$ **d** $-\cos 26°$

2 a $-\dfrac{\sqrt{3}}{2}$ **b** 0
c $-\dfrac{1}{\sqrt{2}}$ **d** 1

3 a

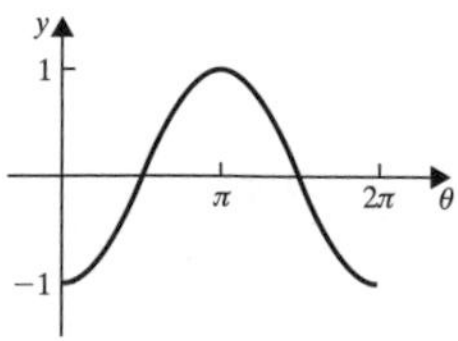

b

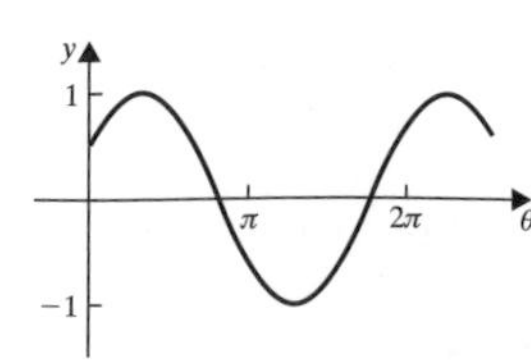

c

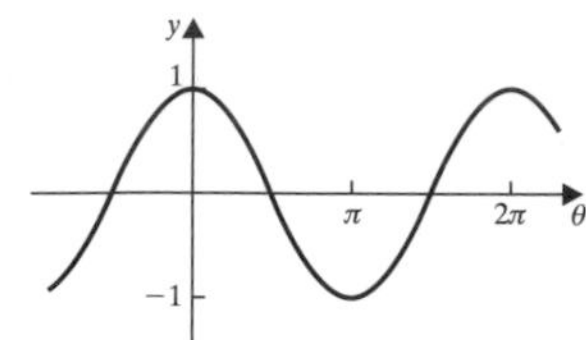

4

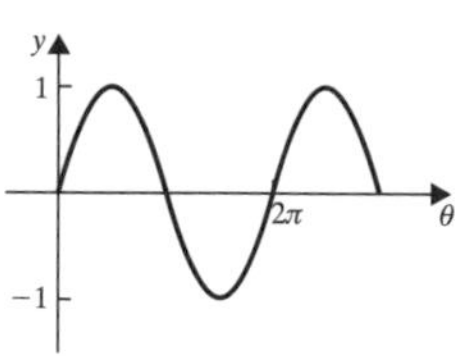

$\sin\theta = \cos(\theta - \frac{1}{2}\pi)$
$\cos\theta = -\sin(\theta - \frac{1}{2}\pi)$

5

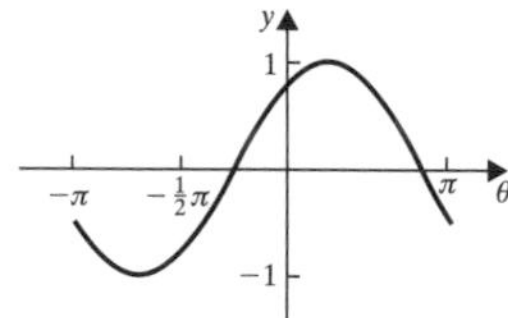

a $\theta = \frac{1}{4}\pi$ **b** $\theta = -\frac{3}{4}\pi$
c $\theta = -\frac{1}{4}\pi$ and $\frac{3}{4}\pi$

6

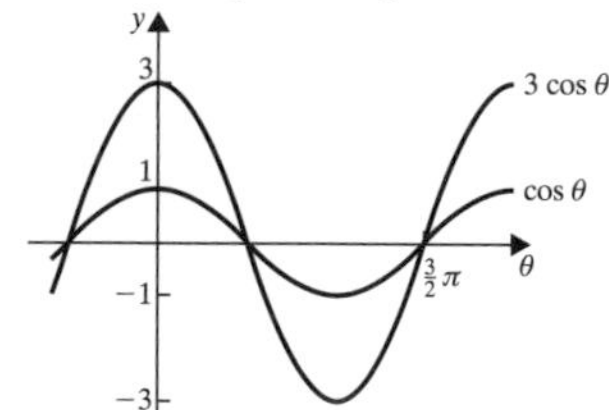

7

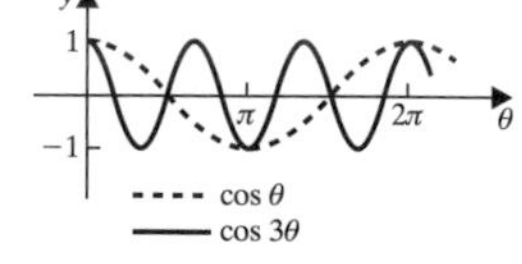

8

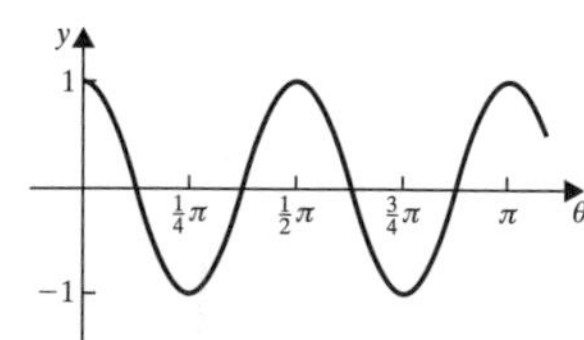

$\frac{1}{8}\pi,\ \frac{3}{8}\pi,\ \frac{5}{8}\pi,\ \frac{7}{8}\pi$

Exercise 15C – p.179

1 a 1 **b** $-\sqrt{3}$
c $\sqrt{3}$ **d** -1

2 a $\tan 40°$ **b** $-\tan\frac{2}{7}\pi$
c $-\tan 50°$ **d** $\tan\frac{2}{5}\pi$

3 a $\frac{1}{4}\pi,\ \frac{5}{4}\pi$ **b** $\frac{3}{4}\pi,\ \frac{7}{4}\pi$
c $0,\ \pi,\ 2\pi$ **d** $\frac{1}{2}\pi,\ \frac{3}{2}\pi$

Exercise 15D – p.180

1 a 0.412 rad, 2.73 rad, −5.87 rad, −3.55 rad
b $-\frac{4}{3}\pi,\ -\frac{2}{3}\pi,\ \frac{2}{3}\pi,\ \frac{4}{3}\pi$
c 0.876 rad, 4.02 rad, −2.27 rad, −5.41 rad

2 a 141.3°, 321.3°, 501.3°, 681.3°
b 191.5°, 348.5°, 551.5°, 708.5°
c 84.3°, 275.7°, 444.3°, 635.7°

3 a 36.9°
b −36.9°
c 0.464 rad

4 $0,\ \pi,\ 2\pi$

5 11.8°, 78.2°, 191.8°, 258.2°

6 $\frac{1}{3}\pi,\ \pi,\ \frac{5}{3}\pi$

Exercise 15E – p.182

1 a 60°, 300°
b 59.0°, 239.0°
c 41.8°, 138.2°

2 a −140.2°, 39.8°
b −131.8°, 131.8°
c −150°, −30°

3 $-\frac{1}{2}\pi,\ \frac{1}{2}\pi$

4 a 1
b $-\sqrt{2}$
c −2

5

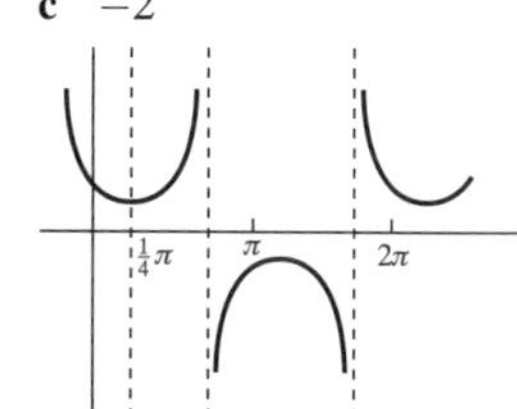

; $\frac{1}{4}\pi$

6

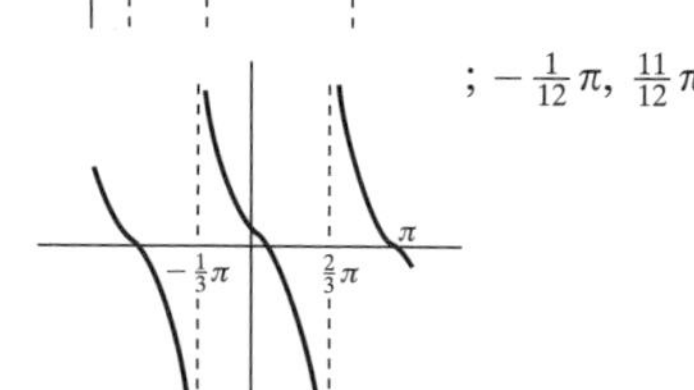

; $-\frac{1}{12}\pi,\ \frac{11}{12}\pi$

Exercise 15F – p.183

1 1.0299 rad

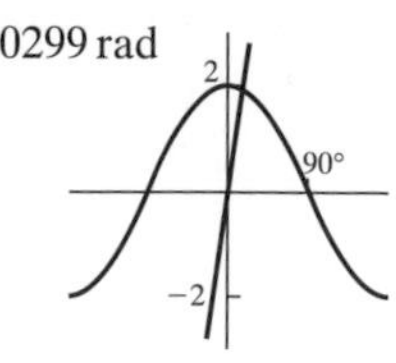

2 a −1.895 49, 0, 1.895 49
b 0, 0.8767 rad
c 1.2834 rad

Chapter 16

Exercise 16A – p.186

		$\sin\theta$	$\cos\theta$	$\tan\theta$
1	**a**	$-\frac{12}{13}$	$-\frac{5}{13}$	$\frac{12}{5}$
	b	$\frac{3}{5}$	$-\frac{4}{5}$	$-\frac{3}{4}$
	c	$\frac{7}{25}$	$\frac{24}{25}$	$\frac{7}{24}$
	d	0	± 1	0

2 $\tan^4 A$
3 1
4 $\sec\theta\operatorname{cosec}\theta$
5 $\sec^2\theta$
6 $\tan\theta$
7 $\sin^3\theta$
8 $x^2 - y^2 = 16$
9 $b^2x^2 - a^2y^2 = a^2b^2$
10 $y^2(4+x^2) = 36$
11 $(1-x)^2 + (y-1)^2 = 1$
12 $y^2(x^2 - 4x + 5) = 4$
13 $x^2(b^2 - y^2) = a^2b^2$

Exercise 16B – p.188

1 60°, 120°
2 90°, 270°
3 120°, 300°
4 194.5°, 345.5°
5 120°, 240°
6 45°, 225°
7 30°, 150°, 210°, 330°
8 57.7°, 122.3°, 237.7°, 302.3°
9 190.1°, 349.9°
10 38.2°, 141.8°
11 30°, 150°
12 30°, 150°
13 ± 0.723 rad
14 -0.315 rad, -2.83 rad
15 $-\frac{3}{4}\pi$, -0.245 rad, $\frac{1}{4}\pi$, 2.90 rad
16 $-\pi, -\frac{1}{3}\pi, \frac{1}{3}\pi, \pi$
17 $-\pi, -\frac{2}{3}\pi, 0, \frac{2}{3}\pi, \pi$
18 $-\pi, -\frac{1}{6}\pi, 0, \frac{1}{6}\pi, \pi$

Exercise 16C – p.189

1 22.5°, 112.5°
2 40°, 80°, 160°
3 none
4 67.5°, 157.5°
5 none
6 60°
7 −149.5°, −59.5°, 30.5°, 120.5°
8 −105.2°, −74.8°, 14.8°, 45.2°, 134.8°, 165.2°
9 ±63.6°
10 $\frac{1}{6}\pi, \frac{5}{12}\pi, \frac{2}{3}\pi, \frac{11}{12}\pi$,
11 $\frac{1}{15}\pi, \frac{1}{3}\pi, \frac{7}{15}\pi, \frac{11}{15}\pi, \frac{13}{15}\pi$
12 none
13 $\frac{1}{12}\pi$
14 $\frac{1}{24}\pi, \frac{13}{24}\pi$

Mixed Exercise 16 – p.190

1 $x^2 + \dfrac{1}{y^2} = 1$
2 $\sin\beta = \pm\dfrac{\sqrt{3}}{2}$, $\tan\beta = \pm\sqrt{3}$
3 $\dfrac{2}{\sin^2\theta}$; $\frac{1}{4}\pi, \frac{3}{4}\pi, \frac{5}{4}\pi, \frac{7}{4}\pi$
4 60°, 109.5°, 250.5°, 300°
6 $-\frac{29}{36}\pi, -\frac{17}{36}\pi, -\frac{5}{36}\pi, \frac{7}{36}\pi, \frac{19}{36}\pi, \frac{31}{36}\pi$
7 **a** $(x-2)^2 + (y+1)^2 = 1$
b $(x+3)^2 = 1 + (2-y)^2$
8 22.5°, 67.5°, 112.5°, 157.5°
10 $\sin^2 A$
11 ±70.5°, ± 180°
12 $\sec^2\theta\tan^2\theta$
13 10°, 110°, 130°
14 −90°, 30°, 150°
15 $\frac{1}{3}\pi, \frac{5}{6}\pi$

Summary C

Multiple Choice Exercise C – p.193

1 C
2 B
3 D
4 C
5 E
6 B
7 B
8 B, C
9 D
10 A, F
11 C
12 D
13 B
14 A, B, C, D
15 B
16 C
17 C
18 A, B
19 B, C
20 B, C
21 B
22 A, C
23 B
24 B, C
25 B, C
26 C
27 F
28 T
29 F
30 T
31 F
32 F

Examination Questions C – p.196

1 0, −1
2 **a** $-4 \leqslant x \leqslant 2$
b

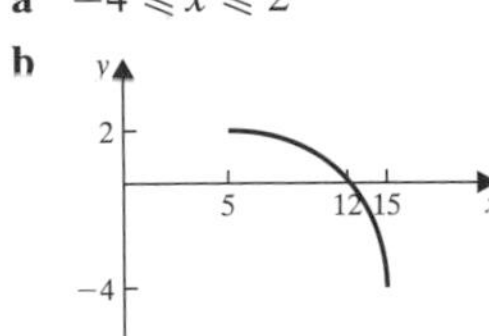

c $-4 \leqslant f'(x) \leqslant 2$
3 $y = 2x + 6$
4 $fg: x \to \dfrac{3}{x+4}, x \in \mathbb{R}, x \geqslant 0$,
$\dfrac{6-3x}{x-1}$
5 **a** **i** no **ii** yes **iii** yes
b

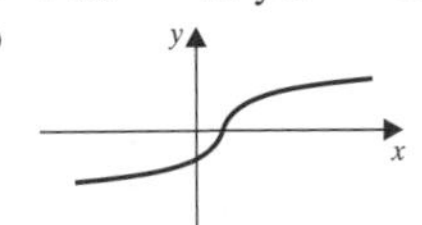

6 −72.6°, −17.4°, 107.4°, 162.6°
7

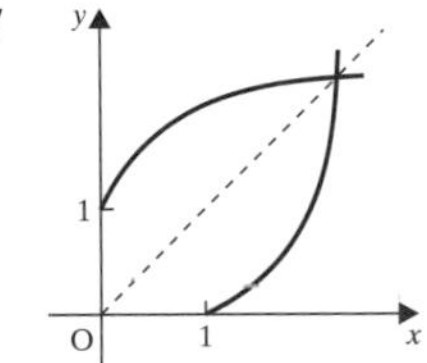

$(x-1)^2, x \geqslant 1$; $f(x)$ and $f^{-1}(x)$ have equal values when $y = x$.
8 **a** $-\sqrt{7}/3$ **b** $\pm\sqrt{14}/2$
9 60°, 300°

10 198°, 342°
11 **a** 31.7°, 121.7°, 211.7°, 301.7°
b 120°, 240°
12 $\frac{6}{7}$
13 **a** $f(x) \geqslant 0, g(x) \geqslant 3$
b Range $g(x) \geqslant 3$ but domain of $f(x), x \geqslant 5$
c **i** $4x - 17$ **ii** $x \geqslant 5, gf(x) \geqslant 3$
d $g(x) \to$ two values of x
e $x^2/4 + 5,\ x \geqslant 0,\ f^{-1}(x) \geqslant 5$
14 **a** $3x^2 + 6x - 9$
b $(-3, 31), (1, -1)$
c

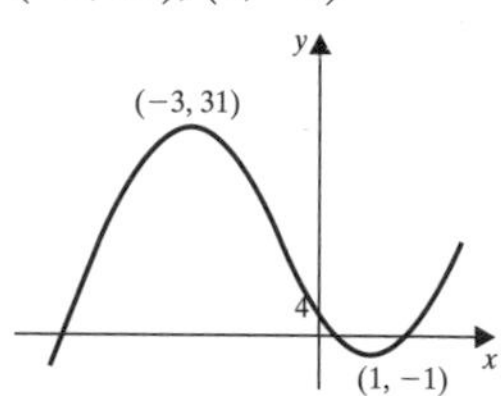

d $-1 < k < 31$
15 **i**

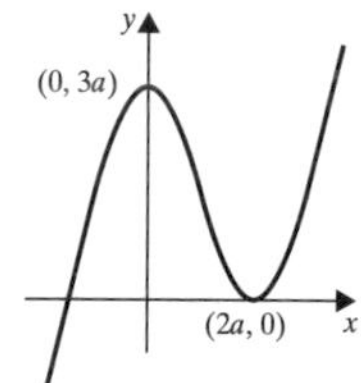

ii

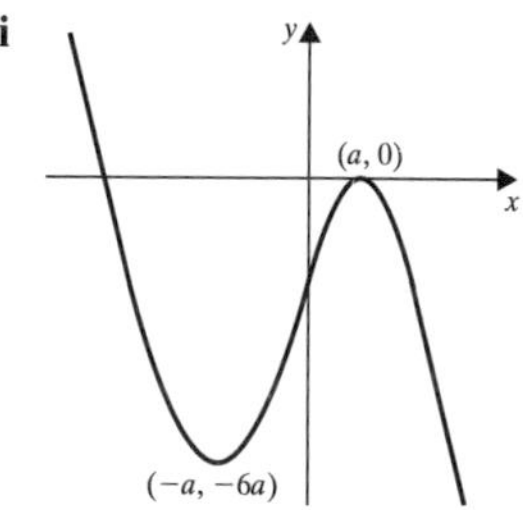

16 **b** $36\,\text{m}^3$
17 **a** $(1 - x)/x$
b **i** translation by 1 unit left
ii translation by 1 unit down
c $x/(1 + x)$ **d** $x - 1$
18 **a** $2x > 21$ **b** $x(x - 5) < 104$
c $10.5 < x < 13$
19 $(2, \frac{1}{4})$ max
i $(2, 5\frac{1}{4})$ **ii** $(3, \frac{1}{2})$

20 −108°, 108°
21 48.6° (1 dp) 131.4° (1 dp)
210° (exact) 330° (exact)
22 **a** 4.176, 8.074
b Feb and Oct
c $L = 6.125 - 2.25\cos(\pi t/6 + \pi/2)$
23 **i**

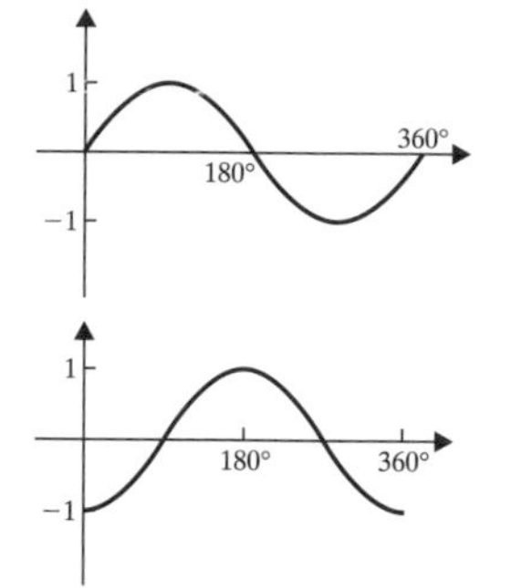

ii P : B, Q : C **iii** 120°, 240°

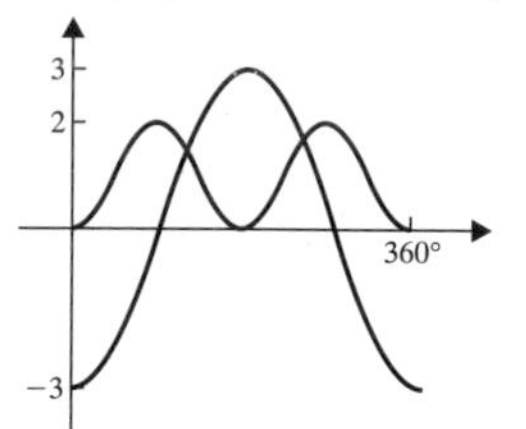

24 **a** **i** $2 - \sqrt{3}$ **ii** $-2 - \sqrt{3}$
b 0.73, 2.41, 4.71
25 **b** $0, \pm 2\sqrt{2}$ **c** 3 units
26 **b** 25 **c** 2 rad **d** $625\,\text{cm}^2$
27 **a** $3x^2 - 1$ **b** 6 **c** 2
d $y = 2x + 8$ **e** $(1, 6)$
f $y = -\frac{1}{2}x + \frac{13}{2}$
28 $-4 < x < 4$
29 **a** −4 **b** 6 **c** 3.50, 5.92
30 **a** $x \geqslant 3$ **b** $0 < k < 4$
c f is one to one, g is not
d $0 \leqslant x \leqslant 4, 0 \leqslant f^{-1}(x) \leqslant 2$
31 $x < -3, x > 4$
32 30°, 150°, 194°, 346°

33 **a** e.g. $x^4 + 1$
b **i** two turning points: looks like a cubic
ii $(0, -4.5), (1, 0), (3, 0)$; $p = 1, q = 3, a = \frac{1}{2}$
iii $t = 2$: model $s = \frac{1}{2}$, graph $s \simeq 0.8$
$t = 4$: model $s = 1\frac{1}{2}$, graph $s \simeq 1$
$t = 5$: model $s = 8$, graph $s \simeq 4$
Not good, very poor for $t > 4$
iv $\frac{1}{10}$
34 **a** $\pi - \alpha$ **b** $3\pi + \alpha, 4\pi - \alpha$
35 **c** $r = 1, \theta = 2$ **d** D
36 **a**
$0 < f(x) \leqslant 4$
b $f^{-1}(x) = \dfrac{8}{x} - 2$ **c** 2
37 $(2, 4)$ min, $(-2, -4)$ max; $x < -2, x > 2$
38 **a** $\frac{1}{3}, -\frac{1}{2}$
b 240°, 289.5°, 430.5°, 480°
39 $3x^2 - 8x + 5$
a 4 **b** 8 **c** $y = 8x - 20$
d $8y + x - 35 = 0;\ 0, \frac{8}{3}$
40 $8.4 < x < 9$
41 45.6°, 91.9°, 268.1°, 314.4°
42 $2(x + \frac{5}{4})^2 + \frac{7}{8}; (-\frac{5}{4}, \frac{7}{8})$
43 **a** $f(x) \geqslant -16$ **b** −2
c $4 - \sqrt{x + 16}$
44 $\cos R = \dfrac{3 + k^2}{4k}$; $1.5 < k < 2$
45 **b** 10 **c** 5
46 **a** $(10x + 1)/(x + 5), x \neq -5$
b $f^{-1}(x) = (2x + 1)/(x - 3), x \neq 3$
47 $8x^3 + 24x^2 + 32x + 14$
48 **a** 0 **b** $x < -3, -1 < x < 2$
49 Least −2, greatest $\frac{1}{4}$

Chapter 17

Exercise 17A – p.209

1 **a** 7.39 **b** 0.368
c 4.48 **d** 0.741
2 **a** $2e^x$
b $2x - e^x$ **c** e^x
3 $e^2 - 2$
4 $2 + 2e$
5 1
6 0
8 **a** $y = 4(2)^{x/6}$
c Any plausible reasons, e.g. (i) inherent errors in measuring instruments (ii) growth rate reduces for $t > 30$ possibly because the dish was left in an unsuitable place after this time.

Exercise 17B – p.211

1 **a** 3.87 **b** 1 **c** 0

2 **a** $2\ln x - \ln(x+1)$
b $\ln(a+b) + \ln(a-b)$
c $\ln\cos x - \ln\sin x$
d $2\ln\sin x$

3 **a** $\ln\cot x$ **b** $\ln ex$
c $\ln(x-1)^{2/3}$

4 **a** $\frac{2}{3}\ln x$ **b** $5\ln x$

5 **a** 2.10 **b** 0
c 1.05 **d** 0

6 **a** $x = 1$ **b** $x = 3$
c $x = 0$

7 **a** Value when new
b £8880, any sensible reason, e.g. depends on what happens to the car in two years.
c $y = Ae^{(-t\ln 1.3)}$

Exercise 17C – p.213

1 **a** $\dfrac{3}{x}$ **b** $\dfrac{1}{x}$ **c** $-\dfrac{2}{x}$
d $-\dfrac{1}{2x}$ **e** $-\dfrac{5}{x}$ **f** $\dfrac{1}{2x}$
g $-\dfrac{3}{2x}$ **h** $\dfrac{5}{2x}$

2 **a** $(1, -1)$ **b** $(2^{1/3}, \{2 - 2\ln 2\})$
c $(4, \{\ln 4 - 2\})$

4 **a**

b $0.3 < x < 2.1$; $f(0.25) = 0.06$ and $f(0.3) = -0.27$, etc.
c $(1, -1.8)$
$(1, 1 - 2\ln 4)$; difficult to be sure that $x = 1$ because the curve is very flat in the region of the minimum point.

Chapter 18

Exercise 18A – p.216

1 $a = 2, b = -3$

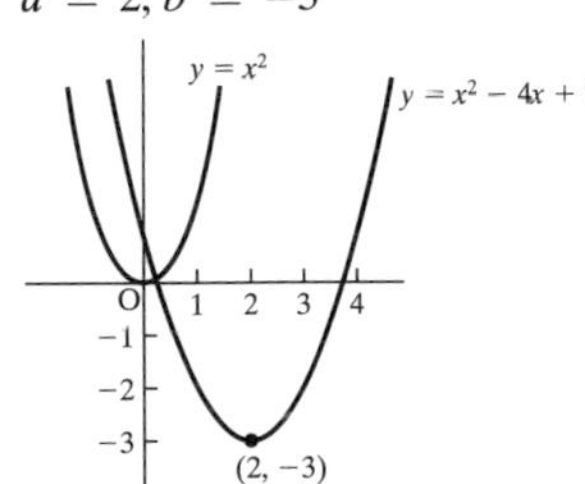

2 **a**

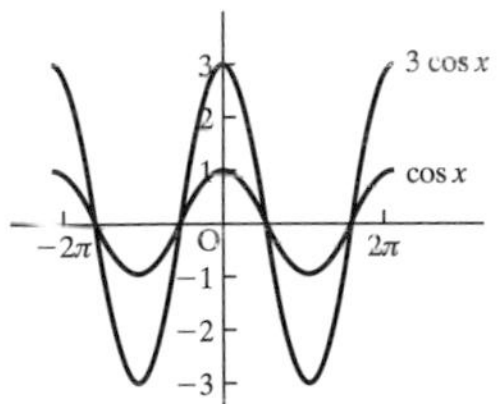

b

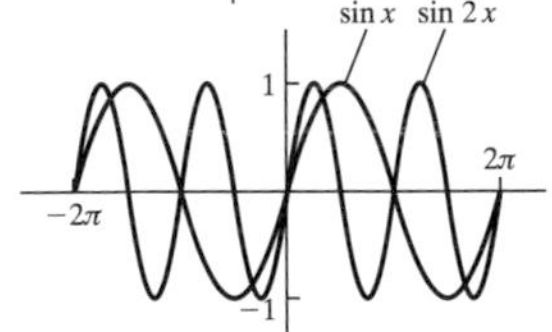

c

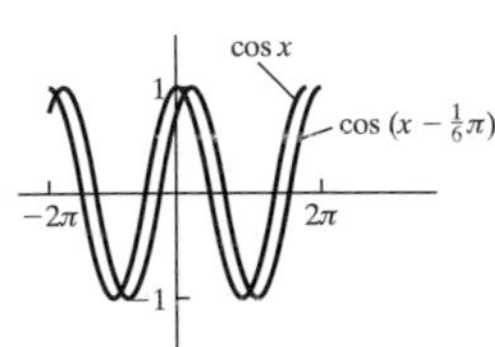

d

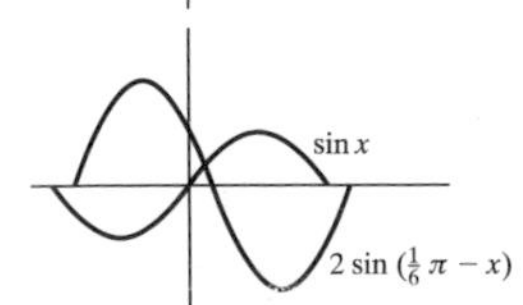

3

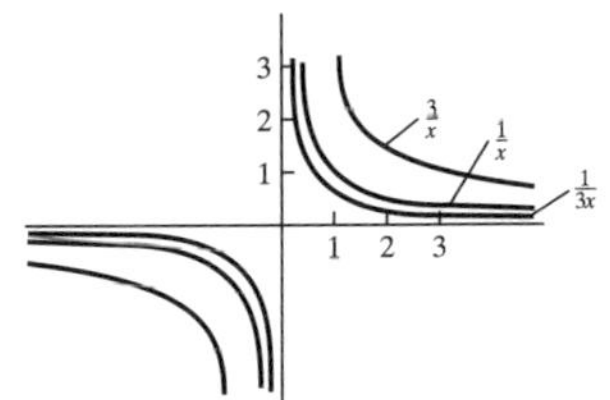

4

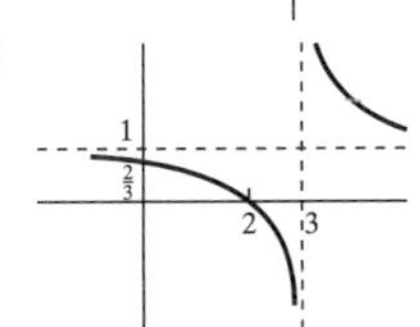

5

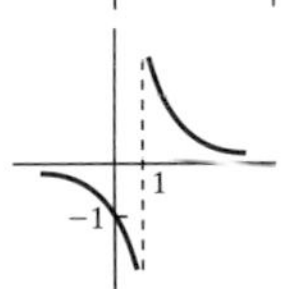

6

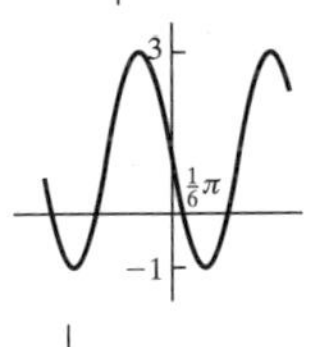

7

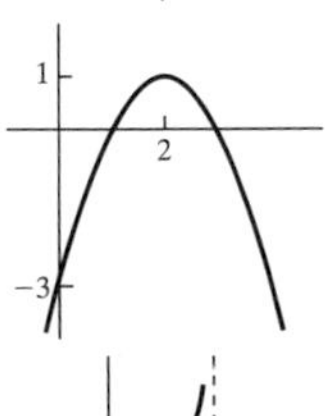

8

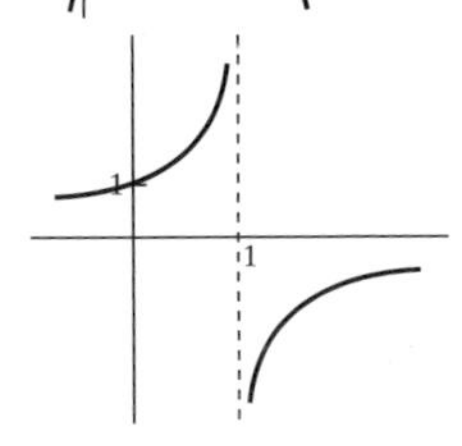

9

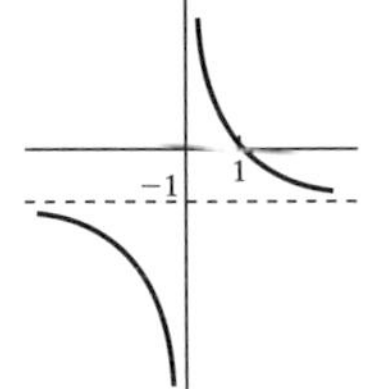

10

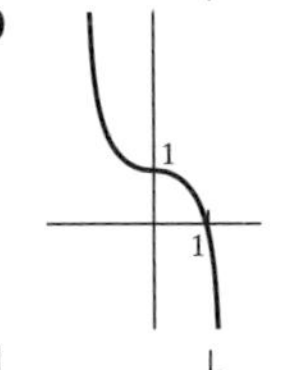

11

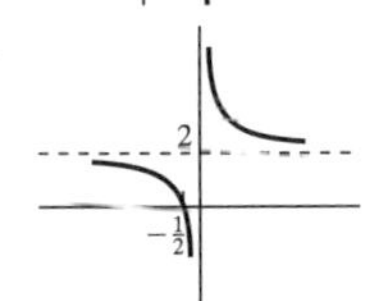

12

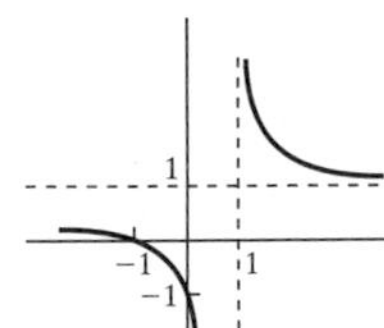

13

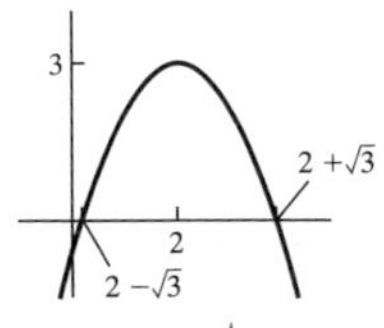

14

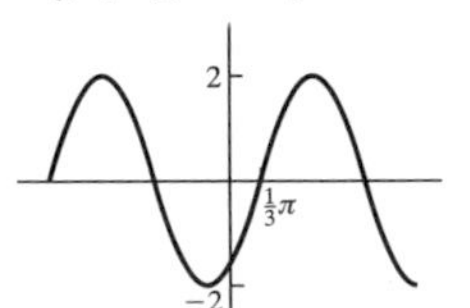

15

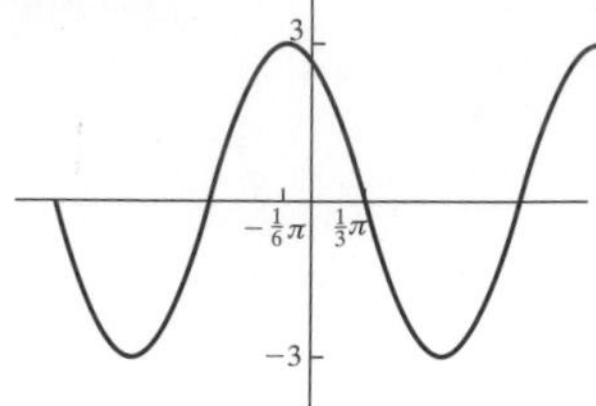

16

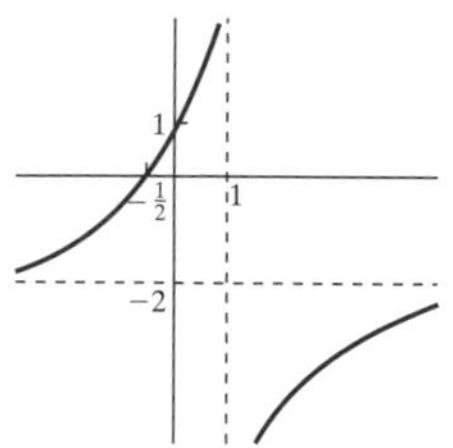

17

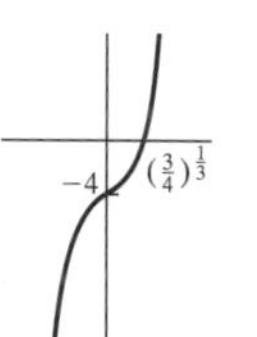

18

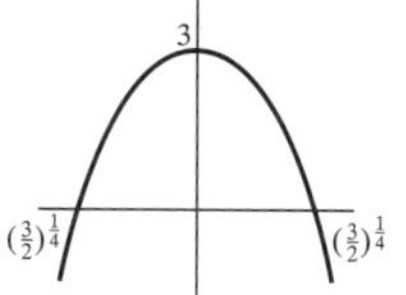

19

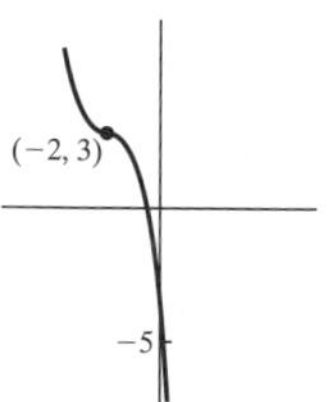

20

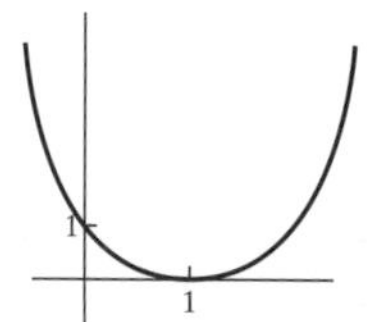

Exercise 18B – p.218

1 **a** even, periodic
b odd, periodic
e odd, periodic
h odd

2

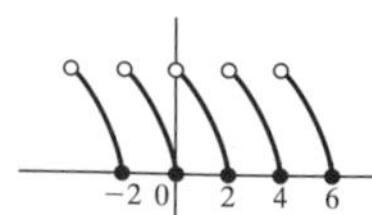

3

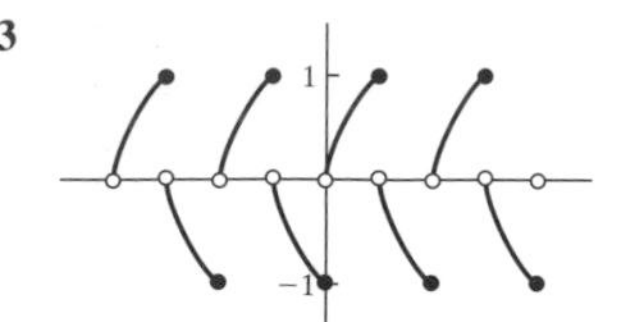

4

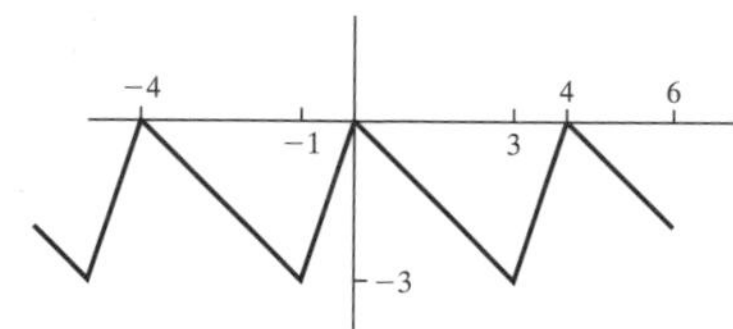

Exercise 18C – p.220

1

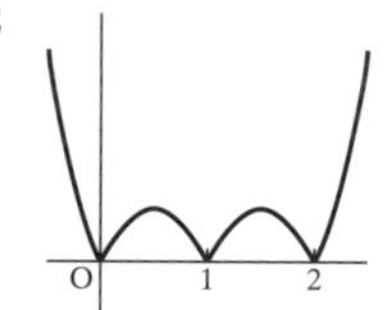

2

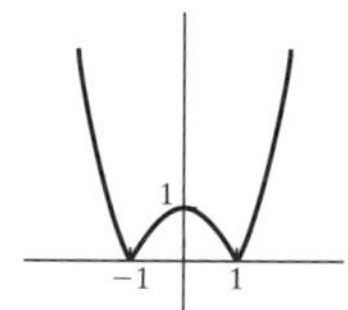

3

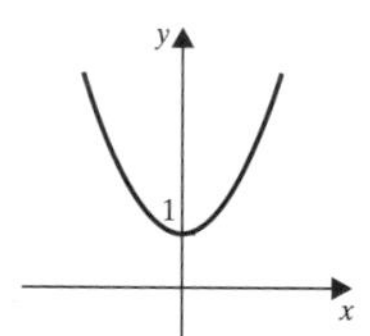

4

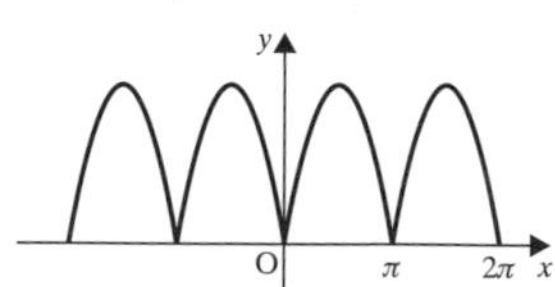

5

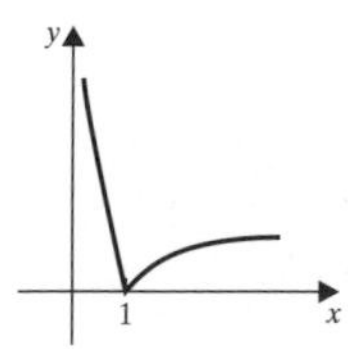

6

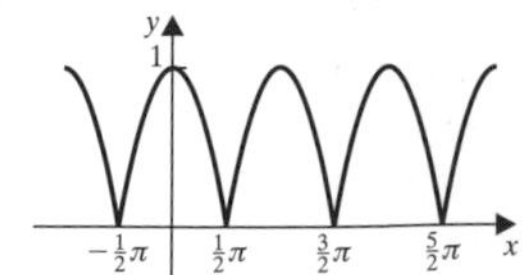

7

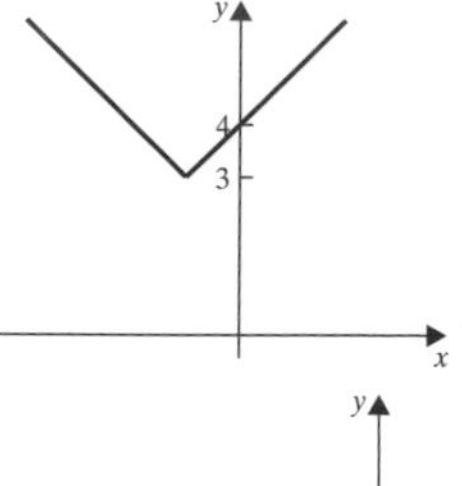

8

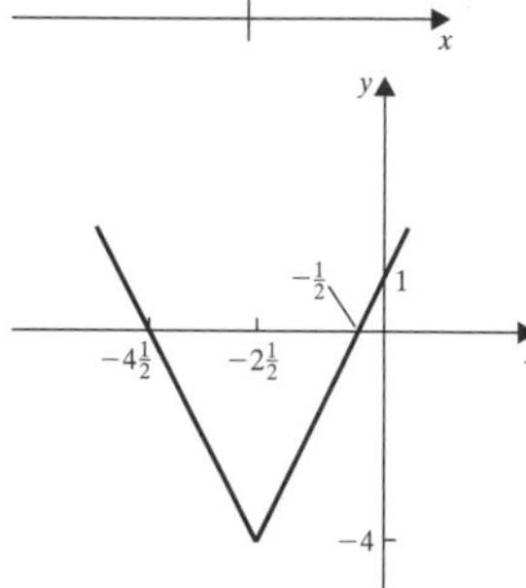

9

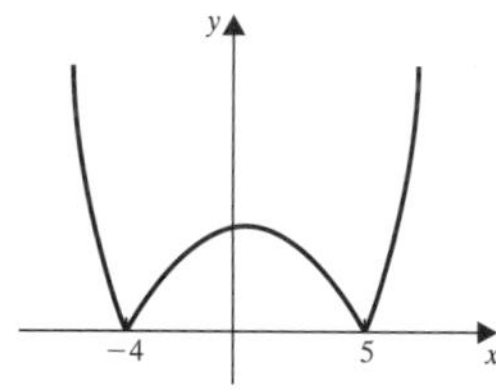

10

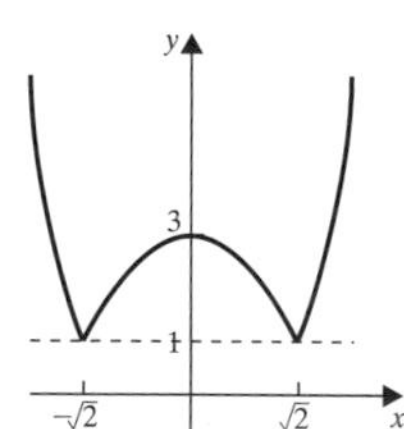

11

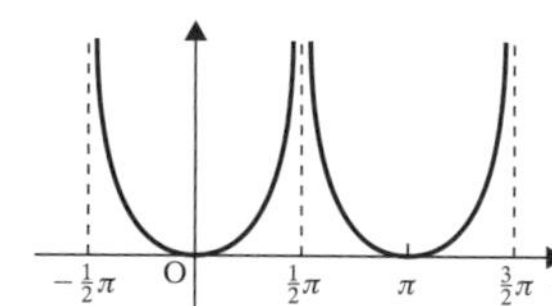

12

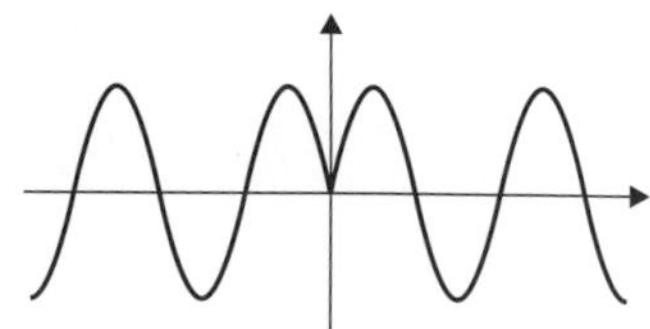

Exercise 18D – p.220

1

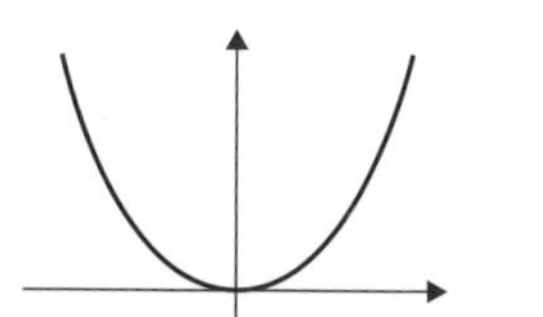

2

3

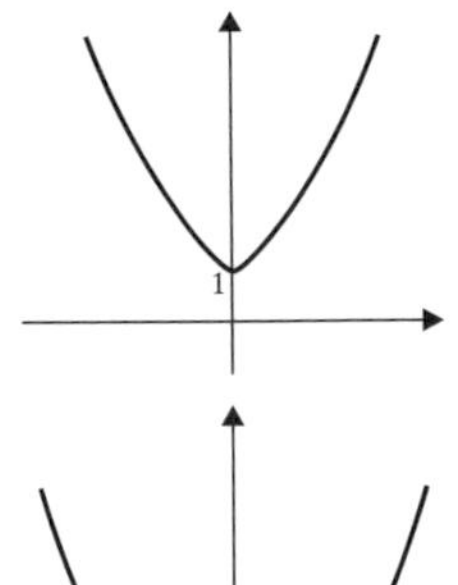

4

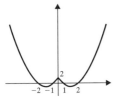

Exercise 18E – p.223

1 $(\frac{1}{2}, \frac{1}{2}), (-\frac{1}{2}, \frac{1}{2})$
2 $(0, 0), (1, 1)$
3 $(\sqrt{3}, 2\sqrt{3}), (2-\sqrt{7}, 2\sqrt{7}-4)$
4 $(1, 1), (-1, 1)$
5 $(1, 3), (1+\sqrt{6}, 3+2\sqrt{6})$
6 3 and $\frac{1}{2}(\sqrt{17}-3)$
7 $\frac{4}{5}$ and $\frac{2}{3}$
8 $\frac{2}{3}$ and 2
9 $-\sqrt{2}-1$ and -1

Exercise 18F – p.224

1 $\frac{1}{2}(1-\sqrt{5}) < x < 1, x > \frac{1}{2}(1+\sqrt{5})$
2 $0 < x < 1$
3 $0 < x < 1$
4 $-1 < x < 0$
5 $-2 < x < -1$
6 $x > 0$
7 $-\frac{1}{2} < x < \frac{1}{2}$
8 $x > -\frac{1}{2}$
9 $-1 < x < \frac{1}{3}$
10 $x > 1$
11 $x < 1$
12 $x < -\frac{5}{4}, x > -\frac{1}{4}$
13 $x < 0, x > 2$
14 $x < 0, x > 2$
15 $0 < x < 1+\sqrt{3}$
16 $-1 < x < 1$
17 $0 < x < \frac{1}{4}\pi, \frac{3}{4}\pi < x < \frac{5}{4}\pi,$
$\frac{7}{4}\pi < x < 2\pi$
18

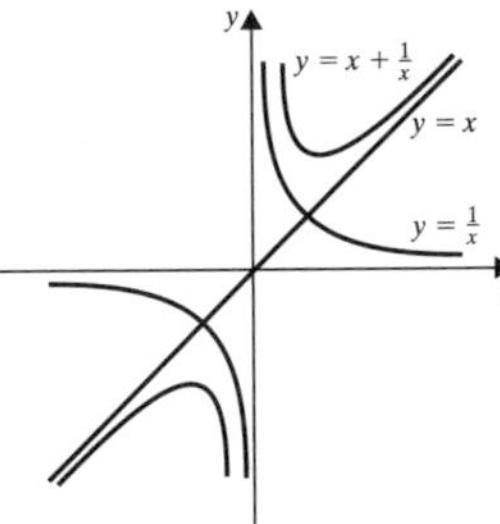

$f(x) \leqslant -2, f(x) \geqslant 2$

19 **a** $f(x) < 0, f(x) > 0$
b $f(x) \leqslant -\frac{1}{4}, f(x) > 0$
c $f(x) \leqslant -\frac{1}{4}$
d $f(x) \leqslant -4, f(x) > 0$

Mixed Exercise 18 – p.225

1

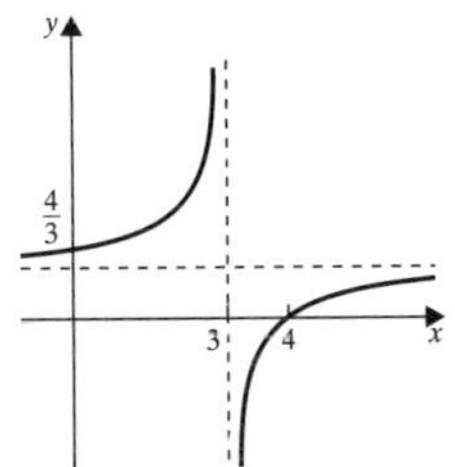

$y = 1$ and $x = 3$; $(4, 0)$ and $(0, \frac{4}{3})$

2

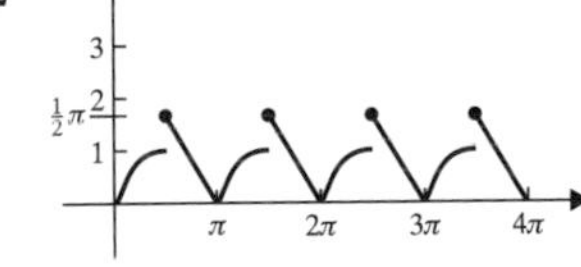

3

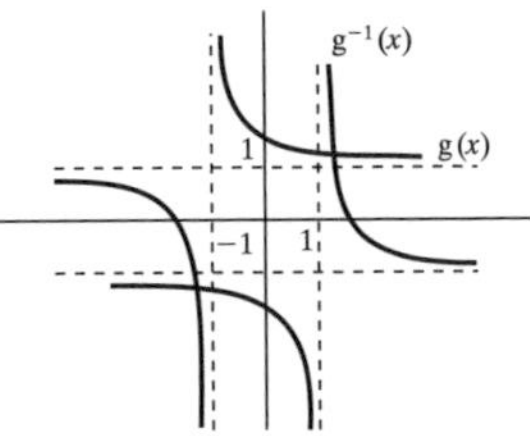

$g^{-1}(x) > -1, g^{-1}(x) < -1$

4 $x < -\frac{1}{2}$
5 $-1 < x < 0$ and $x > 2$
6

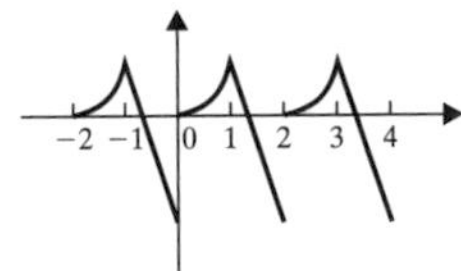

7 **a** odd with period π
b odd
c even
d even with period 2π

8

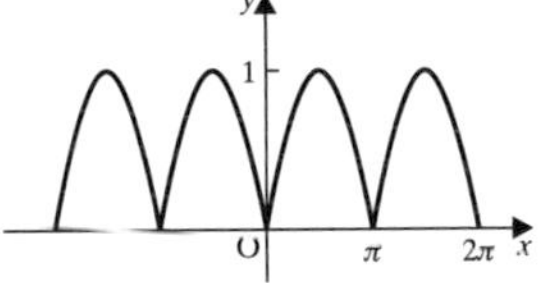

a and c

9

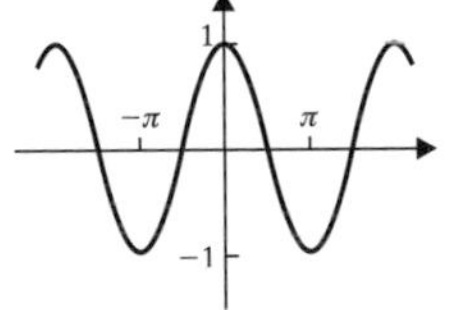

a and c are even.

10 **a** $-\frac{1}{2}$
b 0
c 0, 1

Chapter 19

Exercise 19A – p.229

1 **a** $1, \frac{1}{2^2}, \frac{1}{3^3}, \frac{1}{4^4}, \frac{1}{5^5}, \frac{1}{6^6};$
b $-2, 4, -8, 16, -32, 64;$
c $-\frac{1}{2}, \frac{1}{4}, -\frac{1}{8}, \frac{1}{16}, -\frac{1}{32}, \frac{1}{64};$
d $\frac{1}{2}, \frac{\sqrt{3}}{2}, 1, \frac{\sqrt{3}}{2}, \frac{1}{2}, 0;$

2 **a** 3, 0.57, 0.73, 0.70, 0.71, 0.71;
b $-4, -10, -88, -7654, -5.9 \times 10^7, -3.4 \times 10^{15}$
c 0.5, 1.2, 0.88, 1.1, 0.97, 1.0;
d 1, 0.8, 0.87, 0.85, 0.86, 0.85;

3 **a** Undefined after u_2
b Cycles thro' $\frac{1}{2}, 1\frac{1}{3}, 3, -2$
c Cycles thro' $-2, \frac{1}{2}, 1\frac{1}{3}, 3$

4 **a** converges (to 0.4)
b converges (to 0.4)
c converges (to 0.4)

5 $\frac{1}{2}(3 \pm \sqrt{13})$

Exercise 19B – p.231

1 **a** $\sum_{r=1}^{5} r^3$ **b** $\sum_{r=1}^{10} 2r$
c $\sum_{r=2}^{50} \frac{1}{r}$ **d** $\sum_{r=0}^{\infty} \frac{1}{3^r}$
e $\sum_{r=0}^{7} (-4+3r)$
f $\sum_{r=0}^{\infty} \left(\frac{8}{2^r}\right) = \sum_{r=0}^{\infty} \frac{1}{2^{r-3}}$

2 a $1+\frac{1}{2}+\frac{1}{3}+\ldots$
b $0+2+6+\ldots+30$
c $2+\frac{1}{2}+\frac{4}{15}+\ldots+\frac{22}{861}$
d $1+\frac{1}{2}+\frac{1}{5}+\ldots$
e $0+0+6+\ldots+720$
f $-1+a-a^2+\ldots$
3 a 8; 9 b 9; 10 c −1; 6
d $\frac{1}{420}$; ∞ e $(\frac{1}{2})^n$; ∞
f −48; 23 g 4; 10

Exercise 19C – p.235

1 a 9, $2n-1$ b 16, $4(n-1)$
c 15, $3n$ d 17, $3n+2$
e −2, $8-2n$
f $p+4q, p+(n-1)q$
g 18, $8+2n$ h 17, $4n-3$
i 0, $\frac{1}{2}(5-n)$ j 8, $3n-7$
2 a 100 b 180 c 165
d 185 e −30
f $5(2p+9q)$ g 190
h 190 i $-\frac{5}{2}$ j 95
3 $a = 27.2, d = -2.4$
4 $d = 3$; 30
5 1, $\frac{1}{2}$, 0; $-8\frac{1}{2}$
6 a $28\frac{1}{2}$ b 80 c 400
d 80 e 108 f $3n(1-6n)$
g 40 h $2m(m+3)$
7 4, $2n-4$
9 2, 364
10 39
11 64
12 a 1, 4 b 270
13 a $a = 21, d = -3$
b less than 4 or more than 11

Exercise 19D – p.239

1 a 32, 2^n b $\frac{1}{8}$, $\dfrac{1}{2^{n-2}}$
c 48, $3(-2)^{n-1}$
d $\frac{1}{2}$, $(-1)^{n-1}(\frac{1}{2})^{n-4}$
e $\frac{1}{27}$, $(\frac{1}{3})^{n-2}$
2 a 189 b −255
c $2-(\frac{1}{2})^{19}$ d 781/125
e 341/1024 f 1
3 $\frac{1}{2}$, 2
4 $-\frac{1}{2}$
5 $-\frac{1}{2}$, 1/1024
6 13.21 to 4 s.f.
7 a $(x-x^{n+1})/(1-x)$
b $(x^n-1)/[x^{n-2}(x-1)]$
c $(1+(-1)^{n+1}y^n)/(1+y)$
d $x(2^n-x^n)/[2^{n-1}(2-x)]$
e $[1-(-2)^n x^n]/(1+2x)$
8 $\frac{8}{3}(1-(\frac{1}{4})^n)$, 4
9 62 or 122
10 8.493 to 4 s.f.
11 8 (last repayment is less than £2000)
12 £23.31

Exercise 19E – p.242

1 a yes b no c yes
d yes e no f yes
2 a $-1<x<1$ b $x<-1, x>1$
c $-\frac{1}{2}<x<\frac{1}{2}$ d $0<x<2$
e $-1-a<x<1-a$
f $x<-(1+a), x>1-a$

3 a 6 c $13\frac{1}{3}$ d $\frac{5}{9}$ f $\frac{9}{4}$
4 $\frac{1}{2}$
5 8, 4, 2, 1 or 24, −12, 6, −3

Mixed Exercise 19 – p.243

1 $\frac{1}{2}, \frac{2}{5}, \frac{3}{10}, \frac{4}{17}, \frac{5}{26}, \frac{6}{37}$
converges (to 0)
2 1, −1, 1, −1, 1, −1 cycles 1, −1
3 a 1, 0, 0, 0, 0, 0
b −0.25, 0.3125, −0.215, 0.261, −0.193, 0.230 converges (slowly to 0)
4 a Cycles 2, −1, $\frac{1}{2}$;
b undefined for $r>1$
5 $\frac{2}{3}$
6 $\frac{1}{2}(1+3^{11}) = 88\,574$
7 $\dfrac{ab^4(1-b^{2(n-1)})}{1-b^2}$
8 $2(n+5)(n-4)$
9 $\dfrac{e(1-e^n)}{1-e}$
10 1
11 2.5 or −1
12 1, 7, 19, 37; $3n^2-3n+1$
13 16, −8
14 2
15 1
16 $a = 2\pm\sqrt{2}, r = \frac{1}{4}(2\pm\sqrt{2})$

Chapter 20

Exercise 20A – p.245

1 $\frac{1}{6}x^6+K$
2 $-\frac{1}{4}x^{-4}+K$
3 $\frac{4}{5}x^{5/4}+K$
4 $-\frac{1}{2}x^{-2}+K$
5 $-\frac{2}{3}x^{-3/2}+K$
6 $2x^{1/2}+K$
7 $\frac{1}{2}x^2+K$
8 $\frac{3}{2}x^{2/3}+K$
9 $x+\frac{1}{3}x^3+K$
10 $\frac{1}{2}x^2+e^x+K$
11 $x^2-\frac{2}{3}x^{3/2}+K$
12 $x-1/x+K$
13 x^2-3e^x+K
14 $e^x+(5/2)x^2+2/x+K$
15 $\frac{1}{2}x^2+\frac{1}{3}x^3+K$
16 $2x+(7/2)x^2-5x^3+K$
17 $2\sqrt{x}+\frac{2}{3}x^{3/2}+K$
18 $-\dfrac{1}{2x^2}+\dfrac{2}{x}+K$
19 $\frac{1}{4}x^2-\frac{1}{2}e^x+K$
20 $2\sqrt{x}+\frac{2}{3}x^{3/2}+(2/7)x^{7/2}+K$
21 $x-x^2+\frac{1}{3}x^3+K$
22 $\frac{1}{2}x^2-\frac{1}{4}x^4+K$
23 $x-e^x+K$
24 $-1/x+2/\sqrt{x}+K$

Exercise 20B – p.248

1 4
2 $\frac{2}{7}(8\sqrt{2}-1)$
3 $26\frac{2}{3}$
4 $12\frac{2}{3}$
5 15
6 −2
7 $2\frac{1}{3}$
8 $6\sqrt{2}-4$
9 1
10 $2e^7-2\sqrt{e}+6\frac{1}{2}$

Exercise 20C – p.249

Answers in square units.

1 $5\frac{1}{3}$
2 $12\frac{2}{3}$
3 $2\frac{2}{3}$
4 $13\frac{1}{2}$
5 $5\frac{1}{3}$
6 60
7 $5\frac{1}{3}$
8 $4\frac{7}{8}$
9 24

Exercise 20D – p.250

1 $\frac{4}{3}$
2 $15\frac{1}{4}$
3 $\frac{1}{3}$
4

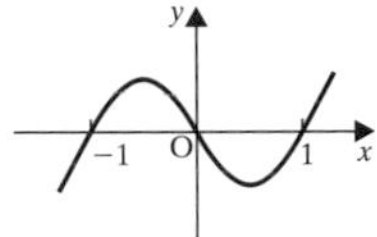

a $\frac{1}{4}$ b $\frac{1}{4}$ c $\frac{1}{2}$

5

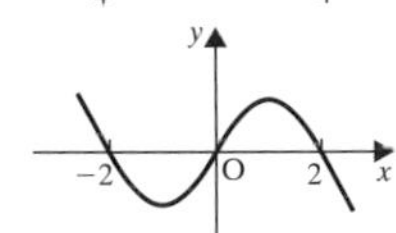

a 4 b 4 c 8

6 a −2 b 2 c 0

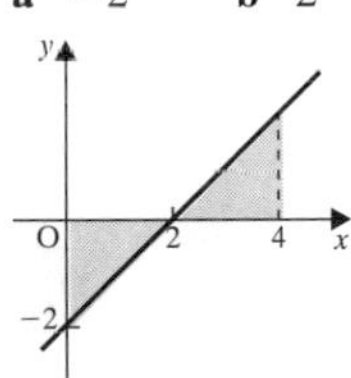

Exercise 20E – p.251

1 a $49\frac{1}{2}$ b $56\frac{1}{3} - \frac{2}{3}\sqrt{2}$
c 27
2 36
3 $2\frac{1}{3}$
4 $5\frac{1}{3}$
5 $4\frac{1}{2}$
6 a $42\frac{2}{3}$ b $42\frac{2}{3}$
7 $A = \int_0^1 x\,dy \quad B = \int_0^1 y\,dx$
$A + B = 1$

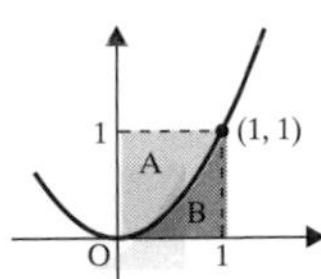

Exercise 20F – p.253

1 9
2 a $\frac{8}{3}$
b $\frac{16}{3}$

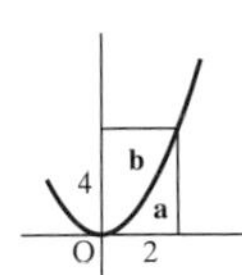

3 $e - 2$
4 $\frac{1}{6}$
5 1
6 18
7 $\frac{4}{3}$
8 a $\frac{4}{3}$ b $\frac{1}{3}$
9 $4\sqrt{3}$

Exercise 20G – p.254

1 $2\ln x + k$
2 $\frac{1}{4}\ln x + k$
3 $(3/2)\ln x + k$
4 $1 + \ln x + k$
5 $\frac{1}{2}x^2 + x - \ln x + k$
6 $\frac{1}{3}\ln 2$
7 $2 - \ln 3$
8 $\ln 2 - 1$
9 $e^3 - e^2 + \ln(2/3)$
10 $\frac{1}{3}(2\ln(5/4) - 1)$

Exercise 20H – p.257

1 22
2 0.705
3 1.625
4 1.282
5

22	0.705
21	0.648
21.33	0.671
21.33	0.667

Mixed Exercise 20 – p.258

1 $\frac{1}{3}x^3 + \frac{1}{x} + K$
2 $\frac{3}{4}x^{4/3} + K$
3 $\frac{2}{3}x^{3/2} + 2x^{1/2} + K$
$= \frac{2}{3}\sqrt{x}(x + 3) + K$
4 $e^x - \ln|x|$
5 $\frac{1}{3}x^3 - \ln|x| + K$
6 $\frac{2}{5}\sqrt{x}(x^2 - 5) + K$
7 9
8 $\frac{297}{10}$
9 $33\frac{3}{4}$
10 $\frac{3}{4}$
11 $10\frac{2}{3}$
12 $\frac{1}{2}$
13 a i 4 square units
ii 4.75 iii 4
b $-4\frac{1}{2}$
14 a $27\frac{1}{2}$ b $27\frac{1}{2}$
c a is exact because the area is that of a trapezium.
15 $e^4 - 1$
16 16/3
17 32/3

Summary D

Multiple Choice Exercise D – p.261

1 D
2 C
3 D
4 B
5 A, C
6 B
7 C
8 A, C
9 A, B
10 C, D
11 D
12 A
13 C, D
14 F
15 T
16 T
17 F
18 T
19 F
20 T
21 F

Examination Questions D – p.264

1 $\frac{1}{2}$, minimum

2

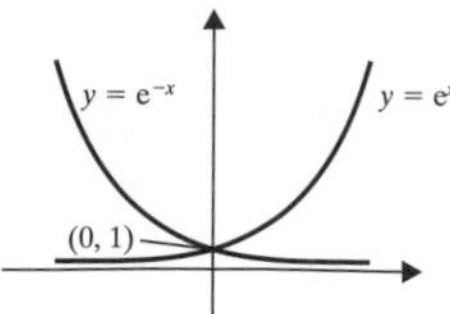

reflection in Oy

3 **b** $f'(x) = e^x - 1 > 0$ for $x > 0$

4 **a** 4 **b** 8 **c** 920

5 **i**

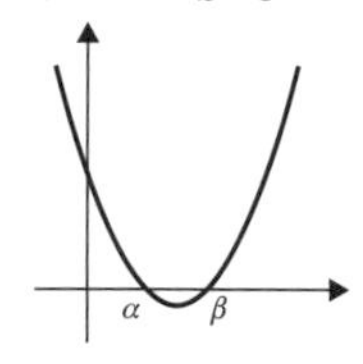

ii

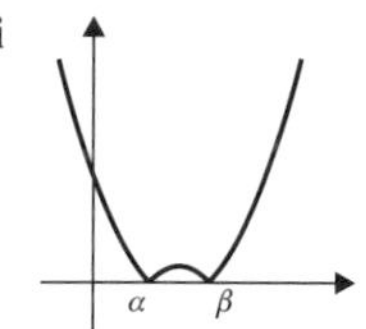

iii

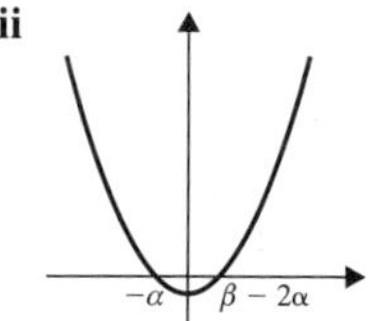

6 **ii** 10

7 $77 + 72\ln 2$

8 **a** 8.4 **b** $f''(x) > 16$

9

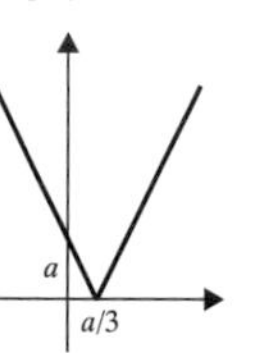

$\frac{a}{4} < x < \frac{a}{2}$

10 $a = -3, b = 5$; $12\frac{2}{3}$

11 **a** $g(x) > 0$

b

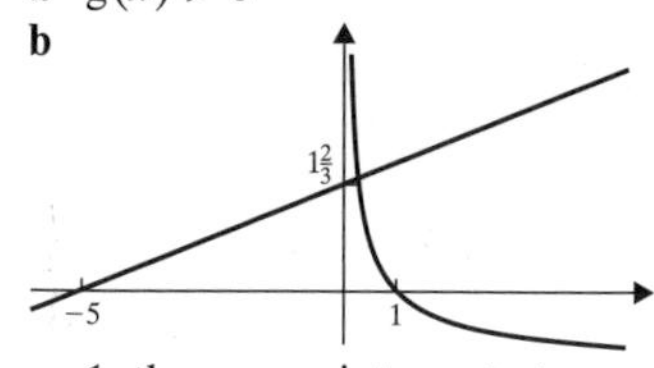

c 1, the curves intersect at 1 point only

d 0.84

12 -8

13 **a** $x + y - 4 = 0$ **b** $\frac{5}{6}$

14 $\frac{3a}{2}, \frac{5a}{2}$

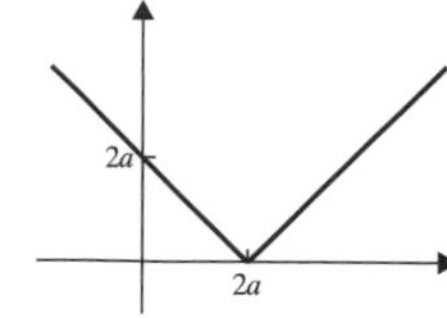

15 **a** 21.17

b $f^{-1} : x \mapsto \frac{x+1}{3}$

c

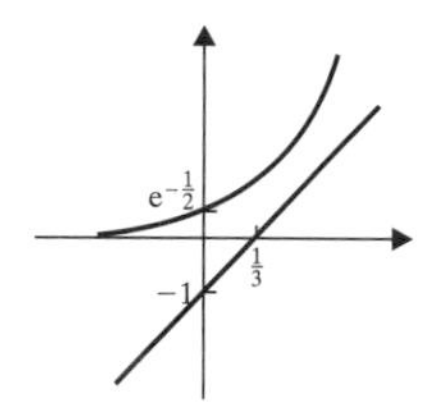

d $-2\frac{2}{3}, 2$

16 **a** $-28\,500$

17 **a** $(\frac{1}{3}, 0)$

b

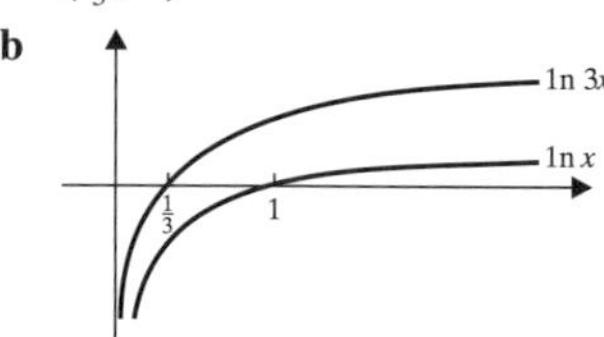

c $\ln 3x$

d

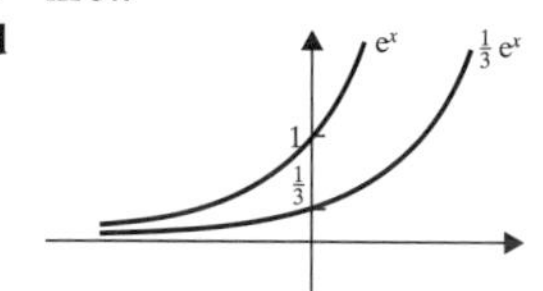

reflections in $y = x$

18 **a** (1, 3), (2, 3) **b** $\frac{1}{6}$

c $x - y + 1 = 0$

19 **a** $p = 0.8, q = 5.2$ **b** 26

20 **a** $f(x) > k$ **b** $2k$

c $f^{-1} : x \mapsto \ln(x - k), x > k$

d

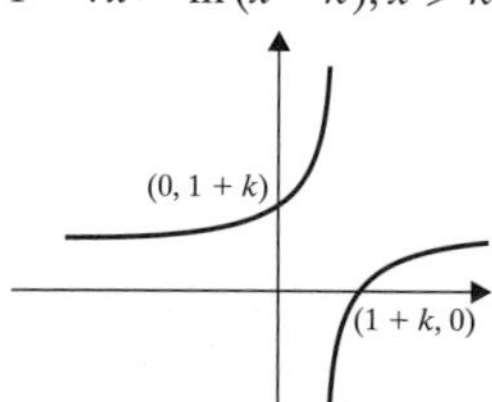

21 **a** £151.20 **b** £54 600

c 4.9%

22 **a** $\ln k$ **b** e

c **i** M is a min **ii** $e^3 - 10$

23 **a** $-3, -37$, **b** 16

24 **a** $4 + 12x^{-1/2} + 9x^{-1}$

b $4x + 24x^{\frac{1}{2}} + 9\ln x + k$

c $44 + 9\ln(9/4)$

25 **b** P(−1, 5), Q(2, 2)

26 **a**

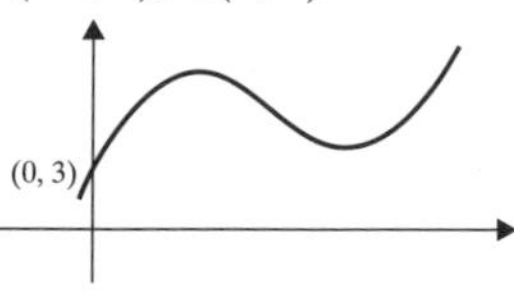

35 sq units

b −4

27 $a = 3, r = -\frac{1}{4}$, 12/5

28 **a** $-\frac{1}{4}x^4 + \frac{27}{2}x^2 - 34x + k$

b 34 **c** 12

29 **a**

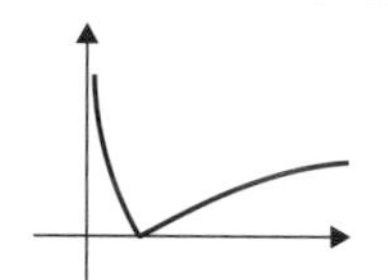

b

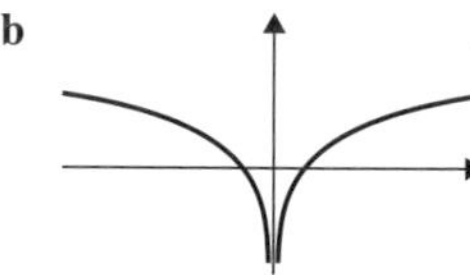

30 $-\frac{1}{3}, -1$

31 £56 007

32 **i** 18 **ii** $4n + 2$

iii $a = 6, d = 4$, 2 004 000

33 **i** $y = 2x - 2$ **ii** $6\frac{3}{4}$

34 **a** £3221 **b** £5187

c $2000(1.1) + 2000(1.1)^2 + 2000(1.1)^3$

d £47 045

35 **a** 33 **b** 1683

36 **a** $x = 4, y = 20$

c $62/5 + 48\ln 4$

37 **a** 0.75, 224 cm^3 **b** 14

38 **b** 8 **c** 8.64 **d** $0 < x < 4$

39 −4, 7

40

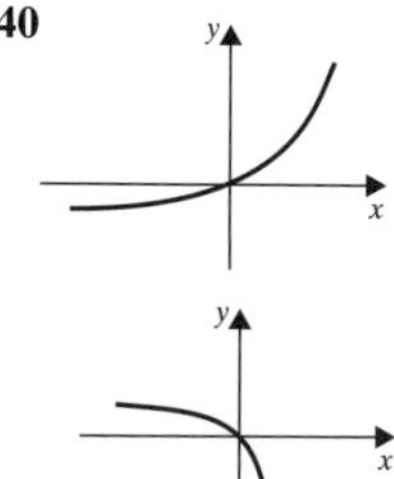

41 **a** £70 620 **b** 1062

42 $20(1.1)^{n-1}$, 18

43 **i** 0, 4 **ii** 1 500 500

44 **a** $g(x) \geqslant 1$ **b** 26

c 0, 1 **d** $3, -2\frac{1}{3}$

45 a $\pm 2, \pm 4$

b ;3, 4

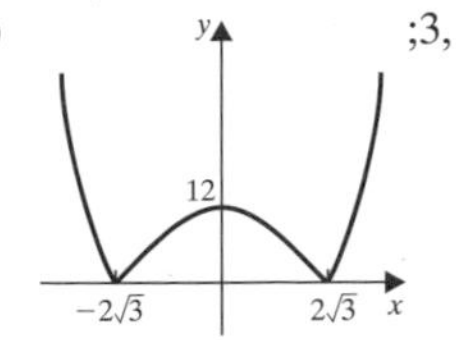

47 i A(0, 2); B($\frac{2}{3}$, 0)

ii

iii $\frac{10}{3}$

48 $x < \frac{1}{2}$

Chapter 21

Exercise 21A – p.275

1 $6(3x+1)$
2 $-4(3-x)^3$
3 $20(4x-5)^4$
4 $6x(x^2+1)^2$
5 $21(2+3x)^6$
6 $-18(2-6x)^2$
7 $4x^3(2x^4-5)^{-1/2}$
8 $-2x(x^2+3)^{-2}$
9 $\dfrac{9x^2}{2\sqrt{3x^3-4}}$
10 $-\left(\dfrac{1}{2\sqrt{x}}+3\right)(\sqrt{x}+3x)^{-2}$
11 $\dfrac{3x}{(4-x^2)^{3/2}}$
12 $\dfrac{-7(x^2+1)}{(x^3+3x)^{4/3}}$
13 $-10(4-2x)^4$
14 $4x(x^2+3)$
15 $21(3x-4)^6$
16 $4x(x^2+4)$
17 $-12x(1-2x^2)^2$
18 $-12x^2(2-x^3)^3$
19 $\frac{3}{2}x(2+x^2)^{-1/4}$
20 $\frac{1}{3}(2x-1)(x^2-x)^{-2/3}$
21 $6x(2-3x^2)^{-2}$
22 $4x(4-x^2)^{-3}$
23 $-\frac{5}{2}x^4(x^5-3)^{-3/2}$
24 $\dfrac{-1}{8\sqrt{x}(6-\sqrt{x})^{3/4}}$

Exercise 21B – p.277

1 $2x(x-3)^2+2x^2(x-3)$
$= 2x(x-3)(2x-3)$
2 $\sqrt{x-6}+\dfrac{x}{2\sqrt{x-6}}$
$= \dfrac{3(x-4)}{2\sqrt{x-6}}$
3 $(x-2)^5+5(x+2)(x-2)^4$
$= (x-2)^4(6x+8)$
4 $(2x+3)^3+6x(2x+3)^2$
$= (2x+3)^2(8x+3)$
5 $4(x+1)^2(x-1)^3$
$+2(x+1)(x-1)^4$
$= 2(x+1)(x-1)^3(3x+1)$
6 $3\sqrt{x}(x-3)^2+\dfrac{1}{2\sqrt{x}}(x-3)^3$
$= \dfrac{(x-3)^2(7x-3)}{2\sqrt{x}}$
7 $\dfrac{4(x-3)(x+5)^3-(x+5)^4}{(x-3)^2}$
$= \dfrac{(x+5)^3(3x-17)}{(x-3)^2}$
8 $\dfrac{(3x+2)^2-6x(3x+2)}{(3x+2)^4}$
$= \dfrac{2-3x}{(3x+2)^3}$
9 $\dfrac{4\sqrt{x}(2x-7)-(2x-7)^2\left(\dfrac{1}{2\sqrt{x}}\right)}{x}$
$= \dfrac{(2x-7)(6x+7)}{2x\sqrt{x}}$
10 $3x^2\sqrt{x-1}+\dfrac{x^3}{2\sqrt{x-1}}$
$= \dfrac{x^2(7x-6)}{2\sqrt{x-1}}$
11 $(x+3)^{-1}-x(x+3)^{-2}$
$= 3(x+3)^{-2}$
12 $2x(2x-3)^2+4x^2(2x-3)$
$= 2x(2x-3)(4x-3)$

Exercise 21C – p.278

1 $\dfrac{2x(x-3)-(x-3)^2}{x^2}$
$= \dfrac{(x-3)(x+3)}{x^2}$
2 $\dfrac{(x+3)(2x)-x^2}{(x+3)^2} = \dfrac{x(x+6)}{(x+3)^2}$
3 $\dfrac{-x^2-2x(4-x)}{x^4} = \dfrac{x-8}{x^3}$
4 $\dfrac{2x^3(x+1)-(x+1)^2(3x^2)}{x^6}$
$= \dfrac{-(x+1)(x+3)}{x^4}$
5 $\dfrac{4(1-x)^3+12x(1-x)^2}{(1-x)^6}$
$= \dfrac{4(1+2x)}{(1-x)^4}$
6 $\dfrac{(x-2)(4x)-2x^2}{(x-2)^2} = \dfrac{2x(x-4)}{(x-2)^2}$
7 $\dfrac{\frac{5}{3}x^{2/3}(3x-2)-3x^{5/3}}{(3x-2)^2}$
$= \dfrac{2x^{2/3}(3x-5)}{3(3x-2)^2}$
8 $\dfrac{-3(1-2x)^2}{x^4}$
9 $\dfrac{(3x-2)(x+1)^{3/2}}{2x^2}$

Exercise 21D – p.279

1 $\dfrac{2}{x}$
2 $4e^{4x-1}$
3 $\dfrac{5}{5x+2}$
4 $2xe^{x^2-3}$
5 $\dfrac{4x}{2x^2-3}$
6 $(9x^2-2)\,e^{3x^3-2x}$
7 $\dfrac{1+2x}{1+x+x^2}$
8 $\dfrac{4}{1+2x}$
9 $(2x+1)\,e^{x(x+1)}$

Exercise 21E – p.280

(pr ≡ product; f of f ≡ function of a function)

1 a pr
b $u = e^x, v = x^2+1, y = uv$
2 a f of f
b $u = x^2+1, y = e^u$

3 **a** pr
b $u = x, v = \ln x, y = uv$
4 **a** f of f
b $u = x+1, y = e^{u/2}$
5 **a** pr
b $u = e^x, v = \ln x, y = uv$
6 **a** f of f
b $u = 3 - x^2, y = \ln u$
7 **a** f of f
b $u = \ln x, y = u^2$
8 **a** f of f
b $u = -2x, y = e^u$
9 **a** f of f
b $u = \ln x, y = \dfrac{1}{u}$
10 $fg(x) = e^{2x}$; $gf(x) = e^{x^2}$
11 **a** $\left(\dfrac{1}{x}\right)^2$ **b** $\ln x^2$
c $\ln\left(\dfrac{1}{x}\right)$ **d** $(\ln x)^2$
e $\ln\left(\dfrac{1}{x}\right)^2$ **f** $\left(\dfrac{1}{x}\right)^2$

Exercise 21F – p.281

1 $e^x(x+1)$
2 $x(2\ln x + 1)$
3 $e^x(x^3 + 3x^2 - 2)$
4 $12x\ln(x-2) + \dfrac{6x^2}{(x-2)}$
5 xe^x
6 $x\ln x + \dfrac{(x^2+4)}{2x}$
7 $\dfrac{4+3x}{2\sqrt{2+x}}$
8 $\frac{1}{2}\ln(x-5) + \dfrac{x}{2(x-5)}$
9 $(x^2 + 2x - 2)e^x$
10 $\dfrac{1-x}{e^x}$
11 $\dfrac{e^x(x-2)}{x^3}$
12 $\dfrac{1 - 3\ln x}{x^4}$
13 $\dfrac{x\ln x - 2(x+1)}{2x\sqrt{x+1}(\ln x)^2}$
14 $\dfrac{e^x(x^2 - 2x - 1)}{(x^2-1)^2}$
15 $\dfrac{-2}{(e^x - e^{-x})^2}$
16 $4e^{4x}$
17 $\dfrac{2x}{x^2 - 1}$
18 $2xe^{x^2}$
19 $-6e^{(1-x)}$ or $-6e(e^{-x})$
20 $2xe^{(x^2+1)}$
21 $\dfrac{1}{2(x+2)}$
22 $\dfrac{2\ln x}{x}$
23 $\dfrac{-1}{x(\ln x)^2}$
24 $\frac{1}{2}\sqrt{e^x}$

Mixed Exercise 21 – p.281

1 $\dfrac{3x+2}{2\sqrt{x+1}}$
2 $6x(x^2-8)^2$
3 $\dfrac{1-x^2}{(x^2+1)^2}$
4 $\dfrac{-4x^3}{3(2-x^4)^{2/3}}$
5 $\dfrac{2x}{(x^2+2)^2}$
6 $\frac{1}{2}x(5\sqrt{x} - 8)$
7 $6x(x^2-2)^2$
8 $\dfrac{1-2x}{2\sqrt{x-x^2}}$
9 $\dfrac{\sqrt{x}+2}{2(\sqrt{x}+1)^2}$
10 $\dfrac{x(5x-8)}{2\sqrt{x-2}}$
11 $\dfrac{-(3x+4)}{2x^3\sqrt{x+1}}$
12 $6x^5(x^2+1)^2(2x^2+1)$
13 $\dfrac{x}{\sqrt{x^2-8}}$
14 $x^2(5x^2 - 18)$
15 $6x(x^2-6)^2$
16 $\dfrac{-(x^2+6)}{(x^2-6)^2}$
17 $-8x^3(x^4+3)^{-3}$
18 $\dfrac{(2-x)^2(2-7x)}{2\sqrt{x}}$
19 $\dfrac{2+5x}{2\sqrt{x}(2-x)^4}$
20 $(x-2)(3x-4)$
21 $30x^2(2x^3+4)^4$
22 $1 + \ln x$
23 $\frac{8}{3}(4x-1)^{-1/3}$
24 $\dfrac{e^x(x-2)}{(x-1)^2}$
25 $\dfrac{-(x^3+4)}{2x^3\sqrt{1+x^3}}$
26 $\dfrac{(x-1)\ln(x-1) - x\ln x}{x(x-1)\{\ln(x-1)\}^2}$
27 $3(\ln 10)10^{3x}$
28 $\dfrac{2x}{(1+x^2)^2}$
29 $\dfrac{2}{x^2 e^{2/x}}$
30 $\dfrac{-e^x}{1-e^x}$ or $\dfrac{e^x}{e^x - 1}$
31 $3x^2e^{3x}(x+1)$
32 $\dfrac{4}{5(2x-1)^2} - \dfrac{6}{5(x-3)^2}$
$= \dfrac{2(3-2x^2)}{(2x-1)^2(x-3)^2}$
33 $\dfrac{e^{x/2}(x-10)}{2x^6}$
34 $\dfrac{2}{x} - \dfrac{1}{x+3} - \dfrac{2x}{x^2-1}$
35 $\dfrac{5x+9}{x(x+3)}$
36 $\dfrac{4}{x}(\ln x)^3$
37 $\dfrac{(x+3)^2(x^2-6x+6)}{(x^2+2)^2}$
38 $\dfrac{e^x - 1}{2\sqrt{e^x - x}}$
39 $\dfrac{8x}{x^2+1}$
40 $\dfrac{dy}{dx} = \dfrac{4}{(1-2x)^2}$; $\dfrac{d^2y}{dx^2} = \dfrac{16}{(1-2x)^3}$
41 $\dfrac{dy}{dx} = \dfrac{1}{x(x+1)}$; $\dfrac{d^2y}{dx^2} = -\dfrac{(2x+1)}{x^2(x+1)^2}$
42 $\dfrac{dy}{dx} = \dfrac{-4e^x}{(e^x-4)^2}$;
$\dfrac{d^2y}{dx^2} = \dfrac{4e^x(e^x+4)}{(e^x-4)^3}$
43 **a** ≈ 1.5; $y = 6.7(15^{x/14})$
b **i** 37 **ii** 79
c ≈ 5.5 per day; larger than the actual increase

Chapter 22

Exercise 22B – p.286

1 0
2 $\frac{1}{2}$
3 $\frac{1}{4}(\sqrt{6}-\sqrt{2})$
4 $-(2+\sqrt{3})$
5 $\frac{1}{4}(\sqrt{6}-\sqrt{2})$
6 $\frac{1}{4}(\sqrt{6}+\sqrt{2})$
7 $\sin 3\theta$
8 0
9 $\tan 3A$
10 $\tan\beta$
11 **a** $\frac{3}{5}$ **b** $-\frac{4}{5}$ **c** $-\frac{3}{4}$
12 **a** 1, 115° **b** 1, 30° **c** 1, 310° **d** 1, 330°
18 67.5°, 247.5°
19 7.4°, 187.4°
20 37.9°, 217.9°
21 15°, 195°
22 0, 60°, 90°, 120°, 180°, 240°, 270°, 300°, 360°

Exercise 22C – p.290

1 $\frac{1}{2}$
2 $\frac{1}{\sqrt{2}}$
3 $\frac{1}{2}\sin 2\theta$
4 $\cos 8\theta$
5 $-\frac{1}{\sqrt{3}}$
6 $\tan 6\theta$
7 $-\frac{1}{\sqrt{2}}$
8 $\frac{1}{\sqrt{2}}$
9 **a** $\frac{24}{25}, -\frac{7}{25}$ **b** $\frac{336}{625}, \frac{527}{625}$ **c** $\frac{120}{169}, -\frac{119}{169}$
10 **a** $-\frac{336}{527}$ **b** $\frac{527}{625}$ **c** $-\frac{336}{625}$ **d** $\frac{164\,833}{390\,625}$
11 **a** $x(1-y^2) = 2y$ **b** $x = 2y^2-1$ **c** $x = 1-\frac{2}{y^2}$ **d** $2x^2y+1 = y$
12 **a** $-\cos 2x$ **b** $3-\cos 2x$ **c** $\frac{1}{2}(\cos 2x+3)$ **d** $\frac{1}{2}(\cos 2x+1)(3+\cos 2x)$ **e** $\frac{1}{4}(1+\cos 2x)^2$ **f** $\frac{1}{4}(1-\cos 2x)^2$
14 **a** $\frac{3}{2}\pi, \frac{1}{6}\pi, \frac{5}{6}\pi$ **b** $\frac{1}{2}\pi, \frac{7}{6}\pi, \frac{11}{6}\pi, \frac{3}{2}\pi$ **c** $0, \frac{2}{3}\pi, \frac{4}{3}\pi, 2\pi$ **d** $\frac{1}{6}\pi, \frac{5}{6}\pi, \frac{1}{2}\pi, \frac{3}{2}\pi$ **e** $\frac{1}{3}\pi, \frac{5}{3}\pi$ **f** $\frac{1}{4}\pi, \frac{1}{2}\pi, \frac{5}{4}\pi, \frac{3}{2}\pi$
15 **a** **i** $\frac{1-2\tan\theta-\tan^2\theta}{1+\tan^2\theta}$ **ii** $\frac{1}{\tan\theta}$ **b** **i** 135° **ii** 45°

Exercise 22D – p.292

1 $3\pi/2+2n\pi$
2 $\pi/6+n\pi, 5\pi/6+n\pi$
3 $\pi/8+n\pi/2$
4 $n\pi, \pi/3+n\pi, 2\pi/3+n\pi$

Mixed Exercise 22 – p.292

1 $y = 1-2x^2$
4 $\frac{56}{65}, -\frac{16}{65}$
5 $x = 2y-1$
6 −155.7°, −114.3°, 24.3°, 65.7°
7 $-\pi, 0, \pi$
9 $\cot^2 x$
10 90°, 270°
11 **a** $2-\cos 2\theta$ **b** $2+2\cos 4A$

Chapter 23

Exercise 23A – p.295

1 **a** 2, 30° **b** $\sqrt{10}$, 71.6° **c** 5, 36.9°
2 $\sqrt{2}\cos(2\theta+\frac{1}{4}\pi)$
3 $\sqrt{29}\sin(3\theta+21.8°)$
4 $-2\sin(\theta-\frac{1}{6}\pi)$; max 2 at $\theta = 300°$, min −2 at $\theta = 106.3°$
5 $25\cos(\theta+73.7°)$; max 28 at $\theta = 286.3°$, min −22 at $\theta = 106.3°$
6 $\sqrt{2}, -\sqrt{2}; -\frac{1}{\sqrt{2}}$ max, $\frac{1}{\sqrt{2}}$ min
7 $-\sqrt{\frac{2}{3}}$ max, $\sqrt{\frac{2}{3}}$ min
8 **a** 45° **b** 118.1°, 323.1° **c** 0, 216.80°, 360° **d** 0, 306.90°, 360°

Exercise 23B – p.297

1 **a** $\frac{1}{3}\pi$ **b** $-\frac{1}{2}\pi$ **c** $\frac{1}{2}\pi$ **d** $-\frac{1}{3}\pi$ **e** $\frac{2}{3}\pi$ **f** $-\frac{1}{4}\pi$
2 **a** $\frac{1}{4}\pi$ **b** $\frac{1}{2}\pi$
3 **a** $\frac{1}{2}\pi$ **b** $\frac{2x}{x^2+1}$ **c** $\pm\frac{1}{2}\pi$ **d** $\frac{7}{11}$

Mixed Exercise 23 – p.298

1 $5\sin(\theta-\alpha)$ where $\tan\alpha = \frac{3}{4}$; $\frac{7}{8}$ min, $-\frac{7}{3}$ max, $\pm\infty$
2 **a** 53.1° **b** 107.5° **c** 26.6°
3 $\sqrt{2}\sin(2\theta-\frac{1}{4}\pi)$; $\frac{3}{8}\pi$
4 $\sqrt{2}\cos(x-\frac{1}{4}\pi)$; $\frac{1}{4}\pi$
5 119.6°, 346.7°
6 0
7 40.2°
9 $5\cos(x+\alpha)$ where $\tan\alpha = \frac{4}{3}$; $4+2\sec(x+\alpha)$

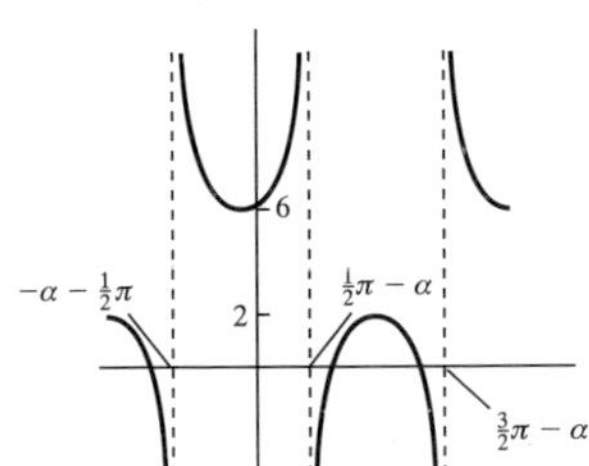

Chapter 24

Exercise 24A – p.299

1 Approximately equal
2 No
3 c $y = -\sin x$
d No; gives different and much smaller values for gradients

Exercise 24B – p.301

1 a $\cos x + \sin x$ b $\cos\theta$
c $-3\sin\theta$ d $5\cos\theta$
e $3\cos\theta - 2\sin\theta$
f $4\cos x + 6\sin x$
2 a -1 b 1 c -1
d 1 e $2(\pi - 1)$
f 4
3 a $\frac{1}{6}\pi$ b $\frac{1}{6}\pi$ c $\frac{1}{4}\pi$
d π
4 a $(\frac{1}{3}\pi, \sqrt{3} - \frac{1}{3}\pi)$, max;
$(\frac{5}{3}\pi, -\sqrt{3} - \frac{5}{3}\pi)$, min
b $(\frac{1}{6}\pi, \frac{1}{6}\pi + \sqrt{3})$, max;
$(\frac{5}{6}\pi, \frac{5}{6}\pi - \sqrt{3})$, min
5 $y + \theta = 3 + \frac{1}{2}\pi$
6 $2\pi y + x = 2\pi^3 - \pi$
7 $(0, 1)$

Exercise 24C – p.303

1 $4\cos 4x$
2 $2\sin(\pi - 2x)$ or $2\sin 2x$
3 $\frac{1}{2}\cos(\frac{1}{2}x + \pi)$ or $-\frac{1}{2}\cos\frac{1}{2}x$
4 $\dfrac{x\cos x - \sin x}{x^2}$
5 $-\dfrac{(\cos x + \sin x)}{e^x}$
6 $\dfrac{\cos x}{2\sqrt{\sin x}}$
7 $2\sin x\cos x$ or $\sin 2x$
8 $\cos^2 x - \sin^2 x$ or $\cos 2x$
9 $\cos x\, e^{\sin x}$
10 $-\tan x$
11 $e^x(\cos x - \sin x)$
12 $x^2\cos x + 2x\sin x$
13 $2x\cos x^2$
14 $-\sin x\, e^{\cos x}$
15 $3\cot x$
16 $\dfrac{\sin x}{\cos^2 x} = \sec x\tan x$
17 $\sec^2 x$
18 $\dfrac{-\cos x}{\sin^2 x} = -\operatorname{cosec} x\cot x$
19 $\dfrac{-1}{\sin^2 x} = -\operatorname{cosec}^2 x$

Exercise 24D – p.304

1 $\dfrac{1}{2x\sqrt{1 + \ln x}}$
2 $-2x\sin(x^2 + 3)$
3 $(3x^2 - 1)e^{(x^3 - x)}$
4 $\dfrac{1}{x}\cos(\ln x)$
5 $-2x\tan(x^2)$
6 $4xe^{x^2}(1 + e^{x^2})$
7 $\dfrac{-\sin x\cos x}{\sqrt{3 - \sin^2 x}}$
8 $\dfrac{2\sec^2 x\ln(\tan x)}{\tan x}$
9 $\dfrac{-xe^{\sqrt{2 - x^2}}}{\sqrt{2 - x^2}}$
10 $-4x\cos(x^2 + 1)\sin(x^2 + 1)$
$= -2x\sin\{2(x^2 + 1)\}$

Exercise 24E – p.305

1 $e^{x^2}(2x\cos x - \sin x)$
2 $-\left(\dfrac{x\sin 2x + \cos^2 x}{x^2}\right)$
3 $\ln\sin x + x\cot x$
4 $\dfrac{2x^2\ln x - x^2 + 1}{x(\ln x)^2}$
5 $\dfrac{e^x(x^2 + x + 2)}{\sqrt{x^2 + 2}}$
6 $\dfrac{2\cos x + x\sin x\ln x}{2x\sqrt{\cos^3 x}}$
7 $e^{\sin x}[1 + (x + 1)\cos x]$
8 $\dfrac{2x(\sin x - x\cos x)}{\sin^3 x}$
9 $\dfrac{e^{x^2}}{x}(1 + 2x^2\ln x)$

Mixed Exercise 24 – p.305

1 a $-4\cos 4\theta$ b $1 + \sin\theta$
c $3\sin^2\theta\cos\theta + 3\cos 3\theta$
2 a $3x^2 + e^x$ b $2e^{(2x+3)}$
c $e^x(\sin x + \cos x)$
3 a $-\dfrac{3}{x}$ b $-\dfrac{2}{x}$ c $\dfrac{1}{2x}$
4 a $3\cos x + e^{-x}$ b $\dfrac{1}{2x} + \frac{1}{2}\sin x$
c $4x^3 + 4e^x - \dfrac{1}{x}$
d $-\frac{1}{2}(e^{-x} + x^{-3/2}) - \dfrac{1}{x}$
5 $1 + \dfrac{1}{x} + \ln x$
6 $3\sin 6x$
7 $\frac{8}{3}(4x - 1)^{-1/3}$
8 $9 - 18\sqrt{x} + 8x$
9 $(x^4 + 4x^3 + 3)/(x + 1)^4$
10 $\dfrac{(x - 1)\ln(x - 1) - x\ln x}{x(x - 1)\{\ln(x - 1)\}^2}$
11 $-1/\sin x\cos x$ or $-2\operatorname{cosec} 2x$
12 $2x\sin x + x^2\cos x$
13 $e^x(x - 2)/(x - 1)^2$
14 $2\cos x/(1 - \sin x)^2$
15 $x(5x - 4)/2\sqrt{x - 1}$
16 $-2(1 - x)^2(2x + 1)$
17 $\dfrac{3}{2(x + 3)} - \dfrac{x}{x^2 + 2}$
18 $\cos^2 x(4\cos^2 x - 3)$
19 $-\sin 2x\, e^{\cos^2 x}$
20 a $x = \ln 3$
b $x = 1$ (*not* -1)
c $x = \frac{1}{4}$
21 a 1 b $y - x = 1 - \frac{1}{2}\pi$
c $y + x = 1 + \frac{1}{2}\pi$
22 a $1 + e$ b $y = x(1 + e)$
c $y(1 + e) + x = (1 + e)^2 + 1$
23 a 2 b $y = 2x + 1$
c $2y + x = 2$
24 a -1 b $x + y = 3$
c $x - y + 1 = 0$
25 a $(\frac{1}{2}\pi, 0)$, min; $(\frac{3}{2}\pi, 2)$, max
b $(\frac{1}{6}\pi, \{\frac{1}{12}\pi + \frac{1}{2}\sqrt{3}\})$, max;
$(\frac{5}{6}\pi, \{\frac{5}{12}\pi - \frac{1}{2}\sqrt{3}\})$, min
c $(\ln 3, \{3 - 3\ln 3\})$, min; only one turning point
26 a $(\frac{1}{6}\pi, \{\frac{1}{2}\pi - \sqrt{3}\})$
b $(1, -1)$

Chapter 25

Exercise 25A – p.309

1 $2x + 2y\frac{dy}{dx} = 0$

2 $2x + y + (x + 2y)\frac{dy}{dx} = 0$

3 $2x + x\frac{dy}{dx} + y = 2y\frac{dy}{dx}$

4 $-\frac{1}{x^2} - \frac{1}{y^2}\frac{dy}{dx} = e^y\frac{dy}{dx}$

5 $-\frac{2}{x^3} - \frac{2}{y^3}\frac{dy}{dx} = 0$

6 $\frac{x}{2} - \frac{2y}{9}\frac{dy}{dx} = 0$

7 $\cos x + \cos y\frac{dy}{dx} = 0$

8 $\cos x \cos y - \sin x \sin y\frac{dy}{dx} = 0$

9 $e^y + xe^y\frac{dy}{dx} = 1$

10 $(1 + x)\frac{dy}{dx} = 2x - 1 - y$

11 $\frac{dy}{dx} = \pm\frac{1}{\sqrt{2x+1}}$

12 $\frac{d^2y}{dx^2} = \pm\frac{x}{\sqrt{(2-x^2)^3}}$

13 $\pm\frac{1}{4}\sqrt{2}$

15 $\frac{dy}{dx} = \frac{1}{1+x^2}$

16 **a** $xx_1 - 3yy_1 = 2(y + y_1)$
b $x(2x_1 + y_1) + y(2y_1 + x_1) = 6$

18 $3x + 12y - 7 = 0$

Exercise 25B – p.312

1 $3^x \ln 3$

2 $2(1.5)^x \ln 1.5$

3 $(2\ln 3)3^{2x}$

4 $3^x(1 + x\ln 3)$

5 $-3^{-x}\ln 3$

6 $5a^x \ln a$

7 $\frac{dy}{dx} = \frac{\ln x - 1}{(\ln x)^2}$

8 $\frac{dy}{dx}\left(\frac{1}{y+1} - \ln x\right) = \frac{y}{x}$

9 $\frac{dy}{dx} = 2x^{2x}(1 + \ln x)$

10 **a** $\frac{-1}{\sqrt{1-x^2}}$ **b** $\frac{2}{\sqrt{1-4x^2}}$
c $\frac{1}{1+x^2}$

Exercise 25C – p.316

1 **a** $\frac{1}{4t}$ **b** $-\cot\theta$ **c** $-\frac{4}{t^2}$

2 **a**

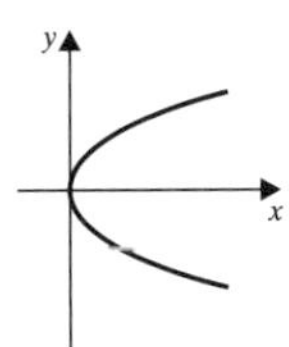

b

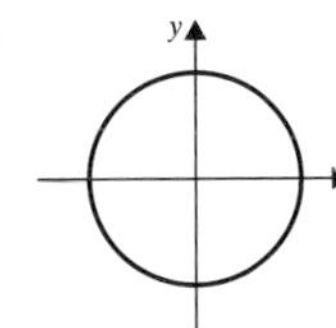

c

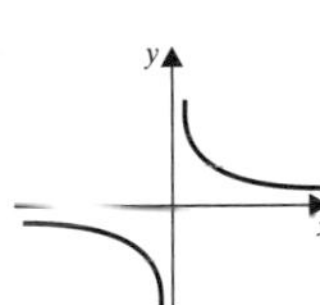

3 $\frac{dy}{dx} = 2t - t^2; \frac{3}{4}$

4 **a** $\frac{3}{2}t$ **b** $\frac{3}{2}\sqrt{x}$
c $x = t^2 \Rightarrow t = \sqrt{x}$

5 **a** $x = 2y^2; \frac{dy}{dx} = \frac{1}{4y} = \frac{1}{4t}$
b $x^2 + y^2 = 1;$
$\frac{dy}{dx} = -\frac{x}{y} = -\cot\theta$
c $xy = 4; \frac{dy}{dx} = -\frac{4}{x^2} = -\frac{4}{t^2}$

6 $(-\frac{1}{3}\sqrt{3}, \frac{2}{9}\sqrt{3})$, max;
$(\frac{1}{3}\sqrt{3}, -\frac{2}{9}\sqrt{3})$, min

7 $\frac{1}{2}\pi$

8 $2x + y + 2 = 0$

9 $6y = 4x + 5\sqrt{2}, (-\frac{137}{97}\sqrt{2}, -\frac{21}{194}\sqrt{2})$

Mixed Exercise 25 – p.316

1 **a** $4y^3\frac{dy}{dx}$ **b** $y^2 + 2xy\frac{dy}{dx}$
c $-\frac{1}{y^2}\frac{dy}{dx}$ **d** $\ln y + \frac{x}{y}\frac{dy}{dx}$
e $\cos y\frac{dy}{dx}$ **f** $e^y\frac{dy}{dx}$
g $\frac{dy}{dx}\cos x - y\sin x$
h $(\cos y - y\sin y)\frac{dy}{dx}$

2 $\frac{x}{2y}$

3 $-\frac{y^2}{x^2}$

4 $-\frac{2y}{3x}$

5 $3(1.1)^x \ln(1.1)$

6 $-\frac{(x-1)^3(3x+13)}{(x+3)^2}$

7 $-\frac{y(y+1)(3x+2)}{x(x+1)(y+2)}$

8 $3t/2$

9 $\frac{t}{t+1}$

10 $-\frac{3}{2}\cos\theta$

11 $-\frac{1}{t^2}$

12 $-\frac{1}{e^t}$

13 $2t - t^2$

14 $\frac{dy}{dx} = -4x$

16 $2y\frac{dy}{dx} - 2x\frac{dy}{dx} - 2y + 3\frac{dy}{dx} = 7$

17 $y = 2x + 2\sqrt{2}$

Chapter 26

Exercise 26A – p.320

1 **a** 10.003 333; 10.003 332
b 2.083 333; 2.080 084
c 3.979 167; 3.979 057

2 **a** 0.857 **b** 0.515 **c** 0.719

3 $\left\{\frac{x}{1+x} + \ln(1+x)\right\}\delta x$; 0.75

4 $(\sec^2 x)\delta x$; $1 + \frac{1}{16}\pi$

5 2.0125

6 $\frac{1}{8}\sqrt{2}; \frac{1}{8}a\sqrt{2}$

ANSWERS

Exercise 26B – p.323
Exercise 26B – p.323

1 0.099 cm/s
2 30 cm^3/s
3 8 m^2/s
4 Decreasing at 0.126 cm/s
5 -2
6 $4a$ cm/minute

Summary E
Summary E

Multiple Choice Exercise E – p.325
Multiple Choice Exercise E – p.325

1 C
2 A, E
3 D, E
4 D
5 A
6 C
7 B, C
8 B, C
9 C
10 C, D
11 B, D
12 F
13 F
14 F
15 F
16 T
17 F
18 T
19 F

Examination Questions E – p.327
Examination Questions E – p.327

1 no; should be $\theta + \tan^{-1} 4/3$
2 $y = 20x + 11$
3 a $\dfrac{-1}{2x^{3/2}}$ b $-e^{-x}$
c $2x\cos 2x - 2x^2 \sin 2x$
4 $t + 2t^2$
5 a 3 b $0 < f''(x) \leqslant 42$
6 27°, 90°, 207°, 270°
7 $4\cos\theta(2\cos^2\theta - 1)$
8 $0, 3/k$
9 0, 120°
10 $R = 5, \alpha = 53.13°$; 6.6°, 120.2°, 186.6°, 300.2°
11 a $\frac{1}{6}\pi$ b $-\dfrac{3}{16\sin t}$
d -1.92
12 b $k = 4$ c $-\left(\dfrac{y}{x}\right)^{1/3}$
13 a 0, 75.5°, 180°, 284.5°, 360°
b 18.4°, 116.6°, 198.4°, 296.6°
14 a $x(1 + 2\ln x)$ b $\dfrac{62x}{(2x^2+7)^2}$
c $30x^2(x^3+5)^9$
15 $\tan\alpha = b/a$; 130°, 342°
16 a $\frac{1}{2}\sqrt{3}$ b $\sqrt{\frac{1}{2} + \sqrt{\frac{3}{16}}}$
17 $2\sec^2 2x$
18 a $2x(\sin 2x + x\cos 2x)$
b $\dfrac{-1}{(x^2-1)\sqrt{x^2-1}}$
19 $\dfrac{2e^{2t}-2}{2e^{2t}-5}$; $\frac{1}{2}\ln 4 (= \ln 2)$
20 a $\cos v$ b $\dfrac{1}{\sqrt{1-u^2}}$
21 a $R = \sqrt{5}, \alpha = 26.6°$; $(63.4°, \sqrt{5})$
b $3 + \sin x - 6\sin^2 x$; 41.8°, 138.2°, 210°, 330°
22 $R = 2\sqrt{2}, \alpha = \pi/4$; $\pi/12$
23 $x + 2x\ln x$
24 b $\theta < 15°$
25 $\dfrac{1-e^{-t}}{1+e^{t}}$
26 a $3y - 2x - 1 = 0$ b $(8, 4)$
27 a $R = 17, A = 61.9°$
b 17, 298.1° c 7.4°, 228.8°
28 $k = -h/3$
29 a $\dfrac{x\cos x - \sin x}{x^2}$ b $\dfrac{-2x}{x^2+9}$
30 a $\pi/2, 7\pi/6, 11\pi/6$
b 0.25, 2.89, 1.57, 4.71
31 a 0.85, 2.29, 4.71
b 1.70, 4.59
32 a $13\sin(x + 67.4°)$
b min 1 where $x = 22.6°$, max $7\frac{1}{2}$ where $x = 202.6°$
33 a $y + xe^{-t} - e^{-t}(1+t) = 0$
c $A = 0$, min $(t = -1)$;
$A = 2/e$, max $(t = 1)$
34 a $3t^2/2$; 6 b $y = \dfrac{(x+1)^3}{8}$
c

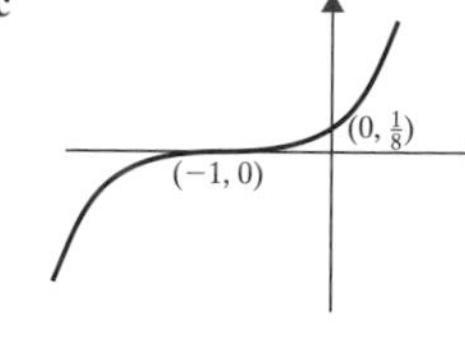

d $y = 6x - 10$; $(-9, -64)$
35 i a $\dfrac{1}{2\sqrt{x}}e^{\sqrt{x}}$
b $\dfrac{2x^2\ln x - (1+x^2)}{x(\ln x)^2}$
ii b $4x - 6y + 1 = 0$
36 a $4\pi r^2$ b 280 cm^{3}s^{-1}
37 a $h = 9$; 5 cm^3 b 0.24 cm^{3}s^{-1}
38 max 25 at 0, 2π, 4π, min 15 at π, 3π
39 c $\frac{32}{81}\pi R^3$
40 a $\dfrac{\ln P - \ln P_0}{\ln 1.09}$ b 8.04
c $P_0(1.09)^T \ln 1.09 = 0.172 P_0$
41 0, 60°, 120°, 180°
42 $k = 0.039, a = 0.90, -0.3$
43 i $\delta P \simeq (0.0008)x\delta x$ iii 1%
44 ii (1, 3), (−1, −3), (3, 1), (−3, −1)
iii $(\sqrt{8}, 0), (-\sqrt{8}, 0), (0, \sqrt{8}), (0, -\sqrt{8})$; 3, $\frac{1}{3}$
45 i 1.65, 65% ii 0.5 iv $\frac{1}{2}$

Chapter 28
Chapter 28

Exercise 28A – p.337
Exercise 28A – p.337

1 $y^2 = 2y + 12x - 13$
2 $x + y = 4$
3 $8x^2 + 9y^2 - 20x = 28$
4 $11y = 3x$ and $99x + 27y + 130 = 0$
5 $x^2 + y^2 - 6x - 10y + 30 = 0$
6 $4x - 3y = 26$ and $4x - 3y + 24 = 0$
7 $x^2 + y^2 = 1$

518

Exercise 28B – p.338

1 **a** $x^2+y^2-2x-4y=4$
b $x^2+y^2-8y+15=0$
c $x^2+y^2+6x+14y+54=0$
d $x^2+y^2-8x-10y+32=0$

2 **a** $(-4, 1); 5$
b $(-\frac{1}{2}, -\frac{3}{2}); 3\sqrt{2}/2$
c $(-3, 0); \sqrt{14}$
d $(\frac{3}{4}, -\frac{1}{2}); \sqrt{5}/4$
e $(0, 0); 2$ **f** $(2, -3); 3$
g $(1, 3); 3$
h $(-1, \frac{1}{2}); \sqrt{69}/6$

3 **a** and **f**

Exercise 28C – p.340

1 **a** $3y+4x=23$
b $4y=x+22$
c $3y=2x+25$

2 $2x^2+2y^2-15x-24y+77=0$

3 $x^2+y^2-8x+6y=0$

4 $x^2+y^2-8x-6y+17=0$

5 $x^2+y^2-4x-14y-116=0$

6 $x^2+y^2-6y=0$

7 $x+2y=0$

Exercise 28D – p.343

1

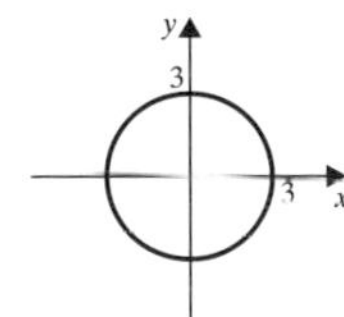

2

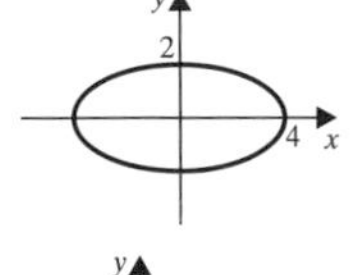

3

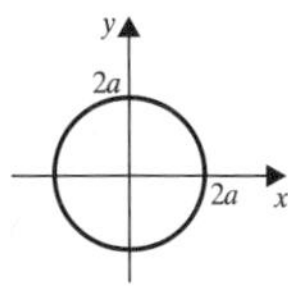

4

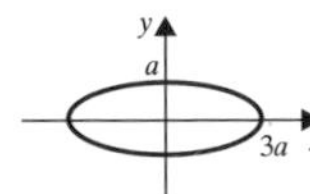

5

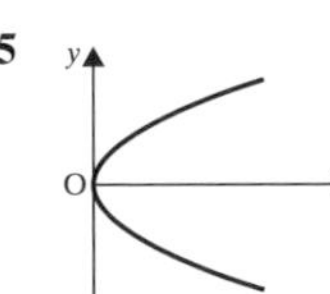

6 **a** $x^2+(y-1)^2=1$

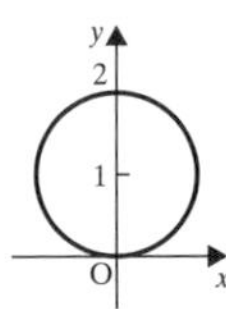

b $y+x=1+\sqrt{2}$

7 **a**

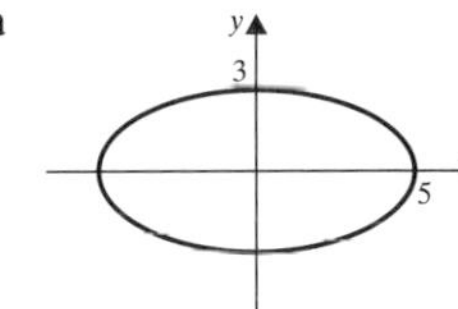

b $x=5$

8 **a** $(x-2)^2+(y-1)^2=4$

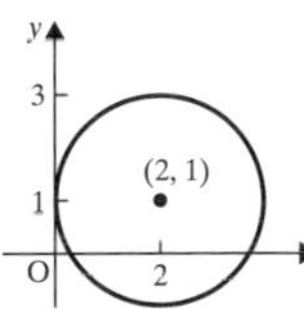

b $\theta=+\frac{\pi}{2}$

9 **a**

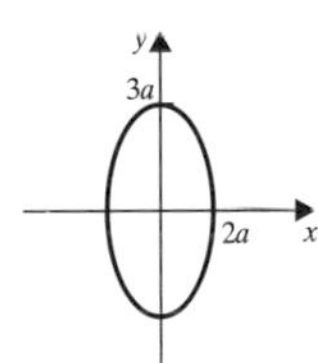

b $2y+3\sqrt{3}x-6\sqrt{3}a=0$

Exercise 28E – p.345

1 $y=\dfrac{1}{T}(xT^3-T^4+1)$

2 $y=\dfrac{2}{p}(x+p^2)$

3 $y\sin\alpha=2-2x\cos\alpha$

4 $\dfrac{-1}{2s_1s_2(s_1+s_2)}$

5 $(0, -1), (1, 0)$

6 **a** $y-\dfrac{2}{s}=s^2(x-2s)$
b $\left(-\dfrac{2}{s^3}, -2s^3\right)$

7 $t^2y+x=2t; (2t, 0), \left(0, \dfrac{2}{t}\right);$
$A=2$

8 $2y+tx=8t+t^3; (8+t^2, 0),$
$(0, \frac{t}{2}(8+t^2)); \frac{t}{4}(8+t^2)^2$

9 **a** $a^3x-ay=a^4-1$
b $a^3b=1$

10 **a** $y^2=8x$
b $x-2y+8=0$
c $2x-y(p+q)+4pq=0$
d $(18, 12), (2, -4)$
e 2
f $(2, 4), (2, -4)$
g $(8, 8)$

11 **a** $\dfrac{\sin\theta_1-\sin\theta_2}{\cos\theta_1-\cos\theta_2}$
b $[\frac{1}{2}(\cos\theta_1+\cos\theta_2),$
$\frac{1}{2}(2+\sin\theta_1+\sin\theta_2)]$
c $\sin\theta_1=\sin\theta_2 \Rightarrow$
$\theta_1=\pi-\theta_2$

12 **b** **i** $-\frac{1}{2}\pi, \frac{1}{2}\pi$ **ii** $-\pi, 0, \pi$
c **i** $-a\leqslant x\leqslant a$ **ii** $-b\leqslant y\leqslant b$
d

y
2
−3
3
x
−2

e a circle

Chapter 29

All indefinite integrals in this chapter require the term $+K$.

Exercise 29A – p.348

1 $\frac{1}{4}e^{4x}$

2 $-4e^{-x}$

3 $\frac{1}{3}e^{(3x-2)}$

4 $-\frac{2}{5}e^{(1-5x)}$

5 $-3e^{-2x}$

6 $5e^{(x-3)}$

7 $2e^{(x/2+2)}$

8 $\dfrac{2^x}{\ln 2}$

9 $\dfrac{4^{(2+x)}}{\ln 4}$

10 $\frac{1}{2}e^{2x}-\dfrac{1}{2e^{2x}}$

11 $\dfrac{a^{(1-2x)}}{-2\ln a}$

12 $\dfrac{2^x}{\ln 2}+\frac{1}{3}x^3$

13 $-(x+3)^{-1}$
14 $\frac{2}{3}(1+x)^{3/2}$
15 $\frac{1}{18}(3x+1)^6$
16 $\frac{1}{25}(5x-2)^5$
17 $\frac{1}{2}\{e^4-1\}$
18 $2\{e^2-1\}$
19 $1-\dfrac{1}{e}$
20 $1-e^2$

Exercise 29B – p.349

Answers can be expressed in other forms also.

1 a $\frac{1}{2}\ln|x|+K$
  b $\frac{1}{2}\ln\{A|x|\}$
2 a $4\ln|x-1|+K$
  b $4\ln\{A|x-1|\}$ or $\ln A(x-1)^4$
3 a $\frac{1}{3}\ln|3x+1|+K$
  b $\frac{1}{3}\ln A|3x+1|$ or $\ln A|\sqrt[3]{(3x+1)}|$
4 a $-\frac{3}{2}\ln|1-2x|+K$
  b $-\frac{3}{2}\ln A|1-2x|$ or $\ln A|(1-2x)^{-3/2}|$
5 a $2\ln|2+3x|+K$
  b $\ln A(2+3x)^2$
6 a $-\frac{3}{2}\ln|4-2x|+K$
  b $\ln A|(4-2x)^{-3/2}|$ or $\ln\dfrac{A}{|(4-2x)^{3/2}|}$ (or each of these results using $\ln|2-x|$)
7 a $-4\ln|1-x|+K$
  b $\ln\dfrac{A}{(1-x)^4}$
8 a $-\frac{5}{7}\ln|6-7x|+K$
  b $\ln\dfrac{A}{|(6-7x)^{5/7}|}$
9 $3\ln 1.5$
10 $\frac{1}{2}\ln 3$
11 $2\ln 2=\ln 4$
12 $\ln 2$

Exercise 29C – p.350

1 $-\frac{1}{2}\cos 2x$
2 $\frac{1}{7}\sin 7x$
3 $\frac{1}{4}\tan 4x$
4 $-\cos(\frac{1}{4}\pi+x)$
5 $\frac{3}{4}\sin(4x-\frac{1}{2}\pi)$
6 $\frac{1}{2}\tan(\frac{1}{3}\pi+2x)$
7 $-\frac{1}{4}\cot 4x$
8 $-\frac{2}{3}\cos(3x-\alpha)$
9 $-10\sin(\alpha-\frac{1}{2}x)$
10 $\frac{5}{4}\sec 4x$
11 $\frac{1}{3}\sin 3x-\sin x$
12 $\frac{1}{2}\tan 2x+\frac{1}{4}\cot 4x$
13 $\frac{1}{3}$
14 $-\frac{1}{4}$
15 0
16 $\frac{1}{2}$

Exercise 29D – p.351

1 $\frac{1}{2}\cos(\frac{1}{2}\pi-2x)$
2 $\frac{1}{4}e^{(4x-1)}$
3 $\frac{1}{7}\tan 7x$
4 $\frac{1}{2}\ln|2x-3|$
5 $\sqrt{2x-3}$
6 $-1/\{3(3x-2)\}$
7 $5^x/\ln 5$
8 $-2\operatorname{cosec}\frac{1}{2}x$
9 $\frac{1}{9}(3x-5)^3$
10 $\frac{1}{4}e^{(4x-5)}$
11 $\frac{1}{6}(4x-5)^{3/2}$
12 $-\frac{1}{3}\cot 3x$
13 $-\frac{3}{2}\ln|1-x|$
14 $10^{(x+1)}/\ln 10$
15 $\frac{1}{3}\sin(3x-\frac{1}{3}\pi)$
16 $\frac{2}{3}x^3-8x$
17 $\frac{1}{4}x^4-2x^3+\frac{9}{2}x^2$
18 $3x-x^2+2x^3$
19 $-\dfrac{1}{3x^2}$
20 $\frac{3}{2}\ln|x|$
21 $\frac{2}{9}(x-3)^3$
22 $\frac{2}{3}\sqrt{2}$
23 $2e(e-1)$
24 $-\frac{1}{2}$

Exercise 29E – p.354

1 e^{x^4}
2 $-e^{\cos x}$
3 $e^{\tan x}$
4 e^{x^2+x}
5 $e^{(1-\cot x)}$
6 $e^{(x+\sin x)}$
7 $e^{(1+x^2)}$
8 $e^{(x^3-2x)}$
9 $\frac{1}{10}(x^2-3)^5$
10 $-\frac{1}{3}(1-x^2)^{3/2}$
11 $\frac{1}{6}(\sin 2x+3)^3$
12 $-\frac{1}{6}(1-x^3)^2$
13 $\frac{2}{3}(1+e^x)^{3/2}$
14 $\frac{1}{5}\sin^5 x$
15 $\frac{1}{4}\tan^4 x$
16 $\dfrac{1}{3(n+1)}(1+x^{n+1})^3$
17 $-\frac{1}{3}\cot^3 x$
18 $\frac{4}{9}(1+x^{3/2})^{3/2}$
19 $\frac{1}{12}(x^4+4)^3$
20 $-\frac{1}{4}(1-e^x)^4$
21 $\frac{2}{3}(1-\cos\theta)^{3/2}$
22 $\frac{1}{3}(x^2+2x+3)^{3/2}$
23 $\frac{1}{2}e^{(x^2+1)}$
24 $\frac{1}{2}(1+\tan x)^2$

Exercise 29F – p.355

1 $\frac{1}{2}(e-1)$
2 $\frac{1}{5}$
3 $\frac{1}{2}(\ln 2)^2$
4 $\dfrac{7^5}{15}$
5 $e-1$
6 9
7 $\frac{1}{2}(e^3-1)$
8 $\frac{13}{24}$
9 $\frac{1}{3}(\ln 3)^3$
10 $\frac{7}{3}$

Exercise 29G – p.357

1 $\frac{1}{2}$
2 does not exist
3 $2\sqrt{2}$
4 does not exist
5 does not exist

Exercise 29H – p.360

1 $x\sin x+\cos x$
2 $e^x(x^2-2x+2)$
3 $\frac{1}{16}x^4(4\ln|3x|-1)$
4 $-e^{-x}(x+1)$
5 $3(\sin x-x\cos x)$
6 $\frac{1}{5}e^x(\sin 2x-2\cos 2x)$
7 $\frac{1}{5}e^{2x}(\sin x+2\cos x)$
8 $\frac{1}{32}e^{4x}(8x^2-4x+1)$
9 $-\frac{1}{2}e^{-x}(\cos x+\sin x)$
10 $x(\ln|2x|-1)$
11 xe^x
12 $\frac{1}{72}(8x-1)(x+1)^8$

13 $\sin(x+\frac{1}{6}\pi)-x\cos(x+\frac{1}{6}\pi)$
14 $\frac{1}{n^2}(\cos nx+nx\sin nx)$
15 $\frac{x^{n+1}}{(n+1)^2}[(n+1)\ln|x|-1]$
16 $\frac{3}{4}(2x\sin 2x+\cos 2x)$
17 $\frac{1}{5}e^x(\sin 2x-2\cos 2x)$
18 $(2-x^2)\cos x+2x\sin x$
19 $\frac{e^{ax}}{a^2+b^2}(a\sin bx-b\cos bx)$
20 $\frac{1}{3}\sin\theta(3\cos^2\theta+2\sin^2\theta)$
21 $\frac{1}{2}e^{x^2-2x+4}$
22 $(x^2+1)e^x$
23 $-\frac{1}{4}(4+\cos x)^4$
24 $e^{\sin x}$
25 $\frac{2}{15}\sqrt{(1+x^5)^3}$
26 $\frac{1}{5}(e^x+2)^5$
27 $\frac{1}{4}e^{2x-1}(2x-1)$
28 $-\frac{1}{20}(1-x^2)^{10}$
29 $\frac{1}{6}\sin^6 x$

Exercise 29I – p.361

1 1
2 $\frac{32}{3}\ln 2-\frac{7}{4}$
3 e
4 $-\frac{1}{2}(e^\pi+1)$
5 $\frac{16}{15}$
6 $\frac{1}{4}\pi^2-2$
7 $2\ln 2-1$
8 $e-2$
9 $\frac{1}{4}$

Mixed Exercise 29 – p.361

1 $\frac{1}{4}e^{2x}(2x^2-2x+1)$
2 e^{x^2}
3 $\frac{1}{6}(3\tan x-4)^2$
4 $\frac{1}{4}(x+1)^2\{2\ln(x+1)-1\}$
5 $\frac{1}{4}\tan^4 x$
6 $(x^2-2)\sin x+2x\cos x$
7 $-e^{\cos x}$
8 $\frac{1}{288}(2x+3)^8(16x-3)$
9 $-\frac{1}{2}e^{(1-x)^2}$
10 $-\frac{1}{9}e^{(2-3x)}(3x+1)$
11 $\frac{1}{5}\sin^5 x$
12 $-\frac{1}{4}(4+\cos x)^4$
13 $\frac{1}{2}e^{(x^2-2x+3)}$
14 $-\frac{1}{30}(1-x^3)^{10}$
15 $\frac{1}{3}(e^9-e^3)$
16 $\frac{1}{4}$
17 $\ln\frac{1}{2}$
18 1
19 $\frac{1}{9}(e^3-1)$
20 $\frac{1}{8}(\pi-2)$
21 $\frac{1}{2}(\ln 2)^2$
22 $\frac{1}{3}\ln 2^8-\frac{7}{9}$
23 does not exist
24 $1/e$
25 $-\frac{1}{4}$
26 $\pi/4$

Chapter 30

Exercise 30A – p.364

1 $\frac{3}{2(x+1)}-\frac{1}{2(x-1)}$
2 $\frac{13}{6(x-7)}-\frac{1}{6(x-1)}$
3 $\frac{4}{5(x-2)}-\frac{4}{5(x+3)}$
4 $\frac{7}{9(2x-1)}+\frac{28}{9(x+4)}$
5 $\frac{1}{x-2}-\frac{1}{x}$
6 $\frac{3}{x-2}-\frac{1}{x-1}$
7 $\frac{1}{2(x-3)}-\frac{1}{2(x+3)}$
8 $\frac{7}{3x}-\frac{1}{3(x+1)}$
9 $\frac{9}{x}-\frac{18}{2x+1}$
10 $\frac{2}{5(x-1)}-\frac{1}{5(3x+2)}$

Exercise 30B – p.365

1 $\frac{1}{x-1}-\frac{1}{x+1}$
2 $\frac{1}{x-2}-\frac{1}{x+1}$
3 $\frac{1}{3(x-3)}-\frac{1}{3x}$
4 $\frac{1}{x-1}-\frac{1}{x+3}$
5 $\frac{1}{2(x-1)}-\frac{1}{2(x+1)}$
6 $\frac{1}{2x-1}-\frac{1}{2x+1}$

Exercise 30C – p.367

1 $\frac{1}{2(x-1)}-\frac{1}{2(x+1)}-\frac{1}{(x+1)^2}$
2 $\frac{3}{2x}-\frac{x}{2(x^2+2)}$
3 $\frac{11}{8(x-3)}+\frac{5}{8(x+1)}-\frac{1}{2(x+1)^2}$
4 $\frac{1}{x}-\frac{x}{2x^2+1}$
5 $\frac{1}{x-1}-\frac{1}{x-2}+\frac{2}{(x-2)^2}$
6 $\frac{2}{x}-\frac{1}{x^2}-\frac{3}{2x+1}$
7 $\frac{7}{16(x+3)}-\frac{1}{4(x-1)^2}+\frac{9}{16(x-1)}$
8 $\frac{3}{(x+2)^2}+\frac{2}{x+2}-\frac{2}{x+1}$
9 $\frac{1}{2(x-2)}-\frac{1}{6(x+2)}-\frac{1}{3(x-1)}$; using **b** goes directly to three partial fractions

Exercise 30D – p.368

1 **a** Q : $x+1$, R : $-6x+3$
b Q : x^3-x^2-4x+4, R : -2
c Q : $3x^2+6x+12$, R : 19
2 **a** $2+\frac{4}{x-2}$ **b** $1+\frac{4}{x^2-1}$
c $x+2+\frac{4}{x-2}$

Exercise 30E – p.368

1 $1+\frac{1}{2(x-1)}-\frac{1}{2(x+1)}$
2 $x+\frac{2}{x-1}-\frac{1}{x+1}$

3 $1 - \frac{7}{4(x+3)} - \frac{1}{4(x-1)}$

4 $1 - \frac{2x+1}{5(x^2+1)} - \frac{8}{5(x+2)}$

Exercise 30F – p.369

1 $\frac{2}{x-2} - \frac{2}{x-1}$; $\frac{-2}{(x-2)^2} + \frac{2}{(x-1)^2}$;
$\frac{4}{(x-2)^3} - \frac{4}{(x-1)^3}$;

2 $\frac{9}{5(x-3)} - \frac{3}{5(2x-1)}$;
$\frac{6}{5(2x-1)^2} - \frac{9}{5(x-3)^2}$;
$\frac{-24}{5(2x-1)^3} + \frac{18}{5(x-3)^3}$

3 $\frac{2}{3(x-4)} + \frac{1}{3(x+2)}$;
$\frac{-1}{3(x+2)^2} - \frac{2}{3(x-4)^2}$;
$\frac{2}{3(x+2)^3} + \frac{4}{3(x-4)^3}$

4 $\frac{1}{x-3} - \frac{1}{x+2}$; $\frac{1}{(x+2)^2} - \frac{1}{(x-3)^2}$;
$\frac{-2}{(x+2)^3} + \frac{2}{(x-3)^3}$;

5 $\frac{3}{2x+3} - \frac{1}{x+1}$;
$\frac{-6}{(2x+3)^2} + \frac{1}{(x+1)^2}$;
$\frac{24}{(2x+3)^3} - \frac{2}{(x+1)^3}$

6 $\frac{3}{2(x-1)} - \frac{9}{2(3x-1)}$;
$\frac{27}{2(3x-1)^2} - \frac{3}{2(x-1)^2}$
$\frac{-81}{(3x-1)^3} + \frac{3}{(x-1)^3}$

Exercise 30G – p.371

1 **a** 3 **b** 18 **c** 47
d $\frac{35}{16}$ **e** $-\frac{16}{27}$
f $a^3 - 2a^2 + 6$ **g** $c^2 - ac + b$
h $\frac{1}{a^4} - \frac{2}{a} + 1$

2 −7

3 $(x-1)(2x-1)(x+1)$;
$-1 < x < \frac{1}{2}, x > 1$

4 $f(x) = (x-2)(x^2+x+1)$
$\Rightarrow \quad f(x) = 0$ only when $x = 2$;
$x < 2$

5 $x = \pm 1, \pm 2$

6 $(x-1)(x+1)(x-2)(x-3)$

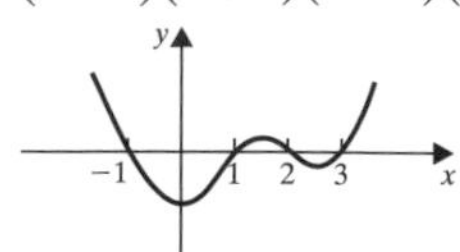

7 **a** $p = 11, q = -5$ **b** 3

Exercise 30H – p.373

1 touch at (1, 2)

2 (1, 1) and (−5, 4)

3 no intersection

4 (1, 0) and (2, 0)

5 (0, 0), (−3, 0) and (−2, 0)

6 touch at (1, 0) and (2, 0)

7 touch at (0, 0)

8 $k = 8$

9 $k = \frac{1}{4}$

10 $k = \pm 2\sqrt{19}$

11 $(\frac{1}{2}, 2)$

12 (1, 2) and (1, −2)

13 touch at (−1, −2); cut at (1, 2) and $(4, \frac{1}{2})$

14 (0, 0) and $)\frac{2}{3}, 2)$

15 **a** $\pm 2\sqrt{3}$ **b** $-2\sqrt{3} < \lambda < 2\sqrt{3}$
c $\lambda < -2\sqrt{3}, \lambda > 2\sqrt{3}$

16

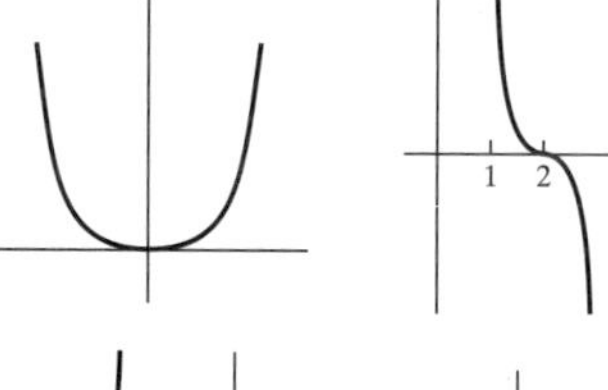

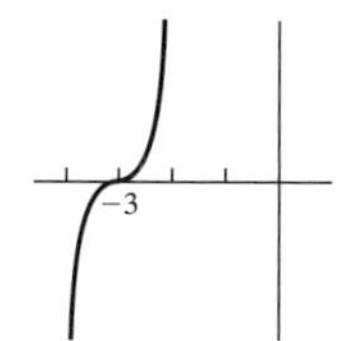

17 **a** **i** $y = \pm(x-1)$
ii $2y + x = 4$
b **i** $y = 36(1-x)$
ii $2y + x = \pm 6\sqrt{2}$
c **i** $3y = 2(x-1), y = 0$
ii $8y + 4x + 3 = 0$

Mixed Exercise 30 – p.374

1 **a** $\frac{4}{7(x-3)} - \frac{8}{7(2x+1)}$
b $\frac{5}{7(x+1)} + \frac{1}{7(4x-3)}$
c $\frac{1}{t-1} + \frac{1}{t+1}$

2 **a** $\frac{3}{x} - \frac{6}{2x+1}$
b $\frac{9}{8(x-5)} - \frac{1}{8(x+3)}$
c $\frac{1}{5(x-2)} + \frac{6}{5(4x-3)}$
d $\frac{1}{2x-3} - \frac{1}{2x+3}$
e $\frac{4}{9(x-8)} - \frac{4}{9(x+1)}$
f $\frac{1}{x-2} + \frac{1}{2(x+1)}$

3 $\frac{4}{9x} - \frac{1}{x^2} + \frac{8}{9(x-3)}$

4 $\frac{1}{x+4} - \frac{x}{x^2+1}$

5 $\frac{1}{2(x+3)} + \frac{2}{(x-1)^2} - \frac{1}{2(x-1)}$

6 $\frac{1}{4(x-1)} - \frac{1}{2(x+1)^2} + \frac{3}{4(x+1)}$

7 $\frac{1}{6(x-1)} + \frac{5-x}{6(x^2+5)}$

8 $\frac{5}{6(x+2)} - \frac{5x-4}{6(x^2+2)}$

9 **a** $1 + \frac{1}{x+1} - \frac{4}{x+2}$
b $1 + \frac{3}{x^2} - \frac{3}{x} + \frac{2}{x+1}$

10 **a** $y = \frac{4}{x-2} - \frac{2}{x-1}$;
$\frac{dy}{dx} = -\frac{4}{(x-2)^2} + \frac{2}{(x-1)^2}$;
$\frac{d^2y}{dx^2} = \frac{8}{(x-2)^3} - \frac{4}{(x-1)^3}$
b $y = \frac{3}{5(x+3)} + \frac{2}{5(x-2)}$;
$\frac{dy}{dx} = -\frac{3}{5(x+3)^2} - \frac{2}{5(x-2)^2}$;
$\frac{d^2y}{dx^2} = \frac{6}{5(x+3)^3} + \frac{4}{5(x-2)^3}$
c $y = \frac{3}{2(x+1)} - \frac{5}{x+2} + \frac{9}{2(x+3)}$;
$\frac{dy}{dx} = -\frac{3}{2(x+1)^2} + \frac{5}{(x+2)^2} - \frac{9}{2(x+3)^2}$;
$\frac{d^2y}{dx^2} = \frac{3}{(x+1)^3} - \frac{10}{(x+2)^3} + \frac{9}{(x+3)^3}$

11 **a** cross at $(0, 2)$ and $(\frac{8}{5}, \frac{14}{5})$
b cross at $(-2, 0)$, touch at $(-3, 0)$

12 **a** $m = \pm 2\sqrt{2}$
b

y
(0, 1)
$x = \frac{1}{2}(3 + \sqrt{5})$
$x = \frac{1}{2}(3 - \sqrt{5})$
x

c $-4 < m < 4$

13 $f(x) = (x - 1)(2x - 1)(2x + 3)$

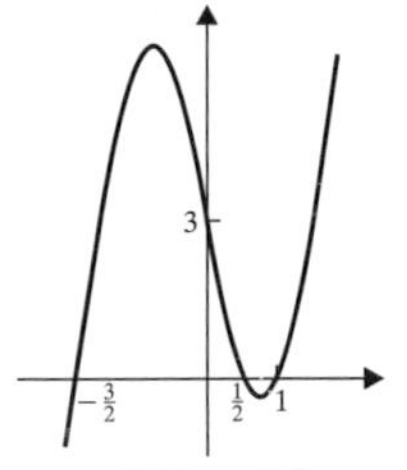

$x < -3/2$ and $\frac{1}{2} < x < 1$

14 $x^3 - 4x^2 + 3x = 0$; 0, 1, 3:

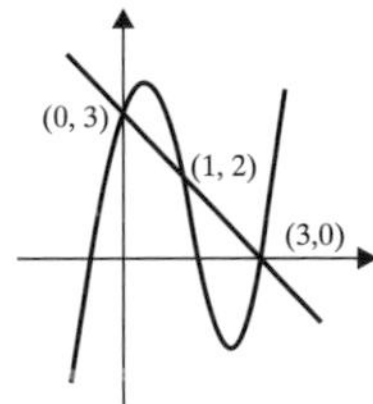

Chapter 31

All indefinite integrals in this chapter require the addition of a constant of integration.

Exercise 31A – p.378

1 $\ln(4 + \sin x)$
2 $\frac{1}{3}\ln|3e^x - 1|$
3 $\dfrac{1}{4(1 - x^2)^2}$
4 $\dfrac{1}{2\cos^2 x}$
5 $\frac{1}{4}\ln(1 + x^4)$
6 $\ln|x^2 + 3x - 4|$
7 $\frac{2}{3}\sqrt{2 + x^3}$
8 $\dfrac{1}{2 - \sin x}$
9 $\ln|\ln x|$
10 $\dfrac{-1}{5\sin^5 x}$
11 $-\ln|1 - x^2|$
12 $-2\sqrt{1 - e^x}$
13 $\frac{1}{6}\ln|3x^2 - 6x + 1|$
14 $\dfrac{-1}{(n - 1)\sin^{n-1} x}$ $(n \neq 1)$
15 $\dfrac{1}{(n - 1)\cos^{(n-1)} x}$ $(n \neq 1)$
16 $\ln|4 + \sec x|$
17 $\frac{1}{2}\ln|x(x - 2)|$
18 $-\dfrac{1}{e^x - x}$
19 $\ln 3$
20 $\ln\sqrt{2}$
21 $\frac{55}{1152}$
22 $\dfrac{e - 1}{2(e + 1)}$
23 0
24 $\dfrac{1}{\ln 4}$

Exercise 31B – p.380

1 $2\ln\left|\dfrac{x}{x + 1}\right|$
2 $\ln\left|\dfrac{x - 2}{x + 2}\right|$
3 $\frac{1}{2}\ln|x^2 - 1|$
4 $\frac{1}{2}\ln\left|\dfrac{(x + 2)^3}{x}\right|$
5 $\ln\dfrac{(x - 3)^2}{|x - 2|}$
6 $\frac{1}{2}\ln\dfrac{|x^2 - 1|}{x^2}$
7 $x - \ln|x + 1|$
8 $x + 4\ln|x|$
9 $x - 4\ln|x + 4|$
10 $\ln\dfrac{|1 - x|}{x^4}$
11 $x - \frac{1}{2}\ln\left|\dfrac{x + 1}{x - 1}\right|$
12 $x + \ln\dfrac{|x + 1|}{(x + 2)^4}$
13 $\frac{1}{2}\ln|x^2 - 1|$
14 $\dfrac{-1}{x^2 - 1}$
15 $\ln\left|\dfrac{x - 1}{x + 1}\right|$
16 $\ln|x^2 - 5x + 6|$
17 $\ln\dfrac{(x - 3)^6}{(x - 2)^4}$
18 $\ln\left|\dfrac{(x - 3)^3}{x - 2}\right|$
19 $4 + \ln 5$
20 $\ln\frac{1}{6}$
21 $\frac{1}{2}\ln\frac{12}{5}$
22 $\ln\frac{5}{3}$
23 $\frac{5}{36}$
24 $1 - \frac{3}{2}\ln\frac{7}{5}$

Exercise 31C – p.383

1 $\frac{1}{4}(2x + \sin 2x)$
2 $\sin x - \frac{1}{3}\sin^3 x$
3 $-\frac{1}{15}\cos x(15 - 10\cos^2 x + 3\cos^4 x)$
4 $\tan x - x$
5 $\frac{1}{32}\{12x - 8\sin 2x + \sin 4x\}$
6 $\frac{1}{2}\tan^2 x - \ln|\sec x|$
7 $\frac{1}{32}\{12x + 8\sin 2x + \sin 4x\}$
8 $\frac{1}{3}\cos x(\cos^2 x - 3)$
9 $\frac{1}{15}\sin^3\theta(5 - 3\sin^2\theta)$
10 $(\sin^{11}\theta)(\frac{1}{11} - \frac{1}{13}\sin^2\theta)$
11 $(\sin^{(n+1)}\theta)\left(\dfrac{1}{n + 1} - \dfrac{1}{n + 3}\sin^2\theta\right)$
$(n \neq -1$ or $-3)$
12 $\frac{1}{32}(4\theta - \sin 4\theta)$
13 $-\frac{1}{16}(\cos 8\theta + 4\cos 2\theta)$
14 $\frac{1}{8}(2\cos 2\theta - \cos 4\theta)$
15 $\frac{1}{12}$
16 $\frac{13}{15} - \frac{1}{4}\pi$
17 $\frac{1}{6}$
18 $\frac{2}{3}$

Chapter 32

All indefinite integrals require the term $+K$.

Exercise 32A – p.385

1 $\frac{1}{6}x^6$
2 $\frac{1}{2}x^2 + e^x$
3 $\frac{1}{2}x^2 + \ln|x|$
4 $2x + \sin x$
5 $\frac{1}{2}e^{2x-3}$
6 $-\cos 5x$
7 $\frac{1}{8}\sin 8x$
8 $-\frac{1}{16}(1-4x)^4$
9 $\dfrac{1}{3(2-3x)}$
10 $\frac{3}{2}\cos(4-2x)$

Exercise 32B – p.387

1 $\frac{1}{21}(x+3)^6(3x+2)$
2 $-\frac{2}{3}(x+6)\sqrt{(3-x)}$
3 $\frac{2}{15}(3x-2)(x+1)^{3/2}$
4 $\dfrac{1-5x}{10(x-3)^5}$
5 $\frac{4}{135}(9x+8)(3x-4)^{3/2}$
6 $-\frac{1}{36}(8x+1)(1-x)^8$
7 $\dfrac{5+4x}{12(4-x)^4}$

Exercise 32C – p.387

1 $\frac{1}{3}x^3 - \frac{1}{2}x^2$
2 $\frac{1}{4}x^4 - \frac{2}{3}x^3$
3 $\frac{1}{2}e^{2x+3}$
4 $\frac{1}{6}(2x^2-5)^{3/2}$
5 $xe^x - x$
6 $x\ln x - x$
7 $\frac{1}{12}(6x - \sin 6x)$
8 $-\frac{1}{2}e^{-x^2}$
9 $\frac{1}{3}\sin^3 x$
10 $\frac{1}{110}(10u-7)(u+7)^{10}$
11 $-\dfrac{1}{12(x^3+9)^4}$
12 $\frac{1}{2}\ln|1-\cos 2y|$ $(y \neq n\pi)$
13 $\frac{1}{2}\ln|2x+7|$
14 $-\frac{2}{9}(1+\cos 3x)^{3/2}$
15 $\frac{1}{16}(\sin 4x - 4x\cos 4x)$
16 $\frac{1}{2}\ln|x^2+4x-5|$
17 $\frac{1}{3}\ln|(x-1)(x+5)^2|$
18 $-\frac{1}{4}(x^2+4x-5)^{-2}$
19 $-(9-y^2)^{3/2}$
20 $\frac{1}{13}e^{2x}(2\cos 3x + 3\sin 3x)$
21 $x(\ln|5x| - 1)$
22 $\frac{1}{6}\sin 2x(3-\sin^2 2x)$
23 $e^{\sin x}$
24 $-2\sqrt{7+\cos y}$
25 $e^x(x^2-2x+2)$
26 $\frac{1}{2}\ln|x^2-4|$
27 $x + \ln\left|\dfrac{x-2}{x+2}\right|$
28 $\frac{1}{4}\ln\left|\dfrac{x-2}{x+2}\right|$
29 $\frac{1}{4}x^2(2\ln|x| - 1)$
30 $\frac{1}{15}\cos^3 u(3\cos^2 u - 5)$
31 $\tan\theta - \theta$
32 $x - \ln|1-x|$
33 $-\ln|1-\tan x|$
34 $\frac{1}{3}(7+x^2)^{3/2}$
35 $-\frac{1}{5}\cos(5\theta - \frac{1}{4}\pi)$
36 $e^{\tan u}$

Exercise 32D – p.390

1 $y^2 = A - 2\cos x$
2 $\dfrac{1}{y} - \dfrac{1}{x} = A$
3 $2y^3 = 3(x^2 + 4y + A)$
4 $x = A\sec y$
5 $(A-x)y = 1$
6 $y = \ln\dfrac{A}{\sqrt{1-x^2}}$
7 $y = A(x-3)$
8 $x + A = 4\ln|\sin y|$
9 $u^2 = v^2 + 4v + A$
10 $16y^3 = 12x^4\ln|x| - 3x^4 + A$
11 $y^2 + 2(x+1)e^{-x} = A$
12 $\sin x = A - e^{-y}$
13 $2r^2 = 2\theta - \sin 2\theta + A$
14 $u + 2 = A(v+1)$
15 $y^2 = A + (\ln|x|)^2$
16 $y^2 = Ax(x+2)$
17 $4v^3 = 3(2+t)^4 + A$
18 $1 + y^2 = Ax^2$
19 $Ar = e^{\tan\theta}$
20 $y^2 = A - \operatorname{cosec}^2 x$
21 $v^2 + A = 2u - 2\ln|u|$
22 $e^{-x} = e^{1-y} + A$
23 $A - \dfrac{1}{y} = 2\ln|\tan x|$
24 $y - 1 = A(y+1)(x^2+1)$

Exercise 32E – p.392

1 $y^3 = x^3 + 3x - 13$
2 $e^t(5 - 2\sqrt{s}) = 1$
3 $3(y^2-1) = 8(x^2-1)$
4 $y = x^2 - x$
5 $y = e^x - 2$
6 $y = 5 - \dfrac{3}{x}$
7 $(0, \sqrt{3+e^{-3}}), (0, -\sqrt{3+e^{-3}})$
8 $(y+1)^2(x+1) = 2(x-1)$
9 $2y = x^2 + 6x$
10 $4y^2 = (y+1)^2(x^2+1)$

Mixed Exercise 32 – p.393

1 $\frac{1}{10}(1+x^2)^5$
2 $-\frac{1}{9}e^{-3x}(3x+1)$
3 $\frac{1}{10}(\sin 5x + 5\sin x)$
4 $x + \ln|x+2|$
5 $\dfrac{-1}{3(x^3+1)}$
6 $\ln\left|\dfrac{x-4}{x-1}\right|$
7 $\ln\left|\dfrac{x}{2x+1}\right|$
8 $\ln\sqrt{x^2+1} - \arctan x$
9 $-2\sqrt{\cos x}$
10 $\sqrt{2}$
11 $\frac{256}{15}$
12 $\frac{1}{2} - \ln\sqrt{2}$
13 $9 - \frac{16}{3}\sqrt{2}$
14 $\ln\frac{9}{4}$
15 $\ln\left\{\dfrac{\ln 3}{\ln 2}\right\}$
16 $\ln\frac{1}{6}$
17 $2\frac{1}{2} + 6\ln\frac{2}{3}$
18 $x^3y = y - 1$
19 $y = \tan\{\frac{1}{2}(x^2-4)\}$
20 $y^2 = 2x$

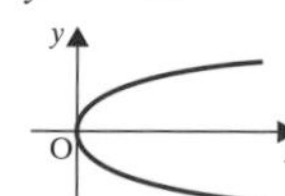

Chapter 33

Exercise 33A – p.395

1 $s\dfrac{ds}{dt} = k$

2 $\dfrac{dh}{dt} = k\ln|H-h|$

3 a $\dfrac{dn}{dt} = 0$ b $\dfrac{dn}{dt} = k\sqrt{n}$

4 a $\dfrac{dn}{dt} = k_1 n$ b $\dfrac{dn}{dt} = \dfrac{k_2}{n}$

c $\dfrac{dn}{dt} = -k_3$

Exercise 33B – p.398

1 $t = 16T/3$
2 $dy/dx = k\sqrt{x}; y = 0.4x^{3/2} + 1.6; 1.2$
3 a $dn/dt = kn$
b $t = T\ln 2/\ln 1.5 = 1.71T$
d e.g. if the colony becomes too great for the amount of liquid to support
4 1000 years (3 s.f.)
5 a $-dm/dt = km; m = 50e^{-kt}$ where $k = 0.002\,554\ldots$
b 26.8 g (3 s.f.)
6 c $k = 0.0357$ (3 .s.f.)
d 44° (nearest degree)

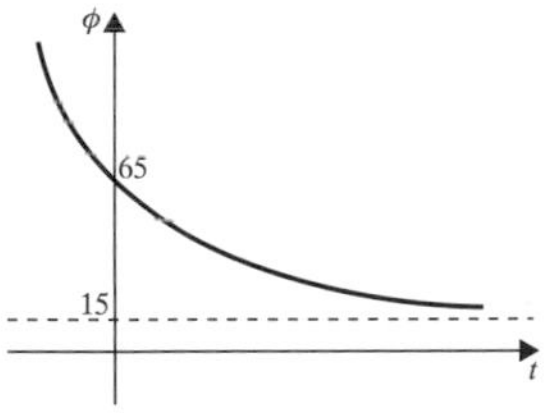

e It may be; $15 + 50e^{-kt} \to 15$ as t gets very large, i.e. the water cools to room temperature. However during 24 hours the room temperature may not have been constant making the model inappropriate.
7 About 4 to $4\frac{1}{2}$ hours before discovery assuming Newton's Law of Cooling and constant ambient temperature of 10 °C.

Exercise 33C – p.400

1 π
2 $(\frac{1}{2}\sqrt{3} - \pi/6)$
3 a

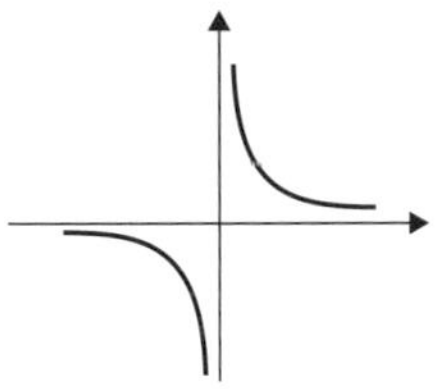

b $8\ln 2$
4 π
5 a

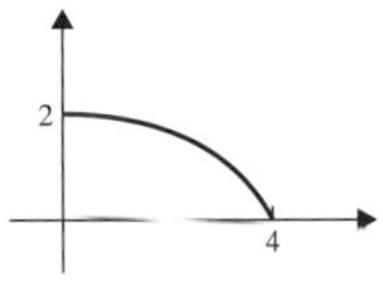

b 2π
6 a $x - 3y + 4 = 0$ b 2.13

Exercise 33D – p.404

1 $\frac{512}{15}\pi$
2 $\frac{1}{2}\pi(e^6 - 1)$
3 $\frac{1}{2}\pi$
4 $\frac{64}{5}\pi$
5 2π
6 8π
7 8π
8 $\frac{3}{5}\pi(\sqrt[3]{32} - 1)$
9 $\frac{1}{2}\pi(e^2 - 1)$
10 $\frac{16}{15}\pi$
11 $\frac{16}{15}\pi$
12 $\frac{1}{2}\pi^2$
13 $\frac{5}{2}\pi$
14 $\frac{3}{10}\pi$

Mixed Exercise 33 – p.405

1 $\frac{1}{3}$
2 $\frac{256}{15}\pi$
3 295 (3 s.f.)
5 $\frac{1}{2}\pi$
6 a $\dfrac{dN}{dt} = kN$ b 33
c E.g. there weren't 200 rabbits on the island
7 a $y = 100e^{-0.105x}$
b E.g. the rate of inflation is very unlikely to remain constant.

Summary F

Multiple Choice Exercise F – p.408

1 C
2 C, D
3 A, B, C
4 B
5 B
6 C
7 B
8 B
9 A
10 B, C
11 A, C
12 B, C
13 B, C
14 B, C
15 A, D
16 F
17 F
18 F
19 T
20 F
21 F

Examination Questions F – p.411

1 $a = -1, b = -2$
2 $2x - e^{-x} + K$
3

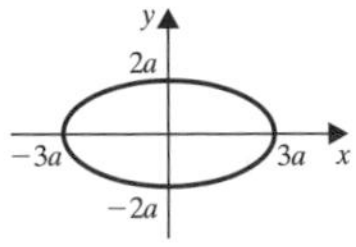

5 $A = 6, B = 1, C = -2$
6 a $\dfrac{2}{x+2} - \dfrac{3}{2x-1}$ b $5\frac{1}{2}$

7 $\pi/12$

8 $\int_0^{\pi} 3\sin x\,dx$; 6 sq units

9 **i** $\frac{1}{2}\ln|A(2x+1)|$

ii $\frac{1}{16}(2x+1)^8+K$

10 (1, 2); (6, 2) and (5, −1)

11 **a** $c = 9, d = -14$

b $-2, -\frac{7}{2}+\frac{1}{2}\sqrt{77}, -\frac{7}{2}-\frac{1}{2}\sqrt{77}$

12 $\dfrac{1}{x+2}+\dfrac{2}{(x+2)^2}-\dfrac{1}{x+1}$

13 **b** 27

14 **a** (1, 4), $\sqrt{17}$ **b** inside

15 $\frac{1}{4}e^4 - e^2 + \frac{7}{4}$

16 **a** false: $n = 4$
$\Rightarrow$ $f(n) = 21$ which is not prime

b true: n odd
$\Rightarrow$ n^2 odd + $(n+1)$ even
$\Rightarrow$ odd
and n even
$\Rightarrow$ n^2 even + $(n+1)$ odd
$\Rightarrow$ odd

17 **a** $3y+2x = 5$ **b** 1.15 sq units

18 $\frac{1}{8}+\frac{1}{16}\pi$

19 $\frac{1}{2}x\sin 2x+\frac{1}{4}\cos 2x+k$

20 $\frac{1}{9}\int u^4(u+1)\,du$; $\dfrac{u^5}{270}(5u+6)+k$

21 **a** (1, 4) **b** 5

c $4x-3y = 17$

23 **a** It shows the formula is correct for $n = 1$, 2 and 3 only, not for all values of n.

b $2S_n = n(2n) \Rightarrow S_n = n^2$

24 $(-2, 5), 4; 2\sqrt{5}$; 2 units

25 1.42 units

26 $\dfrac{1}{x-1}-\dfrac{2}{2x+1}$; $\ln(10/7)$

27 **a** $A = \frac{1}{2}, B = 2, C = -1$

b $\dfrac{1}{x-1}+\ln((x-1)^2\sqrt{x})+K$

28 $7\pi/2$

29 **a** **i** $\ln\left|\dfrac{Ax}{x+1}\right|$ **ii** $\ln\dfrac{Ae^x}{1+e^x}$

b $2\cos x+2x\sin x-x^2\cos x+K$

30 **a** $2\pi/21$

b $\frac{1}{4}\ln|2x+3|+\dfrac{3}{4(2x+3)}+k$

31 **a** $-\frac{1}{2}\cos 2y+k$

b $(x-1)e^x+k$

c $4(x-1)e^x+\cos 2y = k$

32 **a** $k-x^2 = 2/y$

c 1 **e** (−2, −2)

33 (2, 5) (−2, 5); 16π

34 **b** 6.31 hours

35 **a** $3y+x = 33$

b

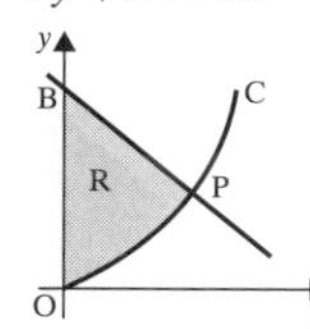

c 186π cubic units

36 **a** $\frac{1}{3}\sin^3 x+k$

b $\frac{1}{4}x^2(2\ln|x|-1)+k$

c $\frac{2}{3}(x+1)^{3/2}-2(x+1)^{1/2}+k$

d $2\frac{2}{3}$

37 **b**

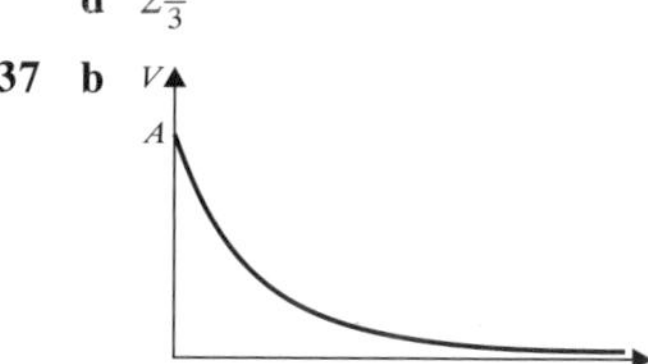

38 **i** $x = kt$; 4 p.m.

ii $\dfrac{dx}{dt} = \dfrac{k}{x}$; $\frac{1}{2}x^2 = kt$;

$x = 2\sqrt{t}$; 6 p.m.

iii temperature remains constant

39 **i** $\frac{1}{2}(1-\cos 2x)$

40 **i** $-e^{-x}(x+1)+k$

41 **a** A(1, 0), B(5, 4)

b 32/3 sq units

c 32π cubic units

42 **a** $\frac{1}{2}x\sin 2x+\frac{1}{4}\cos 2x+k$

b 7/9

43 **a** $\dfrac{2}{x-2}-\dfrac{1}{2x+1}$ **b** $\frac{1}{3}$

c $\dfrac{4}{(x-2)^3}-\dfrac{8}{(2x+1)^3}$; min

d $\ln\left(\dfrac{A(x-2)^2}{\sqrt{2x+1}}\right)$

44 **a** 5, (1, 2)

b $3x+4y-36 = 0$

45 **a** $3-\sqrt{5}$ **b** $\frac{1}{4}(3e^{-2}-7e^{-6})$

46 **a** $\dfrac{dx}{dt} = -ax$ **c** 0.277

47 $\dfrac{dx}{dt} = -kx$

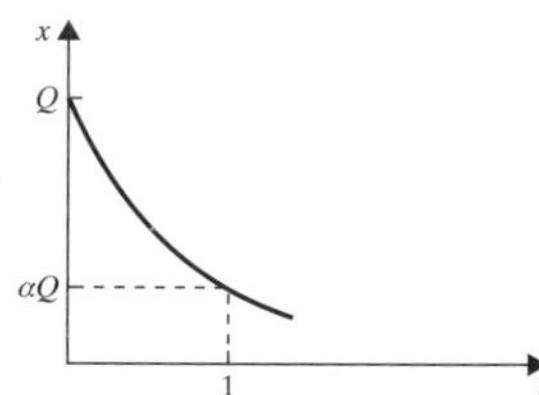

48 **i** $\dfrac{x^2}{9}+\dfrac{y^2}{4} = 1$

ii

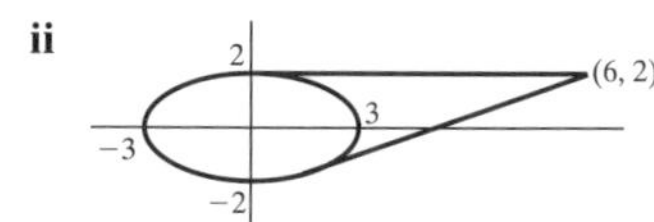

iii $-\frac{2}{3}\cot\theta$ **v** 1.57, 5.64

49 **i** $\dfrac{dr}{dt} = \dfrac{k}{r^2}$ **ii** $k = 5000$; 141 m

iii $\dfrac{dr}{dt} = \dfrac{k}{r^2(t+2)}$; $k = 10^4$

iv 104 m

50 **i** $2x\sin 2x+\cos 2x+k$

iii $y^2 = 4x^2+4x\sin 2x + 2\cos 2x+1$

Chapter 34

Exercise 34A – p.422

1 **a** $1+36x+594x^2+5940x^3$

b $1-18x+144x^2-672x^3$

c $1024+5120x+11\,520x^2 + 15\,360x^3$

d $1-\frac{20}{3}x+\frac{190}{9}x^2-\frac{380}{9}x^3$

e $128-672x+1512x^2-1890x^3$

f $(\frac{3}{2})^9+\dfrac{3^{10}}{27}x+\dfrac{3^9}{8}x^2+\frac{7}{2}(3^7)x^3$

2 **a** $336x^2$ **b** $-10x$

c $-21\,840x^{11}$ **d** $3360p^6q^4$

e $16(3a)^7b$ **f** $7920x^4$

g $63x^5$ **h** $56a^3b^5$

3 **a** $1-8x+27x^2$

b $1+19x+160x^2$

c $2-19x+85x^2$

d $1-68x+2136x^2$

Exercise 34B – p.425

1 0.8171
2 1.062
6 **a** $1-50x$ **b** $256-1024x$
c $1-19x$

Exercise 34C – p.429

1 $1-x-\frac{x^2}{2}-\frac{x^3}{2}, -\frac{1}{2}<x<\frac{1}{2}$
2 $1-10x+75x^2-500x^3, -\frac{1}{5}<x<\frac{1}{5}$
3 $1+\frac{3}{2}x+\frac{3}{2}x^2+\frac{5}{4}x^3, -2<x<2$
4 $1+\frac{3}{2}x+\frac{3}{8}x^2-\frac{1}{16}x^3, -1<x<1$
5 $\frac{1}{3}-\frac{x}{9}+\frac{x^2}{27}-\frac{x^3}{81}, -3<x<3$
6 $1-\frac{x}{4}+\frac{3x^2}{32}-\frac{5x^3}{128}, -2<x<2$
7 $1+2x+3x^2+4x^3, -1<x<1$
8 $1-\frac{1}{2}x+\frac{3}{8}x^2-\frac{5}{16}x^3, -1<x<1$
9 $1+\frac{1}{2}x-\frac{5}{8}x^2-\frac{3}{16}x^3, -1<x<1$
10 $-2-3x-3x^2-3x^3, -1<x<1$
11 $2+2x+\frac{21}{4}x^2+\frac{27}{2}x^3, -\frac{1}{3}<x<\frac{1}{3}$
12 $\frac{1}{2}-\frac{3}{4}x+\frac{13}{8}x^2-\frac{51}{16}x^3, -\frac{1}{2}<x<\frac{1}{2}$
13 $1+x+\frac{1}{2}x^2+\frac{1}{2}x^3, -1<x<1$
14 $1-\frac{1}{9}x^2, -3<x<3$
15 $x+x^2+3x^3, -\frac{1}{2}<x<\frac{1}{2}$
16 $x-x^2+x^3, -1<x<1$
17 $1-3p^{-1}+6p^{-2}-10p^{-3}+15p^{-4}, |p|<1$
18 1.732
19 3.162 28
20 $1+2x+2x^2, -\frac{1}{2}<x<\frac{1}{2}$
22 $1-3x+\frac{7}{2}x^2$
23 $1+2x+5x^2$
24 $\frac{3}{2}+\frac{15}{4}x$
27 $\frac{1}{729}[27+27x+18x^2+20x^3]$

Exercise 34D – p.433

1 $1-x+x^2/2!-x^3/3!+x^4/4!$
2 $1-x^2/2!+x^4/4!-x^6/6!$
3 $x-x^2/2+x^3/3-x^4/4$
4 $1+12x+66x^2+220x^3$
5 $1+2x+2x^2+4x^3/3$
6 $3x-9x^3/2+81x^5/40$
7 $1+x+x^2+x^3+x^4$
8 $1+\frac{1}{2}x-\frac{1}{8}x^2+\frac{1}{16}x^3-\frac{5}{128}x^4$
9 $2x+\frac{1}{3}x^3$
10 $-x-\frac{1}{2}x^2-\frac{1}{3}x^3-\frac{1}{4}x^4$

Exercise 34E – p.435

1 $3x$
2 $3x$
3 $1+2x$
4 $1-\frac{1}{2}x^2$
5 $1-2x+2x^2$
6 x^2
7 1.1
8 $u_{n+1}=\frac{x}{n}u_n$
9 **d** 5 **e** 11
10 **a** $1-\frac{1}{2}x-\frac{1}{8}x^2-\frac{1}{16}x^3$
b $u_{n+1}=(1-2n)u_nx/2(n+1)$

Mixed Exercise 34 – p.436

1 $1+18x+144x^2+\ldots+2^9x^9$; all x
2 $1+15x+90x^2+270x^3+405x^4+243x^5$
$1-15x+90x^2-270x^3+405x^4-243x^5$
4 $1-2x+4x^2-\ldots$
5 $\frac{1}{x-1}-\frac{2}{2x-1}$;
$1+3x+7x^2+15x^3$; $|x|<\frac{1}{2}$
6 $1-x-x^3, |x|<1$
7 $2-x+3x^2, |x|<\frac{1}{2}$
8 $1-x+\frac{1}{2}x^2-\frac{1}{2}x^3$
10 $y=1+4x$
11 $1-2x^2$
13 $1-\frac{1}{6}x^3$
14 $1+x\ln 2+\frac{x^2}{2}(\ln 2)^2$; 1.2599...

Chapter 35

Exercise 35A – p.438

1 $\frac{1}{6}\pi, \frac{2}{3}\pi$
2 $0, \frac{2}{3}\pi$
3 $\tan^{-1}\frac{12}{5}$
4 $0, \pi$
5 $\frac{1}{2}\pi$
6 $\tan^{-1}\frac{3}{4}+\frac{1}{2}\pi$
7 $\frac{1}{2}\pi$
8 $\sin^{-1}(\sqrt{2}-1)$
9 0, 30°, 150°, 180°
10 53.1°, 180°
11 0, 180°
12 201.5°, 338.5°
13 26.6°, 135°, 206.6°, 315°
14 34.6°, 325.4°
15 120°, 300°
16 20°, 100°, 140°, 220°, 260°, 340°

Exercise 35B – p.440

1 $-\frac{3}{4}, -\frac{3}{4}, \frac{2}{5}$
2 $-1, -1, \frac{1}{5}, 5$
3 $\frac{3}{2}, \frac{3}{2}, -\frac{2}{5}$
4 1, 2, 3
5 4
6 $6+2\sqrt{3}$
7 $\frac{1}{2}(3\pm\sqrt{41})$
8 4
9 5
10 −1, 2
11 −2
12 $x=\sqrt{2}, y=\frac{1}{2}\sqrt{2}$ and $x=-\sqrt{2}, y=-\frac{1}{2}\sqrt{2}$
13 $x=2, y=3$ and $x=-2, y=-1$
14 $x=-3, y=4, x=-6, y=3$
15 −2, −2, 1, 4

Exercise 35C – p.442

1 **a** ∞ **b** 2 **c** ∞
d ∞ and all +ve **e** 3
f 1 **g** 1 **h** 2 **i** 1
2 **b** $1<x<1.5$ **e** $2<x<2.5$
f $0.5<x<1$ **g** $1.5<x<2$
h $0.5<x<1$ **i** $0<x<0.5$
3 (0, 1) max, (2, −3) min, 3
4 **a** 2 **b** 3 **c** 3
5 −1 (exact)
6 $-3<x<-2$

Exercise 35D – p.444

1 1.4
2 2.28 rad
3 0.3
4 3.3

Exercise 35E – p.447

1 0.20
2 0.22
3 0.16
4 0.90
5 5.2
6 −1.5

In the remaining questions, the degree of accuracy depends upon the accuracy of the first approximation: these answers are correct to 5 significant figures

7 1.1656
8 −2.6691
9 0 (exact). 1.2564
10 0.510 97
11 0 (exact), 0.746 85
12 0.374 82
13 0 (exact), 1.9038
14 1 (exact)
15 −0.917 35, 0.863 66
16 3.1038
17 −1.4537, 0.539 79

Mixed Exercise 35 – p.448

1 $1/2\log_{10}3$
2 76.7°, 209.6°
3 $x=\frac{1}{2}(1+\sqrt{7}), y=\frac{1}{2}(1-\sqrt{7})$; $x=\frac{1}{2}(1-\sqrt{7}), y=\frac{1}{2}(1+\sqrt{7})$
4 1
5 $-\frac{1}{3}, \frac{1}{3}$
6 $2n\pi$
7 3
8 2.207 07
9 1.3
10 0.7

Chapter 36

Exercise 36A – p.453

1 a $\overrightarrow{AC}$ b $\overrightarrow{BD}$
c $\overrightarrow{AD}$ d $\overrightarrow{DB}$
2 **a**, **b** − **a**, 2**b** − 2**a**
5 **b** − **a**, **a** + **b**, **a** − 3**b**, 2**b**, 2**a** − 2**b**
6 $\overrightarrow{DE}=\overrightarrow{CH}=\overrightarrow{BG}=\mathbf{c}$, $\overrightarrow{DC}=\overrightarrow{EH}=\overrightarrow{FG}=\mathbf{a}$, $\overrightarrow{FE}=\overrightarrow{GH}=\overrightarrow{BC}=\mathbf{b}$
8 **c** − **a**, **b** − **a**, **b** − **c**
9 **b** + **c** − **a**, **b** − **c** − **a**, **a** + **b** + **c**, **a** + **b** − **c**
10 a $\frac{\pi}{2}$ b 0.333 c 1.76
11 a $\frac{1}{2}(\mathbf{a}+\mathbf{c})$ b $\frac{1}{2}(\mathbf{b}+\mathbf{d})$
12 a $\frac{1}{4}(\mathbf{a}+2\mathbf{b}+\mathbf{c})$
b $\frac{1}{4}(\mathbf{b}+2\mathbf{c}+\mathbf{d})$

Exercise 36B – p.458

1 a $3\mathbf{i}+6\mathbf{j}+4\mathbf{k}$ b $\mathbf{i}-2\mathbf{j}-7\mathbf{k}$
c $\mathbf{i}-3\mathbf{k}$
2 a (5, −7, 2) b (1, 4, 0)
c (0, 1, −1)
3 a $\sqrt{21}$ b 5 c 3
4 a 6 b 7 c $\sqrt{206}$
5 a $3\mathbf{i}+4\mathbf{k}$ b $2\mathbf{i}+2\mathbf{j}+2\mathbf{k}$
c $2\mathbf{i}+3\mathbf{j}+3\mathbf{k}$
d $-6\mathbf{i}+12\mathbf{j}-8\mathbf{k}$
6 **b**, **d** and **e**
7 **c**
8 $\lambda=\frac{1}{2}$
9 **b**, **e** and **f**
10 a neither b parallel
c equal
11 a $-2\mathbf{i}-3\mathbf{j}+7\mathbf{k}$
b $-\mathbf{j}+3\mathbf{k}$
c $-2\mathbf{i}-2\mathbf{j}+4\mathbf{k}$
12 $\sqrt{62}, \sqrt{10}, 2\sqrt{6}$
13 $\sqrt{5}$
14 $\overrightarrow{AB}=-\mathbf{i}+3\mathbf{j}$, $\overrightarrow{BD}=-\mathbf{j}-4\mathbf{k}$
$\overrightarrow{CD}=-2\mathbf{i}+2\mathbf{j}-6\mathbf{k}$
$\overrightarrow{AD}=-\mathbf{i}+2\mathbf{j}-4\mathbf{k}$

Exercise 36C – p.459

1 a $\frac{2}{3}\mathbf{i}+\frac{2}{3}\mathbf{j}-\frac{1}{3}\mathbf{k}$
b $\frac{6}{7}\mathbf{i}-\frac{2}{7}\mathbf{j}-\frac{3}{7}\mathbf{k}$
c $\frac{3}{5}\mathbf{i}+\frac{4}{5}\mathbf{k}$ d $\frac{1}{9}\mathbf{i}+\frac{8}{9}\mathbf{j}+\frac{4}{9}\mathbf{k}$
2 a $(\frac{1}{3}, \frac{2}{3}, -\frac{2}{3})$ b $(\frac{3}{7}, \frac{2}{7}, \frac{6}{7})$
c $(\frac{8}{9}, -\frac{1}{9}, -\frac{4}{9})$
d $\left(\frac{1}{\sqrt{3}}, -\frac{1}{\sqrt{3}}, -\frac{1}{\sqrt{3}}\right)$
3 a $4\mathbf{j}+5\mathbf{k}$ b $16\mathbf{i}-16\mathbf{j}-8\mathbf{k}$
c $\pm(\frac{8}{3}\mathbf{i}+\frac{1}{3}\mathbf{j}+\frac{4}{3}\mathbf{k})$
4 a $\frac{1}{\sqrt{3}}\mathbf{i}-\frac{1}{\sqrt{3}}\mathbf{j}+\frac{1}{\sqrt{3}}\mathbf{k}$
b $\frac{5}{13}\mathbf{j}-\frac{12}{13}\mathbf{k}$ c **i**

Exercise 36D – p.461

1 $\sqrt{13}$; 3
2 a no b no c no

Exercise 36E – p.464

1 a 30
b 0; **a** and **b** are perpendicular
c −1
2 a $7, \frac{1}{3}\sqrt{7}$ b $14, \sqrt{\frac{7}{19}}$
c $3, \frac{3}{58}\sqrt{58}$ d $1, \frac{1}{5}$
3 4
4 $-\sqrt{\frac{7}{34}}$; 33.2°
7 a $10\sqrt{3}$ b $41-20\sqrt{3}$
8 67.80°
9 a i $4\mathbf{j}-4\mathbf{k}$ ii $4\mathbf{i}+4\mathbf{j}$
b $\widehat{A}=90°$, $\widehat{E}=19.5°$ $\widehat{H}=70.5°$
10 no; **a** and **b** may be perpendicular
11 b 142° c 4.33 sq units
12 a 17, 40.2°
b $4\mathbf{i}-4\mathbf{j}+3\mathbf{k}$, $\sqrt{206}$
c $-5\mathbf{j}+\mathbf{k}$, $\frac{1}{3}(10\mathbf{i}-7\mathbf{j}+5\mathbf{k})$

Exercise 36F – p.468

1 a $2\mathbf{i}-\mathbf{j}-5\mathbf{k}$, b $3\mathbf{j}-5\mathbf{k}$
c $-2\mathbf{i}+4\mathbf{j}+\mathbf{k}$, d $\mathbf{i}+\mathbf{k}$
e $2\mathbf{i}+3\mathbf{j}+2\mathbf{k}$, f $4\mathbf{i}+3\mathbf{j}+\frac{1}{2}\mathbf{k}$
2 a $\frac{x-1}{1}=\frac{y+5}{3}=\frac{z-2}{1}$
b $x=1, \frac{y-3}{-2}=\frac{z-3}{1}$,
i.e. line is ∥ to the yz-plane.
3 a $\mathbf{r}=3\mathbf{i}-\mathbf{j}+4\mathbf{k}+t(2\mathbf{i}+3\mathbf{j}+2\mathbf{k})$
b $\mathbf{r}=-\mathbf{k}+t(2\mathbf{i}+\mathbf{j}+3\mathbf{k})$
4 a $\mathbf{r}=\mathbf{i}-3\mathbf{j}+2\mathbf{k}+t(5\mathbf{i}+4\mathbf{j}-\mathbf{k})$
b $\mathbf{r}=2\mathbf{i}+\mathbf{j}+t(3\mathbf{j}-\mathbf{k})$
c $\mathbf{r}=t(\mathbf{i}-\mathbf{j}-\mathbf{k})$
5 a no b yes c yes d yes
6 $\mathbf{r}=4\mathbf{i}+5\mathbf{j}+10\mathbf{k}+s(\mathbf{i}+\mathbf{j}+3\mathbf{k})$
$\mathbf{r}=2\mathbf{i}+3\mathbf{j}+4\mathbf{k}+s(\mathbf{i}+\mathbf{j}+5\mathbf{k})$
$\mathbf{r}=4\mathbf{i}+5\mathbf{j}+10\mathbf{k}+s(\mathbf{i}+\mathbf{j}+5\mathbf{k})$
7 a $\mathbf{r}=\mathbf{i}+7\mathbf{j}+8\mathbf{k}+s(3\mathbf{i}+\mathbf{j}-4\mathbf{k})$
(7, 9, 0), $(0, \frac{20}{3}, \frac{28}{3})$,
(−20, 0, 36)
b $\mathbf{r}=\mathbf{i}+\mathbf{j}+7\mathbf{k}+s(2\mathbf{i}+3\mathbf{j}-6\mathbf{k})$
$(\frac{10}{3}, \frac{9}{2}, 0)$, $(0, -\frac{1}{2}, 10)$,
$(\frac{1}{3}, 0, 9)$
8 a $3\mathbf{i}+4\mathbf{j}+5\mathbf{k}$
b $\mathbf{r}=5\mathbf{i}-2\mathbf{j}-4\mathbf{k}+s(3\mathbf{i}+4\mathbf{j}+5\mathbf{k})$,
$(\frac{13}{3}, 0, -\frac{3}{2})$

Exercise 36G – p.470

1 a $\parallel$ b intersect at $\mathbf{i}+2\mathbf{j}$; $27.9°$
c skew

2 -3, $-\mathbf{i}+3\mathbf{j}+4\mathbf{k}$

Exercise 36H – p.474

1 a $x+y-z=2$
b $2x+3y-4z=1$

2 a $\mathbf{r}.(3\mathbf{i}-2\mathbf{j}+\mathbf{k})=5$
b $\mathbf{r}.(5\mathbf{i}-3\mathbf{j}-4\mathbf{k})=7$

3 1: a $\mathbf{i}+\mathbf{j}-\mathbf{k}$
b $2\mathbf{i}+3\mathbf{j}-4\mathbf{k}$
2: a $3\mathbf{i}-2\mathbf{j}+\mathbf{k}$
b $5\mathbf{i}-3\mathbf{j}-4\mathbf{k}$

4 1. a $\dfrac{2}{\sqrt{3}}$ b $\dfrac{1}{\sqrt{29}}$
2. a $\dfrac{5}{\sqrt{14}}$ b $\dfrac{7}{\sqrt{50}}$

5 The normals are parallel, 2 units

6 $\mathbf{r}=s(\mathbf{i}-2\mathbf{j}+\mathbf{k})$

7 $\mathbf{r}=2\mathbf{i}+\mathbf{j}+\mathbf{k}+t(\mathbf{i}+2\mathbf{j}-3\mathbf{k})$

8 $\mathbf{r}.(7\mathbf{i}+2\mathbf{j}-3\mathbf{k})=3$, $\dfrac{3}{\sqrt{62}}$

9 $\mathbf{r}.(\mathbf{i}-\mathbf{j}-3\mathbf{k})=-4$

10 $\mathbf{r}.(-\mathbf{i}+\mathbf{k})=3$

11 $(\frac{5}{2}, -\frac{1}{2}, -\frac{1}{2})$

12 $(1, -1, 4)$

13 a $(\frac{8}{3}, \frac{14}{3}, \frac{2}{3})$ b $10.1°$

14 a $\mathbf{r}.(6\mathbf{i}+4\mathbf{j}+3\mathbf{k})=12$
b $\dfrac{12}{\sqrt{61}}$ c $30.8°$

Summary G

Multiple Choice Exercise g – p.476

1 C
2 B
3 B
4 B
5 C
6 C
7 A, C, D
8 B
9 B
10 B, C
11 C
12 F
13 F
14 T
15 T
16 F

Examination Questions G – p.478

1 i $1+x+x^2+x^3$, $|x|<1$
ii $1-4x+12x^2-32x^3$, $|x|<\frac{1}{2}$
iii $1-3x+9x^2-23x^3$, $|x|<\frac{1}{2}$

2 $142.8°$

3 104.65, $x=100+\ln x$

4 i 5005 ii -455

5 4

6 $x^5-5x^3+10x-\dfrac{10}{x}+\dfrac{5}{x^3}-\dfrac{1}{x^5}$

7 i $-4\mathbf{i}+2\mathbf{j}$, $4\mathbf{i}+8\mathbf{j}$, $\mathbf{PQ}.\mathbf{RQ}=0$, $\therefore \perp$
ii $26.6°$ iii $3\mathbf{i}+7\mathbf{j}$
iv $53.1°$

8 a $1+\dfrac{x^2}{2}+\dfrac{3x^4}{8}+\dfrac{5x^6}{16}$
c 1.060 61

9 $A=8192$, $B=-53\,248$, $C=159\,744$

10 a

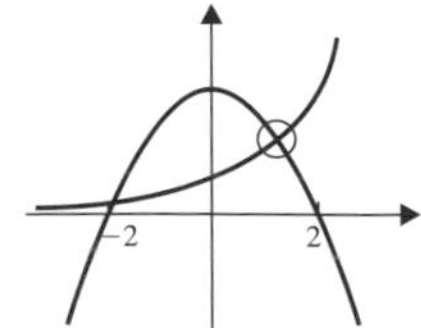

b $x_1=-1.965\,88$,
$x_2=-1.964\,68$,
$x_3=-1.964\,64$,
$x_4=-1.964\,64$; $x\approx-1.9646$
c fails after 3 iterations

11 a $a=6, b=2$ b 2.60

12 a $1-20x+180x^2-960x^3$
b $x=0.01$; 0.817 04

13 a $81+216x+216x^2+96x^3+16x^4$
b $81-216x+216x^2-96x^3+16x^4$
c $x=\sqrt{2}$; 1154

14 a 2.10 b $y^3-3y-3=0$
c $y=2.10 \Rightarrow x=0.68$

15 a $(1,-1)$ b 3.14 c 0.159
f A(0.159, 0), B(3.14, 0)

16 i $16-32x+24x^2-8x^3+x^4$
ii $1-6x+24x^2-80x^3$, $|x|<\frac{1}{2}$
iii $a=-128$, $b=600$

17 $x_1=0.7$, $x_2=0.594$, $x_3=0.622$, $x_4=0.617$, $x_5=0.618$, $\psi=0.62$

18 0.98 rad

19 a i $-6\mathbf{i}-\mathbf{j}$ ii $-\mathbf{i}+6\mathbf{j}+10\mathbf{k}$
b i $\mathbf{r}=-(4+t)\mathbf{i}+(6+6t)\mathbf{j}+(6+10t)\mathbf{k}$

20 a c 3.495

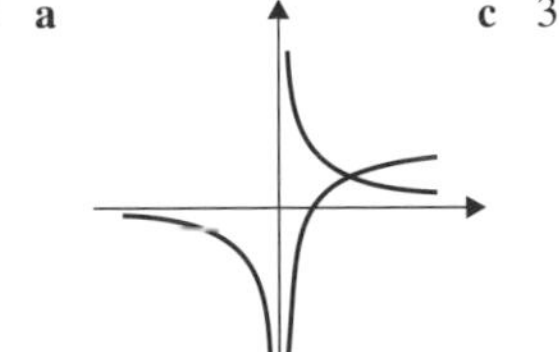

21 a $\dfrac{1}{1-2x}+\dfrac{1}{(1-x)^2}-\dfrac{1}{1-x}$
b $1+3x+6x^2+11x^3$
c $|x|<\frac{1}{2}$

22 a $\mathbf{r}=(2+4t)\mathbf{i}+(1-t)\mathbf{j}+(5+t)\mathbf{k}$
b $85.7°$ c $94.3°$

23 $1+\frac{1}{2}x-\frac{1}{8}x^2$; $a=2, b=-\frac{3}{4}$

24 i $(x\cos x-\sin x)/x^2$
ii 4.49

25 $1+2t-t^2$

26 b $\mathbf{r}=(3-4\mu)\mathbf{i}+(2\mu-1)\mathbf{j}+(2+7\mu)\mathbf{k}$
c $\mathbf{r}=11\mathbf{i}-5\mathbf{j}-12\mathbf{k}$

27 $1-\frac{2}{3}x-\frac{4}{9}x^2-\frac{40}{81}x^3$; 405/188

28 i 3 ii 1.065 iii 188 sq units

29 2.002, $f'(x)=1+\ln(10x)$

30 a 1.733 b 1.620, 1.666

31 240

32 i $\mathbf{i}+3\mathbf{j}+2\mathbf{k}$ ii $-\mathbf{i}+3\mathbf{j}+2\mathbf{k}$; $31°$

33 i $\mathbf{i}+2\mathbf{j}+2\mathbf{k}$, $-2\mathbf{i}+\mathbf{j}+2\mathbf{k}$

34 $\dfrac{3}{4+3x}$; $\ln 4+\frac{3}{4}x-\frac{9}{32}x^2$

35 $-4\cos(\frac{1}{3}\pi+2x)$; $\frac{1}{2}-\sqrt{3}\,x-x^2$

36 $\ln 3+\dfrac{x}{3}-\dfrac{x^2}{18}$

37 a $-\sin x$ b $y=\frac{1}{2}x^2$

38 a $1+6x+27x^2+108x^3$
b i $1+4x+8x^2+\frac{32}{3}x^3$
ii $2x-\frac{4}{3}x^3$
c $a=19, b=98\frac{2}{3}$

39 ii $(\frac{1}{2}\sqrt{2}, \frac{1}{2}\sqrt{2}e^{-1/2})$, $(-\frac{1}{2}\sqrt{2}, -\frac{1}{2}\sqrt{2}e^{-1/2})$
iii 0.0739 iv $x-x^3$ v 0.0736

40 60.5°

41 **i** $a\mathbf{i}+b\mathbf{j}+\mathbf{k}$
ii $2\mathbf{i}-3\mathbf{j}$; $3\mathbf{i}-5\mathbf{j}+\mathbf{k}$
iii $2a-3b=0$; $3a-5b+1=0$
iv $3x+2y+z=6$
v 36.7°
vi $(\frac{11}{3}, -\frac{11}{3}, \frac{7}{3})$

42 **i** $\mathbf{r}=(2\mathbf{i}+3\mathbf{j}+5\mathbf{k}) + \lambda(3\mathbf{i}+\mathbf{j}-2\mathbf{k})$
ii $\lambda=1$; $(5, 4, 3)$
iii $(\frac{19}{2}, \frac{11}{2}, 0)$
iv $(\frac{13}{2}, \frac{9}{2}, 2)$; $\frac{1}{2}\sqrt{14}$
v $\mathbf{i}+2\mathbf{j}-3\mathbf{k}$; 38.2°

43 **i** $\mathbf{r}.(2\mathbf{i}-3\mathbf{j}+7\mathbf{k})=-5$
ii $\mathbf{r}=130\mathbf{i}-40\mathbf{j}+20\mathbf{k}+t(40\mathbf{i}-20\mathbf{j}+5\mathbf{k})$
iii $10\mathbf{i}+20\mathbf{j}+5\mathbf{k}$ **iv** 135 m

44 19°

45 **a** $(0, 4, 3)$ **b** $5\sqrt{2}$
c $\frac{x-5}{-5}=\frac{y}{4}=\frac{z}{3}$

46 **a** $x+z=3$
b $\frac{x-1}{2}=\frac{y}{1}=\frac{z-2}{-2}$

Sine Curve

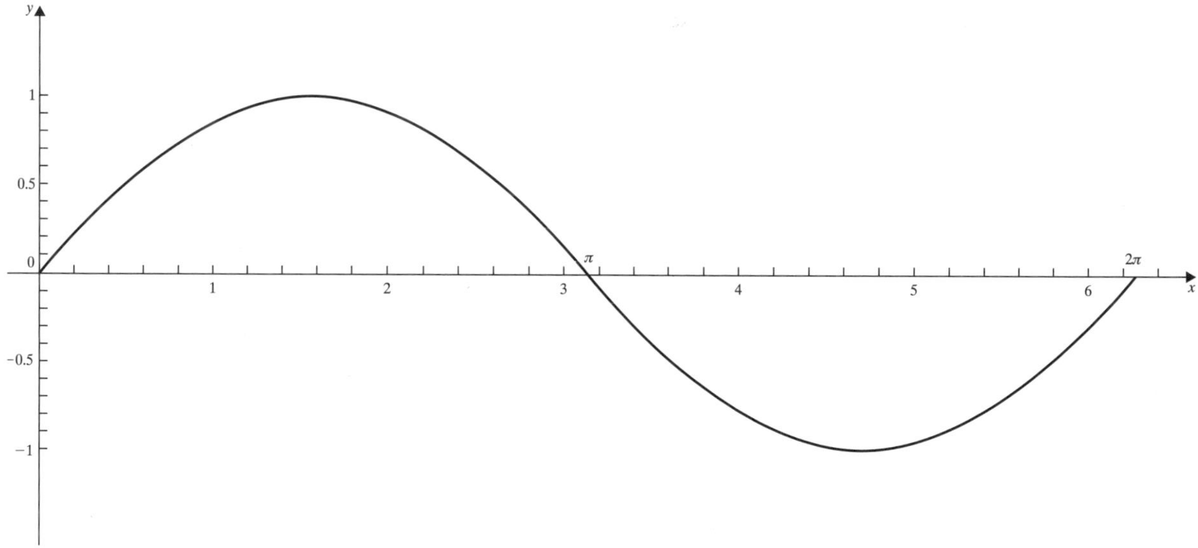

Index